MW00907881

BIOCHEMISTRY AND MOLECULAR BIOLOGY OF FISHES, 6

ENVIRONMENTAL TOXICOLOGY

Biochemistry and Molecular Biology of Fishes

Editors

T.P. Mommsen

Department of Biology
University of Victoria
Victoria, BC V8W 3N5
Canada

and

T.W. Moon

Centre for Advanced Research in Environmental Genomics
and Department of Biology
University of Ottawa
Ottawa, ON K1N 6N5
Canada

Volumes in the Series

1. Phylogenetic and biochemical perspectives
2. Molecular biology frontiers
3. Analytical techniques
4. Metabolic biochemistry
5. Environmental and ecological biochemistry
6. Environmental toxicology

Biochemistry and Molecular Biology of Fishes, 6

Environmental Toxicology

Edited by

T.P. Mommsen

Department of Biology, University of Victoria,
Victoria, BC V8W 3N5, Canada

and

T.W. Moon

Centre for Advanced Research in Environmental Genomics and
Department of Biology, University of Ottawa, Ottawa, ON K1N 6N5, Canada

2005

ELSEVIER

AMSTERDAM - BOSTON - HEIDELBERG - LONDON - NEW YORK - OXFORD
PARIS - SAN DIEGO - SAN FRANCISCO - SINGAPORE - SYDNEY - TOKYO

ELSEVIER B.V.
Radarweg 29
P.O. Box 211, 1000 AE Amsterdam
The Netherlands

ELSEVIER Inc.
525 B Street, Suite 1900
San Diego, CA 92101-4495
USA

ELSEVIER Ltd
The Boulevard, Langford Lane
Kidlington, Oxford OX5 1GB
UK

ELSEVIER Ltd
84 Theobalds Road
London WC1X 8RR
UK

First edition 2005

Library of Congress Cataloging in Publication Data
A catalog record is available from the Library of Congress.

British Library Cataloguing in Publication Data
A catalogue record is available from the British Library.

ISBN (vol. 4): 0 444 51833 9
ISBN (series): 0 444 89185 4

♾ The paper used in this publication meets the requirements of ANSI/NISO Z39.48-1992 (Permanence of Paper).
Printed in The Netherlands

Preface

Volume 1 of Biochemistry and Molecular Biology of Fishes was published in 1991 as a 'forum for state-of-the-art review articles' in this rapidly expanding research area. The editors felt strongly that the significant emphasis on the 'rat-centric' view of biochemistry and molecular biology overlooked the conceptual breakthroughs occurring in the piscine literature. Four additional volumes were published in rapid succession (2–5) to capture key themes in fish research. Volume 3, Analytical Techniques, was a deviation from the norm for this series, and instead was an attempt to standardize many of the common techniques used in the field. Developing new themes and editing such a series is not a trivial task, and with the untimely death of Professor Peter Hochachka, one of the series editors, the series became inactive. However, this does not mean that the series no longer has a raison d'être. To the contrary, our feeling is that more than ever, piscine systems offer unique perspectives and mechanistic advances. For this reason, we approached a number of authorities to contribute to this new volume.

Environmental Toxicology is an unwanted fact of life, with a large anthropogenic component. As standards of living continue to rise worldwide, industrial activities tend to rise in concert. Although by no means unpreventable, a past and current consequence is a continuous input of xenobiotic substances into our waterways. Fish species are numerous and found in most aquatic environments and as such we might expect a fair number of these species to demonstrate unique adjustments to anthropogenic chemicals found in their environment.

The goal of this volume is to capture some of the more important issues affecting the responses of fish to the chemical soup of their environment. Chapters included in this volume identify the systems found in fish to deal with xenobiotics, some of the hormonal interactions initiated in the presence of these chemicals, the unique mechanisms used by fish to adjust to the present chemicals and the new and evolving mixtures of chemicals in their environment. Finally, our authors touch upon the new methods being applied in fish systems to understand the effects of xenobiotics to fish fitness – a key theme in environmental health and crucial to the future of fish populations.

This volume is not just the next book in this series, but we look at it as a new beginning. Future volumes will cover both the obvious and the controversial, in our continued attempt to understand the diversity and unity of fish at molecular and biochemical levels of organization.

Thomas W. Moon, Ottawa
Thomas P. Mommsen, Victoria

Contributors

D. Alsop, *Department of Zoology, University of Guelph, Guelph, ON, Canada N1G 2W1 (Chapter 15),* dalsop@uoguelph.ca

N.C. Bols, *Department of Biology, University of Waterloo, Waterloo, ON, Canada N2L 3G1 (Chapter 2)*, ncbols@sciborg.uwaterloo.ca

Adrienne N. Boone, *Department of Biology, University of Waterloo, Waterloo, ON, Canada N2L 3G1 (Chapter 13)*

S.B. Brown, *Environment Canada, P.O. Box 5050, Burlington, ON, Canada L7R 4A6 (Chapters 14, 15)*, scott.brown@ec.gc.ca

Karen G. Burnett, *Grice Marine Laboratory, The College of Charleston, Charleston, SC 29412, USA (Chapter 8)*, burnettk@cofc.edu

Michael J. Carvan, *Great Lakes WATER Institute and Marine and Freshwater Biomedical Sciences Center, University of Wisconsin-Milwaukee, Milwaukee, WI 53204, USA (Chapter 1)*, carvanmj@uwm.edu

Ann O. Cheek, *Department of Biological Sciences, Southeastern Louisiana University, Box 10736, Hammond, LA 70402, USA (Chapter 16)*, acheek@selu.edu

Tracy K. Collier, *NOAA/NMFS, Northwest Fisheries Science Center, 2725 Montlake Blvd. E., Seattle, WA 98112, USA (Chapter 4)*, tracy.k.collier@noaa.gov

Jean-Pierre Cravedi, *UMR 1089 Xénobiotiques, INRA, BP3 31931 Toulouse Cédex 9, France (Chapter 5)*, jcravedi@toulouse.inra.fr

V.R. Dayeh, *Department of Biology, University of Waterloo, Waterloo, ON, Canada N2L 3G1 (Chapter 2)*

Nancy D. Denslow, *Department of Physiological Sciences and Center for Environmental and Human Toxicology, University of Florida, P.O. Box 10885, Gainesville, FL 32611, USA (Chapter 3)*, denslow@biotech.ufl.edu

J.G. Eales, *Department of Zoology, University of Manitoba, Winnipeg, MB, Canada R3T 3B8 (Chapters 14, 15)*, ealesjg@ms.UManitoba.ca

Adria A. Elskus, *USGS, Department of Biological Sciences, 5751 Murray Hall, University of Maine, Orono, ME 04469-5751, USA (Chapter 4)*, aelskus@pop.uky.edu

Mark E. Hahn, *Biology Department, Woods Hole Oceanographic Institution, Woods Hole, MA 02543 USA (Chapter 7)*, mhahn@whoi.edu

Akihiko Hara, *Division of Marine Biosciences, Graduate School of Fisheries Sciences, Hokkaido University, Hakodate, Hokkaido, 041-8611, Japan (Chapter 16)*, ah011@pop.fish.hokudai.ac.jp

Tisha King Heiden, *Department of Biological Sciences, University of Wisconsin-Milwaukee, Milwaukee, WI 53204, USA (Chapter 1)*

Naoshi Hiramatsu, *Department of Zoology, North Carolina State University, Box 7617, Raleigh, NC 27695-7617, USA (Chapter 16)*, naoshi_hiramatsu@ncsu.edu

Alice Hontela, *Department of Biological Sciences, Water Institute for Semi-arid Ecosystems (WISE), University of Lethbridge, Lethbridge, AB, Canada T1K 3M4 (Chapter 12)*, alice.hontela@uleth.ca

David M. Janz, *Department of Veterinary Biomedical Sciences, University of Saskatchewan, Saskatoon, SK, S7N 5B4, Canada (Chapter 11)*, David.Janz@usask.ca

Sibel I. Karchner, *Biology Department, Woods Hole Oceanographic Institution, Woods Hole, MA, 02543 USA (Chapter 7)*, skarchner@whoi.edu

Peter Kling, *Department of Zoology, University of Göteborg SE-405 30, Göteborg, Sweden (Chapter 10)*, peter.kling@zool.gu.se

Iris Knoebl, *Molecular Ecology Research Branch, EERD/NERL/US EPA, Cincinnati, OH 45268, USA (Chapter 3)*, knoebl.iris@epamail.epa.gov

Patrick Larkin, *EcoArray LLC, Alachua, FL 32615, USA (Chapter 3)*

L.E.J. Lee, *Department of Biology, Wilfrid Laurier University, Waterloo, ON, Canada N2L 3C5 (Chapter 2)*, llee@wlu.ca

Takahiro Matsubara, *Hokkaido National Fisheries Research Institute, 116, Katsurakoi, Kushiro, Hokkaido 085-0802, Japan (Chapter 16)*, sadachan@hnf.affrc.go.jp

Rebeka R. Merson, *Biology Department, Woods Hole Oceanographic Institution, Woods Hole, MA, 02543 USA (Chapter 7)*

Chris D. Metcalfe, *Environmental & Resource Studies Program, Trent University, 1600 West Bank Drive, Peterborough, ON, Canada K9J 7B8 (Chapter 17)*, cmetcalfe@trentu.ca

Caroline Mimeault, *Centre for Advanced Research in Environmental Genomics and Department of Biology, University of Ottawa, Ottawa, ON, Canada K1N 6N5 (Chapter 17)*, cmineaul@science.uottawa.ca

Emily Monosson, *PO Box 329, Montague, MA 01351, USA (Chapter 4)*, emonosson@forwild.umass.edu

Thomas W. Moon, *Centre for Advanced Research in Environmental Genomics and Department of Biology, University of Ottawa, Ottawa, ON, Canada K1N 6N5 (Chapter 17)*, tmoon@science.uottawa.ca

Per-Erik Olsson, *Department of Natural Sciences, Section of Biology, University of Örebro SE-701 82, Örebro, Sweden (Chapter 10)*, per-erik.olsson@nat.oru.se

Gary K. Ostrander, *Departments of Biology and Comparative Medicine, Johns Hopkins University, 3400 North Charles Street, Baltimore, MD 21218, USA (Chapter 9)*, gofish@jhu.edu

Patrick Prunet, *INRA/SCRIBE, Group on Fish Physiology of Adaptation and Stress, Campus de Beaulieu, Rennes Cedex, France (Chapter 13)*, prunet@beaulieu.rennes.inra.fr

Jeanette M. Rotchell, *Centre for Environmental Research, School of Chemistry, Physics and Environmental Sciences, University of Sussex, Falmer, Brighton BN1 9QJ, UK (Chapter 9)*, J.Rotchell@sussex.ac.uk

K. Schirmer, *Junior Research Group – Molecular Animal Cell Toxicology, UFZ-Centre for Environmental Research Leipzig-Halle, 04318 Leipzig, Germany (Chapter 2)*, kristin.schirmer@uoe.ufz.de

Daniel Schlenk, *Department of Environmental Sciences, University of California, Riverside, Riverside, CA 92521, USA (Chapter 6)*, daniel.schlenk@ucr.edu

Helmut Segner, *Centre for Fish and Wildlife Health, University of Bern, Post Box, CH-3001 Bern, Switzerland (Chapter 18)*, helmut.segner@itpa.unibe.ch

Armin Sturm, *King's College London, Division of Life Sciences, Franklin-Wilkins Building, 150 Stamford Street, London SE1 9NN, UK (Chapter 18)*, armin.sturm@kcl.ac.uk

Craig V. Sullivan, *Department of Zoology, North Carolina State University, Box 7617, Raleigh, NC 27695-7617, USA (Chapter 16)*, sull0263@pop-in.ncsu.edu

Henry Tomasiewicz, *Great Lakes WATER Institute and Marine and Freshwater Biomedical Sciences Center, University of Wisconsin-Milwaukee, Milwaukee, WI 53204, USA (Chapter 1)*

Vance L. Trudeau, *Centre for Advanced Research in Environmental Genomics and Department of Biology, University of Ottawa, Ottawa, ON, Canada K1N 6N5 (Chapter 17)*, vtrudeau@science.uottawa.ca

Glen J. Van Der Kraak, *Department of Zoology, University of Guelph, Guelph, ON, Canada N1G 2W1 (Chapters 11, 15)*, gvanderk@uoguelph.ca

Mathilakath M. Vijayan, *Department of Biology, University of Waterloo, Waterloo, ON, Canada N2L 3G1 (Chapter 13)*, mvijayan@sciborg.uwaterloo.ca

Antony W. Wood, *Vincent Center for Reproductive Biology, Massachusetts General Hospital, Harvard Medical School, Bldg 149 13th Street – Room 6607, Charlestown, MA 02129, USA (Chapter 11)*, AWood@partners.org

Daniel Zalko, *UMR 1089 Xénobiotiques, INRA, BP3 31931 Toulouse Cédex 9, France (Chapter 5)*, dzalko@toulouse.inra.fr

Contents

Abbreviations

3MC	3-methylcholanthrene
ACTH	adrenocorticotropin hormone
AFB1	aflatoxinB1
AFC	antibody-forming cell
AHH	aryl hydrocarbon hydroxylase
AHR	aryl hydrocarbon (Ah) receptor
AHRE	AHR-responsive enhancer (= DRE, XRE)
AHRR	AHR repressor
AIF	apoptosis-inducing factor
ANF	α-naphthoflavone
Apaf-1	apoptotic protease-activating factor-1
AR	androgen receptor
ARE	anti-oxidant response element (= EpRE)
ARNT	Ah receptor nuclear translocator
B*a*P	benzo[*a*]pyrene
Bcl-2	B cell lymphoma-2
BH	Bcl-2 homology
bHLH-PAS	basic helix–loop–helix Per-ARNT-SIM
BITEX	benzene, toluenes, ethylbenzene and xylenes
BNF	β-naphthoflavone
bZIP	basic-leucine zipper
CAD	caspase-activated DNase
CAR	constitutive androstane receptor
CBG	corticosteroid binding globulin
Chgs	choriogenins
CL	chemiluminescence
CNC	cap'n'collar
ConA	concanavalin A
COX	cyclooxygenase
CR	corticosteroid receptor
CRH	corticotrophin releasing hormone
CRP	C-reactive protein
CS	corticosteroid
CTL	cytotoxic T lymphocytes
CYP1A1	cytochrome P450 1A1
DEN	diethylnitrosamine
DHEA	dehydroepiandrosterone
DMBA	dimethylbenzanthracene
DMNA	dimethylnitrosamine
DNF	damselfish neurofibromatosis
DR	degradation rates
DRE	dioxin responsive enhancer (= AHRE, XRE)
ECOD	ethoxycoumarin *O*-deethylase
EDC	endocrine disrupting chemical
EE	17α-ethinylestradiol
EGF	epidermal growth factor
EpRE	electrophile response element (= ARE)
ER	estrogen receptor
ERA	environmental risk assessment
EROD	ethoxyresorufin *O*-deethylase
FAA	free amino acids
FADD	Fas-associated death domain
FasL	Fas ligand
FMO	flavin monooxygenase
γ-GCS	γ-glutamylcysteine synthase
GFP	green fluorescent protein
GnRH	gonadotropin-releasing hormone
GR	glucocorticoid receptor
GRE	glucocorticoid response element
GSH	reduced glutathione
GST	glutathione *S*-transferase
HCH	hexachlorohexane
HIF	hypoxia-inducible factor
HO	heme oxygenase
HPA	hypothalamus–pituitary–adrenal
HPI	hypothalamus–pituitary–interrenal
HSE	heat shock element
HSF	heat shock factor
HSP	heat shock protein
ICE	interleukin-1β-converting enzyme
ICP-MS	inductively coupled plasma mass spectrometry
IFN	interferon
IL	interleukin
iNOS	inducible nitric oxide
Keap1	Kelch-like-ECH-associated protein
K_{ow}	octanol:water partitioning coefficient
LOEC	lowest observed effective concentration
LPS	lipopolysaccaride
MAMAc	methylazoxymethanol acetate

MAPK	mitogen-activated protein kinase
MET	methoprene-tolerant
MFO	mixed function oxidase
MHC	major histocompatibility protein
MIH	maturation inducing hormone
MMC	melanomacrophage centers
MNNG	*N*-methyl-*N′*-nitro-*N*-nitrosoguanidine
MO	morpholino-modified oligonucleotide
MRE	metal response element
MT	metallothionein
MTF-1	MRE binding transcription factor
MTI	metallothionein inhibitor
MTT	3-[4,5-dimethylthiazol-2-yl]-2, 5-diphenyltetrazolium bromide
N_3Br_2DD	2-azido-3-iodo-7,8-dibromodibenzo-*p*-dioxin
NCC	non-specific cytotoxic cells
NCoA	nuclear receptor co-activator (= steroid receptor co-activator)
NFAT	nuclear factor of activated T cells
NF-E2	nuclear factor erythroid
NK	natural killer
NLS	nuclear localization signal
NM	nitrosomorpholine
NOEL	no observed effect level
NR	nuclear receptor
Nramp1	natural resistance-associated macrophage protein
OC	organochlorine
OP	organophosphate
OSR	oxidative stress response
PAHs	polycyclic aromatic hydrocarbons
PCB	polychlorinated biphenyl
PCNA	proliferating cell nuclear antigen
P_{CO_2}	carbon dioxide pressure
PCP	pentachlorophenol
PER	period
PHA	phytohemagglutinin
PHAH	planar halogenated aromatic hydrocarbons
P_{O_2}	oxygen pressure
PNEC	predicted no effect concentration
PPAR	peroxisome proliferator-activated receptor
PXR	pregnane X receptor
QSAR	quantitative structure–function relationships
RAPD	random amplified polymorphic DNA
RFLP	restriction fragment length polymorphism
ROS	reactive oxygen species
RT	reverse transcriptase
RT-PCR	reverse transcription-polymerase chain reaction
RXR	retinoid X receptor
SAP	serum amyloid protein
SAR	structure–function relationships
SIM	single-minded
SOD	superoxide dismutase
StAR	steroidogenic acute regulatory protein
TBHQ	*tert*-butylhydroquinone
TBT	tributyl tin
TCB	tetrachlorobiphenyl
TCDD	2,3,7,8-tetrachlorodibenzo-*p*-dioxin
TCE	trichloroethylene
TCIN	chlorothalonil
TCPOBOP	1,4-bis[2-(3,5-dichloropyridyloxy)] benzene
TCR	T cell receptor
T_h	T helper cells
TH	thyroid hormone
TMD	transmembrane domain
TNF	tumour necrosis factor
TRADD	tumour necrosis factor associated death domain
TRAIL	tumour necrosis factor associated apoptosis-inducing ligand
TSH	thyroid stimulating hormone
TUNEL	terminal deoxynucleotidyl-transferase nick end-labelling
UGT	uridine diphosphate glucuronosyl-transferase
UV	ultraviolet
Vg	vitellogenin
XRE	xenobiotic responsive enhancer (= DRE, AHRE)
zVAD-fmk	benzyloxycarbonyl-Val-Ala-Asp-(OMe)-fluoromethylketone

General–Models and Techniques

Biochemistry and Molecular Biology of Fishes, vol. 6
T. P. Mommsen and T. W. Moon (Editors)

CHAPTER 1

The utility of zebrafish as a model for toxicological research

MICHAEL J. CARVAN III*, TISHA KING HEIDEN** AND HENRY TOMASIEWICZ*

**Great Lakes WATER Institute and Marine and Freshwater Biomedical Sciences Center, University of Wisconsin-Milwaukee, Milwaukee, WI 53204, USA, and **Department of Biological Sciences, University of Wisconsin-Milwaukee, Milwaukee, WI 53204, USA*

I. Introduction

Zebrafish (*Danio rerio*) have long been the genetic model of choice for vertebrate developmental biologists, as it provides several advantages for investigating organ and tissue development not available through other model systems. Therefore, it makes logical sense that zebrafish have become a powerful model organism for investigating the molecular and cellular mechanisms by which environmental chemicals disrupt normal developmental processes. The utility of zebrafish as a laboratory model organism also makes it an excellent system for studying processes in juvenile and adult organisms, including reproductive development, carcinogenesis, aging, and the influence of environmental chemicals on these processes. The advent of modern genetic tools and genome sequencing projects has elevated zebrafish as a suitable model to effectively study human disease and pathophysiology, ushering in a new era of comparative biology and medicine.

1. Historical perspective

Zebrafish have historically been the focal point of many investigations into the teratogenic effects of toxicants and the mechanisms underlying cellular responses to toxic chemicals. This work would not be possible without the early work characterizing zebrafish husbandry methods, embryonic development, and genetic manipulation. Zebrafish readily spawn in the laboratory, reach juvenile stage within a few weeks, reach sexual maturity within 3–4 months, produce a large number of offspring that are easy to manipulate and observe throughout embryonic development, and are relatively easy to maintain under laboratory conditions. Husbandry techniques for producing zebrafish eggs for use in embryological research were available as early as 1934[52]. Pioneering work by Roosen-Runge[183] and Lewis and Roosen-Runge[139] describe embryonic development after fertilization, characterizing the early blastodisc stages, cleavage, and differentiation. Another 20 years passed before scientists were able to visualize zebrafish embryogenesis, as embryonic development was fully characterized into comprehensive stages by Hisaoka and Battle[92], which has been succeeded by Kimmel *et al.*[114]. Shortly thereafter, stained and sectioned material became available from selected embryonic stages[93]. In 1970, Snoek[195] characterized the effects of temperature and light on zebrafish embryonic and larval development, providing additional information regarding normal and perturbed development, and the influence of environmental conditions on zebrafish development. Laale[128] provided an invaluable contribution by outlining zebrafish taxonomy and characterizing the general husbandry procedures necessary for using zebrafish for research, which was later expanded by Westerfield[235]. These pioneering scientists provided excellent descriptions of normal embryonic development, and laid the foundation for future investigations of the effects of chemicals on development.

These early studies advanced the use of zebrafish in specialized disciplines such as reproductive biology, ethology, teratology, environmental biology, and genetics. In the following, we will discuss a number of studies that used zebrafish for mechanistic studies in chemical-induced teratogenesis and toxicological research. This is not intended to provide an unabridged history of zebrafish research, as there exists a number of recent reviews describing the utility of zebrafish as a model system[6,29,78,161,197,241,247].

2. *General toxicity screening*

Early toxicological studies utilizing zebrafish were primarily designed to identify potentially harmful substances, to test chemicals for their ability to inhibit cell division, and to delineate the physiological effects of teratogens. Jones *et al.*[104] were among the first to demonstrate both the abilities of thyroxine, thiouracil, and podophyllotoxin to disrupt mitosis, and the utility of the zebrafish embryo for cell cycle experiments. Later, Jones and Huffman[103] were to expand these studies and characterize the immediate effects of several chemicals on mitosis, including estradiol, estriol, diethylstilbestrol, water-soluble and lipid-soluble fractions of extract from an unidentified plant, and equilenin. They exploited the fact that one can directly observe zebrafish embryo development from the earliest stages (one or two cells), and were thus able to carry out rapid toxicity screens and determine the lowest observed effect levels (LOELs) and LC_{egg50} values (concentration of the chemical in water that kills 50% of the eggs in a given time). Battle and Hisaoka[16] were able to demonstrate that exposure of zebrafish embryos to urethane (ethyl carbamate, a known carcinogen) correlates with abnormal development (i.e. disruption in the differentiation of nervous and muscular tissue and eye defects) and that these effects were both time and dose dependent. At the time, studies using invertebrate and amphibian model systems and cultured cell lines had proven inadequate to characterize the effects of urethane on mitosis and cellular differentiation[224]. Hisaoka also used zebrafish embryos to demonstrate that low concentrations of barbiturates delay development, while exposure to high concentrations results in death[94], and to determine the deleterious effects of 2-acetylaminofluorene on development[95].

In the 1960s and 1970s, the efficacy of the zebrafish model system became clear and there was an increase in the number of studies that used zebrafish as bioassays for chemicals that affect normal development and reproductive success. Many focused on the effects of toxicants on early life stages (ELS) in short-term toxicology tests including: zinc sulfate, an industrial waste product[194]; Captan, an agricultural fungicide[1]; and chloramphenicol, a protein synthesis inhibitor[8]. Reproductive success in zebrafish is affected by sublethal exposures to a number of chemicals including: zinc[26,196], mercury[113], cadmium[64,182], selenium[165], and other pollutants[200]. Maternal exposure to certain toxicants directly alters embryonic development via maternal transfer[196], while other toxicants indirectly affect embryonic development by altering parental metabolism[182].

Researchers also used zebrafish to investigate the effects of long-term exposures to xenobiotics on the reproductive system of fishes. Bresch[28] describes a simple method to determine the effect of a substance on reproductive success over several successive cycles in a relatively short period of time, using a combination of acute toxicity tests (96 h pre-tests) to estimate the LC_{50} and embryo–larva survival tests (6-day tests beginning 24 h post-fertilization) to test the effects of several compounds (mercury, copper, cadmium, lead, zinc, potassium dichromate, pentachlorophenol, alachlor, and 4-nitrophenol). Bresch[28] also finds that reproductive tests (adults are exposed to compounds for several days while spawning, using cadmium as an example) in combination with short embryo–larva assays provide more information than extended embryo–larval tests alone, and could replace full life-cycle studies[28]. Landner and colleagues[130] used zebrafish both to determine acute toxic responses of adult fish to xenobiotics and to investigate the adverse transgenerational effects of complex environmental mixtures on embryonic development when the adults were exposed during gametogenesis. In this study, alterations in embryonic development were measured (i.e. survival, hatching success, hatching rate, and stress tolerance of embryos and larvae) following exposure of adult fish to raw pulp mill effluent and bleached kraft mill effluent prior to, but not during, spawning. Their study demonstrates that complete life-cycle assays are not always necessary in the initial stages of investigating toxic effects of chemicals and provides an efficient protocol for maternal exposure in conjunction with embryonic and larval assays for investigating toxic responses to xenobiotics.

3. Carcinogenicity screening

Stanton[202] demonstrated the ability of diethylnitrosamine to induce tumors in zebrafish and provided one of the earliest studies using zebrafish to test and characterize the carcinogenicity of chemical compounds. Early work by Streisinger and co-workers was meant to increase the utility of zebrafish for investigating the effects of genotoxic agents by producing homozygous zebrafish clones to reduce variability associated with heterozygosity[79,80,205,206]. Using these homozygous strains of zebrafish, Streisinger[207] developed a rapid test for genotoxicity (based on studies with mice) to efficiently compare the carcinogenic and genotoxic effects of environmental agents. These strains enabled large-scale animal investigations of the dose–response relationships of potential carcinogens and their ability to induce somatic mosaicism and germline recessive-lethal mutations. At the same time, other laboratories were utilizing the high-throughput capabilities of zebrafish to characterize the carcinogenicity of multiple compounds, including dimethylnitrosamine, diethylnitrosamine, and nitrosomorpholine, and quantify the incidence and morphology of neoplasms, the latency and dynamics of tumor formation, and the influence of age, dose, and environmental temperature on carcinogenesis[111,112,175].

Ethylnitrosourea, the chemical most commonly used in zebrafish to induce mutations for genetic screens, has also been shown to induce epidermal neoplasia[19]. The work of Spitsbergen *et al.*[198,199] reports neoplasia induced in zebrafish treated with 7,12-dimethylbenz[*a*]anthracene or *N*-methyl-*N*′-nitro-*N*-nitrosoguanidine at different developmental stages by multiple exposure routes. Sensitivity varies between the different developmental stages, but zebrafish are a responsive, cost-effective vertebrate model system to study the mechanisms of carcinogenesis[198,199]. The same genetic and genomic tools that have been the foundation of zebrafish developmental biology can be tailored to provide a more complete understanding of the genes that regulate susceptibility to specific tumors, their progression toward metastasis, and the environmental factors that influence these processes.

4. *Comparative toxicology assay system*

While many scientists had already realized the potential for zebrafish as a toxicological model, many remained skeptical. In response to that skepticism, several papers emerged during the 1980s and 1990s that compared the usefulness of zebrafish to other model systems and confirmed that zebrafish are a practical and efficient choice for general toxicology screens, ELS toxicity, and lifetime exposure studies. Dave *et al.*[57] report the toxicity of eight extraction reagents used for the recovery and refinement of chemicals in the nuclear, petrochemical, and pharmaceutical industries. They used *Daphnia*, trout, and zebrafish model systems, and conclude that zebrafish make the most practical choice for toxicology screens. A review by Van Leeuwen *et al.*[223] compare the use of zebrafish, rainbow trout, and white Leghorn chickens for use in ELS toxicity tests for 10 chemicals used in the rubber industry. They describe similar patterns in teratogenic responses to the chemicals and EC_{50} values were similar between the two fish species, and both correlated with that of the white Leghorn. Zebrafish proved to be a good system for ELS toxicity tests because large numbers of embryos were available, year-round, to generate well-supported concentration–response curves. Transparent eggs make observation of the rapid embryonic development feasible allowing for 7-day ELS assays compared to the 60-day ELS assay required for trout. Van Leeuwen *et al.*[223] find strong correlations between zebrafish and other model organisms, such as birds and mammals, indicating that the zebrafish system can be used to make inferences to other species.

A comparative toxicology study by Neilson *et al.*[164] describes the toxic response of several aquatic organisms, including zebrafish, *Daphnia*, copepods, and duckweed, to 4,5,6 trichloroguicol and 3,4,5 trichlorophenol (halogenated phenolic compounds). They find zebrafish to be a particularly amenable system for toxicology studies, while remaining environmentally relevant. Another study by Braunbeck *et al.*[27] compares the toxic responses of rainbow trout and zebrafish following long-term exposures (3 months) to atrazine and finds that zebrafish are less sensitive to the toxic effects of atrazine, making it a good system for continuous and long-term exposure studies. Mizell and Romig[156] report that the zebrafish embryo is particularly effective for investigating the effects of single chemicals and mixtures of chemical toxins on embryonic development and find zebrafish to be effective sentinels for low concentrations of aquatic pollutant mixtures.

Zebrafish are rapidly becoming a mature laboratory-based model system for the toxicological analysis of environmental samples. Murk *et al.*[160] use the zebrafish 8-day ELS test as an *in vivo* validation for the high-throughput CALUX assay, which utilizes rat hepatoma cells transfected with a dioxin-responsive luciferase reporter gene. The results of the zebrafish assay are comparable with those of the CALUX assay following exposure to 2,3,7,8-tetrachlorodibenzo-*p*-dioxin (TCDD); however, the relationship did not transfer to environmental pore water samples where the CALUX assay was specifically responsive to aromatic hydrocarbons, especially those structurally similar TCDD. In contrast, the zebrafish assay is also highly responsive to the heavy metal pollutants Cd, Cu, Zn from the pore water samples and their toxic effects on mortality and malformations. Nagel[162] also discusses the utility of a 48 h zebrafish embryo-based toxicity assay system, DarT, as a substitute for acute fish toxicity tests. Toxicity of a given sample is determined based on a series of criterion with lethal, developmental, or teratogenic endpoints. An international inter-laboratory test validated the repeatability of the DarT assay, which proved to be consistent as long as the laboratories use the same strain of zebrafish. Further discussions and supportive preliminary data on the use of DarT for routine waste water control, quantitative structure–activity relationships, analysis of drug effects on heart beat and pigmentation, and teratogenicity testing[162,173] provide ample evidence for the further development of DarT (and other zebrafish-based assays) for high-throughput screening of municipal and industrial effluent, environmental samples, and preclinical pharmaceuticals for both environmental and medical risk assessment.

'Chemical genetics' exploits the combined power of large and diverse chemical libraries produced through combinatorial chemistry and high-throughput biological assays to carefully dissect biological pathways and explore gene product function. Successful application of chemical genetics depends on the identification of individual compounds that bind to a single protein (or protein family) with high affinity in the target cell or tissue. Peterson *et al.*[173] adopted a 'chemical genetics' approach in zebrafish to gain an understanding of developmental processes. The combinatorial library included 1100 compounds to which zebrafish were exposed in a 96-well format. Developmental defects in the central nervous system, cardiovascular system, pigmentation, and ear were assessed visually with a dissecting microscope. A number of molecules were identified (about 1%) that specifically disrupt the systems examined and warranted further analysis. One of the compounds disrupts normal cardiac development with an ED50 of about 2 nM and does not interfere with other developmental processes at concentrations up to 6 μM[173]. This provides an excellent example of both the specificity of these small molecules and the utility of zebrafish for high-throughput screening – offering access to the whole developing embryo in a manner unmatched by other vertebrate genetic model systems.

II. Mechanistic toxicology

These early studies established the zebrafish as an efficient and reliable system to investigate the toxic effects of chemicals. Researchers continue to use zebrafish to investigate ELS toxicity[67,69,89,223], chronic and lifetime exposure studies[68], and effects of chemicals on reproductive success[168,232]. It is important to remember that oviparous fish, like zebrafish, provide a powerful system for studying interactions between different signaling pathways and their role in the developmental toxicity of chemicals, which cannot be investigated using the mammalian systems due to complex maternal–embryonic interactions. Scientists currently utilizing the zebrafish model system have an array of biochemical and molecular tools at their disposal to investigate the mechanisms by which environmental chemicals cause their toxic effects. We will discuss a few of these chemicals and their mechanisms of toxicity primarily focusing on those classes of environmental chemicals that activate receptor-mediated pathways directly implicated in the mechanisms of toxicity and/or in the activation of detoxification pathways. This receptor dependence is characterized by a latency of response as the newly generated gene products accumulate in the cell to provide the appropriate response.

During the remainder of this chapter, we will discuss a number of genes and/or gene products, which will be identified using standard nomenclature (e.g. human: *AHR1* for the gene, AHR1 for the gene products – mRNA and/or protein, mouse: *Ahr* and Ahr). When specific zebrafish genes and/or gene products are discussed, we will utilize the zebrafish gene nomenclature as much as possible (e.g. *ahr1* and Ahr1, respectively). When discussing multiple species, we default to zebrafish terminology.

1. Estrogen response element

Efforts to increase agricultural productivity through widespread use of insecticides, herbicides, fungicides, and pesticides, in addition to by-products of modern manufacturing processes, have resulted in the release of substantial amounts of toxicants into the environment. Many of these toxicants disturb the development of the endocrine system and the organs that respond to endocrine signals. These compounds are classified as endocrine-disrupting chemicals (EDCs) and are defined as exogenous agents that interfere with the synthesis, storage/release, transport, metabolism, binding action, or elimination of natural blood-borne hormones[106]. Wildlife in aquatic ecosystems is at greater risk from EDCs because these systems act as repositories for such chemicals. EDCs tend to readily bioaccumulate in aquatic species and have been shown to disrupt sexual development, behavior, and fertility as well as alter normal thyroid and immune function in fishes. While physiologic effects of EDCs range from morphologic anomalies to impaired immune systems to behavioral modifications, most investigations of EDCs focus on their effects on the reproductive success of fishes[147,172,208,209,210,244]. There are a number of contributions in this volume that will more fully discuss the effects of EDCs on fishes. We will very briefly discuss research conducted in zebrafish that focuses on the EDCs that impact estrogen receptor function and signaling pathways.

Estrogens are a class of chemicals, both natural and synthetic, that function at the cellular level through a common signaling pathway (Fig. 1). Estrogen is an ovarian steroid hormone that primarily functions in early embryonic development and in establishing and maintaining the female reproductive system. The most potent naturally occurring estrogen is 17β-estradiol (E2). In male and female fish, E2 is important for the coordination of developmental, physiological, and behavioral responses that are essential for reproduction. E2 acts as a ligand for two nuclear estrogen receptors (ERα (ESR1) and ERβ (ESR2)), transcription factors that, once bound to the ligand, induce gene transcription through interaction with an estrogen-responsive DNA element (ERE) (Fig. 2). In the estrogen receptor (ER) knock-out mouse models, alpha ERKO and beta ERKO, the ER (primarily ERα) mediates most of the physiological effects of E2[51,118,119]. In fish, E2 is associated primarily with vitellogenesis and ovarian development. Under the direction of GtHI/FSH, E2 stimulates the liver to produce the yolk protein vitellogenin (Vg), which is sequestered

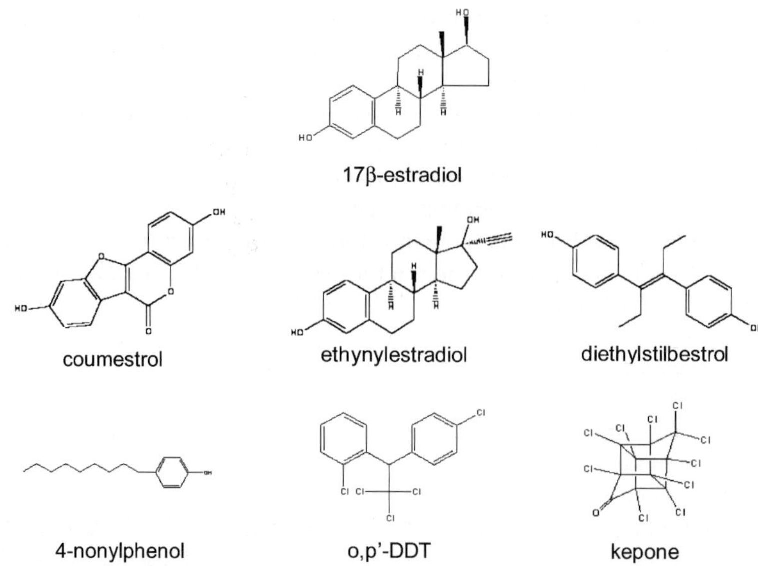

Fig. 1. Estrogen and estrogen mimics. Many natural plant compounds (e.g. coumestrol) and synthetic chemicals (e.g. ethynylestradiol, diethylstilbestrol, 4-nonylphenol, *o,p'* DDT, and kepone) act as endocrine disruptors by mimicking or blocking normal estrogen function via estrogen receptor dependent or independent mechanisms.

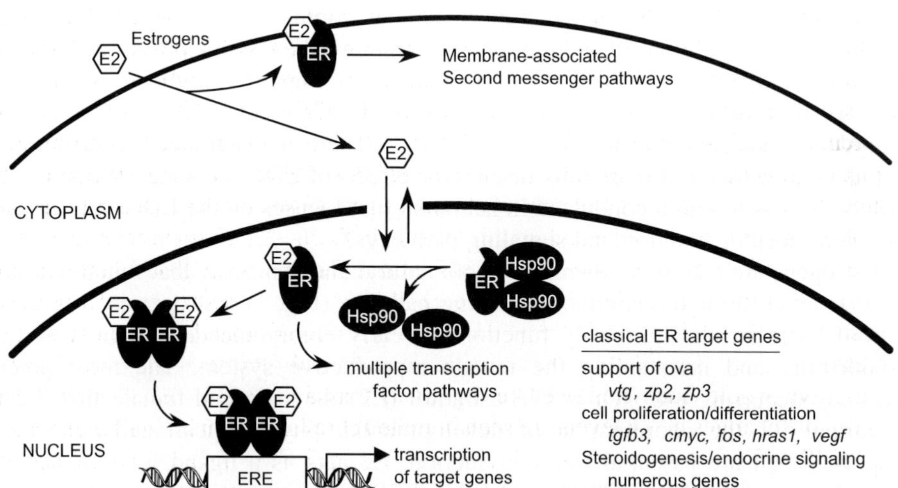

Fig. 2. General estrogen receptor signaling pathway. Binding of a ligand to the ER induces allosteric changes in the receptor, leading to the dissociation of heat-shock proteins and the homodimerazation of ER. The ligand–receptor complex then forms a transcriptional complex at the ERE in the regulatory region of a target gene, inducing gene transcription. Basal transcription factors, including co-activators and co-repressors, and proteins that regulate chromatin remodeling, help regulate target gene expression. Additionally, cell-membrane bound ER can also activate non-nuclear signaling pathways. Black shapes indicate gene products that have been isolated in zebrafish. Abbreviations are given in the text. Hsp90, heat shock protein, 90 kDa.

by the oocyte during oogenesis. As in other vertebrates, E2 also functions to mediate piscine embryonic development, age to maturation, development of secondary sex characteristics, behavior, and immune responses[181,234].

Environmental contaminants that mimic E2 are the best studied of all the endocrine disruptors. The estrogenic compounds responsible for such alterations are believed to be steroids (both natural and synthetic), as well as other compounds previously shown to be ER agonists such as nonylphenol, octylphenol (OP), phthalates, and pesticides such as hexachlorocyclohexane and dieldrin. While these chemicals tend to have similar biochemical activity as estradiol and are known to alter development and sex differentiation in vertebrates by either ER-mediated or ER-independent mechanisms[76], the risks associated with constant low-level exposures are not completely understood. The role of E2 in maintaining reproductive health and regulating development, as well as its effects on growth, metabolism, and immunity, make determination of the biological effects of chronic, low-dose environmental estrogen exposure of particular concern.

Zebrafish have been shown to be sensitive to a number of chemicals that disrupt ER function, including ethynylestradiol (EE2), 4-nonylphenol (NP), OP, and bisphenol A (BPA). In zebrafish embryos, E2 and the estrogen mimics BPA and diethylstilbestrol (DES) influence mortality, hatching, and morphogenesis, in addition to inducing the expression of Cyp19b mRNA (the predominant brain aromatase), but not Cyp19a (the predominant ovarian aromatase) as determined by reverse transcriptase-polymerase chain reaction (RT-PCR)/Southern analysis[116]. In adults, EE2 exposure results in a dose-related reduction in the fertility of both male and female zebrafish, causing significant changes in the gonadosomatic indices and induction of plasma Vg, while OP only has minor effects on gonadosomatic indices[221]. In a follow-up study, Van den Belt *et al.*[222] demonstrate that while the effects of EE2 exposure are dramatic, they are reversible. Exposure to low concentrations of EE2 and NP during sexual differentiation (2–60 days post hatch) induces a concentration-dependent suppression of gametogenesis, skews sex ratios towards females, induces a small percentage of ovotestes in males, causes concentration-dependent kidney, ovarian, and testicular pathology, and induces vitellogenin production in both male and female zebrafish[91,233]. While most of these effects on sexual differentiation and gonad development are reversible, effects on ovarian follicle atresia are latent and persistent. Additionally, exposure to estrogenic chemicals only during the period of sexual differentiation can decrease offspring survival, as well as hatching and swim-up success[91,233]. Using a quantitative whole body homogenate Vg ELISA assay, Rose *et al.*[184] report that quantifiable changes in Vg levels are seen in male zebrafish within 24 h after exposure to E2 (249 ng/l), and increase another 100-fold by day 5. Zebrafish are more sensitive than other small laboratory fish species (medaka, fathead minnow and sheepshead minnow) with an EC50 of 41.2 ng/l for E2 and 2.5 ng/l for EE2[184]. Significant changes in Vg protein levels result from EE2 exposures as low as 1.67 ng/l[71]. Islinger *et al.*[100] identify three different Vg transcripts and demonstrate the sensitivity of zebrafish to EE2 using a semiquantitative RT-PCR assay to measure the induction of two different Vg gene products in male zebrafish livers confirming that increases in Vg protein are the result of increased transcription, which is likely driven by the ER.

Three estrogen receptor cDNAs have been identified from zebrafish: Esr1(ERα), Esr2a (ERβ1/ERβb), and Esr2b (ERβ2/ERβa)[13,132,151], with the *in vitro* translated proteins possessing high affinity binding constants (K_d) of 0.74, 0.75, and 0.42 nM, respectively. The differences in K_d translate to functional differences in heterologous expression systems as demonstrated by the fact that when expressed in CHO cells, Esr2b is able to activate an estrogen-responsive reporter gene at 10^{-11} M E2, compared with 5×10^{-11} for both Esr1 and Esr2a. All three forms are responsive to diethylstilbesterol, estrogen metabolites, and a number of environmental estrogens resulting in the activation of reporter constructs in cultured cells[136,151]. Interestingly, the expression patterns of the three forms of zebrafish ER overlap, being expressed in both reproductive and non-reproductive tissues, predominantly in the gonad, brain, pituitary, and liver[151]. The expression of these ER subtypes has been extensively characterized in the female brain[151]. During development, maternally derived Esr2b is expressed during the first 6 h of development, after which no ER expression is observed until 48 h post fertilization (hpf). After 48 hpf, all three forms are expressed, with Esr1 being the predominant form[13,132]. A number of questions remain regarding the physiological role each ER form plays, the potential for binding of specific estrogens or estrogen mimics, subsequent interactions with specific EREs, and the genes each specific ER form may regulate.

One of the most promising tools for the mechanistic investigation of ER function in fishes is a transgenic zebrafish with a stably integrated luciferase reporter transgene that is regulated by three ERE sequences[136]. This reporter gene is regulated in the same manner as native estrogen-responsive genes, and is very responsive to E2 and EE2, weakly responsive to *o,p*′-DDT (dichlorodiphenyltrichloroethane), and non-responsive to nonylphenol. In comparison, all these chemicals are functional in the ER-CALUX assay in which the same reporter gene has been introduced into human breast cancer cells[135,136]. Since zebrafish ERs show higher transactivation by estrogen mimics relative to E2 than human ER, the incongruence between the *in vivo* and *in vitro* assays likely reflects the differences between whole animal and tissue culture experiments[136,193].

Additionally, endocrine disruptors can also modulate different aspects of steroidogenesis and oogenesis. Several enzymes and proteins involved in both these processes have been characterized in zebrafish, including: steroidogenic acute regulatory protein (StAR)[17], activin[231], 3β-hydroxysteroid dehydrogenase[129], cytochrome P450 side chain cleavage (cyp11a1)[97,129], 20βHSD[230], and aromatase (cyp19a and cyp19b)[31,43,44,110,115,217,220]. The zebrafish model system has great promise to provide insights into the mechanisms that underlie reproductive dysfunction due to exposure to estrogen and estrogen receptor mimics. An in-depth understanding of ER processes in zebrafish will significantly enhance the power of environmental screening tools, like DarT, and provide additional support for its utility in identifying EDCs from environmental samples. Sex differentiation in zebrafish is not strictly genetic and likely has a very strong environmental component[49,96,145]. Analyzing the effects of EDCs on sex differentiation and development in zebrafish could provide insight as to the influence of developmental exposure to EDCs as they may relate to the fetal basis of adult endocrine system dysfunction.

2. *Aromatic hydrocarbon response element*

Aromatic hydrocarbons (AHs; Fig. 3) comprise a diverse group of environmental chemicals, which includes the polycyclic aromatic hydrocarbons (PAHs: e.g. benzo[*a*]pyrene (B[*a*]P), 7,12-dimethylbenzanthrene, *β*-naphthoflavone) and the halogenated aromatic hydrocarbons (HAHs: e.g. dibenzodioxins, dibenzofurans, polychlorinated biphenyls). Within the HAHs, TCDD is often utilized as a 'model' compound and is the most toxic of the aromatic hydrocarbons, although a considerable range exists in sensitivity among different species. The toxicity of specific AHs is determined by a combination of their half-life in the organism and their affinity for the cytoplasmic aromatic hydrocarbon receptor (AHR). The high toxicity of TCDD is primarily due to its stable chemical structure and its high affinity for the AHR[70,152,176,236]. The ligand-activated AHR usually interacts with its dimerization partner, AHR nuclear transporter (ARNT), to form a nuclear complex that induces transcription through interaction with aromatic hydrocarbon responsive DNA elements (AHREs) in the enhancer regions of several genes including *CYP1A1*, *CYP1A2*, and a number of phase II metabolism genes (Fig. 4). Numerous studies suggest that AHR, located in many different tissues, mediates the toxic and biological effects of AHs, the most convincing of which demonstrate that *Ahr* 'knock-out' mice are very resistant to the toxic effects of TCDD and B[*a*]P[72,153,192].

TCDD has the potential to modulate several biological processes that impact growth, immune response, cell cycle regulation and differentiation, and fish are among the most sensitive vertebrates to its developmental toxic effects[172,214,227,228,243]. Fish exhibit similar signs of toxicity following TCDD exposure as described for other

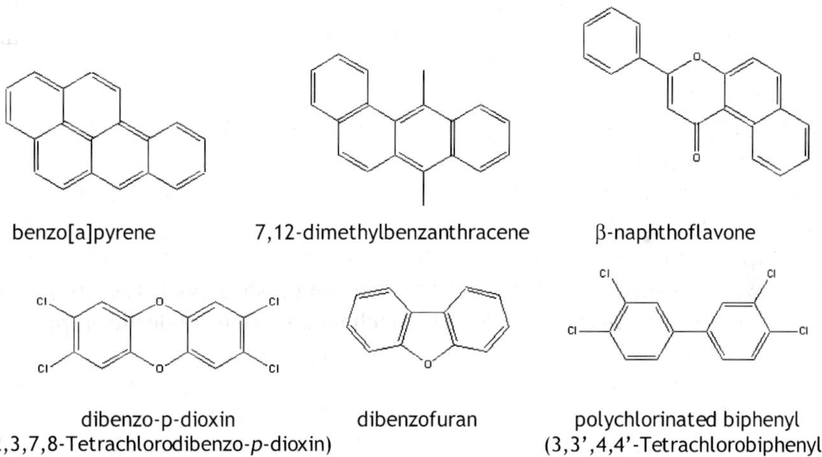

Fig. 3. Representative congeners of aromatic hydrocarbons (AHs). AHs have similar chemical structure, are highly persistent in the environment, produce similar patterns of toxic response that vary in potency, and act via a common mechanism. AHs are, or were, manufactured as commercial products, or result as a by-product from several different manufacturing processes. TCDD is most studied, and often serves as the prototype of the AHs.

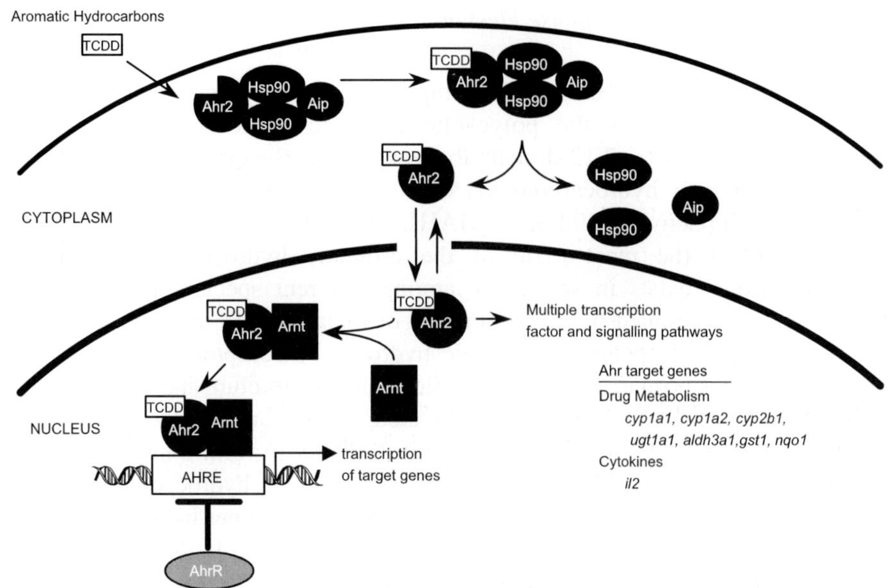

Fig. 4. Aromatic hydrocarbon receptor signaling pathway. Binding of ligand to the cytosolic AHR induces allosteric changes in the receptor, leading to the dissociation of heat-shock proteins and other associated proteins. The ligand-bound receptor then enters the nucleus, and forms a heterodimer with ARNT. The ligand–receptor complex then forms a transcriptional complex at the AHRE located in the regulatory region of several target genes, inducting gene transcription. Black shapes indicate gene products that have been isolated in zebrafish. Abbreviations are given in the text. AhrR, Ahr repressor; Aip, Ahr interacting protein, also known as XAP2 and ARA9.

vertebrates, including decreased food intake, wasting, delayed mortality, lesions in epithelial and lymphomyloid tissue in adults, cardiovascular dysfunction, edema, hemorrhages, craniofacial malformations, impaired reproductive success, and mortality at early stages of development[172,216,227]. While the developmental toxic effects of TCDD are similar among species, sensitivity can vary greatly[67]. Lake and bull trout larvae are the most sensitive fish species studied to date (Fig. 5). Zebrafish larvae are among the least sensitive of the fishes, but much more similar to the most sensitive mammals.

It has been known for decades that exposure of fishes to AHs induces cytochrome P450s (CYPs) and mixed function oxidase activity[150,203]. The mechanistic conservation of the AHR pathway between mammals and fish was always assumed, but has only recently been demonstrated. The AHR-signaling pathway is not well characterized in fish compared to mammalian model systems, but several studies indicate that the AHR signaling pathway is highly conserved between mammals and fish. Hahn *et al.*[83] confirm that several species of fish possess an AH-binding protein with properties similar to the mammalian AHR. The rainbow trout ARNT was the first member of this signaling pathway to be identified and sequenced in fish[177], and other fish AHR sequences followed soon thereafter[84,186]. Using mouse cell lines deficient in specific AHR signaling components, it was demonstrated that AHRE sequences in the

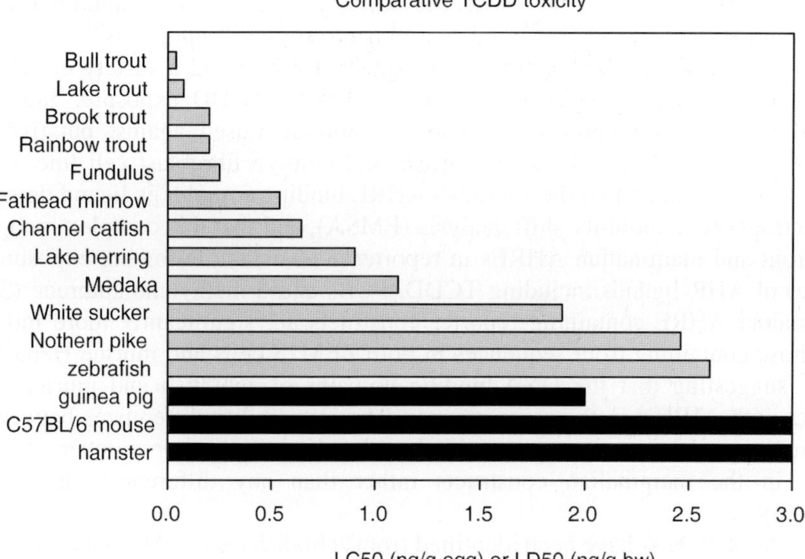

Fig. 5. Comparative sensitivity to the embryotoxic effects of TCDD. Species differences in sensitivity to TCDD-induced mortality among fish and mammalian species. Based on Hahn[141] and Tanguay *et al.*[216], and references therein.

promoter regions of fish CYP1A genes regulate TCDD responsiveness through interaction with AHR and ARNT[34]. Investigations using zebrafish followed logically as a model system that could provide for genetic dissection of the AHR signaling pathway. A number of fish species, besides zebrafish, continue to contribute to our knowledge of fish AHR signaling and our readers are encouraged to read the review by Tanguay *et al.*[216] for a more complete discussion of TCDD toxicity in fish.

The resources available for the zebrafish model system allow scientists to investigate the effects of TCDD at the earliest stages of development and provide the opportunity to investigate the molecular mechanisms of toxicity in the context of a whole organism. Zebrafish embryos respond to TCDD in a manner very similar to other fishes[89] and their sensitivity is comparable to that of mammalian cell culture bioassay systems[159]. Zebrafish are most susceptible to the toxic effects of TCDD when exposed during the first 72 h post-fertilization, and at lower doses, manifestation of the toxic effects will be delayed until the period during or shortly after hatching. Elonen *et al.*[67] and Henry *et al.*[89] report that acute exposure of eggs to TCDD via water affects development of larva, but not egg mortality or hatching time. Both studies suggested that the LC50 for eggs is 2.5 ng TCDD/g egg. Zebrafish exposed to TCDD either via water or maternal transfer present developmental toxic effects resembling 'blue sac disease', including cranial and cardiac edema, uninflated swim bladder, hemorrhage, craniofacial malformation, reduced blood flow, anemia, neuronal apoptosis, and growth retardation[20,60,61,67,89,218,232]. The impact of TCDD on the development of localized circulation correlates with the severity of craniofacial abnormalities[218], cardiovascular system defects[20], and central nervous system apoptosis[61].

Early mechanistic investigations using adult zebrafish and a variety of cell lines report the induction of cytochrome P450 following TCDD exposure, and also a protein that was immunoreactive with an antibody raised against putative trout CYP1A[30,46,48,154]. Using the zebrafish ZEM2S embryo fibroblast cell line, Carvan *et al.*[33,35] demonstrate that the zebrafish AHRE binding complex is ligand dependent by electrophoretic mobility shift analysis (EMSA), and that this complex recognizes both trout and mammalian AHREs in reporter gene assays following exposure to a number of AHR ligands including TCDD, PCBs and 3-methylcholanthrene (3MC). Mammalian AHRE-containing reporter constructs are significantly more inducible than those containing trout sequences in both ZEM2S cells and murine Hepa 1c1c7 cells[34] suggesting that the DNA-binding domains of zebrafish and murine AHRs interact with AHREs in the same manner. The inducibility differences between the mammalian and trout sequences is likely related to the greater number of AHRE motifs in the mammalian constructs rather than any differences in sequence specificity[34].

Two AHR cDNAs have been identified from zebrafish (Ahr1, Ahr2) that show high similarity (>60%) in the deduced amino acid sequence to each other and to the human AHR[10,214]. Using sequence comparisons, Ahr1 is more closely related to the mammalian AHRs and zebrafish Ahr2 is more closely related to other fish Ahr2s[10]. Interestingly, there are striking functional differences between the two zebrafish Ahr proteins. Ahr1 is expressed primarily in the liver, while Ahr2 is expressed in nearly all tissues. Abnet *et al.*[2] reveal that over-expressed zebrafish Ahr2 is able to drive ligand-dependent expression of a luciferase reporter gene regulated by trout AHREs in monkey COS-7 cells. However, the zebrafish Ahr1 is unable to bind ligand[11] and thus does not directly participate in cell signaling in response to AHs. Expression of both zebrafish Ahr mRNAs is detected by RT-PCR at 24 hpf, and both messages are detected throughout development[2,214]. TCDD-dependent induction of zebrafish Cyp1A mRNA and protein are detected in embryo tissues 48 hpf using both *in situ* hybridization and immunohistochemistry[238], in contrast to that previously reported using the less sensitive Western immunoblot technique[148,149].

Three zebrafish Arnt2 cDNAs have been identified that represent alternatively spliced variants with significant homology (>80% deduced amino acid identity) to the murine Arnt2[215]. Only one of the variants (Arnt2c) is able to confer ligand-dependent inducibility of a luciferase reporter gene when over-expressed with Ahr2 in COS-7 cells. ARNT proteins also participate in the signaling pathways that activate transcription through hypoxia response elements (HREs) via dimerization with HIF-1α related proteins[185]. When over-expressed with zebrafish Epas1 (endothelial PAS domain protein 1), all three Arnt2 variants are able to drive activation of an HRE-regulated reporter gene without $CoCl_2$ treatment or hypoxic conditions demonstrating that the function of splice variants may depend on the identity of its dimerization partner[215]. These results illustrate the limitations of studying biochemical interactions in cultured cells using over-expressed proteins, as often these interactions will bypass their normal ligand- or environmental-dependent nature under these conditions.

Zebrafish are one of the least sensitive fish to dioxin toxicity studied to date, yet they exhibit similar signs of toxicity as described for other vertebrates. Zebrafish has proven to be an effective system for investigation into the teratogenic effects of HAHs such as TCDD and the mechanisms that underlie cellular responses to such chemicals. As suggested by Hahn[86], the multiple roles of the original Ahr may be partitioned between the fish Ahr1 and Ahr2. The more ancient functions likely involve developmental regulation, whereas the adaptive responses to AHs were derived more recently[85]. Zebrafish Ahr2 binds dioxin and presumably participates in gene induction via interaction with AHREs. The function of Ahr1 in zebrafish is currently unknown[10], but its lack of TCDD binding suggest that it may provide the ancestral functions of the AHR family without those functions that evolved more recently. The genetic resources of zebrafish and the accessibility of its embryo to genetic manipulation together provide the potential to make significant contributions toward understanding the role of AHR signaling and its potential interactions with other signaling molecules and pathways regulating cell cycle and differentiation.

3. Metal-response element

Only recently have zebrafish been used as a model system to investigate the molecular and biochemical effects of metal exposure. Metallothionein (MT) is induced in fish by a variety of environmental chemicals, including paraquat, Cd, Cu, Hg, and As[37,50,169,191]. The induction of MT in response to metal exposure is mediated by metal-responsive DNA elements (MREs) and is controlled by MRE-binding transcription factor 1 (MTF1; Fig. 6), a Zn-finger transcription factor of the Cys_2His_2 family[90]. The exact relationship between MTF1 binding and MRE activation is unclear, since MTF1 binding of MRE sequences is activated by Zn but not by other heavy metals[21]. However, it has also been shown that both MTF1 and MREs are required for induction of MT in response to Zn and other metals[81]. The first fish *MTF1* cloned was from *Fugu rubripes*, and its gene product is able to mediate MRE-driven reporter gene activity in response to Cd treatment when over-expressed in murine cells at high concentrations of Cd. In lysates from those same cells, pufferfish MTF1 binding to MRE sequences was found by EMSA to increase with Cd treatment[12].

Heavy metals have numerous effects on the growth, reproduction, and survival that vary considerably between different metals, their ionic and organic forms, and in aquatic environments, are influenced by pH, temperature, and the presence of other ions. MT-inducing chemicals have been shown in fish to induce liver necrosis, liver hyperplasia, kidney fibrosis, reproductive abnormalities, and micronucleus formation[88,120,130,133,166,188]. We will focus on cadmium and zinc, which are the most potent inducers of MT in vertebrate systems. In fish, cadmium produces a number of toxic effects including the formation of micronuclei and alteration of protein metabolism[58,188]. As revealed by microangiography and confocal microscopy, cadmium exposure alters embryonic vascular patterning in zebrafish by reducing the branching of cranial vessels[41]. The toxic effects of zinc in fish include the alteration of erythrocyte membrane lipids and suppression of acetylcholinesterase

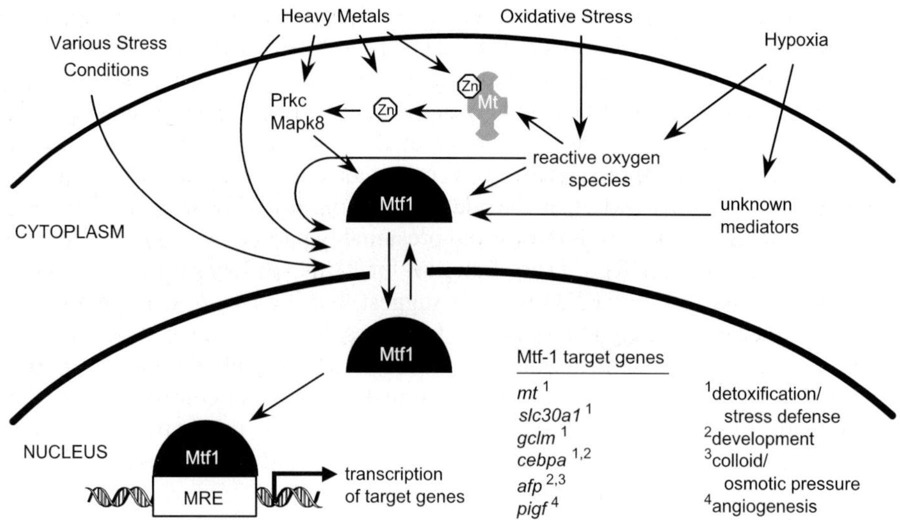

Fig. 6. MTF1 signaling pathway. MTF1 mediates transcriptional responses to a number of stressors including heavy metals, oxidative stress, and hypoxia. The classical pathway involves the activation of MTF1 to its DNA binding form by Zn, Cd and other heavy metals. Recent investigations provide evidence for a significantly more complex regulatory mechanism whereby a number of different cellular pathways influence MTF1 transcriptional activity. Upon entering the nucleus, MTF1 forms a transcriptional complex at the MRE located in the regulatory region of several target genes, inducing gene transcription. Black shapes indicate gene products that have been isolated in zebrafish. Abbreviations are given in the text. Prkc, protein kinase C; Mapk8, mitogen activated protein kinase 8, Jun kinase; *slc30a1*, solute carrier family 30 (zinc transporter), member 1; *gclm*, glutamate cysteine ligase, modifier subunit; *cebpa*, CCAAT/enhancer binding protein (C/EBP), alpha; *afp*, alpha-fetoprotein; *pigf*, phosphatidylinositol glycan, class F. Redrawn after Lichtlen and Schaffner[141].

activity[4,211]. In zebrafish, Zn has been shown to influence inhibitory postsynaptic glycinergic currents in the hindbrain[212].

Initial characterization of the putative zebrafish Mtf1 from ZEM2S cells reveals that it specifically recognizes consensus MRE sequences in EMSA, it is immunologically similar to murine MTF1, and is activated by Zn, but not by Cd[53]. An MRE-driven luciferase reporter gene construct is also highly inducible when transiently transfected into ZEM2S cells following treatment with Cd, while exhibiting less inducibility with Zn and Hg[33]. The zebrafish Mtf1 cDNA has been identified and the deduced amino acid sequence of the DNA-binding domain is highly conserved (>90% identity) when compared with the sequence from other vertebrate MTF1 sequences. Zebrafish Mtf1 is expressed throughout development. It is ubiquitously expressed in the early zebrafish embryo by *in situ* hybridization and the highest levels of expression are associated with neural tissues at about the 14 somite stage[39]. The sequencing of several fish *MT* genes has revealed MRE sequences through which Zn activates the transcription of *MT* genes[167,187]. The zebrafish *Mt* gene promoter also contains four putative MREs and is able to confer responsiveness to Zn, Cd, and Cu in a zebrafish caudal fin cell line[239].

Homozygous disruption of *MTF1* in the mouse is embryonic lethal[81] though not in *Drosophila*[65]. Therefore, zebrafish can make a substantial contribution by revealing the role of Mtf1 in normal development and metal homeostasis. MTF1 is also activated by a number of cellular stressors including oxidative stress and hypoxia, however, the mechanism is poorly understood[141]. Utilization of the genetic and genomic capabilities of the zebrafish model system will allow for the mechanistic dissection of Mtf1 activation and the myriad of pathways that influence its function. This will lead to a greater understanding of the interactions between Cd and MT induction via MREs, the role of accessory factors, and the general effects of metals as transcriptional modulators and initiators of signal transduction cascades. MTF1 plays a key role in a number of processes that are important in development, aging, and carcinogenesis and our exploration of its function in zebrafish has much to contribute.

4. Electrophile response element

A number of enzymes involved in Phase II drug metabolism, glutathione synthesis, and antioxidant activity have been used as biomarkers of pollutant effects in a number of fish species[142,170,190,204]. These biomarkers often correlate with DNA damage, macromolecular adducts, and glutathione depletion, and have been useful as indicators of pollutant-induced ‘oxidative stress’ from the effects of potent electrophilic compounds to which the fish has been directly exposed, or that are generated by cellular metabolic processes. Electrophiles are electron-deficient molecules that attack the electron dense neucleophilic centers of cellular macromolecules, typically carbon, nitrogen, oxygen, or sulfur groups. In proteins, the major reactive sites for electrophiles have tryptophan, tyrosine, methionine and perhaps histidine residues. In DNA, the highly nucleophilic sites include the N-7 and exocyclic N-2 of guanine and the N-3 of adenine, with the affinity for specific nucleophilic sites varying with the type of reactive compound. Quinones, metals, alkylating agents, Michael reaction acceptors, peroxides, and a wide variety of other potent electrophiles, are known to induce a battery of genes by way of a DNA element termed the ‘electrophile response element’ (EPRE; also known as antioxidant response elements) found in the enhancer region of a number of phase II and antioxidant enzymes[53,213]. Regulation of gene expression via EPRE sequences (Fig. 7) has been shown in knock-out mice to be dependent on the Nrf2 transcription factor[101] and Keap1, a cytoskeleton-associated protein that sequesters Nrf2 in the cytoplasm under normal redox conditions[102,124]. Dissociation of Nrf2 from Keap1 is apparently regulated by protein kinase C phosphorylation of specific residues on Nrf2, and by direct interactions between electrophiles and the sulfhydryl groups on Keap1[59,98]. The transcriptional activity of Nrf2 also appears to be influenced by a number of other coactivators that presumably bind to EPRE-specific transcription factors and enhance transcription of the oxidative stress responsive genes by enabling chromatin remodeling[246].

Genes that are regulated by EPRE sequences in mammalian systems are also induced in fish following exposure to a number of different environmental agents[75,204,225]. UV-B induces oxidative stress in zebrafish by a number of biochemical measures[38].

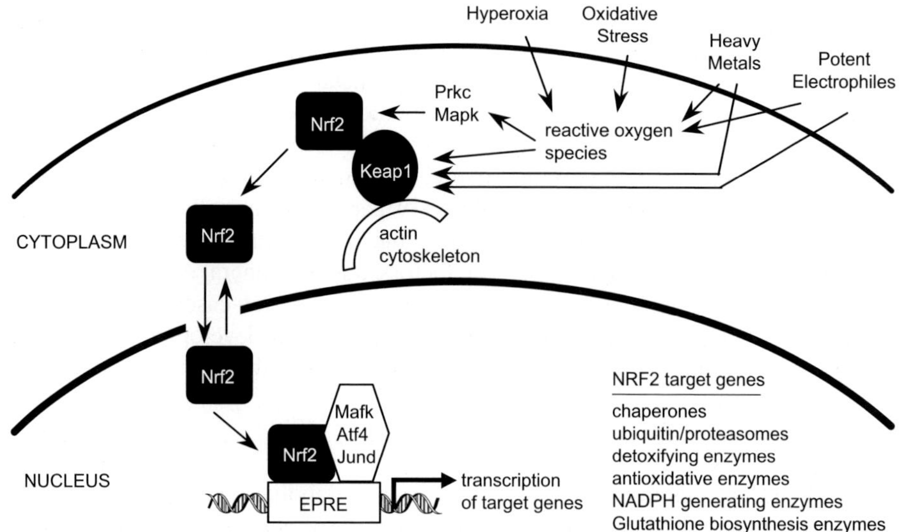

Fig. 7. NRF2 signaling pathway. Potent electrophiles, heavy metals, and reactive oxygen species induce allosteric changes in the cytoskeleton-associated protein Keap1, which causes the dissociation of the transcription factor NRF2. NRF2 enters the nucleus and forms a transcriptional complex at the EPRE located in the regulatory region of several target genes, inducing gene transcription. Black shapes indicate gene products that have been isolated in zebrafish. Abbreviations are given in the text. Mafk, v-maf musculoaponeurotic fibrosarcoma oncogene family, protein K; Atf4, activating transcription factor 4; Jund, jun D proto-oncogene.

The first putative EPRE sequences identified in fish are found in the upstream of the electrophile-responsive *GSTA* gene in plaice (*Pleuronectes platessa*)[134]. Work in zebrafish provided the first functional support for EPRE-dependent transcriptional responses in fish cells or tissues[33,36,117]. ZEM2S cells transfected with an EPRE-driven luciferase reporter gene were highly responsive to a variety of structurally diverse chemicals including TCDD, 3MC, *tert*-butylhydroquinone (tBHQ), Hg, As, Cd, Pb, and Cu[33,36,141]. Kobayashi *et al.*[117] provided elegant molecular and genetic support for the function of the Nrf2–Keap1 system in zebrafish. They verified that tBHQ induction of Gst, Nqo1, and Gclm mRNAs in zebrafish larvae is dependent on both Nrf2 and Keap1, and that zebrafish Nrf2 and Keap1 interact with each other. The functional protein domains of the Nrf2 and Keap1 are well conserved, showing 45–70% identity in the deduced amino acid sequence to the mouse counterparts, and mutation of specific residues in the ETGE (Glu-Thr-Gly-Glu) motif of Nrf2 or the DGR (double glycine repeat) motif of Keap1 abolished their interaction and disrupted normal EPRE-mediated gene regulation[117].

A number of mechanistic questions remain regarding the individual proteins that participate in the EPRE binding complex, their interactions with other signaling pathways, and the influence of tissue-specific or developmental stage-specific factors. Further utilization of the zebrafish model system will prove to be a powerful tool for the identification of accessory factors that may influence the transcriptional activity of Nrf2 and for investigating the toxic mechanisms of electrophiles, which activate numerous cell-signaling pathways. Cultured cells in the laboratory provide an easily manipulated system for the study of oxidative stress and its impact on transcription. However, the impact of oxidative stress on the rapidly changing environment of the developing embryo cannot be adequately modeled using cultured cells. As demonstrated by Kobayashi *et al.*[117], zebrafish embryo provides an organismal platform from which to explore gene–environment interactions and reveal molecular relationships that both cultured cells and 'knock-out' mice are unable to bring to light.

5. Heat shock response element

In virtually all organisms, enhanced synthesis of heat shock proteins (HSPs) occurs in response to environmental, chemical, and physiological stresses. Some members of the evolutionarily conserved HSP gene family are constitutively expressed and function as molecular chaperones. It has been demonstrated that HSP induction results primarily from the activation of a heat shock transcription factor (HSF) with subsequent binding to heat shock response DNA elements (HSEs) in the enhancer regions of the HSP genes[157]. Heat shock responses in fish have been recently reviewed by Basu *et al.*[15] and will be discussed in another chapter of this volume, so we will primarily limit our discussions to heat shock responses in zebrafish.

Krone and Sass[121] characterized the first member of the Hsp gene superfamily in zebrafish when they identified two forms of the Hsp90 chaperone (Hsp90a, Hsp90b). Hsp90b is constitutively expressed throughout development and increases modestly (two to threefold) following heat shock, whereas the Hsp90a expression is barely detectable by Northern analysis in normal embryos and increases dramatically following heat shock at all developmental stages[121]. A number of other zebrafish Hsp cDNAs, including Hsp47 and Hsp 70, have been isolated, sequenced, and characterized with regard to their distinct spatial, temporal, and stress-specific expression during embryonic development, and extensively reviewed by Krone *et al.*[122,123].

Three forms of the zebrafish Hsf1 (Hsf1a, Hsf1b, Hsf1c) have been isolated[178,229] and the deduced amino acid sequence is closely related to Hsf1 from other vertebrates, with high sequence homology in the DNA-binding domain (>90% identity with other vertebrates) and the hydrophobic heptad repeat A/B domain (>75% identity). Disruption of the Hsp1 transcripts in zebrafish embryos using an antisense morpholino (see section III.6) enhances sensitivity to heat shock[229]. The zebrafish Hsf1 mRNA exists as three alternatively spliced transcripts whose ratio varies between tissues. The biochemical consequences for the expression and regulation of the splice variants are not known; however, it is tempting to speculate that the Hsf1 variants may respond to different stressors[3]. The Hsf1b transcript is lost from the adult liver, but not the gills or gonads, following heat shock[178], but not following exposure to cadmium or copper[3].

The role of xenobiotic chemicals in Hsp induction in zebrafish was first described by Lele *et al.*[137], who demonstrated that Hsp47 is significantly induced in pre-cartilagenous cells and other connective tissue cell populations in the developing embryo following ethanol treatment, whereas Hsp70 is only modestly induced by ethanol. Using a transgenic zebrafish in which a GFP transgene is regulated by the

Hsp70 promoter/enhancer, Blechinger *et al.*[24] demonstrated that Hsp70 is constitutively expressed in the developing lens. In this transgenic fish, GFP expression mimics the endogenous Hsp70 gene product, and is induced by heat shock, cadmium, and sublethal laser pulses[24,25,87]. This work shows much promise as zebrafish resources are dedicated toward delineating cellular responses and further research will illuminate the mechanistic convergence of toxic responses at the heat shock pathway and allows for the analysis of cell-specific responses in an intact developing embryo.

6. *Pharmacological agents*

Geldanamycin and its analogs are currently being tested as chemotherapeutic agents[140,155]. Geldanamycin is a potent HSP90 inhibitor and is known to produce significant oxidative stress and cellular lysis. Reduction of the HSP90 cellular content by hammerhead ribozyme in cultured mammalian cells and yeast compromises cellular integrity in a manner similar to, though less severe than, that of geldanamycin[201]. In zebrafish embryos, pharmacological inhibition of Hsp90 by geldanamycin disrupts normal trunk muscle development by specifically suppressing the formation of muscle pioneers and potentially other muscle progenitors. This suggests that during vertebrate development, Hsp90 may perform vital functions beyond its role as a chaperone, and that there may be some redundancy in the specific chaperone functions performed by Hsp90 as there were a number of cell types that were unaffected by its inhibition[138].

Decitabine (2′-deoxy-5-azacytidine) possesses potent antileukaemic activity and is used to induce fetal hemoglobin in sickle cell anemia patients through a mechanism involving DNA hypomethylation[105,189]. Martin *et al.*[146] treated zebrafish embryos with a number of hypomethylating agents, including 2′-deoxy-5-azacytidine and 5-azacytidine, to explore the role of DNA methylation in early development. Embryos treated about 2–3 h after fertilization, just prior to mid-blastula transition, exhibited a number of developmental abnormalities related to gastrulation defects; however, embryos treated during other developmental periods were much more refractory to the toxic effects of hypomethylation demonstrating the temporal specificity of its teratogenic effects.

Alcohol toxicity has not been the primary focus of investigations using zebrafish. Instead, zebrafish have been treated with very high doses of alcohol in order to perturb a particular developmental and/or signaling pathway of interest. Exposure of zebrafish embryos to alcohol causes cyclopia, craniofacial abnormalities, and alters gene expression in the ventral aspects of the fore- and mid-brain[22,23]. Alcohol exposure of zebrafish embryos also induces stress proteins[123,137], developmental abnormalities of the notochord and spinal cord[127], and malformation of the body trunk[18]. A number of genes presumably involved in ethanol metabolism have been cloned, sequenced, and submitted to GenBank, but their involvement in ethanol metabolism is yet to be demonstrated. The ancestral alcohol dehydrogenase gene product (Adh3) metabolizes ethanol in fish[54,179]. Expression of Adh3 mRNA in zebrafish embryos varies during development and is widespread though not homogenously expressed in all tissues[32,56]. It is conceivable that temporal variation in the expression of alcohol dehydrogenase, HSPs, and a number of other factors during development results in a critical window of susceptibility to ethanol teratogenesis, as described by Blader and Strahle[22,23].

III. Advantages of the zebrafish model system: available resources

Although many investigators using zebrafish in their research expound the virtue of zebrafish as a model system for studying vertebrate development, many of the desirable characteristics of zebrafish (clarity of the embryo, small size, fecundity, and short intergeneration time) are not unique to zebrafish at all, but are common to many teleost species, including medaka and fathead minnows. At present, the primary advantage of zebrafish over other fish species as a toxicological model system is the abundance and diversity of the molecular and genetic resources available for investigating mechanistic questions relevant to other vertebrate systems, both fish and human. These biological resources have been rapidly developed due to the utility of zebrafish as a model system for developmental biology and genetics as a complement to the Human Genome Project. These resources allow investigators in toxicology, and many other fields, to address basic mechanistic and physiological questions in ways that no other vertebrate model system can provide.

1. Cell lines

Several cell lines derived from zebrafish are available from the American Type Culture Collection (www.atcc.org) including embryonic fibroblasts, fin-derived fibroblasts, and liver. Many of these cell lines can be used in toxicological assays[46,47,48,154] and can be transfected with reporter constructs or genes of interest to study the role of specific gene products in toxicological responses[35,36]. Zebrafish embryonic fibroblasts are used to create radiation and somatic cell hybrids for gene mapping. One of the first fibroblast cell lines ZF4 cells[62] was selected for its ability to grow without the need for fish serum or extracts, but it contains an excess of 90 chromosomes, almost twice the normal complement of chromosomes in zebrafish[62]. Other cell lines have been developed, including the AB strain derived Lucille's Fish Fibroblasts cells (LFF), and selected for their normal complement of chromosomes[42].

The LFF cells, as well as ZF4 and AB9 cells, have been fused with mouse melanoma or hamster Wg3H cells to generate the somatic cell hybrids used in gene mapping. The resulting cells possess a full complement of rodent chromosomes and a variable number of zebrafish chromosomes. The zebrafish chromosomes are stably maintained in the resulting cell fusions while they are grown under selective conditions. Selectable markers can result either from natural deficiencies inherent in mutant rodent cells that are complemented by the orthologous fish gene, or from selectable markers that can be transfected into the fish fibroblast cells using recombinant vectors containing a selectable marker (e.g. the neomycin resistance gene, *Neo*). Following transfection into zebrafish cells, the *Neo* gene inserts randomly into different chromosomes (usually a single site per cell) and these cells are selected from the non-transfected cells by the addition of G418, a neomycin analog, into the culture medium. As a result, individual G418 resistant clonal fibroblast cell lines are generated with stably integrated *Neo* genes. The different clonal *Neo* lines are then fused with rodent cells and G418 resistant cell fusions selected. Somatic cell hybrids created from different G418-resistant zebrafish fibroblast cell lines will contain different zebrafish chromosomes and are used to map genes to the individual chromosomes by Southern blot analysis or using PCR-based techniques[42,66].

While somatic cell hybrids containing whole zebrafish chromosomes can be used to map genes to specific chromosomes, radiation hybrids can be used to localize genes to specific regions of a chromosome. Rodent cells used in the generation of radiation hybrids have been cultured in medium containing 6-thioguanine to select for cells that are deficient in hypoxanthine phosphoribosyl transferase (*HPRT*). Zebrafish fibroblasts used in the cell fusions have been irradiated with X-rays to break the chromosomes into smaller fragments. During the fusion process, the fish DNA fragments are randomly inserted into the rodent cell chromosomes. Clonal hybrid cell lines that express zebrafish HPRT are then selected using HAT (hypoxanthine, aminopterin, thymidine) containing medium. By selecting for HPRT activity, the hybrid clonal cell lines that are isolated must contain the zebrafish *HPRT* gene, along with a substantial number of other zebrafish genomic DNA fragments. Each hybrid from the Goodfellow T51 panel contains an average 18% of the zebrafish genome[74]. DNA from the resulting hybrid cell lines is then isolated, arrayed, and screened by PCR for the presence of specific genes. This analysis is used to identify several hybrid cell lines containing overlapping pieces of a chromosome to localize genes (markers) to specific chromosomal regions[99,125,126]. These radiation hybrid panels are readily available for use by individual investigators in mapping their gene(s) of interest.

2. *Embryonic stem cells*

While, at present, embryonic stem cells are not available from any fish species, steady progress has been made in developing zebrafish embryonic stem cells for use in gene targeting and chimeric fish experiments. Mixed embryonic stem cell cultures have been generated from zebrafish embryos that have limited use in these capacities because of the difficulty in maintaining them and obtaining pure cultures. Zebrafish embryonic stem cells have been used to generate germline chimeras, which indicate that the embryonic stem cells could be used in gene targeting[144]. In the near future, zebrafish embryonic stem cells will be available, allowing for similar reverse genetic experiments that are common in the mouse to be performed in zebrafish.

3. *Mutagenesis screens*

Several thousand mutant zebrafish lines containing single gene mutations have been generated during two large scale[63,82] and several smaller chemical mutagenesis screens. To date, there is only a single report by Darland and Dowling[55] documenting the use of mutagenesis screens in zebrafish to identify genes that affect an organism's relative sensitivity to a given toxicant or class of toxicants. This pilot screen was able to identify three mutations that enhance resistance to the behavioral effects of cocaine. While genetic screens identifying 'resistant' mutants are relatively simple, screening for 'sensitive' mutants is inherently more complex and requires intensive comparisons between treated and untreated siblings. The large-scale mutagenesis screens conducted in both the Nüsslein-Volhard[82] and Driever laboratories[63] identified hundreds of genes that participate in the developmental process. The mutations in these screens were generated using the point mutagen ethyl nitrosourea (ENU). Since point mutations in a gene can cause either loss or gain of function, mutants from the same zebrafish mutagenesis screen could be analyzed for enhancement of 'sensitivity' and 'resistance' to specific chemicals. Because mutagenic screens for genes that influence 'sensitivity' or 'resistance' are both extremely valuable in mechanistic toxicology, the advancement of zebrafish as a toxicological model system will require large-scale mutagenic screens for genes that will substantially augment the Environmental Genome Project of the National Institute of Environmental Health Sciences.

In addition to ENU-induced mutagenesis, a number of other methods have produced loss of function mutants with altered developmental processes (Table 1). Insertional mutagens disrupt gene function by inserting large DNA fragments into the zebrafish genome, in a nearly random manner. The most common insertional mutagenic techniques in zebrafish involve retroviral- and transposon-mediated insertions, which not only cause mutations, but also serve to tag the affected gene, simplifying the identification and cloning of the mutated gene. The Hopkins lab at MIT reports the cloning and identification of 75 genes isolated in a retroviral insertion screen with an observable developmental phenotype[77]. In comparison, around 50 genes have been isolated and identified by positional cloning from the large-scale ENU-induced mutagenesis screens of the early 1990s. While a number of transposon-based systems have proven to be functional in zebrafish[107,108,109,180], there has not been a published report of a transposon-based mutagenesis screen.

Loss of function mutations can also be generated that involve the loss of chromosomal regions. Gamma and X-rays generate large deletions in the zebrafish genome, resulting in the loss of several genes[73]. The location and extent of the deletion are determined by traditional mapping methods using PCR. Zebrafish mutants possessing a large deletion have also been used in mapping ENU-induced point mutations in mutant fish by complementation analysis[242]. If the ENU-induced mutation lies within the region of the zebrafish genome encompassed by the deletion, then the mutations will not complement each other. Gamma and X-rays cause large deletions involving many genes, thus, the observable phenotype of the deletion mutant may be different from that of the ENU-induced mutation, and most likely more severe. The cross-linking agent trimethyl psoralen (TMP) causes deletions of 3 kb on average, much smaller than those induced by gamma or X-rays when used to generate zebrafish mutants[9]. The smaller deletions result in loss or inactivation of fewer genes and may make these mutants more useful in mapping ENU-induced mutations and the

TABLE 1

Types of mutation generated by commonly used mutagenic agents in zebrafish

Mutagen	Type of mutation	Effect of mutation	Advantages	Disadvantages
Ethylnitroso urea (ENU)	Point mutations, primarily	Gain or loss of function	Can select for either increase or decrease in toxicant sensitivity	Hard to identify mutated gene
X- or gamma-rays	Large deletions, several hundred kilobases in size	Loss of function	Easier to identify mutation location	Can span several genes; rearrangements are common
2,5,8-Trimethyl psoralen	Small deletions, average size of 3 kilobases	Loss of function	Small deletions affecting mostly single genes; fewer rearrangements	Null mutants only
Retrovirus	Insertion	Loss of function	Tag gene for easy identification	Null mutants only
Transposon	Insertion	Loss of function	Tag gene for easy identification	Null mutants only

generation of more precise deletion mutants. TMP could be used to generate a pool of mutants containing small deletions for screening mutants for enhanced sensitivity or resistance to virtually any toxicant.

4. *Genomics*

Once mutants have been generated, the mutations have to be mapped to their location in the genome with subsequent identification by positional cloning. Zebrafish have a big advantage over other fish species in the variety and depth of the genomic resources available for mapping mutations to the genome. There are thousands of markers that can be used for mapping mutations from a number of zebrafish genome maps (see http://zfin.org). These markers consist primarily of PCR-based DNA fragments whose chromosomal locations have been determined. Among the markers are Expressed Sequence Tags (EST), Amplified Fragment Length Polymorphisms (AFLPs), Random Amplified Polymorphic DNAs (RAPDs), Simple Sequence Length Polymorphisms (SSLPs), known genes, large genomic fragments in Bacterial Artificial Chromosomes (BACs), and Single Strand Conformation Polymorphisms (SSCPs). Multiple genetically distinct strains of zebrafish have been identified in which these genomic markers are polymorphic and hence are useful for mapping mutations.

The basic design strategy for a mutagenic screen involves three generations of selective breeding to generate homozygous mutants. First, mutations are made in the experimental strain, which is then crossed with a mapping strain that is highly polymorphic relative to the experimental strain to generate the families. Mutants of interest are identified in the F3 generation and the location of the mutation on a genetic map can be estimated using the appropriate markers. SSLPs or microsatellites are the most frequently used markers because they are widely distributed throughout the genome, can be assayed using traditional agarose gels, and are co-dominant, allowing the alleles from both strains in the crosses to be viewed. The high fecundity of zebrafish allows hundreds to thousands of useful meiotic events to be generated for use in mapping studies with only a few crosses.

The process can be further streamlined by employing a haploid screening strategy, which allows for the screening of F2 homozygotes[40,226]. Haploid zebrafish embryos containing only the maternal or paternal genome will survive for a number of days and can be useful for the identification of mutants in the first few days of development. Half tetrad (homozygous diploid) embryos are derived from haploids by either early pressure or heat shock treatment that disrupts normal meiosis. The resultant embryos have the normal complement of chromosomes and can be raised to adult, which is especially important when examining adult phenotypes. For these reasons it is possible to map genes faster and more economically in zebrafish than in other laboratory vertebrate models.

The zebrafish genome is about 2900 CentiMorgans (cM), therefore, the average distance between markers is much less than 1 cM (approximately 600 kb in the female meiotic map). Once a mutation has been mapped to within 1 cM of a marker, it is generally considered feasible to initiate a positional cloning strategy using the available zebrafish BAC, PAC and YAC genomic libraries to isolate the mutation[7,14,245]. The available genomic libraries contain inserts from 100–300 kb, making the chromosomal walk inherent in positional cloning feasible. While it is possible to create or obtain genomic libraries from any species, it takes several years to develop these resources. The density of genetic markers on the zebrafish genome sets zebrafish apart from other aquatic vertebrates that are used in toxicological studies. Because the density of genetic markers is greater in zebrafish, positional cloning approaches are morel likely to succeed. When the first zebrafish mutation, *one-eyed pinhead* (*oep*), was cloned using a positional approach in 1998, the zebrafish genome had only 1000 markers or one every 3 cM. Since then, positional cloning approaches have identified over 50 developmentally important genes from mutants originally identified in mutagenic screens.

5. *Transgenics*

The utility of zebrafish as a genetic model system has resulted in the development of microinjection techniques to direct gene expression using specific mRNAs or using DNA constructs directing the expression (or over-expression) of specific gene products. Transgenic zebrafish are generated by injection of single-cell zebrafish embryos with DNA constructs or pseudotyped retroviral vectors that direct the expression of a specific gene product. Using specific zebrafish promoters it is possible to create transgenic zebrafish expressing a given protein in a specific tissue or cell type at a defined time during development, providing a powerful tool for studying the role a particular protein plays in a toxic response mechanism. In the near future, as more genes are cloned and their promoters characterized, they can be used to generate fish expressing proteins in defined temporo-spatial patterns. While generation of stable transgenic zebrafish is far from routine, recent advances in the use of recombinant BACs for creating transgenics has increased the success rate. As our knowledge of the zebrafish genome increases our manipulations of the genome will become more common.

Microinjection into mutant one-cell embryos provides the opportunity to rescue a mutant phenotype using either mRNA or DNA constructs directing the expression of a normal copy of the mutated gene. Complementation tests can also be performed using large genomic PAC/BAC/YAC clones, which should have all the genetic components necessary to mimic normal gene expression[240]. Transgenic fish can be generated that express either normal or mutant proteins (e.g. dominant negative mutant zebrafish proteins) or proteins from heterologous systems (e.g. human gene products, reporter genes). The enhancement of green fluorescent protein (GFP) expression systems has resulted in the generation of zebrafish in which internal developmental and physiological process are becoming increasingly accessible. For example, GFP-labeled erythrocytes or endothelial cells can be used to visualize cardiovascular function[143,158], and leukemic lymphoblasts can be used to examine tumor formation and dissemination[131].

A number of toxicologically relevant zebrafish with stably integrated transgenes have been generated, including the ERE-driven luciferase[136] mentioned previously in this chapter, which has utility for detecting environmental estrogen mimics. Also described above is the Hsp70–GFP transgenic zebrafish that responds to Cd concentrations as low as 0.2 μM[24], which could be extremely valuable for the detection of environmental Cd or for the identification of genes that influence Cd toxicity. The expression of transgenic YFP (red shifted GFP variant) under the control of retinoic acid response elements is responsive to exogenous retinoic acid and retinoic acid synthetase inhibitors[171] suggesting that such a transgenic fish has utility for evaluating the impact of environmental samples on retinoid signaling during vertebrate development. Amanuma *et al.*[5] developed a transgenic zebrafish carrying the *rpsL* gene in a bacterial shuttle vector that is capable of detecting induced mutations from water-borne chemicals including ENU, B[*a*]P, and 2-amino-3,8-dimethylimidazo[4,5-f]quinoxaline (MeIQx)[5]. This work suggests that such a transgenic fish could be used for the detection of mutagens in naturally polluted water or utilized as a mutagenesis screening tool similar to the Big Blue transgenic rodent mutagenesis system (developed by Stratagene, La Jolla, CA) or the cII mutant medaka[237] which use viral shuttle vectors as the mutational target gene.

Microinjection of DNA or retroviral particles into the zebrafish embryo is a routine procedure in most zebrafish laboratories. However, a thorough understanding of the processes that lead to the creation of zebrafish with a stably integrated transgene whose function is reliably maintained generation after generation remains elusive. The utilization of transgenic zebrafish technology will not reach its full potential in toxicological sciences until this technology is more completely understood and has been transferred from the basic developmental biology laboratories to those laboratories whose primary focus is toxicology.

6. Morpholinos

Utilization of over-expression techniques and stable transgenics provides tools with which to analyze the function of specific gene products. Recent advances in antisense knock-down techniques now provide the complementary tools necessary to confirm the role of a specific gene product in a given pathway[174]. Morpholinos provide the yin to the transgenics yang. While transgenic fish express specific proteins, morpholino antisense oligos bind and inactivate selected RNA sequences blocking expression of specific proteins[163]. Morpholino antisense oligomers are composed of a backbone of morpholine moieties linked to one of the four bases found in DNA and joined by non-ionic phosphorodiamidate linkages. When injected into zebrafish one-cell embryos, morpholinos can block translation of specific gene products as late as 6–7 days post-fertilization. Usually, the morpholinos are directed against either the splice junction sites of the pre-mRNA or to the translation initiation sites of the mature message resulting in disruption of splicing and translation, respectively. Since morpholinos cause the loss of expression, they can confirm the identification of a mutated gene by blocking expression of a specific protein in an attempt to emulate the mutant phenotype. Several papers published in a special issue of the journal *Genesis* (volume 30, 2001) describe the phenocopying of known zebrafish mutants by injecting morpholinos directed against the normal genes mRNA, supporting the notion that injection of morpholinos directed against the mRNA of genes identified as playing a role in a toxic response will alter the sensitivity of the fish to a toxicant. The power of the morpholino antisense technology in combination with high-throughput potential of the zebrafish model system will be realized as the zebrafish genome sequence nears completion and 'knock-downs' of every known gene product provide invaluable evidence of its function in vertebrate development.

7. Microarrays

The large number of zebrafish cDNA clones available makes it possible to study the expression levels of several thousand genes at the mRNA level using microarrays. Microarrays consist of thousands of unique cDNAs or oligonucleotides anchored to a solid substrate (usually a glass slide), which is then probed with fluorescently labeled cDNAs derived from both toxicant exposed and unexposed fish (see Fig. 8). Microarrays are generally used to identify genes whose relative abundance either increase or decrease upon exposure to a toxicant. A zebrafish cDNA library from pooled 24 h post-fertilization embryos and adult liver has become available for use in generating microarrays. This arrayed, oligonucleotide normalized library contains over 37,000 cDNAs which likely represents about 25,000 different gene products[45]. A number of laboratories, including ours, are developing microarrays from this library to explore a number of pathways. The first publication that examined global gene expression in zebrafish using a 4500 clone cDNA microarray to examine the changes in gene expression associated with embryonic development and the alteration of gene expression by hypoxia[219]. Ton *et al.*[219] were able to identify over 800 cDNAs whose relative abundance is altered in zebrafish under hypoxic conditions. The recent

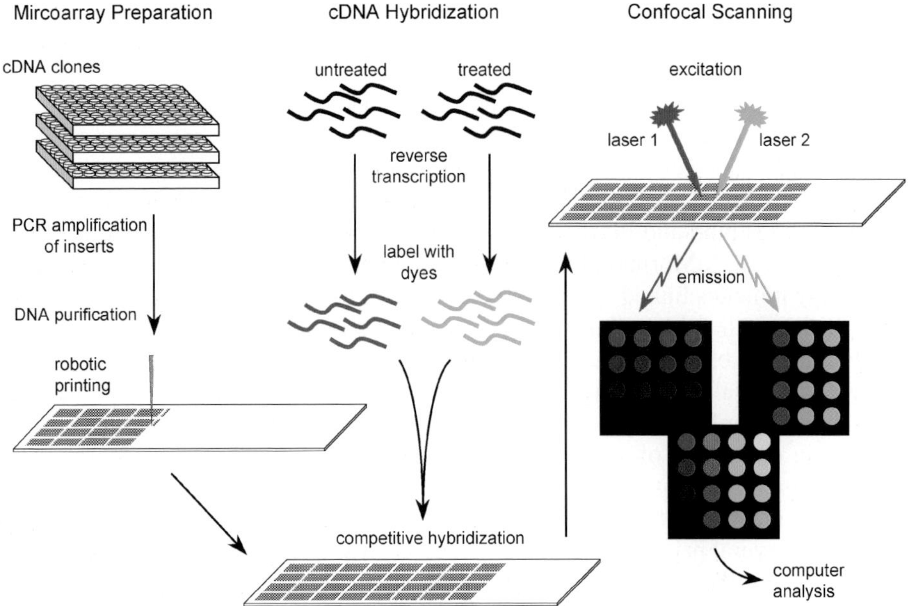

Fig. 8. Diagram of the process for construction and analysis of cDNA microarrays. The cloned inserts from an arrayed cDNA library are used to generate PCR products (probes) that are purified and robotically printed on glass slides. Two mRNA populations to be compared are reverse transcribed and labeled with different fluorescent dyes, generally Cy3 (green) or Cy5 (red). The two labeled cDNA populations are hybridized to the probes immobilized on the glass slide. The competitive nature of the hybridization results in the binding of both Cy3- and Cy5-labeled cDNAs in the proportion they are represented in the original mRNA populations. The Cy3 and Cy5 dyes are detected by laser confocal scanning and quantified by computer image analysis. Spots that appear green represent binding of predominantly Cy3-labeled cDNA. Red spots represent binding of predominantly Cy5-labeled cDNA, and yellow spots contain both Cy3 and Cy5 dyes in approximately equal quantities.

popularity of the zebrafish model system has also induced a number of biotechnology companies to develop oligonucleotide sets for microarray construction based on over 14,000 sequences from the NCBI Unigene database (http://www.ncbi.nlm.nih.gov/).

IV. Conclusions

Obviously, the completed zebrafish genome sequence will expand greatly the ability to map and clone genes, create transgenic fish, and construct zebrafish microarrays for use in identifying genes that directly participate in specific mechanisms of toxicity, or influence the sensitivity/resistance of an individual. With the completion of its genomic sequence, zebrafish will be the first vertebrate model system in which one can query the organism through design of large-scale complementary toxicogenomic, genetic, and complete life-cycle experiments to identify all of the gene products that play a role in specific mechanisms of toxicity. The vertebrate developmental program and the molecular responses to toxicants are dynamic and highly intricate processes.

One must employ a variety of broad-scale, high-resolution techniques to appreciate and understand their complexity, especially when the two processes are superimposed upon one another as in the field of developmental toxicology.

Comparative Toxicogenomics promises that once the complete assemblage of gene products has been revealed in one organism, it will then be possible to derive the complete mechanism of toxicity, which can be tested in other organisms for its universal applicability. The genes, and gene products, identified in zebrafish as participating in specific toxicant pathways can be expected to play the same or similar roles in other organisms, including other fish and humans. Similar approaches can be used to find orthologs in the genomes of mammalian species that may share toxicological response with humans. Once toxicologically relevant genes have been found in humans, it is possible to search for alleles in the gene that may determine the relative sensitivity of individuals to different toxicants. Through the wise use of information derived from toxicogenomic studies initiated in zebrafish, it may well be possible to solve toxicological problems that previously proved intractable in mammalian species.

The zebrafish model system also has much to offer as an instrument for delineating species differences in toxicity. As mentioned above, the genes identified as being part of a toxicant's mode of action in zebrafish can be expected to play the same or similar roles in other organisms, including other fish. Far too often, zebrafish is overlooked as an instrument for extracting relevant information regarding other fish species. As an example, Fig. 5 illustrates the differences in TCDD toxicity between several species of fish. An understanding of the mechanisms influencing these differences in toxicity will lead to better ecological risk assessment, and can be achieved by applying the variety of genomic and genetic tools the zebrafish model system provides. In this way, genomics may begin to fulfill the high hopes that many have predicted for its ability to solve complex mechanistic problems in both normal and stressed biological settings.

V. *References*

1. Abedi, Z.H. and W.P. McKinley. Bioassay of captan by zebrafish larvae. *Nature* 216: 1321–1322, 1967.
2. Abnet, C.C., R.L. Tanguay, W. Heideman and R.E. Peterson. Transactivation activity of human, zebrafish, and rainbow trout aryl hydrocarbon receptors expressed in COS-7 cells: greater insight into species differences in toxic potency of polychlorinated dibenzo-*p*-dioxin, dibenzofuran, and biphenyl congeners. *Toxicol. Appl. Pharmacol.* 159: 41–51, 1999.
3. Airaksinen, S., C.M. Rabergh, A. Lahti, A. Kaatrasalo, L. Sistonen and M. Nikinmaa. Stressor-dependent regulation of the heat shock response in zebrafish, *Danio rerio*. *Comp. Biochem. Physiol.* 134A: 839–846, 2003.
4. Akahori, A., Z. Jozwiak, T. Gabryelak and R. Gondko. Effect of zinc on carp (*Cyprinus carpio* L.) erythrocytes. *Comp. Biochem. Physiol.* 123C: 209–215, 1999.
5. Amanuma, K., H. Takeda, H. Amanuma and Y. Aoki. Transgenic zebrafish for detecting mutations caused by compounds in aquatic environments. *Nat. Biotechnol.* 18: 62–65, 2000.
6. Amatruda, J.F., J.L. Shepard, H.M. Stern and L.I. Zon. Zebrafish as a cancer model system. *Cancer Cell* 1: 229–231, 2002.
7. Amemiya, C.T. and L.I. Zon. Generation of a zebrafish P1 artificial chromosome library. *Genomics* 58: 211–213, 1999.

8. Anderson, P.D. and H.I. Battle. Effects of chloramphenicol on the development of the zebrafish, *Brachydanio rerio*. *Can. J. Zool.* 45: 191–204, 1967.
9. Ando, H. and M. Mishina. Efficient mutagenesis of zebrafish by a DNA cross-linking agent. *Neurosci. Lett.* 244: 81–84, 1998.
10. Andreasen, E.A., M.E. Hahn, W. Heideman, R.E. Peterson and R.L. Tanguay. The zebrafish (*Danio rerio*) aryl hydrocarbon receptor type 1 is a novel vertebrate receptor. *Mol. Pharmacol.* 62: 234–249, 2002.
11. Andreasen, E.A., J.M. Spitsbergen, R.L. Tanguay, J.J. Stegeman, W. Heideman and R.E. Peterson. Tissue-specific expression of AHR2, ARNT2, and CYP1A in zebrafish embryos and larvae: effects of developmental stage and 2,3,7,8-tetrachlorodibenzo-*p*-dioxin exposure. *Toxicol. Sci.* 68: 403–419, 2002.
12. Auf der Maur, A., T. Belser, G. Elgar, O. Georgiev and W. Schaffner. Characterization of the transcription factor MTF-1 from the Japenese pufferfish (*Fugu rubripes*) reveals evolutionary conservation of heavy metal stress response. *Biol. Chem.* 380: 175–185, 1999.
13. Bardet, P.L., B. Horard, M. Robinson-Rechavi, V. Laudet and J.M. Vanacker. Characterization of oestrogen receptors in zebrafish (*Danio rerio*). *J. Mol. Endocrinol.* 28: 153–163, 2002.
14. Barth, A.L., J.C. Dugas and J. Ngai. Noncoordinate expression of odorant receptor genes tightly linked in the zebrafish genome. *Neuron* 19: 359–369, 1997.
15. Basu, N., A.E. Todgham, P.A. Ackerman, M.R. Bibeau, K. Nakano, P.M. Schulte and G.K. Iwama. Heat shock protein genes and their functional significance in fish. *Gene* 295: 173–183, 2002.
16. Battle, H.I. and K.K. Hisaoka. Effects of ethyl carbamate (urethan) on the early development of the teleost, *Brachydanio rerio*. *Cancer Res.* 12: 334–376, 1952.
17. Bauer, M.P., J.T. Bridgham, D.M. Langenau, A.L. Johnson and F.W. Goetz. Conservation of steroidogenic acute regulatory (StAR) protein structure and expression in vertebrates. *Mol. Cell. Endocrinol.* 168: 119–125, 2000.
18. Baumann, M. and K. Sander. Bipartite axiation follows incomplete epiboly in zebrafish embryos treated with chemical teratogens. *J. Exp. Zool.* 230: 363–376, 1984.
19. Beckwith, L.G., J.L. Moore, G.S. Tsao-Wu, J.C. Harshbarger and K.C. Cheng. Ethylnitrosourea induces neoplasia in zebrafish (*Danio rerio*). *Lab. Invest.* 80: 379–385, 2000.
20. Belair, C.D., R.E. Peterson and W. Heideman. Disruption of erythropoiesis by dioxin in the zebrafish. *Dev. Dyn.* 222: 581–594, 2001.
21. Bittel, D., T. Dalton, S.L. Samson, L. Gedamu and G.K. Andrews. The DNA binding activity of metal response element-binding transcription factor-1 is activated in vivo and in vitro by zinc, but not by other transition metals. *J. Biol. Chem.* 273: 7127–7133, 1998.
22. Blader, P. and U. Strahle. Casting an eye over cyclopia. *Nature* 395: 112–113, 1998.
23. Blader, P. and U. Strahle. Ethanol impairs migration of the prechordal plate in the zebrafish embryo. *Dev. Biol.* 201: 185–201, 1998.
24. Blechinger, S.R., T.G. Evans, P.T. Tang, J.Y. Kuwada, J.T. Warren, Jr. and P.H. Krone. The heat-inducible zebrafish hsp70 gene is expressed during normal lens development under non-stress conditions. *Mech. Dev.* 112: 213–215, 2002.
25. Blechinger, S.R., J.T. Warren, Jr., J.Y. Kuwada and P.H. Krone. Developmental toxicology of cadmium in living embryos of a stable transgenic zebrafish line. *Environ. Health Perspect.* 110: 1041–1046, 2002.
26. Bloom, H.D. and A. Perlmutter. Effect of a sublethal concentration of zinc on an aggregating pheromone system in the zebrafish, *Brachydanio rerio*. *Environ. Pollut.* 17: 127–132, 1978.
27. Braunbeck, T.P. Burkhardt-Holm, G. Gorge, R. Nagel, R.D. Negele and V. Storch. Rainbow trout and zebrafish, two models for continuous toxicity tests: relative sensitivity, species and organ specificity in cytopathologic reaction of liver and intestines to atrazine. *Schriftenr Ver Wasser Boden. Lufthyg* 89: 109–145, 1992.
28. Bresch, H. Investigation of the long-term action of xenobiotics on fish with special regard to reproduction. *Ecotoxicol. Environ. Saf.* 6: 102–112, 1982.
29. Briggs, J.P. The zebrafish: a new model organism for integrative physiology. *Am. J. Physiol.* 282: R3–R9, 2002.
30. Buchmann, A., R. Wannemacher, E. Kulzer, D.R. Buhler and K.W. Bock. Immunohistochemical localization of the cytochrome P450 isozymes LMC2 and LM4B (P4501A1) in 2,3,7,8-tetrachlorodibenzo-*p*-dioxin-treated zebrafish (*Brachydanio rerio*). *Toxicol. Appl. Pharmacol.* 123: 160–169, 1993.
31. Callard, G.V., A.V. Tchoudakova, M. Kishida and E. Wood. Differential tissue distribution, developmental programming, estrogen regulation and promoter characteristics of cyp19 genes in teleost fish. *J. Steroid Biochem. Mol. Biol.* 79: 305–314, 2001.
32. Canestro, C., L. Godoy, R. Gonzalez-Duarte and R. Albalat. Comparative expression analysis of Adh3 during arthropod, urochordate, cephalochordate, and vertebrate development challenges its predicted housekeeping role. *Evol. Dev.* 5: 157–162, 2003.

33. Carvan, M.J., III, T.P. Dalton, G.W. Stuart and D.W. Nebert. Transgenic zebrafish as sentinels for aquatic pollution. *Ann. NY Acad. Sci.* 919: 133–147, 2000.
34. Carvan, M.J., III, L.V. Ponomareva, W.A. Solis, R.S. Matlib, A. Puga and D.W. Nebert. Trout CYP1A3 gene: recognition of fish DNA motifs by mouse regulatory proteins. *Mar. Biotechnol.* 1: 155–166, 1999.
35. Carvan, M.J., III, W.A. Solis, L. Gedamu and D.W. Nebert. Activation of transcription factors in zebrafish cell cultures by environmental pollutants. *Arch. Biochem. Biophys.* 376: 320–327, 2000.
36. Carvan, M.J., III, D.M. Sonntag, C.B. Cmar, R.S. Cook, M.A. Curran and G.L. Miller. Oxidative stress in zebrafish cells: potential utility of transgenic zebrafish as a deployable sentinel for site hazard ranking. *Sci. Total Environ.* 274: 183–196, 2001.
37. Castano, A., G. Carbonell, M. Carballo, C. Fernandez, S. Boleas and J.V. Tarazona. Sublethal effects of repeated intraperitoneal cadmium injections on rainbow trout, *Oncorhynchus mykiss*. *Ecotoxicol. Environ. Saf.* 41: 29–35, 1998.
38. Charron, R.A., J.C. Fenwick, D.R. Lean and T.W. Moon. Ultraviolet-B radiation effects on antioxidant status and survival in the zebrafish, *Brachydanio rerio*. *Photochem. Photobiol.* 72: 327–333, 2000.
39. Chen, W.Y., J.A. John, C.H. Lin and C.Y. Chang. Molecular cloning and developmental expression of zinc finger transcription factor MTF-1 gene in zebrafish, *Danio rerio*. *Biochem. Biophys. Res. Commun.* 291: 798–805, 2002.
40. Cheng, K.C. and J.L. Moore. Genetic dissection of vertebrate processes in the zebrafish: a comparison of uniparental and two-generation screens. *Biochem. Cell Biol.* 75: 525–533, 1997.
41. Cheng, S.H., P.K. Chan and R.S. Wu. The use of microangiography in detecting aberrant vasculature in zebrafish embryos exposed to cadmium. *Aquat. Toxicol.* 52: 61–71, 2001.
42. Chevrette, M., L. Joly, P. Tellis and M. Ekker. Contribution of zebrafish–mouse cell hybrids to the mapping of the zebrafish genome. *Biochem. Cell Biol.* 75: 641–649, 1997.
43. Chiang, E.F., Y.L. Yan, Y. Guiguen, J. Postlethwait and B. Chung. Two Cyp19 (P450 aromatase) genes on duplicated zebrafish chromosomes are expressed in ovary or brain. *Mol. Biol. Evol.* 18: 542–550, 2001.
44. Chiang, E.F., Y.L. Yan, S.K. Tong, P.H. Hsiao, Y. Guiguen, J. Postlethwait and B.C. Chung. Characterization of duplicated zebrafish cyp19 genes. *J. Exp. Zool.* 290: 709–714, 2001.
45. Clark, M.D., S. Hennig, R. Herwig, S.W. Clifton, M.A. Marra, H. Lehrach, S.L. Johnson and the WU-GSC EST Group. An oligonucleotide fingerprint normalized and expressed sequence tag characterized zebrafish cDNA library. *Genome Res.* 11: 1594–1602, 2001.
46. Collodi, P., Y. Kamei, T. Ernst, C. Miranda, D.R. Buhler and D.W. Barnes. Culture of cells from zebrafish (*Brachydanio rerio*) embryo and adult tissues. *Cell Biol. Toxicol.* 8: 43–61, 1992.
47. Collodi, P., Y. Kamei, A. Sharps, D. Weber and D. Barnes. Fish embryo cell cultures for derivation of stem cells and transgenic chimeras. *Mol. Mar. Biol. Biotechnol.* 1: 257–265, 1992.
48. Collodi, P., C.L. Miranda, X. Zhao, D.R. Buhler and D.W. Barnes. Induction of zebrafish (*Brachydanio rerio*) P450 *in vivo* and in cell culture. *Xenobiotica* 24: 487–493, 1994.
49. Corley-Smith, G.E., C.J. Lim and B.P. Brandhorst. Production of androgenetic zebrafish (*Danio rerio*). *Genetics* 142: 1265–1276, 1996.
50. Cosson, R.P. Heavy metal intracellular balance and relationship with metallothionein induction in the gills of carp. After contamination by Ag, Cd, and Hg following pretreatment with Zn or not. *Biol. Trace Elem. Res.* 46: 229–245, 1994.
51. Couse, J.F., S.W. Curtis, T.F. Washburn, J. Lindzey, T.S. Golding, D.B. Lubahn, O. Smithies and K.S. Korach. Analysis of transcription and estrogen insensitivity in the female mouse after targeted disruption of the estrogen receptor gene. *Mol. Endocrinol.* 9: 1441–1454, 1995.
52. Creaser, C.W. The technic of handling the zebra fish (*Brachydanio rerio*) for the production of eggs which are favorable for embryological research and are available at any specified time throughout the year. *Copeia* (1934): 159–161, 1934.
53. Dalton, T.P., W.A. Solis, D.W. Nebert and M.J. Carvan, III. Characterization of the MTF-1 transcription factor from zebrafish and trout cells. *Comp. Biochem. Physiol.* 126B: 325–335, 2000.
54. Danielsson, O., H. Eklund and H. Jornvall. The major piscine liver alcohol dehydrogenase has class-mixed properties in relation to mammalian alcohol dehydrogenases of classes I and III. *Biochemistry* 31: 3751–3759, 1992.
55. Darland, T. and J.E. Dowling. Behavioral screening for cocaine sensitivity in mutagenized zebrafish. *Proc. Natl Acad. Sci. USA* 98: 11691–11696, 2001.
56. Dasmahapatra, A.K., H.L. Doucet, C. Bhattacharyya and M.J. Carvan, III. Developmental expression of alcohol dehydrogenase (ADH3) in zebrafish (*Danio rerio*). *Biochem. Biophys. Res. Commun.* 286: 1082–1086, 2001.

57. Dave, G., K. Andersson, R. Berglind and B. Hasselrot. Toxicity of eight solvent extraction chemicals and of cadmium to water fleas, *Daphnia magna*, rainbow trout, *Salmo gairdneri*, and zebrafish, *Brachydanio rerio*. *Comp. Biochem. Physiol.* 69C: 83–98, 1981.
58. De Smet, H. and R. Blust. Stress responses and changes in protein metabolism in carp *Cyprinus carpio* during cadmium exposure. *Ecotoxicol. Environ. Saf.* 48: 255–262, 2001.
59. Dinkova-Kostova, A.T., W.D. Holtzclaw, R.N. Cole, K. Itoh, N. Wakabayashi, Y. Katoh, M. Yamamoto and P. Talalay. Direct evidence that sulfhydryl groups of Keap1 are the sensors regulating induction of phase 2 enzymes that protect against carcinogens and oxidants. *Proc. Natl Acad. Sci. USA* 99: 11908–11913, 2002.
60. Dong, W., H. Teraoka, S. Kondo and T. Hiraga. 2,3,7,8-tetrachlorodibenzo-*p*-dioxin induces apoptosis in the dorsal midbrain of zebrafish embryos by activation of arylhydrocarbon receptor. *Neurosci. Lett.* 303: 169–172, 2001.
61. Dong, W., H. Teraoka, K. Yamazaki, S. Tsukiyama, S. Imani, T. Imagawa, J.J. Stegeman, R.E. Peterson and T. Hiraga. 2,3,7,8-tetrachlorodibenzo-*p*-dioxin toxicity in the zebrafish embryo: local circulation failure in the dorsal midbrain is associated with increased apoptosis. *Toxicol. Sci.* 69: 191–201, 2002.
62. Driever, W. and Z. Rangini. Characterization of a cell-line derived from zebrafish (*Brachydanio rerio*) embryos. *In Vitro Cell. Dev. Biol.-Anim.* 29: 749–754, 1993.
63. Driever, W., L. SolnicaKrezel, A.F. Schier, S.C.F. Neuhauss, J. Malicki, D.L. Stemple, D.Y.R. Stainier, F. Zwartkruis, S. Abdelilah, Z. Rangini, J. Belak and C. Boggs. A genetic screen for mutations affecting embryogenesis in zebrafish. *Development* 123: 37–46, 1996.
64. Eaton, J.G., J.M. McKim and G.W. Holcombe. Metal toxicity to embryos and larvae of seven freshwater fish species – I. Cadmium. *Bull. Environ. Contam. Toxicol.* 19: 95–103, 1978.
65. Egli, D., A. Selvaraj, H. Yepiskoposyan, B. Zhang, E. Hafen, O. Georgiev and W. Schaffner. Knockout of 'metal-responsive transcription factor' MTF-1 in *Drosophila* by homologous recombination reveals its central role in heavy metal homeostasis. *EMBO J.* 22: 100–108, 2003.
66. Ekker, M., M.D. Speevak, C.C. Martin, L. Joly, G. Giroux and M. Chevrette. Stable transfer of zebrafish chromosome segments into mouse cells. *Genomics* 33: 57–64, 1996.
67. Elonen, G.E., R.L. Spehar, G.W. Holcombe, R.D. Johnson, J.D. Fernandez, R.J. Erickson, J.E. Tietge and P.M. Cook. Comparative toxicity of 2,3,7,8-tetrachlorodibenzo-*p*-dioxin to seven freshwater fish species during early life-stage development. *Environ. Toxicol. Chem.* 17: 472–483, 1998.
68. Ensenbach, U. and R. Nagel. Toxicity of binary chemical mixtures: effects on reproduction of zebrafish (*Brachydanio rerio*). *Arch. Environ. Contam. Toxicol.* 32: 204–210, 1997.
69. Ensenbach, U. and R. Nagel. Toxicity of complex chemical mixtures: acute and long-term effects on different life stages of zebrafish (*Brachydanio rerio*). *Ecotoxicol. Environ. Saf.* 30: 151–157, 1995.
70. Exner, J. Perspective on hazardous waste problems related to dioxins. In: *Solving Hazardous Waste Problems*, edited by J. Exner, Washington, DC, American Chemical Society, pp. 1–19, 1987.
71. Fenske, M., R. van Aerle, S. Brack, C.R. Tyler and H. Segner. Development and validation of a homologous zebrafish (*Danio rerio* Hamilton-Buchanan) vitellogenin enzyme-linked immunosorbent assay (ELISA) and its application for studies on estrogenic chemicals. *Comp. Biochem. Physiol.* 129C: 217–232, 2001.
72. Fernandez-Salguero, P., T. Pineau, D.M. Hilbert, T. McPhail, S.S. Lee, S. Kimura, D.W. Nebert, S. Rudikoff, J.M. Ward and F.J. Gonzalez. Immune system impairment and hepatic fibrosis in mice lacking the dioxin-binding Ah receptor. *Science* 268: 722–726, 1995.
73. Fritz, A., M. Rozowski, C. Walker and M. Westerfield. Identification of selected gamma-ray induced deficiencies in zebrafish using multiplex polymerase chain reaction. *Genetics* 144: 1735–1745, 1996.
74. Geisler, R., G.J. Rauch, H. Baier, F. van Bebber, L. Brobeta, M.P. Dekens, K. Finger, C. Fricke, M.A. Gates, H. Geiger, S. Geiger-Rudolph, D. Gilmour, S. Glaser, L. Gnugge, H. Habeck, K. Hingst, S. Holley, J. Keenan, A. Kirn, H. Knaut, D. Lashkari, F. Maderspacher, U. Martyn, S. Neuhauss, C. Neumann, T. Nicolson, F. Pelegri, R. Ray, J.M. Rick, H. Roehl, T. Roeser, H.E. Schauerte, A.F. Schier, U. Schönberger, H.-B. Schönthaler, S. Schulte-Merker, C. Seydler, W.S. Talbot, C. Weiler, C. Nüsslein-Volhard and P. Haffter. A radiation hybrid map of the zebrafish genome. *Nat. Genet.* 23: 86–89, 1999.
75. George, S.G. and P. Young. The time course of effects of cadmium and 3-methylcholanthrene on activities of enzymes of xenobiotic metabolism and metallothionein levels in the plaice, *Pleuronectes platessa*. *Comp. Biochem. Physiol.* 83C: 37–44, 1986.
76. Gillesby, B.E. and T.R. Zacharewski. pS2 (TFF1) levels in human breast cancer tumor samples: correlation with clinical and histological prognostic markers. *Breast Cancer Res. Treat.* 56: 253–265, 1999.
77. Golling, G., A. Amsterdam, Z.X. Sun, M. Antonelli, E. Maldonado, W.B. Chen, S. Burgess, M. Haldi, K. Artzt, S. Farrington, S.Y. Lin, R.M. Nissen and N. Hopkins. Insertional mutagenesis in zebrafish rapidly identifies genes essential for early vertebrate development. *Nat. Genet.* 31: 135–140, 2002.
78. Grunwald, D.J. and J.S. Eisen. Headwaters of the zebrafish – emergence of a new model vertebrate. *Nat. Rev. Genet.* 3: 717–724, 2002.

79. Grunwald, D.J. and G. Streisinger. Induction of mutations in the zebrafish with ultraviolet light. *Genet. Res.* 59: 93–101, 1992.
80. Grunwald, D.J. and G. Streisinger. Induction of recessive lethal and specific locus mutations in the zebrafish with ethyl nitrosourea. *Genet. Res.* 59: 103–116, 1992.
81. Gunes, C., R. Heuchel, O. Georgiev, K.H. Muller, P. Lichtlen, H. Bluthmann, S. Marino, A. Aguzzi and W. Schaffner. Embryonic lethality and liver degeneration in mice lacking the metal-responsive transcriptional activator MTF-1. *EMBO J.* 17: 2846–2854, 1998.
82. Haffter, P., M. Granato, M. Brand, M.C. Mullins, M. Hammerschmidt, D.A. Kane, J. Odenthal, F.J.M. van Eeden, Y.J. Jiang, C.P. Heisenberg, R.N. Kelsh, M. Furutani-Seiki, E. Vogelsang, D. Beuchle, U. Schach, C. Fabian and C. Nusslein-Volhard. The identification of genes with unique and essential functions in the development of the zebrafish, *Danio rerio. Development* 123: 1–36, 1996.
83. Hahn, M.E. and J.J. Stegeman. Regulation of cytochrome P4501A1 in teleosts: sustained induction of CYP1A1 mRNA, protein, and catalytic activity by 2,3,7,8-tetrachlorodibenzofuran in the marine fish *Stenotomus chrysops. Toxicol. Appl. Pharmacol.* 127: 187–198, 1994.
84. Hahn, M.E., S.I. Karchner, M.A. Shapiro and S.A. Perera. Molecular evolution of two vertebrate aryl hydrocarbon (dioxin) receptors (AHR1 and AHR2) and the PAS family. *Proc. Natl Acad. Sci. USA* 94: 13743–13748, 1997.
85. Hahn, M.E. Aryl hydrocarbon receptors: diversity and evolution. *Chem.-Biol. Interact.* 141: 131–160, 2002.
86. Hahn, M.E. Dioxin toxicology and the aryl hydrocarbon receptor: insights from fish and other non-traditional models. *Mar. Biotech.* 3: S224–S238, 2001.
87. Halloran, M.C., M. Sato-Maeda, J.T. Warren, F. Su, Z. Lele, P.H. Krone, J.Y. Kuwada and W. Shoji. Laser-induced gene expression in specific cells of transgenic zebrafish. *Development* 127: 1953–1960, 2000.
88. Hammerschmidt, C.R., M.B. Sandheinrich, J.G. Wiener and R.G. Rada. Effects of dietary methylmercury on reproduction of fathead minnows. *Environ. Sci. Technol.* 36: 877–883, 2002.
89. Henry, T.R., J.M. Spitsbergen, M.W. Hornung, C.C. Abnet and R.E. Peterson. Early life stage toxicity of 2, 3,7,8-tetrachlorodibenzo-*p*-dioxin in zebrafish (*Danio rerio*). *Toxicol. Appl. Pharmacol.* 142: 56–68, 1997.
90. Heuchel, R., F. Radtke, O. Georgiev, G. Stark, M. Aguet and W. Schaffner. The transcription factor MTF-1 is essential for basal and heavy metal-induced metallothionein gene expression. *EMBO J.* 13: 2870–2875, 1994.
91. Hill, R.L., Jr. and D.M. Janz. Developmental estrogenic exposure in zebrafish (*Danio rerio*): I. Effects on sex ratio and breeding success. *Aquat. Toxicol.* 63: 417–429, 2003.
92. Hisaoka, K.K. and H.I. Battle. The normal developmental stages of the zebrafish, *Brachydanio rerio. J. Morphol.* 103: 311–328, 1958.
93. Hisaoka, K.K. and C.F. Firlit. Further studies on the embryonic development of the zebrafish, *Brachydanio rerio. J. Morphol.* 107: 205–225, 1960.
94. Hisaoka, K.K. and A.F. Hopper. Some effects of barbituric acid and diethylbarbituric acid on the development of the zebrafish, *Brachydanio rerio. Anat. Rec.* 129: 297–308, 1957.
95. Hisaoka, K.K. The effects of 2-acetylaminofluorene on the embryonic development of the zebrafish. I. Morphological studies. *Cancer Res.* 18: 527–535, 1958.
96. Horstgen-Schwark, G. Production of homozygous diploid zebrafish (*Brachydanio rerio*). *Aquaculture* 112: 25–37, 1993.
97. Hu, M.C., E.F. Chiang, S.K. Tong, W. Lai, N.C. Hsu, L.C. Wang and B.C. Chung. Regulation of steroidogenesis in transgenic mice and zebrafish. *Mol. Cell. Endocrinol.* 171: 9–14, 2001.
98. Huang, H.C., T. Nguyen and C.B. Pickett. Phosphorylation of Nrf2 at Ser-40 by protein kinase C regulates antioxidant response element-mediated transcription. *J. Biol. Chem.* 277: 42769–42774, 2002.
99. Hukriede, N.A., L. Joly, M. Tsang, J. Miles, P. Tellis, J.A. Epstein, W.B. Barbazuk, F.N. Li, B. Paw, J.H. Postlethwait, T.J. Hudson, L.I. Zon, J.D. McPherson, M. Chevrette, I.B. Dawid, S.L. Johnson and M. Ekker. Radiation hybrid mapping of the zebrafish genome. *Proc. Natl Acad. Sci. USA* 96: 9745–9750, 1999.
100. Islinger, M., D. Willimski, A. Volkl and T. Braunbeck. Effects of 17a-ethinylestradiol on the expression of three estrogen-responsive genes and cellular ultrastructure of liver and testes in male zebrafish. *Aquat. Toxicol.* 62: 85–103, 2003.
101. Itoh, K., T. Chiba, S. Takahashi, T. Ishii, K. Igarashi, Y. Katoh, T. Oyake, N. Hayashi, K. Satoh, I. Hatayama, M. Yamamoto and Y. Nabeshima. An Nrf2/small Maf heterodimer mediates the induction of phase II detoxifying enzyme genes through antioxidant response elements. *Biochem. Biophys. Res. Commun.* 236: 313–322, 1997.
102. Itoh, K., N. Wakabayashi, Y. Katoh, T. Ishii, K. Igarashi, J.D. Engel and M. Yamamoto. Keap1 represses nuclear activation of antioxidant responsive elements by Nrf2 through binding to the amino-terminal Neh2 domain. *Genes Dev.* 13: 76–86, 1999.

103. Jones, R.W. and M.N. Huffman. Fish embryos as bioassay material in testing chemicals for effects on cell division and differentiation. *Trans. Am. Microsc. Soc.* 76: 177–183, 1957.
104. Jones, R.W., W.C. Gibson and C.J. Nickolls. Factors influencing mitotic activity and morphogenesis in embryonic development. I. The effects of thyroxine and thiouracil on the development of *Brachydanio rerio* (zebra fish). *Anat. Rec.* 111: 93, 1951.
105. Kantarjian, H.M., S. O'Brien, J. Cortes, F.J. Giles, S. Faderl, J.P. Issa, G. Garcia-Manero, M.B. Rios, J. Shan, M. Andreeff, M. Keating and M. Talpaz. Results of decitabine (5-aza-2′deoxycytidine) therapy in 130 patients with chronic myelogenous leukemia. *Cancer* 98: 522–528, 2003.
106. Kavlock, R.J. and G.T. Ankley. A perspective on the risk assessment process for endocrine-disruptive effects on wildlife and human health. *Risk Anal.* 16: 731–739, 1996.
107. Kawakami, K. and A. Shima. Identification of the Tol2 transposase of the medaka fish *Oryzias latipes* that catalyzes excision of a nonautonomous Tol2 element in zebrafish *Danio rerio*. *Gene* 240: 239–244, 1999.
108. Kawakami, K., A. Koga, H. Hori and A. Shima. Excision of the tol2 transposable element of the medaka fish, *Oryzias latipes*, in zebrafish, *Danio rerio*. *Gene* 225: 17–22, 1998.
109. Kawakami, K., A. Shima and N. Kawakami. Identification of a functional transposase of the Tol2 element, an Ac-like element from the Japanese medaka fish, and its transposition in the zebrafish germ lineage. *Proc. Natl Acad. Sci. USA* 97: 11403–11408, 2000.
110. Kazeto, Y., S. Ijiri, A.R. Place, Y. Zohar and J.M. Trant. The 5′-flanking regions of CYP19A1 and CYP19A2 in zebrafish. *Biochem. Biophys. Res. Commun.* 288: 503–508, 2001.
111. Khudolei, V.V. The use of the aquarium fishes *Danio rerio* and *Poecilia reticulata* as highly sensitive species for testing the carcinogenicity of chemical compounds. *Eksp. Onkol.* 9: 40–46, 1987.
112. Khudoley, V.V. Use of aquarium fish, *Danio rerio* and *Poecilia reticulata*, as test species for evaluation of nitrosamine carcinogenicity. *Natl Cancer Inst. Monogr.* 65: 65–70, 1984.
113. Kihlstrom, J.E. and L. Hulth. The effect of phenylmercuric acetate upon the frequency of hatching of eggs from the zebrafish. *Bull. Environ. Contam. Toxicol.* 7: 111–114, 1972.
114. Kimmel, C.B., W.W. Ballard, S.R. Kimmel, B. Ullmann and T.F. Schilling. Stages of embryonic development of the zebrafish. *Dev. Dyn.* 203: 253–310, 1995.
115. Kishida, M. and G.V. Callard. Distinct cytochrome P450 aromatase isoforms in zebrafish (*Danio rerio*) brain and ovary are differentially programmed and estrogen regulated during early development. *Endocrinology* 142: 740–750, 2001.
116. Kishida, M., M. McLellan, J.A. Miranda and G.V. Callard. Estrogen and xenoestrogens upregulate the brain aromatase isoform (P450aromB) and perturb markers of early development in zebrafish (*Danio rerio*). *Comp. Biochem. Physiol.* 129B: 261–268, 2001.
117. Kobayashi, M., K. Itoh, T. Suzuki, H. Osanai, K. Nishikawa, Y. Katoh, Y. Takagi and M. Yamamoto. Identification of the interactive interface and phylogenic conservation of the Nrf2–Keap1 system. *Genes Cells* 7: 807–820, 2002.
118. Korach, K.S., J.F. Couse, S.W. Curtis, T.F. Washburn, J. Lindzey, K.S. Kimbro, E.M. Eddy, S. Migliaccio, S.M. Snedeker, D.B. Lubahn, D.W. Schomberg and E.P. Smith. Estrogen receptor gene disruption: molecular characterization and experimental and clinical phenotypes. *Recent Prog. Horm. Res.* 51: 159–186, 1996.
119. Korach, K.S. Estrogen receptor knock-out mice: molecular and endocrine phenotypes. *J. Soc. Gynecol. Invest.* 7: S16–S17, 2000.
120. Kotsanis, N. and J. Iliopoulou-Georgudaki. Arsenic induced liver hyperplasia and kidney fibrosis in rainbow trout (*Oncorhynchus mykiss*) by microinjection technique: a sensitive animal bioassay for environmental metal-toxicity. *Bull. Environ. Contam. Toxicol.* 62: 169–178, 1999.
121. Krone, P.H. and J.B. Sass. HSP 90 alpha and HSP 90 beta genes are present in the zebrafish and are differentially regulated in developing embryos. *Biochem. Biophys. Res. Commun.* 204: 746–752, 1994.
122. Krone, P.H., Z. Lele and J.B. Sass. Heat shock genes and the heat shock response in zebrafish embryos. *Biochem. Cell Biol.* 75: 487–497, 1997.
123. Krone, P.H., J.B. Sass and Z. Lele. Heat shock protein gene expression during embryonic development of the zebrafish. *Cell. Mol. Life Sci.* 53: 122–129, 1997.
124. Kwak, M.K., K. Itoh, M. Yamamoto and T.W. Kensler. Enhanced expression of the transcription factor Nrf2 by cancer chemopreventive agents: role of antioxidant response element-like sequences in the nrf2 promoter. *Mol. Cell Biol.* 22: 2883–2892, 2002.
125. Kwok, C., R. Critcher and K. Schmitt. Construction and characterization of zebrafish whole genome radiation hybrids. *Methods Cell Biol.* 60: 287–302, 1999.
126. Kwok, C., R.M. Korn, M.E. Davis, D.W. Burt, R. Critcher, L. McCarthy, B.H. Paw, L.I. Zon, P.N. Goodfellow and K. Schmitt. Characterization of whole genome radiation hybrid mapping resources for non-mammalian vertebrates. *Nucleic Acids Res.* 26: 3562–3566, 1998.

127. Laale, H.W. Ethanol induced notochord and spinal cord duplications in the embryo of the zebrafish, *Brachydanio rerio*. *J. Exp. Zool.* 177: 51–64, 1971.
128. Laale, H.W. The biology and use of the zebrafish *Brachydanio rerio* in fisheries research. A literature review. *J. Fish Biol.* 10: 121–173, 1977.
129. Lai, W.W., P.H. Hsiao, Y. Guiguen and B.C. Chung. Cloning of zebrafish cDNA for 3beta-hydroxysteroid dehydrogenase and P450scc. *Endocr. Res.* 24: 927–931, 1998.
130. Landner, L., A.H. Neilson, L. Sorensen, A. Tarnholm and T. Viktor. Short-term test for predicting the potential of xenobiotics to impair reproductive success in fish. *Ecotoxicol. Environ. Saf.* 9: 282–293, 1985.
131. Langenau, D.M., D. Traver, A.A. Ferrando, J.L. Kutok, J.C. Aster, J.P. Kanki, S. Lin, E. Prochownik, N.S. Trede, L.I. Zon and A.T. Look. Myc-induced T cell leukemia in transgenic zebrafish. *Science* 299: 887–890, 2003.
132. Lassiter, C.S., B. Kelley and E. Linney. Genomic structure and embryonic expression of estrogen receptor beta a (ERbetaa) in zebrafish (*Danio rerio*). *Gene* 299: 141–151, 2002.
133. Latif, M.A., R.A. Bodaly, T.A. Johnston and R.J. Fudge. Effects of environmental and maternally derived methylmercury on the embryonic and larval stages of walleye (*Stizostedion vitreum*). *Environ. Pollut.* 111: 139–148, 2001.
134. Leaver, M.J., J. Wright and S.G. George. Structure and expression of a cluster of glutathione S-transferase genes from a marine fish, the plaice (*Pleuronectes platessa*). *Biochem. J.* 321: 405–412, 1997.
135. Legler, J., C.E. van den Brink, A. Brouwer, A.J. Murk, P.T. van der Saag, A.D. Vethaak and B. Van der Burg. Development of a stably transfected estrogen receptor-mediated luciferase reporter gene assay in the human T47D breast cancer cell line. *Toxicol. Sci.* 48: 55–66, 1999.
136. Legler, J., L.M. Zeinstra, F. Schuitemaker, P.H. Lanser, J. Bogerd, A. Brouwer, A.D. Vethaak, P. De Voogt, A.J. Murk and B. Van der Burg. Comparison of in vivo and in vitro reporter gene assays for short-term screening of estrogenic activity. *Environ. Sci. Technol.* 36: 4410–4415, 2002.
137. Lele, Z., S. Engel and P.H. Krone. hsp47 and hsp70 gene expression is differentially regulated in a stress- and tissue-specific manner in zebrafish embryos. *Dev. Genet.* 21: 123–133, 1997.
138. Lele, Z., S.D. Hartson, C.C. Martin, L. Whitesell, R.L. Matts and P.H. Krone. Disruption of zebrafish somite development by pharmacologic inhibition of hsp90. *Dev. Biol.* 210: 56–70, 1999.
139. Lewis, W.H. and E.C. Roosen-Runge. The formation of the blastodisc in the egg of the zebrafish, *Brachydanio rerio*. *Anat. Rec.* 85: 326 1943.
140. Liao, Z.Y., S.H. Zhang and Y.S. Zhen. Synergistic effects of geldanamycin and antitumor drugs. *Yao Xue Xue Bao* 36: 569–575, 2001.
141. Lichtlen, P. and W. Schaffner. The 'metal transcription factor' MTF-1: biological facts and medical implications. *Swiss Med. Wkly* 131: 647–652, 2001.
142. Lindesjoo, E., M. Adolfsson-Erici, G. Ericson and L. Forlin. Biomarker responses and resin acids in fish chronically exposed to effluents from a total chlorine-free pulp mill during regular production. *Ecotoxicol. Environ. Saf.* 53: 238–247, 2002.
143. Long, Q., A. Meng, H. Wang, J.R. Jessen, M.J. Farrell and S. Lin. GATA-1 expression pattern can be recapitulated in living transgenic zebrafish using GFP reporter gene. *Development* 124: 4105–4111, 1997.
144. Ma, C., L. Fan, R. Ganassin, N. Bols and P. Collodi. Production of zebrafish germ-line chimeras from embryo cell cultures. *Proc. Natl Acad. Sci. USA* 98: 2461–2466, 2001.
145. Martin, C.C. and R. McGowan. Genotype-specific modifiers of transgene methylation and expression in the zebrafish, *Danio rerio*. *Genet. Res.* 65: 21–28, 1995.
146. Martin, C.C., L. Laforest, M.A. Akimenko and M. Ekker. A role for DNA methylation in gastrulation and somite patterning. *Dev. Biol.* 206: 189–205, 1999.
147. Matthiessen, P. and J.P. Sumpter. Effects of estrogenic substances in the aquatic environment. *EXS* 86: 319–335, 1998.
148. Mattingly, C.J. and W.A. Toscano. Posttranscriptional silencing of cytochrome P4501A1 (CYP1A1) during zebrafish (*Danio rerio*) development. *Dev. Dyn.* 222: 645–654, 2001.
149. Mattingly, C.J., J.A. McLachlan and W.A. Toscano, Jr.. Green fluorescent protein (GFP) as a marker of aryl hydrocarbon receptor (AhR) function in developing zebrafish (*Danio rerio*). *Environ. Health Perspect.* 109: 845–849, 2001.
150. Melancon, M.J., C.R. Elcombe, M.J. Vodicnik and J.J. Lech. Induction of cytochromes P450 and mixed-function oxidase activity by polychlorinated biphenyls and beta-naphthoflavone in carp (*Cyprinus carpio*). *Comp. Biochem. Physiol.* 69C: 219–226, 1981.
151. Menuet, A., E. Pellegrini, I. Anglade, O. Blaise, V. Laudet, O. Kah and F. Pakdel. Molecular characterization of three estrogen receptor forms in zebrafish: binding characteristics, transactivation properties, and tissue distributions. *Biol. Reprod.* 66: 1881–1892, 2002.

152. Mimura, J. and Y. Fujii-Kuriyama. Functional role of AhR in the expression of toxic effects by TCDD. *Biochim. Biophys. Acta* 1619: 263–268, 2003.
153. Mimura, J., K. Yamashita, K. Nakamura, M. Morita, T.N. Takagi, K. Nakao, M. Ema, K. Sogawa, M. Yasuda, M. Katsuki and Y. Fujii-Kuriyama. Loss of teratogenic response to 2,3,7,8-tetrachlorodibenzo-*p*-dioxin (TCDD) in mice lacking the Ah (dioxin) receptor. *Genes Cells* 2: 645–654, 1997.
154. Miranda, C.L., P. Collodi, X. Zhao, D.W. Barnes and D.R. Buhler. Regulation of cytochrome P450 expression in a novel liver cell line from zebrafish (*Brachydanio rerio*). *Arch. Biochem. Biophys.* 305: 320–327, 1993.
155. Mitsiades, C.S., N. Mitsiades, P.G. Richardson, S.P. Treon and K.C. Anderson. Novel biologically based therapies for Waldenstrom's macroglobulinemia. *Semin. Oncol.* 30: 309–312, 2003.
156. Mizell, M. and E.S. Romig. The aquatic vertebrate embryo as a sentinel for toxins: zebrafish embryo dechorionation and perivitelline space microinjection. *Int. J. Dev. Biol.* 41: 411–423, 1997.
157. Morimoto, R.I., K.D. Sarge and K. Abravaya. Transcriptional regulation of heat shock genes. A paradigm for inducible genomic responses. *J. Biol. Chem.* 267: 21987–21990, 1992.
158. Motoike, T., S. Loughna, E. Perens, B.L. Roman, W. Liao, T.C. Chau, C.D. Richardson, T. Kawate, J. Kuno, B.M. Weinstein, D.Y. Stainier and T.N. Sato. Universal GFP reporter for the study of vascular development. *Genesis* 28: 75–81, 2000.
159. Murk, A.J., T.J. Boudewijn, P.L. Menninger, A.T. Bosveld, G. Rossaert, T. Ysebaert, P. Meire and S. Dirksen. Effects of polyhalogenated aromatic hydrocarbons and related contaminants on common tern reproduction: integration of biological, biochemical, and chemical data. *Arch. Environ. Contam. Toxicol.* 31: 128–140, 1996.
160. Murk, A.J., J. Legler, M.S. Denison, J.P. Giesy, C. van de Gucthe and A. Brouwer. Chemical-activated luciferase gene expression (CALUX): a novel *in vitro* bioassay for Ah receptor active compounds in sediments and pore water. *Fundam. Appl. Toxicol.* 33: 149–160, 1996.
161. Myers, D.C., D.S. Sepich and L. Solnica-Krezel. Convergence and extension in vertebrate gastrulae: cell movements according to or in search of identity? *Trends Genet.* 18: 447–455, 2002.
162. Nagel, R. and T. Dar. The embryo test with the zebrafish *Danio rerio* – a general model in ecotoxicology and toxicology. *ALTEX* 19(Suppl. 1): 38–48, 2002.
163. Nasevicius, A. and S.C. Ekker. Effective targeted gene 'knockdown' in zebrafish. *Nat. Genet.* 26: 216–220, 2000.
164. Neilson, A.H., A.S. Allard, S. Fischer, M. Malmberg and T. Viktor. Incorporation of a subacute test with zebra fish into a hierarchical system for evaluating the effect of toxicants in the aquatic environment. *Ecotoxicol. Environ. Saf.* 20: 82–97, 1990.
165. Niimi, A.J. and Q.N. LaHam. Relative toxicity of organic and inorganic compounds of selenium to newly hatched zebrafish (*Brachydanio rerio*). *Can. J. Zool.* 54: 501–509, 1976.
166. Oliveira Ribeiro, C.A., L. Belger, E. Pelletier and C. Rouleau. Histopathological evidence of inorganic mercury and methyl mercury toxicity in the arctic charr (*Salvelinus alpinus*). *Environ. Res.* 90: 217–225, 2002.
167. Olsson, P.E., P. Kling, L.J. Erkell and P. Kille. Structural and functional analysis of the rainbow trout (*Oncorhynchus mykiss*) metallothionein-A gene. *Eur. J. Biochem.* 230: 344–349, 1995.
168. Orn, S., P.L. Andersson, L. Förlin, M. Tysklind and L. Norrgren. The impact on reproduction of an orally administered mixture of selected PCBs in zebrafish (*Danio rerio*). *Arch. Environ. Contam. Toxicol.* 35: 52–57, 1998.
169. Pedrajas, J.R., J. Peinado and J. Lopez-Barea. Oxidative stress in fish exposed to model xenobiotics. Oxidatively modified forms of Cu, Zn-superoxide dismutase as potential biomarkers. *Chem.-Biol. Interact.* 98: 267–282, 1995.
170. Perez, L.M., M.C. Novoa Valinas and M.J. Melgar Riol. Induction of cytosolic glutathione S-transferases from Atlantic eel (*Anguilla anguilla*) after intraperitoneal treatment with polychlorinated biphenyls. *Sci. Total Environ.* 297: 141–151, 2002.
171. Perz-Edwards, A., N.L. Hardison and E. Linney. Retinoic acid-mediated gene expression in transgenic reporter zebrafish. *Dev. Biol.* 229: 89–101, 2001.
172. Peterson, R.E., H.M. Theobald and G.L. Kimmel. Developmental and reproductive toxicity of dioxins and related compounds: cross-species comparisons. *Crit. Rev. Toxicol.* 23: 283–335, 1993.
173. Peterson, R.T., B.A. Link, J.E. Dowling and S.L. Schreiber. Small molecule developmental screens reveal the logic and timing of vertebrate development. *Proc. Natl Acad. Sci. USA* 97: 12965–12969, 2000.
174. Picker, A., S. Scholpp, H. Bohli, H. Takeda and M. Brand. A novel positive transcriptional feedback loop in midbrain–hindbrain boundary development is revealed through analysis of the zebrafish pax2.1 promoter in transgenic lines. *Development* 129: 3227–3239, 2002.

175. Pliss, G.B., M.A. Zabezhinski, A.S. Petrov and V.V. Khudoley. Peculiarities of N-nitramines carcinogenic action. *Arch. Geschwulstforsch* 52: 629–634, 1982.
176. Poland, A. and J.C. Knutson. 2,3,7,8-tetrachlorodibenzo-*p*-dioxin and related halogenated aromatic hydrocarbons: examination of the mechanism of toxicity. *Annu. Rev. Pharmacol. Toxicol.* 22: 517–554, 1982.
177. Pollenz, R.S., H.R. Sullivan, J. Holmes, B. Necela and R.E. Peterson. Isolation and expression of cDNAs from rainbow trout (*Oncorhynchus mykiss*) that encode two novel basic helix–loop–helix/PER-ARNT-SIM (bHLH/PAS) proteins with distinct functions in the presence of the aryl hydrocarbon receptor. Evidence for alternative mRNA splicing and dominant negative activity in the bHLH/PAS family. *J. Biol. Chem.* 271: 30886–30896, 1996.
178. Rabergh, C.M., S. Airaksinen, A. Soitamo, H.V. Bjorklund, T. Johansson, M. Nikinmaa and L. Sistonen. Tissue-specific expression of zebrafish (*Danio rerio*) heat shock factor 1 mRNAs in response to heat stress. *J. Exp. Biol.* 203: 1817–1824, 2000.
179. Ramaswamy, S., M. el Ahmad, O. Danielsson, H. Jörnvall and H. Eklund. Crystal structure of cod liver class I alcohol dehydrogenase: substrate pocket and structurally variable segments. *Protein Sci.* 5: 663–671, 1996.
180. Raz, E., H.G. van Luenen, B. Schaerringer, R.H. Plasterk and W. Driever. Transposition of the nematode *Caenorhabditis elegans* Tc3 element in the zebrafish *Danio rerio*. *Curr. Biol.* 8: 82–88, 1998.
181. Redding, J.M. and R. Patiño. Reproductive physiology. In: *The Physiology of Fishes*, edited by D.H. Evans, Boca Raton, FL, CRS, pp. 503–534, 1993.
182. Rehwoldt, R. and D. Karimian-Teherani. Uptake and effect of cadmium on zebrafish. *Bull. Environ. Contam. Toxicol.* 15: 442–446, 1976.
183. Roosen-Runge, E.C. On the early development – bipolar differentiation and cleavage – of the zebrafish, *Brachydanio rerio*. *Biol. Bull.* 75: 119–133, 1938.
184. Rose, J., H. Holbech, C. Lindholst, U. Norum, A. Povlsen, B. Korsgaard and P. Bjerregaard. Vitellogenin induction by 17beta-estradiol and 17alpha-ethinylestradiol in male zebrafish (*Danio rerio*). *Comp. Biochem. Physiol.* 131C: 531–539, 2002.
185. Rowlands, J.C., I.L. McEwan and J.A. Gustafsson. Trans-activation by the human aryl hydrocarbon receptor and aryl hydrocarbon receptor nuclear translocator proteins: direct interactions with basal transcription factors. *Mol. Pharmacol.* 50: 538–548, 1996.
186. Roy, N.K. and I. Wirgin. Characterization of the aromatic hydrocarbon receptor gene and its expression in Atlantic tomcod. *Arch. Biochem. Biophys.* 344: 373–386, 1997.
187. Samson, S.L. and L. Gedamu. Metal-responsive elements of the rainbow trout metallothionein-B gene function for basal and metal-induced activity. *J. Biol. Chem.* 270: 6864–6871, 1995.
188. Sanchez-Galan, S., A.R. Linde, F. Ayllon and E. Garcia-Vazquez. Induction of micronuclei in eel (*Anguilla anguilla* L.) by heavy metals. *Ecotoxicol. Environ. Saf.* 49: 139–143, 2001.
189. Saunthararajah, Y., C.A. Hillery, D. Lavelle, R. Molokie, L. Dorn, L. Bressler, S. Gavazova, Y.H. Chen, R. Hoffman and J. DeSimone. Effects of 5-aza-2′-deoxycytidine on fetal hemoglobin levels, red cell adhesion, and hematopoietic differentiation in patients with sickle cell disease. *Blood* 2003.
190. Schlenk, D., Y. Sapozhnikova, J.P. Baquirian and A. Mason. Predicting chemical contaminants in freshwater sediments through the use of historical biochemical endpoints in resident fish species. *Environ. Toxicol. Chem.* 21: 2138–2145, 2002.
191. Schlenk, D., Y.S. Zhang and J. Nix. Expression of hepatic metallothionein messenger RNA in feral and caged fish species correlates with muscle mercury levels. *Ecotoxicol. Environ. Saf.* 31: 282–286, 1995.
192. Schmidt, J.V., G.H. Su, J.K. Reddy, M.C. Simon and C.A. Bradfield. Characterization of a murine Ahr null allele: involvement of the Ah receptor in hepatic growth and development. *Proc. Natl Acad. Sci. USA* 93: 6731–6736, 1996.
193. Schreurs, R.H., M.E. Quaedackers, W. Seinen and B. Van der Burg. Transcriptional activation of estrogen receptor ERalpha and ERbeta by polycyclic musks is cell type dependent. *Toxicol. Appl. Pharmacol.* 183: 1–9, 2002.
194. Skidmore, J.F. Resistance to zinc sulphate of the zebrafish (*Brachydanio rerio* Hamilton-Buchanan) at different phases of its life history. *Ann. Appl. Biol.* 56: 47–53, 1965.
195. Snoek, E.F.. A multi-variable study of the interaction of light, temperature, and acridine orange on the embryos and larvae of the zebrafish, *Brachydanio rerio* (Hamilton-Buchanan). *Diss. Abst.* 31: 4520B 1970.
196. Speranza, A.W., R.J. Seeley, V.A. Seeley and A. Perlmutter. The effect of sublethal concentrations of zinc on reproduction in the zebrafish, *Brachydanio rerio* (Hamilton-Buchanan). *Environ. Pollut.* 12: 217–222, 1977.
197. Spitsbergen, J.M. and M.L. Kent. The state of the art of the zebrafish model for toxicology and toxicologic pathology research – advantages and current limitations. *Toxicol. Pathol.* 31(Suppl.): 62–87, 2003.

198. Spitsbergen, J.M., H.W. Tsai, A. Reddy, T. Miller, D. Arbogast, J.D. Hendricks and G.S. Bailey. Neoplasia in zebrafish (*Danio rerio*) treated with *N*-methyl-*N'*-nitro-*N*-nitrosoguanidine by three exposure routes at different developmental stages. *Toxicol. Pathol.* 28: 716–725, 2000.
199. Spitsbergen, J.M., H.W. Tsai, A. Reddy, T. Miller, D. Arbogast, J.D. Hendricks and G.S. Bailey. Neoplasia in zebrafish (*Danio rerio*) treated with 7,12-dimethylbenz[*a*]anthracene by two exposure routes at different developmental stages. *Toxicol. Pathol.* 28: 705–715, 2000.
200. Sprague, J.B.. Current status of sublethal tests of pollutants on aquatic organisms. *J. Fish. Res. Board Can.* 33: 1988–1992, 1976.
201. Sreedhar, A.S., K. Mihaly, B. Pato, T. Schnaider, A. Stetak, K. Kis-Petik, J. Fidy, T. Simonics, A. Maraz and P. Csermely. Hsp90 inhibition accelerates cell lysis: anti-Hsp90 ribozyme reveals a complex mechanism of Hsp90 inhibitors involving both superoxide- and Hsp90-dependent events. *J. Biol. Chem.* 278: 35231–35240, 2003.
202. Stanton, M. Diethylnitrosamine-induced hepatic degeneration and neoplasia in the aquarium fish, *Brachydanio rerio. J. Natl Cancer Inst.* 34: 117–130, 1965.
203. Stegeman, J.J. and R.L. Binder. High benzo[*a*]pyrene hydroxylase activity in the marine fish *Stenotomus versicolor. Biochem. Pharmacol.* 28: 1686–1688, 1979.
204. Stephensen, E., J. Sturve and L. Forlin. Effects of redox cycling compounds on glutathione content and activity of glutathione-related enzymes in rainbow trout liver. *Comp. Biochem. Physiol.* 133C: 435–442, 2002.
205. Streisinger, G., F. Singer, C. Walker, D. Knauber and N. Dower. Segregation analyses and gene-centromere distances in zebrafish. *Genetics* 112: 311–319, 1986.
206. Streisinger, G., C. Walker, N. Dower, D. Knauber and F. Singer. Production of clones of homozygous diploid zebra fish (*Brachydanio rerio*). *Nature* 291: 293–296, 1981.
207. Streisinger, G. Attainment of minimal biological variability and measurements of genotoxicity: production of homozygous diploid zebra fish. *Natl Cancer Inst. Monogr.* 65: 53–58, 1984.
208. Sumpter, J.P. and S. Jobling. Vitellogenesis as a biomarker for estrogenic contamination of the aquatic environment. *Environ. Health Perspect.* 103(Suppl. 7): 173–178, 1995.
209. Sumpter, J.P. Feminized responses in fish to environmental estrogens. *Toxicol. Lett.* 82–83: 737–742, 1995.
210. Sumpter, J.P. Reproductive effects from oestrogen activity in polluted water. *Arch. Toxicol.* (Suppl. 20): 143–150, 1998.
211. Suresh, A., B. Sivaramakrishna, P.C. Victoriamma and K. Radhakrishnaiah. Comparative study on the inhibition of acetylcholinesterase activity in the freshwater fish *Cyprinus carpio* by mercury and zinc. *Biochem. Int.* 26: 367–375, 1992.
212. Suwa, H., L. Saint-Amant, A. Triller, P. Drapeau and P. Legendre. High-affinity zinc potentiation of inhibitory postsynaptic glycinergic currents in the zebrafish hindbrain. *J. Neurophysiol.* 85: 912–925, 2001.
213. Talalay, P., M.J. De Long and H.J. Prochaska. Identification of a common chemical signal regulating the induction of enzymes that protect against chemical carcinogenesis. *Proc. Natl Acad. Sci. USA* 85: 8261–8265, 1988.
214. Tanguay, R.L., C.C. Abnet, W. Heideman and R.E. Peterson. Cloning and characterization of the zebrafish (*Danio rerio*) aryl hydrocarbon receptor. *Biochim. Biophys. Acta* 1444: 35–48, 1999.
215. Tanguay, R.L., E. Andreasen, W. Heideman and R.E. Peterson. Identification and expression of alternatively spliced aryl hydrocarbon nuclear translocator 2 (ARNT2) cDNAs from zebrafish with distinct functions. *Biochim. Biophys. Acta* 1494: 117–128, 2000.
216. Tanguay, R.L., E. Andreasen, M.K. Walker and R.E. Peterson. Dioxin toxicity and aryl hydrocarbon receptor signaling in fish. In: *Dioxins and Health*, 2nd ed., edited by T.A. Gasiewicz and A. Schecter, NewYork, Wiley, pp. 603–628, 2003, Chapter 15.
217. Tchoudakova, A., M. Kishida, E. Wood and G.V. Callard. Promoter characteristics of two cyp19 genes differentially expressed in the brain and ovary of teleost fish. *J. Steroid Biochem. Mol. Biol.* 78: 427–439, 2001.
218. Teraoka, H., W. Dong, S. Ogawa, S. Tsukiyama, Y. Okuhara, M. Niiyama, N. Ueno, R.E. Peterson and T. Hiraga. 2,3,7,8-Tetrachlorodibenzo-*p*-dioxin toxicity in the zebrafish embryo: altered regional blood flow and impaired lower jaw development. *Toxicol. Sci.* 65: 192–199, 2002.
219. Ton, C., D. Stamatiou, V.J. Dzau and C.C. Liew. Construction of a zebrafish cDNA microarray: gene expression profiling of the zebrafish during development. *Biochem. Biophys. Res. Commun.* 296: 1134–1142, 2002.
220. Trant, J.M., S. Gavasso, J. Ackers, B.C. Chung and A.R. Place. Developmental expression of cytochrome P450 aromatase genes (CYP19a and CYP19b) in zebrafish fry (*Danio rerio*). *J. Exp. Zool.* 290: 475–483, 2001.
221. Van den, B.K., R. Verheyen and H. Witters. Reproductive effects of ethynylestradiol and 4t-octylphenol on the zebrafish (*Danio rerio*). *Arch. Environ. Contam. Toxicol.* 41: 458–467, 2001.

222. Van den Belt, K., P.W. Wester, L.T. van der Ven, R. Verheyen and H. Witters. Effects of ethynylestradiol on the reproductive physiology in zebrafish (*Danio rerio*): time dependency and reversibility. *Environ. Toxicol. Chem.* 21: 767–775, 2002.
223. Van Leeuwen, C.J., E.M. Grootelaar and G. Niebeek. Fish embryos as teratogenicity screens: a comparison of embryotoxicity between fish and birds. *Ecotoxicol. Environ. Saf.* 20: 42–52, 1990.
224. Vascotto, S.G., Y. Beckham and G.M. Kelly. The zebrafish's swim to fame as an experimental model in biology. *Biochem. Cell Biol.* 75: 479–485, 1997.
225. Ventura, E.C., L.R. Gaelzer, J. Zanette, M.R. Marques and A.C. Bainy. Biochemical indicators of contaminant exposure in spotted pigfish (*Orthopristis ruber*) caught at three bays of Rio de Janeiro coast. *Mar. Environ. Res.* 54: 775–779, 2002.
226. Walker, C.. Haploid screens and gamma-ray mutagenesis. *Methods Cell Biol.* 60: 43–70, 1999.
227. Walker, M.K. and R.E. Peterson. Toxicity of 2,3,7,8-tetrachlorodibenzo-*p*-dioxin (TCDD) to brook trout (*Salvelinus fontinalis*) during early development. *Environ. Toxicol. Chem.* 113: 817–820, 1994.
228. Walker, M.K., P.M. Cook, B.C. Butterworth, E.W. Zabel and R.E. Peterson. Potency of a complex mixture of polychlorinated dibenzo-*p*-dioxin, dibenzofuran, and biphenyl congeners compared to 2,3,7,8-tetrachlorodibenzo-*p*-dioxin in causing fish early life stage mortality. *Fundam. Appl. Toxicol.* 30: 178–186, 1996.
229. Wang, G., H. Huang, R. Dai, K.Y. Lee, S. Lin and N.F. Mivechi. Suppression of heat shock transcription factor HSF1 in zebrafish causes heat-induced apoptosis. *Genesis* 30: 195–197, 2001.
230. Wang, Y. and W. Ge. Cloning of zebrafish ovarian carbonyl reductase-like 20 beta-hydroxysteroid dehydrogenase and characterization of its spatial and temporal expression. *Gen. Comp. Endocrinol.* 127: 209–216, 2002.
231. Wang, Y. and W. Ge. Involvement of cyclic adenosine $3',5'$-monophosphate in the differential regulation of activin betaA and betaB expression by gonadotropin in the zebrafish ovarian follicle cells. *Endocrinology* 144: 491–499, 2003.
232. Wannemacher, R., A. Rebstock, E. Kulzer, D. Schrenk and K.W. Bock. Effects of 2,3,7,8-tetrachlorodibenzo-*p*-dioxin on reproduction and oogenesis in zebrafish (*Brachydanio rerio*). *Chemosphere* 24: 1361–1368, 1992.
233. Weber, L.P., R.L. Hill, Jr. and D.M. Janz. Developmental estrogenic exposure in zebrafish (*Danio rerio*): II. Histological evaluation of gametogenesis and organ toxicity. *Aquat. Toxicol.* 63: 431–446, 2003.
234. Wendelaar Bonga, S.E. Endocrinology. In: *The Physiology of Fishes*, edited by D.H. Evans, Boca Raton, FL, CRS, pp. 469–502, 1993.
235. Westerfield, M., E. Doerry, A.E. Kirkpatrick, W. Driever and S.A. Douglas. An on-line database for zebrafish development and genetics research. *Semin. Cell Dev. Biol.* 8: 477–488, 1997.
236. Whitlock, J.P., Jr. Induction of cytochrome P4501A1. *Annu. Rev. Pharmacol. Toxicol.* 39: 103–125, 1999.
237. Winn, R.N., M.B. Norris, K.J. Brayer, C. Torres and S.L. Muller. Detection of mutations in transgenic fish carrying a bacteriophage lambda cII transgene target. *Proc. Natl Acad. Sci. USA* 97: 12655–12660, 2000.
238. Yamazaki, K., H. Teraoka, W. Dong, J.J. Stegeman and T. Hiraga. cDNA cloning and expressions of cytochrome P450 1A in zebrafish embryos. *J. Vet. Med. Sci.* 64: 829–833, 2002.
239. Yan, C.H. and K.M. Chan. Characterization of zebrafish metallothionein gene promoter in a zebrafish caudal fin cell-line, SJD 1. *Mar. Environ. Res.* 54: 335–339, 2002.
240. Yan, Y.L., W.S. Talbot, E.S. Egan and J.H. Postlethwait. Mutant rescue by BAC clone injection in zebrafish. *Genomics* 50: 287–289, 1998.
241. Yelon, D. Cardiac patterning and morphogenesis in zebrafish. *Dev. Dyn.* 222: 552–563, 2001.
242. Yoder, J.A., M.G. Mueller, S. Wei, B.C. Corliss, D.M. Prather, T. Willis, R.T. Litman, J.Y. Djeu and G.W. Litman. Immune-type receptor genes in zebrafish share genetic and functional properties with genes encoded by the mammalian leukocyte receptor cluster. *Proc. Natl Acad. Sci. USA* 98: 6771–6776, 2001.
243. Zabel, E.W., M.K. Walker, M.W. Hornung, M.K. Clayton and R.E. Peterson. Interactions of polychlorinated dibenzo-*p*-dioxin, dibenzofuran, and biphenyl congeners for producing rainbow trout early life stage mortality. *Toxicol. Appl. Pharmacol.* 134: 204–213, 1995.
244. Zerulla, M., R. Lange, T. Steger-Hartmann, G. Panter, T. Hutchinson and D.R. Dietrich. Morphological sex reversal upon short-term exposure to endocrine modulators in juvenile fathead minnow (*Pimephales promelas*). *Toxicol. Lett.* 131: 51–63, 2002.
245. Zhong, T.P., K. Kaphingst, U. Akella, M. Haldi, E.S. Lander and M.C. Fishman. Zebrafish genomic library in yeast artificial chromosomes. *Genomics* 48: 136–138, 1998.
246. Zhu, M. and W.E. Fahl. Functional characterization of transcription regulators that interact with the electrophile response element. *Biochem. Biophys. Res. Commun.* 289: 212–219, 2001.
247. Zon, L.I. Zebrafish: a new model for human disease. *Genome Res.* 9: 99–100, 1999.

Biochemistry and Molecular Biology of Fishes, vol. 6
T. P. Mommsen and T. W. Moon (Editors)

CHAPTER 2

Use of fish cell lines in the toxicology and ecotoxicology of fish. Piscine cell lines in environmental toxicology

N.C. BOLS*, V.R. DAYEH*, L.E.J. LEE** AND K. SCHIRMER†

**Department of Biology, University of Waterloo, Waterloo, ON, Canada N2L 3G1, **Department of Biology, Wilfrid Laurier University, Waterloo, ON, Canada N2L 3C5, and †Junior Research Group – Molecular Animal Cell Toxicology, UFZ-Centre for Environmental Research Leipzig-Halle, 04318 Leipzig, Germany*

I. Introduction

Environmental toxicology is an interdisciplinary science dealing with toxicants in the environment. The toxicants usually are contaminants or ecotoxicants, which are a profoundly diverse group of substances. Ecotoxicants can be defined as substances discharged into the environment through human actions and having the potential to impact on ecosystems at relatively low concentrations[61]. Most often, ecotoxicants arise as a result of industrial activities, with polycyclic aromatic hydrocarbons (PAHs) being one example, but pharmaceuticals released through medical and farming practices should also be considered[110].

From the perspective of ecotoxicants, fish are especially important[28]. With approximately 20,000 different species occupying all aquatic niches, fish are the most diverse group of vertebrates. Thus understanding the actions of ecotoxicants on fish assists in evaluating the health of the aquatic environment. In addition, effects on fish are important for what they can say about potential impacts on human health[234]. Ecotoxicants are often released first into aquatic environments, by a variety of routes, humans ultimately can be exposed[4]. As many biological systems have been preserved throughout evolution, effects on the fish can serve as a warning of possible impacts on human health. In the same vein, fish can serve as laboratory models for studying ecotoxicants of concern to human health. Finally, understanding the impact of ecotoxicants on the fish is of economic importance. For a fishery, the health of the target species is of obvious importance but so is the health of other fish species, such as forage species, which might be critical to the maintenance of the fishery.

Toxicology and ecotoxicology might be considered subdisciplines of environmental toxicology, which differ in their level of focus. Toxicology studies the effects of toxicants on individual organisms. Ecotoxicology studies the impact of ecotoxicants on ecosystems. Both toxicology and ecotoxicology try to integrate toxicological information through the various hierarchical levels of biological organization, striving ultimately to explain the impact of toxicants on individuals and ecosystems, respectively[139]. Often the integration process begins with information from the simplest biological organization, the molecular and cellular levels. Toxicological information at these levels can be efficiently acquired with cell cultures. In this chapter, the nature of animal cell cultures is described and the uses of one type, cell lines, in fish environmental toxicology are reviewed. The overall position of cell lines in environmental toxicology is presented in Fig. 1.

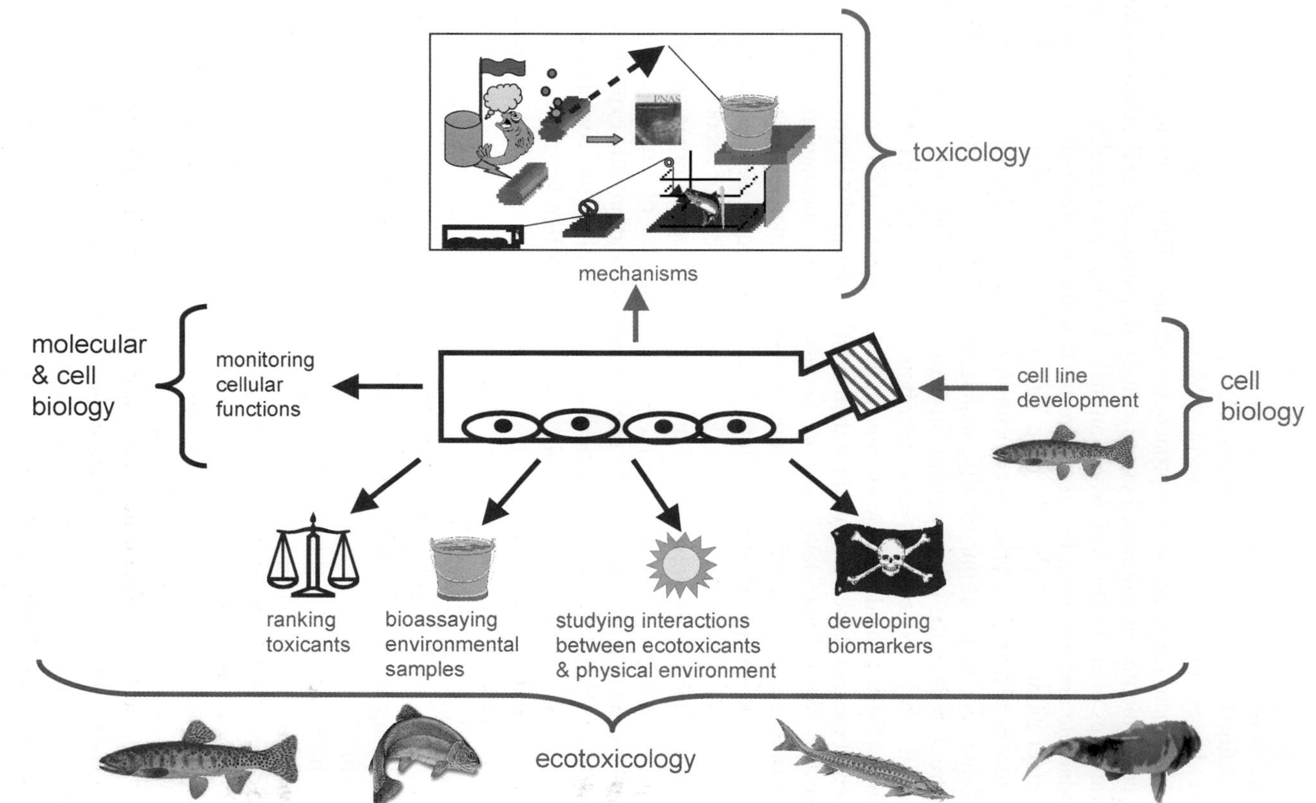

Fig. 1. Four pillars of fish cell line use in environmental toxicology.

II. Advantages of animal cell cultures

Cell cultures offer several advantages as experimental tools. For research in general, cultures allow cellular phenomena to be studied in a controlled and in some cases a completely defined environment, independent of the complexities and variability of systemic or larger physiological controls. Many cellular behaviors, such as movement, and many experimental approaches, such as cellular imaging, can only be studied conveniently in cultures. For toxicology in particular, cell cultures offer several advantages over studies with whole animals. Dosing of cell cultures is easier, more reproducible and produces less toxic waste. Results are obtained more rapidly than with intact animals and with less cost. Interpreting the results of mechanistic studies is often more definitive because cell cultures lack *in vivo* complexities such as bioaccumulation and depuration. Finally, experiments with cell cultures satisfy a societal desire to reduce the use of animals in toxicology testing.

For ecotoxicology, cell cultures can contribute in additional and unique ways. Ecotoxicology has to consider and compare multiple species[35]. Cell cultures allow widely different species to be compared for their relative sensitivity to environmental contaminants at the cellular level and to understand the mechanistic basis of sensitivity differences at the cellular and molecular level. In other words, cell cultures allow species to be compared at the cellular level under equivalent conditions of toxicant exposure. Biomarkers are an additional concern of ecotoxicology. Biomarkers are a measurable response at any level of biological organization that can indicate exposure to ecotoxicants and possible impacts of exposure[121]. Cell cultures can be used to study and identify new biomarkers and provide experimental insight into their basis. Finally, ecotoxicology often deals with complex toxic mixtures, and with animal cell cultures, their toxicity can often be evaluated more easily and with more straightforward results.

1. Cell lines vs. primary cultures

Cell lines make up one of the two general types of cultures that can be used to study animal cells *in vitro*. The other type is the primary culture. The two are interrelated because cell lines are developed from primary cultures. The most obvious difference between the culture types is their life span. Primary cultures are initiated directly from the cells, tissues or organs of fish and typically last for only a few days. The initial step usually involves the preparation of a suspension of cells or clumps of cells in a physiological buffer or a growth medium. As to when this suspension should be considered a primary culture is arbitrary. Often the next step in primary culture preparation is to have the cells attach and spread onto a physical support, such as the glass and plastic surfaces of culture vessels, which can be configured in several ways, including as Petri dishes, flasks and multiwell plates. Attachment can be taken as the start of most primary cultures and can occur as early as 30 min after creation of the cell suspension. However, under some circumstances, the cell suspension is used almost immediately in experiments, without attempts being made to attach or plate cells onto a surface. Such cell preparations likely are best referred to as just cell suspensions rather than primary cultures. However, if the suspensions are maintained for a day or more, the term primary culture is appropriate, especially if the cells proliferate during this time. How long the cultures can be maintained as primary cultures is variable. The extreme is hemopoietic cultures from the rainbow trout spleen and head kidney, which can be maintained for a year or more[84]. By convention, the primary culture ends and the cell line begins upon subcultivation or splitting of the primary culture into new culture vessels[173]. The cell line can continue to be propagated by repeating the cycle of allowing the cell number to increase through cell proliferation followed by splitting the cell population into new culture vessels, usually flasks. This cycle of growth and splitting, which is often referred to as passaging, might be done for only a limited time, usually on the order of months, or done indefinitely, for years or decades.

As to the choice of cell culture type, cell lines have several advantages over primary cultures. For experiments in general, cell lines are a much more reproducible and convenient source of cells because, once established, cell lines are fairly homogeneous and can be cryopreserved indefinitely. Although a single preparation from an organ or pooled organs can yield identical primary cultures, obtaining primary cultures similar to these at another date can sometimes be difficult to accomplish. This is because dissociation procedures are often hard to repeat precisely and the physiological status of the donor fish might be subtly changed at another time. The cells of primary cultures are usually being studied while they are recovering from the trauma of their initiation: e.g. primary culture preparation can elicit heat shock protein synthesis[223]. This recovery might be variable with respect to time and completeness. In addition, unless the specific cell separation techniques are very precise, primary cultures can be more heterogeneous than cell lines. Cell lines also have the advantage of providing an unlimited supply of cells, whereas some organs might be too small to get sufficient material for primary cultures. Some cellular phenomena, such as proliferation, and experimental approaches, such as gene transfer, are more easily studied and applied to cultures of cell lines than primary cultures. Finally, for ecotoxicology, cell lines have special value because the routine preparation of primary cultures from a species of interest might be impossible either because the species is too small, available only seasonally, difficult to collect, and troublesome and/or costly to maintain in the laboratory. Cell lines might be the only regularly available source of biological material for experimentation on these species.

2. *Piscine vs. mammalian cell lines*

The story of animal cell lines is essentially the story of rodent and human cell lines, and procedures and ideas for developing and using cell lines from other vertebrates, such as fish, have been adopted from the experiences and successes with mammalian cells. Some of the similarities and differences between piscine and mammalian cell lines are briefly discussed below to put the properties of fish cell lines in the perspective of the more advanced science of mammalian cell culturing.

Growth is supported by similar media but by different temperatures. Fish cell lines grow well in most basal media, all of which have been optimized for mammalian cells, and with the same supplements as for mammalian cells, usually calf serum (CS) or fetal bovine serum (FBS). Mammalian cells require 34–37°C, whereas fish cells can be grown over a wide temperature range. For example, rainbow trout cells can be grown at temperatures between 5 and 26°C[31], but even at optimal temperatures (18–22°C), they grow more slowly than mammalian cells at 37°C. A reasonable rate of growth and the availability of incubators are practical considerations that usually dictate the propagation temperature for fish cells. If no incubators are available, most fish cell lines can be grown at room temperature (20–24°C). This will necessitate using basal media that do not require an atmosphere of 5% CO_2 for buffering, such as Leibovitz's L-15. If the cells grow above ambient room temperature, as the cells of many warm water fish do, then water-jacketed CO_2 incubators and basal media with a bicarbonate buffering system can be used. Detailed methods for culturing fish cell lines have been described previously[30,85,124,145,146].

Mammalian cell lines are either finite or continuous, whereas most fish cell lines appear to be continuous[29,46]. The number of population doublings is limited in a finite cell line and unlimited in a continuous cell line[173]. The transition from finite to continuous is immortalization, and a variety of techniques immortalize mammalian cells. Many of these direct immortalization methods entail increasing the activity of telomerase, which maintains the length of telomeres and functions as a biological clock[97,199]. By contrast, the cells of many fish appear to immortalize without treatment but instead do so spontaneously. For rainbow trout, this might be explained by the high constitutive telomerase activity in adult organs, where the level of activity was significantly higher than in even immortal human cell lines[116]. Telomerase activity is usually restricted to stem cells in adult humans. The distribution of telomerase activity in fish species and cell lines will be an interesting area of research in the future and should help in understanding spontaneous immortalization and determine whether continuous cell lines will develop as easily from other species. Telomerase activity was found to be present throughout the development of a fibroblast cell line from the rainbow trout hypodermis[152].

Many cell lines from humans and rodents are neoplastic either because they have been developed from tumors or because they have been neoplastically transformed *in vitro*, whereas the neoplastic standing of most fish cell lines is unknown[29]. The neoplastic status of a cell line is suggested by a characteristic set of *in vitro* properties and is proven by the *in vivo* behavior of causing tumors in suitable hosts. For most fish cell lines, knowledge about their capacity to cause tumors is unknown because an appropriate *in vivo* assay has yet to emerge[164].

Heteroploidy, which is a chromosome complement that is not an exact multiple of the haploid chromosome number, is an *in vitro* property of most neoplastically transformed cells and many piscine cell lines. This property is sometimes used to criticize cell line research, but while being recognized, should be put into perspective. The differences between neoplastic and normal cells are not profound, otherwise cancer would not be so difficult to comprehend and treat. All experimental approaches come with caveats, with even fish in a tank being an abnormal situation. In the case of cell lines, whether heteroploid, neoplastic or neither, any differences from cells *in vivo* or in primary cultures rather than being decried should be exploited as a means of gaining insight into the processes under study. In short, cell line research is one of the many ways to build up knowledge of a biological process. Multitiered studies might be beyond the capabilities of a single laboratory but amalgamating the results of a fish research community exploring different approaches should lead to an understanding of the impact of toxicants on fish.

Many mammalian cell lines express functional properties of mature cells or can be triggered to differentiate into more mature cells, whereas the differentiation status or capacity of piscine cell lines is largely unexplored[29,103]. This difference is illustrated with cell lines from the liver, which is commonly studied in toxicology because of the organ's central role in xenobiotic metabolism. For mammals, hepatoma cell lines have long been available and recently several cell lines have been developed from normal hepatocytes[169]. Although all the functions of hepatocytes appear not to be expressed by a single cell line, several hepatoma lines, such as the rat H4IIE and human HepG2, simulate many aspects of hepatocyte metabolism[109,131]. In contrast, although several cell lines have been developed from the liver or hepatomas of fish, knowledge of the extent to which they express liver-specific functions is still rudimentary. Regardless of the tissue of origin, piscinc cell lines are usually described either as fibroblast- or epithelial-like, based simply on cellular morphology.

The intensive background knowledge about many human and rodent cell lines and their more rapid growth has seen them used even in aquatic toxicology, but ultimately piscine cell lines should be superior in fish environmental toxicology for three main reasons. First, the toxicants can be applied to fish cells at temperatures more typical of the temperatures to which fish would be exposed. Second, comparing the susceptibility of different species to toxicants requires working on different species. Finally, the cells of a piscine cell line should better reflect the properties of the fish from which they were derived than the cells of a mammalian cell line. The exception to this might be the situation where the functional properties of the cell line are more important than the species. For example, if toxicants with hepatic actions are being studied, mammalian hepatic cell lines might be appropriate. Indeed, although not reviewed here, several mammalian cell lines have been used in aquatic toxicology, with rat H4IIE being used most often, as illustrated in recent studies on the accumulation of contaminants in fish[115,156].

3. *Sources of piscine cell lines*

Usually extant fish cell lines can be obtained from one of two sources: culture collections and the scientific community. Culture collections usually have only a small list of fish cells relative to the inventory of mammalian cell lines, but the fish list is growing slowly. Table 1 contains the addresses of five repositories with fish cell lines. The number of deposited piscine cell lines likely represents only a small fraction of the cell lines that have been developed from fish by numerous investigators around the world over the last 40 years. Reviews by Wolf and Mann[221] and Fryer and Lannan[82], and to an extent this one, give an overview of the cell lines developed from

TABLE 1

Culture collections that in the fall 2002 had fish cell lines

Culture collection	Address	E-mail and web address	Number of fish cell lines	Represented species
American Type Culture Collection (ATCC)	P.O. Box 1549, Manassas, VA 20108, USA. Tel: (703) 365-2700; fax: (703) 365-2701	E-mail: help@atcc.org; URL: http://www.atcc.org/	18	11
Deutsche Sammlung fur Mikroorganismen und Zellkulturen GmbH, Braunschweig, Germany (DSMZ) (German Collection of Microorganisms and Cell Cultures)	Mascheroder Weg 1b, 38124 Braunschweig, Germany. Tel.: +49 (0) 531-2616-0; fax: +49 (0) 531-2616-418	E-mail: help@dsmz.de; URL: http://www.dsmz.de/index.html	4	2
European Collection of Cell Cultures (ECACC) CAMR Centre for Applied Microbiology & Research	Porton Down, Salisbury, Wiltshire (UK) SP4 0JG, UK. Tel.: +44 (0) 1980 612512; fax: +44 1980 611315	E-mail: ecacc@camr.org.uk; URL: http://www.ecacc.org.uk/	22	20
Istituto Zooprofilattico Sperimentale (IZSBS) Centro Substrati Cellulari	Via A. Bianchi 7, Brescia (BS) 25100, Italy. Tel.: +39-0302290248; fax: +39-030225613	E-mail: substr@bs.izs.it; URL: http://www.biotech.ist.unige.it/cldb/descat6.html	12	8
Riken Cell Bank (RIKEN) The Institute of Physical and Chemical Research	3-1-1 Koyadai, Tsukuba Science City, Ibaraki 305, Japan. Tel.: +81-298-36-3611; fax: +81-298-36-9130	E-mail: cellbank@riken.go.jp; URL: http://www.rtc.riken.go.jp/index.html	12	5

For a general list of culture collections worldwide see http://wdcm.nig.ac.jp/hpcc.html.

TABLE 2

Fish cell lines that have been used in environmental toxicology

Cell line designation	Species	Tissue of origin	Morphology	Culture collection
AB-9	*Brachydanio rerio* (zebrafish)	Caudal fin	Fibroblastic	ATCC CRL-2298
BB	*Ictalurus nebulosus* (bullhead, brown)	Connective tissue and muscle	Fibroblastic	ATCC CCL-59
BF-2	*Lepomis macrochirus* (bluegill)	Caudal trunk	Fibroblastic	ATCC CCL-91
BG/G	*Lepomis macrochirus* (bluegill sunfish)	Gill	Fibroblastic	Not deposited
BG/F	*Lepomis macrochirus* (bluegill sunfish)	Fin	Epithelial	Not deposited
CHSE-214	*Oncorhynchus tshawytscha* (Chinook salmon)	Normal embryo	Fibroblastic	ATCC CRL-1681
CHSE-sp	*Oncorhynchus tshawytscha* (Chinook salmon)	Normal embryo	Suspension	Not deposited
EPC	*Cyprinus carpio* (common carp)	Epithelioma papulosum	Epithelial	ECACC-93120820
FG-9307	*Paralichthys olivaceus* (olive flounder)	Gill	?	Not deposited
FHM	*Pimephales promelas* (fathead minnow)	Connective tissue and muscle	Epithelial	ATCC CCL-42
GFS	*Carassius auratus* (goldfish)	Scale	Fibroblastic	Not deposited
PHL	*Clupea harengus pallasi* (Pacific herring)	Larvae	Epithelial	ATCC CRL-2750
PLHC-1	*Poeciliopsis lucida* (topminnow)	Hepatocellular carcinoma; liver; hepatocyte	Epithelial	ATCC CRL-2406
R1	*Oncorhynchus mykiss* (rainbow trout)	Liver	Fibroblastic	DSMZ ACC 56
RTG-2	*Oncorhynchus mykiss* (rainbow trout)	Mixed gonads	Fibroblastic	ATCC CCL-55
RTgill-W1	*Oncorhynchus mykiss* (rainbow trout)	Gills	Epithelial	ATCC CRL-2523
RTH-149	*Oncorhynchus mykiss* (rainbow trout)	Hepatoma	Epithelial	ATCC CRL-1710
RTL-W1	*Oncorhynchus mykiss* (rainbow trout)	Normal liver	Epithelial	Not deposited
RTP-2	*Oncorhynchus mykiss* (rainbow trout)	Female pituitary	Epithelial	Not deposited
RTP-91E	*Oncorhynchus mykiss* (rainbow trout)	Male pituitary	Epithelial	Not deposited
RTP-91F	*Oncorhynchus mykiss* (rainbow trout)	Male pituitary	Fibroblastic	Not deposited
STE-137	*Oncorhynchus mykiss* (steelhead trout)	Embryo	Epithelial	ECACC 95122020
ZEM2S	*Brachydanio rerio* (zebrafish)	Embryo	Fibroblastic	ATCC CRL-2147
ZF-L	*Brachydanio rerio* (zebrafish)	Normal liver	Epithelial	ATCC CRL-2643

See Fryer and Lannan[82] for a more complete list of cell lines.

fish. The availability of many of these is unknown. Some might be unavailable for a particular purpose because the originating laboratory has similar experimental designs and claim proprietorship, although to date, no fish cell lines appear to have been patented. Many other cell lines likely have been lost, either accidentally through inattention to the maintenance of optimal storage conditions, usually filling up liquid nitrogen, or deliberately through the retirement of the originator and the failure of anyone else to pick up the care of their piscine descendants.

Sometimes no current piscine cell line will appear suitable for the purpose at hand and the researcher will choose to develop the appropriate cell lines. For this, detailed methods for establishing fish cell lines have been presented[30,85,145,146,222], and recent developments in preparing mammalian cell lines should be sought, as well[97]. Although cell line development is encouraged, some difficulties should be mentioned. The major one is time. For cold-water species such as the salmonids, cell lines can take 6 months to 2 years to emerge[26,86]. For a new species, success cannot be certain. Even if obtained, the cell lines might not express the desired functional properties. Defining functional properties can be time consuming and frustrating because of the paucity of reagents that work on fish and distract the researcher from the original goals in environmental toxicology.

If not to be developed, piscine cell lines are best obtained from culture collections, for practical and altruistic reasons. The practical advantages are assurances of quality and delivery. Some quality controls are testing for contamination with fungi and bacteria, including most importantly mycoplasma, and the declaration of their absence. Some culture collections, such as ATCC, have been a dependable source of cell lines for decades, with standard procedures and schedules for shipping, and are likely to go on doing this job into the foreseeable future. Of course, culture collections must charge for their services, and this is perceived as their disadvantage, leading to the distribution of cell lines between individual research laboratories. Although the quality of cell lines from an originating laboratory or from a laboratory using a line might be fine, most individual laboratories do not have the resources to match the continued quality assurances of repositories or to repeatedly send out cultures. Most importantly, this behavior can undercut the role of culture collections, and in the long run, impairs progress in the science of developing and using cell lines from a wide range of animals, which has been referred to as comparative cytotechnology[30,197].

Purchasing piscine cell lines from culture collections will promote the science of comparative cytotechnology, particularly for fish, by encouraging the expansion of collections. For most repositories, the storage of cell lines from animals other than humans and some rodents is a trivial part of their mandates, and an incentive is needed to add cell lines from other species to their collections. The most significant incentive is user demand. In the case of fish, this can be partly forecast from past purchasing records for piscine cell lines already in their collections. The other indication of demand will be the number of requests received by cell line originators from other scientists, but this data will likely be considered softer. However, this informal exchange between research laboratories is important for new or rarely used cell lines because, this is how uses for the cell lines are discovered and a demand created. Once utility and demand have been illustrated, deposition in a culture collection will improve the reliability of the science done with the cell lines and make the cell lines more widely available to scientists from other disciplines, which likely will generate further applications. In other words, the culture collection can be a catalyst to the further use of a cell line. Also, if in the future one or more tests with fish cell lines become part of government required toxicity evaluations, good laboratory practice will necessitate a standard source for the cells. The increased deposition and use of fish cell lines will be incentives for researchers to develop new cell lines from a wider range of species and tissues and to understand their possibly unique culture requirements. Culture collections likely will never store all the cell lines that are developed from fish, but expanding collections should make piscine cell lines a more useful and common source of experimental material for research in fish toxicology and ecotoxicology.

III. Evaluating common cellular responses

The application of fish cell lines for toxicology and ecotoxicology goals often means routinely evaluating some common cellular responses (see Table 2). Four useful parameters are cytotoxicity, cell growth, genotoxicity and xenobiotic metabolism. Although considered separately, these endpoints often encompass overlapping cellular activities. For example, the endpoint of cell growth sometimes reflects a cytotoxic response to a treatment rather than a specific impairment of growth. However, because each response has it own methodology and science, usually developed for mammalian cells and applied with varying degrees of success to piscine cells, they are discussed below separately.

1. Cytotoxicity

Cell lines have been used extensively to study the cytotoxicity of substances to fish cells[12,184]. Most studies have employed tests of general or basal cytotoxicity rather than tests of injury to differentiated cells and their functions. Basal cytotoxicity refers to impairment to cellular activities shared by all or most cells. Evaluating basal cytotoxicity can be done in a variety of ways, which will be referred to as cell viability assays. Usually these tests are performed on cultures after exposure to putative toxicants for 72 h or less and can be described as short-term or acute assays. As a result of this short exposure, toxicants that act by inducing a particular cellular process, such as the xenobiotic metabolism, or by causing cumulative damage might be missed.

Cell viability assays can be profoundly influenced by the exposure medium. Usually less cytotoxicity is observed when the basal medium is supplemented with serum, such as FBS. This is true not only for heavy metals but also for organic ecotoxicants[184]. Although the basis of this protective action likely depends to a degree on the specific toxicant, three general mechanisms to be considered are toxicant availability, toxicant uptake and the protective molecules of serum. For fluoranthene, FBS altered its distribution within the cell cultures and its availability to the cells[175].

The development of serum-free medium for fish cell lines has had some success[3,20] and could be used to overcome the difficulty of toxicant availability[3]. In addition, FBS contains protective molecules, such as antioxidants, which could inhibit death elicited by reactive oxygen species (ROS). This means that basal medium can be more appropriate for acute assays and even simpler exposure solutions might be best. Schirmer *et al.*[175] used an extremely simple one, termed L-15 exposure or L-15/ex. L-15/ex contains only salts, galactose and pyruvate at their concentrations in the basal medium, Leibovitz's L-15. The expression of cytotoxicity appeared to be aided by the absence in L-15/ex of most antioxidants. Also, for photocytotoxicity studies, the lack of vitamins and aromatic amino acids prevented the inadvertent generation of toxicants from some of these compounds during the UV treatment.

Cell viability tests are numerous and have been grouped below into six types based on the cellular process being targeted. Although some tests preferentially measure damage at one site over another, in practice, the results can be due to events at several cellular sites, which can interact, making their relative importance to the overall loss of cell viability difficult to distinguish. For example, neutral red is discussed below as a test measuring lysosomal damage. Yet, under most circumstances neutral red is probably measuring plasma membrane integrity because many toxicants damage membranes generally rather than lysosomes specifically. In fact, most tests of cell viability focus either directly or indirectly on the integrity of the plasma membrane. Despite this, more than one cell viability assay should be performed if possible, in order to ensure a toxicant acting in a specific manner is not missed. The use of multiple tests can potentially reveal mechanism(s) behind cytotoxicity, as well.

1.1. Cell membrane integrity

The integrity of the plasma membrane in cultures of fish cell lines has been assayed in a variety of ways, but most assays can be considered to be one of the two types. Methods that measure the ability of the plasma membrane to exclude large bulky, charged molecules, such as dyes, constitute one type. The classic dye exclusion technique is Trypan blue, which has been applied to fish cells, but can often be tricky and tedious to use because the results must be scored under the light microscope. More recently the fluorescent dye propidium iodide, which has been developed with mammalian cells[225], has been used on RTgill-W1[64]. With the loss of membrane integrity, propidium iodide is taken up and complexes with nuclear DNA, which fluoresces and can be measured with a fluorescent microwell plate reader.

The alternative to dye exclusion as a measure of membrane integrity is the capability of the plasma membrane to retain a marker molecule. The marker can be the appearance in the medium of an intracellular molecule, such as an intracellular enzyme like lactate dehydrogenase (LDH). LDH release has been used for fish cell lines[184], but experience with mammalian cells suggests that the use of LDH can be complicated by several factors[162].

Esterase substrates have long been used as a measure of cell membrane integrity, with the fluorescent product being the marker retained. The one that has been used most often with fish cell lines is 5-carboxyfluorescein diacetate acetoxymethyl ester (CFDA-AM)[174–177]. CFDA-AM diffuses into cells rapidly and is converted by non-specific esterases of living cells from a non-polar, non-fluorescent dye into a polar, fluorescent dye, 5-carboxyfluorescein (CF), which diffuses out of cells slowly. The CFDA-AM assay appears to monitor impairment to plasma membranes, but subtle differences in how the test is performed or how the toxicant acts could result in more complex explanations. When the CFDA-AM is applied to fish cells in microwell plates after having been exposed to ecotoxicants and read sometime later without removing the dye, the fluorescent readings or units (FU) constitute the CF both inside and outside the cells. In this case a decrease in FU with CFDA-AM actually measures a decline in the total esterase activity within a microwell cell culture[66,88]. The decrease in esterase activity with toxicant treatment could be achieved in two general ways: interference with plasma membrane integrity and cellular esterases.

A loss of plasma membrane integrity would decrease esterase activity in two slightly different ways. The first of these would be the complete or partial lysis of the cells upon toxicant exposure so that the esterases are released into the medium and lost when the medium is removed and replaced with the CFDA-AM solution. Another possible cause for the diminution in esterase activity is a change in plasma membrane integrity so that cytoplasmic constituents are lost to the medium but the esterases remain contained within the cells, which are left still attached to the surface of the microwells. This change in the cytoplasmic milieu would be less able to support maximal esterase activity.

Alternatively, the toxicant treatment could leave membrane integrity unimpaired but specifically interfere with cellular esterases, causing the activity to decline. Examples of this would be a toxicant interfering with the uptake of the substrate, CFDA-AM, across the plasma membrane or inhibiting the catalytic activity of the esterases.

1.2. Lysosomal damage

Neutral red (NR) is used frequently to measure plasma membrane integrity in fish cells after exposure to ecotoxicants[12,184], but NR can detect injury specific to lysosomes, as well. The general principle behind the use of this dye is that only viable cells accumulate NR (3-amino-7-dimethylamino-2-methylphenazine hydrochloride)[34]. Accumulation occurs specifically in lysosomes[99]. The dye can be applied before or after the exposure to toxicants, so the endpoint is either the retention or accumulation of NR[34,165], and measurements can be done either spectrophotometrically[34] or fluorometrically[75]. Although accumulating specifically in lyosomes, NR accrual and retention is dependent on an intact plasma membrane, adequate energy metabolism and a functioning lysosome. Under most circumstances, the NR assay likely detects impairment to all three cellular parameters and the results are similar to the results with other viability assays. However, hints of specific lyosomal damage have been seen. For example with the RTgill-W1 cell line, Schirmer *et al.*[177] found that immediately after UV irradiation in the presence of either acenaphthylene, acenaphthene or phenanthrene, photocytotoxicity was detected with NR but not with other indicator dyes, which suggests that lysosomes were being impaired before cell viability was lost.

1.3. Metabolic impairment

The metabolic capability of fish cell cultures has been monitored by measuring their ATP content or their ability to reduce either 3-(4,5-dimethylthizol-2-yl)-2,5 diphenyl tetrazolium bromide (MTT) or resazurin. ATP levels have been measured in cultures after exposure to a variety of ecotoxicants, including heavy metals and phenols[45,46,48]. Less direct are assays that measure the activity of enzymes involved in ATP formation. MTT is reduced by mitochondrial succinate dehydrogenase to a formazan product, which can be quantified spectrophotometrically. MTT has been used successfully with PLHC-1[39,79]. However, Segner[184] found that MTT did not work well with some fish cell lines, possibly due to low succinate dehydrogenase activity in mitochondria. Another dye for measuring metabolism is resazurin, which can be purchased as a commercial solution called Alamar Blue. Originally, Alamar Blue was thought to be reduced by mitochondrial enzymes[67], but now enzymes, such as diaphorases, with both cytoplasmic and mitochondrial locations, are thought to be responsible for dye reduction[148]. Thus a decline in Alamar Blue reduction indicates an impairment of cellular metabolism rather than specific mitochondrial dysfunction. The dye appears to work well on different piscine cell lines[87,176,177]. Alamar Blue has some especially convenient features. Dye reduction can be measured either spectrophotometrically or fluorometrically, and recovery from metabolic impairment can be evaluated by repeatedly applying the dye to the same culture over a period of days[88].

1.4. Cell detachment

Cell detachment can be considered as a cytotoxic endpoint for anchorage-dependent fish cells[11], and in practice, cell detachment contributes to the results of many toxicity experiments regardless of the cell viability assay used. As a cytotoxic endpoint, the principle is that cells with an intact cell membrane and with the capacity for energy metabolism remain attached to the plastic surface of a culture, whereas dead cells do not. In most experiments, the toxicant-containing medium is removed prior to any measurement, an act that divides the culture into remaining adherent cells and removed non-adherent cells. Most cell viability measurements are done only on the remaining adherent cells, and thus removal of the medium is actually enhancing the power of cell viability assays to discriminate between living and non-living cells. However, adherent vs. non-adherent cells do not always equate perfectly with live vs. dead cells: sometimes dead cells remain attached; sometimes viable cells are detached. The importance of this varies with the type of cell viability assay. Viewing cultures with the inverted phase contrast microscope can be an invaluable aid to begin trouble shooting any potential problems.

The first and least troubling scenario is if dead cells remain attached to the culture surface. Viewing cultures will reveal whether cells remain attached to the same extent in control vs. treated cultures and whether treated cultures have apparently dead cells that have 'died standing up'. The appearance of attached dead cells has been described in a variety of ways and will depend to some extent on the toxicant treatment. The cells can have the appearance of being darker, shriveled or empty looking and having blebs and more distinctive nuclear and cellular boundaries[175]. Cultures with such cells always seem to be judged as having lost viability with assays that monitor energy metabolism, such as by Alamar Blue, or cell membrane integrity, such as by dye exclusion. In fact for dye exclusion tests, such as propidium iodide, the assay will really work only if the dead cells do remain attached to the culture surface and can be stained by the dye and their presence to be recorded. Assays that monitor cellular enzymatic activity, such as esterases in the CFDA-AM assay, also usually detect a loss of viability. However, one should be watchful for situations where esterases in the cellular corpses retain their activity during the assay despite cellular disruption. Finally, if protein or DNA content is measured, treated cultures could give very similar values to control cultures, which would be a completely misleading result as to the true impact of the toxicant.

Another potentially vexatious scenario is if viable cells are detached from the growth surface during the period of toxicant exposure and removed along with the toxicant-containing medium. The potential problem can be assessed quickly by viewing cultures at the end of toxicant treatment and before exposure has been terminated by removal of the medium. Viewing should begin by first focusing on the adherent cells. Comparing control wells with treated wells will indicate if cells have been lost from the culture surface of treated cultures. As an aside, the use of a monolayer in the test makes such a visual comparison more obvious. If cells have been lost, the next step is to focus through the culture medium to see whether the medium contains just cellular debris, which would suggest no problem, or round phase bright cells, which is the appearance of viable cells and would suggest a potential overestimation of cytotoxicity. From our experience with PAHs, detachment of viable cells during the period of toxicant exposure does not occur and does not present a problem[171], but this should be judged with each class of cytotoxicants.

The final troublesome scenario is if viable cells detach from the culture surface during removal of the toxicant-containing medium. The likelihood of detachment having occurred can be assessed quickly by viewing cultures just before and after toxicant treatment has been terminated. If a culture had cells that appeared non-viable (see above) prior to termination, the loss of adherent cells from treated cultures would be expected and not be an issue. However, if a culture had only normal appearing cells just prior to termination, the loss of adherent cells after removal of the toxicant-containing medium would raise a concern. The concern would be that removing the medium had inadvertently dislodged viable cells, and assaying only the adherent cells would overestimate the loss of cell viability in the culture. Unfortunately, this issue is not easily resolved on a routine basis. Although viability assays could be done on the dislodged cells, the usual medium-removal procedure of aspiration or dumping of microwell culture plates is not conducive to collect sufficient cells from individual wells for further viability testing. Altering the medium-removal procedure for the purpose of collecting detached cells might no longer dislodge cells, and thus the viability of dislodged cells would still remain unanswered. A more useful approach is to recognize that detachment in toxicant-treated cultures is an indication of cellular impairment. The impairment is in the ability of the cells to remain attached to the substrate under the stress of removing the medium. Basically, this means that test results are due to some combination of two cellular impairments. One is in the ability to remain attached; the second, in the function being monitored by the particular viability test, e.g. metabolism with Alamar Blue.

1.5. Cell attachment

Cell attachment has been used infrequently as a measure of fish cell viability[5,27]. The principle is that only viable anchorage-dependent cells will attach to a growth substrate, such as a plastic surface. In order to attach, cells require an intact cell membrane in which integrins can function to establish attachment and energy metabolism sufficient to support the process. The advantage of cell attachment as an endpoint is the ease with which attached cells can be enumerated, especially now with the development of microwell plate assays[166]. Disadvantages are that toxicants must be applied to cells in suspension and usually the exposure period has to be short because of the difficulty in keeping anchorage-dependent cells in suspension for long periods. In addition, any treatment specifically impeding the functioning of integrins but not impairing plasma integrity could leave viable cells in suspension where they would be scored as dead. To date, ecotoxicants acting this way have yet to be described.

1.6. Cellular morphology and cytoskeleton

Changes in the morphology of fish cells after exposure to ecotoxicants is an easily observable property and presumably reflects underlying changes in the cytoskeleton. Although such changes have been noted on many occasions, more intensive studies of the cytoskeleton have been few[21,213,232,233]. Given the fundamental importance of the cytoskeleton to most cellular processes it probably warrants further studies in the context of ecotoxicants and cell viability.

2. Cell growth

Growth *in vivo* and *in vitro* can be considered as an increase in macromolecule amounts and/or in cell number. Assays to monitor both aspects of growth have been developed for animal cell cultures, and particular assays clearly measure one or the other. Yet, with some methods and under some circumstances, the distinction between the two views of growth is less clear. Also, the impairment of growth is sometimes troublesome to separate from the slow loss of cell viability. Growth experiments require use of medium that supports growth, which usually means a serum supplement. In order for any changes in growth to be realized, the duration of the experiments is usually long, although toxicant exposure can be either at the beginning only or continuous throughout the experiment. A profusion of approaches has been used to evaluate the impact of ecotoxicants on the growth of fish cell lines, and these have been grouped into those monitoring changes in either macromolecules or cell proliferation.

Several endpoints have been used to measure changes in the amounts of specific macromolecules in piscine cell cultures. Most of these were tried early in the use of fish cells in toxicology, as illustrated in a study by Marion and Denizeau[136]. They found that lead decreased total protein, RNA and DNA in RTG-2 cultures, as well as inhibiting radioactive uridine incorporation into RNA. Since then, macromolecular accumulation through biochemical assays or macromolecular synthesis through incorporation of radioactive precursors has been performed in a variety of slightly different ways on several cell lines after exposure to many ecotoxicants. The cell lines have been BF-2, CCL-42, CHSE-214, FHM, FG-9307, PLHC-1, RTG-2 and RTL-W1. The toxicants have included aflatoxin B-1, chlorinated pesticides, heavy metals, PAHs and PCBs[11,17,21,39,45,46,48,69,78,126,129,130].

The second group of growth assays evaluates the capacity of cells to proliferate. Proliferative capacity has been described as being the most comprehensive measure of cellular health because proliferation integrates the soundness of the entire cellular machinery[187]. One way to monitor this is to evaluate cell cycle progression. This has been done with fish cell lines by measuring the incorporation of thymidine or bromodeoxyuridine into DNA[21,39,126,136] or by counting the number of mitotic figures[36,163].

Another method for determining proliferative capacity is to monitor changes in the accumulation of cell number in cultures. Cell number can be derived indirectly by measuring DNA or protein amounts in cell cultures and converting these values to cell number by using a previously established relationship between DNA or protein amounts and cell number[179]. This approach could be misleading if toxicant treatment distorted the relationship between cell number and DNA or protein amounts. Yet, if this happened, the DNA or protein measurements would still suggest a toxicant-induced change in growth, not just in cell number. Proliferation can be measured directly by counting the number of cells in a culture, either with an electronic particle counter or a hemocytometer. Despite being a cumbersome endpoint because of the slow growth of most fish cell lines, cell number can still provide unique insights. For example, long-term exposure to low concentrations of BaP slightly enhanced proliferation of RTL-W1 and R1, whereas high concentrations were cytotoxic to RTL-W1[174].

Finally, proliferation can be measured by determining the proportion of cells able to form colonies. This is sometimes referred to as a reproductive assay, but again is difficult to perform with fish cell lines. In addition to their slow growth, many fish cell lines proliferate poorly at the low density required for the assay and/or form diffuse colonies, which are troublesome to count. Colony formation has been studied with a few fish cell lines after their exposure to ecotoxicants, with results similar to those from other cytotoxic endpoints[17,189].

3. *Genotoxicity*

Several genetic endpoints have been monitored in fish cell lines in order to evaluate the genotoxicity of substances in the aquatic environment[12,184]. Endpoints include the formation of DNA adducts[193] and chromosome aberrations[120] and the induction of mutants[118], micronuclei[13] and sister chromatid exchange[19]. More recently, the comet assay, which detects DNA strand breaks, has received considerable attention. The assay has been used successfully with several cell lines, including RTG-2, RTL-W1,

RTH-149 and EPC[9,112,113,144]. Despite these successes, fish cell lines have some features that make them difficult to use in genotoxicity studies. Their rates of DNA repair appear low relative to mammalian cells, causing assays of unscheduled DNA synthesis difficult to perform[184]. Although a fish cell line, ULF-23HU, from the central mudminnow has a low chromosome number[153], most fish species and cell lines developed from them have large numbers of difficult-to-distinguish chromosomes. As a result, cytogenetic abnormalities can be difficult and tedious to study.

4. Xenobiotic metabolism

Fish cell lines are convenient for studying xenobiotic metabolism and enzymes. Benzo[*a*]pyrene (BaP) is the xenobiotic most often studied[186]. Several cell lines are able to metabolize BaP to water-soluble intermediates and some of these bind DNA and presumably contribute to the cytotoxicity of BaP (Table 3). The most studied xenobiotic enzyme is CYP1A or P4501A, which is a member of the cytochrome P450 superfamily. CYP1A is induced by dioxin-like compounds, such as 2,3,7, 8-tetrachlorodibenzo-*p*-dioxin (TCDD). Induction has been demonstrated by measuring CYP1A catalytic activity as aryl hydrocarbon hydroxylase (AHH) or 7-ethoxyresorufin *O*-deethylase (EROD) activities[127,133], CYP1A protein levels by Western blotting or ELISA[40,96] and CYP1A mRNA by Northern blotting[228]. Strong induction by TCDD has been seen in liver cell lines from the topminnow (PLHC-1)[103], rainbow trout (RTL-W1)[127] and zebrafish (ZF-L)[59], but other liver cell lines, RTH-149 and R1, from rainbow trout had little or no induction[133,174]. Yet, cell lines from other rainbow trout organs, RTG-2 from gonads and RTP-2, RTP-91E, RTP-91F from pituitaries, did show CYP1A induction[197,228].

IV. Toxicology applications of fish cell lines

Toxicology is concerned with mechanisms of toxicity at all organizational levels within an organism. For fish, this research has often been driven by ecotoxicological concerns because basic knowledge of toxicity mechanisms for a particular class of ecotoxicants can provide a better intellectual foundation for assessing their potential impact on fish populations and the aquatic environment. Some examples of how fish cell lines have been used to explore toxicity mechanisms at the molecular and cellular level are discussed below. Many additional mechanisms are likely to be studied in the future as new ecotoxicological problems are identified. One development that would aid the wider use of piscine cell lines for this purpose is the availability of functionally differentiated cell lines, which is discussed in section IV.5

1. Receptor-mediated toxicity

Although receptor-mediated signal transduction pathways will likely be found to be responsible for the toxic actions of many ecotoxicants in the future, for fish,

TABLE 3

Fish cell lines used to study PAH metabolism and metabolic activation to toxic compounds

Species	Cell line	PAH	Evidence of			References
			Metabolite formation	Cytotoxic metabolites	DNA-binding metabolites	
Bluegill	BF-2	BaP	+	+	+	70,118,119,192,193,196
Brown bullhead	BB	BaP	+	+	Not studied	137,193
Chinook salmon	CHSE-214	BaP	Not studied	−	Not studied	126
Fathead minnow	FHM	BaP; DMBA	+	+	Not studied	36,196
Rainbow trout	RTG-2	BaP	+	+	−	52,70,119,137,193,196
Rainbow trout	RTH-149					
Rainbow trout	RTL-W1	BaP	+	+	+	126,144,174
Rainbow trout	R1	BaP	Not studied	−	Not studied	174
Rainbow trout	STE	BaP	Not studied	−	Not studied	119
Topminnow	PLHC-1	BaP	+	+	+	12,191

BaP: benzo[*a*]pyrene; DMBA: 7,12-dimethylbenz(a)anthracene.

environmental estrogens (EEs) and dioxin-like compounds have received the most attention to date. Less successful for EEs, piscine cell lines have been useful for investigating the basic mechanisms by which dioxin-like compounds act.

1.1. Estrogen-receptor: Environmental estrogens

EEs are substances that occur in the environment either naturally or as a result of man's activities and are able to elicit responses normally caused by estrogens, potentially disrupting endocrine circuitry and impairing the reproduction and health of fish[8]. The estrogen receptor (ER) mediates nearly all the significant actions of EE. The most useful *in vitro* system for studying EE has been fish primary hepatocyte cultures, with the induction of vitellogenin (Vg) by estradiol being the most studied endpoint. By contrast, piscine cell lines appear not to be very responsive to estrogens, despite reports of estradiol inducing Vg in RTH-149[83]. The ER was absent from the rainbow trout embryo cell line, STE-137[125], and estradiol did not induce Vg mRNA, Vg protein or ER mRNA in either PLHC-1, RTL-W1 or RTG-2[77]. However, RTG-2 has been genetically engineered to be estrogen responsive[2], which is discussed later under bioassay applications of piscine cell lines.

1.2. Aryl-hydrocarbon receptor: Dioxin-like compounds

Dioxin-like compounds are those chemicals that act as ligands for the aryl hydrocarbon receptor (AhR), which appears to be present in most vertebrate classes[92]. The AhR functions as a ligand-activated transcription factor and is responsible for most of the toxic consequences of dioxin-like compounds[150], which can be diverse and include cardiovascular dysfunctions, immunosuppression and embryotoxicity[24]. Usually, the most potent ligand for the AhR and the most toxic compound is TCDD. Some important classes of ecotoxicants contain members that are AhR active (Table 4).

Several piscine cell lines express the AhR, allowing studies of the components in the fish AhR signal transduction system. The AhR has been identified in at least four cell lines: RTG-2[194], RTH-149[133], RTL-W1[23] and PLHC-1[95]. Two forms of the AhR were expressed in RTG-2 and their expression was induced by TCDD[1]. Upon binding a ligand, AhR moves to the nucleus and forms a dimer with the aryl hydrocarbon nuclear translocator (ARNT), which then binds the xenobiotic response element (XRE) and regulates gene expression. Two forms of ARNT have been identified in RTG-2[157,159] and both forms function in AhR-mediated signaling[158]. Further studies with piscine cell lines on identifying unique features of the AhR signal transduction pathway of fish could help explain why piscine and mammalian cells respond differently to some AhR-active ecotoxicants[55].

Piscine cell lines have been used to study the genes and processes controlled by the AhR signal transduction system and to identify factors that modify signaling. By far the most studied gene is CYP1A[95]. Induction has been monitored as an increase in EROD activity, P4501A protein or CYP1A mRNA[40,95,96,228]. Two forms, CYP1A1 and CYP1A3, have been identified in rainbow trout, but only CYP1A1 was induced by TCDD in RTG-2 and RTH-149[41]. The transcriptional control of CYP1A3 has been studied in a clone of R1[80]. In PLHC-1, CYP1A induction was enhanced by cortisol[49].

TABLE 4

Ecotoxicant classes for which AhR activation as CYP1A induction has been determined with fish cell lines

Ecotoxicant class	# of congeners tested	Inducers[a]	Cell line	References
Azaarenes (*N*-heterocyclic aromatic hydrocarbons)	12	10	PLHC-1	111
Nitrated polycylic aromatic hydrocarbons (NAPs)	12	6	PLHC-1	111
Polycyclic aromatic hydrocarbons (PAHs)				
Unsubstituted PAHs	25	14	RTL-W1	22,32
	1	1	RTP-2	197
	16	7	PLHC-1	211
	18	11	PLHC-1	78,111
Hydroxylated PAHs	2	1	PLHC-1	211
	1	1	RTL-W1	181
Methylated PAHs	7	1	PLHC-1	211
Quinone PAHs	11	5	RTL-W1	181
Polybrominated diphenyl ethers (PBDEs)	12	4	RTL-W1	50
Polychlorinated biphenyls (PCBs)	7	5	RTL-W1	54,55
	11	5	RTG-2	228
	1	1	RTP-2	197
	5	3	PLHC-1	40,94,96
	11	1	ZF-L	101
Polychlorinated dibenzo-*p*-dioxins (PCDDs)	5	5	RTL-W1	57
	2	2	RTP-2	197
	5	5	RTG-2	228
	3	3	ZF-L	101
Polychlorinated dibenzo-*p*-furans (PCDFs)	4	4	RLT-W1	54,55
	1	1	RTP-2	197
	4	4	RTG-2	228
	3	3	ZF-L	101
Polychlorinated naphthalenes (PCNs)	19	2	PLHC-1	210
Tetrachloroxanthenes	1	1	RTG-2	227

[a] CYP1A induction is usually measured as EROD activity, but in some cases as CYP1A mRNA or protein.

In rainbow trout liver and pituitary cell lines, the AhR antagonist, β-naphthoflavone, blocked CYP1A induction as did geldanamycin[126,174,197]. Geldanamycin likely acted by disrupting the interaction of the AhR with heat shock protein 90. Surprisingly, EROD induction has been observed in RTL-W1 after a change in medium and without an exogenous inducer[185]. Perhaps this system could be used to identify an endogenous ligand for the fish AhR. Fish cell lines have also begun to be used to investigate the role of AhR in more complex phenomena. Proliferation of a subclone of PLHC-1 was found to be inhibited by TCDD, suggesting that this clone could be a suitable system for studying the role of AhR in cell cycle progression[102].

2. Xenobiotic metabolism and cytotoxicity

The role of xenobiotic metabolism in the cytotoxicity of an ecotoxicant to fish cell lines has been best studied with BaP, but even with this PAH the story is incomplete (Table 3). Two types of cytotoxic responses to BaP have been documented. One has been a transitory decline in metabolism, but not in cell viability, after BaP exposures of 24–48 h. This was observed in the rainbow trout liver cell lines, RTL-W1 and R1[174]. Metabolism appeared necessary for this effect as the response was blocked in RTL-W1 and R1 by, respectively, the CYP1A inhibitor, α-naphthoflavone (ANF) and the prostaglandin-H-synthase inhibitor, indomethacin. BaP quinones were thought to be the BaP metabolites responsible for this effect.

The second type of response was the loss of cell viability, which has been seen with several fish cell lines upon exposure to BaP in complete growth medium for 72 h or longer[174,184]. This cytotoxicity can be attributed to the generation of toxic metabolites through a pathway(s) in which an essential step(s) is mediated by induced CYP1A because a correlation exists between the expression of inducible CYP1A and the susceptibility of piscine cell lines to BaP. Cell lines expressing inducible CYP1A and BaP cytotoxicity were RTL-W1, RTG-2, BF-2, PLHC-1 and BB[10,16,126,137,174]. Cell lines with little or no inducible CYP1A and no vulnerability to BaP were CHSE-214, R1 and RTH-149[14,127,174]. Generally, phenols, dihydrodiols and quinones have been the major classes of products formed by CYP1A-mediated metabolism of BaP[204]. BaP 7,8-diol and 3-hydroxybenzo[*a*]pyrene were cytotoxic to several different piscine cell lines[119,137].

3. Mechanisms of cell death

Cell death in mammalian cells has been intensively studied in recent years and death is often described as being either apoptosis, which is programmed cell death and a physiological process, or necrosis, which is a pathological response[114]. In fact the term necrosis is perhaps best reserved for the changes occurring after the cells have died. Thus distinguishing mechanisms of cell death involves examining the prelethal events, and different classes of prelethal events have been proposed: usually oncosis vs. apoptosis. Oncosis is characterized by swelling and loss of plasma membrane integrity, whereas apoptosis involves shrinking and nuclear fragmentation caused by activation of caspases[198]. Additional terms have been suggested. The term ultra-fast cell death has been proposed for cell death that occurs quickly in response to strong stimuli and before caspases have been activated[25].

The mechanisms behind the death of fish cell lines in response to ecotoxicants is largely unexplored, despite the many cytotoxicity studies. Studies with viruses suggest that at least some piscine cell lines are capable of undergoing the classic program of apoptosis[108], but few papers have reported on the capacity of ecotoxicants to elicit this response[122,226]. For mammalian cells, Blagosklonny[25] has suggested that the time frame for the development of a decline in cell viability can be used to distinguish ultra-fast cell death (2–16 h), apoptosis (16–36 h) and slow cell death ($<$36 h). As cellular phenomena take longer to develop in fish cells being grown at 18–22°C than in mammalian cells at 37°C, the time frame for these processes in fish cell lines might be increased considerably. If this is indeed the case, many cytotoxicity studies with fish cell lines likely have involved ultra-fast cell death.

4. Metal homeostasis and toxicity

Several classes of molecules are involved in metal homeostasis. Metallothioneins (MTs) are a family of cysteine-rich metal-binding proteins that in mammals appear to function in Zn homeostasis and protect against heavy metal toxicity and oxidative stress[60]. Glutathione (GSH) is a sulfhydryl-rich tripeptide that is generally involved in the protection of cells against toxicants and in the metabolism of xenobiotics[71]. MTs and GSH are found in fish, and their expression and functions have been conveniently studied with piscine cell lines.

The synthesis of MTs has been examined in several cell lines. Most studied have been two rainbow trout cell lines, RTH-149 and RTG-2, and the Chinook embryo cell line, CHSE-214[151,160,229–231]. Heavy metals induced MTs in RTH-149 and RTG-2 but not in CHSE-214. Methylation of the MT genes appeared to account for the failure of MTs to be induced in CHSE-214[161]. MTF-1, the metal response element (MRE)-binding transcription factor-1, has been demonstrated in RTH-149 and RTG-2 and in a zebrafish cell line, ZEM2S[62]. ZEM2S had a single isoform of MTF-1, whereas RTH-149 and RTG-2 contained two isoforms. Zn but not Cd activated MTF-1 binding to MRE in cells from both species. A cell line from at least one marine species has been studied for MT expression. In a turbot fibroblast cell line, Cd, Cu, Hg and Zn but not Pb induced MT mRNA and MT levels[89].

Piscine cell lines have been used to explore the functions of MTs. Several experiments suggest that they function as free radical scavengers. Elevating MT levels in EPC cells by Cd exposure provides protection to the redox cycling toxicants diquat and menadione[224]. Treating RTG-2 with Zn or Cd increased MT levels and raised their resistance to H_2O_2, whereas treating CHSE-214 with Zn or Cd neither induced MT nor enhanced their H_2O_2 tolerance[117]. If MTs were increased in CHSE-214 by transiently transfecting them with an MT expression vector, their resistance to H_2O_2 increased. Adding Zn or Cd did not increase GSH levels in RTG-2 and CHSE-214 cultures, suggesting that MTs alone were responsible for the increase in H_2O_2 tolerance. Prior exposure of PLHC-1 to Cd also provided protection against a subsequent H_2O_2 exposure[183]. However, in this case Cd increased both MTs and GSH.

The role of GSH in metal toxicity has been investigated in another manner with RTG-2[135,185]. When GSH levels were reduced by exposure to buthionine sulfoximine (BSO), which is a specific inhibitor of the rate-limiting enzyme in GSH synthesis, RTG-2 died at lower metal concentrations, particularly for Hg, Cd and Cu. These results suggest that GSH protects fish cells against the toxic actions of high metal concentrations.

5. *Differentiated functions*

From a risk assessment point of view, toxicity studies on differentiated cell functions usually are more valuable than studies of basal cytotoxicity[81]. This is because toxicity *in vivo* is often limited to a small group of organs and even to cells within these organs. Thus specific toxic effects might occur at concentrations well below those causing general cytotoxicity. However, such studies require cell lines expressing differentiated properties or functions. These are available from mammals, making such studies possible. For example, PCBs have been shown to inhibit the differentiation of the rat myoblast cell line L6 into myotubes, which may explain the depression of body mass associated with PCB exposure[58]. By contrast, few differentiated piscine cell lines have been identified[29], although some progress is being made with immune cell lines. The channel catfish has been the source of several lymphocyte cell lines[138], including lines with the properties of NK-like cells[188]. Monocyte/macrophage cell lines have been developed from catfish[200], carp[76], goldfish[216], Atlantic salmon[63] and rainbow trout[86].

V. *Ecotoxicology applications of fish cell lines*

Ecotoxicology has several concerns that can be addressed pragmatically with piscine cell lines[44,68,184]. For the risk assessment process, information is needed on the relative potency of large numbers of diverse ecotoxicants. Also, the toxicity of environmental samples, which can be chemically complex, must be evaluated. Ecotoxicology also has to deal with the changing physical environment, such as rising temperatures and increased exposure to ultraviolet light (UV), and the influence of these parameters on the actions of ecotoxicants. Finally, biomarkers are applied in ecotoxicology in order to assess the impact of environmental contaminants, and the scientific basis for their use needs to be understood.

1. *Ranking compounds for their potency*

Information on the relative toxicity of large numbers of environmental contaminants can be acquired economically and rapidly with cell lines. Several biological activities or potencies can be ranked, and likely more will be evaluated in the future. The information bases are largest for compounds causing cytotoxicity and activation of the AhR.

1.1. *Cytotoxicants*

Fish cell lines have been used to screen and rank ecotoxicants for their basal cytotoxicity. A variety of cytotoxic endpoints as outlined in section III.1 have been used. The number of compounds that have been examined is large and individual compounds are listed in other reviews[184]. Here some of the ecotoxicant classes investigated and the cell lines used are described. Metals have been the most intensively studied and with the largest number of cell lines. These include BB, BF-2,

BG/G, BG/F, CHSE-214, CHSE-sp, EPC, FHM, R1, RTG-2, TF and PLHC-1[12,39,45,69,134,135,141,142,184]. The cytotoxicity of PAHs has been examined with BB, BF-2, CHSE-214, FHM, R1, RTG-2, RTgill-W1 and RTL-W1[7,12,126,137,174,176,177,180]. Naphthalene alone was studied on the Pacific herring cell line, PHL[87]. For pesticides, BF-2, FG9307, EPC and FHM have been the cell lines used[12,15,90,129,224]. Phenolics have been examined with BF-2, CHSE-214, FHM, RTG-2, RTgill-W1 and PLHC-1[12,27,45,48,65,79,167]. BF-2 was used to evaluate the cytotoxicity of biotransformation products of toluene[189]. Cationic softeners were tested on RTG-2[132]; surfactants, on FHM-sp[140]. A large number of antifouling agents have been examined with a strain of CHSE that grows in suspension[149]. These studies have been done for several reasons.

The most general purpose is to identify compounds that have the potential to be acutely toxic *in vivo*. The underlying premise is that all toxic phenomena are fundamentally related to an impairment of some aspect of cellular activity *in vivo*. Therefore, toxicity *in vivo* should be expected if the test agent is available to a target tissue at concentrations that are observed to impair cell viability *in vitro*[81,187]. Applying this approach to humans has had some success. For a diverse range of chemicals, a reasonably good correlation was found between basal cytotoxicity and acute toxicity in humans[53].

Studies with fish have yet to achieve such sophistication, but several reports indicate a strong relationship in the rank order of toxic potencies between fish and fish cells[184]. The cytotoxicity of aromatic hydrocarbons to RTG-2 cells correlated with their acute toxicity to rainbow trout[27]. Likewise, the cytotoxicity of chlorophenols to GFS cells correlated with the LC50 values from *in vivo* experiments with several fish species[171]. However, studies with RTG-2 and either rainbow trout or zebrafish showed that the cells were more tolerant of the absolute chemical concentration than fish[27,45,123]. Possibly lower doses than those causing basal cytotoxicity act *in vivo* to kill sensitive target organ(s) or to cause the malfunction of susceptible physiological systems, which causes a cascade of events leading to death. As a result of these kinds of observations, basal cytotoxicity can be considered as a beginning point in an assessment of potential *in vivo* toxicity.

Another reason for determining the potency of different ecotoxicants is to establish structure–activity relationships. In ecotoxicology, these can be used to predict the toxicity of new compounds that share the molecular or physiochemical properties of chemical classes for which structure–activity relationships have been established. The relationships could be derived with fish but can be done more quickly and at less cost with cells. As a result, fish cell lines have been used to develop structure–cytotoxicity relationships for several chemical categories[184]. The cytotoxicity of lipophilic chemicals was predicted by their octanol/water coefficient[184,176]. The cytotoxicity of metals was predicted by chemical hardness[184].

1.2. AhR-active compounds

As well as being distinguished from inactive compounds, AhR-active compounds must be ranked for their potency in order to apply the toxic equivalency approach to environmental samples. In most environmental samples, dioxin-like compounds are found as complex mixtures, and in order to evaluate the risk of such samples,

Eadon *et al.*[74] devised the toxic equivalency approach. For this, specific dioxin-like compounds are assigned a potency or toxic equivalency factor (TEF) relative to TCDD, which usually has been found to be the most toxic dioxin-like compound and assigned a value of 1.0. The concentration of a specific compound in a sample can then be expressed as a toxic equivalent concentration or quotient (TEQ) by multiplying the concentration of the compound as determined by analytical chemistry techniques by its TEF. Next, the dioxin-like compounds in a sample are assumed to act in an additive manner. Therefore, the TEQ for the sample can be determined by adding together the TEQs for each dioxin-like compound in the sample and the final TEQ can be used in risk assessment.

After considering the results of a variety of *in vivo* and *in vitro* toxicity tests from the literature, an international committee recommends TEFs[201]. In toxicity tests with whole animals, some dioxin-like compounds have been shown to have different potencies in fish than in mammals[155,214]. As a result, different TEFs might have to be used for the same compound depending on the animal group under study. In other words, TEFs derived specifically in fish systems could be necessary for risk assessment. However, the number of compounds that potentially has dioxin-like activity is too large for them all to be tested *in vivo*. Two chemical classes of environmental importance can be used to illustrate this problem: the PCBs consist of 209 congeners and the PAHs contain hundreds of members.

Fish cell lines have been an invaluable alternative to fish for providing relative induction potencies (REPs) that can be used to derive TEFs. REPs are determined in cell lines from CYP1A or EROD induction, which is mediated by the AhR. These can be used as TEFs or used together with other toxicity information to decide upon TEFs. Table 4 lists the cell lines that have been used to evaluate the ability of substances to activate the AhR pathway and induce CYP1A or EROD activity. Also listed are ecotoxicant classes that have been examined and the number of AhR-active compounds identified in each class. In some cases, REPs were developed with the rat H4IIE and rainbow trout RTL-W1 cell lines. Some significant differences were found for a few compounds. The mono-*ortho* PCBs had little or no potency in RTL-W1, although they were inducers in H4IIE[55]. Two non-*ortho* PCBs (126 and 169) were less potent in RTL-W1, whereas two furans were 4–5 times more potent in RTL-W1[54]. These kinds of results illustrate the potential value of using piscine cell lines rather than mammalian cell lines to study dioxin-like compounds in environmental toxicology.

The importance of species-specific TEFs becomes apparent when TEQs are calculated for environmental samples. This has been demonstrated with extracts of livers from lake trout[33,220]. Generally, the TEQs were higher with H4IIE-derived TEFs than with RTL-W1-derived TEFs. However, any differences depended on the combinations of dioxin-like compounds in the sample. For samples where PCBs predominated, the TEQs calculated with mammalian TEFs were higher, whereas for samples where dioxins and furans were most abundant, the relationship was opposite. The value of species-specific TEFs is still hard to judge as studies have been done only on a few samples and with TEFs derived from only a few species for only a fraction of the hundreds of potential dioxin-like compounds. Cell lines would be the most practical approach for studying this further.

2. *Use of piscine cell lines to evaluate environmental samples*

Piscine cell lines are being used as the reporters in bioassays to detect the presence of ecotoxicants in samples from fish, water and sediment. The power of cell bioassays is in providing information and benefits that cannot be achieved by analytical chemistry approaches alone. First, they will detect a biological activity, while chemistry approaches will find only what is being sought and thus might miss an unknown toxicant. Second, depending on whether the sample has been fractionated or not, cell bioassays will integrate the potency of all the compounds in the sample. In other words, the toxicity of the sample rather than of a particular compound will be measured. Of course, sometimes synergism might occur between compounds in a sample and in other cases a toxicant might be masked. These types of interactions could be defined by fractionating the sample and applying individual fractions to the cells. The animal cell bioassays can be done more rapidly and cheaply than analytical techniques. Thus, they can be effective screening tools, and serve as a valuable complement to chemical approaches for identifying toxicants in samples. Although several of these gains could also be achieved with whole organisms, bioassays with cells have advantages over bioassays with animals. Cell bioassays need less sample and can be quicker and less expensive to perform. They can be more sensitive and specific and by avoiding complexities of fish easier to interpret. Finally they reduce the use of whole animals in toxicity testing, which is a general aim in many countries.

Fish cell line bioassays have been developed to detect at least five classes of ecotoxicants and these bioassays and some examples of their applications are briefly reviewed. Some of the additional scientific issues with cell bioassays are how to genetically engineer cells to increase the specificity and sensitivity of the cell lines and how the sample can be prepared for effective presentation to the cells.

2.1. *Bioassays for cytotoxicants*

Piscine cell lines are being used as an alternative to fish in the evaluation of water quality. In some countries, such as Canada, the law requires that certain industrial effluents be tested periodically by the 96 h rainbow trout acute lethality test[65]. Fish bioassays are sometimes used as part of a program of toxicity identification evaluation (TIE) to detect the toxic chemicals in complex effluents, as well. For both purposes, cytotoxicity in cultures of piscine cell lines can replace death of fish in tanks as an indication of water sample toxicity. Two strategies have been used to apply water samples to cell cultures. One is to extract organic contaminants from water and add the extracts to cultures in carrier solvents, such as dimethyl sulfoxide (DMSO)[182]. The other way is to make up culture medium with sample water. This can be done by adding sample water to 2 × tissue culture medium, making the highest water sample concentration 50%[98], or by adding solid components of medium to reconstitute either complete medium or a physiological buffer[65]. In the latter case 100% effluent can be tested.

The water samples that have been tested in piscine cell line bioassays have been quite varied. They include surface water[73], sewage water[206] and seepage water from garbage dumps[233]. Effluents from metal processing plants[98], a fish-canning factory[205], a paper mill[65], a petroleum refinery[182] and an aeronautics industry plant[47] have also been examined. RTG-2 has been the most commonly used cell line, but RTgill-W1 and FHM have also been test cells. Usually, cytotoxicity was evaluated after 24–48 h exposures by different combinations of the endpoints outlined in section III. Cytotoxicity to fish cell lines has been used in several TIE, including ones on river water and sludge from sewage treatment plants[18,143,195,206].

In addition to water samples, piscine lines have been used to test other materials for cytotoxicants. For example, PLHC-1 was used to test the cytotoxicity of wood chips[105].

2.2. Bioassays for AhR-active compounds

Cell lines are increasingly being used in bioassays to screen environmental samples for AhR-active or dioxin-like compounds[33,77,91,93]. The intention can be to determine simply the presence or absence of AhR-active compounds or to quantify the AhR potency of the mixture extracted from samples. Fish cell lines seem appropriate for these bioassays because usually the goal is to evaluate the possible impact of dioxin-like compounds to fish. The most common AhR-mediated endpoint is the induction of CYP1A, which is usually measured as EROD activity. For this response, some fish cell lines seem a little more sensitive than mammalian cell lines, although perhaps less robust. A significant induction of EROD activity was seen in RTL-W1 at a lower TCDD concentration than in H4IIE, resulting in a lower EC50, but the maximum EROD activity was always higher in H4IIE than in RTL-W1[33,54]. The problem with EROD activity as an endpoint is that some sample extracts might contain compounds that interfere with the measurement of EROD activity[56,96]. To overcome this, a fish cell has been genetically engineered to express a reporter enzyme whose activity is induced but not impaired by dioxin-like compounds.

RLT 2.0 cells were created by stably transfecting RTH-149 with a luciferase reporter gene under control of dioxin-responsive elements (DREs)[168,212]. In these cells, the AhR is activated by dioxin-like compounds and binds the DREs, resulting in the production of luciferase. The luciferase activity is measured by adding luciferin and quantifying catalytic activity with a luminometer. Luciferase activity then provides a sensitive measure of potency for AhR-mediated gene expression. In bioassays of river sediments, RLT 2.0 gave results similar to PLHC-1 and H4IIE[104].

A wide range of environmental samples have been evaluated for dioxin-like compounds with piscine cell line bioassays (Table 5). In nearly all studies some samples have been found to be positive. These results have been converted to bioassay-derived TCDD equivalent concentrations[220], and different methods for doing this from the dose–response curves have been developed[38,208]. When the same samples have been applied to bioassays with mammalian and fish cell lines, the results have been broadly similar[104,219,220]. However, a few individual samples have sometimes been observed to differ[33,220]. In most studies, chemical analysis of the

TABLE 5

Use of fish cell lines to detect AhR-active compounds in environmental samples

Organism or material monitored	Some exposure history	Matrix extracted	Cell line	CYP1A endpoint	References
Carp	Saginaw bay	Not stated	RTG-2	mRNA	228
Salmon	Lake Ontario	Not stated	RTG-2	mRNA	228
Sediment	Passaic River	Not stated	RTG-2		228
White suckers	Bleached Kraft mill effluent	Liver	RTL-W1	EROD	202
Lake trout	Three Great Lakes sites	Liver	RTL-W1	EROD	33,220
Rainbow trout	Creosote microcosms	Liver	RTL-W1	EROD	219
SPMD[a]	Creosote microcosms	Triolein	RTL-W1	EROD	219
Water	Petroleum refinery effluent	Water	RTL-W1	EROD	182
	Petroleum refinery effluent	Particulates	RTL-W1	EROD	182
	Petroleum refinery effluent	None[b]	RTL-W1	EROD	182
Ground water	Contaminated industrial site	None[b]	RTL-W1	EROD	178
Sediment	Bitterfeld, Germany	Sediment	RTL-W1	EROD	37
SPMD[a]	Urban stream	Triolein	PLHC-1	EROD	209
Wood chips and cellulose	Bleached and unbleached	Wood chips and cellulose	PLHC-1	EROD/protein	105
Surficial sediments	River Narva, Estonia	Sediments	PLHC-1	EROD/protein	106
SPMD[a]	Oil refinery effluent	Triolein	PLHC-1	EROD	154
Sediment and oil shales	Oil shale processing plants	Sediment and oil shales	PLHC-1	EROD/protein	107
Sediment	Morava and Drevnice Rivers	Sediment	PLHC-1	EROD	104
	Morava and Drevnice Rivers	Sediment	RLT 2.0	–[c]	104

[a] SPMD, semi-permeable membrane device.
[b] Water samples were applied directly to cells.
[c] Luciferase activity.

samples allowed some of the AhR-active compounds in positive samples to be identified.

2.3. *Bioassays for genotoxicants*

Sediment and water samples have been assayed for genotoxicants with fish cell lines. Marine sediment extracts caused chromosomal aberrations in U1-H[215] and DNA damage in EPC[113]. The Kishon River, the most polluted river in Israel, was evaluated for genotoxicants with RTH-149[9,112]. Water samples were collected from different sites over the period of a year and applied to RTH-149 cultures, which were evaluated by the comet assay. Several parameters, comet percentages, average tail lengths and DNA damage levels, suggested that the river water was genotoxic.

2.4. *Bioassays for environmental estrogens*

Although EEs have been detected by vitellogenin induction in fish primary hepatocyte cultures and by several endpoints in mammalian cell lines[77,91], piscine cell lines have proven less successful. To overcome their insensitivity, Ackermann *et al.*[2] genetically engineered RTG-2 in two ways. One method was to permanently transfect RTG-2 with a rainbow trout ERβ (rtERβ) expression vector and select for clones responsive to estradiol. The other was to transiently transfect RTG-2 with rtERβ cDNA, an estrogen-inducible reporter gene, and an internal control reporter gene. The end result of estrogen exposure is the up regulation of luciferase activity. The transiently transfected RTG-2 cells were more sensitive in detecting estrogenic activity than even rainbow trout primary hepatocytes. The engineered RTG-2 cells detected estrogenic activity in an extract from a sewage treatment plant.

2.5. *Bioassays for compounds that generate oxidative stress*

The zebrafish cell line ZEM2S has been transiently transfected to report a response to foreign compounds that act *via* specific DNA motifs, electrophile response elements (EPREs)[42,43]. In mammals EPREs mediate the induction of special protective genes by foreign compounds that generate oxidative stress. ZEM2S were transfected with a plasmid construct that contained a murine EPRE fused to a minimal promoter and the cDNA encoding firefly luciferase (EPRE-LUC). When these cells were exposed to heavy metals and organophosphates, a dose-dependent increase in the expression of EPRE-LUC was observed, suggesting that the cells could be used to detect oxidative stress-inducing chemicals. The cells did respond to aqueous extracts of soil from hazardous waste sites.

3. *Studying interactions between ecotoxicants and physical environment*

The impact of variables, such as temperature and light, on the actions of ecotoxicants can be much more conveniently studied with fish cell cultures than with fish. For some piscine cell lines, the range of temperatures over which cells survive and grow has been defined as has the susceptibility of cells to different UV treatments[31,147]. This information can be used as background to put into context studies on ecotoxicants under different environmental conditions.

The influence of these physical factors on environmental contaminants has been studied most often for PAHs. The metabolism of BaP was found to be near maximal at the optimum temperature for RTG-2 growth[52]. For BF-2 cells, incubation at 35 vs. 23°C increased the proportion of BaP metabolized to an ultimate carcinogenic metabolite[192].

The potential phototoxicity of PAHs to fish has been investigated with piscine cell lines, RTgill-W1 and PLHC-1, by evaluating the capacity of PAHs to be photocytotoxic. Photocytotoxicity refers to the acute damaging or killing of cells that results from the photochemical reactions undergone by aromatic molecules during absorption of UV radiation[175]. The RTgill-W1 cell line was used because prior *in vivo* studies had identified the fish gill epithelium as one of the major target sites of photoinduced PAH toxicity[217]. Although most priority PAHs were found to be photocytotoxic, only six of them were photocytotoxic at concentrations theoretically achievable in water[177]. Four of them, namely fluoranthene, pyrene, anthracene and benzo[*a*]anthracene, were found in a subsequent study to fully explain the photocytotoxic effects to RTgill-W1 cells of a complex mixture, the wood preservative creosote[180]. The general mechanism behind the photocytotoxicity of PAHs appeared to be the impairment of membranes due to oxidative stress[175,177]. Choi and Oris[51] identified reactive oxygen radical-induced lipid peroxidation as an important factor in the PAH photoinduced toxicity to fish, applying fish liver microsomes and the topminnow hepatoma cell line, PLHC-1.

Photocytotoxicity is distinct from the toxicity potentially elicited by stable photomodification products that arise from UV irradiation. For example, creosote that had been UV-irradiated in the absence of RTgill-W1 cells, and subsequently applied to the cells in the dark, was significantly more cytotoxic than intact creosote[180]. In addition, several quinones arising from UV irradiation of the PAH anthracene were, in contrast to the parent compound[32], shown to be able to induce EROD activity in RTL-W1 cells[181]. Although the induction potency of the anthraquinones was comparably low, the importance of this latter finding is in the potential sublethal, chronic effects elicited by products obtained from the UV irradiation of chemical contaminants.

4. Developing and improving biomarkers

Biomarkers have been defined as xenobiotically induced alterations in cellular or biochemical components or processes, structures or functions that can be used to infer exposure and assess untoward biological effects[190]. The simplest uses of biomarkers are as indicators of exposure to ecotoxicants. More difficult to develop but potentially more useful would be biomarkers of outcome. In ecotoxicology, biomarkers are measured in whole organisms, usually to evaluate exposure. For fish, EROD activity[218], stress proteins[128], metallothioneins (MTs) [207] and vitellogenin[203] are some of the biomarkers being used or being considered for use.

Piscine cell lines can be used to improve the utility of these biomarkers in several ways. One is by determining the compounds that the biomarker is responsive to at the cellular level. For example, documenting the xenobiotics that either are or are not capable of inducing EROD activity can be done much more efficiently in fish cell lines than in fish, and as illustrated in Table 3, the number of the compounds that have been examined with cell lines is large. This cell line database can be used to explain potential causes of elevated EROD activity in fish populations and direct the analytical chemistry studies to certain classes of ecotoxicants in the search for the cause of elevated EROD activity in the fish. Cell line databases could also be created for the compounds and treatments capable of inducing stress proteins and metallothioneins. Stress protein induction has been documented in fish cell lines in response to temperature[31], hypoxia[6] and heavy metals[100,170,172]. A second way of using cell lines in biomarker research is to investigate interactions that could potentially confound the use of biomarkers. For example, some combinations of PCBs can impair EROD induction by AhR-active compounds[56]. Third, cell lines can be used to conveniently investigate the molecular and cellular mechanisms responsible for the biomarker responding to particular ecotoxicants or treatments. For example, identifying a receptor-mediated mechanism has a number of implications. Receptor distribution in the whole organism would dictate what tissues should be examined for the biomarker. A receptor-mediated mechanism could lead to structure–activity relationships and predictions about the range of compounds likely to alter the biomarker. Finally, the mechanism might suggest a specific toxic outcome and thus increase the value of the biomarker in risk assessment.

Piscine cell lines can be used to develop new biomarkers. One promising way to do this is through the rapidly emerging technologies of genomics, transcriptomics, proteomics and bioinformatics[127]. In general terms, these technologies promise to allow changes in the flow of information in response to a treatment to be analyzed rapidly and at an unprecedented wide scale, with thousands of genes rather than a few being studied in a single experiment. Cell lines will be valuable subjects for such analysis because exposure conditions can be controlled and varied more easily than *in vivo* experiments. As one example, cells could be exposed to ecotoxicants and changes in the proteins expressed (the proteome) documented. The proteins whose expression changes most profoundly at the cellular level could be identified and methods developed to measure these proteins specifically. From this background information, the potential of these proteins to be biomarkers at the organismal level could be investigated.

VI. Future

Piscine cell lines likely will become increasingly utilized in fish environmental toxicology, but two further developments will be needed to maximize their potential. One is technological; the other, attitudinal. The technological needs are in several interconnected areas in the science of culturing fish cells. These include understanding their nutritional requirements, differentiation capacity, direct immortalization, cell lineage position and transfection. As for attitude, the full value of fish cell lines will be realized when more ecotoxicologists are willing to view cell lines as one of the many complementary approaches to explore the complexity of fish and to place concerns about the normalcy of cell lines in a realistic perspective. Cell lines can never rival the beauty and diversity of whole fish but in conjunction with other approaches they should greatly contribute to the acquisition of knowledge about this marvelous group of vertebrates, and help understand the impact of toxicants on them.

Acknowledgements. The research by the authors was supported by the Natural Sciences and Engineering Research Council of Canada (NSERC).

VII. References

1. Abnet, C.C., R.L. Tanguay, M.E. Hahn, W. Heideman and R.E. Peterson. Two forms of aryl hydrocarbon receptor type 2 in rainbow trout (*Oncorhynchus mykiss*). *J. Biol. Chem.* 274: 15159–15166, 1999.
2. Ackermann, G.E., E. Brombacher and K. Fent. Development of a fish reporter gene system for the assessment of estrogenic compounds and sewage treatment plant effluents. *Environ. Toxicol. Chem.* 21: 1864–1875, 2002.
3. Ackermann, G.E. and K. Fent. The adaptation of the permanent fish cell lines PLHC-1 and RTG-2 to FCS-free media results in similar growth rates compared to FCS-containing conditions. *Mar. Environ. Res.* 46: 363–367, 1998.
4. Adams, S.M. and M.S. Greeley. Establishing possible links between aquatic ecosystem health and human health: an integrated approach. In: *Interconnections Between Human and Ecosystem Health*, edited by R.T. Di Giulio and E. Monosson, London, Chapman & Hall, pp. 91–102, 1999.
5. Ahne, W. Use of fish cell cultures for toxicity determination in order to reduce and replace the fish tests. *Zentbl. Bakteriol. Mikrobiol. Hyg. [B]* 180: 480–504, 1985.
6. Airaksinen, S., C.M.I. Rabergh, L. Sistonen and M. Nikinmaa. Effects of heat shock and hypoxia on protein synthesis in rainbow trout (*Oncorhynchus mykiss*) cells. *J. Exp. Biol.* 201: 2543–2551, 1998.
7. Araújo, C.S.A., S.A.F. Marques, M.J.T. Carrondo and L.M.D. Gonçalves. *In vitro* response of the brown bullhead catfish (BB) and rainbow trout (RTG-2) cell lines to benzo[*a*]pyrene. *Sci. Total Environ.* 247: 127–135, 2000.
8. Arukwe, A. Cellular and molecular responses to endocrine-modulators and the impact on fish reproduction. *Mar. Pollut. Bull.* 42: 643–655, 2001.
9. Avishai, N., C. Rabinowitz, E. Moiseeva and B. Rinkevich. Genotoxicity of the Kishon River, Israel: the application of an *in vitro* cellular assay. *Mutat. Res. – Genet. Toxicol. Environ. Mutagen.* 518: 21–37, 2002.
10. Babich, H. and E. Borenfreund. Polycyclic aromatic hydrocarbon *in vitro* cytotoxicity to bluegill BF-2 cells: mediation by S-9 microsomal fraction and temperature. *Toxicol. Lett.* 36: 107–116, 1987.
11. Babich, H. and E. Borenfreund. Fathead minnow FHM cells for us in *in vitro* cytotoxicity assays of aquatic pollutants. *Ecotoxicol. Environ. Saf.* 14: 78–87, 1987.
12. Babich, H. and E. Borenfreund. Cytotoxicity and genotoxicity assays with cultured fish cells: a review. *Toxicol. In Vitro* 5: 91–100, 1991.
13. Babich, H., S.H. Goldstein and E. Borenfreund. *In vitro* cyto-and genotoxicity of organomercurials to cells in culture. *Toxicol. Lett.* 50: 143–149, 1990.
14. Babich, H., N. Martin-Alquacil and E. Borenfreund. Use of the rainbow trout hepatoma cell line, RTH-149, in a cytotoxic assay. *Altern. Lab. Anim.* 17: 67–71, 1989.
15. Babich, H., M.R. Palace and A. Stern. Oxidative stress in fish cells: *in vitro* studies. *Arch. Environ. Contam. Toxicol.* 24: 173–178, 1993.
16. Babich, H., D.W. Rosenberg and E. Borenfreund. *In vitro* cytotoxicity studies with the fish hepatoma cell line, PLHC-1 (*Poeciliopsis lucida*). *Ecotoxicol. Environ. Saf.* 21: 327–336, 1991.
17. Babich, H., C. Shopsis and E. Borenfreund. *In vitro* cytotoxicity testing of aquatic pollutants (cadmium, copper, zinc, nickel) using established fish cell lines. *Ecotoxicol. Environ. Saf.* 11: 91–99, 1986.
18. Babin, M.M., P. Garcia, C. Fernandez, C. Alonso, G. Carbonell and J.V. Tarazona. Toxicological characterization of sludge from sewage treatment plants using toxicity identification evaluation protocols based on *in vitro* toxicity tests. *Toxicol. In Vitro* 15: 519–524, 2001.
19. Barker, C.J. and B.D. Rackham. The induction of sister-chromatid exchanges in cultured fish cells (*Ameca splendens*) by carcinogenic mutagens. *Mutat. Res.* 68: 381–387, 1979.
20. Barlian, A., R.C. Ganassin, D. Tom and N.C. Bols. A comparison of bovine serum-albumin and chicken ovalbumin as supplements for the serum-free growth of Chinook salmon embryo cells, CHSE-214. *Cell Biol. Int.* 17: 677–684, 1993.

21. Bechtel, D.G. and L.E.J. Lee. Effects of aflatoxin B-1 in a liver-cell line from rainbow trout (*Oncorhynchus mykiss*). *Toxicol. In Vitro* 8: 317–328, 1994.
22. Behrens, A., K. Schirmer, N.C. Bols and H. Segner. Polycyclic aromatic hydrocarbons as inducers of cytochrome P4501A enzyme activity in the rainbow trout cell line, RTL-W1, and in primary cultures of rainbow trout hepatocytes. *Environ. Toxicol. Chem.* 20: 632–643, 2001.
23. Billard, S.M., M.E. Hahn, D.G. Franks, R.E. Peterson, N.C. Bols and P.V. Hodson. Binding of polycyclic aromatic hydrocarbons (PAHs) to teleost aryl hydrocarbon receptors (AHRs). *Comp. Biochem. Physiol.* 133B: 55–68, 2002.
24. Birnbaum, L.S. and J. Tuomisto. Non-carcinogenic effects of TCDD in animals. *Food Addit. Contam.* 17: 275–288, 2000.
25. Blagosklonny, M.V. Cell death beyond apoptosis. *Leukemia* 14: 1502–1508, 2000.
26. Bols, N.C., A. Barlian, M. Chirino-Trejo, S.J. Caldwell, P. Goegan and L.E.J. Lee. Development of a cell line from primary cultures of rainbow trout, *Oncorhynchus mykiss* (Walbaum), gills. *J. Fish Dis.* 17: 601–611, 1994.
27. Bols, N.C., S.A. Boliska, D.G. Dixon, P.V. Hodson and K.L.E. Kaiser. The use of fish cell cultures as an indication of contaminant toxicity to fish. *Aquat. Toxicol.* 6: 147–155, 1985.
28. Bols, N.C., J.L. Brubacher, R.C. Ganassin and L.E.J. Lee. Ecotoxicology and innate immunity in fish. *Dev. Comp. Immunol.* 25: 853–873, 2001.
29. Bols, N.C. and L.E.J. Lee. Technology and uses of cell cultures from the tissues and organs of bony fish. *Cytotechnology* 6: 163–187, 1991.
30. Bols, N.C. and L.E.J. Lee. Cell lines: availability, propagation and isolation, In: *Biochemistry and Molecular Biology of Fishes*, Vol. 3, edited by P.W. Hochachka and T.P. Mommsen, Amsterdam, Elsevier, pp. 145–159, 1994.
31. Bols, N.C., D.D. Mosser and G.B. Steels. Temperature studies and recent advances with fish cells *in vitro*. *Comp. Physiol. Biochem.* 103A: 1–14, 1992.
32. Bols, N.C., K. Schirmer, E.M. Joyce, D.G. Dixon, B.M. Greenberg and J.J. Whyte. Ability of polycyclic aromatic hydrocarbons to induce 7-ethoxyresorufin-*o*-deethylase activity in a trout liver cell line. *Ecotoxicol. Environ. Saf.* 44: 118–128, 1999.
33. Bols, N.C., J.J. Whyte, J.H. Clemons, D.J. Tom, M. van den Heuvel and D.G. Dixon. Use of liver cell lines to develop toxic equivalency factors and to derive toxic equivalent concentrations in environmental samples. In: *Ecotoxicology: Responses, Biomarkers and Risk Assessment*, edited by J.T. Zelikoff, Fair Haven, NJ, SOS Publications, pp. 329–350, 1997.
34. Borenfreund, E. and J.A. Puerner. A simple quantitative procedure using monolayer cultures for cytotoxicity assays. *J. Tissue Cult. Methods* 9: 7–12, 1984.
35. Boudou, A. and F. Ribeyre. Aquatic ecotoxicology: from the ecosystem to the cellular and molecular levels. *Environ. Health Perspect.* 105(Suppl. 1): 21–35, 1997.
36. Bourne, E.W. and R.W. Jones. Effects of 7,12-demethylbenz[a]anthracene (DMBA) in fish cells *in vitro*. *Trans. Am. Microsc. Soc.* 92: 140–142, 1973.
37. Brack, W., K. Schirmer, T. Kind, S. Schrader and G. Schüürmann. Effect-directed fractionation and identification of cytochrome P4501A-inducing halogenated aromatic hydrocarbons in a contaminated sediment. *Environ. Toxicol. Chem.* 21: 2654–2662, 2002.
38. Brack, W., H. Segner, M. Moder and G. Schüürmann. Fixed-effect-level toxicity equivalents – a suitable parameter for assessing ethoxyresorufin-*O*-deethylase induction potency in complex environmental samples. *Environ. Toxicol. Chem.* 19: 2493–2501, 2000.
39. Brüschweiler, B.J., F.E. Würgler and K. Fent. Cytotoxicity *in vitro* of organotin compounds to fish hepatoma cells PLHC-1 (*Poeciliopsis lucida*). *Aquat. Toxicol.* 32: 143–160, 1995.
40. Brüschweiler, B.J., F.E. Würgler and K. Fent. An ELISA assay for cytochrome P4501A in fish liver cells. *Environ. Toxicol. Chem.* 15: 592–596, 1996.
41. Cao, Z., J. Hong, R.E. Peterson and J.M. Aiken. Characterization of CYP1A1 and CYP1A3 gene expression in rainbow trout (*Oncorhynchus mykiss*). *Aquat. Toxicol.* 49: 101–109, 2000.
42. Carvan, M.J., W.A. Solis, L. Gedamu and D.W. Nebert. Activation of transcription factors in zebrafish cell cultures by environmental pollutants. *Arch. Biochem. Biophys.* 376: 320–327, 2000.
43. Carvan, M.J., D.M. Sonntag, C.B. Cmar, R.S. Cook, M.A. Curran and G.L. Miller. Oxidative stress in zebrafish cells: potential utility of transgenic zebrafish as a deployable sentinel for site hazard ranking. *Sci. Total Environ.* 274: 183–196, 2001.
44. Castaño, A., N.C. Bols, T. Braunbeck, P. Dierickx, M. Halder, B. Isomaa, K. Kawahara, L.E.J. Lee, C. Mothersill, P. Part, G. Repetto, J.R. Sintes, H. Rufli, R.C. Smith and H. Segner. The use of fish cells in ecotoxicology. *Altern. Lab. Anim.* 31: 317–351, 2003.

45. Castaño, A., M.J. Cantarino, P. Castillo and J.V. Tarazona. Correlations between the RTG-2 cytotoxicity test EC_{50} and *in vivo* LC_{50} rainbow trout bioassay. *Chemosphere* 32: 2141–2157, 1996.
46. Castaño, A. and J.V. Tarazona. ATP assay on cell monolayers as an index of cytotoxicity. *Bull. Environ. Contam. Toxicol.* 53: 309–316, 1994.
47. Castaño, A., M.M. Vega, T. Blazquez and J.V. Tarazona. Biological alternatives to chemical identification for the ecotoxicological assessment of industrial effluents: the RTG-2 *in vitro* cytotoxicity test. *Environ. Toxicol. Chem.* 13: 1607–1611, 1994.
48. Castaño, A., M.M. Vega and J.V. Tarazona. Acute toxicity of selected metals and phenols on RTG-2 and CHSE-214 fish cell lines. *Bull. Environ. Contam. Toxicol.* 55: 222–229, 1995.
49. Celander, M., M.E. Hahn and J.J. Stegeman. Cytochrome P450 (CYP) in the *Poeciliopsis lucida* hepatocellular carcinoma cell line (PLHC-1): dose- and time-dependent glucocorticoid potentiation of CYP1A induction without induction of CYP3A. *Arch. Biochem. Biophys.* 329: 113–122, 1996.
50. Chen, G., A.D. Konstantinov, B.G. Chittim, E.M. Joyce, N.C. Bols and N.J. Bunce. Synthesis of polybrominated diphenyl ethers and their capacity to induce CYP1A by the Ah receptor mediated pathway. *Environ. Sci. Technol.* 35: 3749–3756, 2001.
51. Choi, J. and J.T. Oris. Anthracene photoinduced toxicity to PLHC-1 cell line (*Poeciliopsis lucida*) and the role of lipid peroxidation in toxicity. *Environ. Toxicol. Chem.* 19: 2699–2706, 2000.
52. Clark, H.F. and L. Diamond. Comparative studies on the interaction of benzo[*a*]pyrene with cells derived from poikilothermic and homeothermic vertebrates. II. Effect of temperature on benzo[*a*]pyrene metabolism and cell multiplication. *J. Cell Physiol.* 77: 385–392, 1971.
53. Clemedson, C., E. McFarlane-Abdulla, M. Andersson, F.A. Barile, M.C. Callega, C. Chesné, R. Clothier, M. Cottin, R. Curren, E. Daniel-Szolgay, P. Dierickx, M. Ferro, G. Fiskesjö, L. Garza-Ocañas, M.J. Gómez-Lechón, B. Isomaa, J. Janus, P. Judge, A. Kahru, R.B. Kemp, G. Kerszman, U. Kristen, M. Kunimoto, G. Persoone, K. Lavrijsen, L. Lewan, H. Lilus, T. Ohno, R. Roguet, L. Romert, T.W. Sawyer, H. Seibert, R. Shrivastava, A. Stammati, N. Tanaka, O. Torres-Alanis, J. Voss, S. Wakuri, E. Walum, X. Wang, F. Zucco and B. Ekwall. MEIC evaluation of acute systemic toxicity – part VII. Prediction of human toxicity by results from testing of the first 30 reference chemicals with 27 further *in vitro* assays. *Altern. Lab Anim.* 28: 161–200, 2000.
54. Clemons, J.H., D.G. Dixon and N.C. Bols. Derivation of 2,3,7,8-TCDD toxic equivalency factors (TEFs) for selected dioxins, furans and PCBs with rainbow trout and rat liver cell lines and the influence of exposure time. *Chemosphere* 34: 1105–1119, 1997.
55. Clemons, J.H., L.E.J. Lee, C.R. Myers, D.G. Dixon and N.C. Bols. Cytochrome P4501A1 induction by polychlorinated biphenyls (PCBs) in liver cell lines from rat and trout and the derivation of toxic equivalency factors (TEFs). *Can. J. Fish. Aquat. Sci.* 53: 1177–1185, 1996.
56. Clemons, J.H., C.R. Myers, L.E.J. Lee, D.G. Dixon and N.C. Bols. Induction of cytochrome P4501A1 by binary mixtures of polychlorinated biphenyls (PCBs) and 2,3,7,8-tetrachlorodibenzo-*p*-dioxin (TCDD) in liver cell lines from rat and trout. *Aquat. Toxicol.* 43: 179–194, 1998.
57. Clemons, J.H., M. van den Heuvel, J.J. Stegeman, D.G. Dixon and N.C. Bols. Comparison of toxic equivalent factors for selected dioxin and furan congeners derived using fish and mammalian liver cell lines. *Can. J. Fish. Aquat. Sci.* 51: 1577–1584, 1994.
58. Coletti, D., S. Palleschi, L. Silvestroni, A. Cannavo, E. Vivarelli, F. Tomei, M. Molinaro and S. Adamo. Polychlorobiphenyls inhibit skeletal muscle differentiation in culture. *Toxicol. Appl. Pharm.* 175: 226–233, 2001.
59. Collodi, P., C.L. Miranda, X. Zhao, D.R. Buhler and D.W. Barnes. Induction of zebrafish (*Brachydanio rerio*) P450 *in vivo* and in cell culture. *Xenobiotica* 24: 487–493, 1994.
60. Colye, P., J.C. Philcox and L.C. Carey. Metallothionein: the multipurpose protein. *Cell. Mol. Life Sci.* 59: 627–647, 2002.
61. Connell, D., P. Lam, B. Richardson and R. Wu. *Introduction to Ecotoxicology*, Oxford, England, Blackwell Science, 1999.
62. Dalton, T.P., W.A. Solis, D.W. Nebert and M.J. Carvan, III. Characterization of the MTF-1 transcription factor from zebrafish and trout cells. *Comp. Biochem. Physiol.* 126B: 325–335, 2000.
63. Dannevig, B.H., B.E. Brudeseth, T. Gjoen, M. Rode, H.I. Wergeland, O. Evensen and C.M. Press. Characterization of a long-term cell line (SHK-1) developed from the head kidney of Atlantic salmon (*Salmo salar* L.). *Fish Shellfish Immunol.* 7: 213–226, 1997.
64. Dayeh, V.R., S.L. Chow, K. Schirmer, D.H. Lynn and N.C. Bols. Evaluating the toxicity of Triton X-100 to protozoan, fish and mammalian cells using fluorescent dyes as indicators of cell viability. *Ecotoxicol. Environ. Saf.* 57: 375–382, 2004.
65. Dayeh, V.R., K. Schirmer and N.C. Bols. Applying whole-water samples directly to fish cell cultures in order to evaluate the toxicity of industrial effluent. *Water Res.* 36: 3727–3738, 2002.

66. Dayeh, V.R., K. Schirmer, L.E.J. Lee and N.C. Bols. The use of fish-derived cell lines for investigation of environmental contaminants. *Curr. Protoc. Toxicol.*: 1.5.1–1.5.17, 2003.
67. de Fries, R. and M. Mitsuhashi. Quantification of mitogen-induced human lymphocytes-proliferation – comparison of alamar Blue^{TM} assay to 3H-thymidine incorporation assay. *J. Clin. Lab. Anal.* 9: 89–95, 1995.
68. Denizeau, F.. The use of fish cells in the toxicological evaluation of environmental contaminants. In: *Microscale Testing in Aquatic Toxicology: Advances, Techniques and Practice*, edited by P.G. Well, K. Lee and C. Blaise, Boca Raton, FL, CRC Press, pp. 113–128, 1998.
69. Devlin, E.W. and B. Clary. *In vitro* toxicity of methyl mercury to fathead minnow cells. *Bull. Environ. Contam. Toxicol.* 61: 527–533, 1998.
70. Diamond, L. and H.F. Clark. Comparative studies on the interaction of benzo(*a*)pyrene with cells derived from poikilothermic and homeothermic vertebrates. I. Metabolism of benzo(*a*)pyrene. *J. Natl Cancer Inst.* 45: 1005–1012, 1970.
71. Dickinson, D.A. and H.J. Forman. Glutathione in defense and signaling: lessons from a small thiol. *Ann. N. Y. Acad. Sci.* 973: 488–504, 2002.
72. Dierickx, P.J.. Increased cytotoxic sensitivity of cultured FHM fish cells by simultaneous treatment with sodium dodecyl sulfate and buthionine sulfoximine. *Chemosphere* 36: 1263–1274, 1998.
73. Dierickx, P.J., C. van der Wielen and S. Lemaire. Toxicological evaluation of surface-water samples in sensitised culture fish cells compared with the Microtox method. *Altern. Lab Anim.* 28: 509–515, 2000.
74. Eadon, G., L. Kaminsky, J. Silkworth, K. Aldous, D. Hilker, P. O'Keefe, R. Smith, J. Gierthy, J. Hawley, N. Kim and A. DeCaprio. Calculation of 2,3,7,8-TCDD equivalent concentrations of complex environmental contaminant mixtures. *Environ. Health Perspect.* 70: 221–227, 1986.
75. Essig-Marcello, J.S. and R.G. Van Buskirk. A double-label in situ cytotoxicity assay using the fluorescent probes neutral red and BCECF-AM. *In Vitro Toxicol.* 3: 219–227, 1990.
76. Faisal, M. and W. Ahne. A cell line (CLC) of adherent peripheral blood mononuclear leucocytes of normal common carp *Cyprinus carpio*. *Dev. Comp. Immunol.* 14: 255–260, 1990.
77. Fent, K. Fish cell lines as versatile tools in ecotoxicology: assessment of cytotoxicity, cytochrome P4501A induction potential and estrogenic activity of chemicals and environmental samples. *Toxicol. In Vitro* 15: 477–488, 2001.
78. Fent, K. and R. Bätscher. Cytochrome P4501A induction potencies of polycyclic aromatic hydrocarbons in a fish hepatoma cell line: demonstration of additive interactions. *Environ. Toxicol. Chem.* 19: 2047–2058, 2000.
79. Fent, K. and J. Hunn. Cytotoxicity of organic environmental chemicals to fish liver cells (PLHC-1). *Mar. Environ. Res.* 42: 377–382, 1996.
80. Ferraris, M., A. Flora, D. Fornasari, S. Radice, L. Marabini, S. Frigerio and E. Chiesara. Response of rainbow trout (*Oncorhynchus mykiss*) D-11 cell line to 3-methylcholanthrene (3MC) exposure. *Toxicol. In Vitro* 16: 365–374, 2002.
81. Flint, O.P. *In Vitro* toxicity testing – purpose, validation and strategy. *Altern. Lab Anim.* 18: 11–18, 1990.
82. Fryer, J.L. and C.N. Lannan. Three decades of fish cell culture: a current listing of cell lines derived from fishes. *J. Tissue Cult. Methods* 16: 87–94, 1994.
83. Gagné, F. and C. Blaise. Evaluation of environmental estrogens with a fish cell line. *Bull. Environ. Contam. Toxicol.* 65: 494–500, 2000.
84. Ganassin, R.C. and N.C. Bols. Development of long-term rainbow trout spleen cell cultures that are hemopoietic and generate dendritic cells. *Fish Shellfish Immunol.* 6: 17–34, 1996.
85. Ganassin, R.C. and N.C. Bols. Development and growth of cell lines from fish: rainbow trout *Oncorhynchus mykiss*. In: *Cell & Tissue Culture: Laboratory Procedures*, edited by J.J. Griffiths, A. Doyle and D.G. Newell, Wilshire, UK, Wiley, pp. 23A:1.1–23A:1.9, 1997.
86. Ganassin, R.C. and N.C. Bols. Development of a monocyte-macrophage-like cell line, RTS11 from rainbow trout spleen. *Fish Shellfish Immunol.* 8: 457–476, 1998.
87. Ganassin, R.C., S.M. Sanders, C.J. Kennedy, E.M. Joyce and N.C. Bols. Development of a cell line from Pacific herring, *Clupea harengus pallasi*, sensitive to both naphthalene cytotoxicity and infection by hemorrhagic septicemia virus. *Cell Biol. Toxicol.* 15: 299–309, 1999.
88. Ganassin, R.C., K. Schirmer and N.C. Bols. Methods for the use of fish cell and tissue cultures as model systems in basic and toxicology research. In: *The Laboratory Fish*, edited by G.K Ostrander, San Diego, CA, Academic Press, pp. 631–651, 2000.
89. George, S., D. Burgess, M. Leaver and N. Frerichs. Metallothionein induction in cultured fibroblasts and liver of a marine flatfish, the turbot, *Scophthalmus maximus*. *Fish Physiol. Biochem.* 10: 43–54, 1992.
90. George, S., C. Riley, J. McEvoy and J. Wright. Development of a fish *in vitro* cell culture model to investigate oxidative stress and its modulation by dietary vitamin E. *Mar. Environ. Res.* 50: 541–544, 2000.

91. Giesy, J.P., K. Hilscherova, P.D. Jones, K. Kannan and M. Machala. Cell bioassays for detection of aryl hydrocarbon (AhR) and estrogen receptor (ER) mediated activity in environmental samples. *Mar. Pollut. Bull.* 45: 3–16, 2002.
92. Hahn, M.E. Dioxin toxicology and the aryl hydrocarbon receptor: insights from fish and other non-traditional models. *Mar. Biotechnol.* 3: S224–S238, 2001.
93. Hahn, M.E. Biomarkers and bioassays for detecting dioxin-like compounds in the marine environment. *Sci. Total Environ.* 289: 49–69, 2002.
94. Hahn, M.E. and K. Chandran. Uroporphyrin accumulation associated with cytochrome P4501A induction in fish hepatoma cells exposed to aryl hydrocarbon receptor agonists, including 2,3,7,8-tetrachlorodibenzo-*p*-dioxin and planar chlorobiphenyls. *Arch. Biochem. Biophys.* 329: 163–174, 1996.
95. Hahn, M.E., T.M. Lamb, M.E. Schultz, R.M. Smolowitz and J.J. Stegeman. Cytochrome P4501A induction and inhibition by 3,3′,4,4′-tetrachlorobiphenyl in an Ah receptor-containing fish hepatoma cell line (PLHC-1). *Aquat. Toxicol.* 26: 185–208, 1993.
96. Hahn, M.E., B.L. Woodward, J.J. Stegeman and S.W. Kennedy. Rapid assessment of induced cytochrome P4501A (CYP1A) protein and catalytic activity in fish hepatoma cells grown in multi-well plates: response to TCDD, TCDF and planar PCBs. *Environ. Toxicol. Chem.* 15: 582–591, 1996.
97. Hahn, W.C. Immortalization and transformation of human cells. *Mol. Cells* 13: 351–361, 2002.
98. Halder, M. and W. Ahne. Evaluation of waste water toxicity with three cytotoxicity tests. *Z. Wasser-Abwasser-Forsch.* 23: 233–236, 1990.
99. Hammond, M.E., J. Goodwin and H.F. Dvorak. Quantitative measurements of neutral red uptake and excretion by mammalian cells. *J. Reticuloendothel. Soc.* 27: 337–346, 1980.
100. Heikkila, J.J., G.A. Schultz, K. Iatrou and L. Gedamu. Expression of a set of fish genes following heat or metal ion exposure. *J. Biol. Chem.* 237: 12000–12005, 1982.
101. Henry, T.R., D.J. Nesbit, W. Heideman and R.E. Peterson. Relative potencies of polychlorinated dibenzo-*p*-dioxin, dibenzofuran, and biphenyl congeners to induce cytochrome P4501A mRNA in a zebrafish liver cell line. *Environ. Toxicol. Chem.* 20: 1053–1058, 2001.
102. Hestermann, E.V., J.J. Stegeman and M.E. Hahn. Relationships among the cell cycle, cell proliferation, and aryl hydrocarbon receptor expression in PLHC-1 cells. *Aquat. Toxicol.* 58: 201–213, 2002.
103. Hightower, L.E. and J.L. Renfro. Recent applications of fish cell culture to biomedical research. *J. Exp. Zool.* 248: 290–302, 1988.
104. Hilscherova, K., K. Kannan, Y.S. Kang, I. Holoubek, M. Machala, S. Masunaga, J. Nakanishi and J.P. Giesy. Characterization of dioxin-like activity of sediments from a Czech river basin. *Environ. Toxicol. Chem.* 20: 2768–2777, 2001.
105. Huuskonen, S.E., M.E. Hahn and P. Lindström-Seppä. A fish hepatoma cell line (PLHC-1) as a tool to study cytotoxicity and CYP1A induction properties of cellulose and wood chip extracts. *Chemosphere* 36: 2921–2932, 1998.
106. Huuskonen, S.E., T.E. Ristola, A. Tuvikene, M.E. Hahn, J.V.K. Kukkonen and P. Lindström-Seppä. Comparison of two bioassays, a fish liver cell line (PLHC-1) and a midge (*Chironomous riparius*), in monitoring freshwater sediments. *Aquat. Toxicol.* 44: 47–67, 1998.
107. Huuskonen, S.E., A. Tuvikene, M. Trapido, K. Fent and M.E. Hahn. Cytochrome P4501A induction and porphyrin accumulation in PLHC-1 fish cells exposed to sediment and oil shale extracts. *Arch. Environ. Contam. Toxicol.* 38: 59–69, 2000.
108. Imajoh, M. and S. Suzuki. Apoptosis induced by a marine birnavirus in established cell lines from fish. *Fish Pathol.* 34: 73–79, 1999.
109. Javitt, N.B.. HepG2 cells as a resource for metabolic studies: lipoprotein, cholesterol, and bile acids. *FASEB J.* 4: 161–168, 1990.
110. Jones, O.A.H.. Human pharmaceuticals in the aquatic environment – a review. *Environ. Technol.* 22: 1383–1394, 2002.
111. Jung, D.K.J., T. Klaus and K. Fent. Cytochrome P450 induction by nitrated polycyclic aromatic hydrocarbons, azaarenes, and binary mixtures in fish hepatoma cell line. *Environ. Toxicol. Chem.* 20: 149–159, 2001.
112. Kamer, I. and B. Rinkevich. *In vitro* application of the comet assay for aquatic genotoxicity: considering a primary culture versus a cell line. *Toxicol. In Vitro* 16: 177–184, 2002.
113. Kammann, U., M. Bunke, H. Steinhart and N. Theobald. A permanent fish cell line (EPC) for genotoxicity testing of marine sediments with the comet assay. *Mutat. Res. – Genet. Toxicol. Environ. Mutagen.* 498: 67–77, 2001.
114. Kanduc, D., A. Mittelman, R. Serpico, E. Sinigaglia, A.A. Sinha, C. Natale, R. Santacroce, M.G. Di Corcia, A. Lucchese, L. Dini, P. Pani, S. Santacroce, S. Simone, R. Bucci and E. Farber. Cell death: apoptosis versus necrosis. *Int. J. Oncol.* 21: 165–170, 2002.

115. Kannan, K., N. Yamashita, T. Imagawa, W. Decoen, J.S. Khim, R.M. Day, C.L. Summer and J.P. Giesy. Polychlorinated naphthalenes and polychlorinated biphenyls in fishes from Michigan waters, including the Great Lakes. *Environ. Sci. Technol.* 34: 566–572, 2000.
116. Klapper, W., K. Heidorn, K. Kuhne, R. Parwaresch and G. Krupp. Telomerase activity in 'immortal' fish. *FEBS Lett.* 434: 409–412, 1998.
117. Kling, P.E. and P. Olsson. Involvement of differential metallothionein expression in free radical sensitivity of RTG-2 and CHSE-214 cells. *Free Radic. Biol. Med.* 28: 1628–1637, 2000.
118. Kocan, R.M., M.L. Landolt, J. Bond and E.P. Benditt. *In vitro* effects of some mutagens/carcinogens on cultured fish cells. *Arch. Environ. Contam. Toxicol.* 10: 663–671, 1981.
119. Kocan, R.M., M.L. Landolt and K.M. Sabo. *In vitro* toxicity of eight mutagens/carcinogens for three fish cell lines. *Bull. Environ. Contam. Toxicol.* 23: 269–274, 1979.
120. Kocan, R.M., M.L. Landolt and K.M. Sabo. Anaphase aberrations: a measure of genotoxicity in mutagen treated fish cells. *Environ. Mutagen.* 4: 181–189, 1982.
121. Lagadic, L. Biomarkers: useful tools for the monitoring of aquatic environments. *Rev. Med. Vet.* 153: 581–588, 2002.
122. Lamache, G. and P. Burkhardt-Holm. Changes in apoptotic rate and cell viability in three fish epidermis cultures after exposure to nonylphenol and to a wastewater sample containing low concentrations of nonylphenol. *Biomarkers* 5: 205–218, 2000.
123. Lange, M., W. Gebauer, J. Markl and R. Nagel. Comparison of testing acute toxicity on embryo of zebrafish, *Brachydanio rerio* and RTG-2 cytotoxicity as possible alternatives to the acute fish test. *Chemosphere* 30: 2087–2102, 1995.
124. Lannan, C.N. Fish cell culture: a protocol for quality control. *J. Tissue Cult. Methods* 16: 99–108, 1994.
125. Le Dréan, Y., L. Kern, F. Pakdel and Y. Valotaire. Rainbow trout estrogen receptor presents an equal specificity but a differential sensitivity for estrogens than human estrogen receptor. *Mol. Cell Endocrinol.* 109: 27–35, 1995.
126. Lee, L.E.J., J.H. Clemons, D.G. Bechtel, S.J. Caldwell, K. Han, M. Pasitschniak-Arts, D.D. Mosser and N.C. Bols. Development and characterization of a rainbow trout liver cell line expressing cytochrome P450-dependent mono-oxygenase activity. *Cell Biol. Toxicol.* 9: 279–294, 1993.
127. Lee, L.E.J., M.M. Vijayan and B. Dixon. Toxicogenomic technologies for *in vitro* aquatic toxicology. In: *In Vitro Aquatic Toxicology*, edited by B. Austin and C. Mothersill, London, Springer, pp. 143–160, 2003.
128. Lewis, S., R.D. Handy, B. Cordi, Z. Billinghurst and M.H. Depledge. Stress proteins (HSPs): methods of detection and their use as an environmental biomarker. *Ecotoxicology* 8: 351–368, 1999.
129. Li, H. and S. Zhang. *In vitro* cytotoxicity of the organophosphorus pesticide parathion to FG-9307 cells. *Toxicol. In Vitro* 15: 643–647, 2001.
130. Li, H. and S. Zhang. *In vitro* cytotoxicity of the organophosphorus insecticide methylparathion to FG-9307, the gill cell line of flounder (*Paralichthys olivaceus*). *Cell Biol. Toxicol.* 18: 235–241, 2002.
131. Li, Z.Q., F.Y. He, C.J. Stehle, Z. Wang, S. Kar, F.M. Finn and B.I. Carr. Vitamin K uptake in hepatocytes and hepatoma cells. *Life Sci.* 70: 2085–2100, 2002.
132. Lopez-Ribas, D., M.C. Riva, M.J. Muñoz and A. Castaño. Acute toxicity of cationic softeners determined *in vitro* on RTG-2 trout cells. *Tenside Surf. Det.* 35: 276–278, 1998.
133. Lorenzen, A. and A.B. Okey. Detection and characterization of [^{3}H]2,3,7,8-tetrachlorodibenzo-*p*-dioxin binding to Ah receptor in a rainbow trout hepatoma cell line. *Toxicol. Appl. Pharmacol.* 106: 53–62, 1990.
134. Magwood, S. and S. George. *In vitro* alternatives to whole animal testing. Comparative cytotoxicity studies of divalent metals in established cell lines derived from tropical and temperate water fish species in a neutral red assay. *Mar. Environ. Res.* 42: 37–40, 1996.
135. Maracine, M. and H. Segner. Cytotoxicity of metals in isolated fish cells: importance of the cellular glutathione status. *Comp. Biochem. Physiol.* 120C: 83–88, 1998.
136. Marion, M. and F. Denizeau. Rainbow trout and human cells in culture for the evaluation of the toxicity of aquatic pollutants: a study with lead. *Aquat. Toxicol.* 3: 47–60, 1983.
137. Martin-Alguacil, N., H. Babich, D.W. Rosenberg and E. Borenfreund. *In vitro* response of the brown bullhead catfish cell line, BB, to aquatic pollutants. *Arch. Environ. Contam. Toxicol.* 20: 113–117, 1991.
138. Miller, N.W., M.A. Rycyzyn, M.R. Wilson, G.W. Warr, J.P. Naftel and L.W. Clem. Development and characterization of channel catfish long term B cell lines. *J. Immunol.* 152: 2180–2189, 1994.
139. Moore, M.N. Biocomplexity: the post-genome challenge in ecotoxicology. *Aquat. Toxicol.* 59: 1–15, 2002.
140. Mori, M., N. Kawakubo and M. Wakabayashi. Cytotoxicity of surfactants to the FHM-sp cell line. *Fish. Sci.* 68: 1124–1128, 2002.
141. Mori, M. and M. Wakabayashi. Cytotoxicity evaluation of chemicals using cultured fish cells. *Water Sci. Technol.* 42: 277–282, 2000.

142. Mori, M. and M. Wakabayashi. Cytotoxicity evaluation of synthesized chemicals using suspension-cultured fish cells. *Fish. Sci.* 66: 871–875, 2000.
143. Muñoz, M.J., A. Castaño, T. Blazquez, M. Vega, G. Carbonell, J.A. Ortiz, M. Carballo and J.V. Tarazona. Toxicity identification evaluations for the investigation of fish kills: a case study. *Chemosphere* 29: 55–61, 1994.
144. Nehls, S. and H. Segner. Detection of DNA damage in two cell lines from rainbow trout, RTG-2 and RTL-W1, using the comet assay. *Environ. Toxicol.* 16: 321–329, 2001.
145. Nicholson, B.L. Techniques in fish cell culture, In: *Techniques in the Life Sciences*, Vol. C1, edited by E. Kurstak, Limmerick, Elsevier, pp. C015/1–C015/16, 1985.
146. Nicholson, B.L. Fish cell culture: an update, In: *Advances in Cell Culture*, Vol. 7, edited by K. Maramorosch, New York, Academic Press, pp. 1–18, 1989.
147. Nishigaki, R., H. Mitani, N. Tsuchida and A. Shima. Effect of cyclobutane pyrimidine dimers on apoptosis induced by different wavelengths of UV. *Photochem. Photobiol.* 70: 228–235, 1999.
148. O'Brien, P.J., I. Wilson, T. Orton and F. Pognan. Investigation of the alamar Blue (resazurin) fluorescent dye for the assessment of mammalian cell cytotoxicity. *Eur. J. Biochem.* 267: 5421–5426, 2000.
149. Okamura, H., T. Watanabe, I. Aoyama and M. Hasobe. Toxicity evaluation of new antifouling compounds using suspension-cultured fish cells. *Chemosphere* 46: 945–951, 2002.
150. Okey, A.B., D.S. Riddick and P.A. Harper. The Ah receptor: mediator of the toxicity of 2,3,7,8-tetrachlorodibenzo-*p*-dioxin (TCDD) and related compounds. *Toxicol. Lett.* 70: 1–22, 1994.
151. Olsson, P.E., S.J. Hyllner, M. Zafarullah, T. Andersson and L. Gedamu. Differences in metallothionein expression in primary cultures of rainbow trout hepatocytes and the RTH-149 cell line. *Biochim. Biophys. Acta* 1049: 78–82, 1990.
152. Ossum, C.G., E.K. Hoffmann, M.M. Vijayan, S.E. Holt and N.C. Bols. Isolation and characterization of a novel fibroblast-like cell line from the rainbow trout (*Oncorhynchus mykiss*) and a study of $p38^{MAPK}$ activation and induction of HSP70 in response to chemically induced ischemia. *J. Fish Biol.* 64: 1103–1116, 2004.
153. Park, E.-H., J.-S. Lee, A.-E. Yi and H. Etoh. Fish cell line (ULF-23HU) derived from the fin of the central mudminnow (*Umbra limi*): suitable characteristics for clastogenicity assay. *In Vitro Cell Dev. Biol.* 25: 987–994, 1989.
154. Parrott, J.L., S.M. Backus, A.I. Borgmann and M. Swyripa. The use of semipermeable membrane devices to concentrate chemicals in oil refinery effluent on the Mackenzie River. *Arctic* 52: 125–138, 1999.
155. Parrott, J.L., P.V. Hodson and M.R. Servos. Relative potency of polychlorinated dibenzo-*p*-dioxins and dibenzofurans for inducing mixed-function oxygenase activity in rainbow trout. *Environ. Toxicol. Chem.* 14: 1041–1050, 1995.
156. Parrott, J.L., M. van den Heuvel, L.M. Hewitt, M.A. Baker and K.R. Munkittrick. Isolation of MFO inducers from tissues of white suckers caged in bleached kraft mill effluent. *Chemosphere* 41: 1083–1089, 2000.
157. Pollenz, R.S. and B. Necela. Characterization of two continuous cell lines from *Oncorhynchus mykiss* for models of aryl-hydrocarbon-receptor-mediated signal transduction. Direct comparison to the mammalian Hepa-1c1c7 cell line. *Aquat. Toxicol.* 41: 31–49, 1998.
158. Pollenz, R.S., B. Necela and K. Marks-Sojka. Analysis of rainbow trout Ah receptor protein isoforms in cell culture reveals conservation of function in Ah receptor-mediated signal transduction. *Biochem. Pharmacol.* 64: 49–60, 2002.
159. Pollenz, R.S., H.R. Sullivan, J. Holmes, B. Necela and R.E. Peterson. Isolation and expression of cDNAs from rainbow trout (*Oncorhynchus mykiss*) that encodes two novel basic helix–loop–helix/PER-ARNT-SIM (bHLH/PAS) proteins with distinct functions in the presence of the aryl hydrocarbon receptor. *J. Biol. Chem.* 271: 30886–30896, 1996.
160. Price-Haughey, J., K. Bonham and L. Gedamu. Heavy metal-induced gene expression in fish and fish cell lines. *Environ. Health Perspect.* 65: 141–147, 1986.
161. Price-Haughey, J., K. Bonham and L. Gedamu. Metallothionein gene expression in fish cell lines: its activation in embryonic cells by 5-azacytidine. *Biochim. Biophys. Acta* 908: 158–168, 1987.
162. Putnam, K.P., D.W. Bombick and D.J. Doolittle. Evaluation of eight *in vitro* assays for assessing the cytotoxicity of cigarette smoke condensate. *Toxicol. In Vitro* 16: 599–607, 2002.
163. Rachlin, J.W. and A. Perlmutter. Fish cells in culture for study of aquatic toxicants. *Water Res.* 2: 409–414, 1968.
164. Rausch, D.M. and S.B. Simpson. *In vivo* test system for tumor production by cell lines derived from lower vertebrates. *In Vitro Cell Dev. Biol.* 24: 217–222, 1988.
165. Reader, S.J., V. Blackwell, R. O'Hara, R.H. Clothier, G. Griffin and M. Balls. Neutral red release from pre-loaded cells as an *in vitro* approach to testing for eye irritancy potential. *Toxicol. In Vitro* 4: 264–266, 1990.

166. Reinhart, B. and L.E.J. Lee. Integrin-like substrate adhesion in RTG-2 cells, a fibroblastic cell line derived from rainbow trout. *Cell Tissue Res.* 307: 165–172, 2002.
167. Repetto, G., A. Jos, M.J. Hazen, M.L. Molero, A. del Peso, M. Saluero, P. de Castillo, M.C. Rodríguez-Vincente and M. Repetto. A test battery for the ecotoxicological evaluation of pentachlorophenol. *Toxicol. In Vitro* 15: 503–509, 2001.
168. Richter, C.A., V.L. Tieber, M.S. Denison and J.P. Giesy. An *in vitro* rainbow trout cell bioassay for aryl hydrocarbon-receptor mediated toxins. *Environ. Toxicol. Chem.* 16: 543–550, 1997.
169. Roberts, E.A., M. Letarte, J. Squire and S. Yang. Characterization of human hepatocyte lines derived from normal liver-tissue. *Hepatology* 19: 1390–1399, 1994.
170. Ryan, J.A. and L.E. Hightower. Evaluation of heavy-metal ion toxicity in fish cells using a combined stress protein and cytotoxicity assay. *Environ. Toxicol. Chem.* 13: 1231–1240, 1994.
171. Saito, H., T. Shigeoka and F. Yamauci. *In vitro* cytotoxicity of chlorophenols to goldfish GS-scale (GFS) cells and quantitative structure–activity relationships. *Environ. Toxicol. Chem.* 10: 235–241, 1991.
172. Sanders, B.M., J. Nguyen, L.S. Martin, S.R. Howe and S. Coventry. Induction and subcellular localization of two major stress proteins in response to copper in the fathead minnow *Pimephales promelas*. *Comp. Biochem. Physiol.* 112C: 335–343, 1995.
173. Schaeffer, W.I. Terminology associated with cell, tissue and organ culture, molecular biology and molecular genetics. *In Vitro Cell Dev. Biol.* 26: 97–101, 1990.
174. Schirmer, K., A.G.J. Chan and N.C. Bols. Transitory metabolic disruption and cytotoxicity elicited by benzo[*a*]pyrene in two cell lines from rainbow trout liver. *J. Biochem. Mol. Toxicol.* 14: 262–276, 2000.
175. Schirmer, K., A.G.J. Chan, B.M. Greenberg, D.G. Dixon and N.C. Bols. Methodology for demonstrating and measuring the photocytotoxicity of fluoranthene to fish cells in culture. *Toxicol. In Vitro* 11: 107–119, 1997.
176. Schirmer, K., A.G.J. Chan, B.M. Greenberg, D.G. Dixon and N.C. Bols. Ability of 16 priority PAHs to be directly cytotoxic to a cell line from the rainbow trout gill. *Toxicology* 127: 129–141, 1998.
177. Schirmer, K., A.G.J. Chan, B.M. Greenberg, D.G. Dixon and N.C. Bols. Ability of 16 priority PAHs to be photocytotoxic to a cell line from the rainbow trout gill. *Toxicology* 127: 143–155, 1998.
178. Schirmer, K., V.R. Dayeh, S. Bopp, S. Russold and N.C. Bols. Applying whole water samples to cell bioassays for detecting dioxin-like compounds at contaminated sites. *Toxicology*, 205: 211–221, 2004.
179. Schirmer, K., R.C. Ganassin, J.L. Brubacher and N.C. Bols. A DNA fluorometric assay for measuring fish cell proliferation in microplates with different well sizes. *J. Tissue Cult. Methods* 16: 133–142, 1994.
180. Schirmer, K., J.S. Herbrick, B.M. Greenberg, D.G. Dixon and N.C. Bols. Use of fish gill cells in culture to evaluate the cytotoxicity and photocytotoxicity of intact and photomodified creosote. *Environ. Toxicol. Chem.* 18: 1277–1288, 1999.
181. Schirmer, K., E.M. Joyce, D.G. Dixon, B.M. Greenberg and N.C. Bols. Ability of the quinones arising from the UV irradiation of anthracene to induce 7-ethoxyresorufin *o*-deethylase activity in a trout liver cell line, In: *Environmental Toxicology and Risk Assessment: Science, Policy, and Standarization – Implications for Environmental Decisions*, Vol. 10, West Conshohocken, PA, ASTM, 2001, pp. 16–26.
182. Schirmer, K., D.J. Tom, N.C. Bols and J.P. Sherry. Ability of fractionated petroleum refinery effluent to elicit cyto- and photocytotoxic responses and to induce 7-ethoxyresorufin-*o*-deethylase activity in fish cell lines. *Sci. Total Environ.* 271: 61–78, 2001.
183. Schlenk, D. and C.D. Rice. Effect of zinc and cadmium treatment on hydrogen peroxide-induced mortality and expression of glutathione and metallothionein in a teleost hepatoma cell line. *Aquat. Toxicol.* 43: 121–129, 1998.
184. Segner, H. Fish cell lines as a tool in aquatic toxicology. In: *Ecotoxicology*, edited by T. Braunbeck, D.E. Hinton and B. Streit, Basel, Birkhäuser, pp. 1–38, 1998.
185. Segner, H., A. Behrens, E.M. Joyce, K. Schirmer and N.C. Bols. Transient induction of 7-ethoxyresorufin-*O*-deethylase (EROD) activity by medium change in the rainbow trout liver cell line, RTL-W1. *Mar. Environ. Res.* 50: 489–493, 2000.
186. Segner, H., D. Lenz, W. Hanke and G. Schüürmann. Cytotoxicity of metals toward rainbow trout R1 cell line. *Environ. Toxicol. Water Qual.* 9: 273–279, 1994.
187. Shaw, A.J. Defining cell viability and cytotoxicity. *Altern. Lab Anim.* 22: 124–126, 1994.
188. Shen, L., T.B. Stuge, H. Zhou, M. Khayat, K.S. Barker, S.M. Quiniou, M. Wilson, E. Bengten, V.G. Chinchar, L.W. Clem and N.W. Miller. Channel catfish cytotoxic cells: a mini-review. *Dev. Comp. Immunol.* 26: 141–149, 2002.
189. Shen, Y., C. West and S.R. Hutchins. *In vitro* cytotoxicity of aromatic aerobic biotransformation products in Bluegill Sunfish BF-2 cells. *Ecotoxicol. Environ. Saf.* 45: 27–32, 2000.
190. Shugart, L.R., J.F. McCarthy and R.S. Halbrook. Biological markers of environmental and ecological contamination. *Risk Anal.* 12: 353–360, 1992.

191. Smeets, J.M.W., A. Voormolen, D.E. Tillitt, J.M. Everaarts, W. Seinen and M. Van Den Berg. Cytochrome P4501A induction, benzo[*a*]pyrene metabolism, and nucleotide adduct formation in fish hepatoma cells: effect of preexposure to 3,3′,-4,4′,5-pentachlorobiphenyl. *Environ. Toxicol. Chem.* 18: 474–480, 1999.
192. Smolarek, T.A., S.L. Morgan and W.M. Baird. Temperature-induced alterations in the metabolic activation of benzo[*a*]pyrene to DNA-binding metabolites in the Blue gill cell line BF-2. *Aquat. Toxicol.* 13: 89–98, 1988.
193. Smolarek, T.A., S.L. Morgan, C.G. Moynihan, H. Lee, R.G. Harvey and W.M. Baird. Metabolism and DNA adduct formation of benzo(*a*)pyrene and 7,12-dimethylbenz(a)anthracene in fish cell lines in culture. *Carcinogenesis* 8: 1501–1509, 1987.
194. Swanson, H.I. and G.H. Perdew. Detection of the Ah receptor in rainbow trout. Use of 2-azido-3-[^{125}I] iodo-7,8-dibromodibenzo-*p*-dioxin in cell culture. *Toxicol. Lett.* 58: 85–95, 1991.
195. Tarazona, J.V., A. Castaño and B. Gallego. Detection of organic toxic pollutants in water and waste-water by liquid -chromatography and *in vitro* cytotoxicity tests. *Anal. Chim. Acta* 234: 193–197, 1990.
196. Thornton, S.C., L. Diamond and W.M. Baird. Metabolism of benzo(*a*)pyrene by fish cells in culture. *J. Toxicol. Environ. Health* 10: 157–167, 1982.
197. Tom, D.J., L.E.J. Lee, J. Lew and N.C. Bols. Induction of 7-ethoxyresorufin-*o*-deethylase activity by planar chlorinated hydrocarbons and polycyclic aromatic hydrocarbons in cell lines from the rainbow trout pituitary. *Comp. Biochem. Physiol.* 128A: 185–198, 2001.
198. Trump, B.F., I.K. Berezesky, S.H. Chang and P.C. Phelps. The pathways of cell death: oncosis, apoptosis, and necrosis. *Toxicol. Pathol.* 25: 82–88, 1997.
199. Tzukerman, M., S. Selig and K. Skorecki. Telomeres and telomerase in human health and disease. *J. Pediatr. Endocrinol. Metab.* 15: 229–240, 2002.
200. Vallejo, A.N., C.F. Ellsaesser, N.W. Miller and L.W. Clem. Spontaneous development of functionally active long-term monocyte like cell lines from channel catfish. *In Vitro Cell Dev. Biol.* 27A: 279–286, 1991.
201. van den Berg, M., L. Birnbaum, A.T. Bosveld, B. Brunstrom, P. Cook, M. Feeley, J.P. Giesy, A. Hanberg, R. Hasegawa, S.W. Kennedy, T. Kubiak, J.C. Larsen, F.X. van Leeuwen, A.K. Liem, C. Nolt, R.E. Peterson, L. Poellinger, S. Safe, D. Schrenk, D. Tillitt, M. Tysklind, M. Younes, F. Waern and T. Zacharewski. Toxic equivalency factors (TEFs) for PCBs, PCDDs, PCDFs for humans and wildlife. *Environ. Health Perspect.* 106: 775–792, 1998.
202. van den Heuvel, M.R., M.R. Servos, K.R. Munkittrick, N.C. Bols and D.G. Dixon. Evidence for a reduction of 2,3,7,8-TCDD equivalent concentrations in white sucker (*Catostomus commersoni*) exposed to bleached kraft pulp mill effluent, following process and treatment improvements. *J. Great Lakes Res.* 22: 264–279, 1996.
203. van der Oost, R., J. Beyer and N.P.E. Vermeulen. Fish bioaccumulation and biomarkers in environmental risk assessment: a review. *Environ. Toxicol. Pharmacol.* 13: 57–149, 2003.
204. Varanasi, U., W.M. Baird and T.A. Smolarek. Metabolic activation of PAHs in subcellular fractions and cell cultures from aquatic and terrestrial species. In: *Metabolism of Polycyclic Aromatic Hydrocarbons in the Aquatic Environment*, edited by U. Varanasi, Boca Raton, FL, CRC Press, 1989, Chapter 6.
205. Vega, M.M., A. Castaño, T. Blazquez and J.V. Tarazona. Assessing organic toxic pollutants in fish-canning factory effluents using cultured fish cells. *Ecotoxicology* 3: 79–88, 1994.
206. Vega, M.M., C. Fernandez, T. Blazquez, J.V. Tarazona and A. Castaño. Biological and chemical tools in the toxicological risk assessment of Jarama River, Madrid, Spain. *Environ. Pollut.* 93: 135–139, 1996.
207. Viarengo, A., B. Dondero, F. Dondero, A. Marro and R. Fabbri. Metallothionein as a tool in biomonitoring programmes. *Biomarkers* 4: 455–466, 1999.
208. Villeneuve, D.L. and A.L. Blankenship. Derivation and application of relative potency estimates based on *in vitro* bioassay results. *Environ. Toxicol. Chem.* 19: 2835–2843, 2000.
209. Villeneuve, D.L., R.L. Crunkilton and W.M. DeVita. Aryl hydrocarbon receptor-mediated toxic potency of dissolved lipophilic organic contaminants collected from Lincoln Creek, Milwaukee, Wisconsin, USA, to PLHC-1 (*Poeciliopsis lucida*) fish hepatoma cells. *Environ. Toxicol. Chem.* 16: 977–984, 1997.
210. Villeneuve, D.L., K. Kannan, J.S. Khim, J. Falandysz, V.A. Nikiforov, A.L. Blankenship and J.P. Giesy. Relative potencies of individual polychlorinated naphthalenes to induce dioxin-like responses in fish and mammalian *in vitro* biossays. *Arch. Environ. Contam. Toxicol.* 39: 273–281, 2000.
211. Villeneuve, D.L., J.S. Khim, K. Kannan and J.P. Giesy. Relative potencies of individual polycylic aromatic hydrocarbons to induce dioxin-like and estrogenic responses in three cell lines. *Environ. Toxicol.* 17: 128–137, 2002.
212. Villeneuve, D.L., C.A. Richter, A.L. Blankenship and J.P. Giesy. Rainbow trout cell bioassay-derived relative potencies for halogenated aromatic hydrocarbons: comparison and sensitivity analysis. *Environ. Toxicol. Chem.* 18: 879–888, 1999.

213. Vosdingh, R.A. and M.J. Neff. Bioassay of aflatoxins by catfish cell cultures. *Toxicology* 2: 107–112, 1974.
214. Walker, M.K. and R.E. Peterson. Potencies of polychlorinated dibenzo-*p*-dioxin, dibenzofuran, and biphenyl congeners, relative to 2,3,7,8-tetrachlorodibenzo-*p*-dioxin, for producing early life stage mortality in rainbow trout (*Oncorhynchus mykiss*). *Aquat. Toxicol.* 21: 219–238, 1991.
215. Walton, D.G., A.B. Acton and H.F. Stich. Chromosome aberrations in cultured central mudminnow heart cells and Chinese hamster ovary cells exposed to polycyclic aromatic hydrocarbons and sediment extracts. *Comp. Biochem. Physiol.* 89C: 395–402, 1988.
216. Wang, R. and M. Belosevic. The *in vitro* effects of estradiol and cortisol on the function of a long-term goldfish macrophage cell line. *Dev. Comp. Immunol.* 19: 327–336, 1995.
217. Weinstein, J.E., J.T. Oris and D.H. Taylor. An ultrastructural examination of the mode of UV-induced toxic action of fluoranthene in fathead minnows, *Pimephales promelas*. *Aquat. Toxicol.* 39: 1–22, 1997.
218. Whyte, J.J., R.E. Jung, C.J. Schmitt and D.E. Tillitt. Ethoxyresorufin-*O*-deethylase (EROD) activity in fish as a biomarker of chemical exposure. *CRC Crit. Rev. Toxicol.* 30: 347–570, 2000.
219. Whyte, J.J., N.A. Karrow, H.J. Boermans, D.G. Dixon and N.C. Bols. Combined methodologies for measuring exposure of rainbow trout (*Oncorhynchus mykiss*) to polycyclic aromatic hydrocarbons (PAHs) in creosote contaminated microcosms. *Polycyclic Aromat. Comp.* 18: 71–98, 2000.
220. Whyte, J.J., M. van den Heuvel, J.H. Clemons, S.Y. Huestis, M.R. Servos, D.G. Dixon and N.C. Bols. Mammalian and teleost cell line bioassay and chemically derived 2,3,7,8-tetrachlorodibenzo-*p*-dioxin equivalent concentrations in lake trout (*Salvelinus namaycush*) from Lake Superior and Lake Ontario, North America. *Environ. Toxicol. Chem.* 17: 2214–2226, 1998.
221. Wolf, K. and J.A. Mann. Poikilotherm vertebrate cell lines and viruses: a current listing for fishes. *In Vitro* 16: 168–179, 1980.
222. Wolf, K. and M.C. Quimby. Fish cell and tissue culture. In: *Fish Physiology*, edited by W.S Hoar and D.J Randall, New York, Academic Press, pp. 253–306, 1969.
223. Wolffe, A.P., J.F. Glover and J.R. Tata. Culture shock: synthesis of heat-shock-like proteins in fresh primary cell cultures. *Exp. Cell Res.* 154: 581–590, 1984.
224. Wright, J., S. George, E. Martinez-Lara, E. Carpene and M. Kindt. Levels of cellular glutathione and metallothionein affect the toxicity of oxidative stressors in an established carp cell line. *Mar. Environ. Res.* 50: 503–508, 2000.
225. Wrobel, K., E. Claudio, F. Segade, S. Ramos and P.S. Lazo. Measurement of cytotoxicity by propidium iodide staining of target cell DNA. *J. Immunol. Methods* 189: 243–249, 1996.
226. Xian, L.X., J.Z. Shao and Z. Meng. Apoptosis induction in fish cells under stress of six heavy metal ions. *Prog. Biochem. Biophys.* 28: 866–869, 2001.
227. Zabel, E.W. and R.E. Peterson. TCDD-like activity of 2,3,6,7-tetrachloroxanthene in rainbow trout early life stages and in a rainbow trout gonadal cell line (RTG-2). *Environ. Toxicol. Chem.* 15: 2305–2309, 1996.
228. Zabel, E.W., R.S. Pollenz and R.E. Peterson. Relative potencies of individual dibenzofuran, and biphenyl congeners and congener mixtures based on induction of cytochrome P4501A mRNA in a rainbow trout gonadal cell line (RTG-2). *Environ. Toxicol. Chem.* 15: 2310–2318, 1996.
229. Zafarullah, M., K. Bonham and L. Gedamu. Structure of the rainbow trout metallothionein B gene and characterization of its metal-responsive region. *Mol. Cell Biol.* 8: 4469–4476, 1988.
230. Zafarullah, M., P.E. Olsson and L. Gedamu. Endogenous and heavy metal ion induced metallothionein gene expression in Salmonid fish tissues and cell lines. *Gene* 83: 85–93, 1989.
231. Zafarullah, M., P.E. Olsson and L. Gedamu. Differential regulation of metallothionein genes in rainbow trout fibroblasts, RTG-2. *Biochim. Biophys. Acta* 1049: 318–323, 1990.
232. Zahn, T., H. Arnold and T. Braunbeck. Cytological and biochemical response of R1 cells and isolated hepatocytes from rainbow trout (*Oncorhynchus mykiss*) to subacute *in vitro* exposure to disulfoton. *Exp. Toxicol. Pathol.* 48: 47–64, 1996.
233. Zahn, T., C. Hauck, J. Holzschuh and T. Braunbeck. Acute and sublethal toxicity of seepage waters from garbage dumps to permanent cell lines and primary cultures of hepatocytes from rainbow trout (*Oncorhynchus mykiss*): a novel approach to environmental risk assessment for chemicals and chemical mixtures. *Zentbl. Hyg. Umweltmed.* 196: 455–479, 1995.
234. Zelikoff, J.T. Biomarkers of immunotoxicity in fish and other non-mammalian sentinel species: predictive value for mammals? *Toxicology* 129: 63–71, 1998.

Biochemistry and Molecular Biology of Fishes, vol. 6
T. P. Mommsen and T. W. Moon (Editors)

CHAPTER 3

Approaches in proteomics and genomics for eco-toxicology

NANCY D. DENSLOW*, IRIS KNOEBL† AND PATRICK LARKIN§

**Department of Physiological Sciences and Center for Environmental and Human Toxicology, University of Florida, Gainesville, FL 32611, USA, †Molecular Ecology Research Branch, EERD/NERL/US EPA, Cincinnati, OH 45268, USA, and §EcoArray LLC, Alachua, FL 32615, USA*

I. Introduction

A new area of scientific investigation, coined toxicogenomics, enables researchers to understand and study the interaction between the environment and inherited genetic characteristics. This understanding will be critical to fully appreciate the response of organisms to environmental stress and toxicants[125]. The availability of whole genome sequences has ushered in this new field, for it is now possible to ask global questions about how physiological systems are perturbed and how metabolic pathways relate to each other. These questions will be important for assessing the risk of exposures to contaminants and for determining specific toxicant pathways and mechanisms of action. It is now clear that classes of chemicals, and perhaps even individual chemicals, will display their own 'chemical signatures,[53,54,79,125] both at the gene transcription and protein expression levels. Once these signatures are identified, it will be easier to determine the risk of exposures to populations of wildlife and even to humans.

Determining the effects at the individual and population levels is important for aquatic toxicology. Multiple studies have begun to relate chemical exposures to phenotypic changes in tissues of affected animals[43,47,55,64,69]. Tissue changes must now be related to both time and dose-dependent changes at the molecular level. This 'phenotypic anchoring' will be indispensable to distinguish changes due to normal environmental stresses from those due to chemical exposure[125]. With these new tools it will be possible to study both toxic and subtoxic doses of chemicals and determine long-range changes that could affect health and populations of wildlife, months or years after the initial exposures.

Fish are excellent vertebrate models for sorting out complex physiological networks. The complete genomes for the zebrafish (*Danio rerio*) and two puffer fish (*Takifugu rubripes* and *Tetraodon nigroviridis*) have been completely sequenced. The sequencing of other fish genomes, including tilapia (*Oreochromis niloticus*), Atlantic salmon (*Salmo salar*), medaka (*Oryzias latipes*), and several other fish species as well as a tunicate (*Ciona intestinalis*), are well underway (see http://wit.integratedgenomics.com/GOLD). These few species represent only a small number of fish of interest in the field of toxicology and certainly do not represent the over 23,000 known species of fish. It will be much more difficult to study models for which genome sequences are not complete. However, numerous ongoing expressed sequence tag (EST) projects are also in progress and information obtained from these efforts will add additional depth to the comparative fish databases. Identifying proteins in these species will also rely heavily on their genome sequences, since the current protein identification procedures by mass spectrometry rely on algorithms based on genetic information. In this chapter we will review current methods to study the transcriptome (the expressed mRNA sequences of the genome) and the proteome (the set of proteins encoded by a genome)[109], and where possible, indicate approaches that can be taken for non-model species.

Global experiments to understand the effects of toxicants on gene expression have been performed successfully at the transcriptome level[6,37]. But, how closely do changes at the mRNA level relate to changes at the protein level or, more importantly, to changes at the protein activity level? To answer these questions, several investigators have begun to compare global transcriptomic experiments with global proteomic experiments. In experiments with yeast, Aebersold's group[49,67] calculated a low correlation between changes at the mRNA level with changes at the protein level ($r^2 = 0.61$). They monitored both changes in transcription (using gene arrays) and changes in protein abundance (using mass spectrometry) and found that the correlation between the two methods was low. While some gene products changed both at the mRNA and protein levels in a predicted pattern, there were others that were unchanged. Several mRNAs had variations in their expression levels without a corresponding change in their protein counterparts. And, some proteins had changes in abundance without a reciprocal modification in mRNA levels. These differences are summarized in Fig. 1 and suggest that both genomic and proteomic measurements must be made to understand the full dynamics of exposure.

To illustrate the importance of measuring both genomic and proteomic changes, one can consider the information obtained both at the mRNA and protein levels after the exposure of fish to estradiol (E_2). Estradiol binds directly to soluble estrogen receptors to turn on gene transcription of estrogen-sensitive genes that are then translated into proteins. One would expect that in such a system, a change at the mRNA level would result in a direct change at the protein level. Vitellogenin (Vg), the egg yolk precursor protein, is an excellent biomarker to illustrate this response since it is regulated by E_2. Several studies have documented both RNA and protein expression time courses after a single injection[14,15,38].

We have performed two studies with sheepshead minnow (*Cyprinodon variegatus*). In one experiment fish were exposed constantly to a low, environmentally relevant level of estradiol and in the second, fish received an injection of 2.5 mg/kg E_2[15,39,60]. In the constant exposure experiment (Fig 2A), male sheepshead minnows were treated with 100 ng/l E_2 in a flow through design[39,60]. Non-exposed fish had about 30 fg Vg mRNA/μg total RNA. This level went up to 50 pg/μg total RNA after 2 days and eventually reached a plateau at 1000 pg/μg total RNA where it remained for the duration of the experiment. In the same time frame, Vg protein levels, not measurable by ELISA at the beginning of the experiment, went up to 30 μg/ml after 2 days and continued to go up at 7, 15, and 21 days before reaching a plateau at about 50 mg/ml. In this experiment, there is good correlation between mRNA expression

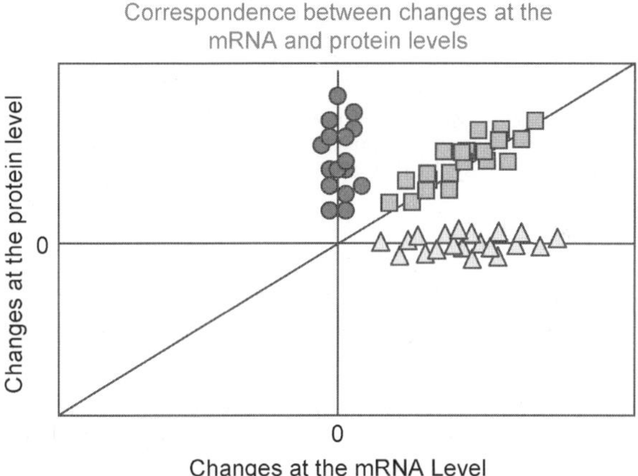

Fig. 1. Cartoon depicting possible correlation between changes in expression at the mRNA and protein levels. This cartoon is based on research results in yeast obtained by Ideker *et al.*[67]. Spots along the diagonal show good correspondence between changes at the mRNA and protein levels, while spots along the *X*-axis show a change in the mRNA level without a corresponding change at the protein level and spots along the *Y*-axis show a change in protein levels, without a corresponding change at the mRNA level.

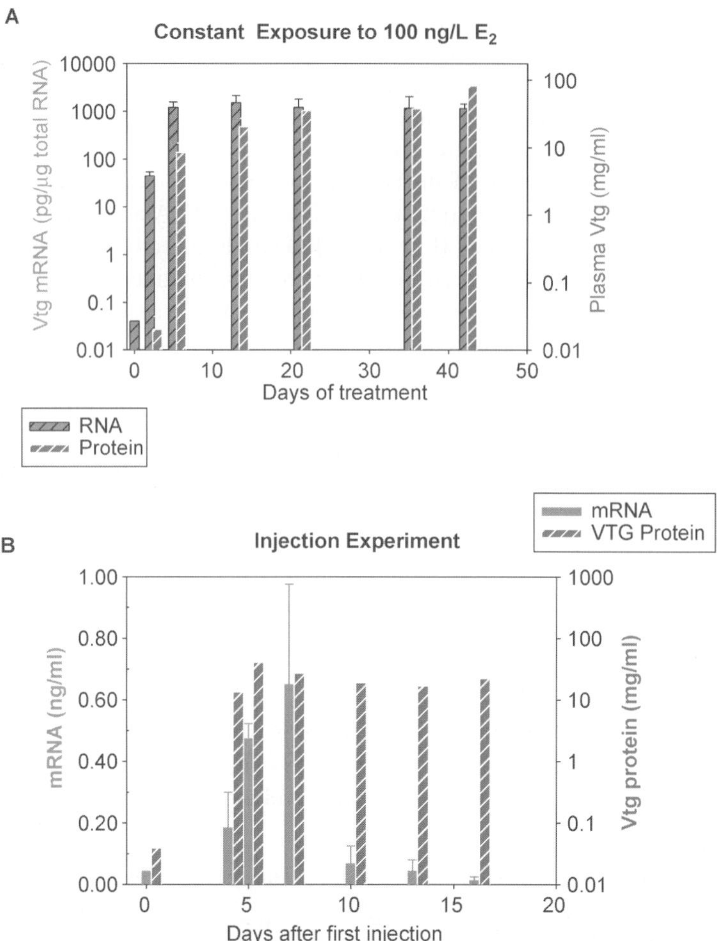

Fig. 2. Exposure of sheepshead minnows to 17-β-estradiol. (A) Exposure to a constant concentration of 100 ng E_2/l in a flow-through design and (B) exposure to two injections (i.p.) of 2.5 mg E_2/kg 4 days apart. Plasma vitellogenin was measured by ELISA and hepatic Vg mRNA was measured by slot blot as described previously[15,39,60].

and protein expression, especially once the fish were acclimated to the constant exposure. However, this is not the case for the second study (Fig. 2B), in which male sheepshead minnows were injected twice with 2.5 mg/kg E_2 with an interlude of 4 days between injections. In this experiment, the expressions of mRNA and protein were monitored beginning immediately after the second injection. The Vg mRNA levels increased at 24 h, maximized by 48 h, and then decreased to background levels after 6 days (day 10 in graph). During the same time frame, protein levels already high from the first injection, increased slightly to 50 mg/ml after 24 h and then declined to a plateau at around 20 mg/ml for the remainder of the experiment. Depending on when the measurements were made, different mRNA and protein expression profiles were observed. These experiments measure the 'steady state' amounts of mRNAs or proteins present in tissues at a particular time, which in turn are a reflection of both *de novo* transcription and translation as well as degradation of the mRNAs and proteins in question. In these particular experiments, it is clear that Vg mRNA has a much shorter half-life than Vg protein. Thus to get a clear view of the dynamics of Vg regulation by E_2, one should measure both the mRNA and the protein, at multiple times and doses. These results illustrate that it is critical to examine both the transcriptome and the proteome in order to gain a better understanding of what is occurring at the cellular level.

II. Genomics

It is now possible to measure global changes in gene expression after exposure to toxicants or other stressors. Rather than studying the effects of a compound, one gene at a time, global measurements of the transcriptome allow a more direct examination of all the pathways affected by exposure to contaminants. Initial results suggest that toxicants in the same group of chemicals will have common targets. However, slight changes in the structures of individual chemicals may cause them to target additional genes, affecting metabolic pathways in unpredictable ways[53,54,79]. It is now clear that transcription factors work together in combination and that competition for factors may be crucial in determining which genes are 'turned on' and which are 'turned off'. For this reason, chemicals that mimic endogenous cellular signal molecules will work in unpredictable ways.

To date, few studies in aquatic toxicology have used genomic methods, but, as in other disciplines, this is likely to change in the near future. The level of sophistication of these techniques will enable toxicologists to determine the risk of exposure of aquatic organisms to anthropogenic compounds in a more objective way with certain knowledge of their mode of action and whether they do or do not cause adverse effects. This approach should yield both biomarkers of exposure and biomarkers of effect. The ultimate goal of genomic experiments is to link induced gene expression patterns to detrimental effects, harmless effects, or even protective effects. To that end, it will be necessary to link gene expression patterns with physiological responses and tissue structure changes. This has been termed 'phenotypic anchoring'[125] by the National Center for Toxicogenomics at NIEHS, and this concept will undoubtedly gain in importance as definitive studies are performed.

To begin to measure the transcriptome, it is first necessary to obtain gene sequences. This is relatively easy to do for fish whose genomes have been studied and for whom sequences exist in the databases. But, for non-model organisms, it is more difficult, since the databases do not contain such information. In this case, researchers must obtain gene transcript information empirically. There are a number of different methods that can be used including random sequencing of cDNA libraries, suppressive subtractive hybridization (SSH), differential display reverse transcriptase polymerase chain reaction (DD RT-PCR), serial analysis of gene expression (SAGE), and direct cloning. Genes obtained by these methods, in turn, can be spotted onto gene arrays for further analysis.

1. Obtaining cDNA sequences from gene libraries for non-model systems

Probably the most efficient method to obtain a large number of genes for gene array analysis is to sequence colonies or phage directly from cDNA libraries. A good cDNA library would contain at least one copy of each mRNA present in the transcriptome of a tissue at a particular time. Many vendors sell kits for constructing cDNA libraries or it is possible to find commercial laboratories that will construct good libraries for a fee. The basic protocol for making a cDNA library is rather simple. First, a poly + RNA fraction must be prepared from the tissues of interest and then this material is converted into cDNA and randomly cloned into phage or plasmid vectors. The library is then plated out on agar plates. If the library is a phage library, it is plated on a bacterial lawn, with each plaque containing a different phage with a gene sequence insert corresponding to a segment of mRNA from the initial mRNA pool. The plaques are picked; the inserts are amplified by PCR and sequenced. If the sequence is of interest, then the phage can be rescued into plasmids. Alternatively, before plating, the phage can be converted into plasmids by mass *in vivo* excision and then directly plated onto clear agar plates. In the case of a plasmid library, the library can be plated directly on agar plates, and individual colonies with inserts can be picked and sequenced directly, without further manipulation. If the library is of high quality, hundreds to thousands of individual colonies can be picked and sequenced and then identified by comparing to sequences in DNA databases. While it is not possible to identify all the transcripts that are expressed within a tissue, one can quickly obtain thousands of unique genes by this method.

A disadvantage of cloning genes from a regular cDNA library, however, is that there is much repetition of highly expressed genes. For example, in our hands vitellogenins, choriogenins, and ribosomal proteins represented more than 56% of the clones that were sequenced from a cDNA library generated from the liver of male largemouth bass (*Micropterus salmoides*) previously treated with estradiol[80]. One way to minimize this redundancy is to prepare normalized libraries[22], which have a more equal representation of the various transcripts. While normalized libraries are indeed superior, they are more difficult and costly to prepare and require a highly trained molecular biologist for success. Another method that can be used to eliminate picking redundant inserts from a regular cDNA library, is to test the colonies first by dot blot with a probe made from the sequences already in hand. In this case, one would only pick plasmids that did not bind with the probe for sequence analysis.

While cDNA libraries can be used to obtain many genes in a short time frame, it may not yield genes of toxicologic importance. To enrich for these genes there are several methods that rely on obtaining mRNAs that are differentially expressed after treatment. These methods, which are described below, include subtractive hybridizations, DD RT-PCR and SAGE.

2. *Subtractive hybridizations*

Subtractive hybridization is a method that enables one to quickly obtain large numbers of cDNAs that are differentially expressed between control and experimental conditions[25,31,34,50,56,57,83,106]. To perform this procedure, one must first isolate poly-A + RNA from both treated and control tissues. The poly-A + RNA is then reverse transcribed into cDNAs and then, by a series of manipulations, the two fractions are hybridized together. cDNAs present in both fractions form duplexes and are removed from consideration. cDNAs that do not hybridize because they are preferentially expressed in either the control or the treated samples can then be PCR amplified, cloned, and subsequently sequenced. The method is normally performed to obtain both up- and down-regulated genes.

A recent refinement of this procedure, called suppressive subtractive hybridization, now enables researchers to obtain both high and low-abundance transcripts[29] by including a normalization step in the procedure. Clontech (Palo Alto, CA) offers a kit for this procedure that works extremely well. This approach enables researchers to quickly find genes that are differentially regulated by toxicants. While the method works well, it is a little 'leaky' in that in addition to the differentially regulated genes, one also obtains genes that are not differentially regulated. Thus, it is important to test all clones by other methods to determine the status of the different genes found. Another protocol for suppressive subtraction uses several more PCR cycles and hybridizations[123] than the Clontech procedure, making it more versatile because genes are highly enriched resulting in fewer false positives. In addition, the procedure subtracts highly expressed genes. Genes that are not differentially regulated are also important because they can be used as normalization controls for arrays. However, this group of genes may be small, therefore additional methods to obtain larger numbers of these genes should be used if the final purpose is to build a gene chip. One possibility is simply to clone genes from a cDNA library as discussed above.

We have used SSH to obtain tissue specific genes from fathead minnows (*Pimephales promelas*), sheepshead minnows, and largemouth bass. By running the SSH procedure one minimizes the presence of abundant gene transcripts regularly expressed in all tissues, such as ribosomal proteins, and maximizes the number of tissue-specific genes. Figure 3 depicts an agarose gel containing fathead minnow cDNA transcripts before and after subtraction from gonad and liver tissue. The subtracted transcripts were then cloned into plasmids and sequenced. The corresponding genes obtained are largely tissue specific.

Subtractive hybridizations have been employed by many research groups to investigate various aspects of fish biology. These include reproduction and development[73,92,104,112,131,132], immunology[7,40,41,75,85,96,105], and other areas of interest. At the time of this writing no papers could be found in the PubMed, National Center for Biotechnology Information (NCBI) database that used subtractive hybridization technology to identify genetic biomarkers for specific classes of environmental contaminants. With the increased use of molecular techniques in the field of fish

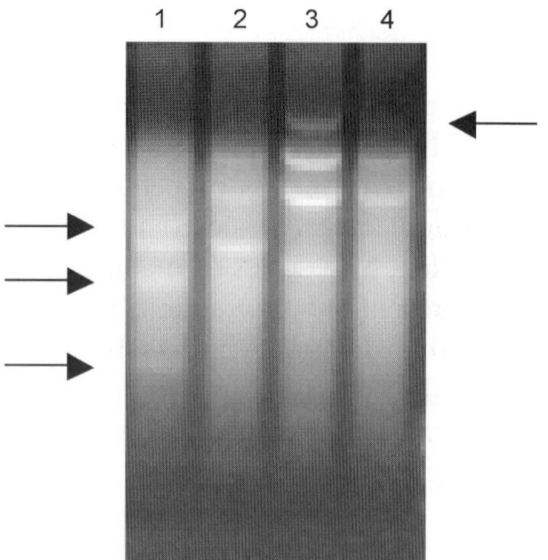

Fig. 3. Subtractive hybridization performed between liver and gonad of male fathead minnows in order to enrich for genes specific to each of these tissues. (1) Subtracted liver tissue, (2) un-subtracted liver tissue, (3) subtracted gonad tissue, and (4) un-subtracted gonad tissue. Arrows represent gene transcripts that are enriched in the subtracted samples compared to the un-subtracted samples.

toxicology, it is only a matter of time before subtractive hybridization will be widely used as a means to identify genes that are differentially regulated by pollutants.

3. Differential display RT-PCR – another open method to analyze differentially expressed genes

Another method that can be used to obtain differentially regulated genes between two tissue populations is called differential display reverse transcriptase (DD RT-PCR), a technology developed by Liang and Pardee[82]. While there are multiple variations of this technology, the basic principle is the same. DD RT-PCR involves the reverse transcription of mRNA with oligo-dT primers anchored to the poly-A tail, followed by a PCR reaction using both the anchor primer and a second short primer with an arbitrary sequence. The amplified cDNA products are separated by size on a DNA sequencing gel and are visualized either by radioactivity or fluorescence, depending on the protocol employed. There are several commercial kits available for this technology. With the kit available from Gen Hunter (Nashville, TN), there are three anchor primers and 80 arbitrary primers for a total of 240 primer pair combinations that must be performed to see the complete transcriptome. Because the anchor primers bind to the poly-A tail, most of the gene fragments amplified with this kit are from the $3'$ untranslated region of the mRNA. This makes it difficult to identify sequences for non-model fish species that have few gene sequences in databases. Kits available from other manufacturers have different primer pair combinations, some of which are designed to preferentially bind to and amplify the coding region of mRNAs, making identification of these genes in the databases more likely.

To discover differences in gene transcription, the PCR-amplified segments obtained from control and treated samples (either whole animals or tissue culture cells) are separated side-by-side on large acrylamide gels. An example of this type of gel can be seen in Fig. 4. Each band on the gel represents a unique fragment of mRNA that was amplified by the primer pair. The length of the fragment depends on the location in the original sequence to which the arbitrary primer was able to bind. In order to minimize false positives, the reactions should be performed at least in triplicate (three control and three treated biological samples). Bands that appear in all the samples that are of equal intensity represent fragments of genes that are not affected by the treatment. Bands that appear in all three experimental samples and are absent from all three controls (or vice versa) represent differentially regulated genes. After identifying differences on the gels, the corresponding gel bands can be cut out, re-amplified with the same primer pairs used in the original PCR, cloned, and sequenced. The genes can then be identified by comparison to sequences in the databases.

The advantages of this method are that it is relatively easy to perform and does not require expensive equipment. It is also possible to identify rare transcripts since these are amplified by PCR. Among the disadvantages are the time it takes to perform the experiments, the relatively high number of false positives, especially if the reactions are not done in triplicate, and the inability to identify certain genes from their 3′

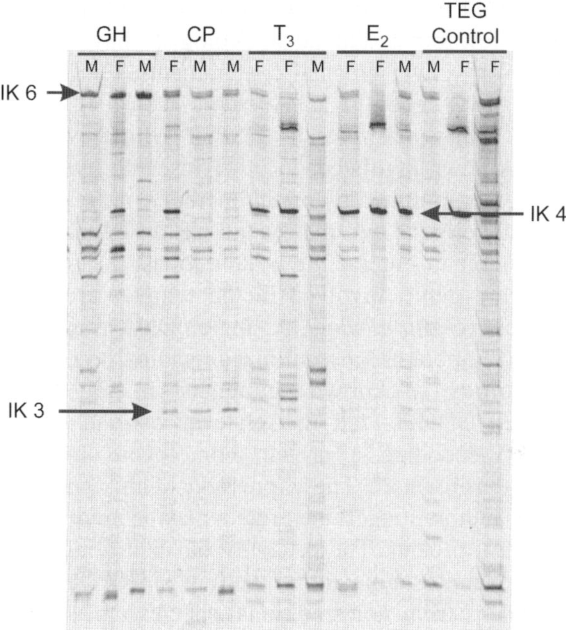

Fig. 4. Representative segment of a differential display (DD) RT-PCR gel of liver mRNA from fish treated either by injection with growth hormone (GH), triiodothyronine (T3), or given an aquatic exposure to estradiol (E_2), chlorpyrifos (CP), or triethylene glycol carrier control (TEG Control). Arrows denote gel bands that were subsequently cloned and sequenced. IK3 represents a gene that appears up-regulated in response to chlorpyrifos (CP) treatment. IK4 represents a gene up-regulated by E_2 treatment and is also present in female fish of other treatment groups. IK6 represents a gene that is present in all treatment groups. The sex of the individual fish is denoted (M or F) at the top of each gel lane.

untranslated sequences. Because DD RT-PCR uses a random approach, important genes may be missed even if many primer combinations are used. Nevertheless, it is possible to quickly determine whether a toxicant has caused changes in the transcriptome by examining differences in banding patterns on the gel.

Differential display technology has been used in our laboratory to determine gene expression patterns in sheepshead minnows[27,28,76] and largemouth bass[14] exposed to natural and anthropogenic estrogens. The gel in Fig. 4 is part of an experiment performed to analyze differences in gene transcription in sheepshead minnows treated with growth hormone, triiodothyronine, estradiol, chlorpyrifos, or triethylene glycol carrier control[76]. Arrows denote gel bands that were differentially expressed as a consequence of treatment. These were subsequently cloned and sequenced. IK3 is a gene that appears to be up-regulated by chlorpyrifos treatment, while IK4 is a female-specific gene that is present in females and E2-treated males. IK6 is a gene that is present in all groups and thus can be used as a normalization gene for arrays. Differentially expressed genes from several gels were isolated, sequenced, and subsequently used in both quantitative real-time PCR assays and microarray technology.

DD RT-PCR was very useful in finding differentially expressed genes. In the case of the sheepshead minnow experiments, we used 50 primer pairs and estimate that we have screened over 2000 genes to obtain those induced by exposure to estrogen mimics[27,28]. To date, few other aquatic toxicologists have used DD RT-PCR to identify genes affected by environmental contaminants. Carginale *et al.*[23] identified several cadmium-sensitive genes in an Antarctic icefish (*Chionodraco hamatus*). These included the heat-shock protein HSP70 which was up-regulated, and transferrin, which was down-regulated. In other research, two genes found by DD RT-PCR were completely inhibited by methylmercury in *Xenopus laevis* embryos[94]. One appears to be a novel gene and the second has a high similarity with a human protein kinase (homeodomain-interacting protein kinase 3). Differential display was also used to investigate genes involved in the temperature-induced gonadal sex differentiation in tilapia (*O. niloticus*)[26]. A single cDNA fragment was found to be highly up-regulated in temperature-masculinized females, however, the authors found no significant homology to known sequences in the available databases. Other researchers used DD RT-PCR to isolate an ovarian mRNA in brook trout (*Salvelinus fontinalis*) encoding a protein homologous to a gene superfamily that includes snake neurotoxins[81]. The expression of this gene was suppressed by phorbol esters and the authors suggest the protein may be involved in the regulation of the complement system in trout. The difficulty in identifying genes that are found to be up- or down-regulated is apparent by the few papers that have been published. As more gene sequences from non-traditional species are added to the databases, identification of novel genes should become easier in the future.

4. *Serial analysis of gene expression*

SAGE was first described in 1995[121]. This technique is another open procedure for finding differentially expressed transcripts and has been used to advantage for mammalian systems. The basic principle involves converting mRNAs to cDNAs and then cutting them specifically with endonucleases into short DNA fragments, or tags of about 10–14 nucleotides, that are then ligated together, interspersed with adaptor sequences into long, serial molecules. These molecules are then cloned and sequenced. Normally, about 20–50 tags are incorporated into each molecule and hundreds to thousands of molecules must be sequenced. This collection of clones is called a SAGE library. Thus, this approach requires high-throughput DNA sequencing. Each tag, representing a different mRNA, is present several times in the sequence in proportion to its concentration in the original mRNA sample and represents the absolute expression level. Thus, when comparing controls and treated samples, the frequency with which each tag is encountered gives information about the differential expression of each gene.

To perform this experiment, poly-A + RNA is bound to oligo(dT) beads and then reverse transcribed into cDNA while bound to the beads. An anchoring enzyme, which is really a specific endonuclease that recognizes sites that should be commonly found in all mRNAs, is used to cut each anchored cDNA. Fragments of different lengths will then be attached to the beads, since the location of the restriction site will vary from one mRNA to another. Then specific linkers are bound to the 5′ ends of each of the fragments with ligase. The linkers are long enough to contain a specific sequence for a rare endonuclease that binds to the site but cleaves the cDNA into 10–14 residues downstream from the 5′ end, thus creating the specific tags for each mRNA. Normally, two tags are connected together by ligase to form di-tags and these are then concatemerized into the long molecules mentioned above. The sequence tags (10–14 bp) are sufficiently long to identify every transcript in the sample, provided the tag is obtained from a unique position within each transcript and there is genomic information for the species of interest in a database.

This method may be useful for those fish species whose genomes are known. It will be much less useful for non-model species because of the lack of information in the databases. In the event that a gene is not in a database, it is possible to identify the gene by performing 5′ rapid amplification of cDNA ends (RACE) (available from several vendors) to determine upstream sequences in the coding region that can match sequences in the databases. While SAGE has been used extensively in human cancer research and cardiovascular biology, no papers have been published to date in which this technique was used to study gene expression in fish or for other applications of environmental toxicology.

5. Directed methods for obtaining genes of toxicologic importance

Another method to obtain cDNAs for genes of interest is to design degenerate primers to conserved gene regions. The conserved gene regions can be determined by aligning protein sequences for multiple species obtained from databases for the gene of interest. Several programs are available that can help with this procedure, but in general the primers are difficult to design and are highly degenerate. Multiple primer sets, typically, must be designed, and fairly extensive PCR optimization performed in order to succeed in cloning genes by this method. However, a definite advantage of the method is that one can specifically target genes of toxicologic importance or genes for a particular pathway.

6. *Expression analysis with gene arrays*

Sequence information obtained by the previously described methods can be used to construct gene arrays (also referred to as microarrays, macroarrays, gene chips, or DNA arrays). An array consists of hundreds to thousands of cDNAs or specific segments arranged specifically in an organized pattern on a glass slide or membrane. The arrays are then probed with mRNA obtained from treated tissues or cells, to obtain a semi-quantitative measure of expression for each of the genes present on the array. Thus, gene arrays can be used to simultaneously measure the expression level of all the genes affixed to the surface. Because thousands of genes can be queried at one time, this is referred to as a global analysis of the effects of toxicologic exposure. Smaller arrays containing fewer genes for only specific pathways can also be constructed, and these would be useful to determine changes in expression profiles for only the pathways of interest. The DNA sequences affixed to the glass or membrane surfaces are called probes and the RNA that is reverse transcribed with fluorescent or radiolabeled markers is called the target[16,108]. In essence DNA arrays are reverse Northern blots, from which the terminologies for 'target' and 'probe' were originally derived.

The first gene arrays constructed used cDNA clones. To make these arrays, inserts from cloned plasmids were amplified by PCR and then they were spotted onto a solid surface – either nylon membranes or glass slides. While the design of arrays has, for the most part, changed little, it is now possible to also use long, gene-specific oligonucleotides (oligos that are 20–70 nucleotides long) that are obtained synthetically to construct arrays. This is very convenient for organisms with many gene entries in databases, since the oligos can be designed directly from these sequences without first having to clone the cDNAs. Affymetrix has led the way with oligonucleotide-based arrays. However for fish, they only provide arrays for zebrafish. The Affymetrix arrays are built directly on the chip by photolithography (http://www.affymetrix.com). The Affymetrix arrays contain tens of thousands of genes; each is represented by several 20mer oligos that span the length of the genes of interest. Immediately below each set of oligos for a specific gene is another set that contains one mismatched nucleotide per 20mer that serve as a control for specificity. The Affymetrix arrays must be used according to the manufacturer's instructions with equipment that is also supplied by the manufacturer.

An alternative to Affymetrix arrays is the use of longer oligos in the 50–70mer range. Designed oligos can be either synthesized directly on glass slides, as done by Agilent by piezoelectric printing or synthesized separately, spotted, and then crosslinked to glass slides. There are a number of vendors who specialize in oligonucleotide design and synthesis, including MWG Biotech Inc., Qiagen Operon Technologies, GmbH (Alameda, CA), and others. Of course, many researchers still spot individual cDNAs, rather than oligos, onto glass slides or membranes.

Consistency and reproducibility of arrays is enhanced when a robot is used to spot the arrays to ensure equal deposition of minute volumes in a high-density format.

However, gene chips can also be constructed manually by using a hand-held spotter. In addition to the genes of interest, several controls should be spotted on the arrays. These include adding genes such as plant genes that are not likely to be found in fish (e.g. genes that encode the photosynthetic pathway in plants). The RNA for these genes is then spiked into the RNAs of both the control and experimental samples as an exogenous control. Other controls used include Cot1- or salmon sperm DNA and poly-A nucleotides which control for non-specific binding.

For expression analysis, total RNA (or poly-A + RNA) is prepared from samples that are to be compared after treatment. It is critical to harvest tissues from animals (or cells) under conditions that will preserve RNA. Methods of choice include freezing the samples in liquid nitrogen and then storing at − 80°C, or using a preservative such as RNA*later* (Ambion, Austin, TX) which is easily transported to the field. Tissue samples should be cut into very small pieces for quick preservation. In the case of RNA*later*, it is important to use the ratio of solution to tissue recommended by the manufacturer. Using less will cause improper preservation of tissue and will result in severe RNA degradation. It is equally important to use a good method for RNA preparation. Some RNases accompany the RNA throughout the isolation procedure, surviving the strong denaturants that are routinely used. These RNases must be removed by treatment of the RNA with RNase inhibitors or with proteinase K, to remove all traces of proteins. An exceptionally good product for this purpose is RNAsecure (Ambion), but several other supply companies have similar products. It is also important to remove all traces of DNA with DNase enzymes (free of RNase activity) also available from many commercial sources. The importance of this initial step of preparing good quality RNA cannot be overstated. Total RNA should be examined by agarose gel electrophoresis to ensure that the rRNAs appear intact and it should be quantified by absorbance at 260 nm. The ratio of absorbances at 260 and 280 nm should be between 1.8 and 2.0. Lower values indicate contamination by proteins. RNA integrity can also be checked by high-throughput machines (Bioanalyzer, Agilent Technologies, Palo Alto, CA).

Once the RNA is prepared, it is converted to cDNA and labeled either with radioactive precursors ([^{33}P]-dCTP or [^{32}P]-dCTP) or with non-radioactive fluorescent labels such as Cy3-dNTP and Cy5-dNTP. Radioactive labels are commonly used for membrane approaches while fluorescent probes are the method of choice for glass slides. When using fluorescent probes, cDNA prepared from controls can be labeled with one dye while cDNA from treated samples can be labeled with the other. These are then mixed and used in competitive binding to probes on the chip. Alternatively, one can compare all control and experimental samples to the same reference RNA sample, labeled with one of the dyes. Reference RNA samples are useful for multiple sample analysis and to account for chip-to-chip variations. In the case of the Affymetrix technique, cDNAs are converted back into fluorescently labeled RNA targets, which bind more tightly than cDNA targets to the short oligonucleotides present on those chips, and the control and experimental samples are applied individually to separate chips.

An important issue with microarrays is the number of samples that must be used to obtain statistically valid data. It is clear that multiple control and treated samples must be analyzed in order to avoid type 1 and type 2 statistical errors. Typically, a minimum of 3–4 arrays should be run per experimental treatment, but this number depends on the expected variability of the fish within a treatment group. For example, one would expect to use a higher number of arrays for field samples compared to samples obtained from a controlled laboratory setting.

6.1. Normalization of data

Once the arrays are imaged, each glass slide or membrane must be normalized before any comparisons can be made between the different treatments. Normalization compensates for any differences in the preparation of RNA, labeling of the targets, probe purification, and hybridization. Several different procedures are commonly used to normalize data. Global normalization is employed when one expects that the overall gene expression levels between the samples to be compared are similar. This method involves adjusting the means (or medians) of the two distributions to equal each other and then scaling the intensity of each spot to the adjusted mean values. This strategy is based on the assumption that the overall level of gene expression is unaffected by the experimental conditions, and any individual changes in gene expression are equally distributed between increases and decreases. This strategy works well for large arrays that are likely to represent an even distribution of expression levels; however, it would not work well for arrays that are predominantly enriched for differentially regulated genes, such as genes picked by DD RT-PCR or SSH.

A second method of data normalization relies on using housekeeping genes. Housekeeping genes have a constant level of expression across a wide range of experimental conditions. It is not safe to assume *a priori*, however, that these genes are not affected by the experimental paradigm. Instead, each control gene must be tested by alternative methods to confirm that it is not affected by the treatment. Several experiments have shown that genes normally thought to be housekeeping genes do in fact vary[11,78,79,107].

A third option to normalize data is to add identical amounts of exogenous RNAs (called spiking RNA) to each RNA pool sample prior to the labeling step (a cDNA complementary to the spiking RNA is also added to the membrane or glass slide). For fish, the best spiking genes come from plants, e.g. genes that are part of the photosynthetic pathway are not likely to be found in vertebrates. These spiking genes can be synthesized in-house or purchased from a vendor (SpotReport 3, Stratagene, La Jolla, CA).

6.2. Data analysis

Once the data have been normalized, comparisons can be made between several different data sets. There are various methodologies and statistical packages available (GeneCluster from Whitehead/MIT Center for Genome Research, http://www.genome.wi.mit.edu/MPR/; Expression Profiler from European Bioinformatics Institute, http://expsrv.ebi.ac.uk/; Cluster from Lawrence Berkeley National Laboratory, http://rana.lbl.gov/EisenSoftware.htm; also see http://www.mpiz-koeln.mpg.de/~weisshaa/Adis/DNA-array-links.html for links to various companies) to mine the wealth of information derived from DNA arrays.

To identify genes that are differentially regulated between two or more treatments, the results should be statistically analyzed. It is not sufficient to simply use a twofold criterion to identify differentially regulated genes. While the *t*-test remains a popular method to test for significant changes in gene expression between paired conditions, it must be used with caution since it can lead to a high degree of false positives. A recently developed program called Significant Analysis of Microarrays (SAM) is a more robust method for analyzing arrays and gives a better estimate of the true number of false positives[119].

Another commonly used method to analyze array data is to perform cluster analysis, which allows for the identification of higher order relationships among the genes by grouping together genes that exhibit similar expression patterns across multiple experiments. The basic assumption underlying this approach to analyzing array data is that genes with similar patterns of expression are likely to be functionally related to each other. Several clustering algorithms used by researchers include hierarchical clustering, self-organizing maps, support vector machine, *k*-means, and Bayesian statistics[2,9,20,32,62,118,126,129]. Another type of analysis, called step-wise discriminate analyses, can be used to determine which group of genes best discriminates between sample types[113].

6.3. Methods for verification of gene array experiments

Results from gene array experiments should be verified by other mRNA quantification methods such as Northern blots[3,4], ribonuclease protection assays[8,36,128] dot and slot blots[72,115], or quantitative real time PCR (Q RT-PCR)[58].

Northern blots involve separating 10–15 μg amounts of RNA on a denaturing gel and then transferring the RNA to a nylon membrane by capillary action or by electroblotting. Probes specific to the genes of interest are then labeled and hybridized to membranes where they bind specifically to the mRNAs of interest. The amount of label that binds is proportional to the amount of specific mRNA present in the sample. In order to control for pipetting errors and other differences in methodology, the expression levels must be compared to a housekeeping gene (known not to change with the treatment) or to rRNA. 18S rRNA is an excellent control for these experiments. Relative abundance of gene transcripts can then be determined. Several advantages of Northern blots are that the technology is easy to implement and the results are easy to analyze, minimal equipment is required, and the size of the mRNA transcript, as well as the number of possible splice variants can be determined. Some disadvantages include the requirement for at least 10 μg of total RNA and the fact that only a limited number of genes can be analyzed by this method at one time. Membranes can be stripped, however, and re-probed several times.

Slot blots are similar to Northern blots, except that the RNA is applied directly to the membrane in a slot created by sandwiching the membrane into a device that contains 96 slots (available from many manufacturers) and is connected to a vacuum. Once the RNA is on the membrane, it is treated exactly like the Northern blot described above.

RNase protection assays are very powerful but more complex to perform than Northern blots. In this case an RNA probe that is complementary to the RNA of interest is made and radiolabeled[91]. It is then allowed to bind to total RNA in solution under conditions that enhance specific binding. Then a single-strand specific nuclease is added to the incubation mixture, which will degrade any RNA that is not duplexed. The protected fragment is then analyzed by electrophoresis on acrylamide gels followed by autoradiography. This experiment is best performed in triplicate, with RNA samples from at least three different control or treated fish. The advantage of this method is that several transcripts can be analyzed on the gel at the same time. A main disadvantage is that the method is difficult to perform.

Quantitative polymerase chain reaction (Q-PCR) is probably the easiest and most accurate method to validate differential transcription. The method involves PCR but in real time, evaluating data through the logarithmic expansion phase of the reaction. For this procedure, primer pairs corresponding to short amplicons of less than 150 nucleotides can easily be designed by software available with the various instruments (e.g. Primer Express, from Applied Biosystems). These primers are used in combination with SYBR green, a dye that fluoresces only when it is bound to double-stranded DNA. Another method to monitor the reaction in real time is to use a specific fluorescent probe (containing a fluorescent dye molecule on one end and a quencher dye on the other) that is complementary to the amplicon. The method works in the following way. Total RNA is prepared and converted into cDNA by reverse transcriptase. The primers (and probe) are then added to the reaction mixture, in a PCR compatible buffer and then the amplification is allowed to proceed. As the amplification occurs the fluorescence increases in real time. In the case of SYBR green, the increased fluorescence is due to the intercalation of the dye into double-stranded DNA. In the case of the probe, fluorescence increases with each cycle because the quencher and dye molecules on either end of the probe are separated from each other as the probe is cleaved during each amplification cycle. In either case, fluorescence is measured in real time and is proportional to the amount of double-stranded DNA that has been amplified in each cycle, and this is ultimately proportional to the amount of mRNA originally present in the sample.

The TaqMan instrument from Applied Biosystems allows evaluation of 96 wells at one time, which enables the validation of several genes at once. Other vendors also have instruments that can monitor multiple wells at the same time. The advantages of this method include the ability to compare multiple genes at once in a very quantitative assay and the ability to design primers that will distinguish and evaluate splice variants or closely related genes. For example, using this method, we were able to design specific primers to distinguish and evaluate the relative abundance of three different largemouth bass estrogen receptor (ER) genes, alpha, beta, and gamma[103]. This is the method of choice for evaluating low-abundance transcripts since an amplification process is used.

6.4. *Incorporation of data into databases*

It is now clear that large databases of toxicologic importance will be constructed to include as much information as possible for experiments performed with these technologies. The data should be made easily available to anyone who wishes to examine or re-analyze the data, for interactions and connections not yet determined.

Likewise, it will be important to standardize methods of isolating RNA, performing experiments, and reporting data. The Minimum Information About a Microarray Experiment (MIAME) protocols developed by the Microarray Gene Expression Database Group (MGED: http://www.mged.org/) aims to describe information that researchers should include in their experiments and also provides guidelines to authors and reviewers of manuscripts describing microarray data[18]. Briefly, the MIAME protocols include guidelines for reporting the experimental design, array design, the minimum number of samples compared, hybridization conditions, and how the data was evaluated.

The National Center for Toxicogenomics (NCT) is establishing a large database, 'Chemical Effects in Biological Systems (CEBS)', to incorporate information that links toxicogenomic biomarkers with chemical effect. Other groups have also come together to pool resources. For example, the Mount Desert Island Biological Laboratory (MDIBL) has started a Comparative Toxicogenomics Database (CTD) that should enable researchers to have access to work done on marine species and to easily compare these results with results obtained with mammals and humans.

For fish that are not model species and therefore have limited genomic information available, it will be important to generate consortia of interested researchers to obtain enough information to utilize these new technologies. However, annotation of genes found for non-mammalian species will continue to be a challenge for years to come.

7. Studies using microarray technology for aquatic toxicology

While array technology is relatively new for aquatic toxicology, several research groups have recently published articles using array technology to study environmental contaminants in fish. Hogstrand and co-workers[63] exposed rainbow trout to a sublethal concentration of zinc and used a cDNA array containing genes from *T. rubripes* to examine gene expression responses in the gill tissue of trout. This study revealed a number of genes that may be differentially regulated by zinc, including genes that encode proteins involved in a variety of functions such as energy production, protein synthesis, paracellular integrity, and inflammatory response. Another research group examined the effects of lithium exposure on gastrula stage zebrafish embryos[61] and found several genes that were differentially regulated in response to lithium exposure. They also found a high correlation of expression profiles for these genes compared to *in situ* hybridization.

Other recent studies using gene array technology in fish include the examination of the development and survival of zebrafish embryos under hypoxic conditions[116], the measurement of the variations in gene expression in natural populations of the killifish, *Fundulus heteroclitus*[97], examination of the effects of hypoxia on different tissues in a goby, *Gillichthys mirabilis*[45], identification of genes that are controlled by nodal signaling in the zebrafish gastrula[30], and the identification of genes that are differentially regulated in the brain of channel catfish in response to the acclimation of these animals to cold temperature[70].

Our research group has examined changes in gene expression profiles in two species of fish (sheepshead minnows and largemouth bass) exposed to endocrine-disrupting chemicals (EDC). In the first experiment, sheepshead minnows were treated by aquatic exposure to environmentally relevant doses of endogenous, pharmaceutical, or suspected estrogenic compounds. Gene expression profiles for livers were examined using a cDNA macroarray[79,80] derived from genes isolated primarily by DD RT-PCR[27,28]. The results of this study indicate that estradiol, the strong synthetic estrogens (diethyl stilbestrol (DES) and ethinylestradiol) and the environmental estrogens (4-nonylphenol and methoxychlor) induce the same overall set of genes that include proteins involved in oogenesis. However, in addition to the overall pattern, the environmental estrogens induced unique patterns of gene expression, suggesting that it may be possible to determine a pattern that is compound specific. Endosulfan, a putative environmental estrogen, on the other hand, only induced estrogen receptor alpha, and not vitellogenins or choriogenins, suggesting that its effect on this pathway is very specific. These data suggest that it may be possible to identify unique expression patterns for individual chemicals. As more expression profiles are determined, they will be useful to identify contaminants in environmental settings. The sheepshead minnow study also showed that the arrays could be used in a semi-quantitative way to determine dose response.

In the second study, largemouth bass were exposed by intraperitoneal injection to estradiol or to two hormonally active agents, 4-nonylphenol (4-NP) and 1,1-dichloro-2,2-bis (*p*-chlorophenyl) ethylene (*p,p*′-DDE)[80]. This study showed that largemouth bass exposed to estradiol and 4-NP had similar, but not identical genetic signatures for the genes examined, some of which are known to be estrogen-regulated genes[79]. Male largemouth bass treated with *p,p*′-DDE also expressed female specific genes, but to a much lower extent in proportion to the relative affinity of *p,p*′-DDE for the estrogen receptor. Interestingly, exposure of vitellogenic females to *p,p*′-DDE had the opposite effect, and actually decreased the expression of the same set of genes. An example of gene arrays that were used in the *p,p*′-DDE experiment for control and treated male largemouth bass is illustrated in Fig. 5.

In summary, gene arrays will play a significant role in determining the response of fish to the vast diversity of environmental contaminants and pharmaceutical drugs. It will be possible to perform time and dose response experiments and obtain an understanding of global changes in gene expression. With time, specific patterns of regulations will emerge which should enhance our knowledge of the mechanisms of toxicity. How arrays will be used for risk assessment remains to be seen, but undoubtedly they will play an important role.

III. Proteomics

The term ‘proteomics’ refers to the large-scale study of proteins in cells or tissues. The term actually has a broader definition, including understanding the complex interactions among proteins that occur within cells. These interactions include formation of functional complexes as well as interactions with other cellular components such as nucleic acids, lipids, and carbohydrates[101]. While studies of protein function and protein–protein interactions are not new, what is new is studying

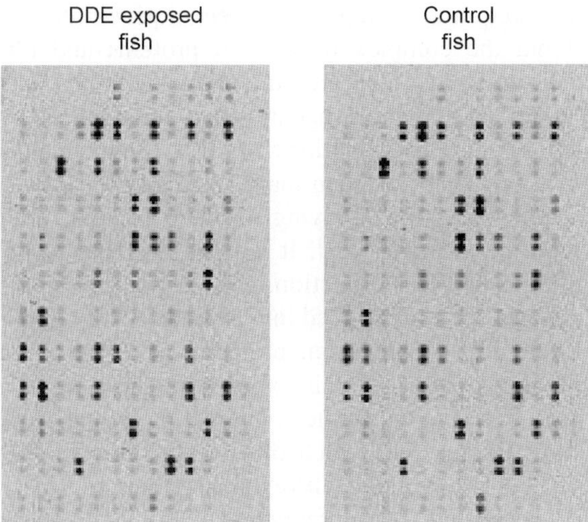

Fig. 5. Representative pattern of gene expression on bass-arrays developed with cDNAs from control or p,p'-DDE-treated male largemouth bass. Groups of fish received a single intraperitoneal (IP) dose of DDE (100 mg/kg) or a vehicle control as previously described[80]. Three separate fish were used per treatment group.

them globally and seeing interactions among proteins that one would not have previously imagined. Large numbers of inter-related proteins and their role(s) in selected physiological or pathological states of cells, tissues, or organs can be studied. This approach will play a large role in understanding the physiological changes brought about by the toxic effects in fish.

Initial results (including results from the human genome project) point to fewer genes than initially expected – in the range of 30,000–40,000 for humans[24,122]. This small number of genes suggests that the intricate pattern of control observed in animals may rely more on control of translation, splice variants, alternative transcription start sites, and protein post-translational modification than originally anticipated. It has been estimated that because of post-translational modifications, there may be over 300,000 different proteins involved in normal physiology. Thus, it will not be sufficient to study changes only at the gene level, but it will be important also to study responses at the protein and activity levels, to understand the physiological consequences of exposure to contaminants or pathogens.

Since the early 1990s, analytical methods to study protein expression levels and identify proteins have improved tremendously, especially with the development of better separation methods and sensitive mass spectrometric techniques. These methods work well for organisms whose genomes are known, but not as well for non-traditional models, where at best, only limited protein sequence information may exist. For these models, it will be important to develop more robust and high-throughput *de novo* sequencing methods by mass spectrometry.

Because of the complexity of the proteome, however, a single method cannot yield full information. Most likely, a combination of different methods, each with its own specific advantages, is required. Using various techniques in combination will allow researchers to sort out the complex mixture of proteins and identify them. It is important to include methods that are amenable to studying post-translational modifications and identifying proteins that are in low abundance or that have atypical characteristics, such as high compositions of basic amino acids or glycosylated residues. While much progress has been made, methods for proteome research still must be refined, especially for studying protein–protein interactions, protein structure, and metabolic pathways[21,42,71]. It will be essential to link the expression or disappearance of a protein to its function in the cell.

Proteome research has been subdivided into two categories: expression proteomics and interaction proteomics[46]. Expression proteomics, as the name implies, is the understanding of factors that affect protein expression, and as such, is complementary to differential mRNA expression studies. Interaction proteomics requires understanding how proteins interact with each other. Proteomics research will generate huge quantities of data. To store and analyze the large data sets that will be produced, bioinformatics will be essential. This cornerstone of biological data manipulation will be described in more detail later in this chapter.

The gold standard for proteomics is still two-dimensional gel electrophoresis (2D-PAGE) followed by mass spectrometry (MS), a method which accurately weighs the trypsin-generated peptides after digestion of proteins by this enzyme[51,98,111]. This two-stage method allows for both hypothesis-driven and discovery-driven research strategies. New methods include separation of proteins by non-gel-based procedures followed by MS, and methods that rely on protein/antibody chips. Development of protein chips is in its infancy, but both antibody-based and protein-based chips are on the horizon. These methods will allow for parallel processing of several proteins at once, and have the potential for increasing the sensitivity of detection, allowing the use of small amounts of material.

1. Two-dimensional polyacrylamide gel electrophoresis followed by mass spectrometry

Analyzing complex protein patterns by 2D gel electrophoresis has been a research tool since the early 1970s[51,68]. With this method it is possible to separate and visualize over 1000 distinct proteins in one experiment. Proteins are separated in the first dimension by isoelectric focusing (in a gel that separates proteins based on their relative amounts of acidic and basic amino acids) and in the second dimension by size. The proteins are visualized by staining and then quantified by densitometry. Figure 6 contains an example of a silver-stained 2D-PAGE analysis of liver proteins obtained from control and E_2-treated largemouth bass. By visual inspection it is clear that there are numerous proteins expressed in the treated sample that are absent in the control, and there are proteins in the control that are not present in the treated sample. These spots would all be candidates for protein identification.

The second part of this method depends on mass spectrometry. Each protein spot of interest is cut from the gel and digested with trypsin[5,111]. The fragments are extracted and their masses analyzed by mass spectrometry using either a matrix-assisted laser

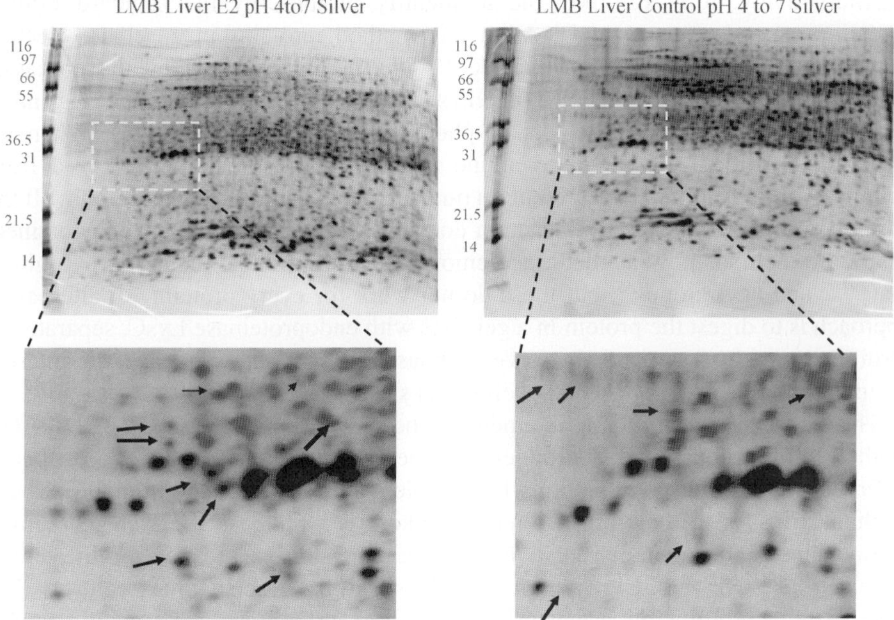

Fig. 6. Two-dimensional polyacrylamide gel separation of hepatic proteins obtained from control and E_2-treated largemouth bass (injected IP with 2.5 mg/kg). The separation in the first dimension was on a pH 4–7 immobilized pH gradient strip and in the second dimension on an 8–16% gradient gel. A small segment of the gel is expanded for visual inspection. Arrows point to proteins that are differentially expressed in either E_2 treated or control fish.

desorption ionization time-of-flight (MALDI TOF) or a liquid chromatography coupled to an electrospray ionization mass spectrometer (LC–ESI–MS)[10,12,89,98]. The peptide fragment masses are used to search databases of the entire predicted proteome for proteins that would generate the same peptide fragments when digested with trypsin[5,111]. With LC–MS/MS, it is possible to further subdivide the fragments and obtain amino acid sequence information, which when added to the database search increases the chance of matching the protein[1,66,133]. This method is particularly interesting for proteins obtained from organisms whose genomes have been completely sequenced, because the entire possible proteome can be interrogated. For species that are not in the databases, this technique will be less efficient. It may be possible to obtain sequence information for proteins that are highly conserved and that have peptides resulting from trypsin digestion that match homologous proteins in the database. But this will be true only for a subset of highly conserved proteins. To identify the majority of proteins from non-model systems, it will be necessary to perform *de novo* protein sequencing on the mass spectrometer. This is a difficult process, as the spectra often must be interpreted manually because few commercially available programs are robust enough to be unequivocal; however, new instruments on the horizon that will measure masses more accurately should help with this approach.

Alternatively, it is still possible to identify proteins by the standard Edman Chemistry sequencing method[65]. With this method, proteins in a gel are electro-transferred to a polyvinylidene difluoride (PVDF) membrane, stained with Coomassie blue, and then applied directly on a protein sequencer. In the sequencer, the free amino terminus of the protein is reacted with phenylisothiocyanate at basic pH. When the protein is acidified, the modified amino acid is cyclized and then cleaved from the protein and identified by its retention time on reverse phase chromatography. If the amino terminus is blocked, the reaction does not work and no sequence is obtained. There are methods now available to remove blocking groups, including pyroglutamate[93,95] and acyl groups[35], but these do not work for every protein. An alternative approach is to digest the protein in a gel slice with endoproteinase LysC, separate the resulting peptide fragments by reverse phase HPLC, and sequence an internal fragment[114]. This approach works well but is slow and expensive.

There have been several improvements in the last 10 years that make the 2D-PAGE method more robust and reproducible. These include the discovery of better detergents and buffer combinations for sample solubilization, new immobilized pH gradient strips for the first dimension electrophoresis[13], pre-cast SDS slab gels for the second dimension, and more sensitive gel stains[33]. In addition, the new 2D-differential gel electrophoresis (DIGE)[120] now marketed as a kit by Amersham (Piscataway, NJ) holds extreme promise, but is expensive. The DIGE system involves a modification of the normal 2D gel method in that the two protein samples to be compared are each pre-labeled with one of the two cyanine dyes, Cy3 or Cy5. The labeled samples are mixed and co-migrated on the same gel in both dimensions, removing imperfections in the separation due to differences in the gel matrix, pH field, or other procedural effects. Because the samples are co-migrated, it is easy to quantify changes in protein expression and pick proteins that are altered by the treatment. The co-migration also reduces the number of gels that must be performed for statistical purposes. The coupling of this method with mass spectrometry to identify proteins that are differentially expressed has recently been validated by Tonge *et al.*[117] and is reviewed by Patton[100]. These advances result in better resolution and visualization.

The 2D gel approach is ideal for identifying proteins with post-translational modifications, since small changes in protein isoelectric points and size result in dramatic changes in the position of the proteins in the gel. For example, proteins modified by phosphorylation move a considerable distance away from the parent, unmodified protein. Despite the improvements, 2D-PAGE remains technically complicated, and requires that at least triplicate samples be processed for statistical purposes. In addition, some classes of proteins are not detectable, including those that are rare (i.e. low abundance), small (i.e. under 2 kDa), glycosylated, or basic.

In general, 2D-PAGE works well for protein discovery and is currently the most widely used technique for detecting proteins. Even though a thousand or more proteins can be visualized on a gel, it is still insufficient for total proteome analysis. There are at least 10,000 proteins per tissue (perhaps more) and most are not visualized on the gel because of low abundance. Sample fractionation procedures are a must, making this system less high throughput than originally perceived. Both

2D-PAGE and MS require highly trained personnel, a requirement that takes the method out of the general laboratory.

Fish toxico-proteomics studies. Few studies using proteomics with fish exist. Somero's group[77] showed differences in protein patterns in gill tissue of the long-jawed mudsucker (*G. mirabilis*) as a consequence of changes in osmotic and thermal conditions. The authors hypothesize that the identified proteins may be involved in acclimation to these conditions. Among the more active research groups, Bradley and his co-workers have performed 2D gel electrophoresis to determine protein expression patterns in fish and marine invertebrates exposed to several toxicants such as metals[90,110], polychlorinated biphenyls[110], and sewage treatment[17] and also to normal stressors such as changing salinities and temperatures. In their work on nonylphenol and sewage treatment effluents, the Bradley group performed 2D gel electrophoresis and showed that several proteins that appeared to be specific to nonylphenol exposure were also present in fish exposed to the effluent, suggesting that it will be possible to detect specific protein expression signatures for contaminants[17]. The next step for Bradley's group is to identify the proteins that were preferentially expressed from the exposures.

Other studies of note are the examination of changes in protein expression patterns in salmon embryos during development[74] and the analysis of changes in proteins in livers from rainbow trout undergoing starvation[88]. Both of these studies also rely entirely on changes in 2D gel patterns of proteins. The proteins with changes in expression are yet to be identified. With the new national focus on proteomics, it is anticipated that many projects employing this technique will be completed in the near future.

2. Non-gel based separations of proteins coupled to mass spectrometry

There are several chromatography-based methods that have been developed to separate proteins prior to identification by mass spectrometry[44,48]. The idea is to use at least two orthogonal systems to increase the separation efficiency, e.g. ion exchange followed by hydrophobic interaction. Both top–down and bottom–up approaches have been developed.

For a top–down approach, intact proteins are separated first, and then fractions are digested with trypsin and mass analyzed. This method can take advantage of specific protein characteristics to separate protein classes, e.g. it is possible to use affinity methods to isolate an entire group of proteins, such as those that are phosphorylated[134] or ubiquitinated[87]. These are then analyzed specifically.

The bottom–up approach differs in that the entire proteome is first digested with trypsin and then the fragments are separated by cation-exchange chromatography followed by reverse phase chromatography[84,124]. The fragments are then analyzed by mass spectrometry and are used to identify the proteins in the sample.

Both methods require complex mass spectrometers that can perform both MS and MS/MS measurements on the tryptic peptides. MS/MS involves the ability to further fragment proteins within the mass spectrometer along the peptide backbone to determine amino acid sequence information. The proteins are identified by comparing both MS and MS/MS data to theoretical masses for proteins in the databases. Both methods have merit, but again are best for organisms whose genomes have been sequenced.

The main innovative technology in this area has been the development of the cleavable Isotope Coded Affinity Tag (cICAT) currently sold by Applied Biosystems (Foster City, CA)[52,67,100]. Briefly, cICAT refers to a pair of affinity-directed reagents, that differ from each other by 9 mass units and that are specific for tagging cysteine residues in proteins. The idea is to tag proteins in the control group with the light reagent (light isotope) and proteins in the experimental group with the heavy reagent (heavy isotope). After tagging, the two protein groups are mixed and digested with trypsin. The reaction is highly specific and occurs with high efficiency. The tags also contain biotin, thus, fragments tagged by the reagents are separated on an avidin column and cleaved at the cleavable site to elute them from the column. The mixture is then further separated by reverse phase liquid chromatography (LC) and analyzed by mass spectrometry (MS/MS). In mass spectrometry lingo this is referred to as LC–MS/MS. Fragments generated from proteins in equal abundance in the two tissues will be only 9 mass units apart (or multiples of 9, if the fragment contains more than 1 Cys) and of equal height in the spectrogram obtained from the initial scan in the mass spectrometer. Fragments from proteins that are differentially expressed will appear to be higher (or lower) than their corresponding partners. These fragments can then be targeted for MS/MS sequencing. Identities of the parent proteins can be obtained by comparing the sequences to databases.

cICAT has the potential to provide excellent data in a short time and is worth considering if the proper mass spectrometer is available. Even though post-translational modifications may not be amenable to this analysis, it can yield information for proteins that are synthesized anew in response to a stressor. This method may be the technique of choice for analyzing proteins in complexes that can be immuno-precipitated with specific antibodies without having to analyze the samples first by gel electrophoresis. It complements the 2D-PAGE/MS method.

3. Protein and antibody chips

The concept of evaluating thousands of proteins at once by a chip-based method is very appealing; however, this technology is still under development for mammalian models and will require considerable work to arrive at a point where it can be used to assess global differential protein expression in fish[68]. Two main types of chips are being investigated: antibody-based and protein-based. In the case of antibody-based chips, antibodies to proteins of interest must be available. Therein lies the problem – few antibodies are commercially available for fish. Similar to methods used for making cDNA microarrays, the antibodies must be fixed to glass slides (or protein-binding membrane, nitrocellulose or polyvinylidene difluoride, PVDF) and probed with the full complement of proteins from the tissue of interest. This method is similar to a reverse Western, where specific antibodies are on the chip and are probed with the proteins. One way to visualize the bound proteins is to pre-label them with biotin and then probe with a streptavidin-coupled alkaline phosphatase linked antibody after the proteins have bound to the chip. Alternatively, one could develop a second set of specific antibodies, one for each captured protein, that could then be detected by conventional methods as used in Sandwich Enzyme-Linked ImmunoSorbent Assays (ELISA). This approach, however, would require having at least two antibodies for every protein, a capture antibody and a detection antibody. New labeling and detection methods are being developed to increase the sensitivity of this approach. The antibody-based chips are used primarily to measure differential expression of proteins.

Protein-based chips, on the other hand, rely on placing proteins directly on the glass slide (or membrane), and are used primarily to test for protein–protein interactions or protein–drug interactions. For these chips, recombinant proteins (or purified proteins) must be available for the species of interest. It may be possible to construct libraries of proteins in easy-to-express recombinant vectors (e.g. the FLEX system used for the human collection at Harvard[19]). The creation of such libraries for fish will require the concerted effort of many scientists.

Advantages and disadvantages exist with chip technology. Advantages include using a single chip to measure multiple proteins simultaneously. Also a single blood or tissue sample can be queried for multiple end points. The main disadvantage is the requirement of an antibody library or protein expression library for the species of interest. This is costly both in time and money.

A different type of chip is available that links the interaction of proteins with an activated surface on a chip with mass spectrometry. Surface Enhanced Laser Desorption Ionization (SELDI) mass spectrometry, a system developed by Ciphergen Biosystems (Palo Alto, CA) over the past 5 years, allows protein profiling for samples deposited onto a special chip containing spots that have specific properties[99]. Complex protein mixtures are essentially sub-fractionated by specific interactions between proteins in the mixture and the various spots on the chip. For example, some proteins have low isoelectric points and will bind preferentially to an anionic surface (weak anion exchanger), whereas other proteins with more basic isoelectric points will bind preferentially to a cationic surface (weak cation exchanger). There are also spots containing a matrix embedded with a metal affinity surface that can be used to capture histidine-tagged proteins (recombinant proteins that have a multiple His sequence (6–10) added to the N- or C-terminus to enable quick purification) or phosphorylated proteins. And, it is also possible to make specific antibody surfaces, to capture specific proteins from complex mixtures. A few researchers are beginning to apply this technology to aquatic toxicology. For example, in one published study, rainbow trout fingerlings were exposed to sublethal concentrations of zinc (3.5 μM) for 1–6 days and then their SELDI profiles were measured. A number of different proteins specific to zinc exposure were identified based on their masses alone[63]. This approach to study protein profiles will undoubtedly gain in popularity as a method to determine toxic changes at the level of protein expression. But, it would still be more usable if the proteins could be unequivocally identified to determine metabolic pathways that are affected by the exposures.

Chip technology holds considerable promise for identifying differentially expressed proteins and would be an excellent complement to cDNA chip technology.

Studies using DNA microarrays could provide valuable information about changes at the mRNA level that may suggest additional proteins to evaluate. While proteomics technology is not currently standardized, a coordinated effort among researchers in this regard would ensure that chips are developed that will be useful to the fish toxicology field.

4. Bioinformatics

Both genomics and proteomics approaches generate extensive data[86]. It will be important to develop methods, or to apply methods developed by others, to interpret and categorize this information in a useful manner. In order to achieve the best information from any technique or method, full and meaningful analysis will be critical.

A recent paper by Wojcik and Schachter[130] describes proteomics databases and software available through the World Wide Web, focusing on current use and applicability. The authors recognize that the evolution of resources is occurring rapidly, and trends and probable changes are discussed wherever applicable[130]. Other investigators have also recognized the need for rapid and accurate computational analysis of protein function and have begun the development of large-scale computational systems for analysis of sequence and structure of proteins[127]. Along these lines, algorithms have been developed to correlate changes in patterns of protein expression determined by SELDI TOF (surface-enhanced laser desorption ionization time of flight) mass spectrometry (Ciphergen Inc.) with human cancer[99,102]. Similar studies with fish protein expression patterns are underway.

In summary, the roles that genomics and proteomics experiments will play in aquatic toxicology are still to be determined. Clearly, there are abundant potential benefits in the ability to determine the exact mechanisms of toxicity of fish exposed to anthropogenic toxicants. It will be important to determine comparative responses of different fish species to the toxicants to determine whether there are different tolerances and adaptabilities in wild fish. Along these lines, it will be important to develop these technologies for those non-model species that are important to evaluate to understand ecosystem health. How these technologies might be used for risk assessment[59] remains to be seen, but the potential benefits in understanding how toxicants work are tremendous.

Acknowledgements. This publication was made possible by grants from the National Institute of Environmental Health Sciences, NIH (P42 ES 07375), NIH (1R43 ES011882-01) and NSF (BES-9906060). The contents are solely the responsibility of the authors and do not necessarily represent the official views of NIEHS or NSF. The authors wish to acknowledge technical help from Marjorie Chow and Ben O'Neal for the 2D gel of largemouth bass, Kevin Kroll for his advice on fish reproduction and Barbara Carter for the subtractive hybridizations.

IV. References

1. Aebersold, R. and D.R. Goodlett. Mass spectrometry in proteomics. *Chem. Rev.* 101: 269–295, 2001.
2. Alon, U., N. Barkai, D.A. Notterman, K. Gish, S. Ybarra, D. Mack and A.J. Levine. Broad patterns of gene expression revealed by clustering analysis of tumor and normal colon tissues probed by oligonucleotide arrays. *Proc. Natl. Acad. Sci. USA* 96: 6745–6750, 1999.
3. Alwine, J.C., D.J. Kemp, B.A. Parker, J. Reiser, J. Renart, G.R. Stark and G.M. Wahl. Detection of specific RNAs or specific fragments of DNA by fractionation in gels and transfer to diazobenzyloxymethyl paper. *Methods Enzymol.* 68: 220–242, 1979.
4. Alwine, J.C., D.J. Kemp and G.R. Stark. Method for detection of specific RNAs in agarose gels by transfer to diazobenzyloxymethyl-paper and hybridization with DNA probes. *Proc. Natl. Acad. Sci. USA* 74: 5350–5354, 1977.
5. Appel, R.D., A. Bairoch and D.F. Hochstrasser. 2D databases on the World Wide Web. In: *2D Proteome Analysis Protocols: Methods in Molecular Biology*, edited by A.J. Link, Totowa, NJ, Humana Press, pp. 383–391, 1999.
6. Bartosiewicz, M., D. Jenkins, S. Penn, J. Emery and A. Buckpitt. Unique gene expression patterns in liver and kidney associated with exposure to chemical toxicants. *J. Pharmacol. Exp. Ther.* 297: 895–905, 2001.
7. Bayne, C.J., L. Gerwick, K. Fujiki, M. Nakao and T. Yano. Immune-relevant (including acute phase) genes identified in the livers of rainbow trout, *Oncorhynchus mykiss*, by means of suppression subtractive hybridization. *Dev. Comp. Immunol.* 25: 205–217, 2001.
8. Beach, L.R. and R.D. Palmiter. Amplification of the metallothionein-I gene in cadmium-resistant mouse cells. *Proc. Natl. Acad. Sci. USA* 78: 2110–2114, 1981.
9. Ben-Dor, A., R. Shamir and Z. Yakhini. Clustering gene expression patterns. *J. Comput. Biol.* 6: 281–297, 1999.
10. Bergquist, J., M. Palmblad, M. Wetterhall, P. Hakansson and K.E. Markides. Peptide mapping of proteins in human body fluids using electrospray ionization Fourier transform ion cyclotron resonance mass spectrometry. *Mass Spectrom. Rev.* 21: 2–15, 2002.
11. Bhatia, P., W.R. Taylor, A.H. Greenberg and J.A. Wright. Comparison of glyceraldehyde-3-phosphate dehydrogenase and 28S-ribosomal RNA gene expression as RNA loading controls for northern blot analysis of cell lines of varying malignant potential. *Anal. Biochem.* 216: 223–226, 1994.
12. Bienvenut, W.V., C. Deon, C. Pasquarello, J.M. Campbell, J.C. Sanchez, M.L. Vestal and D.F. Hochstrasser. Matrix-assisted laser desorption/ionization-tandem mass spectrometry with high resolution and sensitivity for identification and characterization of proteins. *Proteomics* 2: 868–876, 2002.
13. Bjellqvist, B., K. Ek, P.G. Righetti, E. Gianazza, A. Gorg and R. Westermeier. Isoelectric focusing in immobilized pH gradients: principle, methodology and some applications. *J. Biochem. Biophys. Methods* 6: 317–339, 1982.
14. Bowman, C.J., K.J. Kroll, T.G. Gross and N.D. Denslow. Estradiol-induced gene expression in largemouth bass (*Micropterus salmoides*). *Mol. Cell. Endocrinol.* 196: 67–77, 2002.
15. Bowman, C.J., K.J. Kroll, M.J. Hemmer, L.C. Folmar and N.D. Denslow. Estrogen-induced vitellogenin mRNA and protein in sheepshead minnow (*Cyprinodon variegatus*). *Gen. Comp. Endocrinol.* 120: 300–313, 2000.
16. Bowtell, D. Options available – from start to finish – for obtaining expression data by microarray. *Nat. Genet.* 21: 25–32, 1999.
17. Bradley, B.P., E.A. Shrader, D.G. Kimmel and J.C. Meiller. Protein expression signatures: an application of proteomics. *Mar. Environ. Res.* 54: 373–377, 2002.
18. Brazma, A., P. Hingamp, J. Quackenbush, G. Sherlock, P. Spellman, C. Stoeckert, J. Aach, W. Ansorge, C. Ball, H.C. Causton, T. Gaasterland, P. Glenisson, F.C. Holstege, I. Kim, V. Markowitz, J.C. Matese, H. Parkinson, A. Robinson, U. Sarkans, S. Schulze-Kremer, J. Stewart, R. Taylor, J. Vilo and M. Vingron. Minimum information about a microarray experiment (MIAME) – toward standards for microarray data. *Nat. Genet.* 29: 365–371, 2001.
19. Brizuela, L., A. Richardson, G. Marsischky and J. Labaer. The FLEXGene repository. Exploiting the fruits of the genome projects by creating a needed resource to face the challenges of the post-genomic era. *Arch. Med. Res.* 33: 318–324, 2002.
20. Brown, M.P., W.N. Grundy, D. Lin, N. Cristianin, C.W. Sugnet, T.S. Furey, M.J. Ares and D. Haussler. Knowledge-based analysis of microarray gene expression data by using support vector machines. *Proc. Natl. Acad. Sci. USA* 97: 262–267, 2000.
21. Bruno, M.E., C.H. Borchers, M.J. Dial, N. Walker, J.E. Hartis, B.A. Wetmore, J. Barrett, K.B. Tomer and B.A. Merrick. Effects of TCDD upon IkB and IKK subunits localized in microsomes by proteomics. *Arch. Biochem. Biophys.* 406: 153–164, 2002.

22. Caetano, A.R., R.K. Johnson and D. Pomp. Generation and sequence characterization of a normalized cDNA library from swine ovarian follicles. *Mamm. Genome* 14: 65–70, 2003.
23. Carginale, V., C. Capasso, R. Scudiero and E. Parisi. Identification of cadmium-sensitive genes in the Antarctic fish *Chionodraco hamatus* by messenger RNA differential display. *Gene* 299: 117–124, 2002.
24. Claviere, J.M. What if there are only 30,000 human genes? *Science* 291: 1255–1258, 2001.
25. Davis, M.M., D.I. Cohen, E.A. Nielsen, M. Steinmetz, W.E. Paul and L. Hood. Cell-type-specific cDNA probes and the murine I region: the localization and orientation of Ad alpha. *Proc. Natl. Acad. Sci. USA* 81: 2194–2198, 1984.
26. D'Cotta, H., A. Fostier, Y. Guiguen, M. Govoroun and J.F. Baroiller. Search for genes involved in the temperature-induced gonadal sex differentiation in the tilapia, *Oreochromis niloticus*. *J. Exp. Zool.* 290: 574–585, 2001.
27. Denslow, N.D., C.J. Bowman, R.J. Ferguson, H.S. Lee, M.J. Hemmer and L.C. Folmar. Induction of gene expression in sheepshead minnows (*Cyprinodon variegatus*) treated with 17 beta-estradiol, diethylstilbesterol, or ethinylestradiol: the use of mRNA fingerprints as an indicator of gene regulation. *Gen. Comp. Endocrinol.* 121: 250–260, 2001.
28. Denslow, N.D., H.S. Lee, C.J. Bowman, M.J. Hemmer and L.C. Folmar. Multiple responses in gene expression in fish treated with estrogen. *Comp. Biochem. Physiol.* 129B: 277–282, 2001.
29. Diatchenko, L., Y.F. Lau, A.P. Campbell, A. Chenchik, F. Moqadam, B. Huang, S. Lukyanov, K. Lukyanov, N. Gurskaya, E.D. Sverdlov and P.D. Siebert. Suppression subtractive hybridization: a method for generating differentially regulated or tissue-specific cDNA probes and libraries. *Proc. Natl. Acad. Sci. USA* 93: 6025–6030, 1996.
30. Dickmeis, T., P. Aanstad, M. Clark, N. Fischer, R. Herwig, P. Mourrain, P. Blader, F. Rosa, H. Lehrach and U. Strahle. Identification of nodal signaling targets by array analysis of induced complex probes. *Dev. Dyn.* 222: 571–580, 2001.
31. Duguid, J.R. and M.C. Dinauer. Library subtraction of in vitro cDNA libraries to identify differentially expressed genes in scrapie infection. *Nucleic Acids Res.* 18: 2789–2792, 1990.
32. Eisen, M.B., P.T. Spellman, P.O. Brown and D. Botstein. Cluster analysis and display of genome-wide expression patterns. *Proc. Natl. Acad. Sci. USA* 95: 14863–14868, 1998.
33. Epstein, L.B., D.M. Smith, N.M. Matsui, H.M. Tran, C. Sullivan and I. Raineri. Identification of cytokine-regulated proteins in normal and malignant cells by the combination of 2D-PAGE, mass spectrometry, Edman degradation and immunoblotting: approaches to analysis of their functional roles. *Electrophoresis* 17: 1655–1670, 1996.
34. Ermolaeva, O.D. and E.D. Sverdlov. Subtractive hybridization, a technique for extraction of DNA sequences distinguishing two closely related genomes: critical analysis. *Genet. Anal.* 13: 49–58, 1996.
35. Farries, T.C., A. Harris, A.D. Auffret and A. Aitken. Removal of *N*-acetyl groups from blocked peptides with acylpeptide hydrolase: stabilization of the enzyme and its application to protein sequencing. *Eur. J. Biochem.* 196: 679–685, 1991.
36. Favaloro, J., R. Treisman and R. Kamen. Transcription maps of polyoma virus-specific RNA: analysis by two-dimensional nuclease S1 gel mapping. *Methods Enzymol.* 65: 718–749, 1980.
37. Fielden, M.R. and T.R. Zacharewski. Challenges and limitations of gene expression profiling in mechanistic and predictive toxicology. *Toxicol. Sci.* 60: 6–10, 2001.
38. Flouriot, G., F. Pakdel, B. Ducouret, Y. Ledrean and Y. Valotaire. Differential regulation of two genes implicated in fish reproduction: vitellogenin and estrogen receptor genes. *Mol. Reprod. Dev.* 48: 317–323, 1997.
39. Folmar, L.C., M. Hemmer, R. Hemmer, C. Bowman, K. Kroll and N.D. Denslow. Comparative estrogenicity of estradiol, ethinyl estradiol and diethylstibestrol in an *in vivo*, male sheepshead minnow (*Cyprinodon variegatus*), vitellogenin bioassay. *Aquat. Toxicol.* 49: 77–88, 2000.
40. Fujiki, K., C.J. Bayne, D.H. Shin, M. Nakao and T. Yano. Molecular cloning of carp (*Cyprinus carpio*) C-type lectin and pentraxin by use of suppression subtractive hybridisation. *Fish Shellfish Immunol.* 11: 275–279, 2001.
41. Fujiki, K., J. Gauley, N. Bols and B. Dixon. Cloning and characterization of cDNA clones encoding CD9 from Atlantic salmon (*Salmo salar*) and rainbow trout (*Oncorhynchus mykiss*). *Immunogenetics* 54: 604–609, 2002.
42. Gagnon, E., S. Duclos, C. Rondeau, E. Chevet, P.H. Cameron, O. Steele-Mortimer, J. Paiement, J.J. Bergeron and M. Desjardins. Endoplasmic reticulum-mediated phagocytosis is a mechanism of entry into macrophages. *Cell* 11: 119–131, 2002.
43. Gimeno, S., A. Gerritsen, T. Bowmer and H. Komen. Feminization of male carp. *Nature* 384: 221–222, 1996.

44. Goodlett, D.R. and E.C. Yi. Proteomics without polyacrylamide: qualitative and quantitative uses of tandem mass spectrometry in proteome analysis. *Funct. Integr. Genomics* 2: 138–153, 2002.
45. Gracey, A.Y., J.V. Troll and G.N. Somero. Hypoxia-induced gene expression profiling in the euryoxic fish *Gillichthys mirabilis*. *Proc. Natl. Acad. Sci. USA* 98: 1993–1998, 2001.
46. Grant, S.G. and W.P. Blackstock. Proteomics and neuroscience: from protein to network. *J. Neurosci.* 21: 8315–8318, 2001.
47. Gray, M.A. and C.D. Metcalfe. Induction of testis–ova in Japanese medaka (*Oryzias latipes*) exposed to *p*-nonylphenol. *Environ. Toxicol. Chem.* 16: 1082–1086, 1997.
48. Griffin, T.J. and R. Aebersold. Advances in proteome analysis by mass spectrometry. *J. Biol. Chem.* 276: 45497–45500, 2001.
49. Griffin, T.J., S.P. Gygi, T. Ideker, B. Rist, J. Eng, L. Hood and R. Aebersold. Complementary profiling of gene expression at the transcriptome and proteome levels in *Saccharomyces cerevisiae*. *Mol. Cell. Proteomics* 1: 323–333, 2002.
50. Gurskaya, N.G., L. Diatchenko, A. Chenchik, P.D. Siebert, G.L. Khaspekov, K.A. Lukyanov, L.L. Vagner, O.D. Ermolaeva, S.A. Lukyanov and E.D. Sverdlov. Equalizing cDNA subtraction based on selective suppression of polymerase chain reaction: cloning of Jurkat cell transcripts induced by phytohemaglutinin and phorbol 12-myristate 13-acetate. *Anal. Biochem.* 240: 90–97, 1996.
51. Gygi, S.P., G.L. Corthals, Y. Zhang, Y. Rochon and R. Aebersold. Evaluation of two-dimensional gel electrophoresis-based proteome analysis technology. *Proc. Natl. Acad. Sci. USA* 97: 9390–9395, 2000.
52. Gygi, S.P., B. Rist, S.A. Gerber, F. Turecek, M.H. Gelb and R. Aebersold. Quantitative analysis of complex protein mixtures using isotope-coded affinity tags. *Nat. Biotechnol.* 17: 994–999, 1999.
53. Hamadeh, H.K., P.R. Bushel, S. Jayadev, O. DiSorbo, L. Bennett, L. Li, R. Tennant, R. Stoll, J.C. Barrett, R.S. Paules, K. Blanchard and C.A. Afshari. Prediction of compound signature using high density gene expression profiling. *Toxicol. Sci.* 67: 232–240, 2002.
54. Hamadeh, H.K., P.R. Bushel, S. Jayadev, K. Martin, O. DiSorbo, S. Sieber, L. Bennett, R. Tennant, R. Stoll, J.C. Barrett, K. Blanchard, R.S. Paules and C.A. Afshari. Gene expression analysis reveals chemical-specific profiles. *Toxicol. Sci.* 67: 219–231, 2002.
55. Hamadeh, H.K., B.L. Knight, A.C. Haugen, S. Sieber, R.P. Amin, P.R. Bushel, R. Stoll, K. Blanchard, S. Jayadev, R.W. Tennant, M.L. Cunningham, C.A. Afshari and R.S. Paules. Methapyrilene toxicity: anchorage of pathologic observations to gene expression alterations. *Toxicol. Pathol.* 30: 470–482, 2002.
56. Hara, E., T. Kato, S. Nakada, S. Sekiya and K. Oda. Subtractive cDNA cloning using oligo(dT)30-latex and PCR: isolation of cDNA clones specific to undifferentiated human embryonal carcinoma cells. *Nucleic Acids Res.* 19: 7097–7104, 1991.
57. Hedrick, S.M., D.I. Cohen, E.A. Nielsen and M.M. Davis. Isolation of cDNA clones encoding T cell-specific membrane-associated proteins. *Nature* 308: 149–153, 1984.
58. Heid, C.A., J. Stevens, K.J. Livak and P.M. Williams. Real time quantitative PCR. *Genome Res.* 6: 986–994, 1996.
59. Heinrich-Hirsch, B., S. Madle, A. Oberemm and U. Gundert-Remy. The use of toxicodynamics in risk assessment. *Toxicol. Lett.* 120: 131–141, 2001.
60. Hemmer, M.J., B.L. Hemmer, C.J. Bowman, K.J. Kroll, L.C. Folmar, D. Marcovich, M.D. Hoglund and N.A. Denslow. Effects of *p*-nonylphenol, methoxychlor, and endosulfan on vitellogenin induction and expression in sheepshead minnow (*Cyprinodon variegatus*). *Environ. Toxicol. Chem.* 20: 336–343, 2001.
61. Herwig, R., P. Aanstad, M. Clark and H. Lehrach. Statistical evaluation of differential expression on cDNA nylon arrays with replicated experiments. *Nucleic Acids Res.* 29: E117, 2001.
62. Hilsenbeck, S.G., W.E. Friedrichs, R. Schiff, P. O'Connell, R.K. Hansen, C.K. Osborne and S.A. Fuqua. Statistical analysis of array expression data as applied to the problem of tamoxifen resistance. *J. Natl. Cancer Inst.* 91: 453–459, 1999.
63. Hogstrand, C., S. Balesaria and C.N. Glover. Application of genomics and proteomics for study of the integrated response to zinc exposure in a non-model fish species, the rainbow trout. *Comp. Biochem. Physiol.* 133B: 523–535, 2002.
64. Howell, W.M., D.A. Black and S.A. Bortone. Abnormal expression of secondary sex characteristics in a population of mosquitofish (*Gambusia affinis holbrooki*): evidence for environmentally-induced masculanization. *Copeia* 4: 676–681, 1980.
65. Hunkapiller, M.W., R.M. Hewick, W.J. Dreyer and L.E. Hood. High sensitivity sequencing with a gas phase sequenator. *Methods Enzymol.* 91: 399–403, 1983.
66. Hunt, D.F., J.R. Yates, J. Shabanowitz, S. Winston and C.R. Hauer. Protein sequencing by tandem mass spectrometry. *Proc. Natl. Acad. Sci. USA* 83: 6233–6238, 1986.

67. Ideker, T., V. Thorsson, J.A. Ranish, R. Christmas, J. Buhler, J.K. Eng, R. Bumgarner, D.R. Goodlett, R. Aebersold and L. Hood. Integrated genomic and proteomic analyses of a systematically perturbed metabolic network. *Science* 292: 929–934, 2001.
68. Jenkins, R.E. and S.R. Pennington. Arrays for protein expression profiling: towards a viable alternative to two-dimensional gel electrophoresis? *Proteomics* 1: 13–29, 2001.
69. Jobling, S., D. Sheahan, J.A. Osborne, P. Matthiessen and J.P. Sumpter. Inhibition of testicular growth in rainbow trout (*Oncorhynchus mykiss*) exposed to estrogenic alkylphenolic chemicals. *Environ. Toxicol. Chem.* 15: 194–202, 1996.
70. Ju, Z., R.A. Dunham and Z. Liu. Differential gene expression in the brain of channel catfish (*Ictalurus punctatus*) in response to cold acclimation. *Mol. Genet. Genomics* 268: 87–95, 2002.
71. Jung, E., M. Heller, J.C. Sanchez and D.F. Hochstrasser. Proteomics meets cell biology: the establishment of subcellular proteomes. *Electrophoresis* 21: 3369–3377, 2000.
72. Kafatos, F.C., C.W. Jones and A. Efstratiadis. Determination of nucleic acid sequence homologies and relative concentrations by a dot hybridization procedure. *Nucleic Acids Res.* 24: 1541–1542, 1979.
73. Kanamori, A. Systematic identification of genes expressed during early oogenesis in medaka. *Mol. Reprod. Dev.* 55: 31–36, 2000.
74. Kanaya, S., Y. Ujiie, K. Hasegawa, T. Sato, H. Imada, M. Kinouchi, Y. Kudo, T. Ogata, H. Ohya, H. Kamada, K. Itamoto and K. Katsura. Proteome analysis of *Oncorhynchus* species during embryogenesis. *Electrophoresis* 21: 1907–1913, 2000.
75. Khayat, M., T.B. Stuge, M. Wilson, E. Bengten, N.W. Miller and L.W. Clem. Thioredoxin acts as a B cell growth factor in channel catfish. *J. Immunol.* 166: 2937–2943, 2001.
76. Knoebl, I. The use of molecular biological methods to assess the effects of endocrine disrupting chemicals and natural hormones on growth in the sheepshead minnow (*Cyprinodon variegatus*). PhD Thesis. Oregon State University, Corvallis, OR, 2002.
77. Kültz, D. and G.N. Somero. Differences in protein patterns of gill epithelial cells of the fish *Gillichthys mirabilis* after osmotic and thermal acclimation. *J. Comp. Physiol. B* 166: 88–100, 1997.
78. Larkin, P., W. Baehr and S.L. Semple-Rowland. Circadian regulation of iodopsin and clock is altered in the retinal degeneration chicken retina. *Mol. Brain Res.* 70: 253–263, 1999.
79. Larkin, P., L.C. Folmar, M.J. Hemmer, A.J. Poston and N.D. Denslow. Expression profiling of estrogenic compounds using a sheepshead minnow cDNA macroarray. *Environ. Health Perspect. Toxicogenomics* 111: 29–36, 2003.
80. Larkin, P., T. Sabo-Attwood, J. Kelso and N.D. Denslow. Gene expression analysis of largemouth bass exposed to estradiol, nonylphenol, and *p,p′*-DDE. *Comp. Biochem. Physiol.* 133B: 543–557, 2002.
81. Lee, P.H. and F.W. Goetz. Characterization of a novel cDNA obtained through differential-display PCR of phorbol ester-stimulated ovarian tissue from the brook trout (*Salvelinus fontinalis*). *Mol. Reprod. Dev.* 49: 112–118, 1998.
82. Liang, P. and A.B. Pardee. Differential display of eukaryotic messenger RNA by means of the polymerase chain reaction. *Science* 257: 967–971, 1992.
83. Lisitsyn, N. and M. Wigler. Cloning the differences between two complex genomes. *Science* 259: 946–951, 1993.
84. Liu, H., D. Lin and J.R. Yates. Multidimensional separations for protein/peptide analysis in the post-genomic era. *Biotechniques* 32: 898–902, 2002.
85. Liu, L., K. Fujiki, B. Dixon and R.S. Sundick. Cloning of a novel rainbow trout (*Oncorhynchus mykiss*) CC chemokine with a fractalkine-like stalk and a TNF decoy receptor using cDNA fragments containing AU-rich elements. *Cytokine* 17: 71–81, 2002.
86. Maggio, E.T. and K. Ramnarayan. Recent developments in computational proteomics. *Trends Biotechnol.* 19: 266–271, 2001.
87. Marotti, L.A., R. Newitt, Y. Wang, R. Aebersold and H.G. Dohlman. Direct identification of a G protein ubiquitination site by mass spectrometry. *Biochemistry* 41: 5067–5074, 2002.
88. Martin, S.A.M., S. Blaney, A.S. Bowman and D.F. Houlihan. Uniquitin-proteasome-dependent proteolysis in rainbow trout (*Oncorhynchus mykiss*): effect of food deprivation. *Eur. J. Physiol.* 445: 257–266, 2002.
89. Medzihradszky, K.F., H. Leffler, M.A. Baldwin and A. Burlinghame. Protein identification by in-gel digestion, high-performance liquid chromatography and mass spectrometry: peptide analysis by complementary ionization techniques. *J. Am. Soc. Mass Spectrom.* 12: 215–221, 2001.
90. Meiller, J.C. and B.P. Bradley. Zinc concentration effect at the organismal, cellular and subcellular levels in the eastern oyster. *Mar. Environ. Res.* 54: 401–404, 2002.
91. Melton, D.A., P.A. Krieg, M.R. Rebagliati, T. Maniatis, K. Zinn and M.R. Green. Efficient in vitro synthesis of biologically active RNA and RNA hybridization probes from plasmids containing a bacteriophage SP6 promoter. *Nucleic Acids Res.* 12: 7035–7056, 1984.

92. Miura, T., N. Kudo, C. Miura, K. Yamauchi and Y. Nagahama. Two testicular cDNA clones suppressed by gonadotropin stimulation exhibit ZP2- and ZP3-like structures in Japanese eel. *Mol. Reprod. Dev.* 51: 235–242, 1998.
93. Miyatake, N., M. Kamo, K. Satake, Y. Uchiyama and A. Tsugita. Removal of N-terminal formyl groups and deblocking of pyrrolidone carboxylic acid of proteins with anhydrous hydrazine vapor. *Eur. J. Biochem.* 212: 785–789, 1993.
94. Monetti, C., D. Vigetti, M. Prati, E. Sabbioni, G. Bernardini and R. Gornati. Gene expression in *Xenopus* embryos after methylmercury exposure: a search for molecular biomarkers. *Environ. Toxicol. Chem.* 21: 2731–2736, 2002.
95. Moyer, M., A. Harper, G. Payne, H. Ryals and E. Fowler. In situ digestion with pyroglutamate aminopeptidase for N-terminal sequencing of electroblotted proteins. *J. Protein Chem.* 9: 282–283, 1990.
96. O'Farrell, C., N. Vaghefi, M. Cantonnet, B. Buteau, P. Boudinot and A. Benmansour. Survey of transcript expression in rainbow trout leukocytes reveals a major contribution of interferon-responsive genes in the early response to a rhabdovirus infection. *J. Virol.* 76: 8040–8049, 2002.
97. Oleksiak, M.F., G.A. Churchill and D.L. Crawford. Variation in gene expression within and among natural populations. *Nat. Genet.* 32: 261–266, 2002.
98. Panfilov, O. and B. Lanne. Peptide mass fingerprinting from wet and dry two-dimensional gels and its application in proteomics. *Anal. Biochem.* 307: 393–395, 2002.
99. Parveletz, C.P., B. Trock, M. Pennanen, T. Tsangaris, C. Magnant, L.A. Liotta and E.F. Petricoin. Proteomic patterns of nipple aspirate fluids obtained by SELDI-TOF: potential for new biomarkers to aid in the diagnosis of breast cancer. *Dis. Markers* 17: 301–307, 2001.
100. Patton, W.F. Review: detection technologies in proteome analysis. *J. Chromatogr. B* 771: 3–31, 2002.
101. Persidis, A.. Proteomics. *Nat. Biotech.* 18: 45–46, 2001.
102. Petricoin, E.F., A.M. Ardekani, B.A. Hitt, P.J. Levine, V. Fusaro, S.M. Steinberg, G.B. Mills, C. Simone, D.A. Fishman, E. Kohn and L.A. Liotta. Use of proteomic patterns in serum to identify ovarian cancer. *Lancet* 359: 572–577, 2002.
103. Sabo-Attwood, T., K.J. Kroll and N.A. Denslow. Differential expression of estrogen receptor isotypes alpha, beta and gamma by estradiol. *Mol. Cell Endocrinol.* 218: 107–118, 2004.
104. Sagerstrom, C., B.A. Kao, M.E. Lane and H. Sive. Isolation and characterization of posteriorly restricted genes in the zebrafish gastrula. *Dev. Dyn.* 220: 402–408, 2001.
105. Sangrador-Vegas, A., J.B. Lennington and T. Smith. Molecular cloning of an IL-8-like CXC chemokine and tissue factor in rainbow trout (*Oncorhynchus mykiss*) by use of suppression subtractive hybridization. *Cytokine* 17: 66–70, 2002.
106. Sargent, T.D. and I. Dawid. Differential gene expression in the gastrula of *Xenopus laevis*. *Science* 222: 135–139, 1983.
107. Savonet, V., C. Maenhaut, F. Miot and I. Pirson. Pitfalls in the use of several 'housekeeping' genes as standards for quantification of mRNA: the example of thyroid cells. *Anal. Biochem.* 247: 165–167, 1997.
108. Schena, M., R.A. Heller, T.P. Theriault, K. Konrad, E. Lachenmeier and R.W. Davis. Microarrays: biotechnology's discovery platform for functional genomics. *Trends Biotechnol.* 16: 301–306, 1998.
109. Schmidt, C.W. Toxicogenomics. An emerging discipline. *Environ. Health Perspect. Toxicogenomics* 111: A20–A25, 2003.
110. Shepard, J.L. and B.P. Bradley. Protein expression signatures and lysosomal stability in *Mytilus edulis* exposed to graded copper concentrations. *Mar. Environ. Res.* 50: 457–463, 2000.
111. Shevchenko, A., M. Wilm, O. Vorm and M. Mann. Mass spectrometric sequencing of proteins from silver stained polyacrylamide gels. *Anal. Chem.* 68: 850–858, 1996.
112. Shi, Y.H., J. Liu, J.H. Xia and J.F. Gui. Screen for stage-specific expression genes between tail bud stage and heartbeat beginning stage in embryogenesis of gynogenetic silver crucian carp. *Cell Res.* 12: 133–142, 2002.
113. Spanakis, E. and D. Brouty-Boye. Discrimination of fibroblast subtypes by multivariate analysis of gene expression. *Int. J. Cancer* 71: 402–409, 1997.
114. Stone, K.L., M.B. LoPresti, J.M. Crawford, R. DeAngelis and K.R. Williams. Enzymatic digestion of proteins and HPLC peptide isolation. In: *A Practical Guide to Protein and Peptide Purification for Microsequencing*, edited by P.T. Matsudaira, San Diego, CA, Academic Press, pp. 31–47, 1989.
115. Thomas, P. Hybridization of denatured RNA and small DNA fragments transferred to nitrocellulose. *Proc. Natl. Acad. Sci. USA* 77: 5201–5205, 1980.
116. Ton, C., D. Stamatiou, V.J. Dzau and C.C. Liew. Construction of a zebrafish cDNA microarray: gene expression profiling of the zebrafish during development. *Biochem. Biophys. Res. Commun.* 296: 1134–1142, 2002.

117. Tonge, R., J. Shaw, B. Middleton, R. Rowlinson, S. Rayner, J. Young, F. Pognan, E. Hawkins, I. Currie and M. Davison. Validation and development of fluorescence two-dimensional differential gel electrophoresis proteomics technology. *Proteomics* 1: 377–396, 2001.
118. Toronen, P., M. Kolehmainen, G. Wong and E. Castren. Analysis of gene expression data using self-organizing maps. *FEBS Lett.* 451: 142–146, 1999.
119. Tusher, V.G., R. Tibshirani and G. Chu. Significant analysis of microarrays applied to the ionizing radiation response. *Proc. Natl Acad. Sci. USA* 98: 5116–5121, 2001.
120. Unlu, M., M.E. Morgan and J.S. Minden. Difference gel electrophoresis: a single gel method for detecting changes in protein extracts. *Electrophoresis* 18: 2071–2077, 1997.
121. Velculescu, V.E., L. Zhang, B. Vogelstein and K.W. Kinzler. Serial analysis of gene expression. *Science* 270: 484–487, 1995.
122. Venter, J.C., M.D. Adams, E.W. Myers, P.W. Li, R.J. Mural and G.G. Sutton. The sequence of the human genome. *Science* 291: 1304–1351, 2001.
123. Wang, Z. and D.D. Brown. A gene expression screen. *Proc. Natl. Acad. Sci. USA* 88: 11505–11509, 1991.
124. Washburn, M.P., R. Ulaszek, C. Deciu, D.M. Schieltz and J.R. Yates. Analysis of quantitative proteomic data generated via multidimensional protein identification technology. *Anal. Chem.* 74: 1650–1657, 2002.
125. Waters, M., G. Boorman, P. Bushel, M. Cunningham, R. Irwin, A. Merrick, K. Olden, R. Paules, J. Selkirk, S. Stasiewicz, B. Weis, B. Van Houten, N. Walker and R. Tennant. Systems toxicology and the chemical effects in biological systems (CEBS) knowledge base. *Environ. Health Perspect. Toxicogenomics* 111: 15–28, 2003.
126. Weinstein, J.N., T.G. Myers, P.M. O'Connor, S.H. Friend, A.J. Fornace, K. Kohn, T. Fojo, S.E. Bates, L.V. Rubinstein, N.L. Anderson, J.K. Buolamwini, W.W. van Osdol, A.P. Monks, D.A. Scudiero, E.A. Sausville, D.W. Zaharevitz, B. Bunow, V.N. Viswanadhan, G.S. Johnson, R.E. Wittes and K.D. Paull. An information-intensive approach to the molecular pharmacology of cancer. *Science* 275: 343–349, 1997.
127. Weir, M., M. Swindells and J. Overington. Insights into protein function through large-scale computational analysis of sequence and structure. *Trends Biotechnol.* 19: S61–S66, 2001.
128. Williams, D.L., T.C. Newman, G.S. Shelness and D.A. Gordon. Measurement of apolipoprotein mRNA by DNA-excess solution hybridization with single-stranded probes. *Methods Enzymol.* 128: 671–689, 1986.
129. Wittes, J. and H.P. Friedman. Searching for evidence of altered gene expression: a comment on statistical analysis of microarray data. *J. Natl Cancer Inst.* 91: 400–401, 1999.
130. Wojcik, J. and V. Schachter. Proteomic databases and software on the web. *Brief Bioinform.* 1: 250–259, 2000.
131. Xie, J., J.J. Wen, B. Chen and J.F. Gui. Differential gene expression in fully-grown oocytes between gynogenetic and gonochoristic crucian carps. *Gene* 271: 109–116, 2001.
132. Xie, J., J.J. Wen, Z.A. Yang, H.Y. Wang and J.F. Gui. Cyclin A2 is differentially expressed during oocyte maturation between gynogenetic silver crucian carp and gonochoristic color crucian carp. *J. Exp. Zool.* 295A: 1–16, 2003.
133. Yates, J.R., A.L. McCormack and J.K. Eng. Mining genomes with mass spectrometry. *Anal. Chem.* 68: 534A–540A, 1996.
134. Zhou, H., J.D. Watts and R. Aebersold. A systematic approach to the analysis of protein phosphorylation. *Nat. Biotechnol.* 19: 375–378, 2001.

Biotransformations/Toxicokinetics

Biochemistry and Molecular Biology of Fishes, vol. 6
T. P. Mommsen and T. W. Moon (Editors)

CHAPTER 4

Interactions between lipids and persistent organic pollutants in fish

ADRIA A. ELSKUS*, TRACY K. COLLIER** AND EMILY MONOSSON†

**USGS, Department of Biological Sciences, 5751 Murray Hall, University of Maine, Orono, ME 04469-5751, USA, **NOAA/NMFS, Northwest Fisheries Science Center, 2725 Montlake Blvd. E., Seattle, WA 98112, USA, and †PO Box 329, Montague, MA 01351, USA*

I. Introduction

The primary goal of this chapter is to raise awareness of the importance of tissue lipid composition in explaining the disposition of persistent organic pollutants (POPs), and hence POP effects, in fish. Our intention is to stimulate discussion in an area that generally receives little attention in fish toxicity studies with the hope that future experimental designs will consider tissue lipid composition as a critically important determinant of POP dynamics. Toward this end, we present evidence from fish, and in some instances invertebrate studies, supporting our hypothesis that lipid compositional differences among tissues, and over lifecycles and season, can be strong predictors of POP disposition in fish. It is our contention that better understanding of tissue lipid dynamics will dramatically improve our ability to link POP exposure to biological effects.

Despite great strides in elucidating mechanisms of toxicity, and the establishment of tremendous databases on tissue contaminant concentrations, we continue to grapple with our inability to consistently predict toxic effects from either POP exposure or tissue concentrations. This may be the result of how we study exposure and effects, with any given study generally focusing on a single life stage, one or a few tissues and/or cell 'targets', or on a particular endpoint of toxicity. That the bioavailability, uptake, metabolism, and tissue distribution of contaminants may change over an organism's lifetime due to life history factors (e.g. habitat, prey preference, reproductive stage) and the physiological changes that accompany them (e.g. starvation, smoltification, spawning), is seldom considered. In this chapter, we reconsider the role of these factors in the tissue distribution, and hence target site concentrations, of POPs, and hopefully show how better accounting for such factors may provide the missing 'link' in our understanding of why POP toxicities differ among species and life stages.

We begin by defining which compounds are considered POPs in the context of this chapter. The next sections describe the basic mechanisms of POP uptake and bioaccumulation. In a brief review of physiologically based toxicokinetic (PB-TK) modeling, we present evidence that incorporating lipid composition can greatly improve predictions of tissue POP levels relative to the use of total lipid content alone. We then provide basic information on lipid biochemistry and begin our discussion on the influence of lipid composition, fish physiology, and fish life history on POP dynamics and disposition. This is followed by a discussion of the association of POPs with altered lipid composition, with particular attention focused on how POP-mediated alterations in lipids likely underlie some of the classic embryotoxic effects of organochlorines. We conclude with a consideration of the issues involved with linking POP exposure to effects, the factors that make this complicated, and suggestions on the ways to incorporate analyses of lipid composition into experimental designs to better understand, and hence predict, POP effects in fish.

II. Definition of persistent organic pollutants

The POPs we consider in this chapter are non-ionizable, and largely non-polar. These characteristics make them poorly soluble in water and thus hydrophobic. Introduction of an ionizable functional group, such as a sulfate or conjugate as occurs with metabolism, greatly increases the water solubility of hydrophobic compounds (due to hydration of the ionizable moieties), and such modified compounds are rapidly cleared. Halogenated aromatic compounds are often very poorly metabolized by both mammals and fish and thus fit the classification of POPs. Compounds in this group include halogenated aromatic hydrocarbons, particularly the well-studied and widespread toxicants such as polychlorinated biphenyls (PCBs), polychlorinated dibenzo-*p*-dioxins (PCDDs) and polychlorinated dibenzofurans (PCDFs). In contrast, the polynuclear aromatic hydrocarbons (PAHs), which are by definition non-halogenated, are readily metabolized and cleared by fish[106]. Indeed, most parent PAHs are near the limits of detection in fish tissue[32] and evidence of PAH exposure is often confined to biliary concentrations of fluorescent metabolites and/or the presence of DNA adducts in tissue[117]. For these reasons, this chapter will focus on the disposition of un-modified, halogenated organic contaminants.

III. Factors governing the bioavailability of POPs

1. Physicochemical properties of hydrophobic POPs

Several properties govern the bioavailability of a hydrophobic compound, including the compound's physicochemical properties, route of exposure to the organism, and physical transport across the cell membrane. Extensive discussion of the factors influencing the kinetics and disposition of xenobiotics and their bioavailability to aquatic organisms can be found in ref. 25. We briefly discuss on the main factors here.

The physicochemical properties of a compound that most strongly influence its bioavailability are its molecular weight, shape, and degree of hydrophobicity. In aquatic systems, hydrophobic compounds, such as PCBs, preferentially bind to the organic carbon fraction of sediment. Sediments contain varying amounts of organic carbon, and it is the relative hydrophobicity of a compound in addition to the percent organic carbon in the sediment that largely determines the water/sediment partitioning of that compound. Similarly, the presence of high concentrations of dissolved organic matter (DOM) in water decreases the partitioning of the compound from water into biota. Hence, in environments where the DOM levels are high, a smaller fraction of a hydrophobic compound will be available to partition from the water into biological membranes, such as gills. In contrast, high levels of dietary fat enhance dietary absorption of lipophilic xenobiotics, with the lipid solubility of these compounds being directly correlated to intestinal absorption efficiency[28,113].

Persistent organic compounds with similar molecular weights and lipid/water partitioning properties (K_{ow}, discussed below), can differ greatly in their toxic potencies, depending on their reactivity, susceptibility to metabolism, and importantly, upon their ability to cross cell membranes. High molecular weight POPs are generally larger, and thus more sterically hindered than low molecular weight POPs. For these reasons, organisms generally accumulate proportionally lower amounts of high molecular weight POPs than they do of low molecular weight POPs, when compared with the relative abundance of these compounds in sediment or water. POP uptake is also modified by compound shape, with membrane transport of branched compounds more sterically hindered than transport of linear molecules. To further refine the predictions of membrane transport, recent work has focused on molecular cross-sectional diameter[19]. Membrane transport proteins (p-glycoproteins) can also modulate dietary uptake of POPs across the intestinal lumen of fish[27] and are implicated in the development of multi-drug resistance in vertebrates for this reason. However, a discussion of these transporters is outside the scope of this chapter and the reader is referred to recent studies of these transporters and their effects on POP disposition[3,27] (see also chapter 18, this volume).

As discussed above, the bioavailability, and thus to some extent the toxicity, of a compound is linked to its ability to partition between water and lipid (membrane) phases. The relative partitioning of a chemical between water and octanol, a solvent used as a lipid surrogate, is used to determine this characteristic and is known as the octanol–water partition coefficient, K_{ow}. Logarithms of K_{ow} values (log K_{ow} or log p) have been determined empirically for a wide-range of chemicals. Log K_{ow} values are used as standards for predicting partitioning and studying the biological properties (e.g. bioaccumulation) of compounds. Although octanol is used as a surrogate for biological lipid, it cannot simulate barriers to uptake, such as steric hindrance by membranes, and functions instead as a simple measure of linear partitioning. Further, K_{ow} values describe relatively weak (i.e. rapidly reversible) associations between the chemical and lipid and so are not useful in situations where the chemical is known to be tightly bound to a matrix. For example, log K_{ow} values cannot be used to predict dietary uptake of hydrophobic compounds that are tightly bound to food[85]. Indeed, differences in the fatty acid composition of the lipid in which the POP is sequestered dramatically alters POP uptake from the diet[28]. In addition, the degree of water–lipid partitioning is not always predictive of a chemical's potency, as has been demonstrated for xenobiotics with similar log K_{ow} values (hexachlorobenzene and hexabromobenzene) but widely different toxicity[87]. A recent systematic review of published log K_{ow} values demonstrates large inconsistencies in the measurements themselves[30]. These authors found that for even well-studied chemicals such as DDT, log K_{ow} values can differ by 2–4 orders of magnitude. Thus, calculations of chemical partitioning based solely on currently published log K_{ow} values must be considered with caution. Nonetheless, despite these limitations, log K_{ow} values have been successfully employed to explain the toxicity of numerous chemicals and continue to be widely used.

IV. Tissue distribution: Physiologically based toxicokinetic models

Steady-state pharmacokinetic models, generally used to describe the kinetics of uptake and release of lipophilic organic contaminants, consider that partitioning between the environment and the organism is considered to be a single homogenous unit. Empirical data on the relationship of bioconcentration factors (K_b) to octanol–water partition coefficients (K_{ow}), however, often deviate from the predicted values derived from single-compartment models, suggesting the interaction of several factors in determining bioconcentration with the potential for major seasonal differences to occur. Seasonal variation in bioconcentration may be related to dependency on specific metabolic processes involved in the storage and mobilization of lipid reserves *during reproductive events.... Simple lipid (K_{ow}) partitioning phenomena cannot account for some of the second-order fluctuations apparent in contaminant distribution and models to describe selective partitioning within specific tissues are needed*[17].

Physiologically based toxicokinetic (PB-TK) models are used to describe the kinetics of uptake and depuration of hydrophobic contaminants by an organism. The great promise of PB-TK models lies in their potential to link contaminant concentrations in specific tissues with toxic effects in those tissues. We will discuss these models only in relation to the premise that lipid class significantly affects the kinetics of lipophilic chemical uptake and release. For in-depth descriptions of PB-TK models in fish, several excellent reviews and recent articles are available[36,75,84,85].

PB-TK models of hydrophobic chemicals are based on the premise that lipophilic compounds partition between homogenous compartments. In dietary exposure models, chemical distributions among tissues are based on chemical partitioning between blood and tissues, blood and feces, and blood and gut. Tissue concentrations of lipophilic contaminants are measured at steady state when an equilibrium between the environment and the organism with respect to the chemical under study is assumed to have been reached. To simplify the measurements, adjustments are generally not made for variations in relative blood volume or organ size between individual organisms, although recent studies have addressed this issue in fish[69]. Similarly, differences in the lipid composition of various tissues are generally not assessed, lipids being considered a single, homogenous unit for the purpose of simplicity, although it is recognized that lipid composition likely influences contaminant distribution[6,85].

A best-fit model is the one which best reconciles observed assimilation efficiencies with measured chemical concentrations in the tissues. However, contaminant tissue distributions are not always explained on the basis of simple partitioning. As described above, partition coefficients (K_{ow}) are most often derived from experiments using the solvent, octanol, as a surrogate representing all lipids[25]. This approach assumes that lipids are a homogenous class of biomolecules[21] and has led some chemists to consider fish merely as 'bags of lipids' into which chemicals partition based solely on their K_{ow} values. There are several problems with this approach. First, as discussed above, there are large inconsistencies among K_{ow} estimates, which can differ by 2–4 orders of magnitude for a single compound[30]. Second, for very large and/or superhydrophobic chemicals ($K_{ow} > 6.0$)[25] solubilities in octanol and biological lipid are not linearly related[21], making K_{ow} a poor predictor of the partitioning behavior of such compounds. Third, lipids as a group are not homogenous. Some lipids are strongly polar, such as membrane phospholipids, while others are generally classified as neutral, including wax esters, cholesterol, and the ubiquitous storage lipids, the triacylglycerols (TAGs, also commonly referred to as triglycerides). Evidence is accumulating that POPs partition differently among lipid classes and herein lies the foundation for this chapter.

The relative polarity of a POP appears to determine whether it partitions preferentially into polar or non-polar lipid classes[62,95] and may explain, at least in part, deviations from predicted values based on single, homogenous-compartment models.

Tissue concentrations of highly lipophilic, non-polar chemicals, such as PCBs, are more tightly correlated with neutral lipids than with total lipids. Since the neutral storage lipids TAGs can comprise as much as 83% of the total lipids in a given fish tissue[60], strong seasonal and developmental changes in the tissue distribution of TAGs may be a better predictor than total lipid of the tissue partitioning of highly hydrophobic contaminants. Further, lipids are chemically diverse even within a given class, with the length and composition of the fatty acid moieties altering the physicochemical properties of a particular lipid. Models describing selective partitioning within specific tissues have been called for[17], and discussed[6], but none have yet been developed.

A recent study examining the kinetics of an extremely hydrophobic chemical, 2,3,7,8-tetrachlorodibenzo-*p*-dioxin (TCDD), in a fish model provides support for the position that lipid composition, more than total lipids, governs tissue distributions of hydrophobic chemicals. In optimizing their PB-TK model for TCDD kinetics in brook trout, Nichols *et al.*[85] found that rather than explaining concentration differences between tissues, normalizing TCDD to total lipid content (pg TCDD/g total lipid) produced a fourfold variability in TCDD concentration among tissues. These authors speculated that an internal equilibrium condition might exist with respect to TCDD. They used two different models in an attempt to explain these tissue distributions. A diffusion-limited model adequately explained TCDD distribution to adipose tissue, but for non-adipose tissues, even a flow-limited model only fairly approximated TCDD distributions. These models provide insights into how lipid composition may influence tissue distribution patterns of chemicals and are discussed below.

Even though a diffusion-limited model appeared to adequately explain TCDD partitioning into adipose tissue, the physiological mechanisms underlying diffusion limitations of TCDD in brook trout gut are not known. Thus, the authors speculated that complexation with dissolved organic compounds or with organic materials in food (for dietary exposures), limits the bioavailability, and hence assimilation, of highly hydrophobic compounds. Further, because K_{ow} values describe relatively weak (i.e. rapidly reversible) associations between octanol and a hydrophobic chemical, they are likely to be poor predictors of the assimilation efficiency of dietary hydrophobic compounds more tightly associated with tissue lipids. Nichols *et al.*[85] postulate that TCDD distributions between blood and fat may be governed by the perfusion rate of adipose tissue, a rate which is poorly known and for which they suggest their estimate may be inaccurate. Taken together, TCDD levels in fat appear to be largely explained by a diffusion-limited model, with the possibility that blood perfusion rates of adipose also play a role. Since fish adipose tissue is comprised almost entirely of neutral lipids (99%)[6] and TCDD as a non-polar organic compound should preferentially partition into neutral lipids[62,95], a model which normalizes TCDD to total lipids should be a good predictor of TCDD concentrations in adipose tissue, which it was.

Unlike adipose tissues discussed above, however, normalizing TCDD levels in non-adipose tissues to total lipids did not eliminate tissue differences in TCDD concentration, differences which ranged up to fourfold[85]. In an attempt to optimize the model for non-adipose tissues, Nichols *et al.*[85] determined that a blood-flow limited model, rather than a diffusion-limited model, improved predictions of tissue concentrations of TCDD in brook trout liver, ovary, and muscle. Despite this improvement, normalization of TCDD levels to total lipid, even corrected for tissue blood-flow rate differences, still left differences of 0.5–1.5 in blood/tissue ratios of TCDD. However, normalizing tissue TCDD concentrations to the amount of neutral lipid, rather than total lipid, in non-adipose tissues further reduced concentration differences among tissues, providing further evidence that tissue lipid composition is one of, if not the, major factor determining tissue distribution patterns of highly hydrophobic compounds. It should be noted that this is most appropriate for compounds that are not readily metabolized, such as TCDD. For compounds that are readily metabolized by fish, such as the model PAH, benzo(*a*)pyrene, the parent compound is metabolized into more water soluble products within hours of exposure or ingestion. Thus, it is not only parent, but also parent and metabolite compounds, that distribute among lipid and non-lipid/non-water compartments (see ref. 6) according to their relative polarity and lipophilicity. Indeed, for field caught fish, the most reliable indicator of recent exposure to PAHs (compounds which are readily metabolized by fish) is not tissue concentrations but rather the presence of fluorescent aromatic metabolites in the bile[23,24,65,117].

Although as mentioned above, no PB-TK models of chemical partitioning into specific lipid classes have yet been developed that we are aware of, one study has examined the partitioning of hydrophobic compounds among lipid classes extracted from different fish tissues. Six chemicals with moderate hydrophobicity ($K_{ow} = 1.46–4.04$) were partitioned *in vitro* against two lipid classes (polar and non-polar) extracted from seven tissues (blood, liver, kidney, muscle, skin, fat, and viscera) from four fish species (fathead minnows, rainbow trout, channel catfish, and Japanese medaka)[6]. These authors used log-linear equations that incorporate tissue/water partitioning coefficients to take into account both lipid and non-lipid/non-water fractions of the tissue. These authors note that log-linear equations can be derived that take into consideration relative tissue levels of lipid classes with different solubilities for the test chemical. However, the authors feel that this approach is of limited usefulness for PB-TK modeling due to the difficulty of distinguishing the relative contributions of the different lipid classes within a given tissue[6]. Nonetheless, while PB-TK models based on lipid class may generate equations too complex to be practical, the information they provide can still be useful. As discussed above, by using information on tissue lipid composition, Nichols *et al.* found that normalizing to neutral lipid rather than total lipid better described tissue distributions of TCDD in brook trout[85].

V. The role of physiology in POP tissue distribution

Bioconcentration of lipophilic organic contaminants may be influenced by physicochemical properties, such as molecular configuration or steric properties that influence biotransformation and membrane transfer kinetics, and biological factors, such as the partitioning between storage lipids and structural lipids[17].

We begin this section considering why tissue lipid composition, as opposed to total lipids, may be the overlooked linkage between exposure and accumulation of contaminant at the target site. We explore several aspects of lipid biochemistry and POP accumulation, including the concept that tissue differences in POP distributions often closely reflect differences in the lipid composition of tissues, rather than total lipids. We then consider how this concept does, or does not, apply to changes in POP distributions with alterations in the lipid composition of tissues during physiological change, including development, starvation, smoltification, and spawning. For example, changes in lipid composition among tissues as fish develop from embryos to adults may result in changes in POP tissue distributions at different developmental stages. As POP concentrations shift among tissues during development, relative concentrations at target sites may vary, perhaps being high in the liver of juvenile fish, but shifting to the developing gonads during spawning. Hence, liver function may be altered in juveniles, while genetic changes may occur in gametes produced by adults. This section concludes with an examination of how specific biochemical changes, such as the induction of xenobiotic metabolizing enzymes, may affect lipid integrity and POP tissue distributions, and hence effects.

Evidence is accumulating that demonstrates the partitioning of hydrophobic chemicals among and within matrices is governed not simply by the relative level of total lipid in these matrices (as described under section II), but more specifically by the relative proportion of the different lipid classes therein. In this section we will discuss the physicochemical distinctions among and within different lipid classes and provide evidence supporting partitioning based on the relative polarity of chemicals and lipid classes.

1. Effect of lipid class on chemical partitioning

1.1. Lipid classes

Lipids fall into two main groups, polar and non-polar. Polar lipids, such as phospholipids and sphingolipids, orient their non-polar tails inward and their polar heads outward to form the familiar bilayer structure of plasma membranes and lipid micelles. Phosphatidylcholine makes up a significant proportion of membrane lipids. Non-polar lipids generally lack this amphipathic characteristic and are typically found in fat vacuoles in adipose tissue, functioning as storage lipids. Triacylglycerols (TAGs) are the most abundant member of this group comprising the majority of non-polar (i.e. neutral) tissue lipids. TAGs are the main energy reserves, or 'depot lipids'. While adipose is the main storage tissue for TAGs in mammals, several organs serve as TAG storage depots in fish, primarily liver, muscle, and mesenteric fat[102]. In addition to their role as energy reserves, TAGs also function in buoyancy control and as thermal insulators.

In general, studies of fish lipid dynamics indicate that dramatic changes in total lipid are mainly accounted for by changes in non-polar (mainly TAGs) rather than polar lipids (e.g. phospholipids)[12]. The much smaller changes in tissue phospholipid levels may reflect the integral role of these lipids in membrane structure, and thus the necessity to conserve phospholipid concentrations within a relatively narrow range. In contrast, much larger changes are observed in tissue levels of TAGs, the major energy storage macromolecule in fish and the one most readily mobilized under conditions of stress (e.g. starvation)[59,102].

1.2. Contaminant partitioning among lipid classes

1.2.1. Bivalves. Recent evidence indicates that accumulation and partitioning of hydrophobic organic contaminants are strongly affected by lipid class. This contrasts with the vast majority of bioaccumulation studies published to date that normalize tissue contaminant concentrations to total lipid content. Such normalization is based on the simplifying assumption that lipid pools are uniform with relation to chemical lipid solubility[25]. Differential chemical partitioning is based on the concept that chemicals will preferentially partition into lipid classes of like polarity[4,33,95,116]. Studies of PCB congener partitioning in bivalves report that the more polar tri- and tetrachlorinated biphenyl congeners show greater affinity for polar lipids, while those with higher K_{ow} values (penta- and hexachlorinated biphenyls) preferentially accumulate in non-polar lipid pools[62,95]. Bergen *et al.*[4] measured eight lipid components, four polar (membrane) lipids and four non-polar (storage) lipids, in relation to total PCB partitioning. They found total PCB levels in ribbed mussels correlated with neutral lipids, particularly the dominant TAGs but did not correlate with total lipids. From this, they concluded that for total PCBs, normalization to neutral lipids is more appropriate than normalization to total lipids. Others report even further fractionation of some chemicals. The partitioning of chlorobenzene isomers among lipids within a given class was found to be influenced by differences in the fatty acid composition of those lipids[116].

1.2.2. Fish. Studies in fish support the hypothesis that lipophilic contaminants partition into lipid classes based on relative polarity. In flatfish, the more polar, less chlorinated PCB congeners (tri- and tetrachlorinated biphenyls) preferentially partitioned into tissues dominated by polar lipids (kidney and gallbladder)[62]. Less polar, more chlorinated congeners (penta- and hexachlorinated biphenyls, $K_{ow} > 5.7$) accumulated preferentially in tissues with high concentrations of neutral lipids (gill, liver, ovary, stomach)[62]. *In vitro* studies of the partitioning of a tetrachlorinated biphenyl among lipids extracted from the muscle tissue of various fish species further suggest that species-specific differences in PCB tissue distributions are likely related to the relative amounts of polar and neutral lipids in their tissues[33].

In light of these findings on the differential lipid partitioning of POPs based on relative polarity, distributions in fish of other contaminants, such as PAHs, can be re-examined based on probable tissue lipid composition. For example, Neff *et al.*[83] reported tissue differences in rates of accumulation and release of naphthalenes (PAHs) in killifish (*Fundulus similis*). Gall bladder and brain, tissues with high proportions of polar lipids in fish[12,60,62], accumulated the highest concentrations of naphthalenes relative to gut > gill ~ muscle[83]. Similar results were observed in fish early life stages, with pink salmon fry accumulating naphthalene in gut > skeletal muscle > gill[94]. These studies suggest that naphthalene preferentially partitions into polar lipids. Comparison of these two studies could also be used to suggest that, because naphthalene tissue distributions were similar between adult killifish and early life stage salmon, tissue lipid composition may not change over the course of fish development. However, the strong species differences that exist in tissue lipid distribution (discussed below) preclude such conclusions.

The fatty acid chain length of dietary lipids greatly influences the solubility and hence systemic bioavailability of ingested PCBs in vertebrates, including fish. Solubilization of ^{14}C-3,4,3′,4′-tetrachlorobiphenyl (CB77) increased with increasing fatty acid chain length in *in vitro* studies with micelles[28]. This increase in solubilization was associated with increased uptake, as demonstrated in *in vivo* studies with micelles. Blood levels of CB77 were 2.2-fold higher in catfish fed CB77 encapsulated in micelles composed of the long chain length fatty acid, linoleic acid (18:2), relative to those fed a mixture containing both long and short chain fatty acids[28]. As pointed out by these authors, fatty acid chain length and degree of saturation can change seasonally in fish, depending on diet and temperature. Such seasonal changes in lipid composition may produce seasonal changes in POP tissue dynamics. In sum, such studies demonstrate that lipid compositional changes are relevant and important determinants of POP disposition.

In any study, it is important that researchers first establish whether or not their data demonstrate a relationship between POP tissue concentration and tissue lipid levels. This is seldom done, as it is typically assumed that such a relationship *must* exist for lipophilic contaminants. As is compellingly demonstrated by Hebert and Keenleyside[47] in their paper "To normalize or not to normalize? Fat is the question", such assumptions can lead to lipid normalized POP concentrations that are completely at odds with measured wet weight POP values. Further, since factors other than total lipid (such as differences in lipid class, for example) can affect POP levels in organisms, simple ratios (e.g. ng POP/ng lipid) are often inadequate and may actually increase data variability. In many cases, analysis of covariance (ANCOVA) may prove to be a more appropriate method for lipid normalization of POP concentrations[47].

In sum, these studies suggest that tissue concentrations of hydrophobic chemicals are more appropriately normalized to lipid *class* rather than to total lipids[4,95]. Taking this into consideration, the partitioning of hydrophobic pollutants among different lipid classes may help to explain some of the disparate findings amongst studies that normalize tissue contaminant levels to total lipid rather than to lipid class.

1.3. Important analytical considerations regarding lipid normalization

In addition to the problems associated with normalization of contaminants to total lipid discussed above, the problem of lipid normalization is further compounded by differences in the efficiency of various lipid extraction methods. In short, different lipid classes are extracted by different solvents. For example, bluefish liver yielded total lipid levels of 25% when extracted using chloroform/methanol but only 8% when acetonitrile was the extracting solvent[92]. This reflects differences in the lipid class extracted by each solvent: neutral lipids were preferentially extracted by non-polar solvents and more polar membrane lipids were extracted more efficiently by more polar solvents. Since tissues differ in their proportion of neutral/polar lipids, extracting all tissues with a non-polar solvent would yield artificially low lipid levels in tissues with high neutral/polar lipid ratios. Thus, differences in contaminant concentrations or tissue distribution patterns of a given contaminant among different studies may be explained, at least in part, by unwitting normalization to different lipid pools by researchers, depending on the solvents used for lipid extraction. Interestingly, lipid extraction is also strongly affected by another solvent, water. For this reason, the amount of water present in tissue samples must also be taken into account. Because most tissues generally contain about 80% water by weight, this may not be an issue in most situations. However, should samples differ substantially from one another in water content, the analyst should take this additional solvent into consideration. The reader is referred to the seminal work of Bligh and Dyer[9] to understand the importance of this consideration. Finally, it is important to note that lipids degrade in frozen tissues. Triglycerides and phospholipids are particularly susceptible to degradation during freezing[5], which could affect measurements of both total lipids and lipid classes. In sum, predictions and interpretations of contaminant concentration and distribution patterns amongst tissues should improve when contaminant levels are normalized to the appropriate lipid class using appropriate analytical methods. Recent papers comparing lipid extraction methodologies should be consulted when choosing lipid analysis techniques[92,93]. In this regard, the Iatroscan® thin layer chromatography/flame ionization detection method has received particular attention[1,89,101].

1.4. Are lipids protective?

Altered partitioning of contaminants between different lipid classes may affect contaminant toxic potency if the proportion of lipid classes in the total lipid pool of an organism changes over time. For example, if PCBs associate most strongly with storage lipids such as TAGs, as they do in mussels[4], will changes in tissue TAG levels affect tissue susceptibility to PCB toxicity? To our knowledge there are no studies that have addressed this.

Of the two main lipid classes, will the neutral *storage* lipids play a protective role by sequestering toxic POPs away from sensitive target sites, such as membrane lipids and macromolecules (proteins and nucleic acids)? The answer may depend on the mechanism of action of the specific pollutant. Non-polar narcotics, compounds which have no specific mechanism of action, are believed to exert their toxicity non-specifically by inserting into membranes and simply disrupting them[115]. For these toxicants, sequestration into neutral storage lipids is considered protective as it removes these chemicals from interaction with polar membrane lipids, considered the 'target lipids'. This principle is known as 'the survival of the fattest'[66]. This principle does not necessarily hold, however, for toxicants with specific mechanisms of action. This is demonstrated by comparing the body burden of the chemical needed to cause death for a specifically acting and a non-specifically acting toxicant. This lethal body burden (LBB), now commonly called the critical body residue (CBR), differs dramatically between toxicants acting through non-specific vs. specific mechanisms of action. When expressed on a molar concentration basis, the CBR is much higher for narcotics that act non-specifically (2–8 mmol/kg wet weight) than for toxicants with specific mechanisms of action (<0.01 mmol/kg)[73]. The CBR model is predictive only for narcotics that bind reversibly and act non-specifically[67]. For these compounds, sequestration into storage lipids, and thus away from membrane lipids, is protective[37,66]. However, for chemicals that interact irreversibly with specific target sites (e.g. toxicants that covalently bind an enzyme, such as some organophosphorus pesticides[67], or form DNA adducts, such as benzo(*a*)pyrene[70]), increasing storage lipids may have less effect on toxic potency. Some authors have argued that lipids are protective regardless of whether the compound acts via a specific mechanism or not[37]. However, strong species differences have been demonstrated among fish embryos for the toxic potency of TCDD[31], a chemical that exerts its toxicity through a specific mechanism of action[97]. Thus, while highly non-polar POPs may preferentially move into tissues with increased levels of TAGs, whether or not those TAGs are protective may depend largely on whether or not the toxicant acts through a specific mechanism (lipids could be less protective) or acts narcotically (lipids are more protective).

1.5. Lipid metabolism

The issue of whether or not lipids are protective becomes more complicated when one considers that toxicants can affect lipid metabolism. Capuzzo and Leavitt[16] observed increases in non-polar/polar lipid ratios in the tissues of PAH/PCB contaminated mussels and speculated that these contaminants mediate a decrease in the conversion of storage lipids to membrane lipids. Membrane-mediated solute transport, cell signaling, and bioenergetics are all significantly affected by membrane lipid composition[53], and alterations in the proportion of membrane lipids in a tissue could seriously affect membrane integrity and function.

That hydrophobic organic pollutants alter storage lipid metabolism has been demonstrated in both vertebrate and invertebrate organisms. Studies in rats demonstrate that hydrophobic organic contaminants disrupt lipid metabolism in somatic organs. Organochlorine exposure alters profiles of fatty acids, TAGs, phospholipids and plasma lipids in rats (references cited in ref. 28). The chlorinated insecticide toxaphene increased sterols and decreased TAG levels in hepatocytes of yellowtail flounder both *in vivo*[100] and *in vitro*[34]. Since altering membrane levels of sterols is one way ectotherms adjust to thermal stress, organochlorine-mediated alterations in sterols could deleteriously affect fish adaptation to seasonal temperature change[100]. Capuzzo *et al.*[18] found that exposure to petroleum hydrocarbons reduced TAG synthesis in larval lobsters, while PCB exposure dramatically lowered whole body TAG levels in oysters[71]. These studies suggest that similarities may exist between invertebrates and vertebrates in the effects of chlorinated and non-chlorinated toxicants on lipid disposition. Together these studies provide strong evidence that halogenated pollutants alter lipid profiles, which, in turn, could alter the lipid-mediated storage and transport of lipophilic contaminants.

There is evidence from both mammalian and fish studies demonstrating that exposure to organochlorines alters the uptake and tissue disposition of lipophilic toxicants. Dietary pre-treatment with ^{14}C-dieldrin modified the tissue distribution of subsequent doses of dieldrin or PAHs in rainbow trout[26,38,39]. This redistribution of toxicant, from carcass to liver, bile and mesenteric fat, appeared unrelated to xenobiotic metabolizing enzymes, membrane transport proteins, or total body lipid content, none of which were seen to change with treatment. These authors suggest pre-exposure to dieldrin may alter other proteins known to be responsive to lipophilic xenobiotics, such as sterol carrier proteins, plasma lipoproteins (HDL/LDL ratios) or inducible cytosolic lipoprotein complexes, and that these, in turn, may play a role in the observed tissue redistribution of toxicant. Similarly, Doi *et al.*[28] found that dietary pretreatment with CB77 reduced the overall systemic uptake of subsequent dietary doses of ^{14}C-CB77. This effect was not related to intestinal metabolism, which was minor and unaltered by CB77 treatment, and may instead be related to changes in intestinal lipid composition caused by the initial organochlorine exposure. Numerous and extensive studies in mammals also demonstrate altered lipid disposition with xenobiotic exposure, including the effects of organochlorines on lipid metabolism, such as the characteristic induction of 'fatty liver', hypercholesterolaemia, and differential effects on lipid levels in different life stages and with route of exposure[20,82,91,98].

Certain POPs can induce the synthesis of reactive oxygen species (ROS) that mediate lipid peroxidation. By attacking polyunsaturated fatty acids in membranes and producing lipid peroxyl radicals, ROS begin a chain reaction of lipid oxidations leading ultimately to changes in membrane function[88]. Generation of ROS is largely a result of cytochrome P450 monooxygenase-mediated metabolism[10]. In laboratory studies with fish, ROS formation is tightly correlated with induction of CYP1A1 by poorly metabolizable organochlorine inducers[99,121] as well as PAH-mediated photoinduced toxicity[22]. Such relationships have also been reported in field studies. Bioaccumulation of halogenated pesticides, including malathion, dieldrin, aldrin, endrin and methyl parathion, in environmentally exposed silversides correlated with increased lipid peroxidation and decreased acetylcholinesterase activity in gills[35]. Links between induction of CYP1A1, lipid peroxidation, and embryotoxicity are discussed later in this chapter (see section V.3.1.).

Thus, while the majority of these contaminants exert their toxicity through specific mechanisms of action, such as alteration of pollutant-metabolizing monooxygenases (discussed in section V.3), their additional effects on lipid metabolism likely contribute significantly to overall POP bioaccumulation, tissue distribution, and toxic potency.

1.6. Seasonality

There are few studies that have simultaneously addressed seasonal changes in lipid composition and POP levels of aquatic organisms. Capuzzo *et al.*[17] found seasonal distributions of PCB congeners in *Mytilus edulis* correlated with neutral and polar lipid distributions, with individual congeners differentially affected over the spawning season, suggesting differential partitioning of specific congeners in different tissues or different lipid pools. These authors concluded that relative concentrations of individual congeners are modified to some extent by seasonal changes in lipid content. Bergen *et al.*[4] found only a weak relationship between seasonal changes in lipid and tissue levels of PCBs in ribbed mussels, the strongest correlation occurring when tissue TAG and PCB levels were lowest. However, these authors hypothesize that stronger relationships between PCBs and TAGs are likely in species with higher fat content (in ribbed mussels, lipid comprises only 6% of whole body dry weight) such as some fish species. The relationship between TAGs, total lipid and POPs will be discussed in greater detail in section 2.

2. *Effects of physiological changes associated with specific life history stages*

2.1. *From spawning adult to the embryo: Where does the lipid go?*

Our goal is to explore and evaluate the influence of lipid physiology on the disposition of POPs. In our introduction, we explain why this is important for improving the status of linking exposure and effects. In this section, we further explore the influence of a changing lipid physiology on the disposition of POPs. Our interest here is in the fate and disposition of POPs as fish undergo dramatic physiological changes from embryo to spawning adult with concomitant changes in lipid composition. Clearly, one of the greatest difficulties in such an undertaking is trying to make generalizations about changes in lipid physiology of a fish given the enormous variation in life histories amongst species. These include tremendous differences in spawning strategies (from daily to seasonally to once a lifetime), wide ranges in the proportion of time spent as embryos or yolk sac larvae, and the process of smolting in anadromous fish, such as salmon.

Surely, such dramatic differences in life history must influence disposition of xenobiotic chemicals in fish, but what information do we have that this is, or is not, the case? Very little! Since our focus in this chapter is on the relationship between POPs and the main storage lipids (TAGs) we will consider how changes in TAGs during selected life history stages might influence POP distribution. It should be noted that not all studies report TAG concentrations, and in those cases we discuss 'total' lipid rather than TAGs. Although these cases do not directly strengthen our thesis (linking POPs to TAGs), we include them in this section because we believe it is important to consider total lipid distribution within a species (e.g. muscle, liver, gonad, brain). There can be large interspecies differences in lipid content and tissue distribution. Perhaps more studies in the future will provide information on lipid classes in addition to total lipid.

We begin this section with adult fish, noting great interspecies differences in lipid composition and distribution, and how these differences might affect POP distribution. This is followed by a more speculative discussion on how these differences in lipid composition (and therefore POP distribution) might be affected by reproductive strategies such as starvation during migration. Additionally, we will explore how these physiological differences might influence POP deposition into developing oocytes or spawned eggs, and how POP disposition is dependent on lipid content in both females and their developing oocytes. We will then explore how changes in lipid metabolism during embryo/larval development might affect the timing of POP toxicity in developing larvae. Finally, we will briefly address how dramatic changes in lipid composition during the transformation from parr to smolts might influence POP disposition in salmonids.

By integrating what is known about changes in tissue lipid composition and POP distribution over the lifecycle, we endeavor to provide scenarios relevant to environmental situations as well as enough comparative information to improve our understanding of the differences and similarities in POP disposition among species.

2.2. Adult fish: Seasonal changes

It is well known that in addition to great differences in total lipids and lipid distribution amongst species, there are dramatic interspecies differences in life history strategies. Consider, for instance, the many species that fast during spawning migrations, or other differences in reproductive strategies, from semelparous species (spawning just once in a lifetime) to serial spawners with continuous oocyte development over periods ranging from days to months. It is likely that species with such different spawning strategies will have dramatically different tissue lipid profiles at any given time. To understand how this might influence the susceptibility of offspring and parents to POPs, we need to explore changes that occur during the maturation process. Since there are very few data on the relationship between reproductive strategies, lipid distribution and POPs, it must be kept in mind that much of this discussion is speculative. Our main intention in this section is to spark discussion and further thought on the subject.

2.3. Lean fish–fat fish

Relative levels of lipid in muscle tissue underlie dramatic species differences in POP tissue distribution in fish. Adult fish are often categorized as lean or fatty fish based on the lipid content in the fillet or edible muscle. A lean species is one whose muscle tissue generally contains less than 5% lipid (e.g. cod, halibut) while fatty fish are those whose muscle tissue contains 5% or greater lipid (e.g. striped bass, sturgeon). Interspecies differences in muscle lipid can be quite large, ranging from as low as 0.7% lipid to up to 22% lipid (as reported in eel[52]). It is important to realize that characterizations of 'lean' and 'fatty' are for dietary purposes only and can be quite misleading. For example, a lean species such as cod, can contain large amounts of lipid in other tissues such as the liver (the lipid content of different tissues in several species of freshwater fish is reviewed in ref. 52). Total body lipid (all tissues combined) can strongly influence POP deposition into both fatty and lean tissues as will be discussed in section V.2.4. Generally, lipid-rich storage tissues (such as liver, muscle, or mesenteric fat) contain TAGs as the principal lipid, while leaner tissues contain mainly phospholipids[52]. For example, 'fatty fish' contain relatively large amounts of TAGs in their muscle tissues (which serve as storage tissues for those species) compared to lean species, which may have low TAG levels in muscle but high TAG levels in liver (e.g. cod).

How might such interspecies differences in tissue lipid composition play out in terms of POP disposition in lean vs. fatty fish? Examination of POP tissue distributions among species demonstrates that it is inappropriate to compare POP concentrations among species by using a single tissue (e.g. muscle). Hellou *et al.*[50] found that while POP content correlated well with relative total lipid content in liver, gonad, and muscle among four species of wild-caught fish, tissue lipid distributions differed dramatically among these species. For example, in cod, a lean fish, tissue concentrations of most POPs were highest in liver, but below method detection limits in gonads and muscle. In contrast, in turbot and American plaice, whose muscle lipid content exceeds that of cod muscle, liver and muscle POP concentrations were comparable and POP levels in gonad were near detection limits. Comparing across species, relative tissue lipid levels almost appear random, suggesting that the lipid content of a given tissue in one species is not predictive of the relative lipid content of that tissue in other species. While this is not new information, it is very important to bear in mind, especially when measuring POPs. It seems that every so often there are new 'recommendations' of which tissues should be measured to most accurately predict POP effects in fish. Considering these large differences in lipids amongst tissues and that these can vary dramatically among species, selecting one tissue or another to make interspecies comparisons of POP levels could produce very different conclusions as to which species is most 'at risk'. Further, species differences in tissue POP distribution among wild-caught fish can be difficult to interpret because POP accumulation depends not only on lipid composition, but also on age, diet, and spawning status, parameters that can vary widely among field-collected animals.

Because so many factors in addition to lipid composition can affect POP accumulation (discussed above), some researchers currently favor using whole body lipid and POP measurements to estimate risk of effects from POP exposure. This is essentially a lipid-adjusted CBR. A recent example of this approach is shown in detail by Meador *et al.*[122]. These authors used a number of studies from the literature on appropriate life stages of salmon to determine a lipid-adjusted threshold for effects for a wide range of endpoints after PCB exposure. This threshold number was then compared to levels of PCBs measured in wild juvenile salmon from a contaminated estuary, and many of the fish, on the lipid-adjusted basis, were determined to be at risk of effects from environmental exposure to PCBs.

Laboratory studies suggest that whole body lipid may be a strong predictor of whether fatty or lean fish might be at greater 'risk' of POP exposure to specific target organs. Ingebrigsten and colleagues[56,58] explored interspecies differences in susceptible target tissues by evaluating tissue-specific accumulation of radiolabeled organochlorine isomers in a lean species, cod, and a fatty species, rainbow trout. Their findings indicate that given equal doses, cod brain accumulated 6–8 times more radioactivity than trout brain. The authors suggest that since there is no significant difference in the lipid content of the central nervous systems of cod and trout, differences in organochlorine disposition could be attributed to overall differences in body fat between these species[55,58]. In other words, under the same exposure conditions, the lean fish used in this study will accumulate more chemical in their brain relative to fat fish because fat fish have more lipid-rich depots for the chemical to distribute amongst than do lean fish. Similarly, studies with fed ('fatty') and starved ('lean') Arctic charr dosed with octachlorostyrene (OCS) found OCS concentrations were almost fourfold higher in the brains of starved charr relative to fed charr[60]. In that study, brain was the only organ with such a dramatic difference in OCS concentration between the lean and fatty fish, suggesting preferential distribution into the brain. It should be noted, however, that in both lean and fatty charr, brain concentrations of

OCS were lower than in any other tissue measured, including muscle, liver, kidney, gonad, and skin. Follow-up studies by this group using PCB-exposed charr yielded similar results, with a net loss of PCB to carcass and a net gain in liver and brain (up to 10-fold increase in brain PCB) with starvation[61]. These findings have important implications for predicting differences in tissue and target-specific POP toxicity among species.

If TAG is a major factor in POP distribution, as demonstrated for prominent lipophilic organochlorines[4,60,62], any differences in POP concentration between species or among tissues within a species should depend on TAG concentrations as well as total lipid. Unfortunately, there are very few species for which we have all the necessary data (e.g. total body fat and proportion of lipid classes among specific tissues[102]) to test this idea. Nevertheless, studies reporting POP tissue distributions on a total lipid basis can be re-examined based on what is known of relative TAG tissue levels in some species. For example, when exposed to equal amounts of 2,5,2′,5′-tetrachlorobiphenyl (TCB), differences in tissue TCB distributions between yellow perch (a lean species) and rainbow trout (a fatty species), were thought to likely reflect differences in storage lipid (i.e. TAG) levels between these species[41]. These authors also reported that whole body TCB elimination rates were similar between these species, despite the fact that trout have a higher whole body lipid content than yellow perch, suggesting whole body lipid content is not a main factor regulating elimination rates. Though they did not measure all potential target tissues (they did note, however, that neither species was undergoing gonadal development), the data suggest that TCB was distributed based on whole body lipid content so that potential target tissues with equivalent lipid content (e.g. brain) might have higher levels of TCB in one species than the other. A study of PCB-exposed trout also suggests lipid composition may play a role in PCB distribution. In rainbow trout exposed to the PCB mixture A1254 through the diet[68], PCB concentrations on a total lipid weight basis were similar across most tissues (adipose, gill, muscle, stomach) except for the liver which had almost 50% less PCB in lipid than the other organs. The major storage tissue in this species appears to be visceral adipose (with 92.8% lipid compared to 3.5% in liver). We might speculate that if liver had a lower TAG to phospholipid ratio than other tissues, this difference in distribution might be explained by lipid composition. Clearly, predicting interspecies or within species differences in POP tissue distributions is complicated by total lipid and TAG differences amongst tissues and further work in this area is needed.

Lipid, and hence POP, distribution among tissues is also influenced by many dynamic processes, including gonadal maturation and spawning. As discussed in the opening paragraph of this section, fish species not only have different tissue lipid compositions but also very different life histories. Along with relative differences in tissue lipid composition, species may differ in their spawning strategies, ranging from differences in the number of spawns over a lifetime to pre-spawning or spawning behaviors such as fasting or reduced food consumption during spawning migrations (if they migrate at all). Additionally there are large interspecies differences in the lipid composition of eggs. At this point, we should note that these are just some of the potential ‘drivers’ of lipid composition and dynamics; there are many others, including dietary and environmental factors that are beyond the scope of this chapter. Unfortunately, there are very few studies that address these changes from the perspective of lipid composition and POP distribution. Of those that we mentioned, we were able to find sufficient information on the effects of fasting and egg production on lipid composition and POP deposition. Based on these, we discuss below some effects of fasting and egg production on lipid composition and POP deposition.

2.4. Starvation and spawning: Redistribution and use of body lipid

There is evidence that fish undergoing lipid depletion might be at greater risk for contaminant deposition into sensitive target tissues, rather than into storage tissues. Many fish fast during spawning. Anadromous fish, in particular, can undergo drastic changes in lipid composition at this time. Sockeye salmon, for example, may lose 70–80% of their total body lipid during spawning runs. Females lose more body lipid than males, with 8% of depleted body lipid being redeposited into the ovaries compared to only 0.3% into the testes[52]. Seasonal swings related to spawning have also been reported in TAG levels. In Arctic charr, the lowest tissue TAG concentrations occurred in post-spawning, post-fasting fish compared to fish measured just prior to the spawning season[59]. Although not completely unexpected, the teleost brain does not lose weight or lipid during seasonal periods of starvation[52]. Based on reports that brains of lean fish accumulate more POP than brains of fatty fish[56–58], one might predict that as lipid levels drop in other tissues during starvation, POPs are likely to redistribute to brain. This theory is supported by studies with fed and starved sole exposed to PCB[11,12]. While TAG levels in sole changed in several tissues in response to changes in diet, there was little change in brain TAG levels and PCB concentrations in brain increased upon starvation (although on a whole body basis, brain PCB concentrations were still among the lowest of any tissue). After several studies of lipid loss and PCB disposition, these authors concluded that "PCB contents of different organs always resulted from dynamic exchange equilibria. Thus, PCBs were never immobilized in certain organs". Although this seems to state the obvious, it is worth stressing, as it can be quite important, especially for species that fast during migration. Energy metabolism in the brains of fish is strongly affected by both starvation and toxicants, including pesticides, which alter brain utilization of glucose, the main energy source for neural tissue[104]. Taken together, redistribution of POPs into the central nervous system during seasonal starvation (depicted in Fig. 1) could potentially result in neurological damage and behavioral abnormalities that could disrupt migration and/or reproduction.

The importance of relative TAG concentration and total lipid in the distribution of POP amongst tissues is further demonstrated in the Jorgensen *et al.*[60] study in which 'lean' and 'fat' Arctic charr were produced by two different dietary regiments. This study reports distribution differences in a relatively artificial situation (fish were either fed or starved prior to POP exposure), so that it addresses neither active starvation nor interspecies differences since the 'lean' and 'fatty' fish were produced from the same species. Nevertheless, this study still provides insight into the relationship between POP distribution and tissue lipid composition. The total amount of OCS recovered at

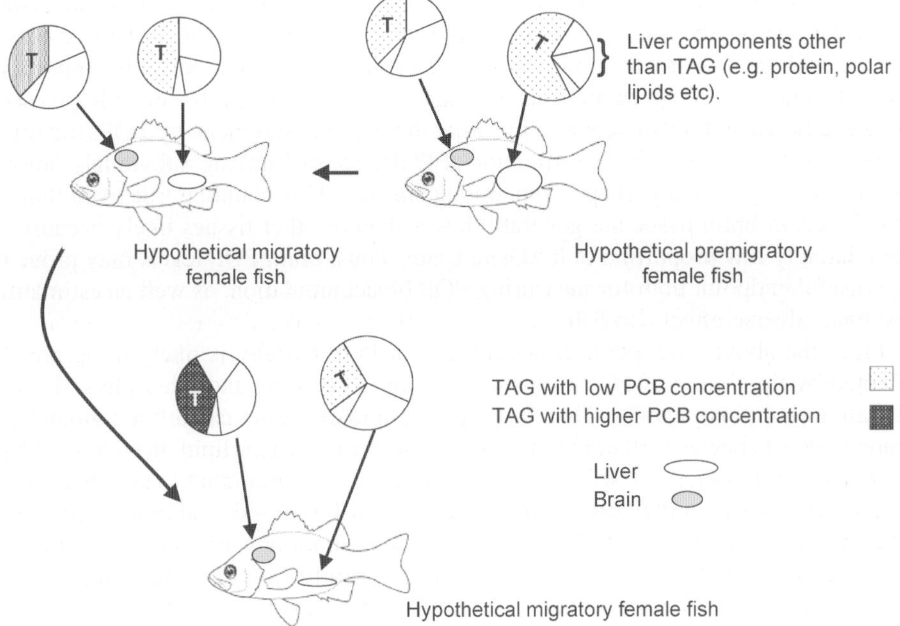

Fig. 1. The hypothetical redistribution of PCBs from liver to brain during a seasonal starvation cycle in fish.

the end of the study was higher in fat charr than in lean charr, indicating that fat fish absorb and retain OCS to a greater extent than lean fish. This is in contrast to the findings of Guiney and Peterson[41] in perch and trout where both species demonstrated similar retention of the administered PCB dose (although in their study, the lean perch were not 'starving'. Rather, they were working with two different species, which might be an important distinction). Most interestingly, OCS accumulation in all organs was relatively equal in both fat and lean charr, except for the brain. The authors report approximately fourfold greater OCS concentrations (on an organ weight basis) in the brain of lean charr compared to fat charr. This finding is particularly interesting since TAG levels were similar in the brains of both fed and starved charr, reflecting the fact that the teleost brain tends to retain lipid during starvation[52]. This was the only organ with such a dramatic difference in concentration between lean and fatty fish, suggesting some kind of preferential distribution into the brain, and is reminiscent of the interspecies differences in POP distribution reported by Ingebrigsten and Solbakken[58] and Ingebrigsten *et al.*[55]. It should be noted, however, that in both groups of fish (lean and fatty), brain concentrations of OCS were lower than in any other tissue measured, including muscle, liver, kidney gonad, and skin. Additionally, Jorgensen *et al.*[60] reported a positive correlation between OCS and TAG in both lean and fat fish and a negative relationship between OCS and phospholipids. They conclude that species undergoing lipid depletion as a normal stage of the life history (such as many of the anadromous salmonids) might be at greater risk for contaminant-associated stress. Particularly, if we consider the potential for many POPs to adversely affect the neuroendocrine system, including induction of apoptosis in teleost neural cells[29], mobilization to the brain during these periods could be quite important. Further, one could speculate that because brain lipid levels are likely more homeostatic than lipid levels in any other tissue, measurement of POPs in brain could provide the best relative estimation of POP exposure among individuals, among life history stages, and perhaps even among species. This could be true even though POP levels in brain tissue are generally lower than in other tissues likely because of the relatively low proportion of TAGs in brain. Thus, brain POP levels may prove to be a useful endpoint both for measuring POP bioaccumulation, as well as estimating resultant adverse effects, in fish.

From the above discussion, it is evident that POP toxicity is likely to be greatly affected by the timing of POP exposure relative to seasonal and other physiological alterations in lipid dynamics. That the timing of lipid dynamics can differ dramatically among teleost species further emphasizes the need to consider lipid disposition when evaluating POP toxicity in wild caught fish. Finally, it is important to note that much of our information on POP distribution in relation to TAGs and total lipid comes from laboratory studies. There are differences between laboratory-induced starvation and natural starvation, and between fish which are naturally lean and which are starved. Comparisons between such studies can only be taken so far.

At this point, it might be tempting to suggest that migrational starvation could lead to increased concentrations of POP in some tissues (such as brain) of pre-spawning or spawning fish. However, during this process, particularly in females, not all lipid is metabolized for energy and a significant proportion is transported to developing oocytes. Thus, rather than simply redistributing to the brain and other tissues, POPs may redistribute to the developing oocytes. However, even this topic is quite complex! As will be discussed below, tissue redistribution of POP is dependent upon both the lipid content of the developing oocyte and the mothers' total lipid content.

2.5. Mom and POP: How might lipid dynamics affect inheritance of POPs?

We shall now explore the dynamics between lipid composition of the mature female and her developing oocytes, with a focus on POP distribution into the developing oocytes. Are the offspring of one species at greater risk of exposure to POP compared to another's because of their mothers' total body lipid?

Lipid redistribution can be quite dramatic in sexually maturing fish. Gonads in immature Arctic charr may account for just 1% of total body mass, but comprise upwards of 15% of total body mass in mature females and 2.5% of total body mass in males. Changes in lipid composition also occur during gonadal development in this species, with TAGs and phospholipids representing up to 60 and 12% of gonadal lipid content, respectively[59]. At the same time when TAGs and total lipid levels are increasing in the ovaries, maturing females can lose over 30% of their body lipid, partly as a result of fasting.

How might such dramatic changes in lipid distribution during female sexual maturation affect POP distribution into the developing oocytes? We might speculate that fish undergoing long periods of 'fasting' during spawning would experience a greater movement of POP to the oocytes due to starvation-induced increases in POP concentration in lipid (or TAG) containing tissues with subsequent transfer of lipid into oocytes. Unfortunately, there are few data available on this dynamic process and studies comparing POP transfer to oocytes in species that fast during spawning with those that do not could be quite informative.

There is strong evidence, however, that tissue distribution (or redistribution) of POP is influenced by percent lipid in the mother and the total lipid deposited into the eggs, and is not driven purely by the amount of lipid transferred into the eggs[86]. A study in yellow perch, a lean species, demonstrated that maturing females transferred up to 50% of their initial PCB body burden to their developing oocytes[118]. In contrast, rainbow trout, a fatty species, transferred only 5–10% of its body burden of a tetrachlorinated PCB to its eggs prior to spawning[40]. Similar findings were reported by Niimi[86] in a study comparing POP distributions in gravid females and their eggs in five species of freshwater fish. Niimi found that in both rainbow trout (12% whole body lipid by weight) and white suckers (6% whole body lipid), concentrations of POP were 10% higher in the whole body of the female (minus eggs) than in their eggs (9 and 6% egg lipid, respectively). In contrast, POP concentrations in smallmouth bass (6% whole body lipid) were 5% lower in females than in their eggs (13% egg lipid), while white bass and yellow perch showed no difference in POP concentrations between whole body (8 and 5% lipid, respectively) and eggs (9 and 6% lipid, respectively).

In addition to species differences in lipid dynamics, POP disposition and potential toxicity to offspring is also likely affected by species differences in spawning strategy. All of the species discussed above have a total body lipid content ranging from approximately 4–15% lipid depending on the species[86], and reproductively mature adults of these species spawn at least once per year. Eels, in contrast, are semelparous, spawn only once in a lifetime, and have very high total body lipid levels (ranging upwards of 22% or higher)[52], with eggs composed of up to 36% lipid[63]. Maturing females fast during migration losing large amounts of total lipid. How might such characteristics affect POP disposition? Hodson *et al.*[54] measured POP concentrations in the maturing gonad and carcass (muscle, skeleton, and skin combined) of migrating eels. POP concentrations were 17–28% higher in gonad than in carcass. This perhaps is a reflection of the high lipid content in the gonads relative to the carcass, which was likely losing lipid as a result of starvation during the spawning migration. It was noted that these gonads were only partially developed at the time of capture and that more mature gonads (after a longer period of maternal fasting and concomitant lipid distribution) might contain even greater proportions of POP. The authors point out that the eel spawns only once in a lifetime, so that there is only one opportunity for such a major redistribution of POP to occur in the eel. This implies that, aside from greater risk of embryonic POP exposure due to the potentially high lipid content of eel eggs relative to female whole body lipid (we do not have data on maternal lipid loss by the time females reach the spawning grounds), embryos may also be at greater risk because the 'opportunity' to deposit POP into the eggs occurs only once in a lifetime. In other words, there will be no subsequent generation with lower embryonic POP concentrations and thus the first, and only, generation may be a highly contaminated one. In comparison to fish that spawn once a year or even more often (such as serial spawners), semelparous species, like the eel, may be at greater risk for experiencing deleterious generational effects of transferred toxicants than species that spawn more frequently.

In short, it is not enough to consider percent lipid in eggs as a potential predictor of POP exposure, but one must also consider the amount of lipid in the eggs relative to lipid concentrations in their mother during the time of oocyte development. Additionally, one must consider whether or not the mother undergoes major physiological changes during maturation (such as fasting and tissue reductions in TAG) or whether she feeds continuously. Finally, there are considerations of spawning strategies (from semelparous to serial spawners). Clearly, the process of maturation and spawning is quite dynamic, particularly in terms of lipid and POP redistribution, and these dynamics are relevant to both maturing females and their eggs. We will next consider the composition of eggs in terms of lipids and POPs.

2.6. Lipid composition of eggs

How might species differences in egg lipid content and the timing of yolk consumption affect POP deposition into developing embryos or larval fish? Before addressing this issue, a brief overview of egg biochemistry and yolk dynamics in fish embryos and larvae is warranted.

The biochemical composition of fish eggs varies dramatically among species. Given the strong influence of both maternal lipids and egg lipids on POP dynamics discussed above, differences in egg composition may help to explain differences in the sensitivity of early life stages to POPs. Fish eggs are composed of protein, lipid, carbohydrate, and ash with relative levels of these components affected by egg size itself, and by age and/or nutritional status of the mother (reviewed in refs. 63,112). In general, protein is the dominant component of dry matter in fish eggs (35–89%, with most eggs being comprised of approximately 55–75% protein)[63,112]. Lipid percentages of dry matter can range from 3 to 54%. Egg yolk is made up of both proteins and lipid (phospholipoproteins), derived from vitellogenin, and provides the main energy source for the developing embryo. In terms of protein and lipid composition, eel eggs fall at one end of the spectrum, having among the lowest protein and highest lipid content of any fish eggs, and cod eggs at the other, containing amongst the highest protein and lowest lipid content of any species. Generally, marine fish eggs tend to have lower percent lipid and lower caloric value than freshwater fish eggs (excluding anadromous and catadromous fishes). Of the different egg components, lipid content is the most labile, and can vary widely both among populations of the same species and within populations. Such great differences in lipid content may perhaps result from an adaptation to various reproductive strategies and/or duration of endogenous nutrition period (reviewed in ref. 51). Some eggs also contain oil droplets (up to 50% of total egg weight in some species) although, as will be discussed below, the oil droplet may or may not be used in a similar manner as yolk lipids during embryo and larval development. Yolk lipids consist of varying amounts of phospholipids, triglycerides, and cholesterol.

In general, eggs with lower total lipid content tend to have relatively high levels of polar lipids (mostly in the form of phospholipids). Eggs with higher lipid content tend to have increased levels of non-polar (TAG) lipids, and/or store lipids in the form of oil (composed of TAGs or wax and steryl esters[51,78]), rather than egg yolk. As would be expected, the amount of egg lipid appears to correlate with timing between spawn and hatch or first feeding so that longer incubation periods tend to have higher egg lipid levels[52]. As the amount and quality of yolk are crucial for healthy development of the embryo, significant deposition of POP into yolk can directly affect embryo development. Given the numerous differences among species in egg composition, and the importance of egg composition on POP distribution between mother and offspring, it would appear that some consideration of egg composition is important when making cross-species comparisons of potential POP exposure during early stages of development. However, we were unable to find many studies relating egg lipid composition to differential POP exposure in developing embryos, suggesting that this is another area for future studies.

Once eggs are fertilized and the embryos begin to develop, carbohydrates, lipids (TAGs and phospholipid) and proteins tend to be consumed prior to hatch, with lipid and protein catabolism dominating after hatch[51]. Yolk absorbed through the periblast is degraded to lower molecular weight compounds and released into blood, providing materials for tissue construction and energy for metabolism. Larger eggs tend to have more rapid yolk absorption rates than smaller eggs (presumably because they have a larger absorptive surface area)[51]. Lipids, especially TAGs, are an important source of energy post-hatch. When the rate of yolk absorption increases during this phase, yolk is consumed often preferentially to oil globules. It is important to consider that there can be interspecies variations in the type of lipid used during embryonic development. For example, a review of lipids in fresh water fish[52] reported that in whitefish, TAGs decreased immediately upon fertilization, while in rainbow trout TAGs were not used until after hatch. These authors also note that phospholipids are an important source of energy in freshwater fish.

After hatching, many fish larvae remain dependent on yolk and oil globules for energy. Although some species seem only to use the oil globules in 'emergency situations', i.e. if hatching is delayed, others (e.g. striped bass) may use yolk and some oil globules, relying more heavily on the oil globule after hatch. In some cases, the energy-rich oil globules may not be absorbed at all even at point of death by starvation[63]. If we follow our thesis that different POP are associated with different types of lipid (TAGs being the most important for non-polar POP), how might species differences in egg lipid content and timing of consumption discussed above affect POP disposition into the developing embryo or larval fish?

Several factors appear to be important in determining POP absorption and susceptibility of early life stages to POPs. Again, there are few, if any, studies that have addressed differences in susceptibility to POP based on embryo/larval use of lipids during development and POP transfer. There is no question that contaminants are transferred to offspring through maternal processes[40,118], and that percent lipid in spawning females and percent of total lipid deposited into eggs influences contaminant levels[86]. Timing of absorption, coupled with the development process, also seems to be important in terms of susceptibility to POP. An early study by Broyles and Noveck[13] found that in both lake trout and chinook salmon eggs exposed to hexachlorobiphenyl, the greatest increase in mortality occurred during the period of greatest yolk absorption. Guiney *et al.*[40] working with rainbow trout, analyzed the uptake of a 2,5,2′,5′-TCB into sac fry. They observed absorption of yolk throughout this period, with absorption of the oil globule just prior to swim-up (or exogenous feeding). The authors report that the greatest absolute amount of TCB was transferred to the fry at the very end of the sac fry stage. They suggest this occurs because rainbow trout fry tend to consume protein and phospholipids first, saving the TAGs (and oil globule), which presumably contain the greatest concentration of TCB, for the final phase of absorption. Interestingly, this 'last bolus' of TCB coincided with the beginning of an increased elimination rate of TCB, which may reflect increases in the activity of xenobiotic metabolizing enzymes over embryolarval development (discussed below). More recently, when developing early life stage LC50s for waterborne exposure to TCDD in several species of freshwater fish, Elonen *et al.*[31] reported that the most sensitive life stage across species tended to be from hatch to swim-up. They speculate that this may result from greater absorption of TCDD at this stage. Unfortunately, TCDD concentrations are reported only for eggs and not larvae, making this impossible to assess. The authors found large interspecies differences in the LC50s, but reported no relationship between total egg lipid and sensitivity to toxicity. They attributed the observed LC50 differences to likely differences in TCDD toxicokinetics mediated via the aryl hydrocarbon receptor (AHR) (further discussed in section V.3). Results from this study suggest that either lipids were not 'protective' in developing larvae and/or that inherent differences in sensitivity to AHR-mediated toxicity in larvae were more important determinants of TCDD toxicity, reminiscent of the earlier discussion on the higher potency of compounds that exert their toxicity through specific mechanisms of action, as TCDD does[96]. Together, these studies demonstrate that several factors are important in determining POP absorption and susceptibility of early life stages to POPs, including maternal transport to eggs, percent lipid in spawning females, timing of yolk absorption, and POP metabolism.

Given the large interspecies differences in toxicity to some POPs, how relevant is it to consider maternal transfer of POP to the offspring? Is it relevant to consider differences in egg-lipid content and absorption of lipid? We think these factors are relevant both for predicting toxicity and for consideration of interspecies differences in exposure (particularly within target tissues) to POP in the field. Understanding the dynamics of POP transfer and uptake could provide insight into susceptibility of certain species such as eel, or on the other extreme, cod, to POP exposure.

2.7. *Smolting*

Many salmonids undergo a process referred to as smolting in preparation to enter the sea. Along with the many morphological and physiological changes that occur during this stage, there is a dramatic depletion of total body lipids. As the body is transformed and prepared for life in the ocean, large amounts of energy are required[52].

Coho salmon may lose up to 60% of their total body lipids. Chinook salmon also undergo significant reductions in total body lipid during smoltification, arriving in the estuary with total body lipid contents that range from 1 to 3% on a wet weight basis (reviewed in ref. 122). During this transformation, large amounts of TAGs are lost relative to other lipid classes (reviewed in ref. 74). Interestingly, Sheridan *et al.*[103] reported loss of TAGs from tissues of steelhead trout during the parr–smolt transformation, including dark muscle (losing up to 91% of TAGs), light muscle and liver, while no seasonal changes were detected in the mesenteric fat, which is mainly TAGs. Thus, pre-smolting parr have higher lipid content, with almost equal amounts of polar and neutral lipids while polar lipids tend to predominate in smolts[52]. How might this time of great lipid loss, which favors loss of TAGs, affect redistribution of POP? Again we can only speculate, given the paucity of studies, but if we liken this phase to starvation, it is possible that a redistribution of POP into potentially sensitive organs (such as the brain) could occur during this critical process. Clearly, further studies are warranted on this topic. Figure 2 depicts whole body lipid content over the life cycle of captive Atlantic salmon, showing that, even in artificially fed fish, a drop in body lipid content is seen during smoltification.

In sum, despite the little work that has been conducted on interspecies differences in total body lipid, lipid distribution and egg lipid, we feel it is a potentially important topic for predicting vulnerabilities to POPs in field-caught fish. Improved understanding of tissue lipid dynamics and lipid composition is likely to be important in any target species selected for field study of POP exposure and effects.

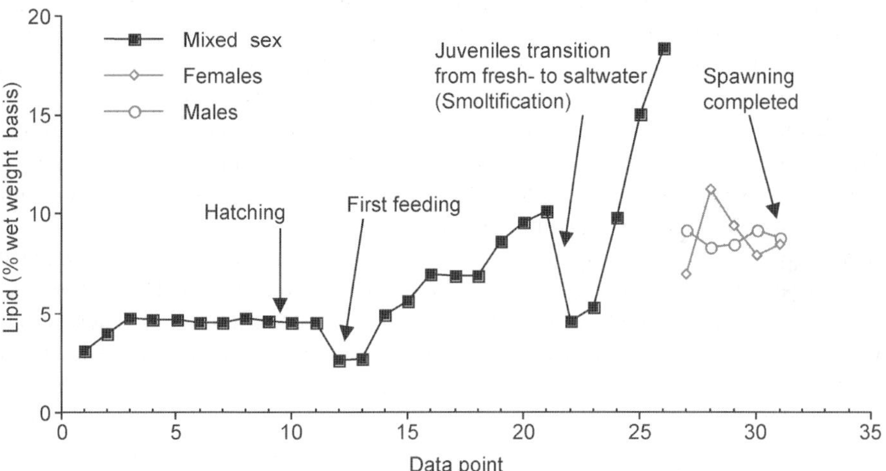

Fig. 2. Total lipid in cultured Atlantic salmon (*Salmo salar*) over their life cycle. Adapted from ref. 123. Data points 1–8 represent doubling of degree days from fertilization until near hatching; 12–26 represent approximate doubling of body weight at each point. Lipids determined gravimetrically after Soxhlet extraction from dried tissues.

3. How changes in metabolizing enzymes during development affect membranes and POP disposition

3.1. Developmental changes in CYP1A1 expression in fish and their relation to POP toxicity

Changes in enzyme activity appear to be more predictive of POP toxicity in embryolarval fish stages than relative lipid levels. CYP1A1 is a monooxygenase enzyme whose induction by xenobiotic compounds is linked to toxicity in early life stage fish. CYP1A1 belongs to a superfamily of enzymes, the cytochrome P450s (CYP450), and is the most prominent xenobiotic metabolizing CYP450 isoform in fish[107]. It is also the most strongly induced. Out of >5000 toxicant-related genes evaluated in a heart DNA array from TCDD-treated zebrafish embryos, CYP1A1 mRNA increased 40-fold in response to an EC10 dose; one other gene increased expression only eightfold[46]. Induction of CYP1A1 is mediated by the binding of a planar ligand, such as TCDD, to a cytosolic transcription factor, the AHR[44,120].

Fish express CYP1A1 very early in embryonic development. Constitutive CYP1A1 activity is first detectable in killifish (*Fundulus heteroclitus*) embryos by 3 days post-fertilization (3 dpf)[81], increasing ninefold immediately post-hatch (14 dpf)[8]. A similar increase in CYP1A1 with hatching is also seen in other teleosts[7,72]. CYP1A1 is also inducible in *F. heteroclitus* embryos as early as 3 dpf[81]. Tissue-specific studies in several teleost species indicate that the cardiovascular system is an early and prevalent location of CYP1A1 expression[42,108]. TCDD exposure induces CYP1A1 in embryo skin, heart, kidney, liver, head, trunk and tail, and can be observed as early as 24–36 hpf[2,14,15,42,43,111].

This early expression of CYP1A1 correlates with the sensitivity of teleost early life stages to POP toxicity. Exposure to the toxic PCB congener, 3,4,5,3′,4′-TCB (CB126) simultaneously induces both CYP1A1 activity (EROD) and developmental abnormalities (pericardial edema) in *F. heteroclitus* embryos from reference sites[81]. Embryos with genetically suppressed CYP1A1 activity, however, are resistant to CB126 toxicity[80]. In larval lake trout, dose–response relationships were shown between TCDD, vascular CYP1A1 expression, and larval mortality[42]. Moreover, induction of CYP1A1 precedes manifestation of xenobiotic-induced vascular toxicity in early life stage teleosts[2,43], strongly suggesting that CYP1A1 expression provides the temporal link between exposure and effect.

There is strong evidence that CYP1A1 expression alters lipid-membrane integrity in embryolarval fish exposed to the model POP, TCDD. Among TCDD's many effects in early life stage fish are edema and cardiovascular dysfunction[48,49]. These effects have been linked to oxidative stress via induction of CYP1A1 by compounds that are poorly metabolized. For example, several organochlorine inducers of CYP1A1, including TCDD and PCBs, are poorly metabolized by CYP1A1 (e.g. CB77 is metabolized 2000–4000 times more slowly than non-halogenated CYP1A1 substrates by scup liver microsomes[119]). The continued presence of these chemicals in the active site uncouples CYP1A-mediated substrate oxidation resulting in the production and cycling of ROS in the form of superoxide and hydroxyl radicals[64,99]. CYP1A1 induction and ROS production increase endothelial membrane permeability[105].

Further, TCDD-induced edema in fish embryos has been linked to increased vascular permeability[43] and vascular collapse[14,15,42], which may be responsible for the reduced blood flow measured in the vasculature of TCDD-exposed embryos[29,110]. Interestingly, TCDD exposure also leads to apoptosis in the neural cells of the embryo midbrain[29], suggesting a basis for neurological effects of TCDD. In sum, studies with fish embryos demonstrate that tissue-specific CYP1A1 expression co-occurs with tissue-specific TCDD toxicity[2,14,15,42,43], suggesting close coupling between CYP1A1 expression and tissue dysfunction. It is clear from these and other studies on larval fish[76,77,80] that CYP1A1 plays a key role in mediating the toxicity of POPs that act through the AHR.

There are differences within a species in the relative toxicant sensitivity of different developmental stages, with the sensitivity to toxicants increasing from embryos to larvae. This is not due to stage differences in metabolic pathways, as larval fish metabolize the model carcinogen, benzo(*a*)pyrene, to the same suite of metabolites as adults[109]. Rather, stage-related sensitivity appears to be due to increased levels of CYP1A expression during development. For example, studies with early life stage medaka demonstrate that larvae are more sensitive to the inhibitory effects of the pesticide, diazinon, on acetylcholinesterase activity than are embryos[45]. Diazinon is activated to its toxic metabolite via CYP450 and there are several studies demonstrating that CYP450 activity increases during embryolarval development in fish[79,90].

It appears that species differences in early life stage sensitivity to POP toxicants are likely related to species differences in early life stage CYP1A1 activity rather than to differences in lipids. For 10 freshwater fish, the LC50 values for egg concentrations of TCDD causing mortality ranged from ~100 pg/g wet weight egg for lake trout to ~2600 pg/g for zebrafish (lake trout > brook trout > rainbow trout > fathead minnow > channel catfish > lake herring > medaka > whitesucker > northern pike > zebrafish; Fig. 3 in ref. 31). Based on the concentration–response curves and the lack of correlation between egg lipid content and relative TCDD sensitivity, the authors speculate that differences in species susceptibility may be related to factors other than lipid, likely species differences in AHR-related toxicodynamics. In support of this hypothesis, zebrafish, the most TCDD-resistant fish species examined to date[31], does not express catalytically active CYP1A1 protein until post-hatch[72], in contrast to other, more POP-sensitive species that express the enzyme embryonically[2,14,15,42,43,79,81,90,111].

In summary, the weight of evidence from embryolarval studies indicates that the expression of CYP1A1, a membrane-bound enzyme, is strongly correlated with exposure to poorly metabolizable chlorinated POPs, resulting in ROS production and decreased membrane integrity in those tissues where CYP1A1 is expressed. Further, changes in CYP1A1 expression as embryolarval development proceeds correlate with changes in POP toxic potency for those toxicants mediating their effects via binding to the AHR. Thus, changes in enzyme activity appear to be more important than lipid levels in predicting the toxicity of POPs in embryolarval fish stages.

VI. Conclusions

As we have described in this chapter, the partitioning of POPs into lipids is a major factor to consider in studies linking exposure to effects. A number of studies demonstrate that toxic responses can be modified by increasing lipid content (reviewed in refs. 66,114). In theory, the higher lipid content serves to sequester the toxicant away from the target site so that it is less available to exert its toxicity. Thus, especially when comparing across species or studies, it is generally accepted that some form of lipid normalization is needed. As we have seen, POPs distribute differently among different lipid classes, raising the question of which lipid class or classes to use for normalization purposes. Most commonly, total lipid levels, rather than individual classes, are used for normalization, total lipid being determined gravimetrically following solvent extraction[9]. As discussed in this chapter, however, POPs accumulate primarily in neutral lipids, especially the TAGs, rather than in polar lipids, such as phospholipids. Thus, when predicting tissue distributions of POPs, especially very hydrophobic POPs such as TCDD, relative tissue levels of TAGs will likely better predict relative tissue distribution than levels of total lipid in tissues. Accordingly, a better approach for lipid normalization would be to quantify TAGs and normalize POP burdens against this class of lipids, at a minimum.

Based on the discussions in this chapter, what predictions can be made, if any, about which species and life stages, and perhaps tissues and organs, are most susceptible to POP toxicity, and how important is the role of lipid dynamics in these differences? First, lipid mobilization leads to dramatic drops in tissue TAG levels and consequently dramatic tissue redistribution of the POPs associated with those TAGs. As adipose levels fall, fish with lower lipid levels on a whole body basis (due to either starvation or genetic makeup), are more likely to partition POPs into central nervous tissue, such as brain, than fatter fish, that can more evenly redistribute adipose-derived POPs among a greater array of fatty tissues. This may make lean fish, or those that are starving during migration or while spawning, more susceptible to POP-mediated neurological, behavioral, and/or reproductive disorders. Certainly the central nervous system may be an important target organ for POP-induced toxicity in situations when storage lipids are being catabolized for energy. While sexually mature females can redistribute some fraction of their body burden of POPs to their developing eggs, the proportion transferred is strongly dependent on the relative level of lipid (mainly TAG) in the female's body vs. the amount of lipid in her eggs. Relatively, little POP is redistributed to testes in any species, and it is likely that changes in nutritional state affect this little, if at all. Larvae appear to be more vulnerable to the adverse effects of POPs than embryos, likely due in large part to increased expression of the POP-metabolizing CYP450 enzymes, and the increased uptake of potentially contaminated TAG over the course of embryo/larval development. In these early life stages, the major POP targets are the endothelial cells of the cardiovascular system, resulting in pericardial edema and vascular collapse. For specifically acting POPs that exert their toxicity via induction of xenobiotic metabolizing enzymes, lipid composition may be less important than relative enzyme activity in determining the relative susceptibility to toxicants of early life stages of different species.

Overall, given the complexity of the relationship between lipid content, lipid class, and their influence on the toxic potency of POPs, we strongly recommend that studies of POP body burdens and effects report lipid analyses in the species of concern. Given that different methods favor the extraction of different lipid classes[92], it is particularly important that methodology for lipid analyses should be provided in considerable detail. Further, we emphasize the importance of establishing whether or not lipid normalization is appropriate for your data set, and using the appropriate normalization methods, of including the determination of lipid classes, especially the ubiquitous storage lipids, the TAGs[47]. Whether the researcher chooses to determine lipid classes in each tissue, or simply total lipid content, will depend largely on the application (e.g. tissue-specific toxicity vs. LC50). In the absence of information on the redistribution of lipids and POPs between tissues in relation to the physiological status of an organism, whole body determinations of lipid and POP concentrations may be highly useful for comparisons between species, life history stages, and exposure histories.

VII. References

1. Ackman, R.G., C.A. McLeod and A.K. Banerjee. An overview of analyses by Chromarod–Iatroscan TLC–FID. *J. Planar Chromatogr.* 3: 450–461, 1990.
2. Andreasen, E.A., J.M. Spitsbergen, R.L. Tanguay, J.J. Stegeman, W. Heideman and R.E. Peterson. Tissue-specific expression of AHR2, ARNT2, and CYP1A in zebrafish embryos and larvae: effects of developmental stage and 2,3,7,8-tetrachlorodibenzo-*p*-dioxin exposure. *Toxicol. Sci.* 68: 403–419, 2002.
3. Bard, S.M. Multixenobiotic resistance as a cellular defense mechanism in aquatic organisms. *Aquat. Toxicol.* 48: 357–389, 2000.
4. Bergen, B.J., W.G. Nelson, J.G. Quinn and S. Jayaraman. Relationships among total lipid, lipid classes, and polychlorinated biphenyl concentrations in two indigenous populations of ribbed mussels (*Geukensia demissa*) over an annual cycle. *Environ. Toxicol. Chem.* 20: 575–581, 2001.
5. Bergen, B.J., J.G. Quinn and C.C. Parrish. Quality assurance study of marine lipid-class determination using Chromarod–Iatroscan thin layer chromatography–flame ionization detector. *Environ. Toxicol. Chem.* 19: 2189–2197, 2000.
6. Bertelsen, S.L., A.D. Hoffman, C.A. Gallinat, C.M. Elonen and J.W. Nichols. Evaluation of log K_{ow} and tissue lipid content as predictors of chemical partitioning into fish tissues. *Environ. Toxicol. Chem.* 17: 1447–1455, 1998.
7. Binder, R.L. and J.J. Stegeman. Basal levels and induction of hepatic aryl hydrocarbon hydroxylase activity during the embryonic period of development in brook trout. *Biochem. Pharmacol.* 32: 1324–1327, 1983.
8. Binder, R.L. and J.J. Stegeman. Microsomal electron transport and xenobiotic monooxygenase activities during the embryonic period of development in the killifish *Fundulus heteroclitus*. *Toxicol. Appl. Pharmacol.* 73: 432–443, 1984.
9. Bligh, E.G. and W.J. Dyer. A rapid method of total lipid extraction and purification. *Can. J. Biochem. Physiol.* 37: 911–917, 1959.
10. Bondy, S.C. and S. Naderi. Contribution of hepatic cytochrome P450 systems to the generation of reactive oxygen species. *Biochem. Pharmacol.* 48: 155–159, 1994.
11. Boon, J.P. and J.C. Duinker. Kinetics of polychlorinated biphenyl (PCB) components in juvenile sole (*Solea solea*) in relation to concentrations in water and to lipid metabolism under conditions of starvation. *Aquat. Toxicol.* 7: 119–134, 1985.
12. Boon, J.P., R.C.H.M. Oudejans and J.C. Duinker. Kinetics of individual polychlorinated biphenyl (PCB) components in juvenile sole (*Solea solea*) in relation to their concentrations in food and to lipid metabolism. *Comp. Biochem. Physiol.* 79C: 131–142, 1984.
13. Broyles, R.H. and M.I. Noveck. Uptake and distribution of 2,4,5,2′,4′,5′-hexachlorobiphenyl in fry of lake trout and Chinook salmon and its effects on viability. *Toxicol. Appl. Pharmacol.* 50: 299–308, 1979.

14. Cantrell, S.M., L. Lutz, D.E. Tillitt and M. Hannink. Embryotoxicity of 2,3,7,8-tetrachlorodibenzo-*p*-dioxin (TCDD): the embryonic vasculature is a physiological target for TCDD-induced DNA damage and apoptotic cell death in medaka (*Orizias latipes*). *Toxicol. Appl. Pharmacol.* 141: 23–34, 1996.
15. Cantrell, S.M., J. Joy-Schlezinger, J.J. Stegeman, D.E. Tillitt and M. Hannink. Correlation of 2,3,7,8-tetrachlorodibenzo-*p*-dioxin-induced apoptotic cell death in the embryonic vasculature with embryotoxicity. *Toxicol. Appl. Pharmacol.* 148: 24–34, 1998.
16. Capuzzo, J. and D. Leavitt. Lipid composition of the digestive glands of *Mytilus edulis* and *Carcinus maenas* in response to pollutant gradients. *Mar. Ecol. Prog. Ser.* 46: 139–145, 1988.
17. Capuzzo, J.M., J.W. Farrington, P. Rantaamaki, C.H. Clifford, B.A. Lancaster, D.F. Leavitt and X. Jia. The relationship between lipid composition and seasonal differences in the distribution of PCBs in *Mytilus edulis* L. *Mar. Environ. Res.* 28: 259–264, 1989.
18. Capuzzo, J.M., B.A. Lancaster and G.C. Sasaki. The effects of petroleum hydrocarbons on lipid metabolism and energetics of larval development and metamorphosis in the American lobster (*Homarus americanus* Milne Edwards). *Mar. Environ. Res.* 14: 201–228, 1984.
19. Cash, G.G. and J.V. Nabholz. Minimum cross-sectional diameter: calculating when molecules may not fit through a biological membrane. *Environ. Toxicol. Chem.* 21: 2095–2098, 2002.
20. Chen, C.-Y., J.T. Hamm, J.R. Hass, P.W. Albro and L.S. Birnbaum. A mixture of polychlorinated dibenzo-*p*-dioxins (PCDDs), dibenzofurans (PCDFs), and non-ortho polychlorinated biphenyls (PCBs) changed the lipid content of pregnant Long Evans rats. *Chemosphere* 46: 1501–1504, 2002.
21. Chessels, M., D. Hawker and D. Connell. Influence of solubility in lipid on bioconcentration of hydrophobic compounds. *Ecotoxicol. Environ. Saf.* 23: 260–273, 1992.
22. Choi, J. and J.T. Oris. Anthracene photoinduced toxicity to PLHC-1 cell line (*Poeciliopsis lucida*) and the role of lipid peroxidation in toxicity. *Environ. Toxicol. Chem.* 19: 2699–2706, 2000.
23. Collier, T.K., C.A. Krone, M.M. Krahn, J.E. Stein, S.-L. Chan and U. Varanasi. Petroleum exposure and associated biochemical effects in subtidal fish after the EXXON Valdez oil spill. *Am. Fish. Soc. Symp.* 18: 671–683, 1996.
24. Collier, T.K. and U. Varanasi. Hepatic activities of xenobiotic metabolizing enzymes and levels of xenobiotics in English sole (*Parophyrus vetulus*) exposed to environmental contaminants. *Arch. Environ. Contam. Toxicol.* 20: 462–473, 1991.
25. Connell, D. *Bioaccumulation of Xenobiotic Compounds*, Boca Raton, FL, CRC Press, 1990, 219pp.
26. Curtis, L.R., M.J. Hemmer and L.A. Courtney. Dieldrin induces cytosolic [^{3}H]7,12-dimethylbenz[*a*]anthracene binding but not multidrug resistance proteins in rainbow trout liver. *J. Toxicol. Environ. Health A* 60: 275–289, 2000.
27. Doi, A.M., E. Holmes and K.M. Kleinow. P-glycoprotein in the catfish intestine: inducibility by xenobiotics and functional properties. *Aquat. Toxicol.* 55: 157–170, 2001.
28. Doi, A.M., Z. Lou, E. Holmes, C.L.J. Li, C.S. Venugopal, M.O. James and K.M. Kleinow. Effect of micelle fatty acid composition and 3,4,3′,4′-tetrachlorobiphenyl (TCB) exposure on intestinal [C-14]-TCB bioavailability and biotransformation in channel catfish in situ preparations. *Toxicol. Sci.* 55: 85–96, 2000.
29. Dong, W., H. Teraoka, K. Yamazaki, S. Tsukiyama, S. Imani, T. Imagawa, J.J. Stegeman, R.E. Peterson and T. Hiraga. 2,3,7,8-Tetrachloridibenzo-*p*-dioxin toxicity in the zebrafish embryo: local circulation failure in the dorsal midbrain is associated with increased apoptosis. *Toxicol. Sci.* 69: 191–201, 2002.
30. Eganhouse, R. and J. Pontolillo. Assessing the reliability of physico-chemical property data (K_{ow}, S_w) for hydrophobic organic compounds: DDT and DDE as a case study. *SETAC Globe* 3: 34–35, 2002.
31. Elonen, G., R. Spehar, G. Holcombe, R. Johnson, J. Fernandez, R. Erickson, J. Tietge and P. Cook. Comparative toxicity of 2,3,7,8-tetrachlorodibenzo-*p*-dioxin to seven freshwater fish species during early life-stage development. *Environ. Toxicol. Chem.* 17: 472–483, 1998.
32. Elskus, A.A. and J.J. Stegeman. Induced cytochrome P-450 in *Fundulus heteroclitus* associated with environmental contamination by polychlorinated biphenyls and polynuclear aromatic hydrocarbons. *Mar. Environ. Res.* 27: 31–50, 1989.
33. Ewald, G. and P. Larsson. Partitioning of a ^{14}C-labeled 2,2′,4,4′-tetrachlorobiphenyl between water and fish lipids. *Environ. Toxicol. Chem.* 13: 1577–1580, 1994.
34. Fahraeus-Van Ree, G.E. and D.R. Spurrell. Effect of toxaphene on isolated hepatocytes of the yellowtail founder, *Pleuronectes ferrugineus* storer. *Ecotoxicol. Environ. Saf.* 46: 289–297, 2000.
35. Favari, L., E. Lopez, L. Martinez-Tabche and E. Diaz-Pardo. Effect of insecticides on plankton and fish of Ignacio Ramirez Reservoir (Mexico): a biochemical and biomagnification study. *Ecotoxicol. Environ. Saf.* 51: 177–186, 2002.
36. Fitzsimmons, P.N., J.D. Fernandez, A.D. Hoffman, B.C. Butterworth and J,W. Nichols. Branchial elimination of superhydrophobic organic compounds by rainbow trout (*Oncorhynchus mykiss*). *Aquat. Toxicol.* 55: 23–34, 2001.

37. Geyer, H., I. Scheunert, R. Bruggemann, M. Matthies, C. Steinberg, V. Zitko, A. Kettrup and W. Garrison. The relevance of aquatic organisms' lipid content to the toxicity of lipophylic chemicals: toxicity of lindane to different fish species. *Ecotoxicol. Environ. Saf.* 28: 53–70, 1994.
38. Gilroy, D.J., H.M. Carpenter, L.K. Siddens and L.R. Curtis. Chronic dieldrin exposure increases hepatic disposition and biliary excretion of [^{14}C]dieldrin in rainbow trout. *Fundam. Appl. Toxicol.* 20: 295–301, 1993.
39. Gilroy, D.J., C.L. Miranda, L.K. Siddens, Q. Zhang, D.R. Buhler and L.R. Curtis. Dieldrin pretreatment alters [C-14]dieldrin and [H-3]7,12-dimethylbenz[*a*]anthracene uptake in rainbow trout liver slices. *Fundam. Appl. Toxicol.* 30: 187–193, 1996.
40. Guiney, P.D., M.J. Melancon, Jr., J.J. Lech and R.E. Peterson. Effects of egg and sperm maturation and spawning on the distribution and elimination of a polychlorinated biphenyl in rainbow trout (*Salmo gairdneri*). *Toxicol. Appl. Pharmacol.* 47: 261–272, 1979.
41. Guiney, P.D. and R.E. Peterson. Distribution and elimination of a polychlorinated biphenyl after acute dietary exposure in yellow perch and rainbow trout. *Arch. Environ. Contam. Toxicol.* 9: 667–674, 1980.
42. Guiney, P.D., R.M. Smolowitz, R.E. Peterson and J.J. Stegeman. Correlation of 2,3,7,8-tetrachlorodibenzo-*p*-dioxin induction of cytochrome P4501A in vascular endothelium with toxicity in early life stages of lake trout. *Toxicol. Appl. Pharmacol.* 143: 256–273, 1997.
43. Guiney, P.D., M.K. Walker, J.M. Spitsbergen and R.E. Peterson. Hemodynamic dysfunction and cytochrome P4501A mRNA expression induced by 2,3,7,8-tetrachlorodibenzo-*p*-dioxin during embryonic stages of lake trout development. *Toxicol. Appl. Pharmacol.* 168: 1–14, 2000.
44. Hahn, M.E., B.R. Woodin, J.J. Stegeman and D.E. Tillitt. Aryl hydrocarbon receptor function in early vertebrates: inducibility of cytochrome P450 1A in agnathan and elasmobranch fish. *Comp. Biochem. Physiol.* 120C: 67–75, 1998.
45. Hamm, J.T., B.W. Wilson and D.E. Hinton. Increasing uptake and bioactivation with development positively modulate diazinon toxicity in early life stage medaka (*Oryzias latipes*). *Toxicol. Sci.* 61: 304–313, 2001.
46. Handley H., M. Grow, M. Fishman and J.J. Stegeman. Generation of zebrafish cDNA microarrays for investigation of cardiovascular embryo toxicity by 2,3,7,8-tetrachlorodibenzo-*p*-dioxin. In: *11th International Symposium on Pollutant Responses in Marine Organisms*, Plymouth, England, July 10–13, Abstract 1204, 2001.
47. Hebert, C.E. and K.A. Keenleyside. To normalize or not to normalize? Fat is the question. *Environ. Toxicol. Chem.* 14: 801–807, 1995.
48. Helder, T. Effects of 2,3,7,8-tetrachlorodibenzo-*p*-dioxin (TCDD) on early life stages of the pike (*Esox lucius* L.). *Sci. Total Environ.* 14: 255, 1980.
49. Helder, T. Effects of 2,3,7,8-tetrachlorodibenzo-*p*-dioxin (TCDD) on early life stages of the rainbow trout (*Salmo gairdneri* Richardson). *Toxicology* 19: 101–112, 1981.
50. Hellou, J., W.G. Warren and G. Mercer. Organochlorines in pleuronectidae: comparison between three tissues of three species inhabiting the Northwest Atlantic. *Arch. Environ. Contam. Toxicol.* 29: 302–308, 1995.
51. Hemming, T.A. and R.K. Buddington. Yolk absorption in embryonic and larval fishes, *Fish Physiology*, Vol. XIA, edited by W.S. Hoar, D.J. Randall and A.P. Farrell, San Diego, CA, Academic Press, pp. 401–446, 1988.
52. Henderson, R.J. and D. Tocher. The lipid composition and biochemistry of freshwater fish. *Prog. Lipid Res.* 26: 281–347, 1987.
53. Hochachka, P.W. and G.N. Somero. *Biochemical Adaptation: Mechanism and Process in Physiological Evolution*, New York, Oxford University Press, 2002, 466pp.
54. Hodson, P.V., M. Castonguay, C.M. Couillard, C. Desjardins, E. Pelletier and R. McLeod. Spatial and temporal variations in chemical contamination of American eels, *Anguilla rostrata* captured in the estuary of the St Lawrence River. *Can. J. Fish. Aquat. Sci.* 51: 464–478, 1994.
55. Ingebrigsten, K., H. Hektoen, T. Andersson, A. Bergman and I. Brandt. Species-specific accumulation of the polychlorinated biphenyl (PCB) 2,3,3′,4,4′-pentachlorobiphenyl in fish brain: a comparison between cod (*Gadus morhau*) and rainbow trout (*Oncorhynchus mykiss*). *Pharmacol. Toxicol.* 67: 344–345, 1990.
56. Ingebrigsten, K., H. Hektoen, T. Andersson, E. Klasson Wehler, A. Bergman and I. Brandt. Enrichment of metabolites in the cerebrospinal fluid of cod (*Gadus morhua*) following oral administration of hexachlorobenzene and 2,4′,5-trichlorobiphenyl. *Pharmacol. Toxicol.* 71: 420–425, 1992.
57. Ingebrigsten, K. and J.U. Skaare. Distribution and elimination of [14C]hexachlorobenzene after single oral exposure in the rainbow trout (*Salmo gairdneri*). *J. Toxicol. Environ. Health* 12: 309–316, 1983.

58. Ingebrigsten, K. and J.E. Solbakken. Distribution and elimination of [^{14}C]hexachlorobenzene after single oral exposure in cod (*Gadus morhua*) and flounder (*Platichthys flesus*). *J. Toxicol. Environ. Health* 16: 197–205, 1985.
59. Jobling, M., S.J.S. Johansen, H. Foshaug, I.C. Burkow and E.H. Jorgensen. Lipid dynamics in anadromous Artic charr *Salvelinus alpinus* (L.): seasonal variations in lipid storage depots and lipid class composition. *Fish Physiol. Biochem.* 18: 225–240, 1998.
60. Jorgensen, E.H., I.C. Burkow, H. Foshaug, B. Killie and K. Ingebrigtsen. Influence of lipid status on tissue distribution of the persistent organic pollutant octachlorostyrene in Arctic charr (*Salvelinus alpinus*). *Comp. Biochem. Physiol.* 118C: 311–318, 1997.
61. Jorgensen, E.H., H. Foshaug, P. Andersson, I.C. Burkow and M. Jobling. Polychlorinated biphenyl toxicokinetics and P4501A responses in anadromous Arctic charr during winter emaciation. *Environ. Toxicol. Chem.* 21: 1745–1752, 2002.
62. Kammann, U., R. Knickmeyer and H. Steinhart. Distribution of polychlorobiphenyls and hexachlorobenzene in different tissues of the Dab (*Limanda limanda* L.) in relation to lipid polarity. *Bull. Environ. Contam. Toxicol.* 45: 552–559, 1990.
63. Kammler, E. *Early Life History of Fish: An Energetics Approach*, London, Chapman & Hall, 1992.
64. Klaassen, C. (Editor). *Casarett and Doull's Toxicology: The Basic Science of Poisons*, New York, McGraw-Hill Medical Publishing Division, 2001, 1236pp.
65. Krahn, M.M., D.G. Burrows, G.M. Ylitalo, D.W. Brown, C.A. Wigren, T.K. Collier, S.L. Chan and U. Varanasi. Mass-spectrometric analysis for aromatic compounds in bile of fish sampled after the Exxon Valdez oil spill. *Environ. Sci. Technol.* 26: 116–126, 1992.
66. Lassiter, R. and T. Hallam. Survival of the fattest: implications for acute effects of lipophylic chemicals on aquatic populations. *Environ. Toxicol. Chem.* 9: 585–595, 1990.
67. Legierse, K., H.J.M. Verhaar, W.H.J. Vaes, J.H.M. De Bruijn and J.L.M. Hermens. Analysis of the time-dependent acute aquatic toxicity of organophosphorus pesticides: the critical target occupation model. *Environ. Sci. Technol.* 33: 917–925, 1999.
68. Lieb, A.J., D.D. Bills and R.O. Sinnhuber. Accumulation of dietary polychlorinated biphenyls (Aroclor 1254) by rainbow trout. *J. Agric. Food Chem.* 22: 638–642, 1974.
69. Lien, G.J., J.M. McKim, A.D. Hoffman and C.T. Jenson. A physiologically based toxicokinetic model for lake trout (*Salvelinus namaycush*). *Aquat. Toxicol.* 51: 335–350, 2001.
70. Maccubbin, A.E. DNA adduct analysis in fish: laboratory and field studies. In: *Aquatic Toxicology: Molecular, Biochemical and Cellular Perspectives*, edited by D.C. Malins and G.K. Ostrander, Ann Arbor, MI, Lewis Publishers, pp. 267–294, 1994.
71. Madueira, M.J., A.M. Picado, A.M. Ferreira, E. Mendonca, Y. Le Gal and C. Vale. PCB contamination in the oyster *Crassotrea angulata*: effects on lipids and adenylic energetic charge. *Sci. Total Environ.* 132: 599–605, 1993.
72. Mattingly, C.J. and W.A. Toscano. Posttranscriptional silencing of cytochrome P4501A1 (CYP1A1) during zebrafish (*Danio rerio*) development. *Dev. Dyn.* 222: 645–654, 2001.
73. McCarty, L. and D. Mackay. Enhancing ecotoxicological modeling and assessment: body residues and modes of toxic action. *Environ. Sci. Technol.* 5: 1071–1080, 1993.
74. McCormick, S.D. and R.L. Saunders. Preparatory physiological adaptations for marine life of salmonids: osmoregulation, growth, and metabolism. *Am. Fish. Soc. Symp.* 1: 211–229, 1987.
75. McKim, J. and J. Nichols. Use of physiologically based toxicokinetic models in a mechanistic approach to aquatic toxicology. In: *Aquatic Toxicology: Molecular, Biochemical and Cellular Perspectives*, edited by D.C. Malins and G.K. Ostrander, Boca Raton, FL, Lewis Publishers, pp. 469–519, 1994.
76. Meyer, J. and R.T. Di Giulio. Patterns of heritability of decreased EROD activity and resistance to PCB 126-induced teratogenesis in laboratory-reared offspring of killifish (*Fundulus heteroclitus*) from a creosote-contaminated site in the Elizabeth River, VA, USA. *Mar. Environ. Res.* 54: 1–6, 2002.
77. Meyer, J., D. Nacci and R.T. Di Giulio. Cytochrome P4501A (CYP1A) in killifish (*Fundulus heteroclitus*): heritability of altered expression and relationship to survival in contaminated sediments. *Toxicol. Sci.* 68: 69–81, 2002.
78. Mommsen, T. and P. Walsh. Vitellogenesis and oocyte assembly, In: *Fish Physiology*, Vol. XIB, edited by W.S. Hoar, D.J. Randall and A.P. Farrell, San Diego, CA, Academic Press, pp. 347–406, 1988.
79. Monod, G., M.A. Boudry and C. Gillet. Biotransformation enzymes and their induction by beta-naphthoflavone during embryolarval development in salmonid species. *Comp. Biochem. Physiol.* 114C: 45–50, 1996.
80. Nacci, D., L. Coiro, D. Champlin, S. Jayaraman, R. McKinney, T.R. Gleason, W.R. Munns, J.L. Specker and K.R. Cooper. Adaptations of wild populations of the estuarine fish *Fundulus heteroclitus* to persistent environmental contaminants. *Mar. Biol.* 134: 9–17, 1999.

81. Nacci, D., L. Coiro, A. Kuhn, D. Champlin, W. Munns, J. Specker and K. Cooper. Nondestructive indicator of ethoxyresorufin-*O*-deethylase activity in embryonic fish. *Environ. Toxicol. Chem.* 17: 2481–2486, 1998.
82. Nagaoka, S., H. Miyazaki, Y. Aoyama and A. Yoshida. Effects of dietary polychlorinated biphenyls on cholesterol catabolism in rats. *Br. J. Nutr.* 64: 161–169, 1990.
83. Neff, J. Accumulation and release of petroleum-derived aromatic hydrocarbons by four species of marine animals. *Mar. Biol.* 38: 279–289, 1976.
84. Nichols, J., P. Rheingans, D. Lothenbach, R. McGeachie, L. Skow and J. McKim. Three-dimensional visualization of physiologically based kinetic model outputs. *Environ. Health Perspect.* 102: 952–956, 1994.
85. Nichols, J.W., K.M. Jensen, J.E. Tietge and R.D. Johnson. Physiologically based toxicokinetic model for maternal transfer of 2,3,7,8-tetrachlorodibenzo-*p*-dioxin in brook trout (*Salvelinus fontinalis*). *Environ. Toxicol. Chem.* 17: 2422–2434, 1998.
86. Niimi, A.J. Biological and toxicological effects of environmental contaminants in fish and their eggs. *Can. J. Fish. Aquat. Sci.* 40: 306–312, 1983.
87. Niimi, A.J. and B.G. Oliver. Influence of molecular weight and molecular volume on dietary absorption efficiency of chemicals by fishes. *Can. J. Fish. Aquat. Sci.* 45: 222–227, 1988.
88. Orrenius, S., G.E.N. Kass, S.K. Duddy and P. Nicotera. Molecular mechanism of oxidative cell damage, In: *Biological Oxidation Systems*, Vol. 2, edited by C.C. Reddy, G.A. Hamilton and K.M. Madyastha, New York, Academic Press, pp. 965–975, 1990.
89. Parrish, C.C. Determination of total lipid, lipid classes and fatty acids in aquatic samples. In: *Lipids in Freshwater Ecosystems*, edited by R.G. Wetzel, M.T. Arts and B.C. Wainmann, New York, Springer, pp. 4–20, 1999.
90. Peters, L.D. and D.R. Livingstone. Studies on cytochrome P4501A in early and adult life stages of turbot (*Scophthalmus maximus* L.). *Mar. Environ. Res.* 39: 5–9, 1995.
91. Pollak, J.K. and W. Harsas. Effects of organochlorine compounds on lipid catabolism of foetal rat liver mitochondria and microsomes. *Bull. Environ. Contam. Toxicol.* 28: 313–318, 1982.
92. Randall, R., H. Lee, II, R. Ozretich, J. Lake and R. Pruell. Evaluation of selected lipid methods for normalizing pollutant bioaccumulation. *Environ. Toxicol. Chem.* 10: 1431–1436, 1991.
93. Randall, R.C., D.R. Young, H.I. Lee and S.F. Echols. Lipid methodology and pollutant normalization relationships for neutral non-polar organic pollutants. *Environ. Toxicol. Chem.* 17: 788–791, 1998.
94. Rice, S., R.E. Thomas and J.W. Short. Effects of petroleum hydrocarbons on breathing and coughing rates and hydrocarbon uptake-depuration in pink salmon fry. In: *Physiological Responses of Marine Biota to Pollutants*, edited by F. Vernberg, A. Calabrese, F. Thurberg and W. Vernberg, New York, Academic Press, pp. 259–278, 1977.
95. Roe, S. and H. MacIsaac. Temporal variation of organochlorine contaminants in the zebra mussel *Dreissena polymorpha* in Lake Erie. *Aquat. Toxicol.* 51: 125–140, 1998.
96. Safe, S. Determination of 2,3,7,8-TCDD toxic equivalent factors (TEFs): support for the use of the in vitro AHH induction assay. *Chemosphere* 16: 791–802, 1987.
97. Safe, S. Polychlorinated biphenyls (PCBs), dibenzo-*p*-dioxins (PCDDs), dibenzofurans (PCDFs), and related compounds: environmental and mechanistic considerations which support the development of toxic equivalency factors (TEFs). *Crit. Rev. Toxicol.* 21: 51–88, 1990.
98. Sanchez, E., M. Santiago, P. Lopez-Aparicio, M. Recio and M. Perez-Albarsanz. Selective fatty acid release from intracellular phospholipids caused by PCBs in rat renal tubular cell cultures. *Chem.-Biol. Interact.* 125: 117–131, 2000.
99. Schlezinger, J.J., R.D. White and J.J. Stegeman. Oxidative inactivation of cytochrome P-450 1A (CYP1A) stimulated by 3,3′,4,4′-tetrachlorobiphenyl: production of reactive oxygen by vertebrate CYP1As. *Mol. Pharmacol.* 56: 588–597, 1999.
100. Scott, K.D., G.E. Fahraeus-Van Ree and C.C. Parrish. Sex differences in hepatic lipids of toxaphene-exposed juvenile yellowtail flounder (*Pleuronectes ferrugineus* Storer). *Ecotoxicol. Environ. Saf.* 51: 168–176, 2002.
101. Sebedio, J.L. and P. Juaneda. Quantitative lipid analyses using the new Iatroscan TLC–FID system. *J. Planar Chromatogr.* 4: 35–41, 1991.
102. Sheridan, M.A. Lipid dynamics in fish: aspects of absorption, transportation, deposition and mobilization. *Comp. Biochem. Physiol.* 90B: 679–690, 1988.
103. Sheridan, M.A., W.V. Allen and T.H. Kersetter. Seasonal variations in the lipid composition of the steelhead trout *Salmo gairdneri* Richardson, associated with the parr–smolt transformation. *J. Fish Biol.* 23: 125–134, 1983.
104. Soengas, J.L. and M. Aldegunde. Energy metabolism of fish brain. *Comp. Biochem. Physiol.* 131B: 271–296, 2002.

105. Stegeman, J., M. Hahn, R. Weisbrod, B. Woodin, J. Joy, S. Najibi and R. Cohen. Induction of cytochrome P4501A1 by aryl hydrocarbon receptor agonists in porcine aorta endothelial cells in culture and cytochrome P4501A1 activity in intact cells. *Mol. Pharmacol.* 47: 296–306, 1995.
106. Stegeman, J.J. Polynuclear aromatic hydrocarbons and their metabolism in the marine environment, In: *Polycyclic Hydrocarbons and Cancer*, Vol. 3, edited by H.V. Gelboin and P.O.P. Ts'o, New York, Academic Press, pp. 1–60, 1981.
107. Stegeman, J.J. and M.E. Hahn. Biochemistry and molecular biology of monooxygenases: current perspectives on forms, functions, and regulation of cytochrome P450 in aquatic species. In: *Aquatic Toxicology: Molecular, Biochemical, and Cellular Perspectives*, edited by D.C. Malins and G.K. Ostrander, Ann Arbor, MI, Lewis Publishers, pp. 87–206, 1994.
108. Stegeman, J.J., M.R. Miller and D.E. Hinton. Cytochrome P450IA1 induction and localization in endothelium of vertebrate (teleost) heart. *Mol. Pharmacol.* 36: 723–729, 1989.
109. Stegeman, J.J., B.R. Woodin and R.L. Binder. Patterns of benzo[*a*]pyrene metabolism by varied species, organs, and developmental stages of fish. *Natl Cancer Inst. Monogr.* 65: 371–377, 1984.
110. Tillitt, D.E. and D.M. Papoulias. Toxicological highlight 2,3,7,8-Tetrachloridibenzo-*p*-dioxin toxicity in the zebrafish embryo: local circulation failure in the dorsal midbrain is associated with increased apoptosis. *Toxicol. Sci.* 69: 1–2, 2002.
111. Toomey, B.H., S. Bello, M.E. Hahn, S. Cantrell, P. Wright, D.E. Tillitt and R.T. Di Giulio. 2,3,7,8-Tetrachlorodibenzo-*p*-dioxin induces apoptotic cell death and cytochrome P4501A expression in developing *Fundulus heteroclitus* embryos. *Aquat. Toxicol.* 53: 127–138, 2001.
112. Trippel, E., O. Kjesbu and P. Solemmial. Effects of adult age and size structure on reproductive output in marine fishes. In: *Early Life History and Recruitment in Fish Populations*, London, Chapman & Hall, pp. 31–55, 1997.
113. Van Veld, P.A. Absorption and metabolism of dietary xenobiotics by the intestine of fish. *Rev. Aquat. Sci.* 2: 185–203, 1990.
114. van Wezel, A., D. de Vries, S. Kostense, D. Sijm and A. Opperhuizen. Intraspecies variation in lethal body burdens of narcotic compounds. *Aquat. Toxicol.* 33: 325–342, 1995.
115. van Wezel, A. and A. Opperhuizen. Narcosis due to environmental pollutants in aquatic organisms: residue-based toxicity, mechanisms and membrane burdens. *Crit. Rev. Toxicol.* 25: 255–279, 1995.
116. van Wezel, A. and A. Opperhuizen. Thermodynamics of partitioning of a series of chlorobenzenes to fish storage lipids, in comparison to partitioning to phospholipids. *Chemosphere* 31: 3605–3615, 1995.
117. Varanasi, U., M. Nishimoto and J. Stover. Analyses of biliary conjugates and hepatic DNA binding in benzo(*a*)pyrene-exposed English sole. In: *Polynuclear Aromatic Hydrocarbons, Eighth Symposium*, edited by M. Cook and A. Dennis, Columbus, OH, Battelle Press, 1986.
118. Vodicnik, M.J. and R.E. Peterson. The enhancing effect of spawning on elimination of a persistent polychlorinated biphenyl from female yellow perch. *Fundam. Appl. Toxicol.* 5: 770–776, 1985.
119. White, R.D., D. Shea and J.J. Stegeman. Metabolism of the aryl hydrocarbon receptor agonist 3,3′,4, 4′tetrachlorobiphenyl by the marine fish scup (*Stenotomus chrysops*) *in vivo* and *in vitro*. *Drug Metab. Dispos.* 25: 564–572, 1997.
120. Whitlock, J.P. Induction of cytochrome P4501A1. *Annu. Rev. Pharmacol. Toxicol.* 39: 103–125, 1999.
121. Wofford, H. and P. Thomas. Effect of xenobiotics on peroxidation of hepatic microsomal lipids from striped mullet (*Mugil cephalus*) and Atlantic croaker (*Micropogonias undulatus*). *Mar. Environ. Res.* 24: 285–289, 1988.
122. Meador, J.P., T.K. Collier and J.E. Stein. Tissue and sediment-based threshold concentrations of polychlorinated biphenyls (PCBs) to protect juvenile salmonids listed under the US Endangered Species Act. *Aquat. Conserv. Mar. Freshw. Ecosyst.* 12: 493–516, 2002.
123. Shearer, K., T. Åsgård, G. Andorsdöttir and G. Aas. Whole body elemental and proximate composition of Atlantic salmon (*Salmo salar*) during the life cycle. *J. Fish Biol.* 44: 785–797, 1994.

Biochemistry and Molecular Biology of Fishes, vol. 6
T. P. Mommsen and T. W. Moon (Editors)

CHAPTER 5

Metabolic fate of nonylphenols and related phenolic compounds in fish

JEAN-PIERRE CRAVEDI AND DANIEL ZALKO

UMR 1089 Xénobiotiques, INRA, BP3 31931 Toulouse Cédex 9, France

I. Introduction

Alkylphenols (APs), particularly nonylphenols (NPs) and to a lesser extent octylphenols (OPs), are extensively used for the production of alkylphenol polyethoxylates (NPEOs), a class of non-ionic surfactants that has been largely employed for more than 40 years in textile and paper processing and in the manufacture of paints, coatings, pesticides, industrial detergents, cosmetics and spermicidal preparations, as well as various cleaning products. NPs are also used in the manufacturing processes of many plastics and as monomers in the production of phenol/formaldehyde resins. Smaller quantities of NPs are employed in the production of tri(4-nonylphenyl) phosphite as an anti-oxidant for rubber and in the manufacture of lubricating oil additives.

NPs cover a large number of isomeric compounds of general formula $C_6H_4(OH)C_9H_{19}$ (Fig. 1). Commercially produced NPs are predominantly 4-nonylphenol, with a varied and undefined degree of branching in the attached alkyl group. Industrial formulations usually comprise less than 10% of the *ortho* isomer. Due to the method of production, very little, if any, straight chain nonylphenol (4-*n*-NP) is present in commercial mixtures. However, in metabolic studies, 4-*n*-NP was often selected as a model for NPs because its chemical structure is defined,

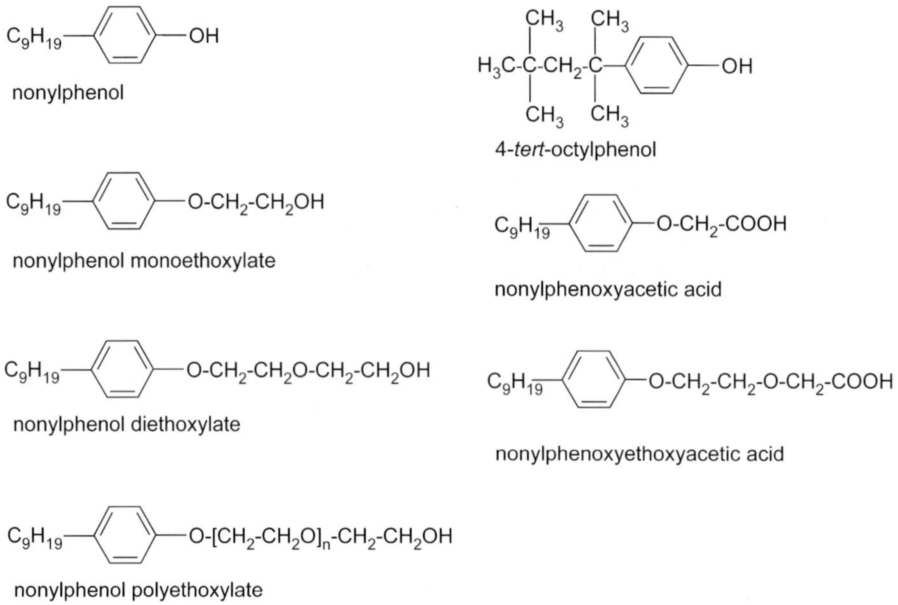

C_9H_{19}–C$_6$H$_4$–OH

nonylphenol

H_3C-C(CH_3)$_2$-CH_2-C(CH_3)$_2$–C$_6$H$_4$–OH

4-*tert*-octylphenol

C_9H_{19}–C$_6$H$_4$–O-CH_2-CH_2OH

nonylphenol monoethoxylate

C_9H_{19}–C$_6$H$_4$–O-CH_2-COOH

nonylphenoxyacetic acid

C_9H_{19}–C$_6$H$_4$–O-CH_2-CH_2O-CH_2-CH_2OH

nonylphenol diethoxylate

C_9H_{19}–C$_6$H$_4$–O-CH_2-CH_2-O-CH_2-COOH

nonylphenoxyethoxyacetic acid

C_9H_{19}–C$_6$H$_4$–O-[CH_2-CH_2O]$_n$-CH_2-CH_2OH

nonylphenol polyethoxylate

Fig. 1. Structures of the alkylphenols and related compounds discussed in this review. Technical nonylphenol consists of up to 22 isomers, most of them having a branched alkyl side-chain in the *para* position.

whereas NPs are a mixture of different isomers, which can make the identification of metabolites difficult.

NPs enter the aquatic environment directly due to their use in plastic production, or through the degradation of NPEOs. When discharged into sewage treatment plants, NPEOs are degraded aerobically to NP di- and mono-ethoxylates (NP2EOs and NP1EOs) and alkylphenol monocarboxylates, which can be further degraded, mostly anaerobically, into the corresponding NP. It has been estimated that about 60% of the NPEOs produced end up in the aquatic environment as NPs, NP2EOs and NP1EOs[8]. Because of the incomplete degradation of these compounds, breakdown products of NPEOs can be detected not only in sewage effluents but also in river systems and coastal environments. The presence of NPs in aquatic ecosystems may be the result of the direct use of this chemical in industry or may originate from plastic leaching[45,62].

Many studies have reported levels of NPs and related compounds in aquatic environments throughout Europe, North America and Japan. Values higher than 100 μg/l have been reported[11,66]. However, this range of concentration is unusual and most of the authors have reported values below 1 μg/l in surface waters, as discussed in recent reviews[10,81]. Due to their hydrophobic properties, NPs tend to be associated with sediments. The analysis of samples collected from the Great Lakes basin and the upper St Lawrence River yielded concentrations between 0.17 and 72 μg/g[9,37], whereas values between 6 and 69 μg/g were found in sediments from the Mediterranean Sea near Barcelona[15]. In Japan, the concentrations of NPs were found to range from 0.03 to 13 μg/g in riverine and bay sediments from the Tokyo metropolitan area[31], whereas OP levels were about one order of magnitude lower.

During the last decade, NPs have attracted interest because of their estrogenic activity. The initial demonstration of the estrogenic properties of NPs was reported by Soto *et al.*[62]. These authors observed that proliferation of estrogen-sensitive breast cancer cells occurred in connection with the leaching of this compound from laboratory plastic-ware. More recently, the estrogenic effects of NPs have been documented in a number of *in vitro* studies conducted in various species of fish. It was found that NPs stimulated vitellogenin production in primary cultures of hepatocytes, including those isolated from male fish, which were not thought to be involved in vitellogenin synthesis[25,32,79]. NPs are shown to competitively bind to the estrogen receptors[25,54,58,76,79] in fish and mammals. Yeast cells transfected with the rainbow trout (*Oncorhynchus mykiss*) estrogen receptor and containing an estrogen responsive element reporter system were used to demonstrate the estrogenicity of NPs as well as that of other structurally related compounds[43,57]. All were less potent than 17β-estradiol. These *in vitro* data are corroborated by several *in vivo* experiments carried out in rainbow trout: both NP (20.3 μg/l) and OP (4.8 μg/l) inhibit testicular growth and induce vitellogenin production after a 3-week exposure period[33]. NPs can also induce the formation of testis–ova and alter the sex ratio in Japanese medaka (*Oryzias latipes*)[28]. Several other experiments in various fish species have revealed elevated plasmatic levels of vitellogenin and/or reproductive abnormalities after exposure to APs[5,16,21,27,36,44,50,65]. Moreover, hormonal imbalances were detected in the offspring of rainbow trout exposed to 10 μg/l NPs, suggesting a transgenerational effect mediated by the endocrine system[59].

In addition to their effect on reproduction, NPs are acutely toxic to fish. For rainbow trout, the LC50 under flow-through conditions for 72 h was calculated at 193.65 μg/l of NP[36]. Similar results are reported for other freshwater as well as saltwater species[63]. In chronic toxicity tests carried out with early life stages of rainbow trout exposed to NP or OP for 90 days, no observable effect concentrations were as low as 6 μg/l for both the compounds[60]. Moreover, it is now established that there is an increase in the toxicity of both NPEOs and OP polyethoxylates with the gradual shortening of the ethoxylate chain and that carboxylated metabolites are less toxic than the corresponding parent compounds.

NPs are hydrophobic organic chemicals with an aqueous solubility of 5.4 mg/l[1] and an octanol:water partition coefficient (log K_{ow}) of 4.48, suggesting a possible bioaccumulation in living organisms and a biomagnification through the aquatic food chain[2]. The main defense mechanism for fish against the concentration of lipophilic chemicals in their tissues is the biotransformation of the foreign compounds into more polar, readily excretable molecules. However, in some cases, biotransformation can result in the formation of products that are much more active or toxic than the parent chemical. For example, rainbow trout can metabolize the pesticide biphenyl into the corresponding hydroxylated and dihydroxylated metabolites[18], which are proven to be considerably more estrogenic than the parent compound[58]. Consequently, to better understand the effects of NPs on fish, to predict the possible impact of these chemicals on the aquatic environment and to assess the risk associated with the presence of NP residues in fish for human consumption, it is essential to know precisely the metabolic fate of these estrogenic compounds in fish species. This article reviews the fate of NPs and OPs, as well as that of closely related phenolic compounds, in fish. When useful, the information will be combined with some available data on the metabolism of these compounds in mammals in order to take into account the comparative aspects of this subject.

II. Bioaccumulation and disposition of NPs and related chemicals in fish

According to several survey programs measuring the concentrations of NPs in fish, a moderate bioaccumulation of these compounds may occur in several species. Chub (*Squalus cephalus*), barbel (*Barbus barbus*) and rainbow trout caught in surface waters from the Glatt Valley (Switzerland) were analyzed for NPs, NP1EOs and NP2EOs levels in various tissues[3]. Concentrations of NPs ranged from <0.03 to 1.6 μg/g, based on dry weight, whereas results for NP1EOs and NP2EOs ranged from 0.15 to 7.0 μg/g and <0.03 to 3.0 μg/g, respectively. In muscle, values ranged from 0.15 to 0.38 μg/g for NPs, from 0.18 to 3.10 μg/g for NP1EOs and from <0.03 to 2.3 μg/g for NP2EOs. Using these data, the authors calculated bioconcentration factors (ratio of the concentration of NPs in fish tissues relative to the concentration of NPs in surrounding water) to be in the range of 13–408, 3–300 and 3–326 for NPs, NP1EOs and NP2EOs, respectively. These values were re-calculated by Staples *et al.*[63] on a wet weight basis (assuming that muscle is 85% water and 15% dry matter) in order to make them comparable with other data; their data indicated a bioconcentration factor in muscle ranging from 0.8 to 37. A 1999 survey of NPs and NPEOs in Kalamazoo River (Michigan, USA) found that residue levels were below the detection limit for these chemicals in 59% of the 183 analyzed fish[34]. Reported concentrations ranged from 3.3 (limit of detection) to 29.1 ng/g wet weight. Another survey in the same area in 2000 found only one fish sample that was above the detection limit. The NP and OP concentrations in fish caught in the rivers flowing into Lake Biwa (Japan) were reported to range from 10 (limit of detection) to 110 ng/g wet weight and from 1 (limit of detection) to 6 ng/g wet weight, respectively[77]. In the same study, analyses were also carried out for water, allowing the calculation of bioconcentration factors for NPs in *Zacco platypus*, *Plecoglossus altivelis*, *Zacco temminckii*, *Carassius carassius*, *Micropterus salmoides*, *Lepomis macrochirus*, and for OPs in *Z. platypus*, *P. altivelis*, *Z. temminckii*, and *M. salmoides*. The values calculated for NPs and OPs were 15–31 and 46–297, respectively.

In addition to these field studies, the bioaccumulation of NPs was investigated in controlled exposure assays. In littoral enclosures, bluegill sunfish (*L. macrochirus*) exposed to NPs exhibited a bioconcentration factor[42] of 87. The bioconcentration factors reported from laboratory experiments vary from 75 to 741, according to the species and the exposure conditions[60,61,63,78]. Radiolabeled NP was used to evaluate the NP concentration in stickleback (*Gasterosteus aculateus*) at steady state in seawater[22]. In these conditions, the residues in tissues amounted to *ca* 6 μg of NP equivalent per gram fresh weight, corresponding to a bioconcentration factor of 1300. The difference between this value and those reported above could be due to the presence of metabolites in the radioactivity extracted from fish tissues, resulting in an overestimation of the quantity of NP in sticklebacks. This hypothesis is in agreement with the bioconcentration factor of 4-*tert*-OP in roach (*Rutilus rutilus*) fry exposed to radiolabeled 4-*tert*-OP in a semi-static system and calculated on the basis of the ratio between radioactivity levels measured in fish and water[23]. In these conditions, apparent bioconcentration factors between 346 and 1134 have been reported, depending on the age of fry and on the length of the exposure period. Nevertheless, the apparent bioconcentration factor estimated in 40–60 g rainbow trout waterborne exposed to ^{14}C-NP, on the basis of radioactivity measurements, was 24 and 98 for carcass and viscera, respectively[41]. This suggests that several parameters such as the species, the exposure conditions, the physiological state of the fish and the structure (or isomeric composition) of the studied AP may influence significantly the bioconcentration factor values. In the Lewis and Lech study[41], both uptake and depuration were quite rapid. A steady state was observed within 12 h during the uptake, and depuration half-lives were 18.6 and 19.6 h for muscle and fat, respectively. In the whole Japanese medaka, the biological half-lives observed by Tsuda *et al.*[78] were 9.9 and 7.7 h for NPs and OPs, respectively. Slower depuration kinetics were obtained by Coldham *et al.*[17] in juvenile rainbow trout (mean weight 122 g) after an intravenous injection of ^{3}H-NP. The depletion kinetics of labeled residues from tissues and plasma were biphasic, corresponding to the distribution (α) and elimination (β) phase half-lives. Prolonged β-phase half-lives of 99 h were reported for muscle and liver residues, whereas the value for plasma was 40 h. These findings concur with earlier studies in Atlantic salmon[48] (*Salmo salar*) estimating that the clearance half-life of NPs was about 4 days.

Levels of AP residues have been shown to vary among tissues with relatively high concentrations generally found in viscera. When rainbow trout were waterborne exposed to 36 μg/l ^{14}C-NP during 14 h, the highest concentration of radioactivity was found in bile[41]. The levels of radioactivity measured in liver, kidney and fat represented less than 1% of those measured in bile, and the concentration of NP residues in muscle represented only 12.6% of the hepatic NP residues. In the Coldham *et al.*[17] study, ^{3}H-4-*n*-NP was administered to rainbow trout as a single dose of 3.1 mg/kg by intravenous injection *via* the caudal vessels. The distribution and elimination of residues were determined by radioactivity measurements in tissues sampled 1, 2, 4, 24, 48, 72 and 144 h after dosing. After 2 h the highest concentration was found in liver, suggesting a rapid biotransformation of NP in this organ, while after 24 h the highest residue levels were detected in bile, feces, intestine and pyloric caeca, indicating a biliary and fecal route of excretion. Feces contained the major part of residues 144 h after dosing. The distribution of residues after oral administration was investigated in immature rainbow trout[67,68,71] and Atlantic salmon[6]. In these studies, a single dose of [R-2,6-^{3}H]-4-*n*-NP was administered by gavage to the fish. In both the species, the results showed that radioactivity was widely distributed in tissues but mainly retained in bile, viscera, liver, fat and kidney. In trout, 2 days after dosing, about 11% of the administered dose was still present in the whole fish (based on the combustion of tissue samples followed by radioactivity determination). Radioactivity was also present in substantial amounts in the gonads. The levels measured in these tissues 48 and 144 h after dosing represented about 25% of the corresponding hepatic residues. These levels are of potential significance since gonads are target tissues for xenoestrogens. In addition, whole body autoradiography has been used by several authors to investigate the disposition of labeled 4-*n*-NP in salmonids[7,17]. The results obtained in these conditions are in accordance with tissue combustion data and indicate high radioactivity levels in bile, pyloric caeca and intestine whatever the exposure route (water exposure, intragastric administration or intravenous injection).

Additionally, the disposition of another AP, ^{14}C-4-*tert*-OP, was investigated in two detailed studies conducted in rainbow trout[24] and rudd[56] (*Scardinius erythrophthalmus*). After 10 days of waterborne exposure to 4 μg/l 4-*tert*-OP, the concentration of residues in rainbow trout was the highest in bile, followed by feces, pyloric caeca, liver and intestine. The apparent bioconcentration factors for liver and muscle were 1020 and 101, respectively. In similar experimental conditions, the majority of radioactivity in rudd was recovered in bile and liver. No data were presented on the amount of residues in pyloric caeca, intestine and feces. When fish were transferred to clean water for 8 days, residue levels in bile and liver were reduced by at least 99 and 81%, respectively.

The toxicokinetics of [^{14}C]-4-heptylphenol were investigated in Atlantic cod (*Gadus morhua*) exposed to a concentration of 3 μg/l sea water[75]. The distribution pattern after an exposure period of 192 h was in close agreement with previously reported information regarding AP residues: the major part of the radioactive material was located in bile, liver and intestine as well as in the content of the digestive tract. This observation was confirmed by whole body autoradiograms. The biological half-life of heptylphenol was estimated at about 13 h, indicating a rapid elimination of this compound. No trace of radioactivity was detected 8 days after fish were transferred to clean seawater.

Residues measured in rainbow trout tissues 72 h after a single oral dose of ^{3}H-4-*n*-NP2EO exhibited a distribution pattern similar to those previously mentioned for NPs[19]. High levels of radioactivity were found in bile, viscera and liver, while the concentration of residues in muscle averaged only 10% of that calculated for liver.

The data from the above studies provide the basis for a better understanding of the disposition of APs in fish. They show that oral administration or waterborne exposure results in an efficient uptake of these chemicals. Radiolabeled NPs are rapidly absorbed from the gastrointestinal tract, as demonstrated by the radioactivity levels detected in blood and tissues 24 h after dosing. It is difficult to determine precisely the oral bioavailability of these compounds on the basis of the published studies. Nevertheless, Thibaut *et al.*[68,69] reported that 48 h after the administration of ^{3}H-4-*n*-NP to rainbow trout, 5.5 and 3.0% of the radioactivity was recovered from bile and urine, respectively. This suggests that the digestive absorption must be at least 8.5%. Nevertheless, very little information is available on the kinetics of residues in fish submitted to a repeated AP oral exposure.

When exposure is carried out using an aqueous solution/suspension, the steady state for NPs may be reached in a few hours, suggesting a ready absorption through the gills, a major route of entrance of chemicals into fish[46]. However, several reports including the OP studies indicate that APs attain steady state conditions after a few days and undergo a widespread body distribution, with the highest concentrations of residues found in the digestive system. For the intravenous route, biphasic kinetics imply an initial phase of distribution from the blood to a second compartment (presumably the lipid compartment and/or the binding of metabolites to liver macromolecules) followed by a slower elimination phase.

Whatever the exposure route, studies show that excretion occurs predominantly in feces and bile. The rapid rate of elimination of these lipophilic compounds, as well as the substantial amount of residues present in liver and bile suggest that NPs could be metabolized in liver prior to biliary excretion.

III. Biotransformation pathways

The large number of existing AP isomers is one of the main challenges faced by scientists wanting to develop a precise understanding of the metabolic fate of APs in fish, as well as in other vertebrates. This research field is key to determine both the nature of AP residues in tissues and the pharmacological (and/or toxicological) properties of AP metabolites. Significant progress has been made over the last few years. The metabolism of NP in fish has been elucidated primarily using the model compound 4-*n*-NP (e.g. the isomer for which the alkyl side-chain of the molecule is linear and is linked to the phenol moiety on the *para* position) (Fig. 1). Conversely, experiments carried out to examine the metabolic fate of OP were achieved using the branched model molecule 4-*tert*-OP (Fig. 1). Based on these data for 4-*n*-NP and 4-*tert*-OP, a general pattern of AP metabolism in fish can be drawn with reasonable certainty (Fig. 2). However, few data are available concerning the biotransformation of branched NP isomers, and additional studies will be necessary in the future to assess some specific points of AP biotransformation.

To date, *in vivo* studies of the fate of APs in fish were carried out in rainbow trout with 4-*n*-NP[17,41,67,68,71], branched NPs[49] and 4-*tert*-OP[24] as models. The metabolic fate of 4-*n*-NP was also investigated in mosquitofish (*Gambusia holbrooki*)[74] and Atlantic salmon[6], whereas 4-*tert*-OP was used for metabolic studies performed in roach (*R. rutilus*)[23] and in rudd[56]. All studies clearly demonstrated that, on a quantitative basis, the predominant metabolic pathways for APs were the conjugation of the phenol group to glucuronic acid (Fig. 2; A, structure I) and the oxidative biotransformations of the alkyl side-chain (Fig. 2; B, structures II to XV).

AP glucuronides are demonstrated not only for the parent compounds NP and OP, but also for most of their metabolites arising from the oxidation of the alkyl side-chain. They are extensively excreted in bile[7,17,23,41,49,56,67,68] and subsequently reach the digestive tract[17,67,68,70]. Some have been characterized in feces[70] and urine[6,69]. Whether the glucuronidation of hydroxylated AP metabolites takes place before or after the hydroxylation step remains unclear and evidence exists to support both

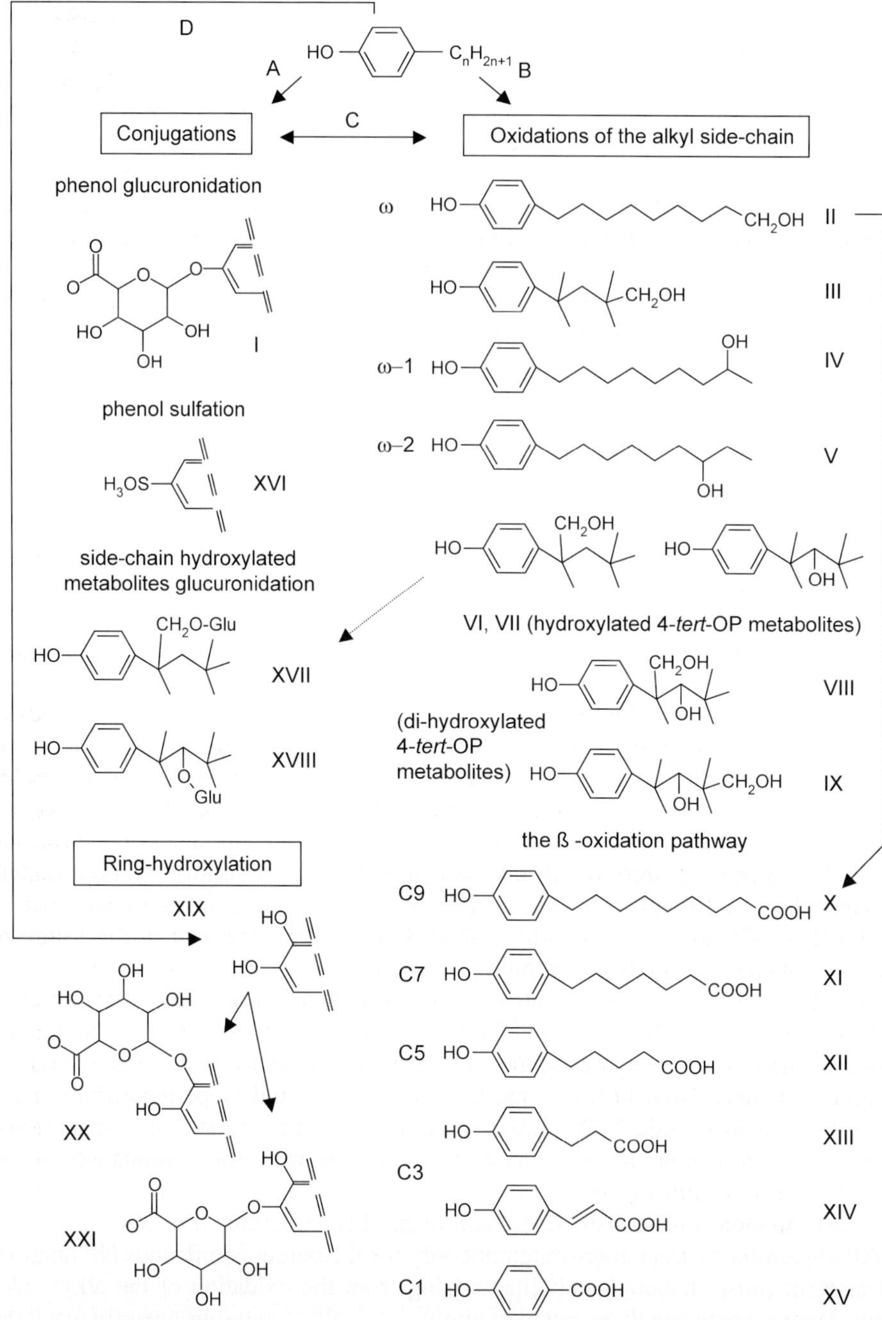

Fig. 2. Synthetic scheme of the biotransformation pathways of alkylphenols in fish, based on the references detailed in the main text. Metabolites II to XV have been solely (or predominantly) identified as glucuronic acid conjugates, as displayed for structure I.

the possibilities (Fig. 2; C). Compared with the glucuronidation pathway, AP sulfation (Fig. 2; XVI) is poorly represented in fish. The sulfation of the phenol moiety of AP was demonstrated *in vivo*, as one 4-*n*-NP sulfated metabolite was isolated from the digestive tract of rainbow trout[71] (Fig. 2; structure XI), and *in vitro* for hydroxylated 4-*n*-NP (Fig. 2; structure II), when incubating the parent compound with isolated trout hepatocytes[17]. Though sulfate conjugation of phenolic compounds is documented in fish and is known to be efficient in cyprinids[35], it is generally expressed at a lower level than it is for mammals, especially in salmonids. Phenyl sulfate kinetics in rainbow trout suggested the existence of a high-affinity/low-capacity pathway, while phenyl glucuronide production was suggestive of a low-affinity/high-capacity formation pathway[47]. Consequently, nearly all AP metabolites have been characterized in fish as glucuronide conjugates, whereas both conjugation pathways have been shown to be equally efficient in the detoxification of 4-*n*-NP in the rat[83].

Oxidative biotransformations of APs in fish are the second major metabolic pathway clearly characterized for APs. The resulting metabolites are shown to be numerous, as listed in Fig. 2 (pathway B). The contribution of this pathway to the metabolic breakdown of APs depends primarily on the structure of alkyl side-chain. The terminal (ω) hydroxylation of the alkyl side-chain was characterized for all APs for which extensive metabolism studies were carried out. For instance, ω-hydroxylated 4-*n*-NP (Fig. 2; metabolite II) was characterized as a glucuronic acid conjugate in Atlantic salmon urine and bile[6], as well as in juvenile rainbow trout bile, liver and feces[17]. Isolated trout hepatocytes were shown to metabolize 4-*tert*-OP into the corresponding ω-hydroxylated metabolite (Fig. 2; metabolite III), detected in incubation supernatants as a glucuronide conjugate[55]. The sub-terminal (ω-1) hydroxylation of APs is also a common metabolic pathway in fish. It was reported for 4-*n*-NP (Fig. 2; structure IV) *in vivo* through the identification of the corresponding glucuronides in bile[6,68] and urine[6], as well as for NP isomers 2-, 3- and 4-(4-hydroxyphenyl)-nonane in rainbow trout bile[49]. However, this biotransformation is structurally impossible for 4-*tert*-OP. The latter AP is predominantly hydroxylated at other locations of the alkyl side-chain, resulting in hydroxylated (Fig. 2; structures VI and VII) and di-hydroxylated (structures VIII and IX) metabolites[56]. It is worth noting that compounds VI and VII are the only hydroxylated AP metabolites for which the occurrence of a glucuronidation on a side-chain hydroxy group was observed (structures XVII and XVIII), as stated by Pedersen and Hill[56]. Some evidence was presented for an ω-2 hydroxylation of 4-*n*-NP following *in vitro* studies carried out with Atlantic salmon liver microsomal fractions[73] (Fig. 2; structure V), but the *in vivo* occurrence of this biotransformation has never been demonstrated.

It is now clearly established that the linear NP isomer 4-*n*-NP can be further oxidized into the corresponding nine-carbon carboxylic acid (Fig. 2; structure X). The glucuronide conjugate of this metabolite has been characterized in mosquitofish bile[74]. This biotransformation is a key-point in the metabolism of linear side-chain APs. Indeed, from this point, linear APs such as 4-*n*-NP enter the β-oxidation metabolic pathway in which each loop results in the loss of two carbon atoms, thereby producing 7-, 5-, 3- and ultimately 1-carbon side-chain carboxylic acid metabolites (Fig. 2;

structures XI to XV). The β-oxidation pathway was extensively characterized for 4-*n*-NP in salmonids *in vivo*[6,67]. In rainbow trout as well as in Atlantic salmon, it results in the complete breakdown of the nonyl chain, the ultimate known metabolite being the *para*-hydroxy benzoic acid (Fig. 2; structure XV), recovered as a glucuronide conjugate. The β-oxidation of 4-*n*-NP may take place in either the mitochondrial and/or the peroxisomal cellular compartment (for a detailed discussion of this subject, see ref. 67). Most of the resulting metabolites have been characterized as glucuronides and are predominantly excreted in bile.

Branched APs such as 4-*tert*-OP cannot undergo a complete shortening of the alkyl side-chain *via* β-oxidation, as this metabolic pathway can only proceed on linear carbon chains. This is probably the reason why other metabolic reactions are favored for branched APs. As mentioned above, various hydroxylation sites of the alkyl side-chain have been characterized in metabolic fate studies of 4-*tert*-OP (Fig. 2; structures VI to IX). However, this AP can also undergo a ring-hydroxylation in the *ortho* position to produce the corresponding catechol[23,55] 4-*tert*-octylcatechol (Fig. 2; structure XIX), a major metabolite found in rainbow trout's digestive tract, feces and liver[23]. This metabolite can be further glucuronidated in two different locations (Fig. 2; structures XX and XXI). Unlike in rat, no methylation of 4-*tert*-octylcatechol seems to occur in fish.

The metabolism of NPEOs in fish, however, is poorly documented. It was demonstrated recently[19] that the major metabolite of 4-*n*-NP2EO in rainbow trout is 4-*n*-NP2EO-glucuronide. Interestingly, no traces of 4-*n*-NP or 4-*n*-NP-glucuronide were detected in the analyzed samples, suggesting that trout cannot metabolize ethoxylated APs into the corresponding APs. This observation suggests that the weak estrogenic activity attributed to NP2EO[32,40,57] is likely due to the effect of the compound *per se* and not to the biotransformation of NP2EO into NP.

The data presented to this point regarding the metabolic fate of APs in fish provide useful indications about their potential toxicity. Several findings point to an efficient detoxification of these xenobiotics. Indeed, AP glucuronides of the parental compounds, which are major *in vivo* metabolites, are unlikely to possess a significant pharmacological activity in terms of estrogenicity, as was demonstrated by Moffat *et al.*[53] for the 4-*n*-NP and 4-*tert*-OP glucuronides. Additionally, the estrogenic activity of the main metabolites produced by the β-oxidation of 4-*n*-NP, again major *in vivo* metabolites, is thought to be impaired by the absence of the alkyl side-chain. We have recently tested *in vitro* several conjugated and unconjugated metabolites originating from 4-*n*-NP *via* the β-oxidation pathway with negative results (Cravedi and Zalko, unpublished data).

Some specific points of AP metabolism deserve additional research. As mentioned above, 4-*n*-NP glucuronide is not estrogenic. However, studies by Thibaut *et al.*[67–71] clearly demonstrated that this metabolite is the major residue of 4-*n*-NP detected in rainbow trout tissues. In some physiological conditions, 4-*n*-NP glucuronide may be released into the blood and hydrolyzed back to the parent compound, and hence may be considered as a metabolite with a potential estrogenic activity.

The terminal and sub-terminal oxidative biotransformations of APs (Fig. 2; structures II to IV) have attracted much attention as the produced metabolites possess two hydroxy groups located at both the extremities of the molecule, and thus may have a closer structure–activity relationship with natural estrogens like estradiol. The estrogenicity of these metabolites might, therefore, be enhanced compared with the parent molecule. As illustrated by the difference between the metabolic fates of 4-*n*-NP and 4-*tert*-OP, branched APs are not expected to undergo a complete breakdown of the alkyl side-chain by β-oxidation. Thus, these isomers are expected to produce more hydroxylated metabolites.

Data concerning the fate of branched NPs are unavailable with the exception of one study[49]. Obviously, our knowledge about these hydroxylations and the estrogenic activity of the corresponding metabolites is limited and it is currently impossible to predict if the hydroxylations of the alkyl side-chain of AP isomers could enhance their estrogenicity. In a study carried out in mosquitofish, Thibaut *et al.*[74] demonstrated that unconjugated (ω-1) hydroxy-4-*n*-NP (Fig. 2; structure IV) was, together with 4-*n*-NP glucuronide, the major metabolite detected in both oocytes and embryos. Since endocrine disruptors such as 4-*n*-NP could play a major role during early life stages, these results clearly point to the need for a better understanding of the biological activity of AP metabolites.

The ring-hydroxylation pathway of AP (Fig. 2; structure XIX) also raises questions. In fish, it has only been demonstrated for 4-*tert*-OP. However, it may also occur for NP, since it was demonstrated previously for 4-*n*-NP[83] and for non-linear NP isomers[20] in the rat. This metabolic pathway is quantitatively minor but may provide some explanations for the relatively high toxicity of APs in fish[63]. Indeed, the resulting metabolites are catechols that are expected to produce reactive intermediates by analogy with the known toxicity of catechol-estrogens[12]. Such intermediates could bind endogenous molecules, resulting in various toxic effects depending on their cellular targets. Therefore, the biological significance of this pathway should be carefully investigated.

IV. Enzymes involved in NP biotransformations and interaction of NPs with these enzymes

An important route of biotransformation of NPs in fish appears to be the formation of hydroxylated metabolites, as reported for other xenobiotics. This reaction is usually mediated by the cytochrome P450-dependent mixed-function oxidase pathway, which is located primarily in the hepatic endoplasmic reticulum. Upon disruption of the cells and fragmentation of the endoplasmic reticulum, these enzymes are localized in the microsomal fraction. Cytochromes P450s (CYPs) constitute a large superfamily of heme thiolated proteins and mediate the NADPH/O_2-dependent oxidation of a broad spectrum of xenobiotics and endogenous compounds. More than 15 CYP isoforms have been described in fish, mainly in rainbow trout[13,80]. Among them, the CYP1A, CYP2K, CYP2M, and CYP3A families are the best documented isoforms.

Meldahl *et al.*[49] reported that the metabolism of NP isomers by trout hepatic microsomes was inhibited when incubations were carried out without NADPH, or when a CYP inhibitor, piperonyl butoxide, was added to the incubations. The *in vivo* treatment of fish with β-naphthoflavone (a known CYP1A1 inducer) prior to the preparation of hepatic microsomes did not result in any major difference in the metabolic profiles obtained after incubating NPs with the microsomal fractions. These findings indicate not only that hepatic CYPs enzymes are involved in NPs metabolism in fish, but also suggest that CYP1A1 is probably not the responsible P450 isoform. Similar conclusions were drawn by Arukwe *et al.*[7] following the incubation of cultured salmon hepatocytes with labeled 4-*n*-NP in the presence or absence of β-naphthoflavone.

Several approaches were undertaken to identify the CYP isoforms involved in the hydroxylation of 4-*n*-NP by Atlantic salmon hepatic microsomes[73]. One experiment involved pre-incubating microsomes with ketoconazole, α-naphthoflavone and ethynylestradiol, which are strong non-specific inhibitors of rainbow trout CYP2K, 1A1 and 3A27, respectively[51]. Ketoconazole was also shown to inhibit laurate ω-hydroxylase in Atlantic salmon, a biotransformation known to be associated with the activity of the CYP2M family[52]. The metabolic profiles obtained in these conditions were compared with those obtained when no inhibitors were used. Ketoconazole inhibited the ω-, ω-1- and ω-2-hydroxylation of 4-*n*-NP, suggesting the involvement of CYP2K-like and possibly CYP2M-like isoforms in NP hydroxylation. Furthermore, ethinyl estradiol specifically inhibited the ω-1- and ω-2-hydroxylation, indicating that CYP3A27 may play a role in the metabolism of the alkyl chain. However, these results were not clearly confirmed in salmon treated with pregnenolone-16α-carbonitrile, a compound reported to be a CYP3A inducer in rainbow trout[14] and tetradecylthioacetic acid, a chemical classified as a peroxisome proliferating agent and a purported inducer of the CYP2K and CYP2M isoforms[73].

Although experimental evidence indicates that fish should possess more than one isoform of UDP-glucuronosyltransferase[26], no information exists on the nature of the isoform(s) involved in the glucuronidation of APs.

NP has been reported to increase the activity of steroid-metabolizing activities such as the 6β-, 16α- and 17α-hydroxylases when injected at low concentrations (1 mg/kg body weight) to juvenile Atlantic salmon, whereas these activities, as well as CYP1A, CYP3A-like and CYP2K-like hepatic proteins were down-regulated at higher concentrations[4]. An up-regulation of the CYP3A-like protein was also observed by Thibaut *et al.*[73] in salmon injected with 5 mg 4-*n*-NP per kg body weight. NP-related compounds can also interfere with the metabolism of steroids. The 6β- and 16β-hydroxylations of testosterone in rainbow trout liver were significantly inhibited by NP2EO[64]. The interference of APs with the expression and activity of steroid biotransformation enzymes is not fish specific. Several investigations in the rat by Lee and co-workers[38,39] demonstrated a down-regulation of the CYP1A hepatic protein and an induction of CYP3A by NPs. In addition, 4-*tert*-OP (and to a lesser extent OP-diethoxylate) are able to inhibit 16β-, 2α- and 6β-testosterone hydroxylases[29]. Conversely, the potential of APs and related compounds to modulate steroid conjugation in fish is difficult to comprehend from the few studies reported in the literature and the apparent conflicting results obtained. The data reported *in vivo* on testosterone glucuronidation indicate that this activity remains unchanged when rainbow trout are exposed to 4-*n*-NP or NP2EO[64,72]. However, NP2EO did inhibit the glucuronidation of testosterone in trout cultured hepatocytes[64]. Although APs are reported to alter the expression of estrogen sulfotransferase in humans[30], whether they modulate the sulfation of steroids in fish remains unknown.

Although the liver is recognized as the major site of xenobiotic-metabolizing enzymes in fish as well as in mammals, other organs may contribute significantly to the biotransformation of chemicals. Incubation of NPs carried out with fish hepatic microsomes and isolated hepatocytes, clearly shows that both the oxidation and the conjugation pathways can take place in the liver. However, it is not known whether extrahepatic tissues could play a role in the biotransformation of APs. In studying the glucuronidation of various environmental estrogens in the carp intestine, Yokota *et al.*[82] were unable to detect the formation of NP-glucuronide or OP-glucuronide. Recently, we incubated labeled 4-*n*-NP with rainbow trout gill epithelial cells in culture (Cravedi, unpublished results). No trace of 4-*n*-NP metabolites was detected in the culture medium or within the cells. Together, these data tend to show a limited contribution, if any, of extrahepatic tissues in the metabolism of NPs in fish.

V. Concluding remarks and future directions

The studies to date on the disposition of NPs in fish have led to the building of a useful body of knowledge, enabling the assessment of the impact of APs in the aquatic environment. Both laboratory and field studies demonstrate that these xenoestrogens are efficiently absorbed by fish through the gills and/or the digestive tract and accumulate moderately in viscera. Current data regarding the metabolic fate of APs in fish indicate that these xenobiotics are efficiently detoxified, mainly through the conjugation with glucuronic acid and the oxidative biotransformation of the alkyl side-chain. The contribution of the latter pathway to the metabolism of APs is highly dependent upon the linear or branched structure of the alkyl chain. Bile and urine are suggested as the main routes of excretion of metabolites. Nevertheless, other routes of elimination including the release of the parent compound and/or related metabolites across the gills, remains to be investigated for APs as this route is known to be efficient for the clearance of xenobiotics in fish. Another important area that requires further attention is the investigation of the biological activity of AP biotransformation products. Although most NP metabolites are unlikely to retain estrogenic activity, this activity has only been examined for a limited number of end products and should be the focus of other studies. In addition, there is no information on the impact of these metabolites on other endocrine functions. Further studies using the tools of molecular biology are needed to investigate the interaction of NP metabolites with thyroid, retinoid and cortisol receptors. At present, most studies on NP metabolism are undertaken with linear alkyl chain isomers and the *in vivo* metabolism of multi-branched NPs, which are the major components of commercially produced NPs, have received little attention comparatively. In addition to playing a key role in the oxidation rates, the conformation of the alkyl chain is known to influence the estrogenic activity of the molecule. Thus, caution must be exercised in extrapolating the published data regarding 4-*n*-NP to real-life situations.

Furthermore, significant differences exist in the levels of xenobiotic metabolizing enzymes between fish species, as well in the sensitivity of these species to NPs toxicity. Accordingly, appropriate comparative metabolic studies may help to identify the NP metabolic steps that play a crucial role in the susceptibility of species to these aquatic pollutants.

Further investigations are also required to assess the sex dependence of NP metabolism as well as the incidence of age and gender on steroid metabolism alteration by APs. Marked differences in xenobiotic metabolizing enzymes between male and female fish are well documented for various species, with higher levels generally noted in males. These differences reflect the combined effect of differential enzyme expression between sexes, together with the relative responsiveness to the predominant gonadal hormones present in mature animals[13]. Consequently, detailed information on sex differences in NP metabolism, or in the modulation of the enzymes involved in steroid biosynthesis or biodegradation, will be required for better insight into the mechanism of endocrine disruption associated with NPs.

Acknowledgements. We thank J. Calabro and L. Dolo for excellent technical support.

VII. References

1. Ahel, M. and W. Giger. Aqueous solubility of alkylphenols and alkylphenol polyethoxylates. *Chemosphere* 26: 1461–1470, 1993.
2. Ahel, M. and W. Giger. Partitioning of alkylphenols and alkylphenol polyethoxylates between water and organic solvents. *Chemosphere* 26: 1471–1478, 1993.
3. Ahel, M., J. McEvoy and W. Giger. Bioaccumulation of the lipophilic metabolites of nonionic surfactants in freshwater organisms. *Environ. Pollut.* 79: 243–348, 1993.
4. Arukwe, A., L. Förlin and A. Goksøyr. Xenobiotic and steroid biotransformation enzymes in Atlantic salmon (*Salmo salar*) liver treated with an estrogenic compound, 4-nonylphenol. *Environ. Toxicol. Chem.* 16: 2576–2583, 1997.
5. Arukwe, A., T. Celius, B.T. Walther and A. Goksøyr. Plasma levels of vitellogenin and plasma zona radiata proteins in 4-nonylphenol and *o,p′*-DDT treated juvenile Atlantic salmon (*Salmo salar*). *Mar. Environ. Res.* 46: 133–136, 1998.
6. Arukwe, A., A. Goksøyr, R. Thibaut and J.P. Cravedi. Metabolism and organ distribution of nonylphenol in Atlantic salmon (*Salmo salar*). *Mar. Environ. Res.* 50: 141–145, 2000.
7. Arukwe, A., R. Thibaut, K. Ingebritsen, T. Celius, A. Goksøyr and J.P. Cravedi. In vivo and in vitro metabolism and organ distribution of nonylphenol in Atlantic salmon (*Salmo salar*). *Aquat. Toxicol.* 49: 289–304, 2000.
8. Bennett, E.R. and C.R. Metcalfe. Distribution of alkylphenol compounds in Great Lakes sediments, United States and Canada. *Environ. Toxicol. Chem.* 17: 1230–1235, 1998.
9. Bennie, D.T., C.A. Sullivan, H.B. Lee, T.E. Peart and R.J. Maguire. Occurrence of alkylphenols and alkylphenol mono- and diethoxylates in natural waters of the Laurentian Great Lakes basin in the upper St. Lauwrence River. *Sci. Total Environ.* 193: 263–275, 1997.
10. Bennie, D.T. Review of the environmental occurrence of alkylphenols and alkylphenol ethoxylates. *Water Qual. Res. J. Can.* 34: 79–122, 1999.
11. Blackburn, M.A. and M.J. Waldock. Concentration of alkylphenols in rivers and estuaries in England and Wales. *Water Res.* 29: 1623–1629, 1995.
12. Bolton, J.L., E. Pisha, F. Zhang and S. Qiu. Role of quinoids in estrogen carcinogenesis. *Chem. Res. Toxicol.* 10: 1113–1127, 1998.

13. Buhler, D.R. and J.L. Wang-Buhler. Rainbow trout cytochrome P450s: purification, molecular aspects, metabolic activity, induction and role in environmental monitoring. *Comp. Biochem. Physiol. C* 121: 107–137, 1998.
14. Celander, M., M. Ronis and L. Förlin. Initial characterization of a constitutive cytochrome P450 isoenzyme in rainbow trout liver. *Mar. Environ. Res.* 28: 9–13, 1989.
15. Chalaux, N., J.M. Bayona and J. Albaigés. Determination of nonylphenols as pentafluorobenzyl derivatives by capillary gas chromatography with electron-capture and mass spectrometric detection in environmental matrices. *J. Chromatogr. A* 686: 275–281, 1994.
16. Christensen, L.J., B. Korsgaard and P. Bjerregaard. The effect of 4-nonylphenol on the synthesis of vitellogenin in the flounder *Platichthys flesus*. *Aquat. Toxicol.* 46: 211–219, 1999.
17. Coldham, N.G., S. Sivapathasundaram, M. Dave, L.A. Ashfield, T.G. Pottinger, C. Goodall and M.J. Sauer. Biotransformation, tissue distribution, and persistence of 4-nonylphenol residues in juvenile rainbow trout (*Oncorhynchus mykiss*). *Drug Metab. Dispos.* 26: 347–354, 1998.
18. Cravedi, J.P., A. Lafuente, M. Baradat, A. Hillenweck and E. Perdu-Durand. Biotransformation of pentachlorophenol and biphenyl in isolated rainbow trout (*Oncorhynchus mykiss*) hepatocytes: comparison with *in vivo* metabolism. *Xenobiotica* 29: 499–509, 1999.
19. Cravedi, J.P., G. Boudry, M. Baradat, D. Rao and L. Debrauwer. Metabolic fate of 2,4-dichloroaniline, prochloraz and nonylphenol diethoxylate in rainbow trout: a comparative *in vivo/in vitro* approach. *Aquat. Toxicol.* 53: 159–172, 2001.
20. Doerge, D.R., N.C. Twaddle, M.I. Churchwell, H.C. Chang, R.R. Newbold and K.B. Delclos. Mass spectrometric determination of *p*-nonylphenol metabolism and disposition following oral administration to Sprague-Dawley rats. *Reprod. Toxicol.* 16: 45–56, 2002.
21. Drèze, V., G. Monod, J.P. Cravedi, S. Biagianti-Risbourg and F. Le Gac. Effects of 4-nonylphenol on sex differentiation in mosquitofish (*Gambusia holbrooki*). *Ecotoxicology* 9: 93–103, 2000.
22. Ekelund, R., A. Bergman, A. Granmo and M. Berggren. Bioaccumulation of 4-*n*-nonylphenol in marine animals. A re-evaluation. *Environ. Pollut.* 64: 107–120, 1990.
23. Ferreira-Leach, A.M.R. and E.M. Hill. Bioconcentration and metabolism of 4-*tert*-octylphenol in roach (*Rutilus rutilus*) fry. *Analusis* 28: 789–792, 2000.
24. Ferreira-Leach, A.M.R. and E.M. Hill. Bioconcentration and distribution of 4-*tert*-octylphenol residues in tissues of the rainbow trout (*Oncorhynchus mykiss*). *Mar. Environ. Res.* 51: 75–89, 2001.
25. Flouriot, G., F. Pakdel, B. Ducouret and Y. Valotaire. Influence of xenobiotics on rainbow trout liver estrogen receptor and vitellogenin gene expression. *J. Mol. Endocrinol.* 15: 143–151, 1995.
26. George, S.G. Enzymology and molecular biology of phase II xenobiotic-conjugating enzymes in fish. In: *Aquatic Toxicology: Molecular, Biochemical and Cellular Perspectives*, edited by D.C. Malins and G.K. Ostrander, Boca Raton, FL, Lewis Publishers, pp. 37–85, 1994.
27. Gimeno, S., A. Gerritsen, T. Bowmer and H. Komen. Feminization in male carp. *Nature* 384: 221–222, 1996.
28. Gray, M.A. and C.D. Metcalfe. Induction of testis–ova in Japanese medadaka (*Oryzias latipes*) exposed to *p*-nonylphenol. *Environ. Toxicol. Chem.* 16: 1082–1086, 1997.
29. Hanioka, N., H. Jinno, Y.S. Chung, T. Tanaka-Kagawa, T. Nishimura and M. Ando. Inhibition of rat hepatic cytochrome P450 activities by biodegradation products 4-*tert*-octylphenol ethoxylate. *Xenobiotica* 29: 873–883, 1999.
30. Harris, R.M., R.H. Waring, C.J. Kirk and P.J. Hughes. Sulfation of ‘estrogenic’ alkylphenols and 17β-estradiol by human platelet phenol sulfotransferases. *J. Biol. Chem.* 275: 159–166, 2000.
31. Isobe, T., H. Nishiyama, A. Nakashima and H. Takada. Distribution and behaviour of nonylphenol, octylphenol, and nonylphenol monoethoxylate in Tokyo metropolitan area: their association with aquatic particles and sedimentary distributions. *Environ. Sci. Technol.* 35: 1041–1049, 2001.
32. Jobling, S. and J.P. Sumpter. Detergent components in sewage effluents are weakly estrogenic to fish: an in vitro study using rainbow trout (*Oncorhynchus mykiss*) hepatocytes. *Aquat. Toxicol.* 27: 361–372, 1993.
33. Jobling, S., D. Sheahan, J.A. Osborne, P. Matthiessen and J.P. Sumpter. Inhibition of testicular growth in rainbow trout (*Oncorhynchus mykiss*) exposed to estrogenic alkylphenolic chemicals. *Eviron. Toxicol. Chem.* 16: 194–202, 1996.
34. Kannan, K., T.L. Keith, C.G. Naylor, C.A. Staples, S.A. Snyder and J.P. Giesy. Nonylphenol and nonylphenol ethoxylates in fish, sediment, and water from the Kalamazoo river, Michigan. *Arch. Environ. Contam. Toxicol.* 44: 77–82, 2003.
35. Kobayashi, K., S. Kimura and Y. Oshima. Sulfate conjugation of various phenols by liver-soluble fraction of goldfish. *Bull. Jpn. Soc. Sci. Fish.* 50: 833–837, 1984.
36. Lech, J.J., S.K. Lewis and L. Ren. In vivo estrogenic activity of nonylphenol in rainbow trout. *Fundam. Appl. Toxicol.* 30: 229–232, 1996.

37. Lee, H.B. and T.E. Peart. Determination of 4-nonylphenol in effluent and sludge from sewage treatment plants. *Anal. Chem.* 67: 1976–1980, 1995.
38. Lee, P.C., S. Chakraborty Patra, C.T. Stelloh, W. Lee and M. Struve. Interaction of nonylphenol and hepatic CYP1A in rats. *Biochem. Pharmacol.* 52: 885–889, 1996.
39. Lee, P.C., S. Chakraborty Patra and M. Struve. Modulation of rat hepatic CYP3A by nonylphenol. *Xenobiotica* 26: 831–838, 1996.
40. Le Gac, F., J.L. Thomas, B. Mourot and M. Loir. *In vivo* and *in vitro* effects of prochloraz and nonylphenol ethoxylates on trout spermatogenesis. *Aquat. Toxicol.* 53: 187–200, 2001.
41. Lewis, S.K. and J.J. Lech. Uptake, disposition and persistence of nonylphenol in rainbow trout. *Xenobiotica* 26: 813–819, 1996.
42. Liber, K., J.A. Gangl, T.D. Corry, L.J. Heinis and F.S. Stay. Lethality and bioaccumulation of 4-nonylphenol on bluegill sunfish in littoral enclosures. *Environ. Toxicol. Chem.* 18: 394–400, 1999.
43. Madigou, T., P. Goff, G. Salbert, J.P. Cravedi, H. Segner, F. Pakdel and Y. Valotaire. Effects of nonylphenol on estrogen receptor conformation, transcriptional activity and sexual reversion in rainbow trout (*Oncorhychus mykiss*). *Aquat. Toxicol.* 53: 173–186, 2001.
44. Magliulo, L., M.P. Schreibman, J. Cepriano and J. Ling. Endocrine disruption caused by two common pollutants at acceptable concentrations. *Neurotoxicol. Teratol.* 24: 71–79, 2002.
45. Maguire, R.J. Review of the persistence of nonylphenol and nonylphenol ethoxylates in aquatic environments. *Water Qual. Res. J. Can.* 34: 37–78, 1999.
46. McKim, J.M. and H.M. Goeden. A direct measure of the uptake efficiency of a xenobiotic chemical across the gills of brook trout (*Salvelinus fontinalis*) under normoxic and hypoxic conditions. *Comp. Biochem. Physiol. C* 72: 65–74, 1983.
47. McKim, J.M., R.C. Kolanczyk, G.L. Lien and A.D. Hoffman. Dynamics of renal excretion of phenol and major metabolites in the rainbow trout (*Oncorhynchus mykiss*). *Aquat. Toxicol.* 45: 265–277, 1999.
48. Mc Leese, D.W., V. Zitko, D.B. Sergeant, L. Burridge and C.D. Metcalfe. Lethality and accumulation of alkylphenols in aquatic fauna. *Chemosphere* 10: 723–730, 1981.
49. Meldahl, A.C., K. Nithipatikom and J.J. Lech. Metabolism of several ^{14}C-nonylphenol isomers by rainbow trout (*Oncorhynchus mykiss*): *in vivo* and *in vitro* microsomal metabolites. *Xenobiotica* 26: 1167–1180, 1996.
50. Miles-Richardson, S.R., S.L. Pierens, K.M. Nichols, V.J. Kramer, E.M. Snyder, S.A. Snyder, J.A. Render, S.D. Fitzgerald and J.P. Giesy. Effects of waterborne exposure to 4-nonylphenol and nonylphenol ethoxylate on secondary sex characteristics and gonads of fathead minnows (*Pimephales promelas*). *Environ. Res. A* 80: S122–S137, 1999.
51. Miranda, C.L., M.C. Henderson and D.R. Buhler. Evaluation of chemicals as inhibitors of trout cytochrome P450s. *Toxicol. Appl. Pharmacol.* 148: 237–244, 1998.
52. Miranda, C.L., J.L. Wang, M.C. Henderson, D.E. Williams and D.R. Buhler. Regiospecificity in the hydroxylation of lauric acid by rainbow trout hepatic cytochrome P450 isozymes. *Biochem. Biophys. Res. Commun.* 171: 537–542, 1990.
53. Moffat, G.J., A. Burns, J. Van Miller, R. Joiner and J. Ashby. Glucuronidation of nonylphenol and octylphenol eliminates their ability to activate transcription via the estrogen receptor. *Regul. Toxicol. Pharmacol.* 34: 182–187, 2001.
54. Mueller, G.C. and U.H. Kim. Displacement of estradiol from estrogen receptors by simple alkylphenols. *Endocrinology* 102: 1429–1435, 1978.
55. Pedersen, R.T. and E.M. Hill. Biotransformation of the xenoestrogen 4-*tert*-octylphenol in hepatocytes of rainbow trout (*Oncorhynchus mykiss*). *Xenobiotica* 30: 867–879, 2000.
56. Pedersen, R.T. and E.M. Hill. Bioconcentration and distribution of 4-*tert*-octylphenol residues in the cyprinid fish (*Scardinius erythrophthalmus*). *Environ. Sci. Technol.* 36: 3275–3283, 2002.
57. Petit, F., P. Le Goff, J.P. Cravedi, Y. Valotaire and F. Pakdel. Two complementary bioassays for screening the estrogenic potency of xenobiotic recombinant yeast for trout estrogen receptor and trout hepatocyte cultures. *J. Mol. Endocrinol.* 19: 321–335, 1997.
58. Petit, F., P. Le Goff, J.P. Cravedi, O. Kah, Y. Valotaire and F. Pakdel. Trout estrogen receptor sensitivity to xenobiotics as tested by different bioassays. *Aquaculture* 177: 353–365, 1999.
59. Schwaiger, J., U. Mallow, H. Ferling, S. Knoerr, T. Braunbeck, W. Kalbfus and R.D. Negele. How estrogenic is nonylphenol? A transgenerational study using rainbow trout (*Oncorhynchus mykiss*) as a test organism. *Aquat. Toxicol.* 59: 177–189, 2002.
60. Servos, M.R. Review of the aquatic toxicity, estrogenic responses and bioaccumulation of alkylphenol and alkylphenol polyethoxylates. *Water Qual. Res. J. Can.* 34: 123–177, 1999.
61. Snyder, S.A., T.L. Keith, S.L. Pierens, E.M. Snyder and J.P. Giesy. Bioconcentration of nonylphenol in fathead minnows (*Pimephales promelas*). *Chemosphere* 44: 1697–1702, 2001.

62. Soto, A.M., H. Justicia, J.W. Wray and C. Sonnenschein. *p*-Nonyl-phenol: an estrogenic xenobiotic released from modified polystyrene. *Environ. Health Perspect.* 92: 167–173, 1991.
63. Staples, C.A., J. Weeks, J. Hall and C.G. Naylor. Evaluation of aquatic toxicity and bioaccumulation of C8- and C9-alkylphenol ethoxylates. *Environ. Toxicol. Chem.* 17: 2470–2480, 1998.
64. Sturm, A., J.P. Cravedi, E. Perdu, M. Baradat and H. Segner. Effects of prochloraz and nonylphenol diethoxylate on hepatic biotransformation enzymes in trout: a comparative *in vitro*/*in vivo*-assessment using cultured hepatocyte. *Aquat. Toxicol.* 53: 229–245, 2001.
65. Tanaka, J.N. and J.M. Grizzle. Effects of nonylphenol on the gonadal differentiation of the hermaphroditic fish. *Rivulus marmoratus*. *Aquat. Toxicol.* 57: 117–125, 2002.
66. Tanghe, T., G. Devriese and W. Verstraete. Nonylphenol and estrogenic activity in aquatic environmental samples. *J. Environ. Qual.* 28: 702–709, 1999.
67. Thibaut, R., L. Debrauwer, D. Rao and J.P. Cravedi. Disposition and metabolism of [^{3}H]-4*n*-nonylphenol in rainbow trout. *Mar. Environ. Res.* 46: 521–524, 1998.
68. Thibaut, R., L. Debrauwer, D. Rao and J.P. Cravedi. Characterization of biliary metabolites of 4*n*-nonylphenol in rainbow trout (*Oncorhynchus mykiss*). *Xenobiotica* 28: 745–757, 1998.
69. Thibaut, R., L. Debrauwer, D. Rao and J.P. Cravedi. Urinary metabolites of 4*n*-nonylphenol in rainbow trout (*Oncorhynchus mykiss*). *Sci. Total Environ.* 233: 193–200, 1999.
70. Thibaut, R., A. Jumel, L. Debrauwer, E. Rathahao, L. Lagadic and J.P. Cravedi. Identification of 4*n*-nonylphenol metabolic pathways and residues in aquatic organisms by HPLC and LC–MS analyses. *Analusis* 28: 793–800, 2000.
71. Thibaut, R., L. Debrauwer and J.P. Cravedi. Tissue residues in rainbow trout dietary exposed to ^{3}H-4*n*-nonylphenol. *Rev. Med. Vet.* 153: 433–436, 2002.
72. Thibaut, R., E. Perdu and J.P. Cravedi. Testosterone metabolism as a possible biomarker of exposure of fish to nonylphenol. *Rev. Med. Vet.* 153: 589–590, 2002.
73. Thibaut, R., L. Debrauwer, E. Perdu, A. Goksøyr, J.P. Cravedi and A. Arukwe. Regio-specific hydroxylation of nonylphenol and the involvement of CYP2K- and CYP2M-like iso-enzymes in Atlantic salmon (*Salmo salar*). *Aquat. Toxicol.* 56: 177–190, 2002.
74. Thibaut, R., G. Monod and J.P. Cravedi. Residues of ^{14}C-4*n*-nonylphenol in mosquitofish (*Gambusia holbrooki*) oocytes and embryos during dietary exposure of mature females to this xenohormone. *Mar. Environ. Res.* 54: 685–689, 2002.
75. Tollefsen, K.E., K. Ingebritsen, A.J. Olsen, K.E. Zachariassen and S. Johnsen. Acute toxicity and toxicokinetics of 4-heptylphenol in juvenile Atlantic cod (*Gadus morhua* L.). *Environ. Toxicol. Chem.* 17: 740–746, 1998.
76. Tollefsen, K.E., R. Mathisen and J. Stenersen. Estrogen mimics bind with similar affinity and specificity to the hepatic estrogen receptor in Atlantic Salmon (*Salmo salar*) and rainbow trout (*Oncorhynchus mykiss*). *Gen. Comp. Endocrinol.* 126: 14–22, 2002.
77. Tsuda, T., A. Takino, M. Kojima, H. Harada, K. Muraki and M. Tsuji. 4-Nonylphenols and 4-tert-octylphenol in water and fish from rivers flowing into Lake Biwa. *Chemosphere* 41: 757–762, 2000.
78. Tsuda, T., A. Takino, K. Muraki, H. Harada and M. Kojima. Evaluation of 4-nonylphenols and 4-tert-octylphenol contamination of fish in rivers by laboratory accumulation and excretion experiments. *Water Res.* 35: 1786–1792, 2001.
79. White, R., S. Jobling, S.A. Hoare, J.P. Sumpter and M.G. Parker. Environmentally persistent alkylphenolic compounds are estrogenic. *Endocrinology* 125: 175–182, 1994.
80. Williams, D.E., J.J. Lech and D.R. Buhler. Xenobiotics and xenoestrogens in fish: modulation of cytochrome P450 and carcinogenesis. *Mut. Res.* 399: 179–192, 1998.
81. Ying, G.G., B. Williams and R. Kookana. Environmental fate of alkylphenols and alkylphenol ethoxylates – a review. *Environ. Int.* 28: 215–226, 2002.
82. Yokota, H., N. Miyashita and A. Yuasa. High glucuronidation of environmental estrogens in the carp (*Cyprinus carpino*) intestine. *Life Sci.* 71: 887–898, 2002.
83. Zalko, D., R. Costagliola, C. Dorio, E. Rathahao and J.P. Cravedi. *In vivo* metabolic fate of the xenoestrogen 4-*n*-nonylphenol in wistar rats. *Drug Metab. Dispos.* 131: 168–178, 2003.

Biochemistry and Molecular Biology of Fishes, vol. 6
T. P. Mommsen and T. W. Moon (Editors)

CHAPTER 6

Pesticide biotransformation in fish

DANIEL SCHLENK

Department of Environmental Sciences, University of California, Riverside, Riverside, CA 92521, USA

I. Introduction

Aquatic organisms, such as fish, that live in and use the aquatic environment are thus frequently exposed to pesticides and their derivatives. Exposures are rare from point source discharges, but are more commonly due to runoff from agricultural fields, domestic uses or specific applications to aqueous environments.

Fish are likely to receive exposure through three specific routes: branchial, dermal and oral primarily through the diet. Typically, dietary exposure to pesticides occurs with compounds that persist in the dietary item consumed by the fish. For example, dietary biomagnification has been noted to occur with several organochlorine insecticides such as DDT in predatory fish (see ref. 60). Environmental persistence and physicochemical properties of the agent will significantly impact uptake by the gills[57]. Lastly, dermal exposure is also possible and is likewise dependent upon the polarity and lipid solubility of the compound[56].

If a xenobiotic is absorbed, distribution may occur to specific target organs where the unchanged chemical causes a direct biological effect (Fig. 1). The agent may also

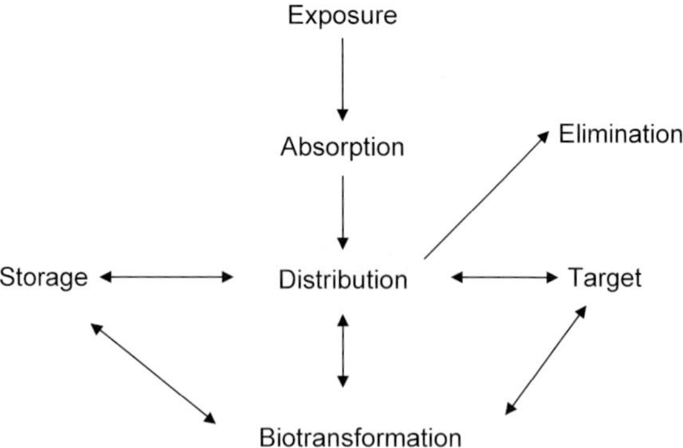

Fig. 1. Dispositional processes of chemicals within organisms.

be transported to storage depots with high lipid content, where it may be mobilized upon lipid utilization by the animal. Certain compounds may be directly excreted without having any interaction with targets or storage depots. However, as most compounds absorbed by fish tend to be somewhat lipophilic, many are biotransformed to more hydrophilic derivatives in an attempt to enhance polarity and subsequent elimination. It is this later dispositional process that will be discussed in this chapter.

1. Biotransformation

In general, biotransformation of an organic chemical can lead to bioactivation and detoxification in most organisms including fish. In an effort to enhance water solubility of the agent, radical alterations are necessary in the initial Phase I reactions (Table 1) either incorporating a polar functional group into the parent compound or splitting the compound such that a polar functional group that is already present in the molecule may become exposed. Monooxygenation is an example of the former process where water or molecular oxygen is utilized as the polar molecule. Examples of monooxygenases include the cytochrome P450 (CYP) and flavin-containing monooxygenases (FMOs). Each of these systems have been characterized in fish[9,69,80]. Oxidation may also occur *via* lesser characterized oxidases or co-oxidation through peroxidase pathways. Exposing polar atoms through cleavage of the xenobiotic is typically brought about through an oxidative dealkylation or a hydrolytic process involving various esterases or amidases. Phase I reactions of this nature often lead to the formation of reactive intermediates or metabolites with greater biological activity than the parent. In an attempt to remedy this potential problem, endogenous macromolecules that are in high cellular concentrations can be used in Phase II reactions to conjugate reactive intermediates further enhancing water solubility and elimination. If a xenobiotic undergoes significant biotransformation, evaluation of a parent compound within an organism may be a fruitless attempt of measuring exposure.

TABLE 1

Subcellular location and representative pesticide substrates of Phase I and Phase II enzymes in fish (see ref. 39)

	Representative enzyme	Subcellular location	Representative pesticide substrate
Phase I			
Oxidation	CYP, FMO, aldehyde oxidase	SER, SER, cytosol	Diazinon, aldicarb, fenthion
Reduction	Nitro-reductase	Cytosol	Parathion
Hydrolysis	Carboxylesterase, epoxide hydrolase	Cytosol/plasma, cytosol/SER	Malathion, carbaryl 3,4 epoxide[a]
Phase II			
Glucuronidation	UDP-glucuronosyl transferase	SER	1-Naphthol (carbaryl)
Sulfation	Sulfotransferase	Cytosol	4-Hydroxy cypermethrin
Glutathione mercapturic acid	Glutathione *S*-transferase[b]	Cytosol, SER	Molinate
Taurine conjugation	Acyl-CoA: amino acid, *N*-acyltransferase	Cytosol/mitochondria	2,4-D

CYP: cytochrome P450 monooxygenase; FMO: flavin-containing monooxygenase; SER: smooth endoplasmic reticulum; 2,4-D: 2,4-dichlorophenoxyacetic acid.
[a] Postulated intermediate.
[b] Other enzymes involved in mercapturic acid formation include gamma-gluatmyl transpeptidase; aminopeptidase M; *N*-acetyltransferase.

2. *Methods to assess biotransformation*

The primary organ of biotransformation is the liver, although significant extrahepatic activities have also been observed in kidney, gut and gill tissues of fish. Enzymes responsible for the catalysis of Phase I and Phase II processes are found throughout the cell, but tend to be either localized in the smooth endoplasmic reticulum (SER) or the cytosol (Table 1). Through differential centrifugation it is possible to isolate the SER (microsomes) and purify the cytosol from other cellular organelles allowing specific evaluation of several of these enzymes. *In vitro* incubations of agents with these cell-free extracts using specific cofactors and optimal buffering conditions can be utilized to measure biotransformation. The most difficult aspect with *in vitro* biotransformation studies is determining the identity of any metabolite present. Turnover rates in fish tend to be relatively low; thus it is suggested that either radiolabeled parent compounds be utilized or compounds with chromophores that allow detection at picomolar concentrations. If comparisons are to be made between individuals or with treatment groups, it is also recommended that proper kinetic experiments be initially performed to evaluate saturation such that rates of turnover may be calculated and compared utilizing substrate concentrations within the linear range of response.

Alternatively, *in vivo* biotransformation of pesticides may also be measured in fish following exposure to the agent either through aqueous or dietary routes of exposure to mimic environmental exposure routes. However, if biotransformation is the ultimate goal of the investigation, fish may be treated with an intraperitoneal injection or an intra-arterial infusion using an arterial cannula. Lastly, tissue residues may be extracted and putative metabolites chromatographically isolated and quantified.

Following exposure, urinary or cloacal catheters may be used to obtain excreta, and several groups have utilized recirculating aqueous extraction methods to obtain branchial metabolites[71,76]. This latter method may also be used following aqueous or dietary exposure as the fish can be moved to a chamber without the parent compound. The water is then pumped over resins which extract metabolites from the water[85]. The resins may be subsequently eluted with specific solvents and the resulting extracts analyzed by various chromatographic methods such as HPLC or LCMS. The major disadvantage using this method is the potential for reuptake of metabolites by the fish in the exposure chamber and the possibility of hydrolysis.

II. Biotransformation of pesticide classes in fish

There are numerous chemicals utilized as pesticides around the world. However, few have been evaluated for the potential of biotransformation in fish. Several reviews of pesticide biotransformation in fish have been previously published[16,39,60].

1. Insecticides

1.1. Organochlorines

Organochlorine pesticides tend to be recalcitrant to biotransformation which is one of the main reasons for the persistence of these compounds in the environment. For example, little excretion of DDT through urine or gill effluate or conversion to metabolites was observed 48 h after administering DDT to dogfish *via* the caudal vein, artery or stomach tube[21]. In fathead minnows exposed to DDT, very low concentrations of DDE and no DDD was observed in the whole body 2 weeks after exposure, but after 266 days of exposure, the DDE concentration was nearly nine times that of DDT; DDD concentration was about one-third that of DDT. The concentrations of DDT, DDE and DDD in embryos obtained from exposed fish were similar to that of the adults[41]. On the basis of extensive studies in trout, Addison and Willis[1] summarized the biotransformation of DDT in fish as shown in Fig. 2. The predominant metabolite appears to be DDE.

A structural analog of DDT which is less persistent is methoxychlor. Demethylation of methoxychlor is required to elicit estrogen receptor binding and activation[10]. This reaction has been observed in green sunfish (*Lepomis cyanellus*) initially[66], and was inhibited by piperonyl butoxide indicating a role of CYP in the oxidative dealkylation. Similar results were observed in channel catfish, where induction with BNF had no effect on the *in vitro* transformation of methoxychlor to ring hydroxylated metabolites as well as demethylation to the desmethyl or bis-desmethyl metabolites[74,75]. However, pretreatment of fish with methoxychlor markedly reduced the *in vitro* biotransformation of methoxychlor and increased serum estradiol concentrations

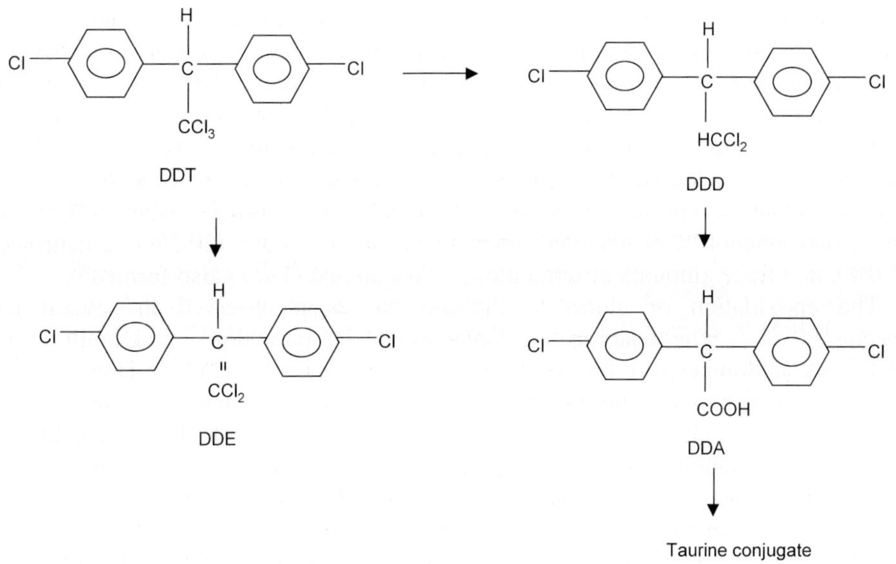

Fig. 2. The biotransformation of DDT within fishes.

indicating CYP3A may be involved in either the ring hydroxylation or dealkylation[74]. It was interesting to note that mono-demethylation of methoxychlor causes the formation of enantiomeric metabolites. However, there was no difference in the binding of the *R* or *S*-desmethyl metabolites to the estrogen receptor in channel catfish[75]. Demethylation and subsequent estrogenic activity was also observed in Japanese medaka (D. Schlenk, unpublished) but the isoform(s) responsible for this reaction were unidentified.

The *in vitro* metabolism of endosulfan was examined in carp microsomes and the 105,000*g* supernatant[50]. NADPH enhanced the formation of endosulfan alpha hydroxyether, endosulfan alcohol and endosulfan ether. There was also evidence of potential glutathione adducts as degradation was enhanced upon addition of glutathione to the incubation mixture. This same group examined the *in vivo* biotransformation of endosulfan in carp which were treated with 4.5 μg/l for 8 h. The hydroxyether, alcohol and ether metabolites were also observed with the additional lactone derivative[50]. In contrast, in *Labeo rohita*[64] and goldfish[65] the sulfate conjugate was observed. There was no sulfate conjugate of endosulfan in *Channa punctata* or carp. In *Macrognathus aculeatum* the sulfate conjugate was observed in the brain, gills, gut, liver and kidney. However, it was suggested by the investigators that the sulfate derivative is an intermediate eventually rearranging to the ether which was observed only in the liver and kidney[65].

Glucuronide conjugates of pentachlorobenzene were observed following the initial phenol formation from hexachlorobenzene in rainbow trout (*Oncorhynchus mykiss*) fry following dietary exposure[27]. Other metabolites included traces of 2,5-dichloro-, 2,3,6-trichloro- and 2,4,5-trichloro-phenol. The biotransformation of pentachlorophenol has been examined in several fish species with fairly similar results. In trout, the glucuronide and sulfate conjugates of pentachlorophenol were observed in isolated cells and *in vivo*[19]. Earlier studies in trout failed to observe the sulfate, but did identify the glucuronide conjugate[34]. In topsmelt (*Atherinops affinis*), pentachlorophenol was primarily excreted unchanged (64.9%), but significant amounts of the sulfate (18.9%) and the glucuronide (16.2%) metabolites were observed in the water[5]. A similar observation was reported in striped bass (*Morone saxatilis*) where 71.5% of pentachlorophenol was excreted unchanged, with sulfate (20.3%), glucuronide (7.0%), and trace amounts of tetrachlorohydroquinone (1.2%) also formed[30].

The epoxidation of aldrin to dieldrin has been observed in several fish species[11,16,54,78]. The reaction was antagonized by parathion[31] and inhibited by SKF525A and piperonyl butoxide indicating catalysis by CYP[54]. However, the specific isoform has yet to be identified. Dieldrin has been shown to be further oxidized in bluegill to dieldrin pentachloroketone as well as *syn*-12-hydroxydieldrin[82]. The major metabolite 4,5-*trans*-dihydroaldrindiol is likely formed by epoxide hydrolase. Chlordene and endrin have also been examined in bluegill and bass fry (*Micropterus dolomeiux*)[78]. Chlordene was converted to the 2,3-epoxide and 1-hydroxy metabolites. Endrin was shown to be a competitive inhibitor of aldrin epoxidation in each fish species. Chlordane biotransformation in fish is likewise complex and has been discussed previously[39].

1.2. Organophosphates

Organophosphate pesticides elicit their toxicity primarily through inhibition of acetylcholinesterases. Activation of P=S to P=O enhances the toxicity of OP compounds such that oxons of parathion, malathion and fenitrothion are more potent inhibitors at the cholinesterase than the parents. Several studies have indicated that this oxidation is primarily carried out by CYP[31]. However, recent studies in mammals suggest that flavin-containing monooxygenases may also bioactivate OPs to more potent inhibitors[51]. Some species of fish can activate and detoxify, *via* hydrolysis, certain compounds at the same time. For example, the lesser toxicity of methyl parathion in sunfish relative to parathion is due to the greater hydrolysis of methyl paraoxon than paraoxon by fish liver homogenates[4]. In addition to sunfish, bullheads, winter flounder and sculpin have been shown to inactivate oxons of malathion, parathion and guthion presumably *via* hydrolytic cleavage[59].

Chlorpyrifos is converted to several metabolites in fish including 3,5,6-trichloro-2-trichloropyridinol (TCP), methoxytrichloropyridine and glucuronide conjugates (Fig. 3)[3]. TCP was the predominant metabolite in goldfish exposed for 5 days to 50 μg/l (see ref. 77). Mosquitofish (*Gambusia affinis*) converted CP to TCP (29%) and unidentified polar metabolites (21%) with 50% of the compound unchanged[2]. In a more recent study in mosquitofish, hepatic A-esterase hydrolyzed chlorpyrifos-oxon and CYP catalyzed dearylation of chlorpyrifos[6]. Similarly, rainbow trout converted CP to TCP and unidentified conjugates[2]. In an attempt to identify specific CYPs involved in the oxidative desulfation and dearylation, catfish were treated with the PCB mixture Aroclor 1254 prior to *in vitro* examination of chlorpyrifos metabolism. Although CYP1A was clearly induced by the PCB treatment, there were no significant

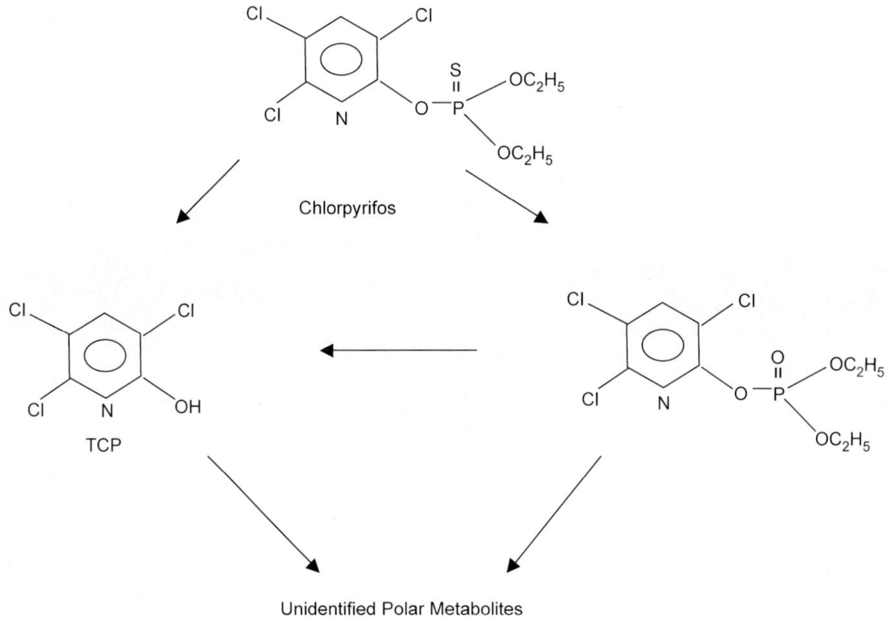

Fig. 3. The biotransformation of chlorpyrifos within fishes.

effects on desulfuration, dearylation or expression of CYP2K1, CYP2M1 or CYP3A immunoreactive proteins from the livers of catfish[81].

Other organophosphates that have been examined in fish include parathion and methyl-parathion[39]. The oxons of parathion and methyl-parathion were not affected by hepatic A esterases in mosquitofish[6]. Methyl-parathion and methyl-parathion oxon were detoxified by glutathione in sunfish hepatic homogenates, but parathion and paraoxon were not affected by glutathione[4].

Fenthion was converted *in vivo* to fenthion sulfoxide and fenthion oxon by goldfish given a 100 mg/kg i.p. injection[45]. Approximately 42–50% of the dose was unchanged after 10 days, and little (i.e. 2.1%) of fenthion sulfoxide was reduced back to the parent compound. Sulfoxidation of fenthion was shown to be catalyzed by CYP and the reduction by a cytosolic aldehyde oxidase[46]. Fenthion was also shown to be converted to the sulfoxide in medaka tissues following an *in vivo* exposure[86]. Fenitrothion will be discussed below (Fig. 11).

Trout, carp, guppy and zebrafish were all capable of converting diazinon to diazoxon and hydrolyzing the oxon to pyrimidinol in liver microsomes, but at differing capacities which significantly influenced toxicity (Fig. 4)[42–44]. Carp was very resistant to diazinon toxicity because of its very low rate of bioactivation and relatively high activity of hydrolyzing esterase activity. The trout was very sensitive to toxicity, although it had low bioactivation potential. Reasons for this response include a lack of esterase activity and a sensitive acetylcholinesterase. Diazinon was very toxic to the guppy because of a relatively high bioactivation enzyme system

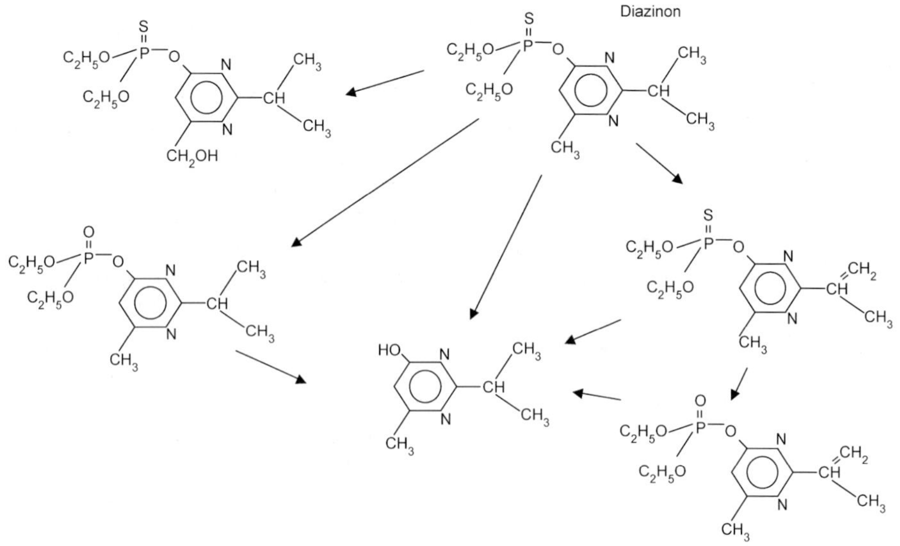

Fig. 4. The biotransformation of diazinon within fishes.

and a relatively sensitive acetylcholinesterase. Like the carp, the zebrafish had the most insensitive acetylcholinesterase and limited bioactivation. *In vivo* biotransformation rates and profiles were consistent with this observation in guppy and zebrafish[44].

More in-depth profiling of the *in vitro* hepatic metabolism of diazinon was carried out in carp, dace (*Tribolodon hakonensis*), channel catfish, rainbow trout and yellowtail (*Seriola quinqueradiata*)[28]. All fish were able to catalyze the desulfuration of diazinon to the oxon. In freshwater fish, the predominant metabolites were isopropenyl diazinon and hydroxypyrimidine. In the carp, dace and catfish, hydroxy diazinon was observed, while hydroxymethyl diazinon was the major metabolite in the yellowtail. In rainbow trout and catfish, isopropenyl diazoxon was observed. In the freshwater fish, more than 70% of diazinon was converted to metabolites with 50% of the metabolites being the nontoxic hydroxypyrimidine. However, in yellowtails, more than 70% of diazinon remained as toxic P-containing esters including the oxon. It was hypothesized that yellowtails had approximately 50% of the CYP levels of carp and rainbow trout and were not capable of converting diazinon to the other metabolites.

Malathion undergoes significant biotransformation in fish with the primary metabolites being malaoxon, malathion monoacid, malathion diacid, *O,O*-dimethyl phosphorodithioate, *O,O*-dimethyl phosphorodithiolate, *O,O*-dimethyl phosphorodithionate and *O,O*-dimethylphosphate (Fig. 5)[26]. No detectable concentrations of malathion or malaoxon was observed in pinfish (*Lagodon* sp.) exposed to 20–75 μg/l. The monoacid and diacid were found in all tissues especially the gut and liver[18]. Intestine and gill accounted for 42.8 and 37.4% of the total metabolite residues in Nile tilapia (*Tiliapia niloticus*) after 24 h of exposure to 200–1000 μg/l. Pretreatment with phenobarbital increased the amounts and percentages of the following metabolites in tilapia: *O,O*-dimethyl phosphorodithioate, *O,O*-dimethyl phosphorodithiolate,

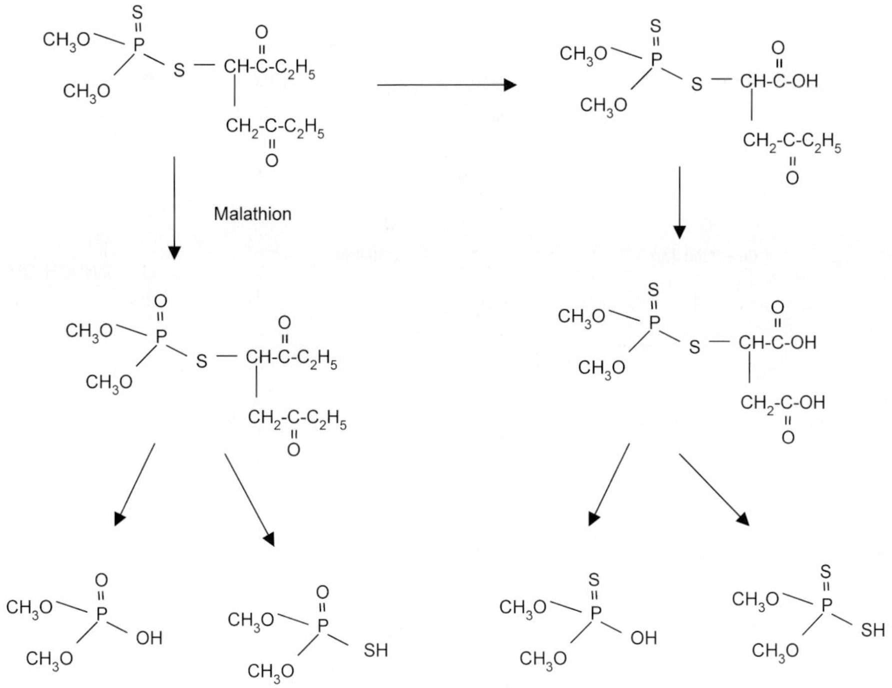

Malathion

Fig. 5. The biotransformation of malathion within fishes.

O,O-dimethyl phosphorodithionate and *O,O*-dimethylphosphate[26]. The investigators suggested that this was due to the enhanced formation of malaoxon. However, malaoxon was not measured, nor was any putative CYP activity, which would likely be responsible for the desulfuration reaction. Malathion was able to enhance the activation of aromatic amines in gilthead seabream (*Sparus aurata*), without inducing CYP1A activity[67].

1.3. Carbamates

One of the most studied carbamate insecticides with regard to biotransformation is that of carbaryl (see ref. 39). Two biliary metabolites were identified in trout exposed to 0.25 mg/l after 24 h (see ref. 79) which were tentatively identified as 5,6-dihydro-5,6-dihydroxy-1-naphthyl-*N*-methylcarbamate and 1-naphthol glucuronide indicating hydrolysis with subsequent conjugation (Fig. 6). In hepatic preparations from bluegill, catfish, perch and goldfish, demethylation, hydrolysis, hydroxylation and oxidation followed by conjugation was observed, with the most common metabolite in all species being 5,6-dihydro-5,6-dihydroxy carbaryl glucuronide[17].

Oxygenation of the sulfur anterior to the carbamyl moiety in aldicarb significantly elevates the toxicity of this compound 40–150 times depending upon species (Fig. 7)[23,24,62]. Aldicarb sulfoxidation has been observed in rainbow trout, Japanese medaka (*Oryzias latipes*), hybrid striped bass (*M. saxatilis* × *chrysops*),

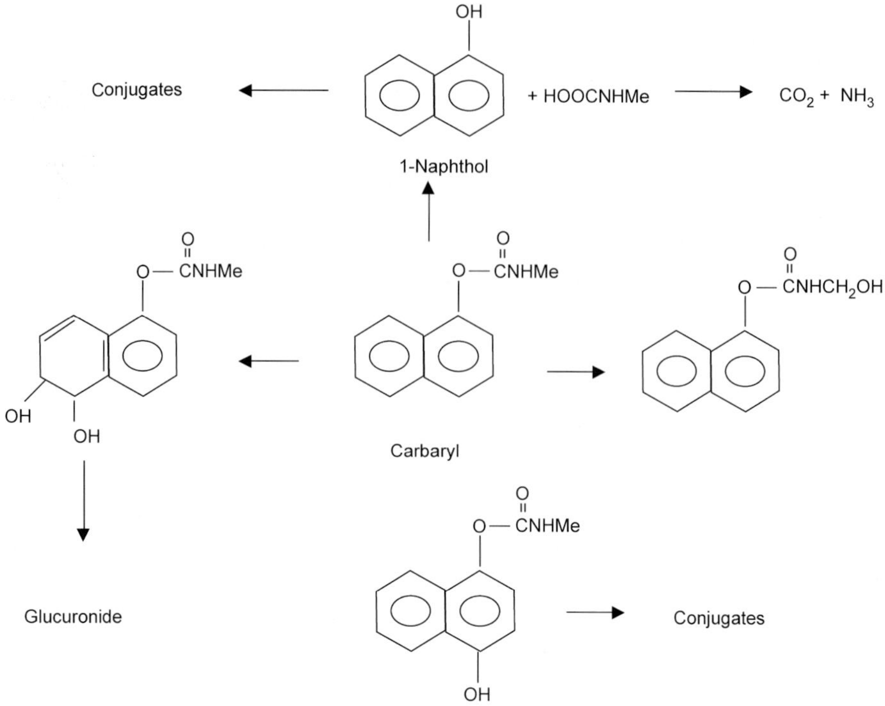

Fig. 6. The biotransformation of carbaryl within fishes.

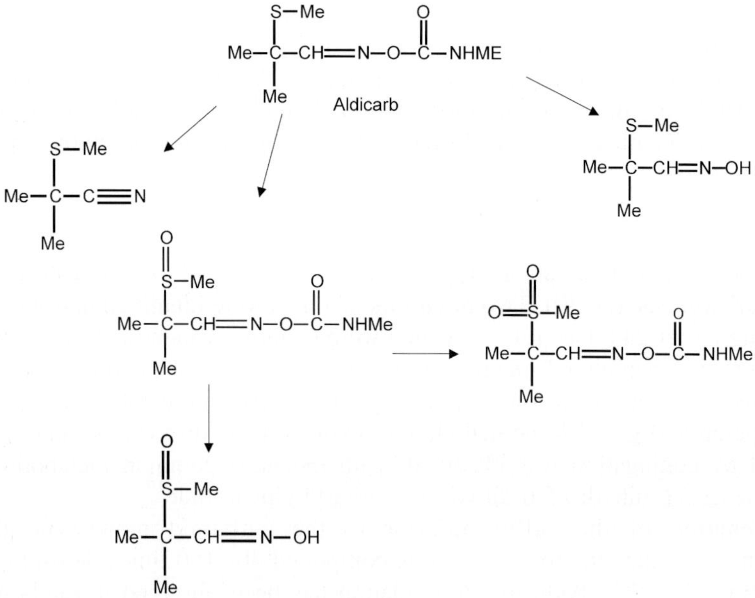

Fig. 7. The biotransformation of aldicarb within fishes.

channel catfish (*Ictalurus punctatus*) and rainbow trout[24,61,62,70]. With the exception of catfish, FMOs catalyzed the S-oxygenation reaction exclusively to the sulfoxide. Since channel catfish and other predominantly freshwater fish do not have active FMO[73], CYP or some other NAPDH-dependent oxygenase appears to be responsible for this reaction[61]. Multiple CYP isoforms apparently catalyze the sulfoxidation of aldicarb as various inhibitors of CYP could not completely inhibit activity[61]. The limited capacity of aldicarb S-oxygenation in channel catfish relative to rainbow trout may explain its resistance against aldicarb toxicity compared to rainbow trout, which readily convert aldicarb to aldicarb sulfoxide[62].

1.4. Pyrethroids

The biotransformation of pyrethroids has been extensively studied in several fish species (see ref. 39). In general, fish are unable to rapidly degrade (by esterase-catalyzed hydrolysis) these compounds and tend to be extremely sensitive to neurotoxicity resulting from exposure (Fig. 8)[35]. In trout, cypermethrin was converted to 4′-hydroxy-*cis*-cypermethrin[58]. In carp and rainbow trout, microsomal esterases hydrolyzed *trans*-permethrin more extensively than *cis*-permethrin[33]. Both isomers of permethrin tend to be converted to hydroxylated derivatives that are conjugated with glucuronic acid and sulfate[32]. Similar pathways were also observed with fenvalerate in trout where the glucuronide of the 4-hydroxylated derivative was observed in the bile[7].

Fipronil is an example of a newer class of insecticides (phenylpyrazole) that act in similar fashion as pyrethroids in targeting the GABA receptor of

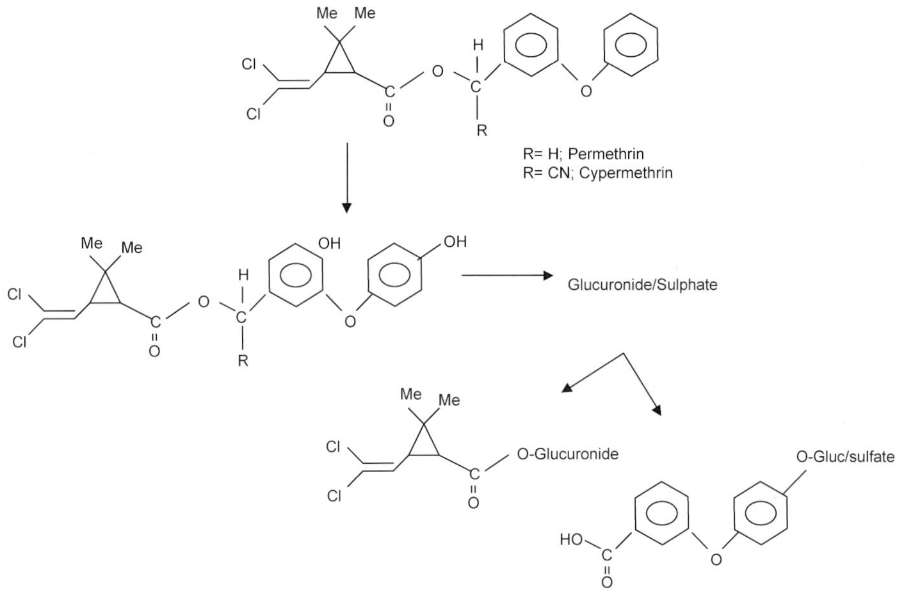

Fig. 8. The biotransformation of select pyrethroids within fishes.

arthropods[37]. A sulfoxide, fipronil was oxygenated to the sulfone, reduced to the sulfide and hydrolyzed in an unidentified fish species exposed to 900 ng/l for 35 days (Fig. 9)[87].

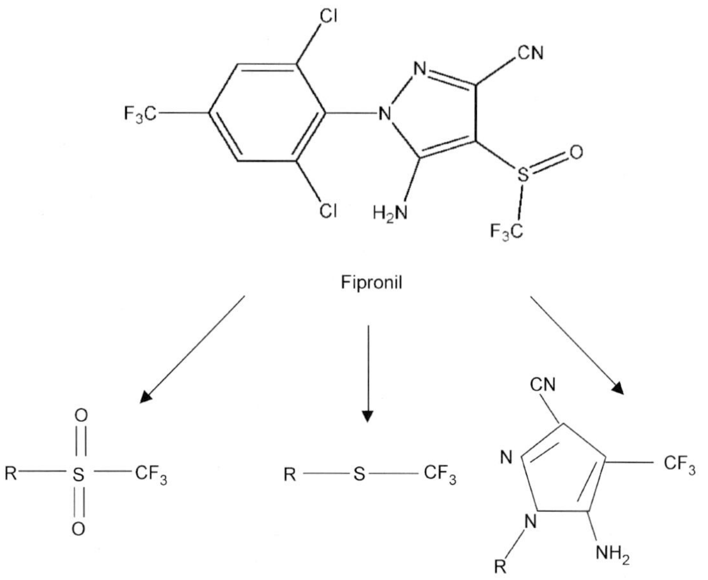

Fig. 9. The biotransformation of fipronil within fishes.

2. Herbicides

The biotransformation of the thiocarbamate, molinate, has been examined in striped bass, white sturgeon (*Acipenser transmontanus*) and two species of carp (Fig. 10)[49,84,85]. Common carp and white sturgeon oxidized molinate to form several products that were hydrolyzed, or conjugated with glutathione, as well as sulfoxide or sulfone metabolites; both fish also form a D-glucuronide. In the bass, metabolites included the sulfoxide, carboxymolinate, 4-hydroxymolinate, mercapturic acid, 4-ketomolinate and hexahydroazepine[84]. Common carp carried out less sulfoxidation and glutathione conjugation of molinate than either white sturgeon or stripped bass which may explain the greater toxicity in carp compared to the other two species[85].

Two other thiocarbamates that have been evaluated in fish include eptam (ethyl *N,N*-dipropylthiocarbamate) and thiobencarb (*p*-chlorobenzyl *N,N*-diethylthiocarbamate)[13]. FMOs and other free-radical mediated co-oxidative pathways catalyzed the S-oxygenation of eptam in hepatic microsomes from fresh and saltwater stripped bass[13]. The *S*-oxide of eptam was shown to be an efficient carbamylating agent of protein thiols indicating potential toxicity of this metabolite to freshwater fish. Similar studies in striped bass were performed with thiobencarb from hepatic microsomes and indicated that FMOs may catalyze S-oxygenation to the *S*-oxide which was also

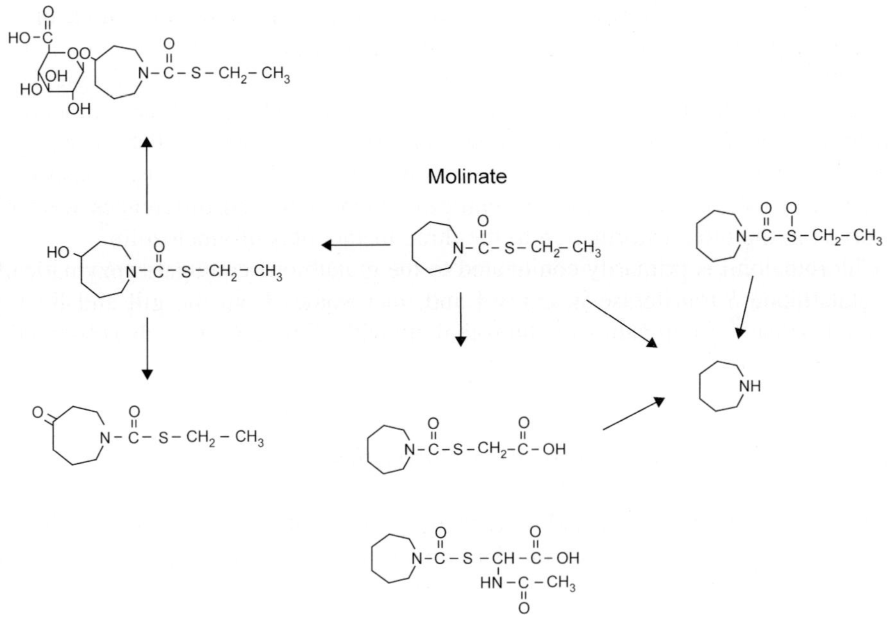

Molinate

Fig. 10. The biotransformation of molinate within fishes.

a more efficient carbamylating agent than the parent compound reacting with thiol and amine nucleophiles[14].

Of the chloroacetanilides, a large concentration of propanil (3′,4′-dichloropropionanilide) was observed in the bile of channel catfish[72]. One metabolite recovered from trout bile following propanil treatment was either 3′,4′-dichloro-2-hydroxypropionanilide or 3′,4′-dichloro-3-hydroxypropionanilide[12].

2,4-Dichlorophenoxyacetic acid (2,4-D) undergoes limited biotransformation in fish with less than 10% converted to primarily taurine acid conjugates *via* the urine[40,63].

Although no specific metabolites have been identified, atrazine has been shown to turnover cytosolic glutathione *S*-transferase indicating conjugation in zebrafish, rainbow trout, starry flounder (*Platichthys stellatus*) and English sole (*Pleuronectes vetulus*)[22,90]. Approximately 50% of atrazine was converted to four uncharacterized metabolites in zebrafish[36].

3. *Fungicides*

Inhibition of cytochrome P450 mediated sterol hydroxylase is one of the most common mechanisms of antifungal compounds. However, within vertebrates inhibition as well as induction of CYP has been observed. Some compounds are converted to nonspecific CYP inhibitors (for review see ref. 39). The imidazole derivative clotrimazole was shown to inhibit benzo(*a*)pyrene elimination in shad (*Dorosoma cepedianum*) presumably through a noncompetitive inhibition of CYP1A which was observed *in vitro* within shad and trout liver microsomes[52]. In contrast, propioconizole was shown to induce the oxygenation and activation of the organophosphate insecticide parathion in fathead minnows[53].

Fathead minnows exposed to 200 and 700 μg/l of the antiandrogenic fungicide for 21 days, vinclozolin, converted the parent compound into three unidentified metabolites *in vivo*[55]. Metabolite M1 was observed in all but one fish primarily in animals exposed to the highest concentrations. M2 and M3 were also consistently observed in fish of the highest concentration. Gender-related differences were not observed in the biotransformation to the three metabolites of vinclozolin[55].

Chlorothalonil is primarily conjugated to the glutathione conjugate enzymatically by glutathione *S*-transferase in cytosol and microsomes from the gill and liver of channel catfish[29]. Conjugation of chlorothalonil with glutathione was also observed in other fish species[20].

III. Effects of mixtures on biotransformation enzymes

Studies in North American and European surface waters have indicated the co-occurrence of a number of steroidal hormone and pesticides[47]. Although several studies have indicated that steroid hormones have significant effects upon biotransformation (primarily Phase I reactions)[8,80], few studies have examined the effects of steroids upon pesticide biotransformation. In Japanese medaka, estradiol diminished the expression and activity of FMOs, which were shown to convert aldicarb to aldicarb sulfoxide in microsomes of gill and liver from male fish as well as liver from female fish. However, estradiol significantly induced expression of FMO1-like protein and aldicarb toxicity in female fish. Testosterone consistently down-regulated FMO expression in male gill and liver, but had no effect on female FMO levels[25].

Metals have been shown to dramatically alter CYP concentrations, but there have not been any studies examining the effects on pesticide biotransformation. In contrast, several studies have examined the effects of enzyme induction on pesticide biotransformation. Pre-exposure of channel catfish to the PCB mixture Aroclor 1254 failed to alter the biotransformation of chlorpyrifos or parathion[81]. As mentioned above, the fungicide propioconizole induced the activation of parathion to the oxon in fathead minnow[53].

IV. Species differences and resistance

Decreased sensitivity to insecticides in resistant strains of mosquitofish from Belzoni, Mississippi is partly based on their ability to metabolize the toxicant rapidly. Resistant strains converted aldrin to dieldrin and water-soluble components to a greater extent than sensitive strains[89]. Resistant strains also had higher activities of methyl-parathion dearylation which appear to be catalyzed by the higher levels of CYP in this strain[15].

Channel catfish were shown to be more resistant to aldicarb toxicity compared to rainbow trout[62]. This species difference is partly explained by the differences in aldicarb activation to aldicarb sulfoxide between the two species, with trout having an active FMO system which is primarily responsible for the oxygenation of aldicarb to the more toxic sulfoxide[73].

Differences in oxidative desulfuration of diazinon have been shown to be responsible for toxicity differences between fish species and developmental stages with susceptible species having greater capacity to form the oxonic metabolite[38,54].

V. Environmental factors that affect pesticide biotransformation

Environmental factors such as temperature and nutrition have been shown to significantly alter the bioconcentration and toxicity of pesticides[39]. However, the effects of these factors on pesticide biotransformation are unclear. In contrast, the effects of salinity upon pesticide biotransformation have been examined in several species of euryhaline fish[68,69]. In mullet and Japanese medaka, fenitrothion biotransformation was unaltered by salinity (Fig. 11)[83]. In contrast, aldicarb biotransformation was significantly altered by salinity in Japanese medaka and rainbow trout with direct correlations observed[23,24,48]. Salinity had no effect on eptam or aldicarb sulfoxidation in striped bass[13,88]. Consistent with these results is the expression and activity of FMOs which are upregulated in trout and medaka by salinity, but not in striped bass[88]. Studies with other thioether pesticides which would

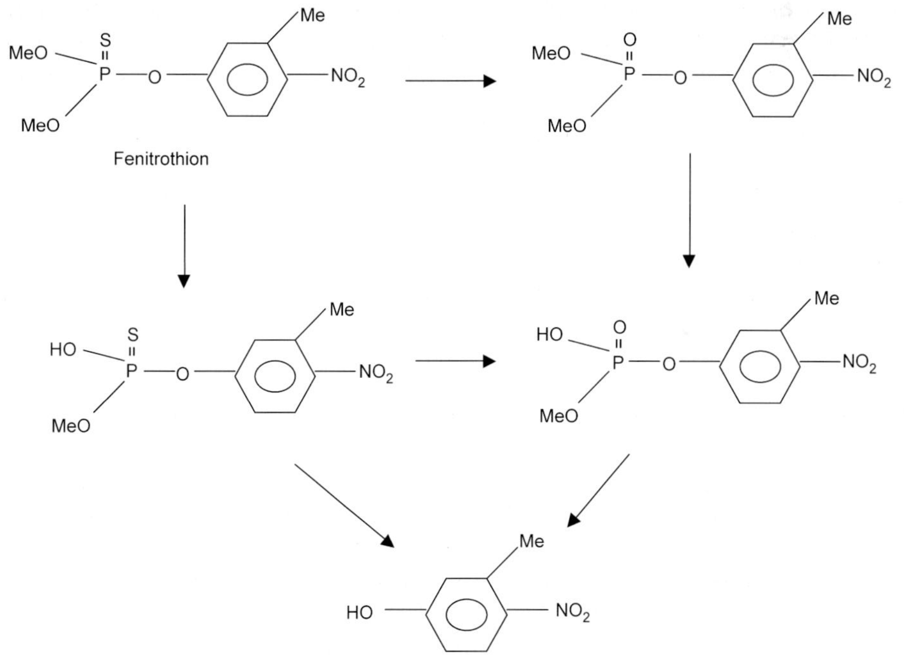

Fig. 11. The biotransformation of fenitrothion within fishes.

serve as FMO substrates are currently underway to determine the consistency of these results.

VI. Summary and conclusions

Fish have active Phase I and Phase II biotransformation pathways that can modify the disposition and toxicity of pesticides. Although more studies are characterizing *in vivo* metabolites of pesticides in fish, there is still a significant lack of knowledge about the ultimate fate of these compounds in the fish. In addition, very little is known about the specific enzymes responsible for the formation of specific metabolites of various pesticides. For example, although multiple CYP isoforms have been identified in fish, the substrate specificities with regard to pesticides are unknown and deserve further study.

Biotransformation plays a significant role in species differences in susceptibility with many resistant species capable of pesticide detoxification, whereas susceptible species tend to bioactivate compounds to more reactive intermediates which may be more toxic. Environmental factors such as salinity and co-exposure to environmental hormones may also significantly affect biotransformation and potential toxicity of pesticides in fish. Clearly, preliminary studies with chemical mixtures have demonstrated a role for biotransformation in subsequent synergistic and antagonistic whole animal endpoints. Further work is necessary in this area as concomitant exposure represents a more realistic setting of xenobiotic exposure to fish since many pesticides are used in combination or may occur in waterways at the same time of the year.

In conclusion, understanding the biotransformation of pesticides is a useful tool in determining species dependent and mechanistic effects of these compounds in fish and helps to reduce uncertainty in ecological risk assessment strategies.

VII. References

1. Addison, R.F. and D.E. Willis. The metabolism by rainbow trout (*Salmo gairdneri*)of *p-p'*-(14-C) DDT and some of its possible degradation products labeled with 14-C. *Toxicol. Appl. Pharmacol.* 43: 303–313, 1978.
2. Baron, M.G. and K.E. Woodburn. Ecotoxicology of chlorpyrifos. *Rev. Environ. Cont. Toxicol.* 144: 1–93, 1995.
3. Barron, M.G., S.M. Plakas and P.C. Wilga. Absorption, tissue distribution and metabolism of chlorpyrifos following intravascular and dietary administration in channel catfish. *Toxicol. Appl. Pharmacol.* 108: 474–482, 1991.
4. Benke, G.M., K.L. Cheever, F.E. Mirer and S.D. Murphy. Comparative toxicity, anticholinesterase action, and metabolism of methyl parathion and parathion in sunfish and mice. *Toxicol. Appl. Pharmacol.* 28: 97–109, 1974.
5. Benner, D.B. and R.S. Tjeerdema. Toxicokinetics and biotransformation of pentachlorophenol in the topsmelt (*Atherinops affinis*). *J. Biochem. Toxicol.* 8: 111–117, 1993.
6. Boone, J.S. and J.E. Chambers. Biochemical factors contributing to toxicity differences among chlorpyrifos, parathion, and methyl parathion in mosquitofish (*Gambusia affinis*). *Aquat. Toxicol.* 39: 333–343, 1997.
7. Bradbury, S.P. and J.R. Coats. Toxicokinetics and toxicodynamics of pyrethroid insecticides in fish. *Environ. Toxicol. Chem.* 8: 373–380, 1989.

8. Buhler, D.R., C.L. Miranda, M.C. Henderson, Y.H. Yang, S.J. Lee and J.L. Wang-Buhler. Effects of 17 beta-estradiol and testosterone on hepatic mRNA/protein levels and catalytic activities of CYP2M1, CYP2K1, and CYP3A27 in rainbow trout (*Oncorhynchus mykiss*). *Toxicol. Appl. Pharmacol.* 168: 91–101, 2000.
9. Buhler, D.R. and J.L. Wang-Buhler. Rainbow trout cytochrome P450s: purification, molecular aspects, metabolic activity, induction, and role in environmental monitoring. *Comp. Biochem. Physiol. C* 121: 107–138, 1998.
10. Bulger, W.H., R.M. Muccitelli and D. Kupfer. Studies on the *in vivo* and *in vitro* estrogenic activities of methoxychlor and its metabolites: role of hepatic monooxygenases in rats and humans. *Biochem. Pharmacol.* 27: 2417–2423, 1978.
11. Burns, K.A. Microsomal mixed function oxidases in an estuarine fish *Fundulus heteroclitus*, and their induction as a result of environmental contamination. *Comp. Biochem. Physiol. B* 53: 443–446, 1976.
12. Call, D.J., L.T. Brooke, R.S. Kent, M.L. Knuth, C. Anderson and C. Moriarity. Toxicity, bioconcentration and metabolism of the herbicide propanil in freshwater fish. *Arch. Environ. Contam. Toxicol.* 12: 175–182, 1983.
13. Cashman, J.R. and L.D. Olsen. S-oxygenation of eptam in hepatic microsomes from fresh and saltwater striped bass *Morone saxatilis*. *Chem. Res. Toxicol.* 2: 392–399, 1989.
14. Cashman, J.R., L.D. Olsen, R.S. Nishioka, E.S. Gray and H.A. Bern. S-oxygenation of thiobencarb (bolero) in hepatic preparations from striped bass (*Morone saxatilis*) and mammalian systems. *Chem. Res. Toxicol.* 3: 433–440, 1990.
15. Chambers, J.E. and J.D. Yarbrough. Organophosphate degradation by insecticide-resistant and susceptible populations of mosquitofish (*Gambusia affinis*). *Pesti. Biochem. Physiol.* 3: 312–322, 1973.
16. Chambers, J.E. and J.D. Yarbrough. Xenobiotic biotransformation systems in fishes. *Comp. Biochem. Physiol. C* 55: 77–84, 1976.
17. Chin, B.H., L.J. Sullivan and J.E. Eldridge. In vitro metabolism of carbaryl by liver explants of bluegill, catfish, perch, goldfish, and kissing gourami. *J. Agricul. Food Chem.* 27: 1395–1398, 1979.
18. Cook, G.H., J.C. Moore and D.L. Coppage. The relation of malathion and its metabolites to fish poisoning. *Bull. Environ. Contam. Toxicol.* 16: 283–290, 1976.
19. Cravedi, J.P., A. Lafuente, M. Baradat, A. Hillenweck and E. Perdu-Durand. Biotransformation of pentachlorophenol, aniline and biphenyl in isolated rainbow trout (*Oncorhynchus mykiss*) hepatocytes: comparison with in vivo metabolism. *Xenobiotica* 29: 499–509, 1999.
20. Davies, P.E. The toxicology and metabolism of chlorothalonil in fish III. Metabolism, enzymatics and detoxification in *Salmo* spp. and *Galaxias* spp. *Aquat. Toxicol.* 7: 277–299, 1985.
21. Dvorchik, B.H. and T.H. Maren. The fate of *p-p′* DDT (2,2-bis (*p*-chlorophenyl),1,1,1-trichloroethane) in the dogfish *Squalus acanthias*. *Comp. Biochem. Physiol. A* 42: 205–215, 1972.
22. Egaas, E., J.U. Skaare, N.O. Svendsen, M. Sandvik, J.G. Falls, W.C. Dauterman, T.K. Collier and J. Netland. A comparative study of effects of atrazine on xenobiotic metabolizing enzymes in fish and insect, and of the *in vitro* phase-I atrazine metabolism in some fish, insects, mammals and one plant species. *Comp. Biochem. Physiol. C* 106: 141–149, 1993.
23. El-Alfy, A., S. Grisle and D. Schlenk. Characterization of salinity-enhanced toxicity of aldicarb to Japanese medaka: sexual and developmental differences. *Environ. Toxicol. Chem.* 20: 2093–2098, 2001.
24. El-Alfy, A. and D. Schlenk. Potential mechanisms of the enhancement of aldicarb toxicity to Japanese medaka, *Oryzias latipes*, at high salinity. *Toxicol. Appl. Pharmacol.* 152: 175–183, 1998.
25. El-Alfy, A.T. and D. Schlenk. Effect of 17 beta-estradiol and testosterone on the expression of flavin-containing monooxygenase and the toxicity of aldicarb to Japanese medaka, *Oryzias latipes*. *Toxicol. Sci.* 68: 381–388, 2002.
26. Eldib, M.A., I.A. Elelaimy, A. Kotb and S.H. Elowa. Activation of in vivo metabolism of malathion in male *Tilapia nilotica*. *Bull. Environ. Contam. Toxicol.* 57: 667–674, 1996.
27. Frankovic, L., M.A.Q. Khan and S.M. Alghais. Metabolism of hexachlorobenzene in the fry of steelhead trout, *Salmo gairdneri* (*Oncorhynchus mykiss*). *Arch. Environ. Contam. Toxicol.* 28: 209–214, 1995.
28. Fujii, Y. and S. Asaka. Metabolism of diazinon and diazoxon in fish liver preparations. *Bull. Environ. Contam. Toxicol.* 29: 455–460, 1982.
29. Gallagher, E.P., G.L. Kedderis and R.T. Digiulio. Glutathione *S*-transferase-mediated chlorothalonil metabolism in liver and gill subcellular fractions of channel catfish. *Biochem. Pharmacol.* 42: 139–145, 1991.
30. Gates, V.L. and R.S. Tjeerdema. Disposition and biotransformation of pentachlorophenol in the striped bass (*Morone saxatilis*). *Pestic. Biochem. Physiol.* 46: 161–170, 1993.
31. Gibson, J.R. and J.L. Ludke. Effect of SKF-525A on brain acetylcholinesterase inhibition by parathion in fishes. *Bull. Environ. Contam. Toxicol* 9: 140–142, 1973.

32. Glickman, A.H., A.A.R. Hamid, D.E. Rickert and J.J. Lech. Elimination and metabolism of permethrin isomers in rainbow trout. *Toxicol. Appl. Pharmacol.* 57: 88–98, 1981.
33. Glickman, A.H., T. Shono, J.E. Casida and J.J. Lech. In vitro metabolism of permethrin isomers by carp and rainbow trout liver microsomes. *J. Agric. Food Chem.* 27: 1038–1048, 1979.
34. Glickman, A.H., C.N. Statham, A. Wu and J.J. Lech. Studies on the uptake, metabolism, and disposition of pentachlorophenol and pentachloroanisole in rainbow trout. *Toxicol. Appl. Pharmacol.* 41: 649–658, 1977.
35. Glickman, A.H., S.D. Weitman and J.J. Lech. Differential toxicity of *trans*-permethrin in rainbow trout and mice. I. Role of biotransformation. *Toxicol. Appl. Pharmacol.* 66: 153–161, 1982.
36. Goerge, G. and R. Nagel. Kinetics and metabolism of carbon-14 lindane and carbon-14 atrazine in early life stages of zebrafish (*Brachydanio rerio*). *Chemosphere* 21: 1125–1137, 1990.
37. Hainzl, D. and J.E. Casida. Fipronil insecticide: novel photochemical desulfinylation with retention of neurotoxicity. *Proc. Natl Acad. Sci. USA* 93: 12764–12767, 1996.
38. Hamm, J.T., B.W. Wilson and D.E. Hinton. Increasing uptake and bioactivation with development positively modulate diazinon toxicity in early life stage medaka (*Oryzias latipes*). *Toxicol. Sci.* 61: 304–313, 2001.
39. Huckle, K.R. and P. Millburn. Metabolism, bioconcentration and toxicity of pesticides in fish. In: *Environmental Fate of Pesticides*, edited by D.H. Hutson and T.R. Roberts, New York, Wiley, pp. 175–243, 1990.
40. James, M.O. and J.R. Bend. Taurine conjugation of 2,4-dichlorophenoxyacetic acid and phenylacetic acid in two marine species. *Xenobiotica* 6: 393–398, 1976.
41. Jarvinen, A.W., M.J. Hoffman and T.W. Thorslund. Long-term toxic effects of DDT food and water exposure on fathead minnows. *J. Fish. Res. Board Can.* 34: 2089–2099, 1977.
42. Keizer, J., G. D'Agostino, R. Nagel, F. Gramenzi and L. Vittozzi. Comparative diazinon toxicity in guppy and zebra fish: different role of oxidative metabolism. *Environ. Toxicol. Chem.* 12: 1243–1250, 1993.
43. Keizer, J., G. D'Agostino, R. Nagel, T. Volpe, P. Gnemi and L. Vittozzi. Enzymological differences of AChE and diazinon hepatic metabolism: correlation of in vitro data with the selective toxicity of diazinon to fish species. *Sci. Total Environ.* 171: 213–220, 1995.
44. Keizer, J., G. D'Agostino and L. Vittozzi. The importance of biotransformation in the toxicity of xenobiotics to fish. I. Toxicity and bioaccumulation of diazinon in guppy (*Poecilia reticulata*) and zebra fish (*Brachydanio rerio*). *Aquat. Toxicol.* 21: 239–254, 1991.
45. Kitamura, S., T. Kadota, M. Yoshida, N. Jinno and S. Ohta. Whole-body metabolism of the organophosphorus pesticide, fenthion, in goldfish, *Carassius auratus*. *Comp. Biochem. Physiol. C* 126: 259–266, 2000.
46. Kitamura, S., T. Kadota, M. Yoshida and S. Ohta. Interconversion between fenthion and fenthion sulfoxide in goldfish *Carassius auratus*. *J. Health Sci.* 45: 266–270, 1999.
47. Kolpin, D.W., E.T. Furlong, M.T. Meyer, E.M. Thurman, S.D. Zaugg, L.B. Barber and H.T. Buxton. Pharmaceuticals, hormones, and other organic wastewater contaminants in US streams, 1999–2000: a national reconnaissance. *Environ. Sci. Technol.* 36: 1202–1211, 2002.
48. Larsen, B.K. and D. Schlenk. Effect of salinity on flavin-containing monooxygenase expression and activity in rainbow trout (*Oncorhynchus mykiss*). *J. Comp. Physiol. B* 171: 421–429, 2001.
49. Lay, M.M. and J.J. Menn. Mercapturic acid occurrence in fish bile. A terminal product of metabolism of the herbicide molinate. *Xenobiotica* 9: 669–673, 1979.
50. Lee, K.B., J.H. Shim and Y.T. Suh. In vivo metabolism of endosulfan in carp (*Cyprinus carpio*). *Agric. Chem. Biotechnol.* 37: 194–202, 1994.
51. Levi, P.E. and E. Hodgson. Metabolism of organophosphorus compounds by the flavin-containing monooxygenase. In: *Organophosphates: Chemistry, Fate and Effects*, edited by J.E. Chambers and P.E. Levi, San Diego, Academic Press, pp. 141–154, 1992.
52. Levine, S.L. and J.T. Oris. Induction of CYP1A mRNA and catalytic activity in gizzard shad (*Dorosoma cepedianum*) after waterborne exposure to benzo[*a*]pyrene. *Comp. Biochem. Physiol. C* 118: 397–404, 1997.
53. Levine, S.L. and J.T. Oris. Enhancement of acute parathion toxicity to fathead minnows following pre-exposure to propiconazole. *Pestic. Biochem. Physiol.* 65: 102–109, 1999.
54. Ludke, J.L., J.R. Bibson and C.I. Lusk. Mixed function oxidase activity in freshwater fishes: aldrin epoxidation and parathion activation. *Toxicol. Appl. Pharmacol.* 21: 89–97, 1972.
55. Makynen, E.A., M.D. Kahl, K.M. Jensen, J.E. Tietge, K.L. Wells, G. Van Der Kraak and G.T. Ankley. Effects of the mammalian antiandrogen vinclozolin on development and reproduction of the fathead minnow (*Pimephales promelas*). *Aquat. Toxicol.* 48: 461–475, 2000.
56. McKim, J.M. and G.L. Lein. Toxic responses of the skin. In: *Target Organ Toxicity in Marine and Freshwater Teleosts*, edited by D. Schlenk and W.H. Benson, London, UK, Taylor & Francis, pp. 151–223, 2001.

57. McKim, J.M. and J.W. Nichols. Use of physiologically based toxicokinetic models in a mechanistic approach to aquatic toxicology. In: *Aquatic Toxicology: Molecular, Biochemical, and Cellular Perspectives*, edited by D.C. Malins and G. Ostrander, Boca Raton, FL, Lewis Publishers, pp. 469–520, 1994.
58. Millburn, E.R. and D.H. Hutson. Comparative toxicity of *cis*-cypermethrin in rainbow trout, frog, mouse, and quail. *Toxicol. Appl. Pharmacol.* 84: 512–522, 1986.
59. Murphy, S.D. Liver metabolism and toxicity of thiophosphate insecticides in mammalian, avian, and piscine species. *Proc. Soc. Exp. Biol. Med.* 123: 392–398, 1966.
60. Murty, A.S. *Toxicity of Pesticides to Fish*, Boca Raton, FL, CRC Press, 1986.
61. Perkins, E.J., A. El-Alfy and D. Schlenk. In vitro sulfoxidation of aldicarb by hepatic microsomes of channel catfish *Ictalurus punctatus*. *Toxicol. Sci.* 48: 67–73, 1999.
62. Perkins, E.J. and D. Schlenk. In vivo metabolism, acetylcholinesterase inhibition, and toxicokinetics of aldicarb in channel catfish (*Ictalurus punctatus*). *Toxicol. Sci.* 53: 308–315, 2000.
63. Plakas, S.M., L. Khoo and M.G. Barron. 2,4-Dichlorophenoxyacetic acid disposition after oral administration in channel catfish. *J. Agric. Food Chem.* 40: 1236–1239, 1992.
64. Ramaneswari, K. and L.M. Rao. Bioconcentration of endosulfan and monocrotophos by *Labeo rohita* and *Channa punctata*. *Bull. Environ. Contam Toxicol.* 65: 618–622, 2000.
65. Rao, D.M.R., A.P. Devi and A.S. Murty. Toxicity and metabolism of endosulfan and its effect on oxygen consumption and total nitrogen excretion of the fish *Macrognathus aculeatum*. *Pestic. Biochem. Physiol.* 15: 282–287, 1981.
66. Reinbold, K.A. and R.L. Metcalf. Effects of the synergist piperonyl butoxide on metabolism of pesticides in green sunfish. *Pestic. Biochem. Physiol.* 6: 401–412, 1976.
67. Rodriguezariza, A., F.M. Diazmendez, J.I. Navas, C. Pueyo and J. Lopezbarea. Metabolic activation of carcinogenic aromatic amines by fish exposed to environmental pollutants. *Environ. Mol. Mutagen.* 25: 50–57, 1995.
68. Schlenk, D. A comparison of endogenous and exogenous substrates of the flavin-containing monooxygenases in aquatic organisms. *Aquat. Toxicol.* 26: 157–162, 1993.
69. Schlenk, D. Occurrence of flavin-containing monooxygenases in non-mammalian eukaryotic organisms. *Comp. Biochem. Physiol. C* 121: 185–195, 1998.
70. Schlenk, D. and D.R. Buhler. Role of flavin-containing monooxygenase in the *in vitro* biotransformation of aldicarb in rainbow trout (*Oncorhynchus mykiss*). *Xenobiotica* 21: 1583–1589, 1991.
71. Schlenk, D., D.A. Erickson, J.J. Lech and D.R. Buhler. The distribution, elimination, and in vivo biotransformation of aldicarb in the rainbow trout (*Oncorhynchus mykiss*). *Fundam. Appl. Toxicol.* 18: 131–136, 1992.
72. Schlenk, D. and C.T. Moore. Distribution and elimination of the herbicide propanil in the channel catfish (*Ictalurus punctatus*). *Xenobiotica* 23: 1017–1024, 1993.
73. Schlenk, D., M.J. Ronis, C.L. Miranda and D.R. Buhler. Channel catfish liver monooxygenases. Immunological characterization of constitutive cytochromes P450 and the absence of active flavin-containing monooxygenases. *Biochem. Pharmacol.* 45: 217–221, 1993.
74. Schlenk, D., D.M. Stresser, J.C. McCants, A.C. Nimrod and W.H. Benson. Influence of beta-naphthoflavone and methoxychlor pretreatment on the biotransformation and estrogenic activity of methoxychlor in channel catfish (*Ictalurus punctatus*). *Toxicol. Appl. Pharmacol.* 145: 349–356, 1997.
75. Schlenk, D., D.M. Stresser, J. Rimoldi, L. Arcand, J. McCants, A.C. Nimrod and W.H. Benson. Biotransformation and estrogenic activity of methoxychlor and its metabolites in channel catfish (*Ictalurus punctatus*). *Mar. Environ. Res.* 46: 159–162, 1998.
76. Seaton, C.L. and R.S. Tjeerdema. Tissue disposition and biotransformation of naphthalene in striped bass (*Morone saxatilis*). *Mar. Environ. Res.* 42: 345–348, 1996.
77. Smith, G.N., B.S. Watson and F.S. Fischer. The metabolism of [14-C] *O,O*-diethyl *O*-(3,5,6-trichloro-2-pyridyl) phosphorothioate (Dursban) in fish. *J. Econ. Entomol.* 59: 1464–1475, 1966.
78. Stanton, R.H. and M.A.Q. Khan. Mixed-function oxidase activity toward cyclodiene insecticides in bass and bluegill sunfish. *Pestic. Biochem. Physiol.* 3: 351–357, 1973.
79. Statham, C.N., S.K. Pepple and J.J. Lech. Biliary excretion products of 1-naphthyl-*N*-methylcarbamate-1-14C (carbaryl) in rainbow trout (*Salmo gairdneri*). *Drug Metab. Dispos.* 3: 400–406, 1975.
80. Stegeman, J.J. and M.E. Hahn. Biochemistry and molecular biology of monooxygenases: current perspectives on forms, functions, and regulation of cytochrome P450 in aquatic species. In: *Aquatic Toxicology: Molecular, Biochemical and Cellular Perspectives*, edited by D.C. Malins and G.K. Ostrander, Boca Raton, FL, Lewis Publishers, pp. 87–206, 1994.
81. Straus, D.L., D. Schlenk and J.E. Chambers. Hepatic microsomal desulfuration and dearylation of chlorpyrifos and parathion in fingerling channel catfish: lack of effect from Aroclor 1254. *Aquat. Toxicol.* 50: 141–149, 2000.

82. Sundershan, P. and M.A.Q. Khan. Metabolism of 14-C-Dieldrin in bluegill fish. *Pestic. Biochem. Physiol.* 15: 192–199, 1981.
83. Takimoto, Y., M. Ohshima and J. Miyamoto. Comparative metabolism of fenitrothion in aquatic organisms I. Metabolism in the euryhaline fish *Oryzias latipes* and *Mugil cephalus*. *Ecotoxicol. Environ. Saf.* 13: 104–117, 1987.
84. Tjeerdema, R.S. and D.G. Crosby. The biotransformation of molinate (Ordram) in the striped bass (*Morone saxatilis*). *Aquat. Toxicol.* 9: 305–317, 1987.
85. Tjeerdema, R.S. and D.G. Crosby. Comparative biotransformation of molinate (Ordram) in the white sturgeon (*Acipenser transmontanus*) and common carp (*Cyprinus carpio*). *Xenobiotica* 18: 831–838, 1988.
86. Tsuda, T., M. Kojima, H. Harada, A. Nakajima and S. Aoki. Accumulation and excretion of fenthion, fenthion sulfoxide and fenthion sulfone by killifish (*Oryzias latipes*). *Comp. Biochem. Physiol. C* 113: 45–49, 1996.
87. USEPA. New pesticide fact sheet, EPA 737-F-96-005, Washington, DC, Office of Pesticide Programs, 1996.
88. Wang, J., S. Grisle and D. Schlenk. Effects of salinity on aldicarb toxicity in juvenile rainbow trout (*Oncorhynchus mykiss*) and striped bass (*Morone saxatilis* × *chrysops*). *Toxicol. Sci.* 64: 200–207, 2001.
89. Wells, M.R., J.L. Ludke and J.D. Yarbrough. Epoxidation and fate of 14-C aldrin in insecticide-resistant and susceptible populations of mosquitofish (*Gambusia affinis*). *J. Agric. Food Chem.* 21: 428–438, 1973.
90. Wiegand, C., S. Pflugmacher, M. Giese, H. Frank and C. Steinberg. Uptake, toxicity, and effects on detoxication enzymes of atrazine and trifluoroacetate in embryos of zebrafish. *Ecotoxicol. Environ. Saf.* 45: 122–131, 2000.

Biochemistry and Molecular Biology of Fishes, vol. 6
T. P. Mommsen and T. W. Moon (Editors)

CHAPTER 7

Xenobiotic receptors in fish: Structural and functional diversity and evolutionary insights

MARK E. HAHN, REBEKA R. MERSON AND SIBEL I. KARCHNER
Biology Department, Woods Hole Oceanographic Institution, Woods Hole, MA 02543 USA

I. Introduction: Xenobiotic receptors

Chemicals interact with biological systems through a variety of mechanisms that give rise to physical, chemical, and biochemical modifications of biological macromolecules or changes in gene expression. Some of the most important – and interesting – mechanisms are those in which the initial steps occur through chemical interactions with intracellular receptor proteins that act as ligand-dependent (or ligand-modified) transcription factors. Many of these receptors appear to have evolved as biological sensors to detect the presence of toxicants (xenobiotic or endogenous) or changes in environmental conditions. Consistent with that notion, these receptors often regulate the expression of genes encoding biotransformation enzymes or small-molecule transporters. The former include Phase I biotransformation enzymes such as cytochromes P450 (CYPs) and flavin monooxygenases (FMOs) as well as Phase II enzymes such as glutathione S-transferases (GSTs) and uridine diphosphate glucuronosyltransferases (UGTs). The induction of such genes can usually be considered an adaptive response. However, in some situations, such as with certain ligands or at specific developmental stages, stimulation of this 'adaptive' response can lead directly to toxicity, for example, through generation of reactive metabolites or interference with endogenous signaling pathways.

Most of the research on xenobiotic receptors has been conducted in mammalian systems. These studies have identified three families of proteins as having important roles in regulating the response to xenobiotic chemicals (Table 1). The aryl hydrocarbon receptor (AHR), a member of the basic helix–loop–helix Per-ARNT-Sim (bHLH-PAS) family of transcription factors, is well known for its role in the altered gene expression and toxicity elicited by chlorinated dioxins and related planar halogenated aromatic hydrocarbons (PHAHs) as well as certain polynuclear aromatic hydrocarbons (PAHs)[114,188,244]. Several members of the nuclear/steroid

TABLE 1

Receptors/sensors involved in regulation of xenobiotic-metabolizing enzymes

Transcription factor	Dimerization partner	Examples of ligands	Genes regulated
AHR	ARNT	Dioxins, non-*ortho* PCBs, some PAHs, bilirubin, etc.	CYP1A, CYP1B GST, UGT, NQO
CAR	RXR	Phenobarbital (PB), TCPOBOP, chlorinated pesticides, *ortho*-PCBs, androstanol/androstenol (inhibits)	CYP2B, CYP3A GST, ABC transporters
PXR (SXR)	RXR	PB, *ortho*-PCBs, organochlorine pesticides, dexamethasone, pregnenalone, corticosterone, bile acids (lithocholic acid)	CYP3A, CYP2B, CYP7A (repression) GST, ABC transporters
PPAR	RXR	Fibrate drugs, phthalate esters, linoleic acid, arachidonic acid	CYP4A, CYP7A (repression), CYP8B, LXR
LXR	RXR	Cholesterol; (24 S)-hydroxycholesterol	CYP7A, ABC transporters, LXR
FXR	RXR	Bile acids, chenodeoxycholic acid	Represses CYP7A, CYP8B, CYP27A
ER	ER	Structurally diverse xenoestrogens	CYP19
NRF	Small Maf proteins	Activated by electrophiles, quinones	GST, γGCS, NQO

For details and references, see text and ref. 52.

receptor (NR) superfamily regulate the expression of multiple CYP and transporter proteins that act on xenobiotic substrates[143,287]. Proteins in the Cap 'n' Collar basic leucine zipper (CnC-bZIP) family, including *N*uclear factor E2-*R*elated *F*actor 2 (NRF2) and related proteins, respond to oxidative stress by enhancing the expression of genes encoding enzymes involved in the anti-oxidant response, including several Phase II biotransformation enzymes[192]. Although not pharmacological 'receptors' (they have not been shown to exhibit high affinity, specific binding to ligands), CnC-bZIP proteins are clearly 'sensors' of environmental chemicals and their effects, and thus are rightly considered together with the bHLH-PAS and NR proteins as part of the transcriptional response to xenobiotic exposure.

All three of these protein families (bHLH-PAS, NR, and CnC-bZIP) exist in fish. In each case, there is strong evidence that the xenobiotic-related roles played by these proteins in fish are similar to those demonstrated in mammals. At the same time, there are important differences in the diversity and function of these proteins in fish as compared to mammals. Such differences complicate attempts to extrapolate findings between these two groups of vertebrates. They also make fish a rich and fascinating group of animals in which to probe the function and evolution of these important signaling pathways. Given the importance of fish both as targets for environmental chemicals[53,54,177] and as model systems for investigating the possible human health impacts of xenobiotics[18,97,254,268] (see Chapter 1), understanding the structural and functional diversity of xenobiotic receptors in fish is essential. In this chapter, we review recent progress in this area and discuss the implications for environmental toxicology and the evolution of gene regulatory pathways.

II. Aryl hydrocarbon receptors and the bHLH-PAS superfamily

1. The bHLH-PAS superfamily and AHR-dependent signaling in mammals

The bHLH-PAS superfamily in vertebrates includes nearly two dozen PAS-domain containing proteins that act as transcription factors to regulate gene expression. Many PAS proteins have important roles during development or act as part of signaling pathways that serve to sense changing environmental conditions (including the internal environment) and initiate homeostatic responses[56,94,95,238]. PAS proteins also occur in invertebrates, plants, bacteria, and archaea, where they also serve functions related to environmental sensing[264]. Metazoan PAS proteins mediate the response to hypoxia (HIF, Trh), serve as co-activators for nuclear receptors (NCoA), form the molecular cogs of circadian clocks (CLOCK, Per), and play important roles in neural development (SIM)[94]. Two PAS proteins interact directly with small molecule ligands: the vertebrate AHR[40] and arthropod MET[16]. Vertebrate bHLH-PAS proteins usually act as heterodimers consisting of one general partner (ARNT or BMAL) and one specific partner (e.g. AHR, HIF, SIM, or CLOCK).

The AHR signaling pathway has been characterized in greatest detail in mammalian (primarily murine and human) systems, and detailed reviews are available[84,95,98,114,186,188,244]. Here, we highlight certain features of the mammalian

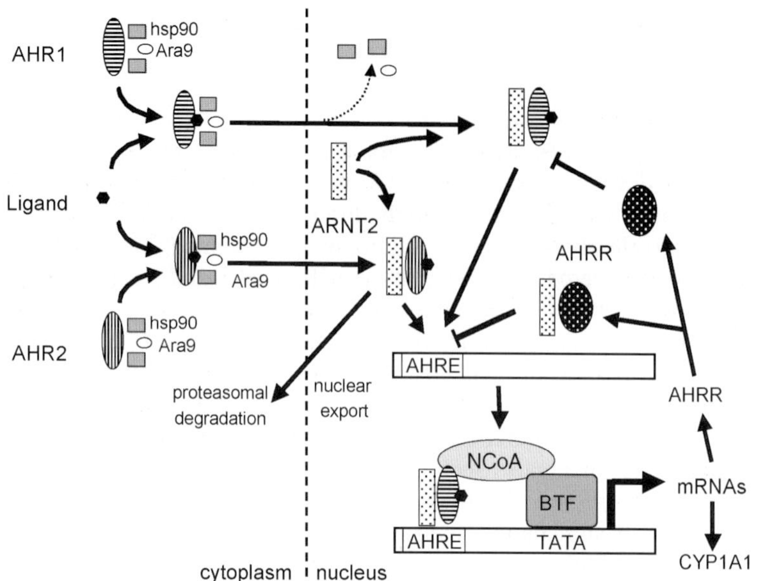

Fig. 1. Outline of the AHR-dependent signaling pathway in Atlantic killifish (*F. heteroclitus*). AHR1 and AHR2 both exhibit high-affinity binding to TCDD and other AHR ligands[137], form dimers with ARNT2[137,213], and bind to AHRE sequences in the promoters of target genes such as *CYP1A*[214] and *AHRR*[135]. Binding to AHREs recruits coactivator proteins (NCoA) and basal transcription factors (BTF) (inferred from studies in mammals) to activate transcription. Signaling is down-regulated by proteasomal degradation of AHRs[206] and through repression by the induced AHRR protein[135], which competes for binding to ARNT and AHREs[175].

AHR pathway as background and for comparison with the fish-specific information that will follow (Fig. 1).

The AHR was first identified[203] as a protein regulating the induction of cytochrome P450-dependent benzo[*a*]pyrene hydroxylase activity – which is catalyzed primarily by CYP1A1 – in inbred mice. Many genes are now known to be regulated by the mammalian AHR; those involved in an autoregulatory loop involving the AHR and CYP1A1 have been classified as the 'Ah gene battery'[187,188]. The AHR regulates gene expression by interacting with a specific DNA sequence motif, referred to variously as a xenobiotic-response element (XRE), dioxin-response element (DRE), or AHR-response element (AHRE), in the promoter of target genes. The AHR binds to AHRE sequences as a dimer with the AHR nuclear translocator (ARNT) protein[121,228]. Our current understanding of AHR–ARNT interactions is that they occur only (or primarily) after activation of the AHR by a small molecule ligand[153], although other activation mechanisms are possible[240]. In most cells, the unliganded AHR exists in the cytoplasm as an inactive complex with chaperone proteins such as hsp90. Ligand binding appears to alter the conformation of this complex, facilitating its movement into the nucleus, the dissociation of the chaperones, and formation of the transcriptionally active AHR–ARNT complex. The AHR signal is terminated when the AHR is exported from the nucleus and subjected to proteasomal degradation[206]. AHR signaling may be negatively regulated by several mechanisms, one of which involves the AHR repressor (AHRR), an AHR-related protein that competes with the AHR for binding to ARNT and AHREs but is transcriptionally inactive[175].

The AHR is required for changes in gene expression (e.g. induction of CYP1A1) after exposure to dioxins and related chemicals, as shown in studies employing mice in which the AHR gene had been inactivated[78,245]. Other studies in these AHR knock-out mice revealed that the AHR also is required for the toxicity of TCDD[77,176,200] and carcinogenicity of benzo[*a*]pyrene[250]. Thus, the toxicity of TCDD and other PHAHs occurs as a result of their interactions with the AHR, most likely through AHR-dependent changes in gene expression (induction or repression) that occur subsequent to ligand binding[39]. It is not yet known which AHR target genes play the most important role in the toxicity of dioxin-like compounds. Nevertheless, the essential role of the AHR in the effects of these compounds has been established, even if the exact mechanisms are not yet clear. In addition to the fact that the AHR is required for toxicity to occur, quantitative differences in the sensitivity of species or strains of animals to the toxicity of AHR ligands are in part the result of differences in AHR functional characteristics such as ligand-binding affinity[96,205].

Although originally thought to exhibit a strict structure–activity relationship for ligand binding, the AHR is now known to be extremely promiscuous in its ligand interactions. In addition to the 'classical' synthetic AHR ligands, which tend to be planar, hydrophobic compounds such as 2,3,7,8-tetrachlorodibenzo-*p*-dioxin (TCDD) and PAHs, the AHR exhibits moderate to high affinity for a variety of flavonoids, indole derivatives, and other structures[59,60]. Some of the latter are candidates for the elusive, endogenous/physiological AHR ligand[34,201,248,251,253].

2. *AHR signaling in fish: Structural diversity*

In contrast to the vast literature on the mammalian AHR signaling pathway, there are just a handful of reports describing AHRs and ARNTs in fish. However, those few studies have revealed a surprising diversity in the structure and function of piscine AHR and ARNT proteins. This diversity includes differences in the fish proteins as compared to their counterparts in mammals, as well as variation among fish species. It occurs through differences in the number of genes, allelic polymorphism, and alternative splicing. Some of the features of AHR signaling in fish have been reviewed previously[95,97–99,263].

One of the most notable features of AHR signaling in fish is the large number of AHR genes that exist as compared to mammals. While all mammalian species examined so far have only a single AHR gene, individual fish species possess from one to at least six AHRs, with two or more AHRs in most species that have been studied (Table 2). For example, two AHRs (AHR1 and AHR2) each have been identified in killifish *Fundulus heteroclitus*[100,104,137] (Fig. 1) and zebrafish *Danio rerio*[10,261], whereas five AHRs (two AHR1 forms and three AHR2s) have been identified in the completed genome of the pufferfish *Takifugu (Fugu) rubripes*[136]. This diversity of AHR genes in fish appears to have arisen by gene or genome duplications prior to the divergence of the fish and tetrapod lineages as well as by lineage-specific duplications

TABLE 2

AHR, AHRR, and ARNT diversity in fish

Species	AHR1	AHR2	AHR3	AHRR	ARNT1	ARNT2	References
Bony fish							
Atlantic killifish (*F. heteroclitus*)	1	1		1		1	100,104,135,137,213
Rainbow trout (*Oncorhynchus mykiss*)	?	3			1		2,115,209
Atlantic salmon (*Salmo salar*)	2	4					115a
Zebrafish (*Danio rerio*)	2	1		2	1	1	10,74,261,262
Medaka (*Oryzias latipes*)	2	2					98,138
Pufferfish (*Takifugu rubripes*)	2	3		1	1	1	136
Atlantic tomcod (*Microgadus tomcod*)		1		1	1		233,236,237
European flounder (*Platichthys flesus*)		1					31
Topminnow (*Poeciliopsis lucida*)		1					95
Cartilaginous fish							
Smooth dogfish (*Mustelus canis*)	1	1					104
Spiny dogfish (*Squalus acanthias*)	1	1	1				32,169
Greenland shark (*Somniosus microcephalus*)	1	1					169
Sandbar shark (*Carcharhinus plumbeus*)		1	1				169
Skate (*Leucoraja erinacea*)			1				104
Jawless fish							
Lamprey (*Petromyzon marinus*)	1*				1*		104

Identification as AHR1, AHR2, or AHR3 was by phylogenetic analyses using maximum parsimony (MP) and neighbor-joining (NJ) methods. Identity of some sequences as AHR1, AHR2, or AHR3 could not be determined or the genes predate the duplications; these are listed under AHR1 and indicated by an asterisk (*). Similarly, ARNT genes not identifiable as ARNT1 or ARNT2 are listed under ARNT1 and indicated by an asterisk (*). A question mark (?) indicates a partial sequence that cannot yet be classified.

in specific groups of fish[98,104,115a,b,136]. The piscine AHR genes have been classified by phylogenetic analysis into three clades, designated AHR1, AHR2, and AHR3. AHR1 genes are orthologous to mammalian AHRs, whereas AHR2 and AHR3 are novel AHR forms first identified in fish[104,137,169]. AHR2 appears to be the predominant AHR form (i.e. most highly and widely expressed) in many teleost fish[2,31,115a,137,237,261]. AHR3 proteins are phylogenetically distinct from those in the AHR1 and AHR2 clades, and have been found so far only in elasmobranchs[32,169]. Fish AHRs also display diversity reflected as polymorphisms. This was first noted in the allelic variants of AHR2 described in tomcod (*Microgadus tomcod*)[237]. Subsequently, it has been shown that the killifish AHR1 gene is highly polymorphic, with numerous single-nucleotide polymorphisms arranged in multiple haplotypes[101,103]. The functional significance of these variants is not yet understood.

AHR splice variants have also been identified in fish (AHR1s)[137,138], as in mammals[48], but again the functional significance of these is unknown.

Despite the diversity in AHR number and structure in fish, all fish AHRs possess N-terminal bHLH and PAS domains that are highly conserved as compared to mammalian AHRs (Fig. 2). In contrast, the C-terminal halves of fish AHRs are quite variable, as seen also for mammalian AHRs and some other PAS proteins[48,137]. Fish AHR2s lack the glutamine-rich transactivation domain found in mammalian AHRs (and killifish AHR1). AHR2s also differ from AHR1 forms in having a modified version (MYCAD) of the LXCXE motif that mediates interactions of the mammalian AHR with pRB, the product of the retinoblastoma tumor suppressor gene[87,137,221] (Fig. 2). However, the functional consequences of such differences are not clear, because fish AHR2s are transcriptionally active[2,3,135,208,261].

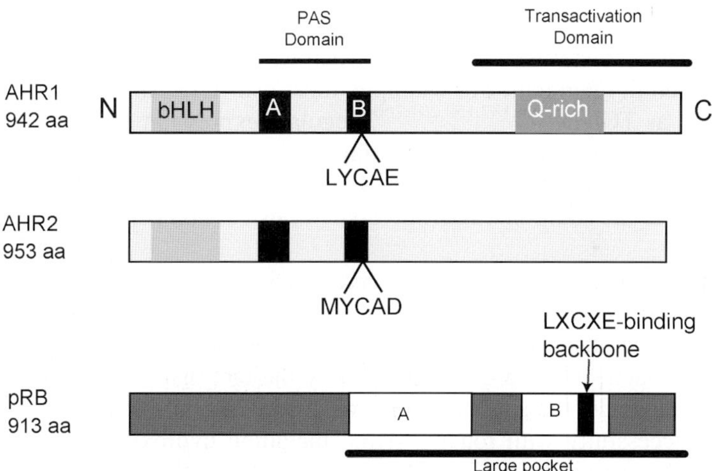

Fig. 2. Domain structure of killifish (*Fundulus heteroclitus*) AHR1, AHR2[104,137], and pRb[170,171]. The bHLH, PAS, and Q-rich domains of the AHRs are indicated, along with the putative pRb-interacting motif of AHR1, LXCXE, which suggests a possible functional difference between fish AHR1 and AHR2. Killifish and other fish AHR2s have MYCAD in place of LXCXE. The functional domains of pRb, including the LXCXE-interacting region, are indicated.

Fish also possess a variety of ARNT forms (Table 2). However, in contrast to AHRs, fish ARNT diversity appears to be generated primarily through alternative splicing, which also occurs in mammalian ARNTs[121,147,222,283]. First identified in rainbow trout[209], fish ARNT splice variants have been also seen in zebrafish[124,262,278] and killifish[212,213]. Although mammals possess two ARNT genes (ARNT1 and ARNT2)[120], the number of ARNT genes in fish is not yet certain. Some of the fish ARNTs that have been characterized to date appear to be more like mammalian ARNT1, whereas others are clearly orthologous to ARNT2[211,213,262]. Tanguay[262] suggested that two ARNT forms may exist in zebrafish. A search of the *Takifugu* genome reveals the presence of one ARNT1 and one ARNT2 (Karchner and Hahn, unpublished; see also Rowatt *et al.*[231]; Table 2).

The AHR repressor (AHRR; Table 2) is an AHR-regulated gene first identified in mice[175] and subsequently also in fish[74,135,233]. AHRR is closely related to AHRs in the primary sequence of the bHLH and PAS-A domain (N-terminal half of the PAS domain), but it is highly divergent from AHRs in the PAS-B domain (location of the AHR ligand-binding domain) and beyond. Consistent with this, fish and mammalian AHRR proteins do not bind typical AHR ligands such as TCDD[135]. The AHRR protein inhibits AHR signaling by competing for binding to ARNT and for AHR–ARNT binding to AHRE sites on target genes[175]. Expression of the AHRR gene is regulated (induced) by the AHR *via* AHRE sequences in the promoter of the mammalian[17,175] and fish[135] AHRR genes. AHRR is inducible by AHR agonists in fish, but its inducibility (like that of CYP1A) is lost in dioxin-resistant fish from highly contaminated sites[135,174].

3. AHR signaling in fish: Functional diversity

3.1. Teleostean fish

The functional characteristics of the AHR pathway in teleost fish are generally like those described in mammals. For example, most fish AHRs display high affinity for PHAHs such as TCDD[2,106,137,159,207,260], regulate expression of CYP1A[216,257], and exhibit ligand structure–activity relationships that are similar (though not identical) to those of mammalian AHRs[3,92,119,276]. However, fish AHRs also exhibit novel features, related in part to the additional AHR diversity in fish as compared to mammals.

3.1.1. AHR ligand-binding characteristics. The set of compounds known to bind mammalian AHRs is large and structurally diverse[59,60] and thus it is of great interest to determine whether fish AHRs exhibit similar promiscuity in their ligand-binding requirements. Such studies are complicated by the fact that fish AHRs appear to be especially labile under the conditions of several batch ligand-binding assays that have been used successfully with rodent AHRs, including hydroxylapatite adsorption[83], protamine sulfate precipitation[58], and filter-binding assays[64]. (Interestingly, the human AHR also is unstable when assayed with these methods[132,184].) However, the more labor-intensive sucrose density gradient competitive binding assay[271] works well with fish AHRs[117,137,159,281] and can be used to compare the structure–binding relationships of fish and mammalian AHRs. Using *in vitro* translated mammalian AHR1 and fish AHR1 and AHR2 forms, we found that the binding of [^{3}H]TCDD could be displaced by planar halogenated compounds (2,3,7,8-tetrachlorodibenzofuran, 3,3′,4,4′-tetrachlorobiphenyl), non-halogenated aromatic hydrocarbon (benzo[*a*]pyrene), a flavonoid (α-naphthoflavone), an indole (indigo), and linear tetrapyrroles (bilirubin, biliverdin)[102]. Although there were some relatively small quantitative differences among receptors in the degree of competition, the overall structure–binding relationships for this set of compounds appears to be similar among fish and mammalian AHRs.

There are, however, some interesting differences in ligand-binding activities of mammalian AHRs and fish AHRs, for certain ligands. For example, fish are less sensitive than mammals to AHR-dependent activation of transcription[3,92] and toxicity[276] caused by mono-*ortho* polychlorinated biphenyls (PCBs) such as PCB-105 (2,3,3′,4,4′-pentachlorobiphenyl). Studies with *in vitro*-expressed proteins and fish cells[119] show that this difference is due not to differences in the relative affinity of AHR binding but rather to differences in intrinsic efficacy for activation of the AHR. Another example of an intriguing difference between mammalian and fish AHRs can be found in the zebrafish (*D. rerio*). This species possesses both AHR1 and AHR2, but zebrafish AHR1 – an ortholog of the mammalian AHR – has lost the ability to bind both halogenated and non-halogenated AHR ligands[10]. Although a second, functional zebrafish AHR1 has recently been discovered (Karchner and Hahn, unpublished), AHR2 appears to be the predominant functional AHR paralog* in most fishes, in contrast to mammals, which possess only AHR1.

3.1.2. AHR role in gene transcription. The transactivation function of mammalian AHRs has been found to reside in the C-terminal half of the protein and three different structural motifs have been implicated: glutamine (Q)-rich, acidic, and proline/serine/threonine (P/S/T)-rich domains[129,232,280]. Q-rich regions are missing or reduced in fish AHR2 forms[2,137,237,261] but are present in at least one fish AHR1[137]. Nevertheless, isolated C-terminal fragments of trout AHR2s exhibit strong transactivation function[208] and most full-length fish AHR1 and AHR2 forms are transcriptionally active in transient transfection assays in mammalian cells[2,3,135,208,261].

Identification of specific amino acid residues of functional importance can be facilitated by a comparison of closely related AHR proteins. Such a comparison was performed for two rainbow trout AHRs – AHR2α and AHR2β – that share 98% amino acid identity. Despite the high degree of relatedness, these two proteins were found to exhibit different, enhancer-specific transactivation activities when transiently expressed in mammalian cells, with AHR2β having reduced activity as compared to AHR2α[2,3]. This reduced activity is reflected in a 10-fold difference in the EC50 for

* Several terms are used to describe the relationships of homologous genes within and among species. Homologous genes in two different species are *orthologous*[79] if they are descended from the same gene in the most recent common ancestor of the two species; thus, the two genes are separated only by a speciation event. *Paralogous* genes[79] are homologs that exist in the same species and resulted from a gene duplication event in that species or an ancestral species. For the case in which a gene duplication has occurred in one lineage but not another, the term *co-ortholog*[86,265] has been used to describe the relationship between each of the two duplicated genes in one species and their single ortholog in the other species.

activation by TCDD[3,208]. Subsequent studies[12] identified amino acid 111 of AHR2β (corresponding to amino acid 110 of AHR2α) as the key difference between these two proteins.

The complete set of AHR target genes is not yet known, even in the well-studied mammalian systems, and only a few have been identified in fish[42,44,45,111,135]. Fish CYP1A genes, which are highly induced by AHR agonists, possess functional AHRE sequences in the 5′-flanking regions[29,44,45,157,194,214,236,282]. AHRR genes in killifish and zebrafish are inducible by TCDD[74,135] and the killifish AHRR promoter includes AHREs that direct TCDD-dependent transactivation in the presence of either AHR1 or AHR2[135]. Interestingly, the expression of AHR2 also has been shown to be inducible by TCDD and other AHR agonists[2,11,174,261,269], but in a species-, tissue-, and developmental stage-specific manner[210,237]. Thus, AHRs and AHRR form a complex feedback loop involving both positive (AHR2 induction) and negative (AHRR induction) components (Fig. 1).

3.1.3. AHR role in sensitivity and resistance to toxicity. From a toxicological perspective, a critical question concerns the role that each AHR plays in the toxicity of AHR agonists such as TCDD. In mice, biochemical and genetic experiments – including studies in AHR knock-out mice – have shown that the AHR is necessary for most forms of TCDD toxicity[77,78,148,176,200,245]. In a few fish studies, the AHR antagonist and CYP1A inhibitor α-naphthoflavone (ANF)* has been used to block TCDD toxicity, providing suggestive evidence for a role of AHRs or CYP1A in this process[65,66,138]. The generation of AHR1- and AHR2-knock-out fish would be enormously valuable in distinguishing the relative roles of these two proteins in the effects of AHR agonists in fish. However, knock-out technology is not yet applicable to fish. Nevertheless, recent studies have utilized the powerful approach of gene knock-down during development using morpholino-modified oligonucleotides (MO)[185] to examine the role of AHR2 and its target gene CYP1A in the early life stage toxicity of TCDD to zebrafish[216,270]. Because zebrafish AHR1 appears to be inactive[10], AHR2 was predicted to control both CYP1A induction and toxicity in this species. Teraoka, Peterson, and colleagues[216,270] found this to be the case. A MO targeted to AHR2 prevented the induction of CYP1A mRNA at 24 h post fertilization in embryos exposed to TCDD. The AHR2 MO also reduced or prevented many of the typical signs[21,118,269] associated with early life stage toxicity of TCDD – pericardial and yolk sac edema, reduced blood flow, craniofacial (jaw) deformities, and the inhibition of definitive erythropoiesis[216,270]. Mortality caused by TCDD exposure of zebrafish embryos was delayed, but not eliminated, by AHR2 MO injection; this could reflect the temporary nature (48–96 h) of the translational block produced by MO treatment[68,185,216].

In zebrafish embryos, CYP1A is the gene most highly induced by TCDD[110,111]. To assess the role of the induced CYP1A in TCDD early life stage toxicity in zebrafish,

* Strictly speaking, ANF is a partial agonist for the AHR. At high concentrations, it has AHR agonist activity, as indicated by its ability to induce CYP1A1 expression. However, at low to moderate concentrations, it antagonizes the ability of full agonists such as TCDD to activate the AHR[85,241].

Teraoka and colleagues[270] used a CYP1A-MO to block expression of CYP1A protein without (presumably) affecting the expression of other TCDD-inducible genes. Knock-down of CYP1A prevented the edema and reduced blood flow caused by TCDD in developing embryos, providing evidence for a direct role for the induced CYP1A in some of the toxic effects of TCDD. However, this finding has been questioned recently. Heideman and co-workers[43] performed similar experiments using a CYP1A–MO to block the induction of CYP1A by TCDD in developing zebrafish, but unlike the earlier study they found that the CYP1A–MO did not prevent the pericardial edema and reduction in blood flow caused by TCDD. Thus, the role of CYP1A in TCDD-induced embryotoxicity remains unresolved.

CYP1A may eventually be found to play an important role in certain aspects of TCDD toxicity. However, it seems unlikely that any one AHR-regulated gene will be identified as responsible for the entire spectrum of TCDD toxic effects. Rather, TCDD toxicity almost certainly involves a complex web of altered gene expression and its sequelae. In light of this, the suggestion by Teraoka *et al.*[270] that AHR2 may play only a 'permissive' role in the mechanism of TCDD toxicity is puzzling. It was hypothesized more than 20 years ago[204] that the role of the AHR in TCDD toxicity occurred through the altered expression of AHR-regulated genes. Our fundamental understanding of that role has not changed and the search for the important AHR target genes continues. However, it appears clear from studies in AHR knock-out mice[77,78,148,176,200,245] and AHR2 knock-down zebrafish[43,216,270] that the AHR plays a necessary and active (not simply 'permissive') role in the toxicity of dioxin-like compounds.

The findings summarized above provide evidence that studies in fish are contributing to our understanding of the roles of AHRs and their target genes in the toxicity of AHR agonists. In addition, fish are serving as valuable models in efforts to understand how variations in AHR structure and function underlie the differential sensitivity to PHAHs and PAHs that is seen when comparing individuals, populations, or species. In at least two species of fish, population-specific differences in sensitivity to PHAHs or PAHs have been documented. Atlantic tomcod (*M. tomcod*) obtained from the Hudson River have reduced sensitivity to CYP1A induction by PHAHs (but not PAHs) as compared to fish from less contaminated reference sites[55,234,235,285,286]. Atlantic killifish inhabiting a variety of PHAH- or PAH-contaminated sites are less sensitive to the biochemical and toxic effects of AHR agonists[23,69,172,173,179–183,198,219,220,273]. This acquired or evolved resistance to AHR agonists appears to involve multiple mechanisms. For example, the relative roles of genetic adaptation vs. physiological acclimation vary between the two species and even among populations of killifish. The resistance is heritable in some populations[22,69,181,218], whereas it is physiological or mixed in others[173,198,235]. The role of altered AHR expression or function has been investigated in some of these populations; specific hypotheses have included (1) down-regulation of AHR1 or AHR2 expression, (2) up-regulation (genetic or physiological) of AHRR expression, and (3) selection for AHR forms with lower ligand-binding affinity or other altered function. An altered pattern of tissue-specific expression of AHR1 was found in PHAH-resistant killifish from New Bedford Harbor, MA, as compared to killifish from a reference site, but unlike the resistance itself, this altered expression pattern was not heritable[210]. Altered AHR expression (AHR1 and/or AHR2) was not seen in killifish from the Elizabeth River[174] or in Hudson River tomcod[233,237]. In the populations of killifish[135,174] and tomcod[233] that have been examined, constitutive expression of AHRR is not altered, suggesting that up-regulation of this protein is not involved in the resistant phenotype. In fact, AHRR appears to be another AHR-regulated gene that – like CYP1A – is refractory to induction in the resistant populations[135,174]. AHR2 polymorphisms, including one unique to Hudson River fish, were identified in tomcod[237]; however, the functional significance of these is not known. Similarly, the killifish AHR1 locus was found to be highly polymorphic, and certain haplotypes (alleles) were under-represented in the PHAH-resistant New Bedford Harbor population[101,103]. Whether the alleles remaining in New Bedford Harbor killifish encode AHR1 proteins with altered functional characteristics is not yet clear. An alternative possibility is that the killifish *AHR1* locus is in linkage disequilibrium with another locus – possibly *AHR2* – bearing alleles that confer reduced sensitivity to dioxin-like compounds. Together, these studies suggest that the mechanisms underlying PHAH and PAH resistance in fish from contaminated sites are complex, likely involving the AHR pathway but possibly other genetic and epigenetic factors as well.

3.1.4. ARNT function. Fish ARNTs are interesting because of the degree of structural diversity, largely generated by alternative splicing. Pollenz and co-workers have provided a detailed structural and functional characterization of two alternatively spliced rainbow trout ARNT forms, designated rtARNT$_a$ and rtARNT$_b$. These two variants are identical over the first 533 amino acids, including the bHLH and PAS domains, but they diverge in the C-terminal portion of the protein[209]. Both forms are able to form dimers with AHR, but they differ in their ability to support transactivation. rtARNT$_b$ is transcriptionally active when complexed with AHR, whereas the AHR–rtARNT$_a$ dimer is inactive due to its poor ability to bind DNA. Moreover, rtARNT$_a$ acts as a dominant negative factor, blocking transcriptional activation mediated by the AHR–rtARNT$_b$ complex[209]. The dominant negative activity of rtARNT$_a$ results from a hydrophobic domain in its C-terminus that blocks DNA binding, not only of the AHR–ARNT complex but of heterologous DNA-binding domains as well[189,190]. Patterns of expression suggest that the two trout ARNT splice variants exhibit distinct functions *in vivo*. rtARNT$_b$ mRNA and protein are widely expressed in a variety of tissues and developmental stages, whereas the expression of rtARNT$_a$ is lower and more restricted[209,252].

Alternatively spliced ARNTs have also been described in zebrafish and killifish. In zebrafish, Tanguay *et al.*[262] identified three such variants. zfARNT2a is a truncated protein resulting from alternative splicing in the C-terminal half of the protein as compared to forms 2b and 2c; the latter two differ only in the lack of an exon just upstream of the basic region of zfARNT2c. Interestingly, only the 2b form is active as a dimerization partner for zfAHR2 in supporting TCDD-dependent transactivation of gene expression from AHREs, whereas all three ARNT variants can support HIF2-dependent transcription from a hypoxia response element (HRE). A zebrafish ARNT2 identical to Tanguay's zfARNT2a was identified by Hu and co-workers[277,278], who showed that this variant, as well as another ARNT splice variant (ARNT2x) could repress TCDD-inducible CYP1A expression in a zebrafish liver cell line[124,278]. Furthermore, microinjection of ARNT2x expression constructs into zebrafish embryos disrupted the development of several organs, presumably through inhibition of processes dependent on full-length ARNT[124].

Killifish express an ARNT2 variant that includes an additional 16 amino acids just upstream of the basic region, as in zfARNT2b[212,213]; this insertion corresponds to a similar region of mammalian ARNT1 forms, encompassing exon 5[212]. The two killifish ARNT2 variants did not differ in their ability to form HRE-binding complexes with killifish HIF-2α[212].

These studies of fish ARNTs have confirmed and extended results in mammalian systems[121,147,222,283] indicating that ARNT genes display considerable transcript diversity, leading to protein variants with distinct functional properties.

3.2. *Elasmobranch fish*

The first evidence for a functional AHR signaling pathway in elasmobranchs came from studies of cytochrome P450 inducibility by PAHs and TCDD[24,25,27,130,202]. For example, benzo[*a*]pyrene (BaP) hydroxylase (also known as aryl hydrocarbon hydroxylase or AHH) and 7-ethoxycoumarin *O*-deethylase (ECOD) activities – both of which are now known to be catalyzed at least in part by CYP1A forms – were significantly increased in Atlantic stingray (*Dasyatis sabina*) injected with 3-methylcholanthrene (3MC) and little skate (*Leucoraja erinacea*) injected with 3MC or 1,2,3,4-dibenzanthracene[26,28,130,202]. Similarly, AHH activity was induced in little skate treated with TCDD[26,202]. Immunochemical confirmation that the induced P450 is a CYP1A was obtained in little skate and spiny dogfish (*Squalus acanthias*) treated with β-naphthoflavone (BNF)[108,255] and smooth dogfish (*Mustelus canis*) captured near a PCB-contaminated site[108]. Recently, a skate CYP1A cDNA has been cloned[168]; it shares 59–61% and 56% identity in deduced amino acid sequence with teleost and human CYP1As, respectively. Thus, elasmobranch fish possess a CYP1A gene that is inducible by typical AHR agonists. Following the demonstration of an inducible CYP1A, studies showed that cartilaginous fish possess a high-affinity AHR. The first direct evidence for this came from the observation that 2-azido-3-[^{125}I]iodo-7,8-dibromodibenzo-*p*-dioxin ([^{125}I]N_3Br_2DD) exhibited specific binding to proteins in cytosol prepared from spiny dogfish and smooth dogfish livers[105,106]. Subsequent cloning efforts using RT-PCR with degenerate primers yielded AHR cDNAs from smooth dogfish and little skate[104]. Two divergent AHR cDNAs were amplified from smooth dogfish; phylogenetic analyses supported the conclusion that these sequences were orthologous to the two bony fish AHRs (AHR1 and AHR2)[104]. Two AHR cDNAs were also obtained from spiny dogfish[32]. One of these is now known to represent a third form of AHR, which has been found only in cartilaginous fish[169] (Table 2). This third AHR paralog is distinct from the AHR repressor identified in bony fish. None of the elasmobranch AHRs has yet been characterized functionally.

3.3. *Jawless fish*

In bony and cartilaginous fish, there is compelling evidence for an AHR–CYP1A pathway that functions like that in mammals. In contrast, the situation in jawless fish is much less clear. At present, the data suggest that in jawless fish the AHR is a low-affinity, transcriptionally competent receptor, but the presence of CYP1 genes and the involvement of the AHR in their regulation is uncertain. Initial studies on the AHR in the jawless fish sea lamprey (*Petromyzon marinus*) and Atlantic hagfish (*Myxine glutinosa*) did not detect specific binding of the photoaffinity ligand [^{125}I]N_3Br_2DD to hepatic cytosol from adult animals[105,106]. Subsequently, an AHR cDNA was cloned from lamprey ammocetes[104]. The lamprey AHR, like other vertebrate AHRs, is highly conserved in the N-terminal half of the protein (bHLH and PAS domains). In radioligand binding experiments, *in vitro*-expressed lamprey AHR has a low affinity for 2,3,7,8-TCDD[107]. When transiently transfected into mammalian cells, the lamprey AHR is able to support TCDD-dependent expression of a luciferase reporter construct driven by AHREs[107].

Immunochemical and enzyme activity data provide suggestive evidence for a CYP1A-like protein in lamprey and the early craniate, hagfish (*M. glutinosa*), but to date typical AHR agonists have failed to induce this protein. In hagfish, hepatic 7-ethoxyresorufin *O*-deethylase (EROD) activity is low and not induced by treatment with β-naphthoflavone (BNF)[90] or 3,3′,4,4′-tetrachlorobiphenyl (TCB)[108]. EROD activity is not detectable in control or TCB-treated sea lamprey[108]. Similarly, in the river lamprey, *Lampetra fluviatilis*, basal levels of hepatic EROD activity are low and not induced by benzo[*a*]pyrene[230]. Immunoblots using polyclonal antibodies against fish CYP1A forms show cross-reactivity with hepatic proteins with sizes expected for a CYP1 homolog, but there is no difference in band intensity with BNF or TCB treatment in hagfish or lamprey. The monoclonal antibody MAb 1-12-3, known for its specificity for CYP1A, did not recognize hepatic proteins from control or treated hagfish or lamprey[90,108].

The many remaining questions concerning the function of AHRs in jawless fish make this a fertile area of future research. It is too soon to know whether studies on AHR pathways in lamprey and hagfish will reveal features of ancestral (ligand-independent?) AHR functions, developmental changes in AHR signaling, or unique modifications of AHR function in these highly specialized fish.

4. *Evolution of AHRs and the AHR signaling pathway*

The findings summarized above clearly indicate that there are differences in the diversity and function of AHRs in fish as compared to mammals. How did these differences arise? The presence of AHR homologs in several invertebrate lineages[41,67,104,215] demonstrates that an AHR existed early in metazoan evolution. We suggest that the diversity of AHR genes in various vertebrate lineages has been shaped by whole genome duplications as well as lineage-specific gene duplications and gene losses (Fig. 3). Along the way the AHR has acquired (and in some cases lost) the ability to bind and be activated by planar aromatic hydrocarbons.

A single AHR existed prior to emergence of the chordates. AHR homologs from arthropods, molluscs, and nematodes do not exhibit specific binding to typical AHR ligands[41,215] and inducible CYP1 homologs do not exist in these groups, suggesting that the original functions of the AHR may not have involved ligand-dependent

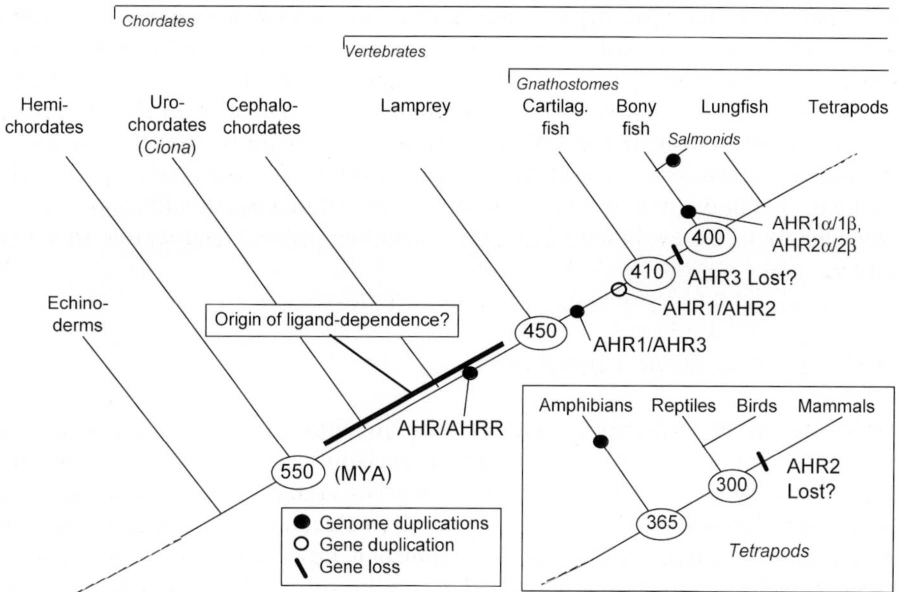

Fig. 3. Evolution of AHR diversity in fish. This updated version of our earlier illustrations[95,98] presents one possible scenario by which AHR diversity in extant vertebrates may have evolved through gene and genome duplications and gene losses. Numbers at nodes represent estimated divergence times, in millions of years ago (MYA). The solid circles indicate the hypothesized times of whole genome duplications occurring in vertebrates[82,195], ray-finned fish[7,265], salmonids[6], and *Xenopus laevis*[33]. The open circle represents an AHR tandem gene duplication, which we infer from the tandem arrangement of AHR1 and AHR2 in the *Takifugu* genome (Karchner and Hahn, unpublished). AHRs have been lost in certain lineages leading to mammals, which possess only AHR1. Based on TCDD-binding data from invertebrate AHRs[41], AHR1/AHR2[137], and AHRR[135], we propose that the ability of AHRs to bind PHAHs and PAHs first evolved after the AHR/AHRR duplication and prior to the AHR1/AHR2 duplication.

regulation of biotransformation enzymes. We cannot at this time, however, rule out a role for invertebrate AHRs in regulation of other CYPs or other xenobiotic-metabolizing enzymes.

Early in the evolution of the chordate lineage, evidence suggests that there were two whole genome duplications (or periods of extensive gene duplication)[82]. One of these might have resulted in the formation of a pair of AHR paralogs that subsequently diverged to form AHR and AHR repressor. Later, another genome duplication may have resulted in two AHRs, which we see today as AHR1 and AHR3 in sharks, with the AHR3 form lost in the bony fish lineage. The presence of two sets of tandem AHR1–AHR2 pairs in the *Fugu* genome suggests that AHR2 originated by a tandem duplication of AHR1. AHR1 and AHR2 both were retained in bony and cartilaginous fish, and possibly also in early tetrapods, up through the bird lineage; however, AHR2 appears to have been lost in mammals[98]. In the bony fish lineage, a fish-specific genome duplication is thought to have occurred, and this could have led to the multiple AHR1 and AHR2 genes that we see today in medaka and *Takifugu*. In addition, salmonids have undergone an additional, more recent genome duplication, which has resulted in up to four AHR2s in this fish family[2,115a,b].

At some point, perhaps early in chordate evolution, CYP1 genes emerged and their regulation became linked to the AHR. This may have coincided with the acquisition of the AHR's ability to bind planar aromatic hydrocarbons with high affinity. A search of the genome of the urochordate *Ciona intestinalis*[57] yields sequences homologous to AHR (Hahn, unpublished) and CYP1[109]. Thus *Ciona* might be used as a model to further understand the evolution of AHR function in early chordates. Additional studies in *Ciona*, lamprey, and other early diverging chordates should improve our understanding of the evolution of the AHR signaling pathway and its role in PHAH toxicity.

III. Nuclear receptor superfamily

Another group of transcription factors with important roles in the response to xenobiotic chemicals is the nuclear receptor superfamily, which includes receptors for steroid and thyroid hormones, sterols, retinoids, fatty acids, and many other endogenous, low-molecular weight signaling molecules[75,197]. Nuclear receptors regulate diverse physiological and developmental processes. Some NRs serve as receptors for known ligands, whereas others are ligand-independent or have ligands that have not yet been identified ('orphan receptors')[143,287].

1. NR signaling in mammals

1.1. CAR and PXR

One of the most exciting developments in pharmacology and toxicology in recent years have been the identification of nuclear receptors controlling the induction of CYPs in families 2, 3, 4, and 7[123,142,143,242,279]. The mechanisms by which these CYPs are induced by structurally diverse xenobiotics have long been of great interest, but until recently had proven elusive.

Research on the mechanism by which xenobiotics induce CYP2 and CYP3 genes in mammals led to the identification of the pregnane X receptor (PXR) and constitutive androstane receptor (CAR) as the ligand-activated transcription factors responsible for induction of these CYPs[93,123,145,223,242,279,287]. PXR, CAR, and their dimerization partners, the retinoid X receptors (RXR), which also are members of the nuclear receptor superfamily[164,227], have been characterized in several mammalian species[30,38,134,144,154,243,289]. PXR and CAR exhibit reciprocal activation of CYP2B and CYP3A gene expression and share certain ligands (Table 1). *Ortho*-PCBs and organochlorine pesticides have been shown to activate the mammalian PXR leading to CYP3A induction[154,246,247]. In addition, the mammalian CYP2Bs can be induced by *ortho*-PCBs and chlorinated pesticides *via* the CAR[35,258].

1.2. PPARs

Peroxisome proliferator-activated receptors (PPARs) comprise another subfamily of nuclear receptors whose members regulate the transcription of genes controlling fatty acid oxidation, including CYP4A (lauric acid hydroxylases). PPAR α, β (also known as δ), and γ homologs bind as heterodimers with RXR to peroxisome proliferator response elements (PPRE) in target genes[5,199]. Fatty acids are the main natural ligands of the PPARs, but these receptors are also activated by synthetic ligands, including fibrate drugs, phthalate ester plasticizers, and herbicides[226]. PPARs have been well characterized in mammalian model systems, especially in the course of hypolipidemic and antidiabetic drug development[126,155].

2. NR diversity in fish

It has long been known that the regulation of CYP2 genes in fish differs from that in mammals[257]. Although fish possess CYP2-family genes[196,288], they have been considered 'non-responsive' to typical inducers of CYP2B such as phenobarbital and *ortho*-substituted PCBs[4,13,70,131,141]. Thus, there has been great interest in understanding the presence and function of the fish orthologs of mammalian receptors that are involved in regulating CYP2B induction.

The diversity and function of CAR- and PXR-related nuclear receptors in fish are not yet well understood. Studies in other non-mammalian species suggest that there are substantial differences among vertebrate classes in the structure and function of PXR and CAR. For example, a chicken nuclear receptor designated 'CXR' appears to be equally related to both mammalian PXR and CAR in its sequence and ligand activation properties[112,113]. Similarly, benzoate X receptor (BXR) is a nuclear receptor identified in the frog *Xenopus laevis* that is activated by endogenous embryonic benzoates and thus has pharmacological properties distinct from those of CAR and PXR[37,178]; its exact relationship to the mammalian PXR and CAR is not clear. A partial cDNA sequence (ligand-binding domain) has been reported for the zebrafish PXR (AF454673)[178]. Ligand-binding assays employing a zebrafish PXR/Gal4 chimeric protein revealed that the fish PXR was activated by a number of xenobiotics, showing a profile similar to that of the chicken CXR. Bainy and Stegeman[19] have cloned and sequenced a full-length zebrafish PXR homolog. A fish protein with immunochemical properties of CAR has been detected in the marine fish scup using anti-human CAR antibodies[128]. Upon treatment with 1,4-bis[2-(3,5-dichloropyridyloxy)]benzene (TCPOBOP, a potent inducer of CYP2B forms in some mammals), no nuclear accumulation of this putative fish CAR was observed, consistent with the idea that there are vertebrate class-specific differences in CAR activation mechanisms[128]. However, despite this immunochemical evidence for a fish CAR, there is yet no definitive evidence that a CAR ortholog exists in fish. An exhaustive search for CAR in the completely sequenced *T. rubripes* (pufferfish) genome[15] did not yield a candidate, whereas *Takifugu* orthologs for several other members of the nuclear receptor family were identified, including PXR and multiple forms of the PPAR (Karchner and Hahn, unpublished) (Fig. 4). Similarly, Maglich *et al.*[162] surveyed the *Takifugu* genome for nuclear receptors and identified a PXR homolog but no CAR. One possibility is that the CAR gene has been lost from some, but not all, species of fish.

There are several reports of fish PPARs. A full-length PPARγ was cloned from the Atlantic salmon (*Salmo salar*)[9]. Multiple variants of PPARγ are expressed in the salmon liver. PPAR ligands caused increased expression of PPARγ and acyl-CoA

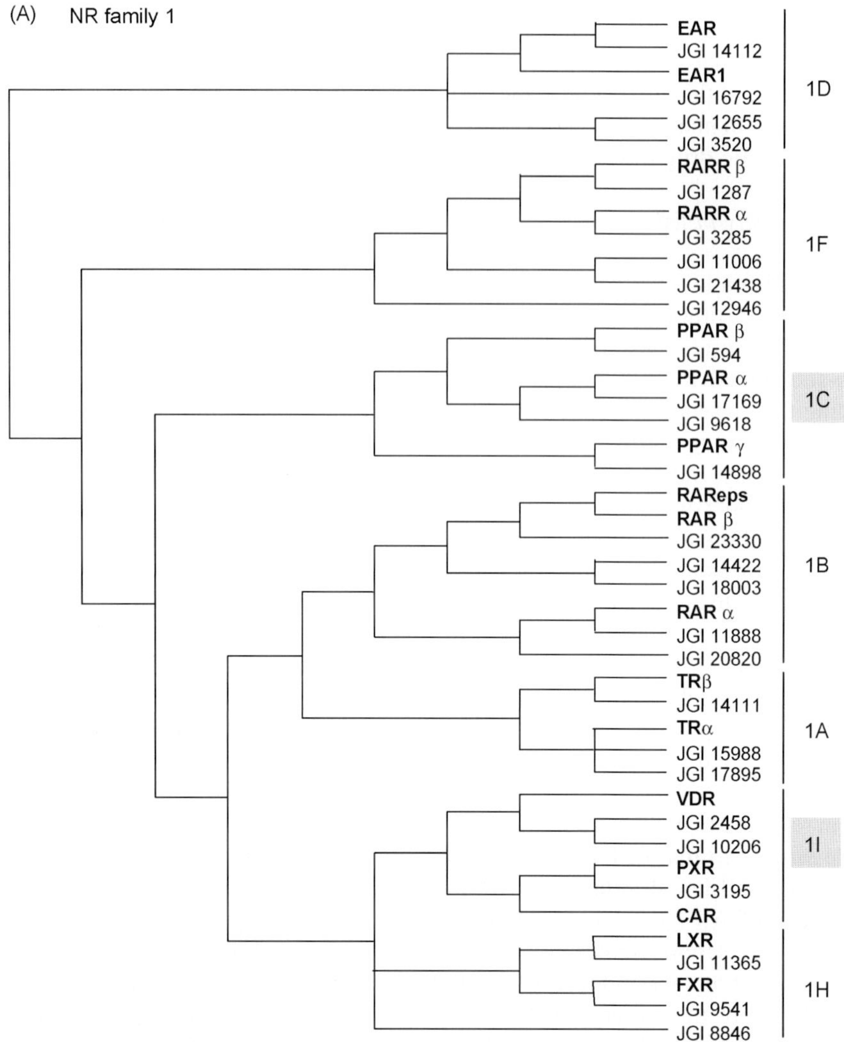

Fig. 4. Phylogenetic analysis of NR genes in the pufferfish (*T. rubripes*) genome[15] as compared to all human NR genes. Amino acid sequences were aligned using ClustalX and a distance tree was constructed using the neighbor-joining algorithm. (A) NR subfamily 1. (B) NR subfamily 2. (C) NR subfamily 3. Human NR genes are named and indicated in bold. *Fugu* predicted NR proteins (see below) are indicated by their JGI number. NR subfamilies involved in the regulation of xenobiotic-metabolizing enzymes or as targets of xenobiotics are shaded: 1C (PPARs), 1I (CAR, PXR, VDR), 2B (RXRs), 3A (ERs), and 3C (AR, GR). *Takifugu* predicted proteins were obtained from the Zon Lab website (http://zfrhmaps.tch.harvard.edu/ZonRHmapper/) in August 2002. Human nuclear receptor protein sequences were used in a blast search against the 'Fugu Genome Sequences' database. The *Fugu* nuclear receptor sequences included in this phylogenetic tree can be retrieved from the same website by accessing the Zon Lab Blast Server followed by the 'Retrieve Sequence in Database' function (http://134.174.23.160/zfBlast/publicRetrieve.htm). Select the 'Takifugu Genome Sequences' database and type in the sequence name as 'JGI_#####', where ##### is the designated protein number in the tree. Alternatively, a text file containing all of these *Fugu* NR sequences is available upon request to the authors. A complete description of NR nomenclature[193] and an independent analysis of NR genes in *Takifugu*[162] can be found in the indicated references.

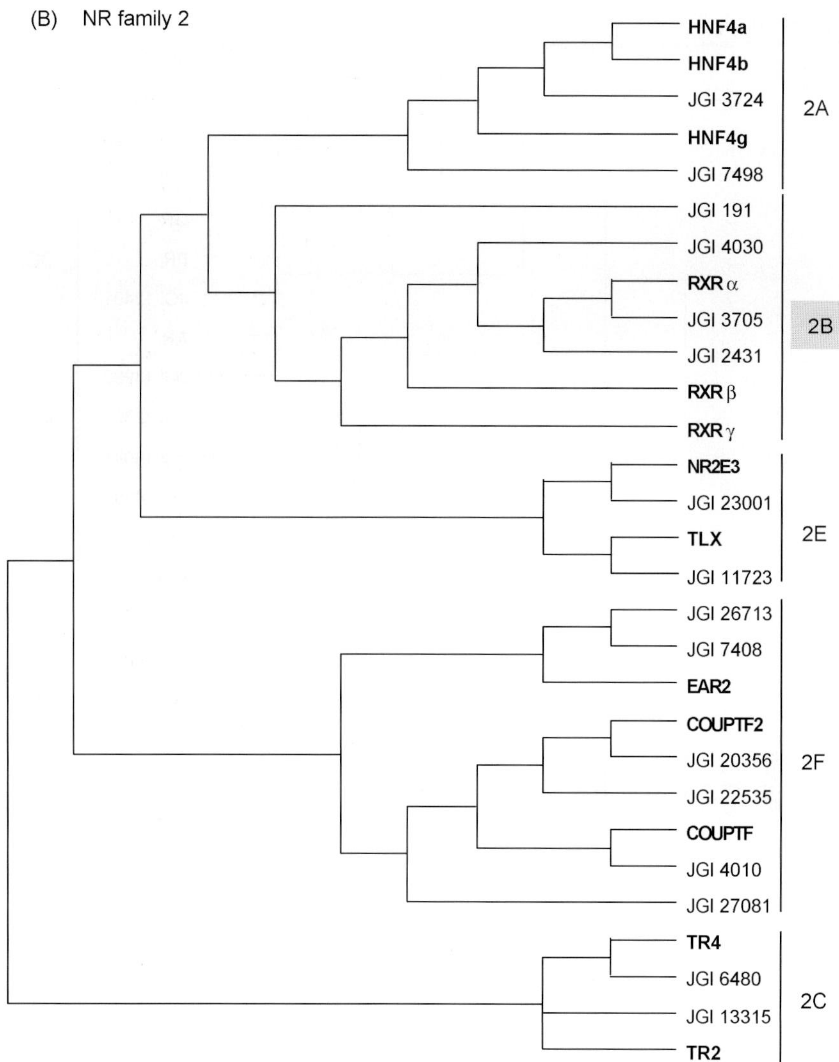

Fig. 4 Continued.

oxidase in cultured salmon hepatocytes, pointing to similarities in the response to peroxisome proliferators in fish and mammals[239]. A full-length plaice (*Pleuronectes platessa*) PPAR gene has been isolated and likely represents the γ form[152]. All three forms of PPAR have been detected in zebrafish (*D. rerio*) tissues *via* immunoblotting[125]. The tissue distribution pattern of the zebrafish PPARs generally matched that of mammalian forms. Partial cDNA sequences for several fish PPARs have been deposited in GenBank; these include PPARs from zebrafish, goldfish, Japanese

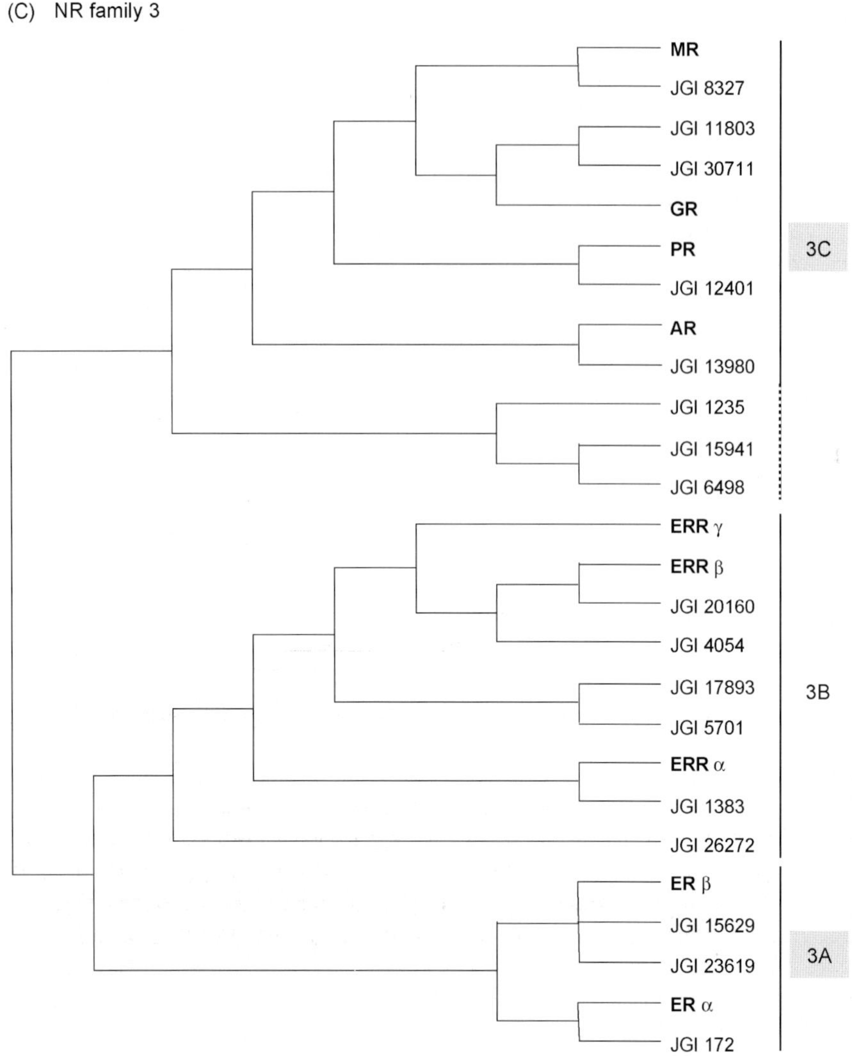

Fig. 4 Continued.

medaka, and fathead minnow. Partial PPAR cDNAs from shark, lamprey, hagfish, and amphioxus have also been reported[72,73].

There is evidence that some fish species possess additional PPAR forms as compared to mammals. Laudet and colleagues[229] reported evidence for a duplicate PPARβalso called PPARδ in several (unnamed) species; they also referred to two PPARα and two PPARβ sequences in zebrafish[72]. In independent surveys of the *Takifugu* genome, four PPAR homologs, including two most closely related to mammalian PPARα, have been identified[162] (Karchner and Hahn, unpublished; see Fig. 4A).

Additional members of the nuclear receptor superfamily, such as the steroid receptors (estrogen receptor (ER), androgen receptor (AR), and glucocorticoid receptor (GR)), are involved in the xenobiotic response or are known to be targets for environmental chemicals. Other NRs have not yet been shown to interact with xenobiotics, but could potentially be targets, based on functional and structural similarities with PXR, CAR, ER, and AR (e.g. LXR, FXR, VDR, RXRs, RARs, ERRs). All of these have been identified in fish, and for some there are more forms in fish than in mammals[162] (Karchner and Hahn, unpublished; see Fig. 4). The estrogen receptors, in particular, have been well characterized in fish. Significantly, there are three ER genes in most fish, as compared to two (ERα and ERβ) in mammals (Fig. 4C). The additional fish ER gene, originally designated ERγ[116], is now known to have originated through a duplication of ERβ, and is thus called ERβ_b or ERβ_2[20,149,167]. These and other nuclear receptors are discussed elsewhere in this volume.

IV. The anti-oxidant response

1. The mammalian oxidative stress response

Research over the past 15 years has shown that mammals possess an inducible enzymatic defense system that acts to detoxify reactive species and replenish small anti-oxidants such as glutathione. Oxidants, electrophiles, and some so-called anti-oxidant chemicals activate this 'anti-oxidant response' or oxidative stress response (OSR) *via* a family of related, cap'n'collar (CnC)-basic-leucine zipper (bZIP) transcription factors that interact with an anti-oxidant response element (ARE; also called the electrophile response element or EpRE) in the promoter region of target genes[81,192,267]. Known target genes in mammals include GSTs, NAD(P)H-quinone oxidoreductase (NQO1), γ-glutamylcysteine synthase (γ-GCS), superoxide dismutase (SOD), and other genes[192].

The best characterized CnC-bZIP protein is nuclear factor erythroid-2 (NF-E2)-related factor 2 (NRF2). NRF2 is normally found in the cytosol as an inactive complex with Kelch-like-ECH-associated protein (Keap1, also called iNRF2). Keap1 both retains NRF2 in the cytoplasm and enhances its proteasomal degradation[61,127,165,191]. Oxidative stress disrupts the interaction between NRF2 and Keap1 through an unknown mechanism that may involve phosphorylation or disruption of sulfhydryl interactions[36,63,192]. Once free of the repressor Keap1, NRF2 protein accumulates in the cell, enters the nucleus and forms a heterodimer with one of several small Maf proteins (MafF, MafG, MafK), which also contain bZIP motifs[133,163]. The NRF2–Maf dimer binds to AREs and activates transcription. A similar mechanism is thought to operate for other NRF isoforms.

The OSR includes induction of genes encoding Phase II biotransformation enzymes (GSTA, NQO1), enzymes involved in glutathione synthesis (γ-glutamylcysteine synthase catalytic and regulatory subunits (γ-GCS-c and γ-GCS-r, respectively)), heme oxygenase (HO-1), ferritin, and other genes. Together, these genes have been referred to as the 'ARE gene battery'[192] in analogy to the '[Ah] gene battery' that is regulated through the XRE as part of an autoregulatory loop involving the AHR and CYP1A1[188].

An OSR involving CNC proteins appears to be conserved in animals. CNC and CNC-bZIP proteins are found in *Drosophila melanogaster*[274] and *Caenorhabditis elegans*[8,275], where they regulate genes encoding Phase II enzymes. However, most of our current understanding of the OSR has been obtained from studies of the four mammalian CNC-bZIP proteins: NF-E2, NRF1, NRF2, and NRF3. NRF proteins are important during development[49,76,156] and in protection against toxicity and carcinogenicity of chemicals that cause oxidative stress[14,50,51,71,224,225].

2. *NRF homologs and the oxidative stress response in fish*

Several laboratories have reported the induction of oxidative stress in fish exposed to TCDD, PCBs, PAHs, and other chemicals, both in the laboratory and in the field; these studies have been reviewed[62,139,284] and will not be discussed in detail here. A common finding in these studies is the induction in fish of catalytic activities or proteins of the same Phase II enzymes that are part of the OSR in mammals[88,256]. Typically, however, the magnitude of induction of these enzymes is much less than that seen for the induction of CYP1A by AHR agonists[88]. The molecular mechanisms underlying induction of Phase II enzymes in fish are just beginning to be uncovered[89,150,151]. Recently, Carvan and colleagues[46,47] demonstrated that reporter gene constructs containing a luciferase gene under the control of mammalian ARE sequences, when transfected into fish cells (ZEM2S), could be induced by exposure to *tert*-butylhydroquinone (*t*BHQ), suggesting that fish oxidant-responsive transcription factors are able to recognize mammalian ARE sequences.

The most direct evidence for an OSR in fish – and its mechanistic similarity to that in mammals – comes from the work of Kobayashi *et al.*[146], who cloned the cDNAs for zebrafish NRF2 and Keap1 and showed that these proteins mediated the induction of GSTP, NQO1, γ-GCS by *t*BHQ in zebrafish embryos. Exposure of zebrafish larvae to *t*BHQ (30 μM) for 6 h at 4- or 7-days post fertilization (dpf) caused the induction of GSTp, NQO1, and γ-GCS. However, exposure of 24-hpf embryos to *t*BHQ for 6 h was not effective at inducing GSTp, suggesting that 24-h old zebrafish embryos are incapable of mounting an OSR. These authors also showed that treatment of embryos with a morpholino anti-sense oligonucleotide directed against NRF2 abolished the response to *t*BHQ at 4 dpf. Additional studies demonstrated the ability of NRF2 to activate reporter gene constructs containing mammalian ARE sequences controlling expression of luciferase, when an NRF2 expression vector and reporter construct were co-injected into early embryos. Thus, the function and target genes of NRF2 appear to be conserved in mammals and fish.

Although four CNC-bZIP members exist in mice and humans (NF-E2, NRF1, NRF2, and NRF3), to date only two of these – NF-E2 and NRF2 – have been reported in fish (zebrafish)[146,217]. However, database searches of the zebrafish genome (still incomplete) and the complete assembly of the pufferfish (*T. rubripes*) genome[15] revealed a predicted

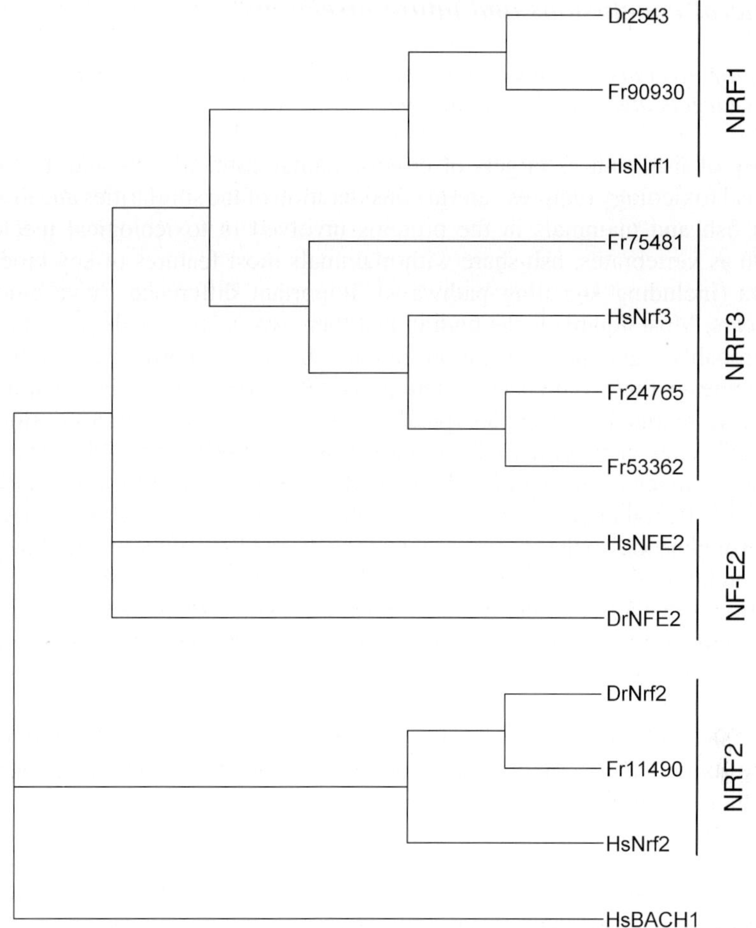

Fig. 5. Phylogenetic analysis of NRF family amino acid sequences. The amino acid sequences of known or predicted members of the NRF family in zebrafish and pufferfish were obtained from the GenBank or genome databases and aligned with the human NRF family sequences. The tree was inferred using maximum parsimony within PAUP * 4b10. The results illustrate the conservation of NRF proteins in mammals and fish as well as the additional diversity in NRF3 forms present in pufferfish. *Danio rerio* (Dr), *Takifugu (Fugu) rubripes* (Fr), *Homo sapiens* (Hs).

zebrafish protein that appears to be an ortholog of mammalian NRF1 (Dr2543) (Fig. 5). The *Takifugu* genome contains evidence for greater NRF diversity in fish as compared to mammals. Thus, *Takifugu* appear to possess one NRF1 form, one NRF2, and three NRF3 forms (Karchner and Hahn, unpublished) (Fig. 5). It will be important to determine the relative roles of these NRF-family proteins in the OSR of fish.

V. General conclusions and future directions

1. Receptor diversity in fish vs. mammals: Implications for investigation of fish as targets and models in toxicology

The study of fish both as targets of environmental contaminants and as models in biomedical toxicology requires careful consideration of the similarities and differences between fish and mammals in the proteins involved in toxicological mechanisms. Although as vertebrates, fish share with mammals most features of key biochemical pathways (including signaling pathways), important differences have emerged in recent years. Most notable is the finding that there was a large-scale gene duplication event (probably a genome duplication) in the fish lineage shortly after its divergence from the lineage that would become tetrapods[7,265]. The result of this is that many fish species have retained extra copies (paralogs[1]) of genes for some transcription factors and other genes, as compared to mammals. Examples of genes involved in toxicological mechanisms include those cited in this chapter as well as CYP19[140,266] and ER[20,116]. For other genes, such as AHRs, fish may possess an extra copy (AHR2) as the result of gene loss in the mammalian lineage. Regardless of the evolutionary origin, such differences raise important questions about the extrapolation of findings between animal groups that differ in such a fundamental way. This remains an open question and one that will only be resolved by obtaining a detailed understanding of signaling pathways in a variety of fish species, especially those proposed as models in toxicology (e.g. zebrafish, medaka, killifish, fathead minnow). The difference in paralog number between fish and mammals also provides an opportunity, because duplicated fish genes might be exploited to obtain new information about the function of their single mammalian counterpart[7,80,160,265]. For example, the duplication, degeneration, complementation model of gene evolution[80,160] predicts that the multiple functions of a mammalian gene may be partitioned between its fish 'co-orthologs'. Thus, mechanistic research in fish may reveal novel functions of vertebrate transcription factors, including xenobiotic receptors.

2. Future directions

Over the next 5 years, research on xenobiotic receptors in fish will be driven primarily by conceptual and technological advances. For example, the emergence of whole genome sequences from several fish species – including representative elasmobranch and agnathan fish – will permit a more complete understanding of xenobiotic receptor diversity in various fish groups. Advances in genomic technologies, including sequencing efficiencies and large-scale gene expression analyses, will provide the tools to better understand the gene targets of these ligand-activated transcription factors. The rapidly expanding toolbox of genetic and genomic techniques for use in fish – especially zebrafish – will foster increasing sophistication in the approaches employed by comparative biologists and toxicologists. For example, the development and application of embryonal stem cells and homologous recombination to generate knock-out fish[161,259] will increase the power of fish models. Adaptation of methods for RNA interference (RNAi)[1,166,249] to modulate gene expression *in vivo* and *in vitro* will be enormously useful. Increased development of transgenic technologies in fish[91,158,272] will provide improved methods for *in vivo* functional analysis of regulatory elements and lead to more sensitive *in vivo* bioassays for the presence and effects of xenobiotics. Finally, recent and continuing advances in theoretical and computational biology[122] will permit more powerful and accurate inferences concerning the evolution and evolutionary relationships of xenobiotic receptors in fish.

Acknowledgements. Preparation of this chapter and the research from our laboratory that is described herein were supported in part by National Institutes of Health grants R01 ES06272, F32 ES05935, and 5P42 ES07381 (Superfund Basic Research Program at Boston University), Grant NA16RG2273 from the National Oceanic and Atmospheric Administration, the Oliver S. and Jennie R. Donaldson Charitable Trust, the Richard B. Sellars Endowed Research Fund and The Andrew W. Mellon Foundation Endowed Fund for Innovative Research. The US Government is authorized to produce and distribute reprints for governmental purposes notwithstanding any copyright notation that may appear hereon. We thank Warren Heideman for discussing data prior to publication, and Tom Moon and two reviewers for helpful suggestions. This is contribution No. 11037 from the Woods Hole Oceanographic Institution.

VI. References

1. Abdelrahim, M., R. Smith, 3rd and S. Safe. Aryl hydrocarbon receptor gene silencing with small inhibitory RNA differentially modulates Ah-responsiveness in MCF-7 and HepG2 cancer cells. *Mol. Pharmacol.* 63: 1373–1381, 2003.
2. Abnet, C.C., R.L. Tanguay, M.E. Hahn, W. Heideman and R.E. Peterson. Two forms of aryl hydrocarbon receptor type 2 in rainbow trout (*Oncorhynchus mykiss*): evidence for differential expression and enhancer specificity. *J. Biol. Chem.* 274: 15159–15166, 1999.
3. Abnet, C.C., R.L. Tanguay, W. Heideman and R.E. Peterson. Transactivation activity of human, zebrafish, and rainbow trout aryl hydrocarbon receptors expressed in COS-7 cells: greater insight into species differences in toxic potency of polychlorinated dibenzo-*p*-dioxin, dibenzofuran, and biphenyl congeners. *Toxicol. Appl. Pharmacol.* 159: 41–51, 1999.
4. Addison, R.F., M.C. Sadler and R.A. Lubet. Absence of hepatic microsomal pentyl- or benzyl-resorufin O-dealkylase induction in rainbow trout (*Salmo gairdneri*) treated with phenobarbitone. *Biochem. Pharmacol.* 36: 1183–1184, 1987.
5. Aldridge, T.C., J.D. Tugwood and S. Green. Identification and characterization of DNA elements implicated in the regulation of CYP4A1 transcription. *Biochem. J.* 306: 473–479, 1995.
6. Allendorf, F.W. and G.H. Thorgaard. Tetraploidy and the evolution of salmonid fishes. In: *Evolutionary Genetics of Fishes*, edited by B.J. Turner, New York, Plenum Press, pp. 1–53, 1984.
7. Amores, A., A. Force, Y.-L. Yan, L. Joly, C. Amemiya, A. Fritz, R.K. Ho, J. Langeland, V. Prince, Y.-L. Wang, M. Westerfield, M. Ekker and J.H. Postlethwait. Zebrafish hox clusters and vertebrate genome evolution. *Science* 282: 1711–1714, 1998.
8. An, J.H. and T.K. Blackwell. SKN-1 links *C. elegans* mesendodermal specification to a conserved oxidative stress response. *Genes Dev.* 17: 1882–1893, 2003.
9. Andersen, O., V.G. Eijsink and M. Thomassen. Multiple variants of the peroxisome proliferator-activated receptor (PPAR) gamma are expressed in the liver of Atlantic salmon (*Salmo salar*). *Gene* 255: 411–418, 2000.
10. Andreasen, E.A., M.E. Hahn, W. Heideman, R.E. Peterson and R.L. Tanguay. The zebrafish (*Danio rerio*) aryl hydrocarbon receptor type 1 (zfAHR1) is a novel vertebrate receptor. *Mol. Pharmacol.* 62: 234–249, 2002.

11. Andreasen, E.A., J.M. Spitsbergen, R.L. Tanguay, J.J. Stegeman, W. Heideman and R.E. Peterson. Tissue-specific expression of AHR2, ARNT2, and CYP1A in zebrafish embryos and larvae: effects of developmental stage and 2,3,7,8- tetrachlorodibenzo-*p*-dioxin exposure. *Toxicol. Sci.* 68: 403–419, 2002.
12. Andreasen, E.A., R.L. Tanguay, R.E. Peterson and W. Heideman. Identification of a critical amino acid in the aryl hydrocarbon receptor. *J. Biol. Chem.* 277: 13210–13218, 2002.
13. Ankley, G.T., R.E. Reinert, R.T. Meyer, M.D. Burke and M. Agosin. Metabolism of alkoxyphenoxazones by channel catfish liver microsomes: effects of phenobarbital, Aroclor 1254, and 3-methylcholanthrene. *Biochem. Pharmacol.* 36: 1379–1381, 1987.
14. Aoki, Y., H. Sato, N. Nishimura, S. Takahashi, K. Itoh and M. Yamamoto. Accelerated DNA adduct formation in the lung of the Nrf2 knockout mouse exposed to diesel exhaust. *Toxicol. Appl. Pharmacol.* 173: 154–160, 2001.
15. Aparicio, S., J. Chapman, E. Stupka, N. Putnam, J.M. Chia, P. Dehal, A. Christoffels, S. Rash, S. Hoon, A. Smit, M.D. Gelpke, J. Roach, T. Oh, I.Y. Ho, M. Wong, C. Detter, F. Verhoef, P. Predki, A. Tay, S. Lucas, P. Richardson, S.F. Smith, M.S. Clark, Y.J. Edwards, N. Doggett, A. Zharkikh, S.V. Tavtigian, D. Pruss, M. Barnstead, C. Evans, H. Baden, J. Powell, G. Glusman, L. Rowen, L. Hood, Y.H. Tan, G. Elgar, T. Hawkins, B. Venkatesh, D. Rokhsar and S. Brenner. Whole-genome shotgun assembly and analysis of the genome of *Fugu rubripes*. *Science* 297: 1301–1310, 2002.
16. Ashok, M., C. Turner and T.G. Wilson. Insect juvenile hormone resistance gene homology with the bHLH-PAS family of transcriptional regulators. *Proc. Natl Acad. Sci. USA* 95: 2761–2766, 1998.
17. Baba, T., J. Mimura, K. Gradin, A. Kuroiwa, T. Watanabe, Y. Matsuda, J. Inazawa, K. Sogawa and Y. Fujii-Kuriyama. Structure and expression of the Ah receptor repressor gene. *J. Biol. Chem.* 276: 33101–33110, 2001.
18. Bailey, G.S., D.E. Williams and J.D. Hendricks. Fish models for environmental carcinogenesis: the rainbow trout. *Environ. Health Perspect.* 104(Suppl. 1): 5–21, 1996.
19. Bainy, A.C.D. and J.J. Stegeman. Cloning and identification of a full length pregnane X receptor and expression in vivo in zebrafish (*Danio rerio*). *Mar. Environ. Res.* 58: 133–134, 2004.
20. Bardet, P.L., B. Horard, M. Robinson-Rechavi, V. Laudet and J.M. Vanacker. Characterization of oestrogen receptors in zebrafish (*Danio rerio*). *J. Mol. Endocrinol.* 28: 153–163, 2002.
21. Belair, C.D., R.E. Peterson and W. Heideman. Disruption of erythropoiesis by dioxin in the zebrafish. *Dev. Dyn.* 222: 581–594, 2001.
22. Bello, S.M. Characterization of resistance to halogenated aromatic hydrocarbons in a population of *Fundulus heteroclitus* from a marine superfund site. PhD Thesis. Woods Hole Oceanographic Institution/ Massachusetts Institute of Technology, 1999.
23. Bello, S.M., D.G. Franks, J.J. Stegeman and M.E. Hahn. Acquired resistance to aryl hydrocarbon receptor agonists in a population of *Fundulus heteroclitus* from a marine Superfund site: in vivo and in vitro studies on the induction of xenobiotic-metabolizing enzymes. *Toxicol. Sci.* 60: 77–91, 2001.
24. Bend, J.R. and M.O. James. Xenobiotic metabolism in marine and freshwater species. In: *Biochemical and Biophysical Perspectives in Marine Biology*, edited by D.C. Malins and J.R. Sargent, New York, Academic Press, pp. 125–188, 1978.
25. Bend, J.R., M.O. James and P.M. Dansette. In vitro metabolism of xenobiotics in some marine animals. *Ann. NY Acad. Sci.* 298: 505–521, 1977.
26. Bend, J.R., R.J. Pohl, E. Arinc and R.M. Philpot. Hepatic microsomal and solubilized mixed-function oxidase systems from the little skate, *Raja erinacea*, a marine elasmobranch. In: *Microsomes and Drug Oxidations*, edited by V. Ullrich, I. Roots, A. Hildebrandt, R.W. Estabrook and A.H. Conney, Oxford, Pergamon Press, pp. 160–169, 1977.
27. Bend, J.R., R.J. Pohl, N.P. Davidson and J.R. Fouts. Response of hepatic and renal microsomal mixed-function oxidases in the little skate, *Raja erinacea*, to pretreatment with 3-methylcholanthrene or TCDD (2,3,7,8-tetrachlorodibenzo-*p*-dioxin). *Bull. Mt Desert Isl. Biol. Lab.* 14: 7–12, 1974.
28. Bend, J.R., R.J. Pohl and J.R. Fouts. Further studies of the microsomal mixed-function oxidase system of the little skate, *Raja erinacea*, including its response to some xenobiotics. *Bull. Mt Desert Isl. Biol. Lab.* 13: 9–13, 1973.
29. Berndtson, A.K. and T.T. Chen. Two unique CYP1 genes are expressed in response to 3-methylcholanthrene treatment in rainbow trout. *Arch. Biochem. Biophys.* 310: 187–195, 1994.
30. Bertilsson, G., J. Heidrich, K. Svensson, M. Asman, L. Jendeberg, M. Sydow-Backman, R. Ohlsson, H. Postlind, P. Blomquist and A. Berkenstam. Identification of a human nuclear receptor defines a new signaling pathway for CYP3A induction. *Proc. Natl Acad. Sci. USA* 95: 12208–12213, 1998.
31. Besselink, H.T., M.S. Denison, M.E. Hahn, S.I. Karchner, A.D. Vethaak, J.H. Koeman and A. Brouwer. Low inducibility of CYP1A activity by polychlorinated biphenyls (PCBs) in flounder (*Platichthys flesus*): characterization of the Ah receptor and the role of CYP1A inhibition. *Toxicol. Sci.* 43: 161–171, 1998.

32. Betka, M., A. Welenc, D.G. Franks, M.E. Hahn and G.V. Callard. Characterization of two aryl hydrocarbon receptor (AhR) mRNA forms in *Squalus acanthias* and stage-specific expression during spermatogenesis. *Bull. Mt Desert Isl. Biol. Lab.* 39: 110–112, 2000.
33. Bisbee, C.A., M.A. Baker, A.C. Wilson, I. Haji-Azimi and M. Fischberg. Albumin phylogeny for clawed frogs (*Xenopus*). *Science* 195: 785–787, 1977.
34. Bittinger, M.A., L.P. Nguyen and C.A. Bradfield. Aspartate aminotransferase generates protagonists of the aryl hydrocarbon receptor. *Mol. Pharmacol.* 64: 550–556, 2003.
35. Blizard, D., T. Sueyoshi, M. Negishi, S.S. Dehal and D. Kupfer. Mechanism of induction of cytochrome P450 enzymes by the proestrogenic endocrine disruptor pesticide-methoxychlor: interactions of methoxychlor metabolites with the constitutive androstane receptor system. *Drug Metab. Dispos.* 29: 781–785, 2001.
36. Bloom, D.A. and A.K. Jaiswal. Phosphorylation of Nrf2S40 by PKC in response to antioxidants leads to the release of Nrf2 from INrf2 but not required for Nrf2 stabilization/accumulation in the nucleus and transcriptional activation of ARE-mediated NQO1 gene expression. *J. Biol. Chem.* 278: 44675–44682, 2003.
37. Blumberg, B., H. Kang, J.J. Bolado, H. Chen, A.G. Craig, T.A. Moreno, K. Umesono, T. Perlmann, E.M. De Robertis and R.M. Evans. BXR, an embryonic orphan nuclear receptor activated by a novel class of endogenous benzoate metabolites. *Genes Dev.* 12: 1269–1277, 1998.
38. Blumberg, B., W.J. Sabbagh, H. Juguilon, J.J. Bolado, C.M. van Meter, E.S. Ong and R.M. Evans. SXR, a novel steroid and xenobiotic-sensing nuclear receptor. *Genes Dev.* 12: 3195–3205, 1998.
39. Bunger, M.K., S.M. Moran, E. Glover, T.L. Thomae, G.P. Lahvis, B.C. Lin and C.A. Bradfield. Resistance to 2,3,7,8-tetrachlorodibenzo-*p*-dioxin toxicity and abnormal liver development in mice carrying a mutation in the nuclear localization sequence of the aryl hydrocarbon receptor. *J. Biol. Chem.* 278: 17767–17774, 2003.
40. Burbach, K.M., A. Poland and C.A. Bradfield. Cloning of the Ah receptor cDNA reveals a distinctive ligand-activated transcription factor. *Proc. Natl Acad. Sci. USA* 89: 8185–8189, 1992.
41. Butler, R.B., M.L. Kelley, W.H. Powell, M.E. Hahn and R.J. Van Beneden. An aryl hydrocarbon receptor homologue from the soft-shell clam, *Mya arenaria*: evidence that invertebrate AHR homologues lack TCDD and BNF binding. *Gene* 278: 223–234, 2001.
42. Cao, Z., R.L. Tanguay, D. McKenzie, R.E. Peterson and J.M. Aiken. Identification of a putative calcium-binding protein as a dioxin-responsive gene in zebrafish and rainbow trout. *Aquat. Toxicol.* 63: 271–282, 2003.
43. Carney, S.A., W. Heideman and R.E. Peterson. 2,3,7,8-Tetrachlorodibenzo-*p*-dioxin activation of the aryl hydrocarbon receptor/aryl hydrocarbon receptor nuclear translocator pathway causes developmental toxicity through a CYP1A-independent mechanism in zebrafish. *Mol. Pharmacol.* 66: 512–521, 2004.
44. Carvan, M.J., III, L.V. Ponomareva, W.A. Solis, R.S. Matlib, A. Puga and D.W. Nebert. Trout CYP1A3 gene: recognition of fish DNA motifs by mouse regulatory proteins. *Mar. Biotechnol.* 1: 155–166, 1999.
45. Carvan, M.J., III, W.A. Solis, L. Gedamu and D.W. Nebert. Activation of transcription factors in zebrafish cell cultures by environmental pollutants. *Arch. Biochem. Biophys.* 376: 320–327, 2000.
46. Carvan, M.J., III, W.A. Solis, L. Gedamu and D.W. Nebert. Activation of transcription factors in zebrafish cell cultures by environmental pollutants. *Arch. Biochem. Biophys.* 376: 320–327, 2000.
47. Carvan, M.J., III, D.M. Sonntag, C.B. Cmar, R.S. Cook, M.A. Curran and G.L. Miller. Oxidative stress in zebrafish cells: potential utility of transgenic zebrafish as a deployable sentinel for site hazard ranking. *Sci. Total Environ.* 274: 183–196, 2001.
48. Carver, L.A., J.B. Hogenesch and C.A. Bradfield. Tissue specific expression of the rat Ah-receptor and ARNT mRNAs. *Nucleic Acids Res.* 22: 3038–3044, 1994.
49. Chan, J.Y., M. Kwong, R. Lu, J. Chang, B. Wang, T.S. Yen and Y.W. Kan. Targeted disruption of the ubiquitous CNC-bZIP transcription factor, Nrf-1, results in anemia and embryonic lethality in mice. *EMBO J.* 17: 1779–1787, 1998.
50. Chan, K., X.D. Han and Y.W. Kan. An important function of Nrf2 in combating oxidative stress: detoxification of acetaminophen. *Proc. Natl Acad. Sci. USA* 98: 4611–4616, 2001.
51. Chan, K. and Y.W. Kan. Nrf2 is essential for protection against acute pulmonary injury in mice. *Proc. Natl Acad. Sci. USA* 96: 12731–12736, 1999.
52. Chiang, J.Y. Bile acid regulation of gene expression: roles of nuclear hormone receptors. *Endocrinol. Rev.* 23: 443–463, 2002.
53. Cook, P.M., J.A. Robbins, D.D. Endicott, K.B. Lodge, P.D. Guiney, M.K. Walker, E.W. Zabel and R.E. Peterson. Effects of aryl hydrocarbon receptor-mediated early life stage toxicity on lake trout populations in Lake Ontario during the 20th century. *Environ. Sci. Technol.* 37: 3864–3877, 2003.

54. Cooper, K.R. Effects of polychlorinated dibenzo-*p*-dioxins and polychlorinated dibenzofurans on aquatic organisms. *CRC Crit. Rev. Aquat. Sci.* 1: 227–242, 1989.
55. Courtenay, S.C., C.M. Grunwald, G.-L. Kreamer, W.L. Fairchild, J.T. Arsenault, M. Ikonomou and I.I. Wirgin. A comparison of the dose and time response of CYP1A1 mRNA induction in chemically treated Atlantic tomcod from two populations. *Aquat. Toxicol.* 47: 43–69, 1999.
56. Crews, S.T. Control of cell lineage-specific development and transcription by bHLH-PAS proteins. *Genes Dev.* 12: 607–620, 1998.
57. Dehal, P., Y. Satou, R.K. Campbell, J. Chapman, B. Degnan, A. De Tomaso, B. Davidson, A. Di Gregorio, M. Gelpke, D.M. Goodstein, N. Harafuji, K.E. Hastings, I. Ho, K. Hotta, W. Huang, T. Kawashima, P. Lemaire, D. Martinez, I.A. Meinertzhagen, S. Necula, M. Nonaka, S. Putnam, S. Rash, H. Saiga, M. Satake, A. Terry, L. Yamada, H.G. Wang, S. Awazu, K. Azumi, J. Boore, M. Branno, S. Chin-Bow, R. DeSantis, S. Doyle, P. Francino, D.N. Keys, S. Haga, H. Hayashi, K. Hino, K.S. Imai, K. Inaba, S. Kano, K. Kobayashi, M. Kobayashi, B.I. Lee, K.W. Makabe, C. Manohar, G. Matassi, M. Medina, Y. Mochizuki, S. Mount, T. Morishita, S. Miura, A. Nakayama, S. Nishizaka, H. Nomoto, F. Ohta, K. Oishi, I. Rigoutsos, M. Sano, A. Sasaki, Y. Sasakura, E. Shoguchi, T. Shin-i, A. Spagnuolo, D. Stainier, M.M. Suzuki, O. Tassy, N. Takatori, M. Tokuoka, K. Yagi, F. Yoshizaki, S. Wada, C. Zhang, P.D. Hyatt, F. Larimer, C. Detter, N. Doggett, T. Glavina, T. Hawkins, P. Richardson, S. Lucas, Y. Kohara, M. Levine, N. Satoh and D.S. Rokhsar. The draft genome of *Ciona intestinalis*: insights into chordate and vertebrate origins. *Science* 298: 2157–2167, 2002.
58. Denison, M.S., J. Fine and C.F. Wilkinson. Protamine sulfate precipitation: a new assay for the Ah receptor. *Anal. Biochem.* 142: 28–36, 1984.
59. Denison, M.S. and S.R. Nagy. Activation of the aryl hydrocarbon receptor by structurally diverse exogenous and endogenous chemicals. *Annu. Rev. Pharmacol. Toxicol.* 43: 309–334, 2003.
60. Denison, M.S., A. Pandini, S.R. Nagy, E.P. Baldwin and L. Bonati. Ligand binding and activation of the Ah receptor. *Chem. Biol. Interact.* 141: 3–24, 2002.
61. Dhakshinamoorthy, S. and A.K. Jaiswal. Functional characterization and role of INrf2 in antioxidant response element-mediated expression and antioxidant induction of NAD(P)H:quinone oxidoreductase1 gene. *Oncogene* 20: 3906–3917, 2001.
62. Di Giulio, R.T., P.C. Washburn, R.J. Wenning, G.W. Winston and C.S. Jewell. Biochemical responses in aquatic animals: a review of determinants of oxidative stress. *Environ. Toxicol. Chem.* 8: 1103–1123, 1989.
63. Dinkova-Kostova, A.T., W.D. Holtzclaw, R.N. Cole, K. Itoh, N. Wakabayashi, Y. Katoh, M. Yamamoto and P. Talalay. Direct evidence that sulfhydryl groups of Keap1 are the sensors regulating induction of phase 2 enzymes that protect against carcinogens and oxidants. *Proc. Natl Acad. Sci. USA* 99: 11908–11913, 2002.
64. Dold, K.M. and W.F. Greenlee. Filtration assay for quantitation of 2,3,7,8-tetrachlorodibenzo-*p*-dioxin (TCDD) specific binding to whole cells in culture. *Anal. Biochem.* 184: 67–73, 1990.
65. Dong, W., H. Teraoka, S. Kondo and T. Hiraga. 2,3,7,8-tetrachlorodibenzo-*p*-dioxin induces apoptosis in the dorsal midbrain of zebrafish embryos by activation of arylhydrocarbon receptor. *Neurosci. Lett.* 303: 169–172, 2001.
66. Dong, W., H. Teraoka, K. Yamazaki, S. Tsukiyama, S. Imani, T. Imagawa, J.J. Stegeman, R.E. Peterson and T. Hiraga. 2,3,7,8-tetrachlorodibenzo-*p*-dioxin toxicity in the zebrafish embryo: local circulation failure in the dorsal midbrain is associated with increased apoptosis. *Toxicol. Sci.* 69: 191–201, 2002.
67. Duncan, D.M., E.A. Burgess and I. Duncan. Control of distal antennal identity and tarsal development in *Drosophila* by spineless-aristapedia, a homolog of the mammalian dioxin receptor. *Genes Dev.* 12: 1290–1303, 1998.
68. Ekker, S.C. and J.D. Larson. Morphant technology in model developmental systems. *Genesis* 30: 89–93, 2001.
69. Elskus, A.A., E. Monosson, A.E. McElroy, J.J. Stegeman and D.S. Woltering. Altered CYP1A expression in *Fundulus heteroclitus* adults and larvae: a sign of pollutant resistance? *Aquat. Toxicol.* 45: 99–113, 1999.
70. Elskus, A.A. and J.J. Stegeman. Further consideration of phenobarbital effects on cytochrome P-450 activity in the killifish, *Fundulus heteroclitus*. *Comp. Biochem. Physiol.* 92C: 223–230, 1989.
71. Enomoto, A., K. Itoh, E. Nagayoshi, J. Haruta, T. Kimura, T. O'Connor, T. Harada and M. Yamamoto. High sensitivity of Nrf2 knockout mice to acetaminophen hepatotoxicity associated with decreased expression of ARE-regulated drug metabolizing enzymes and antioxidant genes. *Toxicol. Sci.* 59: 169–177, 2001.
72. Escriva, H., L. Manzon, J. Youson and V. Laudet. Analysis of lamprey and hagfish genes reveals a complex history of gene duplications during early vertebrate evolution. *Mol. Biol. Evol.* 19: 1440–1450, 2002.
73. Escriva, H., R. Safi, C. Hanni, M.-C. Langlois, P. Saumitou-Laprade, D. Stehelin, A. Capron, R. Pierce and V. Laudet. Ligand binding was acquired during evolution of nuclear receptors. *Proc. Natl Acad. Sci. USA* 94: 6803–6808, 1997.

74. Evans, B.R., S.I. Karchner and M.E. Hahn. Zebrafish AHR repressor: cloning, regulatory interactions and inducibility by TCDD. *Toxicol. Sci.* 78(Suppl. 1): 122 2004, Abstract #591.
75. Evans, R.M. The steroid and thyroid hormone receptor superfamily. *Science* 240: 889–895, 1988.
76. Farmer, S.C., C.W. Sun, G.E. Winnier, B.L. Hogan and T.M. Townes. The bZIP transcription factor LCR-F1 is essential for mesoderm formation in mouse development. *Genes Dev.* 11: 786–798, 1997.
77. Fernandez-Salguero, P., D.M. Hilbert, S. Rudikoff, J.M. Ward and F.J. Gonzalez. Aryl-hydrocarbon receptor-deficient mice are resistant to 2,3,7,8-tetrachlorodibenzo-*p*-dioxin-induced toxicity. *Toxicol. Appl. Pharmacol.* 140: 173–179, 1996.
78. Fernandez-Salguero, P., T. Pineau, D.M. Hilbert, T. McPhail, S.S.T. Lee, S. Kimura, D.W. Nebert, S. Rudikoff, J.M. Ward and F.J. Gonzalez. Immune system impairment and hepatic fibrosis in mice lacking the dioxin-binding Ah receptor. *Science* 268: 722–726, 1995.
79. Fitch, W.M. Distinguishing homologous from analogous proteins. *Syst. Zool.* 19: 99–113, 1970.
80. Force, A., M. Lynch, F.B. Pickett, A. Amores, Y.-L. Yan and J.H. Postlethwait. Preservation of duplicate genes by complementary, degenerative mutations. *Genetics* 151: 1531–1545, 1999.
81. Friling, R.S., A. Bensimon, Y. Tichauer and V. Daniel. Xenobiotic-inducible expression of murine glutathione S-transferase Ya subunit gene is controlled by an electrophile-responsive element. *Proc. Natl Acad. Sci. USA* 87: 6258–6262, 1990.
82. Furlong, R.F. and P.W. Holland. Were vertebrates octoploid? *Philos. Trans. R. Soc. London B Biol. Sci.* 357: 531–544, 2002.
83. Gasiewicz, T.A. and R.A. Neal. The examination and quantitation of tissue cytosolic receptors for 2,3,7,8-tetrachlorodibenzo-*p*-dioxin using hydroxylapatite. *Anal. Biochem.* 124: 1–11, 1982.
84. Gasiewicz, T.A. and S.-K. Park. The Ah receptor: involvement in toxic responses, Dioxins and Health, 2nd edition, edited by A. Schecter and T.A. Gasiewicz, London, UK, Taylor & Francis, pp. 491–532, 2003.
85. Gasiewicz, T.A. and G. Rucci. α-Naphthoflavone acts as an antagonist of 2,3,7,8-tetrachlorodibenzo-*p*-dioxin by forming an inactive complex with the Ah receptor. *Mol. Pharmacol.* 40: 607–612, 1991.
86. Gates, M.A., L. Kim, E.S. Egan, T. Cardozo, H.I. Sirotkin, S.T. Dougan, D. Lashkari, R. Abagyan, A.F. Schier and W.S. Talbot. A genetic linkage map for zebrafish: comparative analysis and localization of genes and expressed sequences. *Genome Res.* 9: 334–347, 1999.
87. Ge, N.L. and C.J. Elferink. A direct interaction between the aryl hydrocarbon receptor and retinoblastoma protein – linking dioxin signaling to the cell cycle. *J. Biol. Chem.* 273: 22708–22713, 1998.
88. George, S.G.. Enzymology and molecular biology of phase II xenobiotic-conjugating enzymes in fish. In: *Aquatic Toxicology: Molecular, Biochemical and Cellular Perspectives*, edited by D.C. Malins and G.K. Ostrander, Boca Raton, FL, CRC/Lewis, pp. 37–85, 1994.
89. George, S.G. and B. Taylor. Molecular evidence for multiple UDP-glucuronosyltransferase gene familes in fish. *Mar. Environ. Res.* 54: 253–257, 2002.
90. Goksøyr, A., T. Andersson, D.R. Buhler, J.J. Stegeman, D.E. Williams and L. Forlin. Immunochemical cross-reactivity of β-naphthoflavone-inducible cytochrome P450 (P450IA) in liver microsomes from different fish species and rat. *Fish Physiol. Biochem.* 9: 1–13, 1991.
91. Gong, Z., B. Ju and H. Wan. Green fluorescent protein (GFP) transgenic fish and their applications. *Genetica* 111: 213–225, 2001.
92. Gooch, J.W., A.A. Elskus, P.J. Kloepper-Sams, M.E. Hahn and J.J. Stegeman. Effects of *ortho* and non-*ortho* substituted polychlorinated biphenyl congeners on the hepatic monooxygenase system in scup (*Stenotomus chrysops*). *Toxicol. Appl. Pharmacol.* 98: 422–433, 1989.
93. Goodwin, B., M.R. Redinbo and S.A. Kliewer. Regulation of cyp3a gene transcription by the pregnane X receptor. *Annu. Rev. Pharmacol. Toxicol.* 42: 1–23, 2002.
94. Gu, Y.-Z., J.B. Hogenesch and C.A. Bradfield. The PAS superfamily: sensors of environmental and developmental signals. *Annu. Rev. Pharmacol. Toxicol.* 40: 519–561, 2000.
95. Hahn, M.E. The aryl hydrocarbon receptor: a comparative perspective. *Comp. Biochem. Physiol.* 121C: 23–53, 1998.
96. Hahn, M.E. Mechanisms of innate and acquired resistance to dioxin-like compounds. *Rev. Toxicol.* 2: 395–443, 1998.
97. Hahn, M.E. Dioxin toxicology and the aryl hydrocarbon receptor: insights from fish and other non-traditional models. *Mar. Biotechnol.* 3: S224–S238, 2001.
98. Hahn, M.E. Aryl hydrocarbon receptors: diversity and evolution. *Chem.-Biol. Interact.* 141: 131–160, 2002.
99. Hahn, M.E. Evolutionary and physiological perspectives on Ah receptor function and dioxin toxicity, In: *Dioxins and Health*, 2nd edition, edited by A. Schecter and T.A. Gasiewicz, New York, Wiley, pp. 559–602, 2003, Chapter 14.

100. Hahn, M.E. and S.I. Karchner. Evolutionary conservation of the vertebrate Ah (dioxin) receptor: amplification and sequencing of the PAS domain of a teleost Ah receptor cDNA. *Biochem. J.* 310: 383–387, 1995.
101. Hahn, M.E., S.I. Karchner and D.G. Franks. Aryl hydrocarbon receptor polymorphisms and dioxin resistance in Atlantic killifish (*Fundulus heteroclitus*). *Mar. Environ. Res.* 54(409) 2002, Abstract.
102. Hahn, M.E., S.I. Karchner and D.G. Franks. The Ah receptor and its ligands: a comparative perspective. *Organohal. Comp.* 65: 110–113, 2003.
103. Hahn, M.E., S.I. Karchner, D.G. Franks and R.R. Merson. Aryl hydrocarbon receptor polymorphisms and dioxin resistance in Atlantic killifish (*Fundulus heteroclitus*). *Pharmacogenetics* 14: 131–143, 2004.
104. Hahn, M.E., S.I. Karchner, M.A. Shapiro and S.A. Perera. Molecular evolution of two vertebrate aryl hydrocarbon (dioxin) receptors (AHR1 and AHR2) and the PAS family. *Proc. Natl Acad. Sci. USA* 94: 13743–13748, 1997.
105. Hahn, M.E., A. Poland, E. Glover and J.J. Stegeman. The Ah receptor in marine animals: phylogenetic distribution and relationship to P4501A inducibility. *Mar. Environ. Res.* 34: 87–92, 1992.
106. Hahn, M.E., A. Poland, E. Glover and J.J. Stegeman. Photoaffinity labeling of the Ah receptor: phylogenetic survey of diverse vertebrate and invertebrate species. *Arch. Biochem. Biophys.* 310: 218–228, 1994.
107. Hahn, M.E., J. Sakai, D. Greninger, R.R. Merson and S.I. Karchner. Structural and functional characterization of the aryl hydrocarbon receptor in an early diverging vertebrate, the lamprey *Petromyzon marinus*. *Mar. Environ. Res.* 58: 137–138, 2004.
108. Hahn, M.E., B.R. Woodin, J.J. Stegeman and D.E. Tillitt. Aryl hydrocarbon receptor function in early vertebrates: inducibility of cytochrome P4501A in agnathan and elasmobranch fish. *Comp. Biochem. Physiol.* 120C: 67–75, 1998.
109. Handley, H.H., J.V. Goldstone, A.M. Morrison, A.M. Tarrant, J.Y. Wilson, C.A. Godard, B.R. Woodin and J.J. Stegeman. Evolution and diversity of the cytochrome P450 1 (CYP1) family: CYP1 genes in urochordates. *Mar. Environ. Res.* 58: 131–132, 2004.
110. Handley, H.M. Zebrafish cardiovascular cDNA microarrays: expression profiling and gene discovery in embryos exposed to 2,3,7,8-tetrachlorodibenzo-*p*-dioxin. PhD Thesis. Woods Hole Oceanographic Institution (WHOI)/Massachusetts Institute of Technology (MIT) Joint Graduate Program in Oceanography, 2003.
111. Handley, H.M., J.J. Stegeman and M.C. Fishman. Characterization of TCDD-responsive genes identified by cDNA microarray analysis. *Toxicol. Sci.* 72(1S): 1284S 2003.
112. Handschin, C., M. Podvinec and U.A. Meyer. CXR, a chicken xenobiotic-sensing orphan nuclear receptor, is related to both mammalian pregnane X receptor (PXR) and constitutive androstane receptor (CAR). *Proc. Natl Acad. Sci. USA* 97: 10769–10774, 2000.
113. Handschin, C., M. Podvinec, J. Stockli, K. Hoffmann and U.A. Meyer. Conservation of signaling pathways of xenobiotic-sensing orphan nuclear receptors, chicken xenobiotic receptor, constitutive androstane receptor, and pregnane X receptor, from birds to humans. *Mol. Endocrinol.* 15: 1571–1585, 2001.
114. Hankinson, O. The aryl hydrocarbon receptor complex. *Annu. Rev. Pharmacol. Toxicol.* 35: 307–340, 1995.
115a. Hansson, M.C., H. Wittzell, K. Persson and T. von Schantz. Characterization of two distinct aryl hydrocarbon receptor (AhR2) genes in Atlantic salmon (*Salmo salar*) and evidence for multiple AhR2 gene lineages in salmonid fish. *Gene* 303: 197–206, 2003.
115b. Hansson, M.C., H. Wittzell, K. Persson and T. von Schantz. Unprecedented genomic diversity of AhR1 and AhR2 genes in Atlantic salmon (*Salmo salar* L.). *Aquat. Toxicol.* 68: 219–232, 2004.
116. Hawkins, M.B., J.W. Thornton, D. Crews, J.K. Skipper, A. Dotte and P. Thomas. Identification of a third distinct estrogen receptor and reclassification of estrogen receptors in teleosts. *Proc. Natl Acad. Sci. USA* 97: 10751–10756, 2000.
117. Heilmann, L.J., Y.Y. Sheen, S.W. Bigelow and D.W. Nebert. The trout P450IA1: cDNA and deduced protein sequence, expression in liver, and evolutionary significance. *DNA* 7: 379–387, 1988.
118. Henry, T.R., J.M. Spitsbergen, M.W. Hornung, C.C. Abnet and R.E. Peterson. Early life stage toxicity of 2,3,7,8-tetrachlorodibenzo-*p*-dioxin in Zebrafish (*Danio rerio*). *Toxicol. Appl. Pharmacol.* 142: 56–68, 1997.
119. Hestermann, E.V., J.J. Stegeman and M.E. Hahn. Relative contributions of affinity and intrinsic efficacy to aryl hydrocarbon receptor ligand potency. *Toxicol. Appl. Pharmacol.* 168: 160–172, 2000.
120. Hirose, K., M. Morita, M. Ema, J. Mimura, H. Hamada, H. Fujii, Y. Saijo, O. Gotoh, K. Sogawa and Y. Fujii-Kuriyama. cDNA cloning and tissue-specific expression of a novel basic helix–loop–helix/PAS

factor (Arnt2) with close sequence similarity to the aryl hydrocarbon receptor nuclear translocator (Arnt). *Mol. Cell. Biol.* 16: 1706–1713, 1996.

121. Hoffman, E.C., H. Reyes, F.-F. Chu, F. Sander, L.H. Conley, B.A. Brooks and O. Hankinson. Cloning of a factor required for activity of the Ah (dioxin) receptor. *Science* 252: 954–958, 1991.
122. Holder, M. and P.O. Lewis. Phylogeny estimation: traditional and Bayesian approaches. *Nat. Rev. Genet.* 4: 275–284, 2003.
123. Honkakoski, P. and M. Negishi. Regulation of cytochrome P450 (CYP) genes by nuclear receptors. *Biochem. J.* 347: 321–337, 2000.
124. Hsu, H.J., W.D. Wang and C.H. Hu. Ectopic expression of negative ARNT2 factor disrupts fish development. *Biochem. Biophys. Res. Commun.* 282: 487–492, 2001.
125. Ibabe, A., M. Grabenbauer, E. Baumgart, H.D. Fahimi and M.P. Cajaraville. Expression of peroxisome proliferator-activated receptors in zebrafish (*Danio rerio*). *Histochem. Cell Biol.* 118: 231–239, 2002.
126. Issemann, I., R.A. Prince, J.D. Tugwood and S. Green. The peroxisome proliferator-activated receptor:retinoid X receptor heterodimer is activated by fatty acids and fibrate hypolipidaemic drugs. *J. Mol. Endocrinol.* 11: 37–47, 1993.
127. Itoh, K., N. Wakabayashi, Y. Katoh, T. Ishii, T. O'Connor and M. Yamamoto. Keap1 regulates both cytoplasmic-nuclear shuttling and degradation of Nrf2 in response to electrophiles. *Genes Cells* 8: 379–391, 2003.
128. Iwata, H., K. Yoshinari, M. Negishi and J.J. Stegeman. Species-specific responses of constitutively active receptor (CAR)–CYP2B coupling: lack of CYP2B inducer-responsive nuclear translocation of CAR in marine teleost, scup (*Stenotomus chrysops*). *Comp. Biochem. Physiol.* 131C: 501–510, 2002.
129. Jain, S., K.M. Dolwick, J.V. Schmidt and C.A. Bradfield. Potent transactivation domains of the Ah receptor and Ah receptor nuclear translocator map to their carboxyl termini. *J. Biol. Chem.* 269: 31518–31524, 1994.
130. James, M.O. and J.R. Bend. Polycyclic aromatic hydrocarbon induction of cytochrome P-450-dependent mixed-function oxidases in marine fish. *Toxicol. Appl. Pharmacol.* 54: 117–133, 1980.
131. James, M.O. and P.J. Little. Polyhalogenated biphenyls and phenobarbital: evaluation as inducers of drug metabolizing enzymes in the sheepshead, *Archosargus probatocephalus*. *Chem.-Biol. Interact.* 36: 229–248, 1981.
132. Jensen, B.A. and M.E. Hahn. cDNA cloning and characterization of a high affinity aryl hydrocarbon receptor in a cetacean, the beluga, *Delphinapterus leucas*. *Toxicol. Sci.* 64: 41–56, 2001.
133. Johnsen, O., N. Skammelsrud, L. Luna, M. Nishizawa, H. Prydz and A.B. Kolsto. Small Maf proteins interact with the human transcription factor TCF11/Nrf1/LCR-F1. *Nucleic Acids Res.* 24: 4289–4297, 1996.
134. Jones, S.A., L.B. Moore, J.L. Shenk, G.B. Wisely, G.A. Hamilton, D.D. McKee, N.C. Tomkinson, E.L. LeCluyse, M.H. Lambert, T.M. Willson, S.A. Kliewer and J.T. Moore. The pregnane X receptor: a promiscuous xenobiotic receptor that has diverged during evolution. *Mol. Endocrinol.* 14: 27–39, 2000.
135. Karchner, S.I., D.G. Franks, W.H. Powell and M.E. Hahn. Regulatory interactions among three members of the vertebrate aryl hydrocarbon receptor family: AHR repressor, AHR1, and AHR2. *J. Biol. Chem.* 277: 6949–6959, 2002.
136. Karchner, S.I. and M.E. Hahn. Pufferfish (*Fugu rubripes*) aryl hydrocarbon receptors: unusually high diversity in a compact genome. *Mar. Environ. Res.* 58: 139–140, 2004.
137. Karchner, S.I., W.H. Powell and M.E. Hahn. Identification and functional characterization of two highly divergent aryl hydrocarbon receptors (AHR1 and AHR2) in the teleost *Fundulus heteroclitus*. Evidence for a novel subfamily of ligand-binding basic helix–loop–helix Per-ARNT-Sim (bHLH-PAS) factors. *J. Biol. Chem.* 274: 33814–33824, 1999.
138. Kawamura, T. and I. Yamashita. Aryl hydrocarbon receptor is required for prevention of blood clotting and for the development of vasculature and bone in the embryos of medaka fish, *Oryzias latipes*. *Zool. Sci.* 19: 309–319, 2002.
139. Kelly, K.A., C.M. Havrilla, T.C. Brady, K.H. Abramo and E.D. Levin. Oxidative stress in toxicology: established mammalian and emerging piscine model systems. *Environ. Health Perspect.* 106: 375–384, 1998.
140. Kishida, M. and G.V. Callard. Distinct cytochrome P450 aromatase isoforms in zebrafish (*Danio rerio*) brain and ovary are differentially programmed and estrogen regulated during early development. *Endocrinology* 142: 740–750, 2001.
141. Kleinow, K.M., M.L. Haasch, D.E. Williams and J.J. Lech. A comparison of hepatic P450 induction in rat and trout (*Oncorhynchus mykiss*): delineation of the site of resistance of fish to phenobarbital-type inducers. *Comp. Biochem. Physiol.* 96C: 259–270, 1990.
142. Kliewer, S.A., J.M. Lehmann, M.V. Milburn and T.M. Willson. The PPARs and PXRs: nuclear xenobiotic receptors that define novel hormone signaling pathways. *Recent Prog. Horm. Res.* 54: 345–367, 1999.

143. Kliewer, S.A., J.M. Lehmann and T.M. Willson. Orphan nuclear receptors: shifting endocrinology into reverse. *Science* 284: 757–760, 1999.
144. Kliewer, S.A., J.T. Moore, L. Wade, J.L. Staudinger, M.A. Watson, S.A. Jones, D.D. Mckee, B.B. Oliver, T.M. Willson, R.H. Zetterstrom, T. Perlmann and J.M. Lehmann. An orphan nuclear receptor activated by pregnanes defines a novel steroid signaling pathway. *Cell* 92: 73–82, 1998.
145. Kliewer, S.A. and T.M. Willson. Regulation of xenobiotic and bile acid metabolism by the nuclear pregnane X receptor. *J. Lipid Res.* 43: 359–364, 2002.
146. Kobayashi, M., K. Itoh, T. Suzuki, H. Osanai, K. Nishikawa, Y. Katoh, Y. Takagi and M. Yamamoto. Identification of the interactive interface and phylogenic conservation of the Nrf2-Keap1 system. *Genes Cells* 7: 807–820, 2002.
147. Korkalainen, M., J. Tuomisto and R. Pohjanvirta. Identification of novel splice variants of ARNT and ARNT2 in the rat. *Biochem. Biophys. Res. Commun.* 303: 1095–1100, 2003.
148. Lahvis, G.P., S.L. Lindell, R.S. Thomas, R.S. McCuskey, C. Murphy, E. Glover, M. Bentz, J. Southard and C.A. Bradfield. Portosystemic shunting and persistent fetal vascular structures in aryl hydrocarbon receptor-deficient mice. *Proc. Natl Acad. Sci. USA* 97: 10442–10447, 2000.
149. Lassiter, C.S., B. Kelley and E. Linney. Genomic structure and embryonic expression of estrogen receptor beta a (ERbetaa) in zebrafish (*Danio rerio*). *Gene* 299: 141–151, 2002.
150. Leaver, M., D. Clarke and S. George. Molecular studies of the phase II xenobiotic conjugative enzymes of marine Pleuronectid flatfish. *Aquat. Toxicol.* 22: 265–278, 1992.
151. Leaver, M.J., J. Wright and S.G. George. Structure and expression of a cluster of glutathione S-transferase genes from a marine fish, the plaice (*Pleuronectes platessa*). *Biochem. J.* 321: 405–412, 1997.
152. Leaver, M.J., J. Wright and S.G. George. A peroxisomal proliferator-activated receptor gene from the marine flatfish, the plaice (*Pleuronectes platessa*). *Mar. Environ. Res.* 46: 75–79, 1998.
153. Lees, M.J. and M.L. Whitelaw. Multiple roles of ligand in transforming the dioxin receptor to an active basic helix–loop–helix/PAS transcription factor complex with the nuclear protein Arnt. *Mol. Cell. Biol.* 19: 5811–5822, 1999.
154. Lehmann, J.M., D.D. McKee, M.A. Watson, T.M. Willson, J.T. Moore and S.A. Kliewer. The human orphan nuclear receptor PXR is activated by compounds that regulate CYP3A4 gene expression and cause drug interactions. *J. Clin. Invest.* 102: 1016–1023, 1998.
155. Lehmann, J.M., L.B. Moore, T.A. Smith-Oliver, W.O. Wilkison, T.M. Willson and S.A. Kliewer. An antidiabetic thiazolidinedione is a high affinity ligand for peroxisome proliferator-activated receptor gamma (PPAR gamma). *J. Biol. Chem.* 270: 12953–12956, 1995.
156. Leung, L., M. Kwong, S. Hou, C. Lee and J.Y. Chan. Deficiency of the Nrf1 and Nrf2 transcription factors results in early embryonic lethality and severe oxidative stress. *J. Biol. Chem.* 278: 48021–48029, 2003.
157. Lewis, N., T.D. Williams and K. Chipman. Functional analysis of xenobiotic response elements (XRES) in CYP 1A of the European Flounder (*Platichthys flesus*). *Mar. Environ. Res.* 58: 101–105, 2004.
158. Linney, E. and A.J. Udvadia. Construction and detection of fluorescent, germline transgenic zebrafish. In: *Methods in Molecular Medicine, vol. 254: Germ Cell Protocols, Volume 2: Molecular Embryo Analysis, Live Imaging, Transgenesis, and Cloning*, edited by H. Schatten, Totowa, NJ, Humana Press, 2004, pp. 271–288.
159. Lorenzen, A. and A.B. Okey. Detection and characterization of [^{3}H]2,3,7,8-tetrachlorodibenzo-*p*-dioxin binding to Ah receptor in a rainbow trout hepatoma cell line. *Toxicol. Appl. Pharmacol.* 106: 53–62, 1990.
160. Lynch, M. and A. Force. The probability of duplicate gene preservation by subfunctionalization. *Genetics* 154: 459–473, 2000.
161. Ma, C., L. Fan, R. Ganassin, N. Bols and P. Collodi. Production of zebrafish germ-line chimeras from embryo cell cultures. *Proc. Natl Acad. Sci. USA* 98: 2461–2466, 2001.
162. Maglich, J.M., J.A. Caravella, M.H. Lambert, T.M. Willson, J.T. Moore and L. Ramamurthy. The first completed genome sequence from a teleost fish (*Fugu rubripes*) adds significant diversity to the nuclear receptor superfamily. *Nucleic Acids Res.* 31: 4051–4058, 2003.
163. Marini, M.G., K. Chan, L. Casula, Y.W. Kan, A. Cao and P. Moi. hMAF, a small human transcription factor that heterodimerizes specifically with Nrf1 and Nrf2. *J. Biol. Chem.* 272: 16490–16497, 1997.
164. McKenna, N.J. and B.W. O'Malley. Combinatorial control of gene expression by nuclear receptors and coregulators. *Cell* 108: 465–474, 2002.
165. McMahon, M., K. Itoh, M. Yamamoto and J.D. Hayes. Keap1-dependent proteasomal degradation of transcription factor Nrf2 contributes to the negative regulation of antioxidant response element-driven gene expression. *J. Biol. Chem.* 278: 21592–21600, 2003.
166. McManus, M.T. and P.A. Sharp. Gene silencing in mammals by small interfering RNAs. *Nat. Rev. Genet.* 3: 737–747, 2002.

167. Menuet, A., E. Pellegrini, I. Anglade, O. Blaise, V. Laudet, O. Kah and F. Pakdel. Molecular characterization of three estrogen receptor forms in zebrafish: binding characteristics, transactivation properties, and tissue distributions. *Biol. Reprod.* 66: 1881–1892, 2002.
168. Merson, R.R., D. Gilbert, J.J. Stegeman and M.E. Hahn. Xenobiotic metabolizing enzymes in elasmobranchs: cloning of the cytochrome P450-1A (CYP1A) cDNA in little skate *Raja erinacea. Mar. Environ. Res.* 58: 538–539, 2004.
169. Merson, R.R. and M.E. Hahn. Are sharks susceptible to dioxins? cDNA cloning and characterization of elasmobranch aryl hydrocarbon receptors. *Mar. Environ. Res.* 54: 410, 2002.
170. Merson, R.R., S.I. Karchner and M.E. Hahn. Cloning and characterization of the Atlantic killifish retinoblastoma tumor supressor cDNA and interaction with the aryl hydrocarbon receptor. *Toxicol. Sci.* 66(1S): 45, 2002, Abstract.
171. Merson, R.R., S.I. Karchner and M.E. Hahn. Interaction of the aryl hydrocarbon receptor and the retinoblastoma tumor suppressor gene product in the Atlantic killifish (*Fundulus heteroclitus*). *Mar. Environ. Res.* 54: 409–410, 2002, Abstract.
172. Meyer, J. and R. Di Giulio. Patterns of heritability of decreased EROD activity and resistance to PCB 126-induced teratogenesis in laboratory-reared offspring of killifish (*Fundulus heteroclitus*) from a creosote-contaminated site in the Elizabeth River, VA, USA. *Mar. Environ. Res.* 54: 621–626, 2002.
173. Meyer, J.N., D.E. Nacci and R.T. Di Giulio. Cytochrome P4501A (CYP1A) in killifish (*Fundulus heteroclitus*): heritability of altered expression and relationship to survival in contaminated sediments. *Toxicol. Sci.* 68: 69–81, 2002.
174. Meyer, J.N., D.M. Wassenberg, S.I. Karchner, M.E. Hahn and R.T. Di Giulio. Expression and inducibility of aryl hydrocarbon receptor (AHR) pathway genes in wild-caught killifish (*Fundulus heteroclitus*) with different contaminant exposure histories. *Environ. Toxicol. Chem.* 22: 2337–2343, 2003.
175. Mimura, J., M. Ema, K. Sogawa and Y. Fujii-Kuriyama. Identification of a novel mechanism of regulation of Ah (dioxin) receptor function. *Genes Dev.* 13: 20–25, 1999.
176. Mimura, J., K. Yamashita, K. Nakamura, M. Morita, T. Takagi, K. Nakao, M. Ema, K. Sogawa, M. Yasuda, M. Katsuki and Y. Fujii-Kuriyama. Loss of teratogenic response to 2,3,7,8-tetrachlorodibenzo-*p*-dioxin (TCDD) in mice lacking the Ah (dioxin) receptor. *Genes Cells* 2: 645–654, 1997.
177. Monosson, E. Reproductive and developmental effects of PCBs in fish: a synthesis of laboratory and field studies. *Rev. Toxicol.* 3: 25–75, 2000.
178. Moore, L.B., J.M. Maglich, D.D. McKee, B. Wisely, T.M. Willson, S.A. Kliewer, M.H. Lambert and J.T. Moore. Pregnane X receptor (PXR), constitutive androstane receptor (CAR), and benzoate X receptor (BXR) define three pharmacologically distinct classes of nuclear receptors. *Mol. Endocrinol.* 16: 977–986, 2002.
179. Mulvey, M., M.C. Newman, W. Vogelbein and M.A. Unger. Genetic structure of *Fundulus heteroclitus* from PAH-contaminated and neighboring sites in the Elizabeth and York Rivers. *Aquat. Toxicol.* 61: 195–209, 2002.
180. Mulvey, M., M.C. Newman, W.K. Vogelbein, M.A. Unger and D.R. Ownby. Genetic structure and mtDNA diversity of *Fundulus heteroclitus* populations from polycyclic aromatic hydrocarbon-contaminated sites. *Environ. Toxicol. Chem.* 22: 671–677, 2003.
181. Nacci, D., L. Coiro, D. Champlin, S. Jayaraman, R. McKinney, T. Gleason, W.R. Munns, Jr., J.L. Specker and K. Cooper. Adaptation of wild populations of the estuarine fish *Fundulus heteroclitus* to persistent environmental contaminants. *Mar. Biol.* 134: 9–17, 1999.
182. Nacci, D.E., D. Champlin, L. Coiro, R. McKinney and S. Jayaraman. Predicting the occurrence of genetic adaptation to dioxinlike compounds in populations of the estuarine fish *Fundulus heteroclitus. Environ. Toxicol. Chem.* 21: 1525–1532, 2002.
183. Nacci, D.E., M. Kohan, M. Pelletier and E. George. Effects of benzo[*a*]pyrene exposure on a fish population resistant to the toxic effects of dioxin-like compounds. *Aquat. Toxicol.* 57: 203–215, 2002.
184. Nakai, J.S. and N.J. Bunce. Characterization of the Ah receptor from human placental tissue. *J. Biochem. Toxicol.* 10: 151–159, 1995.
185. Nasevicius, A. and S.C. Ekker. Effective targeted gene 'knockdown' in zebrafish. *Nat. Genet.* 26: 216–220, 2000.
186. Nebert, D.W. The Ah locus: genetic differences in toxicity, cancer, mutation, and birth defects. *CRC Crit. Rev. Toxicol.* 20: 137–152, 1989.
187. Nebert, D.W., D.D. Petersen and A.J. Fornace. Cellular responses to oxidative stress: the [Ah] gene battery as a paradigm. *Environ. Health Perspect.* 88: 13–25, 1990.

188. Nebert, D.W., A.L. Roe, M.Z. Dieter, W.A. Solis, Y. Yang and T.P. Dalton. Role of the aromatic hydrocarbon receptor and [Ah] gene battery in the oxidative stress response, cell cycle control, and apoptosis. *Biochem. Pharmacol.* 59: 65–85, 2000.
189. Necela, B. and R.S. Pollenz. Functional analysis of activation and repression domains of the rainbow trout aryl hydrocarbon receptor nuclear translocator (RtARNT) protein isoforms. *Biochem. Pharmacol.* 57: 1177–1190, 1999.
190. Necela, B. and R.S. Pollenz. Identification of a novel C-terminal domain involved in the negative function of the rainbow trout Ah receptor nuclear translocator protein isoform a (rtARNTa) in Ah receptor-mediated signaling. *Biochem. Pharmacol.* 62: 307–318, 2001.
191. Nguyen, T., P.J. Sherratt, H.C. Huang, C.S. Yang and C.B. Pickett. Increased protein stability as a mechanism that enhances Nrf2-mediated transcriptional activation of the antioxidant response element. Degradation of Nrf2 by the 26 S proteasome. *J. Biol. Chem.* 278: 4536–4541, 2003.
192. Nguyen, T., P.J. Sherratt and C.B. Pickett. Regulatory mechanisms controlling gene expression mediated by the antioxidant response element. *Annu. Rev. Pharmacol. Toxicol.* 43: 233–260, 2003.
193. Nuclear Receptors Nomenclature Committee, A unified nomenclature system for the nuclear receptor superfamily. *Cell* 97: 161–163, 1999.
194. Ogino, Y., T. Itakura, H. Kato, J. Aoki and M. Sato. Functional analysis of promoter region from eel cytochrome P450 1A1 gene in transgenic medaka. *Mar. Biotechnol.* 1: 364–370, 1999.
195. Ohno, S.. *Evolution by Gene Duplication*, New York, Springer, 1970.
196. Oleksiak, M.F., S. Wu, C. Parker, S.I. Karchner, J.J. Stegeman and D.C. Zeldin. Identification, functional characterization and regulation of a new cytochrome P450 subfamily, the CYP2Ns. *J. Biol. Chem.* 275: 2312–2321, 2000.
197. Owen, G.I. and A. Zelent. Origins and evolutionary diversification of the nuclear receptor superfamily. *Cell Mol. Life Sci.* 57: 809–827, 2000.
198. Ownby, D.R., M.C. Newman, M. Mulvey, W.K. Vogelbein, M.A. Unger and L.F. Arzayus. Fish (*Fundulus heteroclitus*) populations with different exposure histories differ in tolerance of creosote-contaminated sediments. *Environ. Toxicol. Chem.* 21: 1897–1902, 2002.
199. Palmer, C.N., M.H. Hsu, A.S. Muerhoff, K.J. Griffin and E.F. Johnson. Interaction of the peroxisome proliferator-activated receptor alpha with the retinoid X receptor alpha unmasks a cryptic peroxisome proliferator response element that overlaps an ARP-1 binding site in the CYP4A6 promoter. *J. Biol. Chem.* 269: 18083–18089, 1994.
200. Peters, J.M., M.G. Narotsky, G. Elizondo, P.M. Fernandez-Salguero, F.J. Gonzalez and B.D. Abbott. Amelioration of TCDD-induced teratogenesis in aryl hydrocarbon receptor (AhR)-null mice. *Toxicol. Sci.* 47: 86–92, 1999.
201. Phelan, D., G.M. Winter, W.J. Rogers, J.C. Lam and M.S. Denison. Activation of the Ah receptor signal transduction pathway by bilirubin and biliverdin. *Arch. Biochem. Biophys.* 357: 155–163, 1998.
202. Pohl, R.J., J.R. Fouts and J.R. Bend. Response of hepatic microsomal mixed-function oxidases in the little skate, *Raja erinacea*, and the winter flounder, *Pseudopleuronectes americanus* to pretreatment with TCDD (2,3,7,8-tetrachlorodibenzo-*p*-dioxin) or DBA (1,2,3,4-dibenzanthracene). *Bull. Mt Desert Isl. Biol. Lab.* 15: 64–66, 1975.
203. Poland, A., E. Glover and A.S. Kende. Stereospecific, high-affinity binding of 2,3,7,8-tetrachlorodibenzo-*p*-dioxin by hepatic cytosol. *J. Biol. Chem.* 251: 4936–4946, 1976.
204. Poland, A. and J.C. Knutson. 2,3,7,8-Tetrachlorodibenzo-*p*-dioxin and related halogenated aromatic hydrocarbons: examination of the mechanism of toxicity. *Annu. Rev. Pharmacol. Toxicol.* 22: 517–554, 1982.
205. Poland, A., D. Palen and E. Glover. Analysis of the four alleles of the murine aryl hydrocarbon receptor. *Mol. Pharmacol.* 46: 915–921, 1994.
206. Pollenz, R.S. The mechanism of AH receptor protein down-regulation (degradation) and its impact on AH receptor-mediated gene regulation. *Chem. Biol. Interact.* 141: 41–61, 2002.
207. Pollenz, R.S. and B. Necela. Characterization of two continuous cell lines derived from *Oncorhynchus mykiss* for models of Ah-receptor mediated signal transduction. Direct comparison to the mammalian Hepa-1c1c7 cell line. *Aquat. Toxicol.* 41: 31–49, 1998.
208. Pollenz, R.S., B. Necela and K. Marks-Sojka. Analysis of rainbow trout Ah receptor protein isoforms in cell culture reveals conservation of function in Ah receptor-mediated signal transduction. *Biochem. Pharmacol.* 64: 49–60, 2002.
209. Pollenz, R.S., H.R. Sullivan, J. Holmes, B. Necela and R.E. Peterson. Isolation and expression of cDNAs from rainbow trout (*Oncorhynchus mykiss*) that encode two novel basic helix–loop–helix/PER-ARNT-SIM (bHLH/PAS) proteins with distinct functions in the presence of the aryl hydrocarbon receptor.

Evidence for alternative mRNA splicing and dominant negative activity in the bHLH/PAS family. *J. Biol. Chem.* 271: 30886–30896, 1996.

210. Powell, W.H., R. Bright, S.M. Bello and M.E. Hahn. Developmental and tissue-specific expression of AHR1, AHR2, and ARNT2 in dioxin-sensitive and -resistant populations of the marine fish, *Fundulus heteroclitus. Toxicol. Sci.* 57: 229–239, 2000.
211. Powell, W.H. and M.E. Hahn. The evolution of aryl hydrocarbon signaling proteins: diversity of ARNT isoforms among fish species. *Mar. Environ. Res.* 50: 39–44, 2000.
212. Powell, W.H. and M.E. Hahn. Identification and functional characterization of hypoxia-inducible factor 2α from the estuarine teleost, *Fundulus heteroclitus*: interaction of HIF-2α with two ARNT2 splice variants. *J. Exp. Zool. (Mol. Dev. Evol.)* 294: 17–29, 2002.
213. Powell, W.H., S.I. Karchner, R. Bright and M.E. Hahn. Functional diversity of vertebrate ARNT proteins: identification of ARNT2 as the predominant form of ARNT in the marine teleost, *Fundulus heteroclitus. Arch. Biochem. Biophys.* 361: 156–163, 1999.
214. Powell, W.H., H.G. Morrison, E.J. Weil, S.I. Karchner, M.L. Sogin, J.J. Stegeman and M.E. Hahn. Cloning and analysis of the CYP1A promoter from the Atlantic killifish (*Fundulus heteroclitus*). *Mar. Environ. Res.* 58: 119–124, 2004.
215. Powell-Coffman, J.A., C.A. Bradfield and W.B. Wood. *Caenorhabditis elegans* orthologs of the aryl hydrocarbon receptor and its heterodimerization partner the aryl hydrocarbon receptor nuclear translocator. *Proc. Natl Acad. Sci. USA* 95: 2844–2849, 1998.
216. Prasch, A.L., H. Teraoka, S.A. Carney, W. Dong, T. Hiraga, J.J. Stegeman, W. Heideman and R.E. Peterson. Aryl hydrocarbon receptor 2 mediates 2,3,7,8-tetrachlorodibenzo-*p*-dioxin developmental toxicity in zebrafish. *Toxicol. Sci.* 76: 138–150, 2003.
217. Pratt, S.J., A. Drejer, H. Foott, B. Barut, A. Brownlie, J. Postlethwait, Y. Kato, M. Yamamoto and L.I. Zon. Isolation and characterization of zebrafish NFE2. *Physiol. Genomics* 11: 91–98, 2002.
218. Prince, R. Comparisons of the effects of 2,3,7,8-tetrachlorodibenzo-*p*-dioxin on chemically impacted and nonimpacted subpopulations of *Fundulus heteroclitus*. PhD Thesis. Rutgers/Robert Wood Johnson Medical School, 1993.
219. Prince, R. and K.R. Cooper. Comparisons of the effects of 2,3,7,8-tetrachlorodibenzo-*p*-dioxin on chemically impacted and nonimpacted subpopulations of *Fundulus heteroclitus*: I. TCDD toxicity. *Environ. Toxicol. Chem.* 14: 579–587, 1995.
220. Prince, R. and K.R. Cooper. Comparisons of the effects of 2,3,7,8-tetrachlorodibenzo-*p*-dioxin on chemically impacted and nonimpacted subpopulations of *Fundulus heteroclitus*: II. Metabolic considerations. *Environ. Toxicol. Chem.* 14: 589–595, 1995.
221. Puga, A., S.J. Barnes, T.P. Dalton, C. Chang, E.S. Knudsen and M.A. Maier. Aromatic hydrocarbon receptor interaction with the retinoblastoma protein potentiates repression of E2F-dependent transcription and cell cycle arrest. *J. Biol. Chem.* 275: 2943–2950, 2000.
222. Qin, C., C. Wilson, C. Blancher, M. Taylor, S. Safe and A.L. Harris. Association of ARNT splice variants with estrogen receptor-negative breast cancer, poor induction of vascular endothelial growth factor under hypoxia, and poor prognosis. *Clin. Cancer Res.* 7: 818–823, 2001.
223. Quattrochi, L.C. and P.S. Guzelian. Cyp3A regulation: from pharmacology to nuclear receptors. *Drug Metab. Dispos.* 29: 615–622, 2001.
224. Ramos-Gomez, M., P.M. Dolan, K. Itoh, M. Yamamoto and T.W. Kensler. Interactive effects of nrf2 genotype and oltipraz on benzo[*a*]pyrene-DNA adducts and tumor yield in mice. *Carcinogenesis* 24: 461–467, 2003.
225. Ramos-Gomez, M., M.K. Kwak, P.M. Dolan, K. Itoh, M. Yamamoto, P. Talalay and T.W. Kensler. Sensitivity to carcinogenesis is increased and chemoprotective efficacy of enzyme inducers is lost in nrf2 transcription factor-deficient mice. *Proc. Natl Acad. Sci. USA* 98: 3410–3415, 2001.
226. Rao, M.S. and J.K. Reddy. Peroxisome proliferation and hepatocarcinogenesis. *Carcinogenesis* 8: 631–636, 1987.
227. Rastinejad, F. Retinoid X receptor and its partners in the nuclear receptor family. *Curr. Opin. Struct. Biol.* 11: 33–38, 2001.
228. Reyes, H., S. Reisz-Porszasz and O. Hankinson. Identification of the Ah receptor nuclear translocator protein (Arnt) as a component of the DNA binding form of the Ah receptor. *Science* 256: 1193–1195, 1992.
229. Robinson-Rechavi, M., O. Marchand, H. Escriva, P.L. Bardet, D. Zelus, S. Hughes and V. Laudet. Euteleost fish genomes are characterized by expansion of gene families. *Genom. Res.* 11: 781–788, 2001.
230. Rotchell, J.M., G.B. Steventon and D.J. Bird. Catalytic properties of CYP1A isoforms in the liver of an agnathan (*Lampetra fluviatilis*) and two species of teleost (*Pleuronectes flesus*, *Anguilla anguilla*). *Comp. Biochem. Physiol.* 125C: 203–214, 2000.

231. Rowatt, A.J., J.J. DePowell and W.H. Powell. ARNT gene multiplicity in amphibians: characterization of ARNT2 from the frog *Xenopus laevis*. *J. Exp. Zool. Part B. Mol. Dev. Evol.* 300: 48–57, 2004.
232. Rowlands, J.C., I.J. McEwan and J.A. Gustafsson. Trans-activation by the human aryl hydrocarbon receptor and aryl hydrocarbon receptor nuclear translocator proteins: direct interactions with basal transcription factors. *Mol. Pharmacol.* 50: 538–548, 1996.
233. Roy, N.K., E. Carlson, R.C. Chambers, D.S. Cerino, D.A. Whitting and I. Wirgin. Comparison of AHR, ARNT, and AHRR expression in chemically-treated and environmentally-exposed Atlantic tomcod from three populations. *Mar. Environ. Res.* 58: 134–135, 2004.
234. Roy, N.K., S. Courtenay, G. Maxwell, Z. Yuan, R.C. Chambers and I. Wirgin. Cytochrome P4501A1 is induced by PCB 77 and benzo[*a*]pyrene treatment but not by exposure to the Hudson River environment in Atlantic tomcod (*Microgadus tomcod*) post-yolk sac larvae. *Biomarkers* 7: 162–173, 2002.
235. Roy, N.K., S. Courtenay, Z. Yuan, M. Ikonomou and I. Wirgin. An evaluation of the etiology of reduced CYP1A1 messenger RNA expression in the Atlantic tomcod from the Hudson River, New York, USA, using reverse transcriptase polymerase chain reaction analysis. *Environ. Toxicol. Chem.* 20: 1022–1030, 2001.
236. Roy, N.K., B. Konkle and I. Wirgin. Characterization of CYP1A1 regulatory elements in cancer-prone Atlantic tomcod. *Pharmacogenetics* 6: 273–277, 1996.
237. Roy, N.K. and I. Wirgin. Characterization of the aromatic hydrocarbon receptor gene and its expression in Atlantic tomcod. *Arch. Biochem. Biophys.* 344: 373–386, 1997.
238. Rutter, J., M. Reick and S.L. McKnight. Metabolism and the control of circadian rhythms. *Annu. Rev. Biochem.* 71: 307–331, 2002.
239. Ruyter, B., O. Andersen, A. Dehli, A.K. Ostlund Farrants, T. Gjoen and M.S. Thomassen. Peroxisome proliferator activated receptors in Atlantic salmon (*Salmo salar*): effects on PPAR transcription and acyl-CoA oxidase activity in hepatocytes by peroxisome proliferators and fatty acids. *Biochim. Biophys. Acta* 1348: 331–338, 1997.
240. Santiago-Josefat, B., E. Pozo-Guisado, S. Mulero-Navarro and P.M. Fernandez-Salguero. Proteasome inhibition induces nuclear translocation and transcriptional activation of the dioxin receptor in mouse embryo primary fibroblasts in the absence of xenobiotics. *Mol. Cell. Biol.* 21: 1700–1709, 2001.
241. Santostefano, M., H. Liu, X.H. Wang, K. Chaloupka and S. Safe. Effect of ligand structure on formation and DNA binding properties of the transformed rat cytosolic aryl hydrocarbon receptor. *Chem. Res. Toxicol.* 7: 544–550, 1994.
242. Savas, U., K.J. Griffin and E.F. Johnson. Molecular mechanisms of cytochrome P-450 induction by xenobiotics: an expanded role for nuclear hormone receptors. *Mol. Pharmacol.* 56: 851–857, 1999.
243. Savas, U., M.R. Wester, K.J. Griffin and E.F. Johnson. Rabbit pregnane X receptor is activated by rifampicin. *Drug Metab. Dispos.* 28: 529–537, 2000.
244. Schmidt, J.V. and C.A. Bradfield. Ah receptor signaling pathways. *Annu. Rev. Cell Dev. Biol.* 12: 55–89, 1996.
245. Schmidt, J.V., G.H.-T. Su, J.K. Reddy, M.C. Simon and C.A. Bradfield. Characterization of a murine Ahr null allele: involvement of the Ah receptor in hepatic growth and development. *Proc. Natl Acad. Sci. USA* 93: 6731–6736, 1996.
246. Schuetz, E.G., C. Brimer and J.D. Schuetz. Environmental xenobiotics and the antihormones cyproterone acetate and spironolactone use the nuclear hormone pregnenolone X receptor to activate the CYP3A23 hormone response element. *Mol. Pharmacol.* 54: 1113–1117, 1998.
247. Schuetz, E.G., S.A. Wrighton, S.H. Safe and P.S. Guzelian. Regulation of cytochrome P-450p by phenobarbital and phenobarbital-like inducers in adult rat hepatocytes in primary monolayer culture and in vivo. *Biochemistry* 25: 1124–1133, 1986.
248. Seidel, S.D., G.M. Winters, W.J. Rogers, M.H. Ziccardi, V. Li, B. Keser and M.S. Denison. Activation of the ah receptor signaling pathway by prostaglandins. *J. Biochem. Mol. Toxicol.* 15: 187–196, 2001.
249. Shi, Y. Mammalian RNAi for the masses. *Trends Genet.* 19: 9–12, 2003.
250. Shimizu, Y., Y. Nakatsuru, M. Ichinose, Y. Takahashi, H. Kume, J. Mimura, Y. Fujii-Kuriyama and T. Ishikawa. Benzo[*a*]pyrene carcinogenicity is lost in mice lacking the aryl hydrocarbon receptor. *Proc. Natl Acad. Sci. USA* 97: 779–782, 2000.
251. Sinal, C.J. and J.R. Bend. Aryl hydrocarbon receptor-dependent induction of Cyp1a1 by bilirubin in mouse hepatoma Hepa 1c1c7 cells. *Mol. Pharmacol.* 52: 590–599, 1997.
252. Sojka, K.M. and R.S. Pollenz. Expression of aryl hydrocarbon receptor nuclear translocator (ARNT) isoforms in juvenile and adult rainbow trout tissues. *Mar. Biotechnol.* 3: 416–427, 2001.
253. Song, J., M. Clagett-Dame, R.E. Peterson, M.E. Hahn, W.M. Westler, R.R. Sicinski and H.F. DeLuca. A ligand for the aryl hydrocarbon receptor isolated from lung. *Proc. Natl Acad. Sci. USA* 99: 14694–14699, 2002.

254. Spitsbergen, J.M. and M.L. Kent. The state of the art of the zebrafish model for toxicology and toxicologic pathology research – advantages and current limitations. *Toxicol. Pathol.* 31(Suppl.): 62–87, 2003.
255. Stegeman, J.J. Cytochrome P450 forms in fish: catalytic, immunological and sequence similarities. *Xenobiotica* 19: 1093–1110, 1989.
256. Stegeman, J.J., M. Brouwer, R.T. DiGiulio, L. Forlin, B.M. Fowler, B.M. Sanders and P. Van Veld. Molecular responses to environmental contamination: enzyme and protein systems as indicators of contaminant exposure and effect. In: *Biomarkers for Chemical Contaminants*, edited by R.J. Huggett, Boca Raton, FL, CRC Press, pp. 237–339, 1992.
257. Stegeman, J.J. and M.E. Hahn. Biochemistry and molecular biology of monooxygenases: current perspectives on forms, functions, and regulation of cytochrome P450 in aquatic species. In: *Aquatic Toxicology: Molecular, Biochemical and Cellular Perspectives*, edited by D.C. Malins and G.K. Ostrander, Boca Raton, FL, CRC/Lewis, pp. 87–206, 1994.
258. Sueyoshi, T., T. Kawamoto, I. Zelko, P. Honkakoski and M. Negishi. The repressed nuclear receptor CAR responds to phenobarbital in activating the human CYP2B6 gene. *J. Biol. Chem.* 274: 6043–6046, 1999.
259. Sun, L., C.S. Bradford, C. Ghosh, P. Collodi and D.W. Barnes. ES-like cell cultures derived from early zebrafish embryos. *Mol. Mar. Biol. Biotechnol.* 4: 193–199, 1995.
260. Swanson, H.I. and G.H. Perdew. Detection of the Ah receptor in rainbow trout. Use of 2-azido-3-[^{125}I]iodo-7,8-dibromodibenzo-*p*-dioxin in cell culture. *Toxicol. Lett.* 58: 85–95, 1991.
261. Tanguay, R.L., C.C. Abnet, W. Heideman and R.E. Peterson. Cloning and characterization of the zebrafish (*Danio rerio*) aryl hydrocarbon receptor. *Biochim. Biophys. Acta* 1444: 35–48, 1999.
262. Tanguay, R.L., C.C. Abnet, W. Heideman and R.E. Peterson. Identification and expression of alternatively spliced aryl hydrocarbon nuclear translocator 2 (ARNT2) cDNAs from zebrafish with distinct functions. *Biochim. Biophys. Acta* 1494: 117–128, 2000.
263. Tanguay, R.L., E.A. Andreasen, M.K. Walker and R.E. Peterson. Dioxin toxicity and aryl hydrocarbon receptor signaling in fish. In: *Dioxins and Health*, 2nd ed., edited by A. Schecter and T.A. Gasiewicz, New York, Wiley, pp. 603–628, 2003, Chapter 15.
264. Taylor, B.L. and I.B. Zhulin. PAS domains internal sensors of oxygen, redox potential, and light. *Microbiol. Mol. Biol. Rev.* 63: 479–506, 1999.
265. Taylor, J.S., Y. Van de Peer, I. Braasch and A. Meyer. Comparative genomics provides evidence for an ancient genome duplication event in fish. *Phil. Trans. R. Soc. London B* 356: 1661–1679, 2001.
266. Tchoudakova, A. and G.V. Callard. Identification of multiple CYP19 genes encoding different cytochrome P450 aromatase isozymes in brain and ovary. *Endocrinology* 139: 2179–2189, 1998.
267. Telakowski-Hopkins, C.A., R.G. King and C.B. Pickett. Glutathione S-transferase Ya subunit gene: identification of regulatory elements required for basal level and inducible expression. *Proc. Natl Acad. Sci. USA* 85: 1000–1004, 1988.
268. Teraoka, H., W. Dong and T. Hiraga. Zebrafish as a novel experimental model for developmental toxicology. *Congenit. Anom. Kyoto* 43: 123–132, 2003.
269. Teraoka, H., W. Dong, S. Ogawa, S. Tsukiyama, Y. Okuhara, M. Niiyama, N. Ueno, R.E. Peterson and T. Hiraga. 2,3,7,8-Tetrachlorodibenzo-*p*-dioxin toxicity in the zebrafish embryo: altered regional blood flow and impaired lower jaw development. *Toxicol. Sci.* 65: 192–199, 2002.
270. Teraoka, H., W. Dong, Y. Tsujimoto, H. Iwasa, D. Endoh, N. Ueno, J.J. Stegeman, R.E. Peterson and T. Hiraga. Induction of cytochrome P450 1A is required for circulation failure and edema by 2,3,7,8-tetrachlorodibenzo-*p*-dioxin in zebrafish. *Biochem. Biophys. Res. Commun.* 304: 223–228, 2003.
271. Tsui, H.W. and A.B. Okey. Rapid vertical tube rotor gradient assay for binding of 2,3,7,8-tetrachlorodibenzo-*p*-dioxin to the Ah receptor. *Can. J. Physiol. Pharmacol.* 59: 927–931, 1981.
272. Udvadia, A.J. and E. Linney. Windows into development: historic, current, and future perspectives on transgenic zebrafish. *Dev. Biol.* 256: 1–17, 2003.
273. Van Veld, P.A. and D.J. Westbrook. Evidence for depression of cytochrome P4501A in a population of chemically resistant mummichog (*Fundulus heteroclitus*). *Environ. Sci.* 3: 221–234, 1995.
274. Veraksa, A., N. McGinnis, X. Li, J. Mohler and W. McGinnis. Cap 'n' collar B cooperates with a small Maf subunit to specify pharyngeal development and suppress deformed homeotic function in the *Drosophila* head. *Development* 127: 4023–4037, 2000.
275. Walker, A.K., R. See, C. Batchelder, T. Kophengnavong, J.T. Gronniger, Y. Shi and T.K. Blackwell. A conserved transcription motif suggesting functional parallels between *Caenorhabditis elegans* SKN-1 and Cap'n'Collar-related basic leucine zipper proteins. *J. Biol. Chem.* 275: 22166–22171, 2000.
276. Walker, M.K. and R.E. Peterson. Potencies of polychlorinated dibenzo-*p*-dioxin, dibenzofuran, and biphenyl congeners, relative to 2,3,7,8-tetrachlorodibenzo-*p*-dioxin, for producing early life stage mortality in rainbow trout (*Oncorhynchus mykiss*). *Aquat. Toxicol.* 21: 219–238, 1991.

277. Wang, W.-D., Y.-M. Chen and C.-H. Hu. Detection of Ah receptor and Ah receptor nuclear translocator mRNAs in the oocytes and developing embryos of zebrafish (*Danio rerio*). *Fish Physiol. Biochem.* 18: 49–57, 1998.
278. Wang, W.-D., J.-C. Wu, H.-J. Hsu, Z.-L. Kong and C.-H. Hu. Overexpression of a zebrafish ARNT2-like factor represses CYP1A transcription in ZLE cells. *Mar. Biotechnol.* 2: 376–386, 2000.
279. Waxman, D.J. P450 gene induction by structurally diverse xenochemicals: central role of nuclear receptors CAR, PXR, and PPAR. *Arch. Biochem. Biophys.* 369: 11–23, 1999.
280. Whitelaw, M., J.A. Gustafsson and L. Poellinger. Identification of transactivation and repression functions of the dioxin receptor and its basic helix–loop–helix/PAS partner factor ARNT: inducible versus constitutive modes of regulation. *Mol. Cell. Biol.* 14: 8343–8355, 1994.
281. Willett, K., M. Steinberg, J. Thomsen, T.R. Narasimhan, S. Safe, S. McDonald, K. Beatty and M.C. Kennicutt. Exposure of killifish to benzo[*a*]pyrene: comparative metabolism, DNA adduct formation and aryl hydrocarbon (Ah) receptor agonist activities. *Comp. Biochem. Physiol.* 112B: 93–103, 1995.
282. Williams, T.D., J.S. Lee, D.L. Sheader and J.K. Chipman. The cytochrome P450 1A gene (CYP1A) from European flounder (*Platichthys flesus*), analysis of regulatory regions and development of a dual luciferase reporter gene system. *Mar. Environ. Res.* 50: 1–6, 2000.
283. Wilson, C.L., J. Thomsen, D.J. Hoivik, M.T. Wormke, L. Stanker, C. Holtzapple and S.H. Safe. Aryl hydrocarbon (Ah) nonresponsiveness in estrogen receptor-negative MDA-MB-231 cells is associated with expression of a variant Arnt protein. *Arch. Biochem. Biophys.* 346: 65–73, 1997.
284. Winston, G.W. and R.T. Di Giulio. Prooxidant and antioxidant mechanisms in aquatic organisms. *Aquat. Toxicol.* 19: 137–161, 1991.
285. Wirgin, I., G.L. Kreamer, C. Grunwald, K. Squibb, S.J. Garte and S. Courtenay. Effects of prior exposure history on cytochrome P4501A mRNA induction by PCB congener 77 in Atlantic tomcod. *Mar. Environ. Res.* 34: 103–108, 1992.
286. Wirgin, I. and J.R. Waldman. Altered gene expression and genetic damage in North American fish populations. *Mutat. Res.* 399: 193–219, 1998.
287. Xie, W. and R.M. Evans. Orphan nuclear receptors: the exotics of xenobiotics. *J. Biol. Chem.* 276: 37739–37742, 2001.
288. Yang, Y.H., J.L. Wang, C.L. Miranda and D.R. Buhler. CYP2M1: cloning, sequencing, and expression of a new cytochrome P450 from rainbow trout liver with fatty acid (omega-6)-hydroxylation activity. *Arch. Biochem. Biophys.* 352: 271–280, 1998.
289. Zhang, H., E. LeCluyse, L. Liu, M. Hu, L. Matoney, W. Zhu and B. Yan. Rat pregnane X receptor: molecular cloning, tissue distribution, and xenobiotic regulation. *Arch. Biochem. Biophys.* 368: 14–22, 1999.

Cellular Changes

Biochemistry and Molecular Biology of Fishes, vol. 6
T. P. Mommsen and T. W. Moon (Editors)

CHAPTER 8

Impacts of environmental toxicants and natural variables on the immune system of fishes

KAREN G. BURNETT

Grice Marine Laboratory, The College of Charleston, Charleston, SC 29412, USA

I. Introduction

Broadly defined, the immune system is the set of mechanisms that protects the host from microorganisms. Suppression of the immune system renders the host susceptible to infection with pathogenic microorganisms or unable to recognize and destroy host cells that have undergone neoplastic transformation. Environmental toxicants such as metals, pesticides and organics may induce immunosuppression. Alternatively, these substances may cause inappropriate responses against host tissues (autoimmunity) or hyperactivity against environmental targets producing allergic disorders. The extent of these impacts can be modulated in the face of natural variation in the environment. This overview will present some of the evidence for direct impacts of individual environmental toxicants on cells of the immune system in fish. The body of relevant literature is large. The chapter will begin with a description of the fish immune system emphasizing those aspects most relevant to current immunotoxicology testing schemes. Next, the effects of metals and organometals, pesticides and aromatic hydrocarbons will be addressed, focusing on intracellular signaling events triggered by toxicant exposures. Finally, some evidence for substantial effects of oxygen and pH on the fish immune system will be presented, arguing that the immunotoxic impacts of xenobiotics must be considered in light of variations in these, as well as other, natural stressors.

II. Immune defenses of fish

Immune defenses of vertebrate organisms are generally grouped into three categories that differ in the speed with which the defense can be mobilized and the specificity of the response. The first line of defense is the external barrier dividing the host from its external environment. In fish, the integument, scales and a robust mucous layer along with linings of the gill and digestive tract constitute major barriers against the aquatic milieu that teems with potential pathogens, including bacteria, viruses, fungi and parasites. Connective tissue also serves as a barrier to the external milieu as does the stroma of organs and tissues of the immune system. The role of this external innate immune system and its sensitivity to environmental toxicants have recently been reviewed[19,93].

When these external barriers are penetrated by an invading microorganism, the host mounts a multi-faceted immune response involving lymphoid and myeloid cell types. These cell lineages that arise from a single stem cell precursor populate the primary lymphoid organs (thymus and anterior kidney in bony fish, or epigonal tissues and Leydig organs in most elasmobranchs) and the secondary organs (spleen). Myeloid cells, along with epithelial cells and stroma, are also organized into a network called the reticuloendothelial system (RES) which serves as a surveillance mechanism for invading pathogens or 'non-self'[32,93].

Most multicellular organisms, including fish, can rapidly respond to the presence of a potential pathogen. This innate immune response is triggered by recognition of molecular structures that are common to many microorganisms, such as the yeast glucans and bacterial lipopolysaccharide (LPS), and does not require prior exposure to the pathogen. Major cellular effectors of the innate immune response include granulocytes (largely neutrophils and eosinophilic granular cells), macrophages and non-specific cytotoxic cells (NCC). The innate immune response is usually initiated when tissue macrophages at the site of an invading pathogen emit chemical signals (chemokines) to activate and attract other cells of the immune system[108], setting off a cascade of events called the inflammatory response. Neutrophils move from the vascular compartment into the tissue, while acquiring enhanced ability to bind, engulf (phagocytose) and kill foreign particles, such as invading microorganisms. Monocytes circulating in the blood also respond to the chemokines by moving into the tissues where they mature into macrophages that can phagocytose, kill and digest potential pathogens (see reviews, refs. 40,80). A third group of cellular effectors of innate immunity is the NCC. NCC have been described from a variety of fish species, including *Ictalurus punctatus* (channel catfish), *Oreochromis niloticus* (tilapia) and *Oncorhynchus mykiss* (rainbow trout). These cells recognize and kill a wide variety of larger cell targets, including xenogeneic and allogeneic tumor cells, virus-transformed syngeneic cells and protozoan parasites[42]. NCC also produce cytokines that promote the inflammatory response. The molecular mechanisms of recognition and cell killing used by NCC have been the subject of particularly intense scrutiny[15,56,85].

As part of the early response to binding a particulate target, neutrophils and macrophages convert molecular oxygen into superoxide (O_2^-), a process catalyzed by NADPH oxidase. This highly reactive oxygen radical is spontaneously or enzymatically converted into a variety of other reactive oxygen species (ROS) such as hydrogen peroxide (H_2O_2), hydroxyl radical (OH^-) and hypochlorous acid (OCl^-) through the action of superoxide dismutase and myeloperoxidase. Oxygen is consumed during ROS production, so this process is more generally referred to as the respiratory burst. Molecular and biochemical approaches have confirmed that fish phagocytic cells also possess an inducible form of nitric oxide synthase (iNOS) whose endproduct, nitric oxide (NO), can combine with O_2^- to form the potent microbicidal peroxynitrite ($ONOO^-$). Expression of fish iNOS is upregulated by LPS, extracellular and intracellular protozoans, Gram-positive bacteria and microsporidians[80,105,131]. When activated by LPS and the anti-viral cytokine interferon-gamma (IFN-γ), mammalian macrophages also upregulate the expression of the natural resistance-associated macrophage protein (Nramp1). Nramp1 functions as a proton-gradient coupled divalent cation transporter that removes redox-active metals from the phagosomal compartment[2]. Depletion of divalent ions from the phagosome is believed to have a generalized suppressive effect on a wide variety of microbial pathogens. Nramp cDNA sequences have been documented in rainbow trout[35], common carp *Cyprinus carpio*[106] and channel catfish[31] and, at least in the catfish, their expression was induced by LPS. The extent to which fish Nramp can be induced by cytokines or infection remains to be demonstrated.

Eosinophilic granulocytes, neutrophils, tissue macrophages and NCC all synthesize and store a broad array of degradative enzymes, such as lipases, esterases and nucleases, along with antimicrobial peptides that are immediately available following phagocytosis. In contrast, fish granulocytes may act like mammalian mast cells, releasing their toxic products to act against larger protozoan or parasitic pathogens that cannot be effectively engulfed[101]. Fish phagocytic cells, NCC and endothelial cells of the RES also produce cytokines that increase vascular permeability and adhesiveness and promote other components of the inflammatory response. Among these pivotal cytokines are interleukin-1 (IL-1)[14,41,108] and tumor necrosis factor-alpha (TNF-α)[52,64]. For example, the expression of cyclooxygenase-2 (COX-2) is induced by these early cytokines and LPS. COX-2, which has recently been cloned from rainbow trout and brook trout[47,140], catalyzes the synthesis of prostaglandins which in turn regulate multiple points of the inflammatory response.

Another group of cytokines produced by many cell types, including macrophages, in response to viral challenge is the interferons (IFN). In mammalian cells, IFN induces the synthesis of Mx proteins that can inhibit the translation of viral mRNA. Induction of Mx proteins has been used as an indirect measure of IFN activity in fish[40,68]. Recently, Altmann *et al.*[3] cloned, sequenced and characterized the first teleost interferon gene from the zebrafish, *Danio rerio*, so it should soon become feasible to quantify IFN production directly as a measure of anti-viral response.

Other soluble (humoral) factors play key roles in the innate immunity of fish. Lysozyme is an enzyme that hydrolyzes glycosidic linkages in the peptidoglycan layer of Gram-positive bacteria, and on the inner peptidoglycan layer of Gram-negative bacteria, after complement and other enzymes have disrupted the outer wall. In both bony and jawless fish, as in the higher vertebrates, products of the complement cascade induce death in cellular targets and support the inflammatory response[81,120,133]. Other soluble factors with a role in innate immunity are produced by the liver including lectins such as mannan-binding lectin and the pentraxins, C reactive protein (CRP) and serum amyloid protein (SAP). These lectins bind discrete polysaccharide structures in the presence of Ca^{2+} ions. In mammals lectins and pentraxins activate complement and serve as receptors for phagocytes, and in fish they are presumed to play similar roles in immune defense[40].

The third line of immune defense is adaptive immunity, the set of humoral and cellular responses that arise following exposure of the host to an invading microorganism. Adaptive immune responses develop more slowly than innate immunity, often taking weeks to mature and target discrete structures (epitopes) on the invading microorganism. Adaptive, also called specific acquired, immunity arose among the vertebrates and is associated primarily with functions of B- and T-lymphocytes (see ref. 77 for a recent review of molecular and functional characteristics of teleost lymphocytes). Like their mammalian counterparts, fish B-lymphocytes produce antibodies against foreign epitopes by splicing and transcribing immunoglobulin genes that are part of the genetic repertoire. These antibodies can be used to mobilize complement fixation or trigger phagocytic cell activity. Following a second exposure to the same pathogen, specific antibodies may have higher affinity for the target and are produced more rapidly than the first (primary) response. The increased speed and the change in antibody specificity/affinity associated with this secondary immune response has been documented to occur in fish[60], but is relatively limited when compared to the secondary response in higher vertebrates. This ability to recall prior exposure to a potential pathogen is called memory and is a unique feature of the adaptive immune system. The production of antigen-specific antibodies and maintenance of memory is supported by a second type of T-lymphocytes, called T helper cells (T_h). Functional fish T_h were described some years ago[76]. Only recently have Stuge *et al.*[119], Nakanishi *et al.*[79] and Fischer *et al.*[43] used both functional and genetic approaches to demonstrate that channel catfish, ginbuna crucian carp (*Carassius auratus langsdorfii*) and rainbow trout also produce specific cytotoxic T-lymphocytes (CTL) against virus-infected syngeneic or allogeneic cells with phenotypic and gene expression patterns similar to those of CTL in higher vertebrates. Fish CTL activities were greatly magnified by prior exposure to virus, consistent with the existence of immunologic memory.

Fish B- and T-lymphocytes express antigen-specific receptors[69]. Ligation of B cell antigen receptors (BCR) triggers tyrosine phosphorylation and other intracellular signaling events leading to cellular proliferation[74,125]. In mammals T cell antigen receptors (TCR) on T_h recognize antigenic peptides bound to Class II major histocompatibility complex (MHC) proteins, while TCR on CTL recognize peptides presented by MHC Class I proteins. The functional importance of the MHCs to the fish immune response is only beginning to be understood[49]. Activation of the mammalian TCR triggers a signal transduction cascade, leading from tyrosine phosphorylation of several cytosolic proteins and a concomitant increase in second messenger generation and intracellular free Ca^{2+}, to the activation of several transcription factors *via* the Ras and calcium/calmodulin signaling pathways, production of interleukin-2, and cell division[1]. Activation of channel catfish T cells triggers phosphorylation and nuclear translocation of the nuclear factor of activated T cells (NFAT) transcription factor that is involved in cytokine gene expression in mammalian cells[86]. This suggests that pathways involved in lymphocyte activation are highly conserved among the vertebrates.

1. General considerations for immunotoxicology

The impact of environmental toxicants on immune function cannot be separated from the larger biology of the organism, and as such it is important to note fundamental differences in lymphoid and myeloid organizations between mammals and fishes, as well as the diversity of lymphoid organ structures among the fishes[93]. In mammals, stem cells of the immune system are located in the bone marrow and during differentiation migrate to other organs. Fish have no bone marrow. In teleosts, lymphoid and myeloid stem cells are believed to arise within the anterior kidney. This organ ranges from being a discrete, encapsulated structure in the channel catfish to being a diffuse tissue in which immune and renal cell types are interspersed with neuroendocrine cells. This suggests that stem cells and their immune cell progeny might be highly accessible to the environmental toxicant taken up into tissues and blood and to products of the neuroendocrine system. In contrast, the liver of fish is not believed to be as important to immune function as it is in higher vertebrates. Tissue macrophages in the liver do function as part of the RES and, along with spleen and kidney, the liver is major site for the development of melanomacrophage centers (MMC). MMC appear to be sites of inflammation and deposition of melanin, lipofuscin and hemosiderin and have been used as bioindicators of oxidative stress[90,93].

Substantial evidence supports the concept that the fish immune system participates in a complex communication network with the neuroendocrine system. This network, called the hypothalamus–pituitary–interrenal (HPI) axis, is comparable to the hypothalamus–pituitary–adrenal (HPA) axis in mammals. For example, under a variety of stressful conditions, the fish endocrine system releases substantial amounts of cortisol into the bloodstream. High levels of cortisol negatively depress numbers of B-lymphocytes and antibody production, but appear to rescue neutrophilic granulocytes, thereby favoring the innate immune response[41,132]. In turn, mammalian cytokines, especially IL-1, TNF-α and interleukin-6 (IL-6), participate in immune signaling to the neuroendocrine system. Holland *et al.*[54] recently provided evidence for the comparable effects of immune cytokines on the fish neuroendocrine system. In this study intraperitoneal injections of trout recombinant IL-1β significantly elevated plasma cortisol levels in a dose-dependent manner, conclusively showing that the HPI in trout both produces and responds to IL-1. The clear significance of the HPI axis to this review is that toxicant impacts on endocrine or neurological system, which are themselves highly complex and sensitive to environmental perturbations, could indirectly impact immune function. It is a difficult, and perhaps impossible, task to attribute immunotoxic effects in whole animals independent of toxic effects on other tissues, organs and cells. The HPI axis is dealt with in greater detail elsewhere in this volume (see Chapters 12 and 13).

Numerous reviews have documented immune modulation in response to metal, pesticide and organic contaminants in fish[4,19,37,135]. Assays traditionally used to assess perturbations of immune function in fish fall into three broad categories: pathogen challenge models, assays that monitor immune suppression/activation or immunopathology. Immune suppression/activation is usually measured with assays of phagocytosis, respiratory burst, cell proliferation, as well as quantification of soluble factors such as lysozyme, serum antibody, CRP or complement.

Few studies have addressed the ability of xenobiotics to induce autoimmunity or hypersensitivity in fish as they can in higher vertebrates[93]. Allergens or autoimmunogens can be generated when structures that are too small to be recognized by the immune system (the so-called 'haptens') react with normal 'self' antigens to produce new structures that provoke an immune response against self. Metals can form protein–metal complexes by binding to several amino acids or oxidizing proteins. As pointed out elsewhere in this volume, metabolism of organic xenobiotics in fish can occur in the liver, gut[126] and anterior kidney[100], and the latter two organs play a critical role in immune surveillance and maturation of the immune system. Reactive metabolites of organic compounds can bind covalently to hydroxyl, amino or thiol groups to form hapten–self protein conjugates that trigger allergic or autoimmune responses. Rose *et al.*[100] demonstrated that anterior kidney preparations from the mummichog, *Fundulus heteroclitus*, could metabolize benzo(*a*)pyrene (B*a*P) to a mutagen and carcinogenic diol-epoxide structure similar to those formed by hepatic parenchymal cells. These metabolites were able to form DNA adducts, a possible first step in generating auto-DNA antibodies, a common target for autoimmune responses in humans. COX-2, the potent co-oxidation-peroxidation enzyme that catalyzes the production of prostaglandins, may also act on xenobiotic substrates to generate reactive haptens capable of triggering autoimmune responses. As discussed above, the adaptive immune system in fish is remarkably similar to those of mammals, suggesting that tools to examine autoimmunity and allergy may prove to be critical to fish immunotoxicologists in the near future.

2. *Metals and organometals*

There is a substantial literature supporting the idea that heavy metals and organometals, including mercury, cadmium, zinc, chromium (VI) and tributyltin (TBT), can modulate both innate and adaptive immunities in fish as well as humans. However, the direction of change and effective dose at which these changes are observed vary greatly among reports. These differences have been attributed to mode, dose and length of exposure to the metal, the species examined and the assays employed to examine immunomodulation[19]. Discrepancies among these reports might also be explained by the changes in the cell redox status caused by accumulated heavy metals, leading to additional ROS production. In addition, activated phagocytes normally show enhanced expression of the protective antioxidant enzymes such as catalase, superoxide dismutase, glutathione *S*-transferase and glutathione reductase.

Many immunotoxicology studies of heavy metals in fish assess cellular innate immunity by measuring phagocytosis and/or the respiratory burst; some include NCC activity as well as plasma lysozyme activity. Measurements of adaptive immunity are usually directed at proliferative responses of lymphocytes to the T-lymphocyte mitogens Concanavalin A (ConA) or phytohemagglutinin (PHA) and/or to the B-lymphocyte mitogen LPS. Some studies also follow the induction of specific antibody following exposure to a target molecule (antigen) as a measure of acquired immunity.

One of the metals most extensively examined with regard to immunotoxicity in fish and humans is mercury in its inorganic (Hg^{2+}) and organic forms, e.g. methylmercury (CH_3Hg^{2+}) (see review 121). Both enhancement and suppression of innate immunity have been reported, particularly with regard to the activity of phagocytic cells. By way of example, *in vitro* exposure to $HgCl_2$ significantly suppressed respiratory burst activity of leukocytes from rainbow trout[127] as well as from tilapia *O. aureus* and blue gourami *Trichogaster trichopterus*[70–72], but the effects coincided with the onset of cytotoxicity. The production of ROS in rainbow trout leukocytes was inhibited by *in vivo* exposure to 1.8×10^{-3} μM Hg^{2+} (ref. 107) but increased in tilapia *O. aureus* and blue gourami exposed to 3.3×10^{-1} μM Hg^{2+} (refs. 70–72). These two observations are difficult to reconcile, but it should be noted that trout O_2^- production was monitored by fluorescence flow-cytometry while Low and Sin[72] employed a chemiluminescence assay (CL) for ROS that may differ in sensitivity to oxyradical species. In fact, Zelikoff *et al.*[136] observed increased production of O_2^-, but not H_2O_2 by head kidney leukocytes of Japanese medaka, *Oryzias latipes*, exposed *in vivo* to even lower waterborne doses of 2.2×10^{-2} μM $HgCl_2$ using a spectrophotometric assay. Low and Sin[72] attributed increased ROS production in their *in vivo* Hg^{2+}-treated fish to potential increases in IL-1 production similar to those observed in mice at 1–10 μM $HgCl_2$[134]. Although ROS production was decreased, another parameter of innate immunity, plasma lysozyme activity, was significantly elevated in rainbow trout exposed for 30 days to 1.8×10^{-3} μM Hg^{2+} (ref. 107), while blue gourami exposed for 2 weeks to a 100-fold higher dose (3.3×10^{-1} μM) displayed no significant change in plasma lysozyme, although the activity in kidney was significantly increased[73]. An altered form of CRP was reported in freshwater carp *Catla catla* exposed to very low doses of Hg^{2+} (10^{-4} μM for 6 h); but the significance of this altered form to immunocompetence remains unclear[87] (Table 2).

When examining parameters of adaptive immunity Low and Sin[73] reported slight, but significant decreases in levels of agglutinating antibody to *Aeromonas hydrophila* L37 in fish exposed to 3.3×10^{-1} μM $HgCl_2$ prior to immunization with the bacterin; comparable decreases were observed by Sanchez-Dardon *et al.*[107] in rainbow trout at 1.8×10^{-3} μM Hg^{2+}. But, while Low and Sin[73] found that kidney leukocyte proliferation to the T cell mitogen ConA was significantly increased with *in vitro* exposure to 3.7×10^{-2}–1.7×10^{-1} μM Hg^{2+} and decreased at concentrations $>3.3 \times 10^{-1}$ μM Hg^{2+} in the blue gourami, only decreases of ConA, PHA or

TABLE 1

Immune cell subsets and assays mentioned in this review that are commonly used to assess immunotoxicity in fishes

Soluble factors not produced by myeloid or lymphoid tissues	Myeloid tissues: granulocytes, neutrophils, macrophages	Lymphoid tissues		
		NCC	B-lymphocytes	T-lymphocytes
Lysozyme	Cell number	Cell number	Cell number	Cell number
C-reactive protein	Bactericidal activity	Tumor cell killing	Antigen-specific AFC	Proliferation to ConA, PHA
	Migration	cAMP	Antigen-specific antibodies	Calcium flux
	Phagocytosis	Calcium flux	AFC	Apoptosis/ cytotoxicity
	ROS production	Apoptosis/ cytotoxicity	Proliferation to LPS	Receptor phosphorylation
	iNOS induction	Receptor phosphorylation	Calcium flux	
	Calcium flux		Apoptosis/ cytotoxicity	
	Apoptosis/ cytotoxicity		Receptor phosphorylation	
	Receptor phosphorylation			

LPS-stimulated proliferation were observed for rainbow trout thymocytes and head kidney leukocytes by Voccia *et al.*[127] when tested over the range of 10^2–10^{-3} μM Hg^{2+} or CH_3Hg^{2+}. The biphasic effect on mitogen-stimulated proliferation is supported in part by MacDougal *et al.*[74] who reported that mercuric chloride and methylmercury (at 10-fold lower doses) had a biphasic effect on *in vitro* proliferation driven by phorbol ester in peripheral blood lymphocytes of the red drum *Sciaenops ocellatus*. Significant increases in proliferation were induced at very low doses (0.1–1.0 μM) $HgCl_2$ coincident with a slow sustained rise in intracellular Ca^{2+}. Concentrations of inorganic mercury ≥ 10 μM suppressed DNA synthesis and induced rapid influx of radiolabeled Ca^{2+}, as well as tyrosine phosphorylation of numerous cellular proteins. In marked contrast, thymocytes isolated from lake trout *Salvelinus namaycush* displayed dose-dependent, Hg^{2+}-mediated induction of cell death by apoptosis following a 6 h exposure to the metal[75].

In mammals, mercury is generally immunosuppressive, and apoptosis has been suggested as a possible mechanism for immunosuppression. Apoptosis may be induced in human $CD4^+$ T cells at concentrations as low as 0.5 μM $HgCl_2$ (ref. 114), comparable to the results with lake trout thymocytes[75]. The target organelle for this apoptotic effect in human T cells is the mitochondrion, and the induction of oxidative stress is critical to the activation of death-signaling pathways[115,116]. In contrast, $HgCl_2$ is also well known for its ability to induce autoimmunity in susceptible (H-2s) mice[8]. The autoimmune response is characterized by the proliferation of T and B cells, increases in selected immunoglobulin classes and production of antinucleolar antibodies. Consistent with the finding of Hg^{2+}-induced autoimmunity in mice and activation of fish leukocytes[73,74] $HgCl_2$ has also been shown to induce DNA synthesis in $CD4^+$ T cells from some murine strains, but not from others[57]. Furthermore, memory and resting T cells differ in their responses to Hg, Cd and Pb[114]. Thus, mercuric compounds suppress or enhance immune response, depending on the species, cell type, dose, along with the dose and route of administration.

Other heavy metals and organometals, including the butyltins, cadmium, nickel, lead, cause immunomodulation of one or more parameters of immune function in fish and, as is the case for mercury, assays of phagocyte function provide the greatest variety of response. In the case of TBTs, the most toxic form of the butyltins in mammals, low dose, and short-term exposures activated ROS production in cultured toadfish (*Opsanus tau*) and channel catfish macrophages. Longer term *in vitro* or *in vivo* exposures lead to decreased respiratory burst activity. Early activation was associated with an increase in Ca^{2+} (refs. 94–96,98). *In vitro* exposure to TBT induced iNOS in channel catfish macrophages, further suggesting that this toxicant induces an inflammatory response in phagocytic cells. Induced macrophages also became resistant to the cytotoxic effects of PCB 104 and TBT[131] (Table 2).

For Cd Sanchez-Dardon[107] reported suppression of phagocytosis and ROS production by head kidney leukocytes in rainbow trout, while Zelikoff *et al.*[136] reported enhancement of ROS production in medaka leukocytes at approximately 5×10^{-2} μM Cd and inhibition at higher doses. Sanchez-Dardon[107] showed that Cd decreased ConA-mediated proliferation of head kidney leukocytes in trout. Robohm[99] also reported that exposure to cadmium enhanced the antibody response to immunization with sheep red blood cells (SRBC) in striped bass (*Morone saxatilis*), but the same Cd exposure regime suppressed anti-SRBC antibody response in cunners (*Tautogolabrus adspersus*). Such direct comparisons reported from the same laboratory suggest that fish species may differ in the magnitude and direction of heavy metal effects on immune function.

Like Hg, cadmium belongs to the same group II B in the periodic table suggesting that they may have a similar mechanism of toxicity. Avidity of these metals to SH-radicals allows them to bind indiscriminately to SH groups in proteins. Based on low dose immunostimulation reported by Zelikoff *et al.*[137] it was reasonable to postulate that Cd should have a biphasic effect on PMA-driven proliferation, as we observed for mercury[74]. We found that $CdCl_2 \geq 10$ μM significantly suppressed *in vitro* PMA-mediated proliferation of peripheral blood leukocytes of the red drum (*S. ocellatus*), but enhancement was not observed at any $CdCl_2$ concentration over the test range of $10^{-3} - 10^2$ μM. Cadmium also failed to activate tyrosine phosphorylation or mitogen-activated protein kinase, intracellular signaling events that are linked to cell proliferation[24,25]. It is possible that the atomic radius of Hg, which is larger than that of Cd, would be able to cross-link SH groups on extracellular receptors across greater spans of the cell membrane. Further, it has been often suggested that because it has an

TABLE 2

Modulation of selected immune functions by individual chemical contaminants and natural variables as detailed in this review

	Serum lysozyme/ C-reactive protein	Phagocytosis/ ROS production	NCC killing	T/B-lymphocyte function
Metals and organometals				
Hg	△	△▼		△▼
Cd		△▼		△▼
Cr (VI)		▼		
Ni		△		
Mn		△	△	△
Cu	△▼	▼	▼	
Zn		▼	▼	
TBT		△▼	▼	
Pesticides				
Organochlorine				
Lindane		△▼		
Endosulfan				▼
Organophosphate				
Malathion		▼		▼
Trichlorfon	▼	▼		
Dichlorvos	▼	▼		
Pyrethroid				
Esfenvalerate				▼
AFB1				
PAH				
TCDD	△			
DMBA		▼		▼
B*a*P		▼		▼
3MC	△			
Creosote	▼	▼		
PCB				
PCP		▼		
PCB126		▼		▼
Aroclor 1254	△			
Chlorothalonil		△		
Natural variables				
Temperature				△▼
Low pH		▼		
Low P_{O_2}		▼		

Activities are reported as △ (enhancement), ▼ (suppression). Reports of no change are not included in this summary.

ionic radius similar to calcium, cadmium might be able to displace calcium in intracellular signaling pathways. Resting cells have relatively low intracellular and high extracellular calcium levels. A temporary, controlled influx of calcium is a common early event in cell signaling, mediating calmodulin/calcineurin-directed systems. Larger sustained increases in intracellular Ca^{2+} are associated with apoptosis. In human T_h cells Cd required a concentration of 100 μM to induce apoptosis, comparable to the dose at which the metal suppressed proliferation of red drum leukocytes. At lower doses that inhibited human T_h proliferation and IL-2 production, Cd enhanced the production of IL-3; mercury did not display this enhancement of IL-3. This result further emphasizes that Cd and Hg may have distinctive low dose effects on the immune system[65]. Another mechanism by which Cd and other heavy metals may exert multi-phasic effects on the immune system involves the ability of metallothionein (MT) to bind to Cd and ameliorate its toxicity. Phagocytic cells can express MT. However, NO in an activated macrophage might displace cadmium from MT, thereby enhancing the metal's toxic effects[78]. Metals and organic contaminants can induce iNOS in fish phagocytes[131] and this response may contribute to the complex pattern of Cd toxicity to the immune system.

Impacts of chromium, nickel and manganese on the fish immune system have been investigated less extensively, and these studies have focused primarily on phagocyte and NCC effects. *In vivo* and *in vitro* exposure to chromium suppressed ingestion of SRBC by spleen or head kidney leukocytes of the freshwater catfish *Saccobranchus fossilis*[62]. Nickel enhanced random migration, but had no effect on the respiratory burst in rainbow trout peritoneal macrophages[20]. Relatively low concentrations of manganese (6.26 μg/ml) stimulated ROS production of carp macrophages following *in vitro* exposure for 2 h. Mn also enhanced NCC activity in fish. Whether treated *in vivo* by injection or *in vitro* in culture medium, carp's (*C. carpio*) NCC activity increased with exposure to $MnCl_2$ (ref. 45).

In contrast to the wide range of effects on phagocytic cell function, NCC activity is either suppressed or unaffected by most heavy metals, with the exception of Mn (see ref. 19). One possible explanation for this uniformity of response is suggested by the action of TBT on NCC and human NK cells. Grinwis *et al.*[48] reported that *in vivo* waterborne exposure of flounder to TBT significantly suppressed NCC activity. Similarly, brief exposure (5 min) to 300 nM TBT significantly decreased intracellular cAMP in human natural killer (NK) cells, followed by significantly suppressed *in vitro* cytotoxic activity after 1 h exposure. The loss of activity could be restored with IL-2 and IL-12, two cytokines that upregulate the expression of granzyme and perforin, enzymes which mediate NK killing activity against cellular targets[129]. This suggests the possibility that heavy metals may have a discrete target available in NCC as in NK cells that can alter intracellular signaling which mediates killing activity. The effects of Mn vs. other heavy metals on signaling processes that control NCC activity bear further study.

Transition metals such as copper and zinc are essential for health in teleost fish, forming integral components of proteins involved in all aspects of biological function. However, in excess, these metals are also potentially toxic, and to maintain metal homeostasis organisms must tightly coordinate metal acquisition and excretion. Copper suppressed the innate immunity in zebrafish as measured by cellularity of the kidney, head kidney macrophage phagocytosis and NCC activity in fish exposed to 0.05, 0.10 and 0.15 mg/l copper for 7 days at 22°C[104]. Over approximately the same dose range, Japanese carp *Puntius gonionotus* (Bleeker) exposed to copper for 61 days and challenged with *A. hydrophila* on day 56 also displayed a significant decrease in

ROS production as well as a decrease in serum lysozyme[112], while a 96 h exposure caused an increase in head kidney and plasma lysozyme activity in carp (*C. carpio*)[33]. Zebrafish exposed for 7 days to 0.05, 0.15 and 0.25 mg/l zinc displayed reduced NCC activity, enhanced macrophage responses and hypocellularity of the kidney[104]. Interestingly, Sanchez-Dardon *et al.*[107] provided data suggesting that zinc can protect against immunosuppressive effects of Hg and Cd on phagocytic cell function (phagocytosis and respiratory burst) and mitogenic assays in trout.

3. *Pesticides*

Fish species vary in their susceptibility to the immunotoxic effects of organochlorine (OC) and organophosphate (OP) pesticides[39], but where effects have been documented these compounds appear to suppress innate immunity. Levels of serum lysozyme decreased slightly in *C. carpio* treated for 15 min with 20,000 ppm of the OP insecticide trichlorfon, an effect which occurred rapidly (3 days) and persisted for 56 days. Ceruloplasmin, a copper transport protein that may compete with microbial pathogens for available iron, showed a transitory increase at 3 days, but decreased to control levels at 56 days[117]. The same group reported a small reduction in serum lysozyme 7 days after exposing carp for 30 min to 30 mg/l dichlorvos, a degradation product of trichlorfon[38]. No change was observed in serum CRP levels of rainbow trout exposed to two OCs, mirex or kepone[130].

In vitro and *in vivo* exposure of NCC to pesticides appears to have little effect (reviewed in ref. 19), while alterations in phagocytic cell function have been reported frequently. When medaka were exposed to waterborne malathion (OP) at 0.2 or 0.8 mg/l for 7 or 14 days there was a slight reduction in phagocytic cells and ROS production at the highest malathion dose. In addition, medaka exposed for up to 21 days to either 0.1 or 0.3 mg malathion/l were more susceptible than controls to lethal infection with *Yersinia ruckeri*[12]. In rainbow trout exposure *in vivo* concentrations of lindane (OC) over the range of $17.2 \times 10^{-10} - 17.2 \times 10^{-6}$ M led to a large depression of ROS production by head kidney macrophages[39]. *In vitro* exposure of carp (*C. carpio*) head kidney phagocytes to trichlorfon or dichlorvos also inhibited the respiratory burst in a dose-dependent manner[38].

Pesticides also generally have suppressive effects on acquired immunity in fish. In the Beaman *et al.*[12] study in medaka described above, malathion reduced the development of antibody-forming cells (AFC), but T cell proliferation to ConA was unaffected. Mitogen-stimulated proliferation of T- and B-lymphocytes by rainbow trout leukocytes was suppressed by *in vitro* exposure to endosulfan (OC), esfenvalerate (pyrethroid) or malathion, with little accompanying cytotoxicity. The doses of malathion and esfenvalerate that induced suppression were above expected tissue levels in contaminated environments, but the effective endosulfan dose was relevant to aquatic habitats[83].

Despite the relatively uniform *in vivo* impacts of this group of compounds, there is a suggestion from the work by Betoulle *et al.*[13] that at least one OC pesticide, lindane, may have biphasic effects on fish macrophages. At low concentrations (2.5–25 μM), *in vitro* exposure to lindane stimulated ROS production by Ca^{2+}-independent mechanisms in trout macrophages, suggesting a direct or indirect effect of lindane on NADPH oxidase. Slightly higher doses of lindane (25–50 μM) were associated with increases in ROS production and influx of Ca^{2+} from the extracellular medium. At concentrations above 50 μM, lindane increased intracellular Ca^{2+} in macrophages by release of intracellular calcium stores and was accompanied by increasing cytotoxicity. The tissue concentrations of lindane that caused depression of ROS production *in vivo*[39] were less than those used *in vitro* by Betoulle *et al.*[13], but the latter's observations were conducted over a range associated with acute intoxication and death in fish. In general, this observation supports the contention that ROS are beneficial in protection of the organism but when ROS are produced in excess, they can be toxic to cells, tissues and to the organism. It also supports a bi- or multi-phasic response to this pesticide, which should be considered in evaluating other OC pesticides.

4. Polycyclic and polychlorinated aromatic hydrocarbons

In sampling from the available literature regarding the immunotoxic impacts of polycyclic and polychlorinated aromatic hydrocarbons (PAHs and PCBs) on fish, focus is given to studies that employ individual or well-defined mixtures of these compounds. The intent of this perspective is to aid in the identification of mechanism(s) underlying common patterns of immune modulation. Changes in lysozyme have been most well-studied in response to substances of environmental interest such as oil-contaminated sediments, creosote and sewage sludge. Where significant impacts are reported, serum lysozyme is usually decreased[61,109,123]. Levels of CRP were increased in rainbow trout exposed to Aroclor 1254, 2,3,7,8-tetraclorodibenzo-*p*-dioxin (TCDD) or 3-methylcholanthrene (3MC)[130].

Patterns of responses by phagocytic cells to PAH and PCBs are more complex. Particle ingestion is usually suppressed. Injecting TCDD into rainbow trout did not change the number of SRBC engulfed by phagocytes[118]. However, injection of DMBA into oyster toadfish reduced phagocytic activity of peritoneal leukocytes in a dose-dependent fashion[111]. Similarly, injection of B*a*P in rainbow trout decreased uptake of yeast particles by phagocytic cells, while the exposure in sea bass (*Dicentrarchus labrax*) and tilapia (*O. niloticus*) had no effect on phagocytic function[53,66,127,128]. *In vitro* exposure to pentachlorophenol (PCP) inhibited phagocytosis of yeast particles by macrophages and eosinophils of the mummichog[103]. Chlorothalonil (TCIN), structurally similar to PCP, had no such effect in striped bass (*M. saxatilis*) macrophages[9].

Exposures to high doses of single PAH and PCB species generally have a suppressive effect on respiratory burst. In recent reports a sublethal concentration (940 nM) of TCIN modulated ROS (H_2O_2/OCl^- and O_2^-) production in striped bass macrophages[10,11]. PCB126 suppressed superoxide production by phagocytes in medaka, with age playing a significant role in the magnitude and duration of the response[36]. Immersion exposure to liquid creosote, comprised predominantly of PAHs, caused a transitory suppression of superoxide production by rainbow trout macrophages[61]. Holladay *et al.*[53] found that tilapia (*O. niloticus*) exposed by intraperitoneal injection to 5, 25 or 50 mg/kg B*a*P displayed significant suppression of phorbol ester-stimulated respiratory burst in head kidney leukocytes, but only at the highest dose employed. At this high dose, total head kidney cell counts were reduced and the tissues displayed evidence of vacuolation and immune cell apoptosis. High concentrations of B*a*P also reduced phagocyte-mediated superoxide production in the medaka[30]. Although the study employed a PAH mixture, it is interesting to note that Hutchinson *et al.*[55] found that sediment exposure to PAHs, but not PCBs, for 7 days could mediate a suppression of phagocytic respiratory burst in the female dab (*Limanda limanda*), detected as a reduction in H_2O_2, but not superoxide production. This observation stands in contrast to the previously reported effects of PCP on mummichog and medaka phagocytes[30,36,101,102] and of sewage sludge, also in dab[109,110]. The latter exposures depressed superoxide formation but had no effect on hydrogen peroxide. They were also associated with decreased bactericidal activity against *A. salmonicida* in the dab and *Listonella anguillarum* in the mummichog. Hutchinson *et al.*[55] did not report whether changes in bactericidal activity were associated with the reduction in hydrogen peroxide production. However, Powell *et al.*[89] concluded that dietary exposure to Aroclor 1254 did not alter growth or immunocompetence of juvenile chinook salmon (*Oncorhynchus tshawytscha*), as measured by resistance to challenge with *L. anguillarum* bacteria.

Components of the acquired immune response in fish are also sensitive to PAHs and PCBs. Although not technically a PAH, aflatoxin B1 (AFB1) has been commonly employed by immunotoxicologists as a model for studying the general group of aromatic hydrocarbons. Exposure to AFB1 is associated with hepatic carcinogenesis and immunomodulation in a wide variety of vertebrates. *In vitro* exposure to AFB1 suppressed proliferation and immunoglobulin secretion of rainbow trout leukocytes in response to the polyclonal B cell mitogen LPS. These cells displayed up to a 1000-fold greater sensitivity to the immunosuppressive effects of AFB1 *in vitro* than is observed with murine leukocytes[84]. However, exposure of trout embryos to AFB1 produced a long-term immune dysfunction in adults, characterized by enhanced proliferation and immunoglobulin secretion of peripheral blood leukocytes, but not anterior kidney leukocytes, to LPS[59]. These results suggest that AFB1 set in motion a persistent imbalance among components of the acquired immune system.

In a classic set of studies to explore the impact of environmental toxicants on acquired immunity, juvenile chinook salmon were exposed in the laboratory to either a technical PCB mixture or to DMBA. Immune function was analyzed in these fish by examining the ability of their head kidney and splenic leukocytes to produce a primary and secondary *in vitro* antibody-forming cell (AFC) response to the hapten, trinitrophenyl and by determining their susceptibility to a marine pathogen, *Vibrio anguillarum*. Animals exposed to the PCB mixture or to DMBA produced a lower secondary AFC response and were more susceptible to disease compared to unexposed controls, suggesting that memory cells are particularly vulnerable to PAH and PCB toxicity in salmonids[5,6]. In a more recent experiment, PCB126 suppressed the production of AFC to SRBC in medaka[36]. A second medaka study by Carlson *et al.*[29] explored the importance of metabolic degradation to immunotoxic effects on acquired immunity. Medaka were given a single exposure to a relatively low dose of B*a*P (2 μg/g body mass). This dose of B*a*P suppressed mitogen-stimulated T- and B-lymphocyte proliferation without inducing hepatic CYP1A expression or activity. At higher concentrations, B*a*P also reduced AFC number and T-dependent antibody responses. B*e*P failed to induce immunotoxicity, probably because it did not effectively induce CYP1A.

5. Subject and environmental variables

Application of immune function assays to field studies is complicated by many factors related to the test organism and to the environment. Genetic variation in resident fish populations is undoubtedly a contributing factor. Duffy *et al.*[36] have also provided clear evidence that age can be an important variable in determining adverse outcome. In this study there were significant differences between 4–6 month juveniles and 12–15 month aged fish in the production of O_2^- by phagocytes and induction of AFC following immunization with SRBC-aged medaka. Seasonal influences must also be considered. The sensitivity of salmonids to AFB1[84] appeared to be seasonal, exhibiting an acute phase during the July–December period of the year, with the sensitivity falling to lower, more murine-like levels from January to June. A winter-associated decrease in immune reactivity has been observed in a number of ectothermic species. Seasonal effects can be profound in ectotherms, including the onset of substantial changes in the testosterone, estradiol and cortisol, all of which been found to alter immune cell function (reviewed in refs. 41,50). Crowding and hierarchy stress may also influence the fish neuroendocrine system (reviewed in ref. 17). These age-related and environmental influences, largely driven and coordinated by the HPI, have tremendous implications for the interpretation of field studies involving measurements of immune function.

Other naturally occurring variations to which the fish immune system may be sensitive include temperature, salinity, dissolved oxygen and pH. Among these factors, temperature has been particularly well studied[17,67]. Temperature is likely to have pleiotropic effects on the immune system, altering both acquired and innate immunities. For example, Concanavalin A-mediated proliferation of channel catfish T cells is particularly temperature sensitive, correlated at least in part to the distinctive phospholipid composition of their membranes[16]. Low temperatures also altered the distribution of mucosal cells in the channel catfish rendering the animal more susceptible to fungal infections[91]. The direction and magnitude of change induced by a natural stressor may also vary dramatically among fish species. Robohm[99] found that serum antibody responses against immunization with SRBC increased in striped bass exposed to 9°C water compared to 14°C water, while in the same study the anti-SRBC antibody response of cunners was decreased at the lower temperature.

Another natural variable in aquatic habitats less frequently considered in immune function studies is dissolved oxygen (hypoxia). Hypoxia occurs naturally as a result of aquatic respiration in excess of photosynthesis and oxygen can fluctuate both seasonally and diurnally[26]. The magnitude and duration of hypoxic events may increase due to eutrophication[34]. When excess nutrients enter the water they support the production of large algal blooms and the decomposition of organic matter in the sediments. Both these events severely deplete oxygen levels in the water column[58]. Low oxygen is almost always accompanied by an increase in carbon dioxide pressure

(P_{CO_2}) or hypercapnia[26]. These elevated levels of CO_2 derived from respiration drive a decrease in pH, or acidosis, in hypoxic waters. Aquatic organisms have only a limited ability to regulate internal levels of dissolved gasses and pH by behavioral and physiological mechanisms[27], therefore blood levels of P_{O_2} and P_{CO_2} will reflect changes in water P_{O_2}. To assess whether the respiratory burst of fish macrophages was altered by physiologically relevant levels of oxygen and pH, head kidney cells of the mummichog were incubated under moderate conditions simulating blood oxygen and pH of animals in hypercapnic hypoxia. Control cells were held under blood gas and pH conditions of animals from air-saturated water. Superoxide production stimulated by yeast zymosan or live *Vibrio parahaemolyticus* was measured by NBT and chemiluminescence. Bactericidal activity against *V. parahaemolyticus* was monitored by quantifying the reduction in bacterial colony forming units. Moderate conditions of hypercapnic hypoxia significantly suppressed both superoxide production and bactericidal activity as compared to normoxic conditions. Hypoxia alone caused a significant reduction in bactericidal activity, but the magnitude of the suppression was greater in the presence of low pH[18]. This is consistent with the increased mortalities among tilapia hybrids (*O. niloticus*) and carp (*C. carpio*) infected with *Streptococcus* spp. and exposed to hypoxia[23]. Low dissolved oxygen levels also increased mortality in yellowtail flounder, *Seriola quinqueradiata*, challenged with *Enterococcus seriolicida*[44]. It is likely that hypercapnic hypoxia alters more than one enzymatic and/or transport activity associated with ROS production and bactericidal activity. Boyd and Burnett[21] suggested that the pH-mediated reduction in O_2^- production may be focused at a point prior to activation of NADPH oxidase. The very high affinity of NADPH oxidase for O_2 suggests that this enzyme does not serve as a regulatory point in the respiratory burst over the range of physiologically relevant oxygen pressures. An alternative possibility is the involvement of a low-affinity cell surface oxygen receptor that coordinates the response to hypoxia through the hypoxia-inducible factor (HIF)-1 transcription factor[92].

III. Future considerations

By its nature the immune system is designed to be sensitive to changes or so-called 'danger signals' in the host organism. It is therefore not surprising that toxicologists have been able to use different assays in a variety of fish species to reach the same conclusion: that the immune system can serve as a sensitive indicator of environmental changes, including the presence of xenobiotics, toxins and alterations in natural water quality parameters. Furthermore, these assays in appropriate fish models can be used as biological indicators of human and environmental health risks[7,97,138,138]. As new reagents become available, new assays and new assay formats have been proposed for inclusion in immunotoxicological assessments, including profiles of leukocyte populations with subset-specific antibodies, activation status of leukocyte with fluorescent dyes responsive to calcium flux, antigen-specific cell-mediated cytotoxicity[63] or analysis of apoptosis/necrosis[46,122]. New assays for phagocyte activation, such as induction of iNOS, Nramp, IL-1, TNFα and IFN may be included in future immunological assessments. With the recent cloning of the zebrafish IFN gene, direct assays for this cytokine should become an important tool in evaluating toxicant effects on anti-viral immunity. Although not yet in routine use, rapid, inexpensive profiling of gene transcripts and expressed proteins/peptides in fish models, such as medaka, rainbow trout, zebrafish and channel catfish, is underway[28,51]. These approaches have shown promise in delineating the immune responses of rainbow trout to rhabodovirus infection[82]. Functional genomic and proteomic-based techniques are already being used to address questions related to the effects of toxicants on immune function in humans[88] and in fish[22,124]. Perhaps the most exciting potential of these high-throughput technologies is the ability to 'capture' profiles of host:pathogen interactions across broad stretches of the genome and proteome. The ability to simultaneously examine a large array of genes or proteins that are modulated in response to stress or toxicants should be a powerful tool in dissecting the complex interactions among the neural, endocrine and immune systems and the integration of the host response *in toto* to its environment.

Existing and soon-to-be-developed assays for immunotoxic effects in fish will increasingly be used to protect natural populations as well as human health, to enhance productivity in aquaculture and to monitor restoration of aquatic ecosystems. Interpretations of these data have been limited by the great variation in the nature, sensitivity, magnitude and direction of specific immunotoxic responses to any given stressor. It is also clear that over a range of concentrations, environmental toxicants or stressors may activate or then suppress key pathways involved in the complex network of immune defense. Another layer of complexity is added by the sensitivity of the fish immune system to natural stressors, including temperature, salinity, oxygen and pH, which may directly impact cells of the immune system or intervene indirectly through other physiological mechanisms. We are only beginning to unravel the biochemical and molecular mechanisms by which these individual natural and xenobiotic factors alter disease resistance in fish. Delineation of these mechanisms would significantly enhance our ability to harness the sensitivity of the fish immune system to monitor aquatic ecosystems, protect human health and assure the safety of natural populations in our common environment.

Acknowledgements. Contribution No. 235 from the Grice Marine Laboratory.

IV. References

1. Acuto, O. and D. Cantrell. T cell activation and the cytoskeleton. *Annu. Rev. Immunol.* 18: 165–184, 2000.
2. Agranoff, D., I.M. Monahan, J.A. Mangan, P.D. Butcher and S. Krishna. *Mycobacterium tuberculosis* expresses a novel pH-dependent divalent cation transporter belonging to the Nramp family. *J. Exp. Med.* 190: 717–724, 1999.
3. Altmann, S.M., M.T. Mellon, D.L. Distel and C.H. Kim. Molecular and functional analysis of an interferon gene from the zebrafish, *Danio rerio*. *J. Virol.* 77: 1992–2002, 2003.

4. Anderson, D.P. and M.G. Zeeman. Immunotoxicology in fish, In: *Fundamentals of Aquatic Toxicology*, Vol. 2, edited by G.M. Rand, London, Taylor & Francis, pp. 371–404, 1995.
5. Arkoosh, M.R., C. Casillas, E. Clemons, P. Huffman, A.N. Kagley, E. Casillas, N. Adams, H.R. Sanborn, T.K. Collier and J.E. Stein. Increased susceptibility of juvenile Chinook salmon to vibriosis after exposure to chlorinated and aromatic compounds found in contaminated urban estuaries. *J. Aquat. Anim. Health* 13: 257–268, 2001.
6. Arkoosh, M.R., E. Clemons and M. Myers. Suppression of B-cell-mediated immunity in juvenile Chinook salmon (*Oncorhynchus tshawytscha*) after exposure to either a polycyclic aromatic hydrocarbon or to polychlorinated biphenyls. *Immunopharmacol. Immunotoxicol.* 16: 293–314, 1994.
7. Arkoosh, M.R. and T.K. Collier. Ecological risk assessment paradigm for salmon: analyzing immune function to evaluate risk. *Hum. Ecol. Risk Assess.* 8: 265–276, 2002.
8. Bagenstose, L.M., P. Salgame and M. Monesteir. IL-12 down-regulates autoantibody production in mercury-induced autoimmunity. *J. Immunol.* 160: 1612–1617, 1998.
9. Baier-Anderson, C. and R.S. Anderson. Chlorothalonil inhibits reactive oxygen species production, but not phagocytosis in fish (*Morone saxatilus*) phagocytes and oyster (*Crassostrea virginica*) hemocytes. *Dev. Comp. Immunol.* 21: 127 1997.
10. Baier-Anderson, C. and R.S. Anderson. Characterization of the immunotoxicity of chlorothalonil to striped bass phagocytes following in vitro exposure. *Mar. Environ. Res.* 46: 337–340, 1998.
11. Baier-Anderson, C. and R.S. Anderson. Suppression of superoxide production by chlorothalonil in striped bass (*Morone saxatilus*) macrophages: the role of cellular sulfhydryls and oxidative stress. *Aquat. Toxicol.* 50: 85–96, 2000.
12. Beaman, J.R., R. Finch, H. Gardner, F. Hoffmann, A. Rosencrance and J.T. Zelikoff. Mammalian immunoassays for predicting the toxicity of malathion in a laboratory fish model. *J. Toxicol. Environ. Health* 56A: 523–542, 1999.
13. Betoulle, S., C. Duchiron and P. Deschaux. The organochlorine insecticide γ-hexachlorocyclohexane (γ-HCCH), lindane increases in vitro respiratory burst activity and intracellular calcium levels in rainbow trout (*Oncorhynchus mykiss*) head kidney phagocytes. *Aquat. Toxicol.* 48: 211–221, 2000.
14. Bird, S., J. Zou, T. Wang, B. Munday, C. Cunningham and C.J. Secombes. Evolution of interleukin-1β. *Cytokine Growth Factor Rev.* 13: 483–502, 2002.
15. Bishop, G.R., S. Taylor, L. Jaso-Friedmann and D.L. Evans. Mechanisms of nonspecific cytotoxic cell regulation of apoptosis: cytokine-like activity of Fas ligand. *Fish Shellfish Immunol.* 13: 47–67, 2002.
16. Bly, J.E., T.M. Buttke and L.W. Clem. Differential effects of temperature and exogenous fatty acids on mitogen induced proliferation in channel catfish T and B lymphocytes. *Comp. Biochem. Physiol.* 95A: 417–424, 1990.
17. Bly, J.E., S.M. Quiniou and L.W. Clem. Environmental effects on fish immune mechanisms. *Dev. Biol. Stand.* 90: 33–43, 1997.
18. Boleza, K.A., L.E. Burnett and K.G. Burnett. Hypercapnic hypoxia compromises bactericidal activity of fish anterior kidney cells against opportunistic environmental pathogens. *Fish Shellfish Immunol.* 11: 593–610, 2001.
19. Bols, N.C., J.L. Brubacher, R.C. Ganassin and L.E.J. Lee. Ecotoxicology and innate immunity in fish. *Dev. Comp. Immunol.* 25: 853–873, 2001.
20. Bowser, D.H., K. Frenkel and J.T. Zelikoff. Effects of in vitro nickel exposure on the macrophage-mediated immune functions of rainbow trout (*Oncorhynchus mykiss*). *Bull. Environ. Contam. Toxicol.* 52: 367–373, 1994.
21. Boyd, J.N. and L.E. Burnett. Reactive oxygen intermediate production by oyster hemocytes exposed to hypoxia. *J. Exp. Biol.* 202: 3135–3143, 1999.
22. Bradley, B.P., E.A. Shrader, D.G. Kimmel and J.C. Meiller. Protein expression signatures; an application of proteomics. *Mar. Environ. Res.* 54: 373–377, 2002.
23. Bunch, E.C. and I. Bejerano. The effect of environmental factors on the susceptibility of hybrid tilapia, *Oreochromis niloticus* × *Oreochromis aureus* to *Streptococcosis*. *Isr. J. Aquacult.* 49: 67–76, 1997.
24. Burnett, K.G. Evaluating intracellular signaling pathways as biomarkers for environmental contaminant exposures. *Am. Zool.* 37: 585–594, 1997.
25. Burnett, K.G., A. Karlsson and K. Kohlberg. Differential activation of mitogen-activated protein kinase (MAPK) pathways in teleost leukocyte by environmental metals. *Mar. Environ. Res.* 50: 466 2000.
26. Burnett, L.E. The challenges of living in hypoxic and hypercapnic aquatic environments. *Am. Zool.* 37: 633–640, 1997.
27. Burnett, L.E. and W.B. Stickle. Physiological responses to hypoxia. In: *Effects of Hypoxia on Living Resources, with Emphasis on the Northern Gulf of Mexico*, edited by N.N. Rabalais and R.E. Turner, Washington, DC, American Geophysical Union, pp. 101–114, 2001.

28. Cao, D., A. Kocabas, Z. Ju, A. Karsi, P. Li, A. Pataterson and Z. Liu. Transcriptome of channel catfish (*Ictalurus punctatus*): initial analysis of genes and expression profiles of the head kidney. *Anim. Genet.* 32: 169–188, 2001.
29. Carlson, E.A., Y. Li and J.T. Zelikoff. The Japanese medaka (*Oryzias latipes*) model: applicability for investigating immunosuppressive effects of the aquatic pollutant benzo[*a*]pyrene (B*a*P). *Mar. Environ. Res.* 54: 565–568, 2002.
30. Carlson, E.A., Y. Li and J.T. Zelikoff. Exposure of Japanese medaka (*Oryzias latipes*) to benzo(*a*)pyrene suppresses immune function and host resistance against bacterial challenge. *Aquat. Toxicol.* 56: 289–301, 2002.
31. Chen, H., G.C. Waldbieser, C.D. Rice, B. Elibol and L.A. Hanson. Isolation and characterization of a channel catfish natural resistance associated macrophage protein (Nramp) gene. *Dev. Comp. Immunol.* 26: 517–531, 2002.
32. Dalmo, R.A., K. Ingebrigste and J. Bogwald. Non-specific defence mechanisms in fish, with particular reference to the reticuloendothelial system (RES). *J. Fish Dis.* 20: 241–273, 1997.
33. Dautremepuits, C., S. Betoulle and G. Vernet. Antioxidant response modulated by copper in healthy or parasitized carp (*Cyprinus carpio* L.) by *Ptychobothrium* sp. (Cestoda). *Biochim. Biophys. Acta* 1573: 4–8, 2002.
34. Diaz, R.J. and R. Rosenberg. Marine benthic hypoxia: a review of its ecological effects and the behavioural responses of benthic macrofauna. *Oceanogr. Mar. Biol. Annu. Rev.* 33: 245–303, 1995.
35. Dorschner, M.O. and R.B. Phillips. Comparative analysis of two Nramp loci from rainbow trout. *DNA Cell Biol.* 18: 573–583, 1999.
36. Duffy, J.E., E. Carlson, Y. Li, C. Prophete and J.T. Zelikoff. Impact of polychlorinated biphenyls (PCBs) on the immune function of fish: age as a variable in determining adverse outcome. *Mar. Environ. Res.* 54: 559–563, 2002.
37. Dunier, M. and A.K. Siwicki. Effects of pesticides and other organic pollutants in the aquatic environment on immunity of fish: a review. *Fish Shellfish Immunol.* 3: 423–438, 1993.
38. Dunier, M., A.K. Siwicki and A. Demael. Effects of organophosphorus insecticides: effects of trichlorfon and dichlorvos on the immune response of carp (*Cyprinus carpio*) III. In vitro effects on lymphocyte proliferation and phagocytosis and in vivo effects on humoral response. *Ecotoxicol. Environ. Saf.* 22: 79–87, 1991.
39. Dunier, M., A.K. Siwicki, J. Scholtens, S. Dal Molin, D. Vergnet and M. Studnicka. Effects of lindane exposure on rainbow trout (*Oncorhynchus mykiss*) immunity. III. Effect on nonspecific immunity and B lymphocyte functions. *Ecotoxicol. Environ. Saf.* 27: 324–334, 1994.
40. Ellis, A.E. Innate host defense mechanisms of fish against viruses and bacteria. *Dev. Comp. Immunol.* 25: 827–840, 2001.
41. Engelsma, M.Y., M.O. Huising, W.G. van Muiswinkel, G. Flik, J. Kwang, H.F. Savelkoul and B.M. Verburg-van Kemenade. Neuroendocrine-immune interactions in fish: a role for interleukin-1. *Vet. Immunol. Immunopathol.* 87: 467–479, 2002.
42. Evans, D.L. and L. Jaso-Friedmann. Nonspecific cytotoxic cells as effectors of immunity in fish. *Annu. Rev. Fish Dis.* 1: 109–121, 1992.
43. Fischer, U., K. Utke, M. Ototake, J.M. Dijkstra and B. Köllner. Adaptive cell-mediated cytotoxicity against allogeneic targets by CD8-positive lymphocytes of rainbow trout (*Oncorhynchus mykiss*). *Dev. Comp. Immunol.* 27: 323–337, 2003.
44. Fukuda, Y., M. Maita, K. Satoh and N. Okamoto. Influence of dissolved oxygen concentration on the mortality of yellowtail experimentally infected with *Enterococcus seriolicida*. *Fish Pathol.* 32: 129–130, 1997.
45. Ghanmi, Z., M. Rouabhia, M. Alifuddin, D. Troutaud and P. Deschaux. Modulatory effect of metal ions on the immune response of fish: in vivo and in vitro influence of $MnCl_2$ on NK activity of carp pronephros cells. *Ecotoxicol. Environ. Saf.* 20: 241–245, 1990.
46. Gogal, R.M., B.J. Smith, J. Kalnitsky and S.D. Holladay. Analysis of apoptosis of lymphoid cells in fish exposed to immunotoxic compounds. *Cytometry* 39: 310–318, 2000.
47. Grayson, T.H., L.F. Cooper, A.B. Wrathmell, J. Roper, A.J. Evenden and M.L. Gilpin. Host responses to *Renibacterium salmoninarum* and specific components of the pathogen reveal the mechanisms of immune suppression and activation. *Immunology* 106: 273–283, 2002.
48. Grinwis, G.C.M., A.D. Vethaak, P.W. Wester and J.G. Vos. Toxicology of environmental chemicals in the flounder (*Platichthy flesus*) with emphasis on the immune system: field semi-field (mesocosm) and laboratory studies. *Toxicol. Lett.* 113: 289–301, 2000.
49. Hansen, J.D. and S. La Patra. Induction of the rainbow trout MHC class I pathway during acute IHNV infection. *Immunogenetics* 54: 654–661, 2002.

50. Harris, J. and D.J. Bird. Modulation of the fish immune system by hormones. *Vet. Immunol. Immunopathol.* 77: 163–176, 2000.
51. Henrich, T., M. Ramialison, R. Quirin, B. Wittbrodt, M. Furutani-Seiki, J. Wittbrodt and H. Kondoh. MEPD: a medaka gene expression pattern database. *Nucleic Acids Res.* 31: 72–74, 2003.
52. Hirono, I., B.H. Nam, T. Kurobe and T. Aoki. Molecular cloning, characterization, and expression of TNF cDNA and gene from Japanese flounder *Paralychthys olivaceus*. *J. Immunol.* 165: 4423–4427, 2000.
53. Holladay, S.D., S.A. Smith, E.G. Besteman, A.S.M.I. Deyab, R.M. Gogal, T. Hrubec, J.L. Robertson and S.A. Ahmed. Benzo[*a*]pyrene-induced hypocellularity of the pronephros in tilapia (*Oreochromis niloticus*) is accompanied by alterations in stromal and parenchymal cells and by enhanced immune cell apoptosis. *Vet. Immunol. Immunopathol.* 64: 69–82, 1998.
54. Holland, J.W., T.G. Pottinger and C.J. Secombes. Recombinant interleukin-1 beta activates the hypothalamic-pituitary-interrenal axis in rainbow trout, *Oncorhynchus mykiss*. *J. Endocrinol.* 175: 261–267, 2002.
55. Hutchinson, T.H., M.D.R. Field and M.J. Manning. Evaluation of non-specific immune functions in dab, *Limanda limanda* L., following short-term exposure to sediments contaminated with polyaromatic hydrocarbons and/or polychlorinated biphenyls. *Mar. Environ. Res.* 55: 193–202, 2003.
56. Jaso-Friedmann, L., J.H. Leary, III and D.L. Evans. The non-specific cytotoxic cell receptor (NCCRP-1): molecular organization and signaling properties. *Dev. Comp. Immunol.* 25: 701–712, 2001.
57. Jiang, Y. and G. Moller. In vitro effects of $HgCl_2$ on murine lymphocytes. I. Preferable activation of $CD4^+$ T cells in a responder strain. *J. Immunol.* 154: 3138–3146, 1995.
58. Justic, D., N.N. Rabalais, R.E. Turner and W.J. Wiseman, Jr. Seasonal coupling between riverborne nutrients, net productivity and hypoxia. *Mar. Pollut. Bull.* 26: 184–189, 1993.
59. Kaattari, S.L. and C.A. Ottinger. Seasonality of trout leukocytic sensitivity to toxicity: implications for immunotoxic biomonitoring. *Mar. Environ. Res.* 50: 465 2000.
60. Kaattari, S.L., H.L. Zhang, I.M. Khor and D.A. Shapiro. Affinity maturation in trout: clonal dominance of high affinity antibodies late in the immune response. *Dev. Comp. Immunol.* 26: 191–200, 2002.
61. Karrow, N.A., H.J. Boermans, D.G. Dixon, A. Hontella, K.R. Solomon, J.J. Whyte and N.C. Bols. Characterizing the immunotoxicity of creosote to rainbow trout (*Oncorhynchus mykiss*): a microcosm study. *Aquatic Toxicol.* 45: 223–239, 1999.
62. Khangarot, B.S., R.S. Rathore and D.M. Tripathi. Effects of chromium on humoral and cell-mediated immune responses and host resistance to disease in a freshwater catfish, *Saccobranchus fossilis* (Bloch). *Ecotoxicol. Environ. Saf.* 43: 11–20, 1999.
63. Köllner, B., B. Wasserrab, G. Kotterba and U. Fischer. Evaluation of immune function of rainbow trout (*Oncorhynchus mykiss*) – how can environmental influences be detected? *Toxicol. Lett.* 1–2: 83–95, 2002.
64. Laing, K.J., T. Wang, J. Zou, J. Holland, S. Hong, N. Bols, I. Hirono, T. Aoki and C.J. Secombes. Cloning and expression analysis of rainbow trout *Oncorhynchus mykiss* tumour necrosis factor-alpha. *Eur. J. Biochem.* 268: 1315–1322, 2001.
65. Lee, K., X. Shen and R. Konig. Effects of cadmium and vanadium ions on antigen-induced signaling in CD4(+) T cells. *Toxicology* 169: 53–65, 2001.
66. Lemaire-Gony, S., P. Lemair and A.L. Pulsford. Effects of cadmium and benzo[*a*]pyrene on the immune system, gill ATPase and EROD activity of European sea bass *Dicentrarchus labrax*. *Aquat. Toxicol.* 31: 297–313, 1995.
67. Le Morvan, C., D. Troutaud and P. Deschaux. Differential effects of temperature on specific and nonspecific immune defences in fish. *J. Exp. Biol.* 201: 165–168, 1998.
68. Leong, J.-A.C., G.D. Trobridge, C.H.Y. Kim, M. Johnson and B. Simon. Interferon-inducible Mx proteins in fish. *Immunol. Rev.* 166: 349–363, 1998.
69. Litman, G.W., M.K. Anderson and J.P. Rast. Evolution of antigen binding receptors. *Annu. Rev. Immunol.* 17: 109–147, 1999.
70. Low, K.W. and Y.M. Sin. Effects of mercuric chloride on chemiluminescent response of phagocytes and tissue lysozyme activity on tilapia, *Oreochromis aureus*. *Bull. Environ. Contam. Toxicol.* 54: 302–308, 1995.
71. Low, K.W. and Y.M. Sin. In vitro effect of mercuric chloride and sodium selenite on chemiluminescent response of pronephros cells isolated from tilapia, *Oreochromis aureus*. *Bull. Environ. Contam. Toxicol.* 55: 909–915, 1995.
72. Low, K.W. and Y.M. Sin. In vivo and in vitro effects of mercuric chloride and sodium selenite on some non-specific immune responses of blue gourami, *Trichogaster trichopterus* (Pallus). *Fish Shellfish Immunol.* 6: 351–362, 1996.
73. Low, K.W. and Y.M. Sin. Effects of mercuric chloride and sodium selenite on some immune responses of blue gourami, *Trichogaster trichopterus* (Pallus). *Sci. Total Environ.* 214: 153–164, 1998.

74. MacDougal, K.C., M.D. Johnson and K.G. Burnett. Exposure to mercury alters early activation events in fish leukocytes. *Environ. Health Perspect.* 104: 1102–1106, 1996.
75. Miller, G.G., L.I. Sweet, J.V. Adams, G.M. Omann, D.R. Passino-Reader and P.G. Meier. In vitro toxicity and interactions of environmental contaminants (Arochlor 1254 and mercury and immunomodulatory agents (lipopolysaccharide and cortisol) on thymocytes from lake trout (*Salvelinus namaycush*). *Fish Shellfish Immunol.* 13: 11–26, 2002.
76. Miller, N.W., R.C. Sizemore and L.W. Clem. Phylogeny of lymphocyte heterogeneity: the cellular requirements for in vitro antibody responses of channel catfish leukocytes. *J. Immunol.* 134: 2884–2888, 1985.
77. Miller, N., M. Wilson, E. Bengten, T. Stuge, G. Warr and L.W. Clem. Functional and molecular characterization of teleost leukocytes. *Immunol. Rev.* 166: 187–197, 1998.
78. Misra, R.R., J.F. Hochadel, G.T. Smith, J.C. Cook, M.P. Waalkes and D.A. Wink. Evidence that nitric oxide enhances cadmium toxicity by displacing the metal from metallothionein. *Chem. Res. Toxicol.* 9: 326–332, 1996.
79. Nakanishi, T., U. Fischer, J.M. Dijkstra, S. Hasegawa, T. Somamoto, N. Okamoto and M. Ototake. Cytotoxic T cell function in fish. *Dev. Comp. Immunol.* 26: 131–139, 2002.
80. Neumann, N.F., J.L. Stafford, D. Barreda, A.J. Ainsworth and M. Belosevic. Antimicrobial mechanisms of fish phagocytes and their role in host defense. *Dev. Comp. Immunol.* 25: 807–825, 2001.
81. Nonaka, M.. Evolution of the complement system. *Curr. Opin. Immunol.* 13: 69–73, 2001.
82. O'Farrell, C., N. Vaghefi, M. Cantonnet, B. Buteau, P. Boudinot and A. Benmansour. Survey of transcript expression in rainbow trout leukocytes reveals a major contribution of interferon-responsive genes in the early response to a rhabdovirus infection. *J. Virol.* 76: 8040–8049, 2002.
83. O'Halloran, K., J.T. Ahokas and P.F.A. Wright. In vitro responses of fish immune cells to three classes of pesticides. In: *Modulators of Immune Responses: The Evolutionary Trail*, Breckenridge Series 2, edited by J.S. Stolen, T.C. Fletcher, C.J. Bayne, C.J. Secombes, J.T. Zelikoff, L.E. Twerdok and D.P. Anderson, Fair Haven, NJ, SOS Publications, pp. 535–537, 1996.
84. Ottinger, C.A. and S.L. Kaattari. Sensitivity of rainbow trout leukocytes to aflatoxin B1. *Fish Shellfish Immunol.* 8: 515–530, 1998.
85. Oumouna, M., L. Jaso-Friedmann and D.L. Evans. Activation of nonspecific cytotoxic cells (NCC) with synthetic oligodeoxynucleotides and bacterial genomic DNA: binding, specificity and identification of unique immunostimulatory motifs. *Dev. Comp. Immunol.* 26: 257–269, 2002.
86. Park, H., H. Zhou, E. Bengten, M. Wilson, V.G. Chinchar, L.W. Clem and N.W. Miller. Activation of channel catfish (*Ictalurus punctatus*) T cells involves NFAT-like transcription factors. *Dev. Comp. Immunol.* 26: 775–784, 2002.
87. Paul, I., C. Mandal and C. Mandal. Effect of environmental pollutants on the C-reactive protein of a freshwater major carp, *Catla catla*. *Dev. Comp. Immunol.* 22: 519–532, 1998.
88. Pennie, W.D. Custom cDNA microarrays: technologies and applications. *Toxicology* 181–182: 551–554, 2002.
89. Powell, D.B., R.C. Palm, Jr., A. Skillman and K. Godtfredsen. Immunocompetence of juvenile Chinook salmon against *Listonella anguillarum* following dietary exposure to Aroclor 1254. *Environ. Toxicol. Chem.* 22: 285–295, 2003.
90. Press, C.McL. and Ø Evensen. The morphology of the immune system in teleost fishes. *Fish Shellfish Immunol.* 9: 309–318, 1999.
91. Quiniou, S.M.-A., S. Bigler, L.W. Clem and J.E. Bly. Effects of water temperature on mucous cell distribution in channel catfish epidermis: a factor in winter saprolegniasis. *Fish Shellfish Immunol.* 8: 1–11, 1998.
92. Ratcliffe, P.J., J.E. O'Rourke, P.H. Maxwell and C.W. Pugh. Oxygen sensing, hypoxia-inducible factor-1 and the regulation of mammalian gene expression. *J. Exp. Biol.* 201: 1153–1162, 1998.
93. Rice, C.D. Fish immunotoxicology: understanding mechanisms of action, In: *Target Organ Toxicity in Marine and Freshwater Teleosts*, Vol. 2, edited by D. Schlenk and W.H. Benson, London, Taylor & Francis Publishers, pp. 96–138, 2001.
94. Rice, C.D. and B.A. Weeks. Influcence of tributyltin on in vitro activation of oyster toadfish macrophages. *J. Aquat. Anim. Health* 1: 62–68, 1989.
95. Rice, C.D. and B.A. Weeks. The influence of in vivo exposure to tributyltin on reactive oxygen formation in oyster toadfish macrophages. *Arch. Environ. Contam. Toxicol.* 19: 854–857, 1990.
96. Rice, C.D. and B.A. Weeks. Tributyltin stimulates reactive oxygen formation in toadfish macrophages. *Dev. Comp. Immunol.* 15: 431–436, 1991.
97. Rice, C.D. and M.R. Arkoosh. Immunological indicators of environmental stress and disease susceptibility in fishes. In: *Biological Indicators of Stress in Aquatic Ecosystems, Symposium 9*, edited by S.M. Adams, Bethesda, MD, American Fisheries Society Publications, pp. 187–220, 2002.

98. Rice, C.D., M.M. Banes and T.C. Ardelt. Immunotoxicity in channel catfish, *Ictalurus punctatus*, following acute exposure to tributyltin. *Arch. Environ. Contam. Toxicol.* 28: 464–470, 1995.
99. Robohm, R.A. Paradoxical effects of cadmium exposure on antibacterial antibody responses in two fish species: inhibition in cunners (*Tautogolabrus adspersus*) and enhancement in striped bass (*Morone saxatilis*). *Vet. Immunol. Immunopathol.* 12: 251–262, 1986.
100. Rose, W.L., B.L. French, W.L. Reichert and M. Faisal. DNA adducts in hematopoietic tissues and blood of the mummichog, *Fundulus heteroclitus*, from a creosote-contaminated site in the Elizabeth River, Virginia. *Mar. Environ. Res.* 50: 581–589, 2000.
101. Roszell, L.E. and R.S. Anderson. Inhibition of phagocytosis and superoxide production by pentachlorophenol in two leukocyte subpopulations from *Fundulus heteroclitus*. *Mar. Environ. Res.* 38: 195–206, 1994.
102. Roszell, L.E. and R.S. Anderson. Effect of in vivo pentachlorophenol exposure on *Fundulus heteroclitus*: modulation of bactericidal activity. *Dis. Aquat. Org.* 26: 205–211, 1996.
103. Roszell, L.E. and C.D. Rice. Innate cellular immune function of anterior kidney leucocytes in the gulf killifish, *Fundulus grandis*. *Fish Shellfish Immunol.* 8: 129–142, 1998.
104. Rougier, F., D. Troutaud, A. Ndoye and P. Deschaux. Non-specific immune response of Zebrafish, *Brachydanio rerio* (Hamilton-Buchanan) following copper and zinc exposure. *Fish Shellfish Immunol.* 4: 115–127, 1994.
105. Saeij, J.P., R.J. Stet, A. Groenveld, L.B. Verburg-van Kemenade, W.B. van Muiswinkel and G.F. Wiegertjes. Molecular and functional characterization of a fish inducible-type nitric oxide synthase. *Immunogenetics* 51: 339–346, 2000.
106. Saeij, J.P., G.F. Wiegertjes and R.J.M. Stet. Identification and characterization of a fish natural resistance-associated macrophage protein (Nramp) cDNA. *Immunogenetics* 50: 60–66, 1999.
107. Sanchez-Dardon, J., I. Voccia, A. Hontela, P. Anderson, P. Brousseau, B. Blakely, H. Boermans and M. Fournier. Immunotoxicity of cadmium, zinc and mercury after in vivo exposure, alone or in mixture in rainbow trout (*Oncorhynchus mykiss*). *Dev. Comp. Immunol.* 21: 133 1997.
108. Secombes, C.J.. Cytokines and innate immunity of fish. *Dev. Comp. Immunol.* 25: 713–724, 2001.
109. Secombes, C.J., T.C. Fletcher, J.A. O'Flynn, M.J. Costello, R. Stagg and D.F. Houlihan. Immunocompetence as a measure of the biological effects of sewage sludge pollution in fish. *Comp. Biochem. Physiol.* 100C: 133–136, 1991.
110. Secombes, C.J., T.C. Fletcher, A. White, M.J. Costello, R. Stagg and D.F. Houlihan. Effects of sewage sludge on immune responses in the dab, *Limanda limanda*. *Aquat. Toxicol.* 23: 217–230, 1992.
111. Seeley, K.R. and B.A. Weeks-Perkins. Altered phagocytic activity of macrophages in oyster toadfish from a highly polluted subestuary. *J. Aquat. Anim. Health* 3: 224–227, 1991.
112. Shariff, M., P.A. Jayawardena, F.M. Yusoff and R. Subasinghe. Immunological parameters of Javanese carp *Puntius gonionotus* (Bleeker) exposed to copper and challenged with *Aeromonas hydrophila*. *Fish Shellfish Immunol.* 11: 281–291, 2001.
113. Shen, L., T.B. Stuge, H. Zhou, M. Khayat, K.S. Barker, S.M. Quiniou, M. Wilson, E. Bengten, V.G. Chinchar, L.W. Clem and N.W. Miller. Channel catfish cytotoxic cells: a mini-review. *Dev. Comp. Immunol.* 26: 141–149, 2002.
114. Shen, X., K. Lee and R. König. Effects of heavy metal ions on resting and antigen-activated $CD4^+$ T cells. *Toxicology* 169: 67–80, 2001.
115. Shenker, B.J., T.L. Guo and I.M. Shapiro. Low-level methylmercury exposure causes human T-cells to undergo apoptosis: evidence of mitochondrial dysfunction. *Environ. Res.* 77: 149–159, 1998.
116. Shenker, B.J., L. Pankoski, A. Zekavat and I.M. Shapiro. Mercury-induced apoptosis in human lymphocytes: caspase activation is linked to redox status. *Antioxid. Redox. Signal.* 4: 379–389, 2002.
117. Siwicki, A.K., M. Cossarini-Dunier, M. Studnicka and A. Demael. In vivo effect of the organophosphorus insecticide trichlorphon on immune response of carp (*Cyprinus carpio*). II. Effect of high doses of trichlorophon on nonspecific immune response. *Ecotoxicol. Environ. Saf.* 29: 99–105, 1990.
118. Spitsbergen, J.M., K.A. Schat, J.M. Kleeman and R.E. Peterson. Interactions of 2,3,7,8-tetrachlorodibenzo-*p*-dioxin (TCDD) with immune responses of rainbow trout. *Vet. Immunol. Immunopathol.* 12: 263–280, 1986.
119. Stuge, T.B., M.R. Wilson, H. Zhou, K.S. Barker, E. Bengten, G. Chinchar, N.W. Miller and L.W. Clem. Development and analysis of various clonal alloantigen-dependent cytotoxic cell lines from channel catfish. *J. Immunol.* 15: 2971–2997, 2000.
120. Sunyer, J.O., I.K. Zarkadis, A. Sahu and J.D. Lambris. Multiple forms of complement C3 in trout that differ in binding to complement activators. *Proc. Natl Acad. Sci. USA* 93: 8546–8551, 1996.
121. Sweet, L.I. and J.T. Zelikoff. Toxicology and immunotoxicology of mercury: a comparative review in fish and humans. *J. Toxicol. Environ. Health B Crit. Rev.* 4: 161–205, 2001.

122. Sweet, L.I., D.R. Passino-Reader, P.G. Meier and G.M. Omann. Xenobiotic-induced apoptosis: significance and potential application as a general biomarker of response. *Biomarkers* 4: 237–253, 1999.
123. Tahir, A. and C.J. Secombes. The effects of diesel oil-based drilling mud extracts on immune responses of rainbow trout. *Arch. Environ. Contam. Toxicol.* 29: 27–32, 1995.
124. Thorgaard, G.H., G.S. Bailey, D. Williams, D.R. Buhler, S.L. Kaattari, S.S. Ristow, J.D. Hansen, J.R. Winton, J.L. Bartholomew, J.J. Nagler, P.J. Walsh, M.N. Vijayan, R.H. Devlin, R.W. Hardy, K.E. Overturf, W.P. Young, B.D. Robison, C. Rexroad and Y. Palti. Status and opportunities for genomics research with rainbow trout. *Comp. Biochem. Physiol.* 133B: 609–646, 2002.
125. van Ginkel, F.W., N.W. Miller, M.A. Cuchens and L.W. Clem. Activation of channel catfish B cells by membrane immunoglobulin cross-linking. *Dev. Comp. Immunol.* 18: 97–107, 1994.
126. van Veld, P.A., R.D. Vetter, R.F. Lee and J.S. Patton. Dietary fat inhibits the intestinal metabolism of the carcinogen benzo[*a*]pyrene in fish. *J. Lipid Res.* 28: 810–817, 1987.
127. Voccia, I., K. Krzystyniak, M. Dunier, D. Flipo and M. Fournier. In vitro mercury-related cytotoxicity and functional impairment of the immune cells of rainbow trout (*Oncorhynchus mykiss*). *Aquat. Toxicol.* 29: 37–48, 1994.
128. Walczak, B.Z., B.R. Blunt and P.V. Hodson. Phagocytic function of monocytes and haematological changes in rainbow trout injected intraperitoneally with benzo(*a*)-pyrene (B*a*P) and benzo(*a*)anthracene (B*a*A). *J. Fish Biol.* 31: 251–253, 1987.
129. Whalen, M.M., T.B. Williams, S.A. Green and B.G. Loganathan. Interleukins 2 and 12 produce recovery of cytotoxic function in tributyltin-exposed human natural killer cells. *Environ. Res.* 88: 199–209, 2002.
130. Winkelhake, J.L., M.J. Vodicnik and J.L. Taylor. Induction in rainbow trout of an acute phase (C-reactive) protein by chemicals of environmental concern. *Comp. Biochem. Physiol.* 74C: 55–58, 1983.
131. Xiang, Y. and C. Rice. Expression of fish iNOS is increased by pro-inflammatory signals and xenobiotics. *Mar. Environ. Res.* 50: 466–467, 2000.
132. Yada, T. and T. Nakanishi. Interaction between endocrine and immune systems in fish. *Int. Rev. Cytol.* 220: 35–92, 2002.
133. Zarkadis, I.K., D. Mastellos and J.D. Lambris. Phylogeneic aspects of the complement system. *Dev. Comp. Immunol.* 25: 713–724, 2001.
134. Zdolsek, J.M., O. Soder and P. Hultman. Mercury induces in vivo and in vitro secretion of interleukin-1 in mice. *Immunopharmacology* 28: 201–208, 1994.
135. Zelikoff, J.T.. Fish immunotoxicology. In: *Immunotoxicology and Immunopharmacology*, edited by J.H. Dean, M.I. Luster, A.E. Munson and I. Kimber, New York, Raven Press, pp. 72–95, 1994.
136. Zelikoff, J.T., D. Bowser, K.S. Squibb and K. Frenkel. Immunotoxicity of low level cadmium exposure in fish: an alternative animal model for immunotoxicological studies. *J. Toxicol. Environ. Health* 45: 235–248, 1995.
137. Zelikoff, J.T., A. Raymond, E. Carlson, Y. Li, J.R. Beaman and M. Anderson. Biomarkers of immunotoxicity in fish: from the lab to the ocean. *Toxicol. Lett.* 112–113: 325–331, 2000.
138. Zelikoff, J.T., E. Carlson, Y. Li, A. Raymond, J. Duffy, J.R. Beaman and M. Anderson. Immunotoxicity biomarkers in fish: development, validation and application for field studies and risk assessment. *Hum. Ecol. Risk Assess.* 8: 253–263, 2002.
139. Zou, J., N.F. Neumann, J.W. Holland, M. Belosevic, C. Cunningham, C.J. Secombes and A.K. Rowley. Fish macrophages express a cyclo-oxygenase-2 homologue after activation. *Biochem. J.* 340: 153–159, 1999.

Biochemistry and Molecular Biology of Fishes, vol. 6
T. P. Mommsen and T. W. Moon (Editors)

CHAPTER 9

Fish models of carcinogenesis

GARY K. OSTRANDER* AND JEANETTE M. ROTCHELL**

**Departments of Biology and Comparative Medicine, Johns Hopkins University, 3400 North Charles Street, Baltimore, MD 21218, USA, and **Centre for Environmental Research, School of Chemistry, Physics and Environmental Sciences, University of Sussex, Falmer, Brighton BN1 9QJ, UK*

I. Introduction

Experimental and environmental oncology in fish has been under study for more than 40 years, and a number of 'favorites' have made it to the forefront: rainbow trout (*Oncorhynchus mykiss*), *Xiphophorus* spp. and medaka (*Oryzias latipes)* in the realm of tank-held experimental models. Likewise, there are naturally occurring epizootics of cancer that have proven to be worthy models of study including the English sole (*Parophrys vetulus*), mummichog (*Fundulus heteroclitus*) and shad (*Dorosoma *cepedianum*) to name only a few. There are a number of reasons why piscine models have received so much attention. In some instances, their value may be related to application in human cancer studies where, for instance, rodent models have failed to provide an alternate model (such as the case with retinoblastoma eye tumors) or where reduced costs in carrying out carcinogenic/toxicological tests are desired. Alternatively, their value may be related to understand the mechanisms that underpin the carcinogenesis process in feral fish populations, or relating long-term effects of cancer development on transgenerational levels and, ultimately, fish population decline. In either case, fish offer much to understand the carcinogenesis process and recently, offer increasing potential in medical application (with the generation of transgenic and knockout models), and in answering the 'big questions' of ecotoxicology such as the relevance of molecular level damage to population level repercussions.

Herein, we examine a variety of models of carcinogenesis. Our review is not intended to be exhaustive. Instead, we focus on models that have received considerable attention and in doing so have contributed much to our understanding. Likewise, we also highlight fish model systems that offer the potential to provide unique insights upon further study.

II. Laboratory models of carcinogenesis

1. Response of fishes to carcinogens

Laboratory fish models of carcinogenesis complement and attempt to provide an explanation for observations made in the field. Historically, the fish species selected for such studies differ: in the lab, small domestic freshwater fish are opted for; in the field, estuarine species living in close association with contaminated sediments are selected. Extrapolating across species has its limits, though it has nevertheless been possible to determine much of the universal mechanisms relating to the carcinogenesis process in fish species.

It is well established that environmental contaminants, such as hydrocarbons, can 'initiate' a cancer, but additional events, generated by subsequent exposure to another 'promoting' agent, are also required in order for a cancer to develop. Initiation, in the case of polycyclic aromatic hydrocarbons (PAHs), involves the formation of a covalent adduct of an epoxide metabolite of a specific hydrocarbon to a guanine nucleotide base in DNA, ultimately resulting in mutations. The changes brought about by the initiating agent require cell proliferation, in which the promoting agent triggers transformation of a single cancer cell into a multicellular tumor. Further, molecular level changes are required to produce a cancer; in the final stages of carcinogenesis 'progression' occurs involving changes in growth regulatory mechanisms. Fish have thus been employed to determine a variety of initiating and (lesser examples of) promoting agents that are present in the aquatic environment.

Among the chemical carcinogens, dimethylbenzanthracene (DMBA) is one that is widely used in experimental studies. A large body of literature gathered over several decades exists from studies of DMBA carcinogenicity in fish (see review 13).

Benzo[*a*]pyrene (BaP) has been tested for carcinogenicity in a number of fish species including trout, medaka and guppy (*Poecilia reticulata*) (see review 13). For instance, Japanese medaka and guppy have been exposed to water-borne BaP in several doses. Following high-dose exposure (of 270 ppb and not based on environmental concentrations) hepatocellular foci, adenomas and carcinomas were observed in both species[72]. In medaka, the lesions appeared to develop sequentially with the appearance of foci followed by adenomas then hepatocellular carcinomas.

The carcinogenic response of medaka, guppies and trout to the PAHs, BaP and DMBA indicates that these, and probably other fish species, are capable of incorporating PAHs and metabolizing them to intermediates that initiate hepatic tumor formation. Studies have also shown that a complex mixture of PAHs and other compounds can induce hepatocarcinomas in a rainbow trout carcinogenesis assay using contaminant-loaded sediment extracts[116].

Laboratory studies have begun to investigate the causal relationship between a number of carcinogens, especially the environmental contaminants PAHs, and neoplasia in a number of fish species. Research has involved tank-held fish and exposures to defined initiating and promoting agents, such as BaP and DMBA and chemical mixes of various PAHs, in order to investigate the role and mechanism of PAH-induced cancer in fish. Such studies aim to address progression in terms of histopathology, but also describe the underlying molecular etiology and, in particular, the specific genetic changes occurring during tumor development.

2. *Medaka: An introduction*

There are currently four different common models of development in human cancer research: (1) animal experiments of the carcinogenesis steps including initiation, promotion and progression; (2) cell biology investigations of cell immortalization and anchorage independence; (3) molecular biology investigations that pinpoint DNA changes in the progressive steps of precancer, cancer and metastasis; and (4) clinical data. Medaka have provided the opportunity for studying two such models combined: the carcinogenesis process and their underpinning molecular etiology. Accordingly, medaka, along with trout, have received a lot of recent attention as models for various forms of cancer[30,75,112], providing a surrogate role for wild fish populations in assessing suspected carcinogens in an aquaria setting, a surrogate role for humans in assessing carcinogenic activity of compounds, as well as a model with which to study the carcinogenesis process.

2.1. *Medaka as a surrogate for wild fish populations*

Feral fish populations that display a high incidence of various tumors are often not suitable for use in controlled laboratory exposure regimes. Such fish are often large and require specialized husbandry. It is also a challenge to obtain control fish from 'clean' reference sites for use in parallel exposure experiments. Additionally, the carcinogenesis process in such fish is usually a long, gradual process, increasing the time required and, thus, the cost involved to conduct exposure experiments. In contrast, medaka are small enough to be exposed in aquaria in large numbers and also exhibit a relatively short time to tumor production, weeks or months as opposed to months or years, as is often the case with feral fish. There is also a wealth of available information regarding medaka husbandry. Recent advances have provided a highly novel 'see-through' medaka, in which most of the pigments have been genetically removed from the body, potentially making it an extremely useful tool in non-invasive studies of morphological and molecular events that occur in internal organs following contaminant exposure[183].

Exposure experiments have shown that medaka are sensitive to a variety of carcinogen-initiating agents. A variety of promoting agents, environmental contaminants that act as mitogens and encourage cellular proliferation following initiation events, have also been investigated using medaka. Such compounds include endogenous and environmental estrogens, specifically 17β-estradiol, β-hexachlorocyclohexane (both in liver tumors)[41], nonylphenol and quercetin[186] as well as other classes of common aquatic environment contaminants[59]. Following exposure, be it one step initiation or two step initiation and promotion, medaka develop assorted neoplasms, mainly in the liver[145,167], though also in other organs, including kidney[73], eye[71], gall bladder[76], muscle[172], thyroid[65], ovary and skin[172]. Liver tumors have been found in medaka following exposure to a broad variety of chemical carcinogens. These include diethylnitrosamine (DEN)[88], methylazoxymethanol acetate (MAMAc)[7], PAHs[72] (cf. Table 1) and preneoplastic lesions using polychlorinated biphenyls (PCBs) and DDT[70]. The sensitivity of medaka to such a broad range of direct and indirect carcinogens thus makes it an ideal model species for determining whether suspect environmental contaminants are indeed carcinogenic both quickly (within months instead of years, as with rodents and common feral flatfish species) and cheaply. Practically, all published scientific data can be found on two website resources now available for medaka – Medakafish Homepage: http://biol1.bio.nagoya-u.ac.jp:8000/ and Medaka ToxiNet: http://medaka.fish.agr.kyushu-u.ac.jp/.

Natural variation related to life stage and sex have also been investigated in medaka. The growth and survival of medaka at several life stages have also been examined following tolerated doses of carcinogens[109]. Results indicate that growth is a sensitive indicator of carcinogenic effect in medaka 13 days or fewer post-hatch. Others report a gender-specific response to carcinogen exposure, indicating that tumor incidence and time to tumor development are positively affected in females following inducing agent exposure[167]. Such an effect is probably due to the promotion role of endogenous estrogens in the carcinogenesis process. Although some have reported in the literature that medaka display a low level of spontaneous neoplasia, recent evidence contradicts this. The occurrence of spontaneous neoplasia in medaka includes examples of thymic lymphoma, swim bladder and ovarian neoplasia[15,55,64,112]. Such frequently occurring spontaneous neoplasm in medaka, as well as natural variation factors, need to be considered when assessing the results of laboratory carcinogenesis investigations.

Cumulatively, laboratory studies involving controlled exposures with tank-held medaka demonstrate that environmental contaminants, such as PAHs, present in the environment have carcinogenic effects on small fish species, following brief exposures to low concentrations and support the argument that such contaminants

TABLE 1

Tumors found in Japanese medaka (*Oryzias latipes*) tissues following exposure to chemical carcinogens

Initiating agent	Dose	Organ/pathology	Reference
DEN	Larvae, 350/500 ppm, 48 h; Adults, 50 ppm, 5 week	Hepatocellular carcinoma, biliary tumors	130
DEN + heavy metals + PAHs	10 mg/l DEN + contaminated groundwater	Liver neoplasm + ovary, skeletal muscle, skin, swim bladder, thymus and thyroid neoplasms	172
DEN			41
DEN	Larvae, 10–100 ppm, 48 h	Hepatocellular carcinoma	27
DEN			167
DEN			134
DEN			171
DEN	Adult, 50 ppm, 5 week	Hepatocellular carcinoma	83
DEN			25
DEN			115
DEN			29
MNNG	Larvae, 0.5–1 mg/l	Branchioblastoma, thyroid adenoma, adenocarcinoma, fibrosarcoma	37
MNNG	Fry, 30 mg/l, 1 h	Blood vascular neoplasm, various sarcomas	31
MAMAc	Fry, 0–100 mg/l, 2 h	Intraocular neoplasms	71
PAH (DMBA + BaP)	30–250 μg/l, 6 h	Hepatic neoplasms (and extra hepatic DMBA only)	74
TCE	All life stages, 0–40 mg/l	Survival and growth rates	109

PAHs, polycyclic aromatic hydrocarbons; DMBA, dimethylbenzanthracene; BaP, benzo[*a*]pyrene; DEN, diethylnitrosamine; MAMAc, methylazoxymethanol acetate; MNNG, *N*-methyl-*N*′-nitro-*N*-nitrosoguanidine; TCE, trichloroethylene.

are at least partially responsible for the occurrence of certain cancers in wild fish populations. Determining the precise steps in carcinogenesis development in feral fish populations is an impossible task given the heterogeneity of types and levels of common environmental contaminants present, yet the use of medaka in a controlled laboratory exposure regime has made considerable progress in elucidating such steps. Initiation agents (nitrosamines and hydrocarbons, for instance) and promoting agents (such as environmental and endogenous estrogens) are now well characterized and others will likely be revealed following controlled exposure regimes using tank-held medaka as surrogates for feral fish.

2.2. *Medaka as a model for determining the molecular mechanisms of carcinogenesis*

The underlying molecular steps of carcinogenesis in terms of mutations in DNA caused by initiator exposure, in particular damage to human ‘cancer genes’ (oncogenes and tumor suppressor genes), have been the focus of intensive research in the last decade. One oncogene that has received considerable attention is the *ras* gene. *Ras* genes encode proteins that play a central role in cell growth signaling cascades, promoting cell growth and differentiation (Fig. 1). The fish *ras* genes characterized to date have a high degree of nucleotide sequence and deduced amino acid similarity with the human *ras* gene counterparts. A large proportion and wide variety of human tumors possess mutant forms of *ras*. In such cases, the localization of *ras* mutations

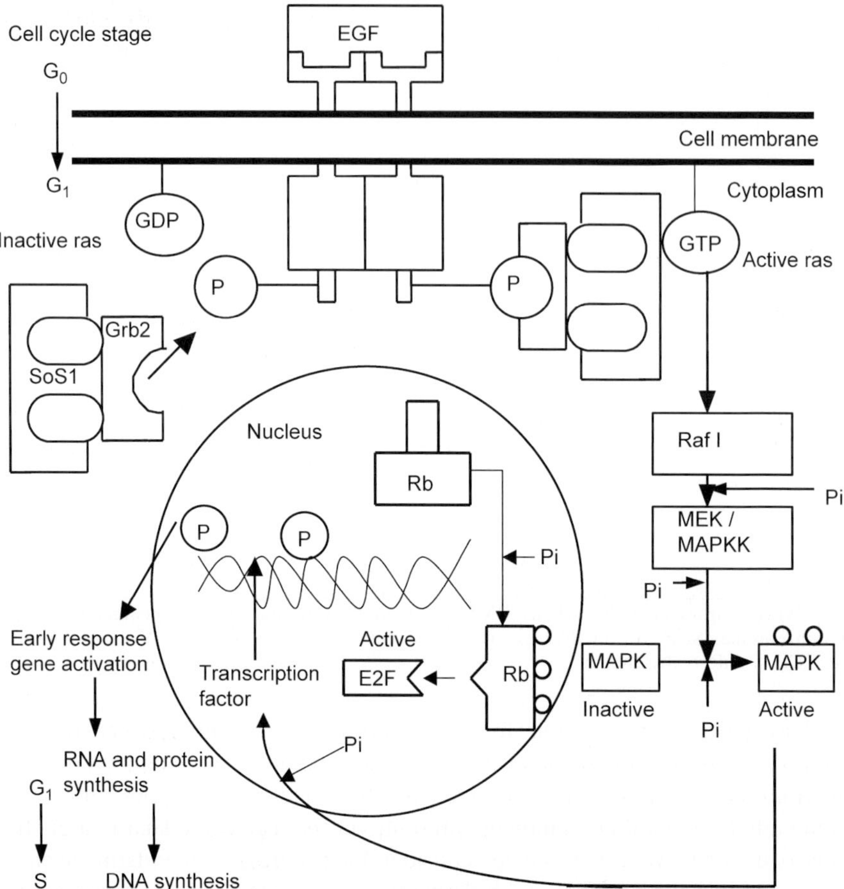

Fig. 1. An overview of the roles of ras and Rb in cell signaling. Ras: an activated receptor tyrosine kinase (EGF in mammals, Xmrk in fish) binds to an adaptor protein (Grb2) that links to the nucleotide-releasing factor SoS, leading to ras activation and kinase cascade to the nucleus. At the nucleus activation of early response genes, such as *jun*, *myc* and *fos*, result in mRNA synthesis and progression of the cell growth cycle to S phase (DNA synthesis). Rb: phosphorylation of Rb releases the transcription factor, in this case E_2F. In its unphosphorylated state Rb inhibits the activity of transcription factors and thus prevents progression of the cell cycle. Abbreviations: SoS1, son of sevenless/nucleotide exchange factor; Grb2, nucleotide exchange factor; EGF, epidermal growth factor; MAPK, mitogen-activated protein kinase; G_0, 'resting' phase of cell cycle; G_1, the gap period of the cell cycle after mitosis and before DNA replication; P, phosphates.

has been restricted to exons I and II of the gene, and specifically to codons 12, 13 and 61. Experimental exposure of fish to a range of genotoxic compounds has similarly led to the production of a *ras* mutational profile for selected species (see review 146).

Two closely related *ras* genes, assigned K-*ras*-1 and K-*ras*-2, have been reported in medaka (Genbank identifiers: AF030545, AF032713). In addition to the characterization of the medaka K-*ras*-1 and K-*ras*-2 genes, others have identified and sequenced three medaka partial *ras* gene sequences. The three partial medaka *ras* sequences were found to diverge predominantly at the third base position of each codon and, thus, they represent polymorphic sequences of a medaka K-*ras* gene. This is the first reported incidence of fish *ras* polymorphic variation[2]. Subsequent studies have investigated point mutations in exons I and II of the medaka K-*ras* genes in tissue isolated from DEN-induced liver tumors[171]. A significant proportion (3/11) of the DEN-initiated tumor DNAs showed evidence of activating point mutations in the medaka K-*ras* gene. Among these, the predominant lesion was a GGT to GAT transversion in codon 12[171], a mutation common in human cancers. The remaining mutant *ras* genotype was a codon 16 AAA to ACA transition, representing the first reported incidence of a lesion within this codon[171] and not present in human cancers (Table 2).

The role of *ras* genes in tumor formation in feral fish has been investigated using several species collected from areas of high hydrocarbon contamination. Tomcod (*Microgadus tomcod*), winter flounder (*Pleuronectes americanus*) and dragonet (*Callionymus lyra*) liver samples display evidence of *ras* gene mutations, though for the latter species the codon affected is not characteristic of *ras* gene mutational

TABLE 2

Summary of mutational alterations in fish *ras* genes

	Codon									
	Normal sequence					Mutated sequence				
	11	12	13	16	61	11	12	13	16	61
Rainbow trout		GGA	GGT		CAG		AGA[a,b,c]	GTT[c]		CTG[b,c]
							AGA[d,f,g]	CGT[b,g]		
							GTA[b,c,g]			
							CGA[b]			
							TGA[g]			
							GAA[a]			
Medaka		GGT		AAA			GAT[d]		ACA[d]	
Pink salmon embryos		GGA	GGT		CAG		GAA[e]	AGT[e]		AAG[e]
Winter flounder		GGT	GGT				AGT			
Dragonets	GCT	GGT	GGT		CAG	ACT				

Chemically induced samples for rainbow trout, medaka, pink salmon embryos and field samples for winter flounder and dragonets. Superscript characters: a, *N*-methyl-*N'*-nitro-*N*-nitrosoguanidine (MNNG)/nitrosomethylurea (NMU)-induced[13,26]; b, dimethylbenzanthracene (DMBA)-induced[13,26,51,107]; c, aflatoxin (AFB_1)-induced[13,35]; d, *N*-nitrosodiethylamine (DEN)-induced[82,107,171]; e, Exxon Valdez oil-exposed[194]; f, dehydroepiandrosterone (DHEA)-induced[131]; g, dibenzo[*a,l*]pyrene (DBP)-induced[13]. McMahon *et al.*[195] and Vincent *et al.*[196] characterized the winter flounder and dragonet field-induced mutations.

profiles (Table 2). English sole (*Parophrys vetulus*) and European flounder (*Platichthys flesus*) liver tumor samples so far examined, however, do not display *ras* gene mutations. Thus, the pattern and incidence of *ras* gene mutations in environmentally induced tumors also appear to be species specific although medaka, acting as a surrogate, do seem to be among the susceptible species.

Where oncogenes promote inappropriate cell growth and differentiation, tumor suppressor genes provide a regulatory braking mechanism to inhibit cell growth. It is well established in mammalian models that both proto-oncogene activation and tumor suppressor gene loss are both required before uncontrolled cell growth and cancer to develop. To date, two key tumor suppressor genes have been isolated and characterized using medaka, *p53* and *Rb*[98,147].

P53 is a critical regulator of cell cycle progression in mammals and is conserved across species including trout and medaka[45,98]. The genomic organization of the *p53* gene is similar in all species. It consists of 11 exons, interrupted by 10 introns. Exons 2, 4, 5, 7 and 8 code for five clusters of amino acid sequences called domains I–V, which are highly conserved. Mutations in *p53* are the most common genetic defects found in human cancers[62]. The mutations are not distributed randomly, most cluster within the homology boxes and are restricted to four 'hot spots', between amino acids 129–146, 171–179, 234–260 and 270–287[84]. As with *ras* oncogene, some carcinogens appear to leave characteristic mutation 'fingerprints' in the *p53* gene. In contrast to human cancer studies, a preliminary investigation of carcinogen-induced tumors in medaka has revealed no *p53* mutations in hot spots within the conserved domains IV and V of the gene[98]. Results, following an investigation of ultraviolet light inducibility of medaka *p53*, also suggest that the p53 protein has a different function in lower vertebrates[38]. Consequently, interest is now mounting in the role of alternate tumor suppressor genes, particularly *Rb*, in the development of cancers in medaka and other fish.

The *retinoblastoma* gene (*Rb*) was the first tumor suppressor gene to be characterized in any species[57]. It plays a key role in transcription as it progressively phosphorylated during the cell cycle (Fig. 1). Among humans, the loss of function of the *Rb* gene occurs by mutation and deletion. Loss of function of the encoded gene product, pRb, results in a diverse set of cancers including osteosarcoma[57], lung cancer[66], breast cancer[166] and bladder cancer[87], as well as a high proportion of retinoblastomas[86]. Consequently, considerable research effort has been directed towards understanding the role of the *Rb* gene and its gene product in various types of cancer. A medaka *Rb* homolog has been isolated and DNA sequence analysis with other vertebrate *Rb* sequences demonstrates that it is highly conserved in regions of functional importance[147]. Structural alterations in the coding region of the *Rb* gene in methylene chloride-induced medaka liver tumors have also been reported[145] and include point mutations and a deletion (Table 3). Such results suggest that the molecular etiology of the medaka hepatocellular carcinoma model appears to be similar to that reported in humans. As such, the medaka appears to be a valid model for the study of *Rb*-implicated tumorigenesis, where it had previously failed with *p53*-implicated tumorigenesis.

TABLE 3

Summary of mutational alterations in the medaka *Rb* cDNA

Sample	Mutation	Position	Putative consequence
Normal + liver tumor			Polymorphic variation?
	C addition	Exon 9 (codon 313)	Frameshift
	C addition	Exon 9 (codon 316)	Frameshift
	C addition	Exon 9 (codon 318)	Frameshift
	T addition	Exon 9 (codon 320)	Frameshift
	G addition	Exon 9 (codon 321)	Inappropriate termination codon
Liver tumor	G → A	Exon 18 (codon 588)	Glu → Lys
Liver tumor	C → A	Exon 18 (codon 593)	His → Asn
Liver tumor	A → G	Exon 18 (codon 609)	Ser → Glu
Liver tumor	A → G	Exon 18 (codon 612)	Silent
Normal + liver tumor	T → G	Exon 20 (codon 686)	Polymorphic variation? Ser → Ala
Normal + liver tumor	A → G	Exon 23 (codon 812)	Polymorphic variation? Silent
Liver tumor	T → C	Exon 23 (codon 823)	Val → Ala
Liver tumor	C → T	Exon 23 (codon 834)	Silent
Liver tumor	23 bp deletion	Exon 19 (codons 621–628)	Causes frameshift and inappropriate termination codon at codon 667
Eye tumor	A → T	Exon 15 (codon 474)	Lys → Stop
Eye tumor	A → G	Exon 21 (codon 747)	Lys → Glu
Eye tumor	G → C	Exon 23 (codon 838)	Ala → Pro
Eye tumor	C → T	Exon 8 (codon 288)	Silent
Eye tumor	A → G	Exon 23 (codon 812)	Silent
Eye tumor	C → T	Exon 22 (codon 759)	His → Tyr
Eye tumor	C → T	Exon 23 (codon 808)	Pro → Leu
Eye tumor	G → A	Exon 16 (codon 498)	Silent
Eye tumor	141 bp deletion	Exons 22–23 (codons 777–824)	Partial loss of binding domains for *Rb* Chain B and *myc*

2.3. Medaka as a human surrogate: The retinoblastoma example

A number of investigations using animal models have been conducted in order to determine the significance of *Rb* gene mutations, associated altered expression and corresponding cellular effects of the altered protein[39,89,101]. There are limitations, however, associated with studying pRb, especially its role in retinoblastoma tumor formation, using rodent models since tumors cannot be induced with chemical carcinogens. Moreover, mice, which are heterozygous for the *Rb* mutations, are not predisposed to retinoblastoma formation, tending instead to develop brain or pituitary tumors[89,101]. The utilization of transgenic rodent models has other limitations as animals tend to develop uncharacteristic patterns of tumor formation and retinoblastomas cannot be induced through chemical carcinogenesis regimes[21]. Researchers therefore continue to rely on human surgery and autopsy materials to characterize mutations in the *Rb* gene and investigate their potential involvement in oncogenesis.

An incidence of naturally occurring retinoblastoma has been reported in fish[53,67]. Subsequently, studies demonstrated that these tumors could also be induced experimentally in medaka using the compound MAMAc[135]. The histopathology of the induced eye tumors in medaka corresponds with the human condition[71]. The sequence reported for the medaka *Rb* cDNA, combined with the high degree of conservation observed in critical domains, facilitated an investigation of the molecular etiology of the chemically induced retinoblastoma. The medaka tumor samples examined included several mutational alterations within the *Rb* cDNA sequence, including silent and missense mutations, as well as a large deletion. Such mutational alterations suggest that medaka may provide a novel model and, thus, provide additional insight into the human retinoblastoma condition[147]. There are potentially other human cancers for which no alternative suitable vertebrate model exists, but where medaka or other small fish can be utilized. This is particularly notable given the apparent differences in the importance of *ras* oncogene and *p53* tumor suppressor gene involvement in rodent cancers when compared with humans.

2.4. Medaka transgenic and knockout models: Advantages over rodents

Medaka have been chosen for transgenic and knockout applications for several reasons. First, there is a wealth of information regarding cancer studies using this species, and importantly many of the key players in control of growth and the cell cycle have been characterized. Second, the groups of Winn and Wakamatsu have provided both the techniques to create transgenic medaka and also provided the groundwork to demonstrate the efficacy of using the medaka for knockout research. Medaka have already been used successfully in many transgenic applications; both medical and environmental. For instance, transgenic medaka containing the *Xiphophorus Xmrk* gene were used to determine the importance of mutational activation of the INV-*Xmrk* to ONC-*Xmrk* in melanoma formation[46], whereas transgenic medaka have also been developed to determine the mutation frequency following exposure to environmental mutagens[158,190–192].

For knockout applications, medaka are easily reared in the laboratory, spawn daily and produce many eggs. Furthermore, the eggs have a clear chorion, allowing easy manipulation compared with rodents. Embryos can also be reared in Petri dishes from fertilization through hatching and early development, allowing easy monitoring and timing of experiments. Critically, the embryonic stem cells required to create the *Rb* medaka knockout are only available for this fish species[85]. In diploid organisms mutational alteration of one copy of a tumor suppressor gene is known to increase the likelihood of acquiring more point mutation damage, thus destabilizing the entire genome. Knockouts allow the investigation of mutator phenotype by creating medaka with both copies of a gene altered and inactive (knockouts). It is then possible to investigate the role of specific genes and their protein products in cellular signaling and cell growth checkpoint control. Also, such fish could be used in controlled exposure experiments in order to determine the rate of further mutational inactivation and its effect on tumor development. Such studies are underway in a number of laboratories. Recent advances have also been made in nuclear transplantation techniques using medaka[184], offering another mode of genetic modification and cloning.

3. Zebrafish: An introduction

Even though humans and zebrafish (*Danio rerio*) are separated by approximately 300 million years of evolution, both organisms contain roughly 30,000 genes and many of these genes are sequentially homologous and appear to have similar function. Cancer presents in both species and they share many of the same oncogenes and tumor suppressor genes, which are similarly mutated during oncogenesis. Historically, the zebrafish has been optimized as a model for developmental biology and much has been learned (e.g. see special issue of *Development* 123, 1996). However, though few relevant studies have been completed to date, this is another small fish that holds tremendous potential as a cancer model and will likely mature into the preferred model for studies of the genetics of cancer among non-mammalian models. Finally, in a manner similar to medaka, the fish produces externally fertilized eggs with transparent embryos. Their small size results in easy maintenance and 150 + progeny can be produced per female on a weekly basis. They are amenable to production of transgenic animals as well as large-scale chemical and genetic screens. A high-resolution genetic/physical map is available and sequencing of the genome is underway. Finally, there appears to be a strongly conserved phylogenetic relationship between zebrafish, mice and humans.

3.1. Zebrafish cancer

Among the early papers demonstrating the efficacy of the zebrafish as a cancer model was the work of Khudoley[94], which evaluated the carcinogenic effects of various does, length of exposures and temperatures of dimethylnitrosamine (DMNA), diethylnitrosamine (DEN) and nitrosomorpholine (NM) during short-term exposures. Increasing doses of DMNA (20–100 ppm) lead to an increased incidence of tumors (31–76%) and the average latent period to tumor development decreased with increasing concentration of DMNA. Temperature was also found to impact tumor development with fish kept at a lower temperature (17 ± 1°C) exhibiting a tumor incidence of 38% compared to 88% among fish maintained in water 10° warmer. Similar results were reported among fish exposed to DENA and neoplastic lesions included hepatocellular carcinomas, adenomas, cholangiomas, cholangiocarcinomas and esophageal papillomas. Exposure of zebrafish to NM also resulted in the induction of the intestinal adenocarcinomas.

Subsequent studies have demonstrated that zebrafish are sensitive to other carcinogens (e.g. DMBA and MNNG) from varying routes of exposure (e.g. refs. 18,161,162). These and similar studies have clearly demonstrated that although the zebrafish may not be as sensitive to carcinogen exposure as medaka or the rainbow trout (discussed below), it clearly is susceptible to cancer from exposure to environmentally relevant carcinogens. These studies are largely redundant with what has been previously accomplished with other fish models and they do not really exploit the novelty or wealth of available information from the zebrafish.

A potential approach that appears to exploit unique aspects of the zebrafish, and in particular their genetic accessibility, was recently reported by Langenau and colleagues[100]. The investigators created transgenic zebrafish that expressed the mouse c-myc oncogene. Clonally derived t-cell acute lymphoblastic leukemia was induced in zebrafish using c-myc and chimeric *EGFP*-mMyc transgenes under the control of the zebrafish *Rag2* promoter which is responsible for targeting genes expression specifically to lymphoid cells. At the one-cell stage of development, 215 fish were injected with appropriate constructs, and depending on the construct, 5–6% (18 fish) progressed to tumors. The authors concluded that their transgene model provides a platform for drug and genetic screen aimed at identifying and understanding mutations involved in *c-myc*-induced carcinogenesis. These types of studies point to the promise for this model.

Finally, it is clear that the use of zebrafish as a model for carcinogenesis studies is at its infancy. Nonetheless, as Amatruda *et al.*[8] recently observed "The zebrafish system, with its combination of forward genetics and vertebrate biology, has great potential as a cancer model system." Currently, studies along these lines are underway at various laboratories around the world.

4. *Rainbow trout: An introduction*

Many of the advantages of employing medaka in carcinogenesis bioassays apply equally well to rainbow trout (*O. mykiss*). Trout incur low rearing costs (though not as low as the former) and provide an early life history stage that is highly sensitive in bioassay. They are also sensitive to a wide variety of carcinogens, as well as responsive to promoting and inhibiting agents. The tumor pathology is well described in the literature and the mechanistic etiology at the molecular level appears comparable to mammalian models. Their limitations as a human surrogate are the same as medaka, in that, there is no complete organ homology and the extra facet of gene duplication must be considered, though most scientists view the diploid condition in trout and other salmonids as being functionally tetraploid. This view is based on chromosome number and is likely due to a relatively recent genome-wide duplication event.

An extensive amount of data regarding exposure regimes, including different types of carcinogens, doses and mode of exposure for trout is available (e.g. Bailey *et al.*[13]). The sensitivity of trout to a wide variety of carcinogens is attributed to a high capacity for cytochrome P450-mediated bioactivation and a corresponding low capacity for Phase 2 enzyme-mediated detoxification of the resultant epoxide. Confirmed carcinogen-initiating agents include MAMAc[79], *N*-methyl-*N*′-nitro-*N*-nitrosoguanidine (MNNG)[132], dehydroepiandrosterone (DHEA)[131,133], aflatoxins[10,12,14], nitrosamines[82] and PAHs[51,81].

In the case of the latter, trout embryos exposed to water-borne DMBA (the most potent carcinogen of this class of compounds) at 1 ppm for 24 h, 5 ppm for 2 h and 5 ppm for 24 h resulted in an increasing incidence of liver tumors of 3.8, 23.0 and 85.0%, respectively[51]. This contradicted suggestions at that time, that high doses of hydrocarbon were critical in inducing tumors. Early studies of BaP-induced carcinogenesis in trout have shown BaP induces the trout MFO system, as with mammalian systems[81] although a few notable differences between fish and mammalian systems have been determined. For example, the initiation of hepatocellular carcinomas does not require partial hepatectomy or infant exposure to BaP as has been the case in rodents. Mode of administration was also investigated and it was found that trout given intraperitoneal injections, although receiving less BaP overall than fish in dietary groups, had a higher prevalence of hepatocellular carcinoma[81]. This was attributed to the comparative insolubility of BaP when mixed with the diet and for feeding hierarchies within the tank. In addition to hepatocellular carcinoma, BaP also induced two other identifiable tumor types: fibrosarcoma of the liver and swim bladder adenomas.

Building on the information gathered regarding carcinogen initiators, subsequent studies have demonstrated that trout also respond to a wide range of tumor-modulating compounds including promoters, inhibitors and cocarcinogens. Promoters of AFB1-initiated hepatocarcinogenesis include indole-3-carbinol (I3C), at low levels only, otherwise the Ah pathway in addition to the estrogen metabolic pathway is induced and the promotion effect is lost[129]. DHEA also acts as a promoter and is thought to act *via* cell cycle proteins p53 and p34cdc2-mediated mechanisms[130,132]. Hydrogen peroxide is a promoter of MNNG-induced hepatocarcinogenesis, possibly as a result of enhancing oxidative stress[93]. Other promoters identified include fumonisin B1, a mycotoxin[37], Aroclors-1242 and 1254, I3C and β-naphthoflavone following DEN initiation[50,157]. Established inhibitors of aflatoxin-induced carcinogenesis include chlorophyllin[77], α- and β-naphthoflavone[165], each of which modulate DNA adduct formation. Tumor modulators of PAH-induced tumorigenesis have also been characterized and examples include dieldrin and chlordecone[48]. Like many modulators, the effects of such compounds (whether they inhibit or promote) are dependent on dose and timing of administration[48]. For instance, pre-treatment with chlordecone prior to DMBA initiation will not promote tumor development, whereas exposure post-initiation and at low dose will promote tumor development[48]. In summary, the extent of any modulation observed is protocol dependent.

Trout, therefore, respond to a wide variety of tumor-modulating compounds and, as a result, can offer both (i) comparative mechanistic information for mammalian studies as well as (ii) provide an insight into the synergistic or antagonistic effects following exposure of feral fish to complex mixtures of contaminants in the aquatic environment.

4.1. Rainbow trout as a model for determining the molecular mechanisms of carcinogenesis

The underlying molecular steps of carcinogenesis in terms of mutations in DNA caused by initiator exposure, in particular damage to oncogenes, have been studied in trout as with medaka. Again, one oncogene, the *ras* gene, has received considerable attention. Approximately 20–30% of human tumors contain some form of an activated *ras* oncogene making them the most frequently detected oncogenes in human tumors. Such a high incidence of activated *ras* among tumors of different etiologies points to the critical function of the normal cellular proteins. The *ras* gene superfamily contains over 40 related genes classified into three families: *ras*, *rho* and *rab*. Genes from the *ras* encode small (21 kDa) membrane-bound proteins involved in signal transduction. Activation of c-*ras* is a growth signal for eukaryotic cells and is similar to other G proteins in that it binds GTP. Its GTPase activity catalyzes the conversion of GTP to GDP. The pathway from GTP to GDP and then to the nucleus is more complicated, involving MAP kinases and raf, passing on the signal to cytoplasmic receptors and an ultimate endpoint in the nucleus. Protein alterations can occur *via* point mutations, deletions, insertions or rearrangement through chromosomal translocation, gene amplification or proviral insertion. In addition, and highly novel compared with medaka, is the implication of gene duplication and triploidy in tumor suppressor gene function in trout and other salmonids.

Two *ras* genes have been isolated from trout[108]. The trout *ras* genes have more homology to their mammalian counterparts (76.8–87.1%) than the trout *myc* gene (52.9% over exons II and III)[174]. Extensive studies have investigated point mutations in exons I and II of the trout *ras* genes in tissue isolated from experimentally induced tumors[13,36,51,82]. Initially, the effects of aflatoxin B_1 exposure on *ras* genes in trout liver tumor samples were investigated[36]. Later, experimental exposures have been repeated with a variety of genotoxic compounds and their effects on trout *ras* genes studied in several tissues including liver, stomach and swim bladder[13,51,82].

Chang *et al.*[36] was the first study to demonstrate *ras* gene activation by a known carcinogen in any fish species. A high proportion (10/14) of the aflatoxin B_1-initiated tumor DNAs showed evidence of activating point mutations in the trout *ras*-1 gene. Among these, the predominant lesion (7/10) was a GGA to GTA transversion in codon 12, the most commonly found molecular lesion in rodent carcinogenesis models and many human tumors. Of the remaining mutant *ras* genotypes (3/10), two were codon 13 GGT to GTT transversions and one was a codon 12 GGA to AGA transition.

In addition to the investigations on aflatoxin B_1-induced *ras* gene mutations, others have subsequently exposed trout embryos to several genotoxic compounds, including DMBA, DEN and MNNG, and also successfully induced *ras* gene mutations[13,51,82]. In 11 DMBA-induced tumors, nine displayed *ras* gene mutations, four affected *ras*-1 codon 12 GGA to AGA, while another four tumors displayed a *ras*-1 codon 12 GGA to GTA mutation. A single tumor sample displayed a codon 61 CAG to CTG point mutation in the *ras*-1 gene[51]. In subsequent DMBA exposures carried out by the same workers, additional *ras* gene mutations were characterized and the incidences were dissimilar to those found in mammalian models. More than 90% of DMBA-initiated mouse skin papillomas display changes at H-*ras* codon 61 A to T transversions compared with 0, 44 and 22% of trout liver, stomach and swim bladder tumors, respectively[13]. Also, more than 80% of DMBA-induced hepatic tumors in CD-1 mice contain H-*ras* codon 13 GGC to CGC transversions[106], whereas only 11% (1/9) contained mutations at this codon in the trout DMBA-initiated (and Aroclor-promoted) hepatic tumors[13].

Trout experimentally exposed to DEN, administered in the diet, also display K-*ras* mutations[82]. Seven DEN-induced liver tumors were examined for evidence of K-*ras*-activating mutations. Of these, a high proportion (6/7) carried a codon 12 GGA to AGA mutation[82]. The pattern of mutational spectrum for this compound, again, varies compared with rodents, where the predominant *ras* lesion is a codon 61 change. The effect of MNNG exposure on *ras* has also been investigated using trout[13]. In this case, MNNG-induced tumors revealed *ras* codon 12 GGA to AGA and GGA to GAA changes in 10 and 83% of samples, respectively. The latter *ras* mutation is characteristic of MNNG exposure in rodent models[26].

The data accumulated for trout clearly indicates that both the type of inducing compound and the tissue investigated affect the *ras* mutational spectrum and incidence observed (Table 2). Furthermore, these extensive studies using trout also show that the *ras* mutational profile differs from that observed in rodent models. There are no reports as yet of *p53*, or any other tumor suppressor gene, mutations in trout tumor tissue despite readily available methods[99]. While an apparent lack of similarity in mutational profiles may exclude trout as a viable model of mammalian carcinogenesis at the molecular mechanistic level, there still remains the enormous potential by virtue of their induced triploidy status in the development of trout as a model of heritable disease where tumor suppressor genes are implicated.

Thorgaard and colleagues[170] investigated tumor suppression in triploid trout. Diploid and triploid trout were exposed to three model carcinogens (DMBA, MNNG and aflatoxin B_1) and the incidence of tumors recorded. Triploid trout displayed significantly fewer tumors in many organ systems. The mechanistic explanation for the lower tumor incidence is that triploid trout have more copies of the tumor suppressor genes and are therefore less likely to possess all copies that have suffered mutational inactivation. Since tumor suppressor genes act in a recessive way, triploid trout are more likely to retain a normal non-mutated working form of the tumor suppressor protein product. Knockout animals provide a means to study tumor suppressor gene and oncogene function by targeting and removal, polyploid animals may offer the reverse – a means to study the role of tumor suppressor genes in excess to determine their function.

In addition to triploidy, isolated gene duplication events in trout also have a potential application in determining the molecular etiology underpinning carcinogenesis. Two pertinent examples of duplication are the WT1 tumor suppressor gene[28] and the aryl hydrocarbon receptor gene[1]. Studies of the two forms in these cases may help to reveal the multiple functions of the similar lone homolog in mammalian species. Spontaneous and induced nephroblastoma resembling Wilms' tumors have been characterized in fish including Japanese eels (*Anguilla japonica*)[69] and trout[141]. In humans, it is thought that there are three *WT* genes, though current evidence implicates *WT1* gene mutations alone, in the formation of Wilms' tumors[47]. Trout appear to have three distinct *WT1* as well as two *WT2* genes as a result of gene duplication events (Genbank accession numbers: AAK52719 to AAK52723)[28]. A future analysis of the trout *WT1* genes in chemically induced tumors will elucidate the role of mutational inactivation of such genes in the development of Wilms' tumors. Trout *WT* provides a medical application, similar to that of *Rb* and medaka, but again such models also have environmental application – the lack of *p53* involvement implies that other fish tumor suppressor genes are pivotal in carcinogenesis. *WT* and *Rb* are involved in a wide variety of cancers in mammals, possibly in fish cancer as well.

Trout have thus become adopted and established as a model of carcinogenesis for a number of reasons. In summary, they (i) have low rearing costs compared to rodents, (ii) provide a highly sensitive early life-stage bioassay, (iii) are sensitive to a wide variety of carcinogens, (iv) are responsive to promoters and inhibitors, (v) have a well-defined tumor pathology, (vi) are mechanistically comparable to mammalian models and (vii) provide more tissues for analysis compared with smaller aquarium fish.

5. *Xiphophorus spp.: An introduction*

Research on the role of oncogenes in the carcinogenesis process in *Xiphophorus* began more than 70 years ago (pre-dating rodent models) with systematic crossings between populations, races and species and with mutagenesis studies in purebred and hybrid fish (see reviews 106,127,148). Hybrids derived from crossings between different wild populations developed neoplasms spontaneously or following treatment with carcinogens (chemicals or radiation). The most studied model is the Gordon–Kosswig hybrid, in which melanomas form spontaneously in all individuals of a subset of backcross hybrids between platyfish and swordtail species[61]. Linkage studies found that melanoma and a large number of other neoplasms could be assigned to a particular locus located on particular sex chromosomes designated '*Tu*'[5].

5.1. *Xiphophorus spp. and molecular mechanisms of carcinogenesis*

In a search for molecular markers within the *Tu* locus, a probe from the v-*erb* B gene, containing the conserved kinase domain of an oncogenic version of the avian epidermal growth factor receptor gene, identified a restriction fragment length polymorphism (RFLP) closely linked with *Tu*[3]. The sequence detected by the v-*erb* B probe turned out to be a critical constituent of the *Tu* locus. The fragment sequenced was used to isolate a full-length cDNA clone from a *Xiphophorus* melanoma cell line. Translation of the isolated cDNA predicted an epidermal growth factor receptor-related receptor, subsequently termed X*mrk* (for *Xiphophorus* melanoma receptor kinase)[193]. *Tu* has also been referred to as X-*erb* B (for Xiphorine epidermal growth factor receptor gene)[113].

There are two versions of the X*mrk* gene: one a melanoma-inducing oncogene, ONC-X*mrk*; the other a proto-oncogene, INV-X*mrk*. The oncogene was produced by translocation of the INV-X*mrk* gene downstream of an, as yet uncharacterized, donor gene locus[52]. ONC-X*mrk* thus obtained a new promoter from the donor gene that led to gene duplication. Loss-of-function investigations, using disrupted ONC-X*mrk* mutant hybrids, also confirm its role as a tumor-inducing gene[149]. It is now well established that the development of malignant melanomas results from abnormal regulation and over-expression of the ONC-X*mrk* gene. The ONC-X*mrk* alleles contain a high concentration of retroelements not observed in the proto-oncogene version, resulting in an unstable composition and frequent removal by deletion or disruption of transposable elements[58]. The encoded protein also differs from that of the INV-X*mrk*-encoded product in a number of amino acid changes. The resulting conformational change in the tyrosine kinase protein leads to strong ligand-independent tyrosine phosphorylation[58].

Mutational activation, in addition to over-expression of the tyrosine kinase receptor, is now believed to be necessary to achieve full tumorigenic potential[46].

The mechanism by which over-expressed ONC-X*mrk* signals cell proliferation has yet to be fully determined. Studies have looked at relative binding affinities of a general receptor tyrosine kinase (PLCγ) as well as *Xiphophorus* cytoplasmic X*src*, X*fyn* and X*yes* and found that each, X*fyn* in particular, is enhanced[187]. Binding motifs for STAT5, GRB2 and Shc have also been characterized on the X*mrk* gene[188,189]. These provide evidence of adapter proteins that link X*mrk* to the ras/MAP kinase-signaling pathway. Consistent with the theory of multistage development of carcinogenesis, other oncogenes are also implicated as operating either downstream of X*mrk* induction of melanomas, or independently. For instance, Maueler *et al.*[114] looked at X*src* expression in hereditary melanomas and in carcinogen-induced tumors, finding it expressed at high levels, especially in non-sporadic tumors. The location of the *Xiphophorus yes1* and *fyn* genes have also been mapped to LG VI, in close proximity to X*mrk*[119]. These encode non-receptor tyrosine kinase proteins that possibly operate downstream of Xmrk in the signaling cascade that leads to melanoma progression.

ONC-X*mrk* acts as a dominant tumor gene under the control of regulatory genes (located at a specific locus), termed 'R', which in turn act as tumor suppressor genes[16]. The R genes have been categorized into three inter-related R gene systems. The first category includes *Tu*-linked tissue-specific R genes: R-*mel*, R-*neu*, R-*epi* and R-*mes*. If impaired or lost, these allow *Tu*-encoded tumor formation in the pigment cell system and neurogenic, epithelial and mesenchymal tissues, respectively[6]. The second category includes *Tu*-linked compartment-specific R genes. These restrict spots and melanomas to specific compartments of the platyfish body. A total of 14 have been identified (collectively called R-co) and if impaired correspond to sites of the body where the melanomas in the hybrids occurred[6]. The third category includes *Tu*-non-linked modifying genes (e.g. '*Diff*'), which control proliferation and differentiation of transformed pigment cells[6]. *Diff* maps to linkage group V and certain hybrids that do not inherit a copy are susceptible to melanoma development[148]. A *CDKN2*-related gene has been suggested as a candidate for the *Diff* gene[92,126]. In mammals, such genes encode proteins critical to the regulation of the cell cycle and are also mutationally inactivated in several types of tumor including melanoma and liver[11]. The *Xiphophorus p53* tumor suppressor homolog (another key player in control of the cell cycle in mammals) has also been characterized, and even though this is not linked to *Tu*, it may be involved in the development of other, more diverse, tumor types[91]. A Xiphorine homologue of the tumor suppressor *retinoblastoma* (*Rb*), which is evolutionarily conserved[147] and most likely also critical in cell cycle control, has yet to be characterized. Repeated backcrossing thus eliminates the regulatory control of the R tumor suppressor genes and allows increased expression of the dominant oncogenes, such as X*mrk*, resulting in the development of malignant melanoma (or other tumor types) in hybrids[4,148].

DNA repair mechanisms in different *Xiphophorus* hybrids have also recently been investigated[117,185]. There is already considerable interest in understanding the molecular basis of the defects in human syndromes associated with deficiencies in

relationships between anthropogenic chemicals and observed alterations. Identification of etiological agents often remains uncertain, owing largely to a lack of controlled tumor studies with appropriate organisms exposed to suspect compounds. Further, the complexity of chemical profiles in these field samples and the lack of information on the interactions between such chemicals in experimental carcinogenesis make interpretation difficult. The hepatic lesions characterized in feral flatfish do, however, closely resemble those experimentally induced by single carcinogen exposure (reviewed in ref. 118).

Laboratory studies, in parallel to field studies, have started to investigate the causal relationship between a number of carcinogens, especially the environmental contaminants, PAHs, and hepatic neoplasia in a number of fish species. For instance, hepatic lesions in feral flatfish also resemble those characterized in mummichog, an estuarine fish (discussed below), following controlled exposure to PAH-contaminated sediments and diet[182]. There is also a statistically significant relationship between environmental contaminant levels, especially PAHs, and lesion incidence[105,118]. Hepatic lesions have thus become accepted biomarkers of contaminant exposure. Future emphasis may be placed in relating observed histopathological stages in development of the liver lesion with the underlying molecular mechanisms and multistage polygenic development of carcinogenesis.

1.1. Feral flatfish cancer genes

The underlying molecular steps of carcinogenesis in terms of mutations in DNA caused by initiator exposure, in particular damage to oncogenes, have been studied in feral flatfish as with aquaria fish. The precise genetic epidemiology of multistage carcinogenesis in feral fish has yet to be tackled. Since the neoplasms themselves are heterogeneous, it is likely that the causes, in terms of genetic alterations and pathways, are also likely to be heterogeneous and thus complex. The current opinion among human cancer epidemiologists is that 2–4 genetic alterations may be required before cancer development has passed through all the necessary stages. Such stages must include a flood of growth signals (oncogene activation), loss of growth inhibitory signals (tumor suppressor gene inactivation), escape from regulation by programmed cell death (*p53* or similar inactivation?), limitless replicative ability (promotion step?), maintained angiogenesis (promotion step?), tissue invasion and metastasis (promotion steps?) (see review 78). Such stages can potentially be related to histopathological stages, though at present few studies have linked the two models. Attempts have been made to link metabolic events, such as proliferating cell nuclear antigen (PCNA): a marker of cell proliferation and potentially the promotion step, with histopathological stages[23,97]. However, only rudimentary details are thus far known in terms of genetic alterations in feral fish displaying hepatic tumors.

At present, one oncogene (*ras*) and one tumor suppressor gene (*p53*) have been examined in flatfish species. Two closely related *ras* genes, *ras*-1 and *ras*-2, have been reported in European flounder (Genbank identifiers: Y17187, Y17188)[180,181]. European flounder have been identified as a cancer-prone species[178,179], showing population variation in contaminant-metabolizing enzymes[20]. However, preneoplastic and neoplastic livers collected from 14 flounder did not display any mutation within the *ras*-2 gene[181]. A second analysis of flounder liver tumors also failed to find *ras* or *p53* mutations that changed the amino acid sequence of either encoded protein[56]. In contrast, a third study has reported evidence of a *p53* mutation, that changes the amino acid sequence and thus alters protein conformation, in hyperplastic foci from a single flounder captured from the Seine Estuary[32]. The latter two studies also highlight the considerable polymorphic variation observed in the flounder *p53* gene. A single *ras* homolog, K-*ras*-B, has thus far been characterized from English sole (*P. vetulus*)[139]. Again, the English sole *ras* gene displays a high predicted (97%) amino acid sequence identity with mammalian and fish counterparts. The authors report that each of the functional domains is conserved. However, no mutations in codons 12, 13 or 61 of K-*ras* have been found in hepatic lesions from English sole, although the authors suggest that mutations could exist at levels below the detection limits of analysis[139]. Partial dab *ras* genes have also been characterized[144] but, as yet, no mutations have been reported in tumor tissue. The apparent lack of *ras* and *p53* gene involvement in the etiology of flatfish liver tumors suggests that other genes, such as *Rb*, are involved.

2. *Mummichog*

Mummichog (*F. heteroclitus*) spend much of their life in close association with contaminated coastal areas of the Eastern US and Canada, moving only limited distances toward freshwater to spawn. Spawning behavior brings gametes into direct contact with contaminated surface micro-layer and sediment. Eggs remain in contact with the sediment for several days before hatching. Consequently, there is the possibility of contaminant-induced damage early in their life history. There is substantial literature concerning feral adult mummichog, their utilization in controlled contaminant exposure[63,125] and various field studies[54,182]. The main research interest in mummichog is in their apparent resistance to the toxic action of certain classes of environmental contaminants, including PCBs and PAHs and the mechanism that allows such resistance.

Populations of *Fundulus* sp. have seemingly developed resistance to a number of classes of contaminants including dioxins, PCBs and PAHs. The two most cited subpopulations are those at New Bedford Harbor (NBH), MA, a heavily PCB-contaminated environment, and Elizabeth River (ELZR), VA, a heavily PAH-contaminated environment. NBH has extremely high PCB contaminant levels in both the sediments (~2100 μg/g dry weight sediment) and biota (~320 μg/g dry weight tissue)[123]. Despite such high levels of PCBs, individual fish appear 'healthy' as measured using general health parameters[124]. One suggestion offered, to explain the apparent resistance to PCB toxicity in the short term, is that such individuals are relatively unresponsive to PCB exposure, particularly in terms of CYP activity[123]. In the longer term, the resistance may come at a cost in terms of reduced metabolism and elimination of xenobiotics generally. Evidence suggests that such costs are real: for instance, high PAH concentrations in tissues of exposed fish have been reported[63], and future consequences may include an increase in maternal transfer to offspring as well as increased susceptibility to photoactivation toxic effects.

In contrast to the NBH PCB-resistant individuals, the ELZR PAH-tolerant individuals appear to be cancer prone[54,182]. Isolated regions of the ELZR have PAH concentrations 13-fold higher compared with contaminated NBH sediments[182]. In such areas, the mummichog are apparently insensitive to CYP1A-inducing contaminants[176], yet display a high incidence of hepatic (up to 93% of individuals in localized areas) and exocrine neoplasms[54,182]. The mechanism of tolerance is thought to involve an alteration in glutathione *S*-transferase activity and/or increased P-glycoprotein levels (a general xenobiotic transporter)[9,42]. Further work has suggested that such tolerance is heritable[138]. The combination of apparent tolerance, yet increased tumor incidence, would, however, suggest that such mummichog populations in areas of high contamination are not suitable as biomonitoring indicator species, since they fail to provide an early indication of a toxic effect. They do, however, present an opportunity to understand the mechanisms that underpin the apparent tolerance or resistance to environmentally induced carcinogenesis.

3. *Gizzard shad*

In the early 1990s, Ostrander and colleagues put out a call to scientists and fishermen in the central North America requesting any fish collected with apparent tumors. In addition to receiving many fish with injuries and parasites, an interesting epizootic of pigmented subcutaneous spindle cell tumors, initially among gizzard shad (*D. cepedianum*), emerged.

Initially, nearly 25% of the adult gizzard shad from Lake of the Arbuckles (located in central Oklahoma, USA) collected over a 2-year period presented with tumors primarily distributed over the head, trunk and fish[136]. The tumors were almost always darkly pigmented and appeared as large, raised masses. Histologically, they were located in the dermis, had a variable amount of connective tissues and consisted of cells in a variety of forms and arrangement. The majority of the tumors were comprised of fusiform or spindle cells arranged in wavy bundles, whirling patterns or interwoven fascicles. Large dense deposits of melanin or scattered individual melanin-containing cells caused the pigmentation. Staining with PCNA verified a high mitotic activity for the spindle cells. Although the cell of origin of the tumors was not identified, evidence points toward melanocytes or possibly nerve sheath cells, though it is also possible that some other poorly differentiated cell is the cell of origin.

To date, the etiology of the tumors has not been determined, though numerous avenues have been investigated. Initially, water and sediment samples were fractionated and subjected to gas chromatography–mass spectrometry. Obtained spectra were matched to an on-line NIST library of 40,000 known spectra. No significant matches were observed[136]. Simultaneously, EPA-approved method 800 for BITEX (benzene, toluenes, ethylbenzene and xylenes) analysis of water samples was negative as well.

As described elsewhere herein, oncogenic viruses have been implicated in a variety of cancers in fish. To this end the potential for retroviral involvement in the shad tumors was investigated. A method was modified from published materials for assaying for reverse transcriptase activity (RT), which is an appropriate marker for retroviral activity. No RT activity was detected in any of the homogenates of tumor tissues examined. Attempts to concentrate potential RT activity were accomplished *via* sucrose gradient centrifugation. Appropriate fractions were examined and again no RT activity was detected. Finally, viral particles were not detected by electron microscopy and attempts to transmit the potential virus likewise proved negative[90,136].

Subsequent studies[90] focused on the population structure of fish presenting tumors as well as the role of heavy metals. Overall neoplasm prevalence appears to be stable at about 22% of the population. No juvenile fish, out of 2000 + examined, exhibited tumors. Likewise, tumor prevalence did not appear to be seasonal or site specific within the Lake of the Arbuckles. Water, sediment and fish tissues were collected from the Lake of the Arbuckles, a reference lake outside the drainage, and were analyzed for total recoverable metals (cadmium, chromium, copper, nickel and lead) by graphite furnace atomic absorption. Chromium, copper and nickel were found in the water samples at concentrations of $>1-8.6\ \mu g/l$. Low concentrations ($>1-13.6\ \mu g/g$ wet weight) of all the metals were found in the sediment and liver tissues from both sites. Based on these concentrations, heavy metal contamination does not appear linked to neoplasm occurrence.

Further attempts to identify the etiological agent responsible for the gizzard shad tumors included inductively coupled plasma mass spectrometry (ICP-MS). The ICP-MS analysis was conducted on sediment, water and shad liver and muscle tissues. Potentially carcinogenic trace elements examined included beryllium, chromium, nickel, arsenic, selenium, cadmium, mercury and lead. Detectable levels of some compounds were observed in water and sediment samples, however, they were all present at levels well below the EPA guidelines. Analysis of tumor-bearing and non-tumor-bearing tissues showed statistical differences between beryllium (<0.05 vs. $0.79\ \mu g/g$) and nickel (<0.05 vs. $21.25\ \mu g/g$) in liver and nickel (10.35 vs. $4.48\ \mu g/g$) in muscle[60]. Again, these levels are not likely to be causative.

Naturally occurring radiation can harm aquatic systems by producing a range of syndromes, from reduced vigor to lethality, shortened life span, diminished reproductive rate and genetic transmission of radiation-altered genes[173]. Forty-five water samples, collected from various locations, were examined for environmental radiation[62]. Radioactivity levels ranged from less than 0.07 to 0.51 Bq/l for alpha, from less than 0.40 to 1.6 Bq/l for beta and <100 pCi/l for radon-222 radiation. All levels detected were well below EPA guidelines for alpha, beta and radon radiation.

Tumor-bearing and non-tumor-bearing shad populations have also been examined for genetic markers produced by random amplified polymorphic DNA (RAPD) and double-stringency polymerase chain reaction (DS-PCR) to determine if population differences could be distinguished[60]. No differences were observed between tumor-bearing and non-tumor-bearing shad by genetic marker comparison. Band-sharing analysis did not reveal any differences between the two populations with RAPD ($p = 0.294$) or DS-PCR ($p = 0.236$) markers.

In recent years, the tumor-bearing fish have been collected from gizzard shad residing in other lakes within the drainage to include Lake Murray and Lake Texoma with a similar neoplasm prevalence (ca. 17–20%)[60,137]. Interestingly, the epizootic of tumors does not extend to other species in these lakes. To date only two similar-appearing tumors have been observed in threadfin shad (*D. petenense*) and a single white bass (*Morone chrysops*) presented with hemagiopericyoma. The latter is interesting in that it involves cells that surround blood vessels which is analogous to the shad tumors which appear to arise from cells surrounding nerves.

The etiology of these tumors remains unresolved, though it may be noteworthy that the tumors have been around for some time as examination of archived fish collection from various locations in Oklahoma and Texas revealed tumor-bearing shad as early as the 1970s[137]. Moreover, it is not clear why the tumor incidence is being maintained at about 22%. Sampling by Ostrander and his colleagues, taking place nearly 10 years after the initial discovery, still reveals an incidence of tumor greater than 20%[137].

4. *Damselfish*

Malignant tumors in the bicolor damselfish (*Stegastes partitus*) were initially observed in South Florida (USA) coral reefs in the early 1980s. Subsequent characterization[152–154] revealed a naturally occurring transmissible tumor involving neuroectodermal cell types and was named damselfish neurofibromatosis (DNF). The disease presents as nerve sheath and pigment cell tumors and is ultimately fatal[35]. The primary cell types expressed in the tumors included Schwann cells and chromatophores. Intramuscular injections of cell-free tumor homogenates and injection of cells from long-term cell cultures resulted in tumor transmission/formation and indicates a sub-cellular infectious agent is likely the causative factor in tumor formation[153].

Schmale and colleagues[151] have previously demonstrated the presence of several retroviruses in damselfish cell lines, though they have been unable to consistently recognize retroviral genomes in damselfish tumors or other tissues and as such retroviruses are not likely the agent causing this disease. Recently, they have reported the occurrence of a group of extrachromosomal DNAs in tumors from fish affected with DNF but not in healthy individuals[155]. They also determined the distribution of extrachromosomal DNA in tumor-bearing fish, healthy fish and in cells lines derived from these fish. Analysis of the extrachromosomal DNA revealed a distinct DNA fragment with mobility approximately equivalent to that of 1.4 and 2.6 kb double-stranded DNA. Cell lines derived from non-tumor-bearing healthy fish and fibroblasts infected with damselfish retroviruses did not contain detectable DNA sequences homologous to the 1.4 kb fragment or PCR fragments derived from it. The potential of a variety of damselfish cell lines to induce tumors *in vivo* relative to extrachromosomal DNA distribution was examined. Some lines were derived from tumors and others from healthy damselfish tissues. Among the 12 cell lines examined, the six containing various forms of extrachromosomal DNA led to the formation of neurofibromas *in vivo* independent of the expression of retroviral genomes in the cell lines. Cell line from healthy fish or embryos and cell line fibroblasts infected with a previously identified damselfish retrovirus did not possess DNA sequences homologous to the 1.4 kb fragment or PCR products derived from it[155]. The pattern of expression of extrachromosomal DNA was similar in experimentally induced tumors. Moreover, a DNase-resistant component of the DNA was isolated from both tumor cells and conditioned media of tumor cell lines leading to the suggestion that these sequences were encapsulated viral particles. Thus, the data gathered so far supports the hypothesis that one or more of the extrachromosomal DNA forms may be the genome of a novel virus, which is the causative agent of DNF[33,34,155].

5. *Chondrichthyan fishes*

The first reported case of neoplasia in a chondrichthyan fish occurred in 1853 and was a large (30 cm × 8 cm) fibroma at the base of the tail in a thornback skate, *Raja clavata* (Deslongchamps, 1853, cited in ref. 169). A second tumor was described from the liver of a blue shark *Prionace glauca* in 1908 and was likely a hepatocellular carcinoma rather than an adenoma as originally reported[156]. Fourteen more neoplasms were reported among chondrichthyans over the next 40 years including six malignant tumors and a single case of a metastatic melanoma (see review 150). Ostrander and colleagues[137] recently reported on three new cases and provided a tabulation of the 42 cases reported to date among chondrichthyan fishes.

The claims by the shark cartilage industry and others that sharks do not get cancer are inaccurate and cannot be attributed to a lack of published information. Nonetheless, there are several explanations for the relatively low number of documented neoplasms among sharks and their allies. First, sharks limit carcinogen exposure by their pelagic life history. There is limited exposure to toxicants in open water and this trend is consistent with what is observed with pelagic marine bony fish that likewise are nearly devoid of known neoplasms. In comparison, benthic bony fish that feed on the flora and/or fauna in polluted waterways frequently have epizootic skin and liver neoplasms at rates that can exceed 50%. In fact, the approximately 150 epizootic neoplasms known in fish have all occurred in inland or coastal waters – none has occurred in fish from open waters[68] (Harshbarger, personal communication). In addition, it is logical to assume that cancer-weakened pelagic animals are much more likely than inshore animals to be removed by predators before they are caught by fishermen.

Second, systematic tumor surveys have not been completed in chondrichthyans which is a sharp contrast to the frequent tumor surveys among bony fish that have yielded a large part of the known fish tumor cases. Moreover, far fewer chondrichthyan specimens are available for examination from sportsmen and commercial fishermen compared to bony fish and shellfish. Among fish (e.g. refs. 22,104,136,182) and shellfish[175] that have been systematically studied, neoplasia has often been found. Neoplasms have also been documented among diverse invertebrate groups that include coral[140] and flatworms[160].

Third, low susceptibility to tumor induction may be due to a variety of innate factors including carcinogen metabolism and DNA repair. Differential susceptibility to carcinogens is well established in fishes. In bony fish, e.g. for many years investigators have reported on high incidences of liver neoplasms among English sole (*P. vetulus*), residing in contaminated waterways in Puget Sound[104,105]. However, in these same waters the incidence of liver lesions in starry flounder (*P. stellatus*), a fish from the same family (Pleuronectidae) as the English sole, is comparatively low, attributed to species-specific differences in hepatic xenobiotic metabolizing enzymes[40]. Differences in detoxification mechanisms have also been found among some chondrichthyans that could contribute to overall lower tumor prevalence[24]. This issue clearly needs further study.

It is significant to note that malignant tumors have been reported in the cartilaginous jawless fish more primitive than sharks including a metastatic melanoma in a lamprey and epizootic hepatocellular carcinoma in the hagfish[49]. Likewise, neoplasms have been reported in a variety of evolutionarily advanced, cartilaginous fish, relative to sharks, to include lungfish (*Protopterus aethiopicus*)[110,111], paddlefish (*Polyodon spathula*)[102], sturgeon (Acipenseridae)[143] and bowfin (*Amia calva*)[103]. These are not all isolated incidences as paddlefish from the Detroit River exhibit epizootic hepatocellular carcinoma[102].

Sharks as well as skates and rays, like nearly all other vertebrates that have been closely studied, do develop cancer. In general, these tumors are analogous to their counterparts among other vertebrates including fishes, rodents and humans.

IV. Summary

We have reviewed a number of interesting examples of carcinogenesis among the fishes. Such models hold the promise to reveal unique mechanism of carcinogenesis among all species as well as focus studies on other piscine models.

V. References

1. Abnet, C.C., R.L. Tanguay, M.E. Hahn, W. Heideman and R.E. Peterson. Two forms of aryl hydrocarbon receptor type 2 in rainbow trout (*Oncorhynchus mykiss*). Evidence for differential expression and enhancer specificity. *J. Biol. Chem.* 274: 15159–15166, 1999.
2. Abrams, E.S., S.E. Murdaugh and L.S. Lerman. Comprehensive screening of the human K*RAS2* gene for sequence variants. *Genes Chromosomes Cancer* 6: 73–85, 1993.
3. Adam, D., J. Wittbrodt, A. Telling and M. Schartl. RFLP for an EGF-receptor related gene associated with the melanoma oncogene locus of *Xiphophorus maculatus*. *Nucl. Acids Res.* 16: 7212–7214, 1998.
4. Adam, D., N. Dimitrijevic and M. Schartl. Tumour suppression in *Xiphophorus* by an accidentally acquired promoter. *Science* 259: 816–819, 1993.
5. Anders, F. and A. Anders. Aetiology of cancer as studied in the platyfish–swordtail system. *Biochim. Biophys. Acta* 516: 61–65, 1978.
6. Anders, F., M. Schartl, A. Barnekow and A. Anders. *Xiphophorus* as an *in vivo* model for studies on normal and defective control of oncogenes. *Adv. Cancer Res.* 42: 191–275, 1984.
7. Aoki, K. and H. Matsudaira. Induction of hepatic tumors in a teleost (*Oryzias latipes*) after treatment with methylazoxymethanol acetate: brief communication. *J. Natl Cancer Inst.* 59: 1747–1749, 1977.
8. Amatruda, J.F., J.L. Shepard, H.M. Stern and L.I. Zon. Zebrafish as a cancer model system. *Cancer Cells* 1: 229–231, 2002.

9. Armknecht, S.L., S.L. Kaatari and P.A. Van Veld. An elevated glutathione *S*-transferase in creosote-resistant mummichog (*Fundulus heteroclitus*). *Aquat. Toxicol.* 41: 1–16, 1998.
10. Ayres, J.L., D.J. Lee, J.H. Wales and R.O. Sinnhuber. Aflatoxin structure and hepatocarcinogenicity in rainbow trout (*Salmo gairdneri*). *J. Natl. Cancer Inst.* 46: 561–564, 1971.
11. Baek, M.J., Z. Piao, C. Park, E.C. Shin, J.H. Park, H.J. Jung, C.G. Kim and H. Kim. p16 is a major inactivation target in hepatocellular carcinoma. *Cancer* 89: 60–68, 2000.
12. Bailey, G.S., P.M. Loveland, C. Pereira, D. Pierce, J.D. Hendricks and J.D. Groopman. Quantitative carcinogenesis and dosimetry in rainbow trout for aflatoxin B1 and aflatoxicol, 2 aflatoxins that form the same DNA adduct. *Mutat. Res. Environ. Mutagen.* 313: 25–38, 1994.
13. Bailey, G.S., D.E. Williams and J.D. Hendricks. Fish models for environmental carcinogenesis: the rainbow trout. *Environ. Health Perspect.* 104: 5–21, 1996.
14. Bailey, G.S., R. Dashwood, P.M. Loveland, C. Pereira and J.D. Hendricks. Molecular dosimetry in fish: quantitative target organ DNA adduction and hepatocarcinogenicity for four aflatoxins by two exposure routes in rainbow trout. *Mutat. Res. Fund. Mol. Mech. Mutagen.* 399: 233–244, 1998.
15. Battalora, M.S., W.E. Hawkins, W.W. Walker and R.M. Overstreet. Occurrence of thymic lymphoma in carcinogenesis bioassay specimens of the Japanese medaka (*Oryzias latipes*). *Cancer Res.* 50: 5675S–5678S, 1990.
16. Baudler, M., J. Duschl, C. Winkler, M. Schartl and J. Altschmied. Activation of transcription of the melanoma inducing *Xmrk* oncogene by a GC box element. *J. Biol. Chem.* 272: 131–137, 1997.
17. Baumann, P.C. Epizootics of cancer in fish associated with genotoxins in sediment and water. *Mutat. Res.* 411: 227–233, 1998.
18. Beckwith, L.G., J.L. Moore, G.S. Tsao-Wu, J.C. Harshbarger and K.C. Cheng. Ethylnitrosourea induces neoplasms in zebrafish (*Danio rerio*). *Lab. Invest.* 80: 379–385, 2000.
19. Berneburg, M. and A.R. Lehmann. Xeroderma pigmentosum and related disorders: defects in DNA repair and transcription. *Adv. Genet.* 43: 71–102, 2001.
20. Besselink, H.T., M.S. Denison, M.E. Hahn, S.I. Karchner, A.D. Vethaak, J.H. Koeman and A. Brouwer. Low inducibility of CYP1A activity by polychlorinated biphenyls (PCBs) in flounder (*Platichthys flesus*): characterization of the Ah receptor and the role of CYP1A inhibition. *Toxicol. Sci.* 43: 161–171, 1998.
21. Bignon, Y.-J., Y. Chen, C.-Y. Chang, D.J. Riley, J.J. Windle, P.L. Mellon and W.-H. Lee. Expression of a *RB* transgene; results in dwarf mice. *Genes Dev.* 7: 1654–1662, 1993.
22. Black, J.J. and P.C. Baumann. Carcinogens and cancers in freshwater fishes. *Environ. Health Perspect.* 90: 27–33, 1991.
23. Blas-Machado, U., H.W. Taylor and J.C. Means. Apoptosis, PCNA, and p53 in *Fundulus grandis* fish liver after in vivo exposure to *N*-methyl-*N'*-nitro-*N*-nitrosoguanidine and 2-aminofluorene. *Toxicol. Pathol.* 28: 601–609, 2000.
24. Bodine, A.B., C.A. Luer, S.A. Gangjee and C.J. Walsh. In vitro metabolism of the pro-carcinogen aflatoxin B1 by liver preparations of the calf, nurse shark and clearnose skate. *Comp. Biochem. Physiol.* 94C: 447–453, 1989.
25. Braunbeck, T.A., S.J. Teh, S.M. Lester and D.E. Hinton. Ultrastructural alterations in the liver of medaka (*Oryzias latipes*) exposed to diethylnitrosamine. *Toxicol. Pathol.* 20: 179–196, 1992.
26. Brown, K., A. Buchmann and A. Balmain. Carcinogen-induced mutations in the mouse c-Ha-*ras* gene provide evidence of multiple pathways for tumour progression. *Proc. Natl. Acad. Sci. USA* 87: 538–542, 1990.
27. Brown-Peterson, N.J., R.M. Krol, Y.L. Zhu and W.E. Hawkins. *N*-nitrosodiethylamine initiation of carcinogenesis in Japanese medaka (*Oryzias latipes*): hepatocellular proliferation, toxicity, and neoplastic lesions resulting from short term, low level exposure. *Toxicol. Sci.* 50: 186–194, 1999.
28. Brunelli, J.P., B.D. Robison and G.H. Thorgaard. Ancient and recent duplications of the rainbow trout Wilms' tumor gene. *Genome* 44: 455–462, 2001.
29. Bunton, T.E. Diethylnitrosamine (DEN)-induced degenerative, proliferative and neoplastic lesions in the liver of the medaka (*Oryzias latipes*) following short-term exposure. *Mar. Environ. Res.* 28: 369–374, 1989.
30. Bunton, T.E. Experimental chemical carcinogenesis in fish. *Toxicol. Pathol.* 24: 603–618, 1996.
31. Bunton, T.E. and M.J. Wolfe. *N*-methyl-*N'*-nitro-*N*-nitrosoguanidine-induced neoplasms in medaka (*Oryzias latipes*). *Toxicol. Pathol.* 24: 323–330, 1996.
32. Cachot, J., Y. Cherel, F. Galgani and F. Vincent. Evidence of *p53* mutation in an early stage of liver cancer in European flounder, *Platichthys flesus*. *Mutat. Res.* 464: 279–287, 2000.
33. Campbell, C.E., P.D.L. Gibbs and M.C. Schmale. Progression of infection and tumor development in damselfish. *Mar. Biotechnol.* 3: S107–S114, 2001.

34. Campbell, C.E. and M.C. Schmale. Distribution of a novel infectious agent in healthy and tumored bicolor damselfish in Florida and the Caribbean. *Mar. Biol.* 139: 777–786, 2001.
35. Carlson, D.B., D.E. Williams, J.M. Spitsbergen, P.F. Ross, C.W. Bacon, F.I. Meredith and R.T. Riley. Fumonisin B-1 promotes aflatoxin B-1 and *N*-methyl-*N*′-nitro-nitrosoguanidine-initiated liver tumors in rainbow trout. *Toxicol. Appl. Pharmacol.* 172: 29–36, 2001.
36. Chang, Y.-J., C. Mathews, K. Mangold, K. Marien, J. Hendricks and G.S. Bailey. Analysis of *ras* gene mutations in rainbow trout liver tumors initiated by aflatoxin B1. *Mol. Carcinog.* 4: 112–119, 1991.
37. Chen, H.C., I.J. Pan, W.J. Tu, W.H. Lin, C.C. Hong and M.R. Brittelli. Neoplastic response in Japanese medaka and channel catfish exposed to *N*-methyl-*N*′-nitro-*N*-nitrosoguanidine. *Toxicol. Pathol.* 24: 696–706, 1996.
38. Chen, S., Y. Hong, S.J. Scherer and M. Schartl. Lack of ultraviolet-light inducibility of the medakafish (*Oryzias latipes*) tumor suppressor gene. *Gene* 264: 197–203, 2001.
39. Clarke, A.R., E.R. Maandag, M.V. Roon, N.M.T. Lugt, M. Valk, M.L. Hooper, A. Berns and H. te Riele. Requirement of a functional Rb-1 gene in murine development. *Nature* 359: 328–330, 1992.
40. Collier, T.K., S.V. Singh, Y.C. Awasthi and U. Varanasi. Hepatic xenobiotic metabolizing enzymes in two species of benthic fish showing different prevalences of contaminant-associated liver neoplasms. *Toxicol. Appl. Pharmacol.* 113: 319–324, 1992.
41. Cooke, J.B. and D.E. Hinton. Promotion by 17 beta-estradiol and beta-hexachlorocyclohexane of hepatocellular tumors in medaka, *Oryzias latipes*. *Aquat. Toxicol.* 45: 127–145, 1999.
42. Cooper, P.S., W.K. Vogelbein and P.A. Van Veld. Altered expression of the xenobiotic transporter P-glycoprotein in liver and liver tumours of the mummichog (*Fundulus heteroclitus*) from a creosote-contaminated environment. *Biomarkers* 4: 48–58, 1999.
43. Cormier, S.M., R.N. Racine, C.E. Smith, W.P. Dey and T.H. Peck. Hepatocellular carcinoma and fatty infiltration in the Atlantic tomcod. *J. Fish Dis.* 12: 105–116, 1989.
44. Couch, J.A. and L.A. Courtney. *N*-nitrosodiethylamine-induced hepatocarcinogenesis in estuarine sheepshead minnow (*Cyprinodon variengatus*): neoplasms and related lesions compared with mammalian lesions. *J. Natl. Cancer Inst.* 79: 297–321, 1987.
45. DeFromentel, C.C., F. Pakdel, A. Chapus, C. Baney, P. May and T. Soussi. Rainbow trout p53: cDNA cloning and biochemical characterisation. *Gene* 112: 241–245, 1992.
46. Dimitrijevic, N., C. Winkler, C. Wellbrock, A. Gomez, J. Duschl, J. Altschmied and M. Schartl. Activation of the *Xmrk* proto-oncogene of *Xiphophorus* by overexpression and mutational alterations. *Oncogene* 16: 1681–1690, 1998.
47. Dome, J.S. and M.J. Coppes. Recent advances in Wilms tumor genetics. *Curr. Opin. Pediatr.* 14: 5–11, 2002.
48. Donohoe, R.M., Q.A. Zhang, L.K. Siddens, H.M. Carpenter, J.D. Hendricks and L.R. Curtis. Modulation of 7,12-dimethylbenz[*a*]anthracene disposition and hepatocarcinogenesis by dieldrin and chlordecone in rainbow trout. *J. Toxicol. Environ. Health A* 54: 227–242, 1998.
49. Falkmer, S., S.O. Emdin, Y. Östberg, A. Mattisson, M.-L. Sjöbeck and R. Fänge. Tumor pathology of the hagfish, *Myxine glutinosa* and the river lamprey, *Lampetra fluviatilis*. A light-microscopical study with particular reference to the occurrence of primary liver carcinoma, islet-cell tumors, and epidermoid cysts of the skin. *Prog. Exp. Tumor Res.* 20: 217–250, 1976.
50. Fong, A.T., J.D. Hendricks, R.H. Dashwood, S. Vanwinkle, B.C. Lee and G.S. Bailey. Modulation of diethylnitrosamine-induced hepatocarcinogenesis and *O*-6-ethylguanine formation in rainbow trout by indole-3-carbinol, beta-naphthoflavone and aroclor-1254. *Toxicol. Appl. Pharmacol.* 96: 93–100, 1988.
51. Fong, A.T., R.H. Dashwood, R. Cheng, C. Mathews, B. Ford, J.D. Hendricks and G.S. Bailey. Carcinogenicity, metabolism and Ki-*ras* proto-oncogene activation by 7,12-dimethylbenz[*a*]anthracene in rainbow trout embryos. *Carcinogenesis* 14: 629–635, 1993.
52. Fornzler, D., J. Altschmied, I. Nanda, R. Kolb, M. Baudler, M. Schmid and M. Schartl. The *Xmrk* oncogene promoter is derived from a novel amplified locus of unusual organisation. *Genome Res.* 6: 102–113, 1996.
53. Fournie, J.W. and R.M. Overstreet. Retinoblastoma in the spring cavefish, *Chologaster agassizi*. *J. Fish Dis.* 8: 377–381, 1985.
54. Fournie, J.W. and W.K. Vogelbein. Exocrine pancreatic neoplasms in the mummichog (*Fundulus heteroclitus*) from a creosote-contaminated site. *Toxicol. Pathol.* 22: 237–247, 1994.
55. Fournie, J.W., W.E. Hawkins and W.W. Walker. Proliferative lesions in swimbladder of Japanese medaka *Oryzias latipes* and guppy *Poecilia reticulata*. *Dis. Aquat. Org.* 38: 135–142, 1999.
56. Franklin, T.M., J.S. Lee, A. Kohler and J.K. Chipman. Analysis of mutations in the *p53* tumour suppressor gene and Ki- and Ha-*ras* proto-oncogenes in hepatic tumours of European flounder (*Platichthys flesus*). *Mar. Environ. Res.* 50: 251–255, 2000.

57. Friend, S.H., R. Bernards, S. Rogeij, R.A. Weinberg, J.M. Rapaport, D.M. Alberts and R.P. Dryja. A human DNA segment with properties of the gene that predisposes to retinoblastoma and osteosarcoma. *Nature* 323: 643–646, 1986.
58. Froschauer, A., C. Korting, W. Bernhardt, I. Nanda, M. Schmid, M. Schartl and J.-N. Volff. Genomic plasticity and melanoma formation in the fish *Xiphophorus*. *Mar. Biotechnol.* 3: S72–S80, 2001.
59. Gardner, H.S., L.M. Brennan, M.W. Toussaint, A.B. Rosencrance, E.M. Boncavage-Hennessey and M.J. Wolfe. Environmental complex mixture toxicity assessment. *Environ. Health Perspect.* 106(Suppl. 6): 1299–1305, 1998.
60. Geter, D.R., W.E. Hawkins, J.C. Means and G.K. Ostrander. Pigmented skin tumors in gizzard shad (*Dorosoma cepedianum*) from the south-central United States: range extension and further etiological studies. *Environ. Toxicol. Chem.* 17: 2282–2287, 1998.
61. Gordon, M. Hereditary basis of melanosis in hybrid fishes. *Am. J. Cancer* 15: 1495–1523, 1931.
62. Greenblatt, M.S., W.P. Bennett, M. Hollstein and C.C. Harris. Mutations in the *p53* tumor suppressor gene – clues to cancer etiology and molecular pathogenesis. *Cancer Res.* 54: 4855–4878, 1994.
63. Gutjahr-Gobell, R.E., D.E. Black, L.J. Mills, R.J. Pruell, B.K. Taplin and S. Jayaraman. Feeding the mummichog (*Fundulus heteroclitus*) a diet spiked with non-*ortho* and mono-*ortho*-substituted polychlorinated biphenyls: accumulation and effects. *Environ. Toxicol. Chem.* 18: 699–707, 1999.
64. Harada, T., N. Okazaki, S.S. Kubota, J. Hatanaka and M. Enomoto. Spontaneous ovarian tumor in a medaka (*Oryzias latipes*). *J. Comp. Pathol.* 104: 187–193, 1991.
65. Harada, T., H. Itoh, J. Hatanaka, S. Kamiya and M. Enomoto. A morphological study of a thyroid carcinoma in a medaka, *Oryzias latipes*. *J. Fish Dis.* 19: 271–277, 1996.
66. Harborn, J.M., S.L. Lai, J. Whang-Peng, A.F. Gadzar, J.D. Minna and F.J. Kaye. Abnormalities in structure and expression of the human retinoblastoma gene in SCLC. *Science* 241: 353–357, 1988.
67. Harshbarger, J.C. *RTLA Supplements: Registry of Tumors in Lower Animals*, Washington, DC, Smithsonian Institute, 1965–1980.
68. Harshbarger, J.C. and J.B. Clark. Epizootiology of neoplasms in bony fish of North America. *Sci. Total Environ.* 94: 1–32, 1990.
69. Harshbarger, J.C. and M.S. Slatick. Lesser known aquarium fish tumor models. *Mar. Biotechnol.* 3: S115–S129, 2001.
70. Hatanaka, J., N. Doke, T. Harada, T. Aikawa and M. Enomoto. Usefulness and rapidity of screening for the toxicity and carcinogenicity of chemicals in medaka, *Oryzias latipes*. *Jpn. J. Exp. Med.* 52: 243–253, 1982.
71. Hawkins, W.E., J.W. Fournie, R.M. Overstreet and W.W. Walker. Intraocular neoplasms induced by methylazoxymethanol acetate in Japanese medaka (*Oryzias latipes*). *J. Natl Cancer Inst.* 76: 453–465, 1986.
72. Hawkins, W.E., W.W. Walker, R.M. Overstreet, J.S. Lytle and T.F. Lytle. Dose-related carcinogenic effects of waterborne benzo(*a*)pyrene on livers of two small fish species. *J. Ecotox. Environ. Saf.* 16: 219–231, 1988.
73. Hawkins, W.E., R.M. Overstreet and W.W. Walker. Carcinogenicity tests with small fish species. *Aquat. Toxicol.* 11: 113–128, 1988.
74. Hawkins, W.E., W.W. Walker, R.M. Overstreet, J.S. Lytle and T.F. Lytle. Carcinogenic effects of some polycyclic aromatic hydrocarbons on the Japanese medaka and guppy in waterborne exposures. *Sci. Total Environ.* 94: 155–167, 1990.
75. Hawkins, W.E., W.W. Walker and R.M. Overstreet. Carcinogenicity tests using aquarium fish. *Toxicol. Methods* 5: 225–263, 1995.
76. Hawkins, W.E., W.W. Walker, M.O. James, C.S. Manning, D.H. Barnes, C.S. Heard and R.M. Overstreet. Carcinogenic effects of 1,2-dibromoethane (ethylene dibromide; EDB) in Japanese medaka (*Oryzias latipes*). *Mutat. Res. Fund. Mol. Mech. Mutagen.* 399: 221–232, 1998.
77. Hayashi, T., M. Schimerlik and G. Bailey. Mechanisms of chlorophyllin anticarcinogenesis: dose-responsive inhibition of aflatoxin uptake and biodistribution following oral co-administration in rainbow trout. *Toxicol. Appl. Pharmacol.* 158: 132–140, 1999.
78. Hemminki, K. and P. Mutanen. Genetic epidemiology of multistage carcinogenesis. *Mutat. Res. Fund. Mol. Mech. Mutagen.* 473: 11–21, 2001.
79. Hendricks, J.D., D.W. Shelton, J.L. Casteel, T.R. Meyers and R.O. Sinnhuber. Carcinogenicity of methylazoxymethanol acetate (MAMA) to rainbow trout (*Salmo gairdneri*) embryos, with and without prior exposure to Aroclor-1254 (PCB). *Proc. Am. Assoc. Cancer Res.* 24: 54 1983.
80. Hendricks, J.D., T.R. Meyers, J.L. Casteel, J.E. Nixon, P.M. Loveland and G.S. Bailey. Rainbow trout embryos: advantages and limitations for carcinogenesis research. *Natl. Cancer Inst. Monogr.* 65: 129–137, 1984.

81. Hendricks, J.D., T.R. Meyers, D.W. Shelton, J.L. Casteel and G.S. Bailey. Hepatocarcinogenicity of benzo[*a*]pyrene to rainbow trout by dietary exposure and intraperitoneal injection. *J. Natl Cancer Inst.* 74: 839–851, 1985.
82. Hendricks, J.D., R.S. Cheng, D.W. Shelton, C.B. Pereira and G.S. Bailey. Dose-dependent carcinogenicity and frequent Ki-ras protooncogene activation by dietary *N*-nitrosodiethylene in rainbow trout. *Fund. Appl. Toxicol.* 23: 53–62, 1994.
83. Hinton, D.E., S.J. Teh, M.S. Okihiro, J.B. Cooke and L.M. Parker. Phenotypically altered hepatocyte populations in diethylnitrosamine-induced medaka liver carcinogenesis – resistance, growth, and fate. *Mar. Environ. Res.* 34: 1–5, 1992.
84. Hollstein, M., D. Sidransky, B. Vogelstein and C.C. Harris. P53 mutations in human cancers. *Science* 253: 49–53, 1991.
85. Hong, Y.H., C. Winkler and M. Schartl. Production of medakafish chimeras from a stable embryonic stem cell line. *Proc. Natl. Acad. Sci. USA* 95: 3679–3684, 1998.
86. Horowitz, J.M., S.H. Par, E. Bogenmann, J.C. Cheng, D.W. Yandell, F.J. Kaye, T.P. Dryja and R.A. Weinberg. Frequent inactivation of retinoblastoma anti-oncogene is restricted to a subset of human tumor cells. *Proc. Natl. Acad. Sci. USA* 87: 2775–2790, 1990.
87. Ishikawa, J., H.J. Xu, S.X. Hu, D.W. Yandell, S. Maeda, S. Kamidono, W.F. Benedict and R. Takahashi. Inactivation of the retinoblastoma gene in human bladder and renal cell carcinomas. *Cancer Res.* 51: 5736–5743, 1991.
88. Ishikawa, T., T. Shimamine and S. Takayama. Histologic and electron microscope observations on diethylnitrosamine-induced hepatomas in small aquarium fish (*Oryzias latipes*). *J. Natl Cancer Inst.* 55: 909–916, 1975.
89. Jacks, T., A. Fazeli, E.M. Schmitt, R.T. Bronson, M.A. Goodell and R.A. Weinberg. Effects of an Rb mutation in the mouse. *Nature* 359: 295–300, 1992.
90. Jacobs, A.D. and G.K. Ostrander. Further investigation of the etiology of subcutaneous neoplasms in native gizzard shad (*Dorosoma cepedianum*). *Environ. Toxicol. Chem.* 14: 1781–1787, 1995.
91. Kazianis, S., L. Gan, L. Della Coletta, B. Santi, D.C. Morizot and R.S. Nairn. Cloning and comparative sequence analysis of TP53 in *Xiphophorus* fish hybrid melanoma models. *Gene* 212: 31–38, 1998.
92. Kazianis, S., D.C. Morizot, L. Della Coletta, D.A. Johnston, B. Woolcock, J.R. Vielkind and R.S. Nairn. Comparative structure and characterisation of a CDKN2 gene in a *Xiphophorus* fish melanoma model. *Oncogene* 18: 5088–5099, 1999.
93. Kelly, J.D., G.A. Orner, J.D. Hendricks and D.E. Williams. Dietary hydrogen-peroxide enhances hepatocarcinogenesis in trout – correlation with 8-hydroxy-2′-deoxyguanosine levels in liver DNA. *Carcinogenesis* 13: 1639–1642, 1992.
94. Khudoley, V.V. Use of the aquarium fish, *Danio rerio* and *Poecilia reticulata* as a test species for the evaluation of nitrosamine carcinogenicity. In: *Use of Small Fish Species in Carcinogenicity Testing*, NCI Monograph 65, edited by K.L. Hoover, pp. 65–70, 1984.
95. Köhler, A. Cellular effects of environmental contamination in fish from the River Elbe and the North Sea. *Mar. Environ. Res* 28: 417–424, 1989.
96. Köhler, A., H. Deissemann and B. Lauritzen. Histological and cytochemical indices of toxic injury in the liver of dab *Limanda limanda. Mar. Ecol. Prog. Ser.* 91: 141–153, 1992.
97. Köhler, A., S. Bahns and C.J.F. Van Noorden. Determination of kinetic properties of G6PDH and PGDH and the expression of PCNA during liver carcinogenesis in coastal flounder. *Mar. Environ. Res.* 46: 179–183, 1998.
98. Krause, M.K., L.D. Rhodes and R.J. Van Beneden. Cloning of the p53 tumor suppressor gene from the Japanese medaka (*Oryzias latipes*) and evaluation of mutational hotspots in MNNG-exposed fish. *Gene* 189: 101–106, 1997.
99. Kusser, W.C., R.L. Parker and X.L. Miao. Polymerase chain reaction and DNA sequence of rainbow trout tumor suppressor gene P53 exon 5, exon 6 and exon 7 to exon 9. *Aquat. Living Resour.* 7: 11–16, 1994.
100. Langenau, D.M., D. Traver, A.A. Ferrando, J.L. Kutok, J.C. Aster, J.P. Kanki, S. Lin, E. Prochownik, N.S. Trede, L.I. Zon and A.T. Look. Myc-induced T cell leukemia in transgenic zebrafish. *Science* 299: 887–890, 2003.
101. Lee, E.Y.-H.P., C.-Y. Chang, N. Hu, Y.-C.J. Wang, C.-C. Lai, K. Herrup, W.-H. Lee and A. Bradley. Mice deficient for RB are nonviable and show defects in neurogenesis and hematopoiesis. *Nature* 359: 288–294, 1992.
102. Maccubin, A.E. and N. Ersing. Tumors in fish from the Detroit River. *Hydrobiologia* 219: 301–306, 1991.
103. Maccubbin, A.E., J. Black and J.C. Harshbarger. A case report of hepatocellular neoplasia in bowfin, *Amia calva. J. Fish Dis.* 10: 329–331, 1987.

104. Malins, D.C., M.M. Krahn, D.W. Brown, L.D. Rhodes, M.S. Myers, B.B. McCain and S.L. Chan. Toxic chemicals in marine sediment and biota from Mukilteo, Washington: relationships with hepatic neoplasms and other hepatic lesions in English sole (*Parophyrs vetulus*). *J. Natl. Cancer Inst.* 74: 487–494, 1985.
105. Malins, D.C., B.B. McCain, M.S. Myers, D.W. Brown, M.M. Krahn, W.T. Roubal, M.H. Schiewe, J.T. Landahl and S.L. Chan. Field and laboratory studies of the etiology of liver neoplasms in marine fish from Puget Sound. *Environ. Health Perspect.* 71: 5–16, 1987.
106. Malitschek, B., D. Fornzler and M. Schartl. Melanoma formation in *Xiphophorus*: a model system for the role of receptor tyrosine kinases in tumorigenesis. *Bioessays* 17: 1017–1023, 1995.
107. Manam, S., R.D. Storer, S. Prahalada, K.R. Leander, A.R. Kraynak, B.J. Ledwith, M.J. Van Zwieten, M.O. Bradley and W.W. Nichols. Activation of the Ha-, Ki-, and N-*ras* genes in chemically-induced liver tumors from CD-1 mice. *Cancer Res.* 52: 3347–3352, 1992.
108. Mangold, K., Y.-J. Chang, C. Mathews, K. Marien, J. Hendricks and G. Bailey. Expression of *ras* genes in rainbow trout liver. *Mol. Carcinog.* 4: 97–102, 1991.
109. Manning, C.S., W.E. Hawkins, D.H. Barnes, W.D. Burke, C.S. Barnes, R.M. Overstreet and W.W. Walker. Survival and growth of Japanese medaka (*Oryzias latipes*) exposed to trichlorethylene at multiple life stages: implications of establishing the maximum tolerated dose for chronic aquatic carcinogenicity bioassays. *Toxicol. Methods* 11: 147–159, 2001.
110. Masahito, P., T. Ishikawa and S. Takayama. Spontaneous spermatocytic seminoa in African lungfish, *Protopterus aethiopicus* Heckel. *J. Fish Dis.* 7: 169–172, 1984.
111. Masahito, P., T. Ishikawa, H. Sugano, H. Uchida, T. Yasuda, T. Inaba, Y. Hirosaki and A. Kasuga. Spontaneous hepatocellular carcinomas in lungfish. *J. Natl. Cancer Inst.* 77: 291–298, 1986.
112. Masahito, P., K. Aoki, N. Egami, T. Ishikawa and H. Sugano. Life span studies on spontaneous tumor-development in the medaka (*Oryzias latipes*). *Jap. J. Cancer Res.* 80: 1058–1065, 1989.
113. Maueler, W., F. Raulf and M. Schartl. Expression of proto-oncogenes in embryonic, adult, and transformed tissue of *Xiphophorus*. *Oncogene* 2: 421–430, 1988.
114. Maueler, W., A. Schartl and M. Schartl. Different expression patterns of oncogenes and proto-oncogenes in hereditary and carcinogen-induced tumors of *Xiphophorus*. *Int. J. Cancer* 55: 288–296, 1993.
115. McCarthy, J.F., H. Gardner and M.J. Wolfe. DNA alterations and enzyme activities in Japanese medaka (*Oryzias latipes*) exposed to diethylnitrosamine. *Neurosci. Biobehav.* 15: 99–102, 1991.
116. Metcalfe, C.D., V.W. Cairns and J.D. Fitzsimons. Experimental induction of liver tumors in rainbow trout by contaminated sediment from Hamilton Harbour, Ontario. *Can. J. Fish. Aquat. Sci.* 45: 2161–2167, 1988.
117. Mitchell, D.L., J.A. Meador, M. Byrom and R.B. Walter. Resolution of UV-induced DNA damage in *Xiphophorus* fishes. *Mar. Biotechnol.* 3: S61–S71, 2001.
118. Moore, M.J. and M.L. Myers. Pathology of chemical-associated neoplasia in fish. In: *Aquatic Toxicology: Molecular, Biochemical and Cellular Perspectives*, edited by D.C. Malins and G.K. Ostrander, Boca Raton, FL, CRC Press/Lewis Publishers, pp. 327–386, 1994.
119. Morizot, D.C., B.B. McEntire, L. Della Coletta, S. Kazianis, M. Schartl and R.S. Nairn. Mapping of tyrosine kinase gene family members in *Xiphophorus* melanoma model. *Mol. Carcinog.* 22: 150–157, 1998.
120. Myers, M.L., J.T. Landahl, M.M. Krahn, L.L. Johnson and B.B. McCain. Overview on studies on liver carcinogenesis in English Sole from Puget Sound, evidence for a xenobiotic chemical aetiology. I Pathology and epizootiology. *Sci. Total Environ.* 94: 33–50, 1990.
121. Myers, M.L., J.T. Landahl, M.M. Krahn and B.B. McCain. Relationships between hepatic neoplasms and related lesions and exposure to toxic chemicals in marine fish from the US West Coast. *Environ. Health Perspect.* 90: 7–15, 1991.
122. Myers, M.L., L.L. Johnson, P. Olson, C.M. Stehr, B.H. Horness, T.K. Collier and B.B. McCain. Toxicopathic hepatic lesions as biomarkers of chemical contaminant exposure and effects in marine bottomfish species from the Northeast and Pacific Coasts, USA. *Mar. Pollut. Bull.* 37: 92–113, 1998.
123. Nacci, D.E., L. Coiro, D. Champlin, S. Jayaraman, R. McKinney, T.R. Gleason, W.R. Munns, J.L. Specker and K.R. Cooper. Adaptations of wild populations of estuarine fish *Fundulus heteroclitus* to persistent environmental contaminants. *Mar. Biol.* 134: 9–17, 1999.
124. Nacci, D.E., S. Jayaraman and J.L. Specker. Stored retinoids in populations of the estuarine fish *Fundulus heteroclitus* indigenous to PCB-contaminated and reference sites. *Arch. Env. Contam. Toxicol.* 40: 511–518, 2001.
125. Nacci, D.E., M. Kohan, M. Pelletier and E. George. Effects of benzo[*a*]pyrene exposure on a fish population resistant to the toxic effects of dioxin-like compounds. *Aquat. Toxicol.* 57: 203–215, 2002.

126. Nairn, R.S., D.C. Morizot, S. Kazianis, A.D. Woodhead and R.B. Setlow. Nonmammalian models for sunlight carcinogenesis: genetic analysis of melanoma formation in *Xiphophorus* hybrid fish. *Photochem. Photobiol.* 64: 440–448, 1996.
127. Nairn, R.S., S. Kazianis, L. Della Coletta, D. Trono, A.P. Butler, R.B. Walter and D.C. Morizot. Genetic analysis of susceptibility to spontaneous and UV-induced carcinogenesis in *Xiphophorus* hybrid fish. *Mar. Biotechnol.* 3: S24–S36, 2001.
128. Nunez, O., J.D. Hendricks and J.R. Duimstra. Ultrastructure of hepatocellular neoplasms in aflatoxin B_1-initiated rainbow trout. *Toxicol. Pathol.* 19: 11–23, 1991.
129. Oganesian, A., J.D. Hendricks, C.B. Pereira, G.A. Orner, G.S. Bailey and D.E. Williams. Potency of dietary indole-3-carbinol as a promoter of aflatoxin B-1-initiated hepatocarcinogenesis: results from a 9000 animal tumor study. *Carcinogenesis* 20: 453–458, 1999.
130. Okihiro, M.S. and D.E. Hinton. Progression of hepatic neoplasia in medaka (*Oryzias latipes*) exposed to diethylnitrosamine. *Carcinogenesis* 20: 933–940, 1999.
131. Orner, G.A., C. Mathews, J.D. Hendricks, H.M. Carpenter, G.S. Bailey and D.E. Williams. Dehydroepiandrosterone is a complete hepatocarcinogen and potent tumor promoter in the absence of peroxisome proliferation in rainbow trout. *Carcinogenesis* 16: 2893–2898, 1995.
132. Orner, G.A., J.D. Hendricks, D. Arbogast and D.E. Williams. Modulation of *N*-methyl-*N*′-nitro-nitrosoguanidine multiorgan carcinogenesis by dehydroepiandrosterone in rainbow trout. *Toxicol. Appl. Pharmacol.* 141: 548–554, 1996.
133. Orner, G.A., J.D. Hendricks, D. Arbogast and D.E. Williams. Modulation of aflatoxin-B-1 hepatocarcinogenesis in trout by dehydroepiandrosterone: initiation/post-initiation and latency effects. *Carcinogenesis* 19: 161–167, 1998.
134. Ortego, L.S., W.E. Hawkins, Y. Zhu and W.W. Walker. Chemically-induced hepatocyte proliferation in the medaka (*Oryzias latipes*). *Mar. Environ. Res.* 42: 75–79, 1996.
135. Ostrander G.K., J.-K. Shim, W.E. Hawkins and W.W. Walker. A vertebrate model for investigation of retinoblastoma. Proceedings of the 83rd Annual Meeting of the American Association for Cancer Research, p. 109, 1992, Abstract 652.
136. Ostrander, G.K., W.E. Hawkins, R.L. Kuehn, A.D. Jacobs, K.D. Berli and J. Pigg. Pigmented subcutaneous spindle cell tumors in native gizzard shad (*Dorosoma cepedianum*). *Carcinogenesis* 16: 1529–1535, 1995.
137. Ostrander, G.K., K. Cheng, M.J. Wolfe and J. Wolf. Shark cartilage, cancer and the growing threat of pseudoscience. Cancer Research. In press, 2004.
138. Ownby, D.R., M.C. Newman, M. Mulvey, W.K. Vogelbein, M.A. Unger and L.F. Arzayus. Fish (*Fundulus heteroclitus*) populations with different exposure histories differ in tolerance of creosote-contaminated sediments. *Environ. Toxicol. Chem.* 21: 1897–1902, 2002.
139. Peck-Miller, K.A., M. Myers, T. Collier and J.E. Stein. Complete cDNA sequence of the Ki-*ras* proto-oncogene in the liver of wild English sole (*Pleuronectes vetulus*) and mutation analysis of hepatic neoplasms and other toxicopathic liver lesions. *Mol. Carcinog.* 23: 207–216, 1998.
140. Peters, E.C., J.C. Halas and H.B. McCarty. Calicoblastic neoplasms in *Acropora palmata*, with a review of reports on anomalies of growth and form in corals. *J. Natl. Cancer Inst.* 76: 895–912, 1986.
141. Rehulka, J. Spontaneous nephroblastoma in a hatchery rainbow trout *Oncorhynchus mykiss*. *Dis. Aquat. Org.* 14: 75–79, 1992.
142. Reichert, W.L., M.S. Myers, K. Peck-Miller, B. French, B.F. Anulacion, T.K. Collier, J.E. Stein and U. Varanasi. Molecular epizootiology of genotoxic events in marine fish: linking contaminant exposure, DNA damage, and tissue-level alterations. *Mutat. Res. Rev.* 411: 215–225, 1998.
143. Romanov, A.A. and Yu.V. Altufev. Tumors in the sex glands and liver of sturgeons (Acipenseridae) of the Caspian Sea. *J. Ichthyol.* 31: 44–49, 1991.
144. Rotchell, J.M., R.M. Stagg and J.A. Craft. Chemically-induced genetic damage in fish: isolation and characterization of the dab (*Limanda limanda*) *ras* gene. *Mar. Pollut. Bull.* 31: 457–459, 1995.
145. Rotchell, J.M., E. Unal, R.J. Van Beneden and G.K. Ostrander. Induction of retinoblastoma tumor suppressor gene mutations in chemically-induced liver tumors in medaka (*Oryzias latipes*). *Mar. Biotechnol.* 3: S44–S49, 2001.
146. Rotchell, J.M., J.-S. Lee, J.K. Chipman and G.K. Ostrander. Structure, expression and activation of fish *ras* genes. *Aquat. Toxicol.* 55: 1–21, 2001.
147. Rotchell, J.M., J. Shim, J.B. Blair, W.E. Hawkins and G.K. Ostrander. Cloning of the retinoblastoma cDNA from the Japanese medaka (*Oryzias latipes*) and preliminary evidence of mutational alterations in chemically-induced retinoblastomas. *Gene* 263: 231–237, 2001.
148. Schartl, M. Platyfish and swordtails: a genetic system for the analysis of molecular mechanisms of tumor formation. *Trends Genet.* 11: 185–189, 1995.

149. Schartl, M., U. Hornung, H. Gutbrod, J.-N. Volff and J. Wittbrodt. Melanoma loss-of-function mutants in *Xiphophorus* caused by X*mrk*-oncogene deletion and gene disruption by a transposable element. *Genetics* 153: 1385–1394, 1999.
150. Schlumberger, H.G. and B. Lucké. Tumors of fishes, amphibians, and reptiles. *Cancer Res.* 8: 657–753, 1948.
151. Schmale, M.C. Experimental induction of neurofibromatosis in the bicolor damselfish. *Dis. Aquat. Org.* 23: 201–212, 1995.
152. Schmale, M.C. and G.T. Hensley. Transmissibility of a neurofibromatosis-like disease in bicolor damselfish. *Cancer Res.* 48: 3828–3833, 1988.
153. Schmale, M.C., G.T. Hensley and L.R. Udey. Multiple schwanomas in the biocolor damselfish, *Pomacentrus partitus*: a possible model of von Rocklinghausen neurofibromatosis. *Am. J. Pathol.* 112: 238–241, 1983.
154. Schmale, M.C., G.T. Hensley and L.R. Udey. Neurofibromatosis in the biocolor damselfish (*Pomacentrus partitus*) as a model of von Rocklinghausen neurofibromatosis. *Ann. NY Acad. Sci.* 486: 386–402, 1986.
155. Schmale, M.C., P.D.L. Gibbs and C.E. Campbell. A virus-like agent associated with neurofibromatosis in damselfish. *Dis. Aquat. Org.* 49: 107–115, 2002.
156. Schroeders, V.D. Tumors of fishes. Dissertation in Russian. Translation in Army Medical Library, Washington, DC, St Petersburg, 1908.
157. Shelton, D.W., J.D. Hendricks and G.S. Bailey. The hepatocarcinogenicity of diethylnitrosamine to rainbow trout and its enhancement by aroclor-1242 and aroclor-1254. *Toxicol. Lett.* 22: 27–31, 1984.
158. Shih, P.M.T., R. Winn, M. Norris, K. Brayer and B.S. Shane. Mutagenesis of ethylnitrosourea at the lacI locus in a transgenic fish (*Oryzias latipes*). *Environ. Mol. Mutagen.* 35(Suppl. 31): 186, 2000.
159. Simpson, M.G. Histopathological changes in the liver of dab (*Limanda limanda*) from a contamination gradient in the North Sea. *Mar. Environ. Res.* 34: 39–43, 1992.
160. Sparks, A.K. Tumors and tumorlike lesions in invertebrates. *Synopsis of Invertebrate Pathology Exclusive of Insects*, Amsterdam, Elsevier Science Publishers, pp. 91–131, 1985.
161. Spitsbergen, J.M., H. Tsai, A.R. Reddy, D. Arbogast, J.D. Hendricks and G.S. Bailey. Neoplasia in zebrafish treated with 7,12-dimethylbenz[*a*]anthracene by two exposure routes at different developmental stages. *Toxicol. Pathol.* 28: 705–715, 2000.
162. Spitsbergen, J.M., H. Tsai, A.R. Reddy, D. Arbogast, J.D. Hendricks and G.S. Bailey. Neoplasia in zebrafish treated with *N*-methyl-*N*′-nitro-*N*-nitrosoguanidine by three exposure routes at different developmental stages. *Toxicol. Pathol.* 28: 716–725, 2000.
163. Stein, J.E., W.L. Reichert, M. Nishimoto and U. Varanasi. Overview of studies on liver carcinogenesis in English sole from Puget Sound; evidence for a xenobiotic chemical aetiology II: Biochemical studies. *Sci. Total Environ.* 94: 51–69, 1990.
164. Stentiford, G.D., M. Longshaw, B.P. Lyons, G. Jones, M. Green and S.W. Feist. Histopathological biomarkers in estuarine fish species for the assessment of biological effects of contaminants. *Mar. Environ. Res.* 55: 137–159, 2003.
165. Takahashi, N., C.L. Miranda, M.C. Henderson, D.R. Buhler, D.E. Williams and G.S. Bailey. Inhibition of in vitro aflatoxin B1 DNA binding in rainbow trout by CYP1A inhibitors – alpha-naphthoflavone, beta-naphthoflavone and trout CYP1A1 peptide antibody. *Comp. Biochem. Physiol.* 110C: 273–280, 1995.
166. T'Ang, A., J.M. Varlay, S. Chakraborty, A.L. Murphree and Y.T.K. Fung. Structural rearrangement of the retinoblastoma gene in human breast carcinoma. *Science* 242: 263–266, 1988.
167. Teh, S.J. and D.E. Hinton. Gender-specific growth and hepatic neoplasia in medaka (*Oryzias latipes*). *Aquat. Toxicol.* 41: 141–159, 1998.
168. Thiyagarajah, A. and M.E. Bender. Lesions in the pancreas and liver of an oyster toadfish collected from the Lower York River, Virginia. *J. Fish Dis.* 11: 359–364, 1988.
169. Thomas, L. Les tumeurs des poissons (étude anatomique et pathogénique). *Bull. Assoc. Fr. Etude Cancer* 20: 703–760, 1931.
170. Thorgaard, G.H., D.N. Arbogast, J.D. Hendricks, C.B. Pereira and G.S. Bailey. Tumor suppression in triploid trout. *Aquat. Toxicol.* 46: 121–126, 1999.
171. Torten, M., Z. Liu, M.S. Okihiro, S.J. Teh and D.E. Hinton. Induction of ras oncogene mutations and hepatocarcinogenesis in medaka (*Oryzias latipes*) exposed to diethylnitrosamine. *Mar. Environ. Res.* 42: 93–98, 1996.
172. Toussaint, M.W., M.J. Wolfe, D.T. Burton, F.J. Hoffmann, T.R. Shedd and H.S. Gardner. Histopathology of Japanese medaka (*Oryzias latipes*) chronically exposed to a complex environmental mixture. *Toxicol. Pathol.* 27: 652–663, 1999.
173. US Environmental Protection Agency. Effects of radiation on aquatic organisms and radiobiological methodologies for effects assessment. EPA 520/1 – 85-016. Washington, DC, 1986.

174. Van Beneden, R.J., D.K. Watson, T.T. Chen, J.A. Lautenberger and T.S. Papas. Cellular *myc* (c-*myc*) in fish: its relationship to other vertebrate *myc* genes and to the transforming genes of the MC29 family of viruses. *Proc. Natl Acad. Sci. USA* 83: 3698–3702, 1986.
175. Van Beneden, R.J., G.R. Gardner, N.J. Blake and D.G. Blair. Implications for the presence of transforming genes in gonadal tumors in two bivalve mollusk species. *Cancer Res.* 53: 2976–2979, 1993.
176. Van Veld, P.A. and D.J. Westbrook. Evidence for a depression of cytochrome P450IA in a population of chemically-resistant mummichog (*Fundulus heteroclitus*) from a creosote-contaminated environment. *Environ. Sci.* 3: 221–224, 1995.
177. Vethaak, A.D., D. Bucke, T. Lang, P.W. Wester, J. Jol and M. Carr. Fish disease monitoring along a pollution transect: a case study using dab (*Limanda limanda*) in the German Bight. *Mar. Ecol. Prog. Ser.* 91: 173–192.
178. Vethaak, A.D. and J.G. Jol. Diseases of flounder (*Platichthys flesus*) in Dutch coastal and estuarine waters, with particular reference to environmental stress factors. *Dis. Aquat. Org.* 26: 81–97, 1996.
179. Vethaak, A.D. and P.W. Wester. Diseases of flounder (*Platichthys flesus*) in Dutch coastal and estuarine waters, with particular reference to environmental stress factors. II Liver histopathology. *Dis. Aquat. Org.* 26: 99–116, 1996.
180. Vincent, F., S. Jaunet, F. Galgani, H. Besselink and J. Koeman. Expression of ras gene in flounder (*Platichthys flesus*) and red mullet (*Mullus barbatus*). *Biochem. Biophys. Res. Commun.* 215: 659–665, 1995.
181. Vincent-Hubert, F. cDNA cloning and expression of two Ki-*ras* genes in the flounder, *Platichthys flesus*, and analysis of hepatic neoplasms. *Comp. Biochem. Physiol.* 126C: 17–27, 2000.
182. Vogelbein, W.K., J.W. Fournie, P.A. Van Veld and R.J. Huggett. Hepatic neoplasms in the mummichog *Fundulus heteroclitus* from a creosote-contaminated site. *Cancer Res.* 50: 5978–5986, 1990.
183. Wakamatsu, Y., S. Pristyazhnyuk, M. Kinoshita, M. Tanaka and K. Ozato. The see-through medaka: a fish model that is transparent throughout life. *Proc. Natl Acad. Sci. USA* 98: 10046–10050, 2001.
184. Wakamatsu, Y., B.S. Ju, I. Pristyaznhyuk, K. Niwa, T. Ladygina, M. Kinoshita, K. Araki and K. Ozato. Fertile and diploid nuclear transplants derived from embryonic cells of a small laboratory fish, medaka (*Oryzias latipes*). *Proc. Natl Acad. Sci. USA* 98: 1071–1076, 2001.
185. Walter, R.B., H.-M. Sung, R.D. Obermoeller, D.L. Mitchell, G.W. Intano and C.A. Walter. Relative base excision repair in *Xiphophorus* fish tissue extracts. *Mar. Biotechnol.* 3: S50–S60, 2001.
186. Weber, L.P., Y. Kiparissis, G.S. Hwang, A.J. Niimi, D.M. Janz and C.D. Metcalfe. Increased cellular apoptosis after chronic aqueous exposure to nonylphenol and quercetin in adult medaka (*Oryzias latipes*). *Comp. Biochem. Physiol.* 131C: 51–59, 2002.
187. Wellbrock, C., R. Lammers, A. Ullrich and M. Schartl. Association between the melanoma-inducing receptor kinase X*mrk* and the *src* family tyrosine kinases in *Xiphophorus*. *Oncogene* 10: 2135–2143, 1995.
188. Wellbrock, C., E. Geissinger, A. Gomez, P. Fischer, K. Friedrich and M. Schartl. Signalling by the oncogenic receptor tyrosine kinase X*mrk* leads to activation of STAT5 in *Xiphophorus* melanoma. *Oncogene* 16: 3047–3056, 1998.
189. Wellbrock, C. and M. Schartl. Multiple binding sites in the growth factor receptor X*mrk* mediate binding to $p59^{fyn}$, GRB2 and Shc. *Eur. J. Biochem.* 260: 275–283, 1999.
190. Winn, R., M. Norris, K. Brayer, S. Muller and C. Torres. Bacteriophage lambda and plasmid pUR 288 transgenic fish models for detecting in vivo mutations. *Mar. Biotechnol.* 3: S185–S195, 2001.
191. Winn, R., M. Norris, K. Brayer, C. Torres and S. Muller. Detection of mutations in transgenic fish carrying a bacteriophage l cII transgene target. *Proc. Natl Acad. Sci. USA* 97: 12655–12660, 2000.
192. Winn, R.N., R.J. Van Beneden and J.G. Burkhart. Transfer, methylation and spontaneous mutation frequency of PHI-X174AM3CS70 sequences in medaka (*Oryzias latipes*) and mummichog (*Fundulus heteroclitus*) – implications for gene transfer and environmental mutagenesis in aquatic species. *Mar. Environ. Res.* 40: 247–265, 1995.
193. Wittbrodt, J.D., B. Adam, W. Malitschek, F. Maueler, F. Raulf, A. Telling, S.M. Robertson and M. Schartl. Novel putative receptor tyrosine kinase encoded by the melanoma-inducing *Tu* locus in *Xiphophorus*. *Nature* 341: 415–421, 1989.
194. Roy, N.K., J. Stabile, J.E. Seeb, C. Habicht and I. Wirgin. High frequency of K-ras mutations in pink salmon embryos experimentally exposed to Exxon Valdez oil. *Environ. Toxicol. Chem.* 18: 1521–1528, 1999.
195. McMahon, G., L.J. Huber, M.J. Moore and J.J. Stegeman. Mutations in c-Ki-ras oncogenes in diseased livers of winter flounder from Boston Harbor. *Proc. Natl Acad. Sci. USA* 87: 841–845, 1990.
196. Vincent, F., J. De Boer, A. Pfohl-Leszkowicz, Y. Cherrel and F. Galgani. Two cases of ras mutation associated with liver hyperplasia in dragonets (*Callionymus lyra*) exposed to polychlorinated biphenyls and polycyclic aromatic hydrocarbons. *Mol. Carcinog.* 21: 121–127, 1998.

Biochemistry and Molecular Biology of Fishes, vol. 6
T. P. Mommsen and T. W. Moon (Editors)

CHAPTER 10

Metallothionein: Structure and regulation

PETER KLING* AND PER-ERIK OLSSON**

**Department of Zoology, University of Göteborg SE-405 30, Göteborg, Sweden, and **Department of Natural Sciences, Section of Biology, University of Örebro SE-701 82, Örebro, Sweden*

I. Introduction

Metallothionein (MT) is an ubiquitous cytosolic, cysteine-rich, metal- and free radical-binding protein. Since the discovery of MT four decades ago, extensive efforts have been directed to define the physiological role of MT. The first extensive chemical characterization of MT demonstrated that it contained 5.9% Cd^{2+}, 2.2% Zn^{2+} and 8.5% S (w/w), and that 95% of the S was present as sulfhydryl groups of cysteine residues[40]. The sulfhydryl groups were all found to be involved in cation binding. Aromatic amino acids and histidine were absent and the protein had little or no absorbance at 280 nm.

Initially, MT was proposed to protect cells from Cd toxicity[40]. The finding that large amounts of MT were formed in the liver of rabbits in response to repeated doses of Cd supported this suggestion[72]. Thus, Piscator proposed that once thionein synthesis had been initiated, the animal develops resistance to Cd, and will consequently tolerate higher doses of this toxic metal[72]. At about this time environmental cadmium pollution was identified as a causative factor in the Japanese Itai-itai disease[32]. These observations promoted biochemical and chemical research into the toxicological importance of MT production as a defense mechanism against low levels of cadmium exposure. However, it soon became apparent that other metals, zinc and copper in particular, could also induce MT synthesis. Currently, the primary function of MT is believed to be the maintenance of homeostasis of these essential elements[35].

While the primary physiological role of MT involves the homeostasis of zinc and copper, it remains that MT also has a role in the cellular defense against cadmium and mercury. In addition, being a thiol containing protein MT has the potential to be an effective free radical scavenger, therefore, important in regulating the cellular redox-state.

During the processes of transcription, translation and replication, many Zn requiring enzymes are required. This suggests that MT could be involved in the control of development and differentiation[28]. Recently, studies have revealed a role for MT in preventing oxidative stress. Disturbances in trace metal metabolism and generation of free radicals is often implicated in the development of pathological states and aging, and MT may be a link in these processes.

The present review is focused on the regulation of teleost fish MT by heavy metals and free radicals. The conservation of regulatory mechanisms between fish and mammals is discussed and the unique characteristics of fish MT is highlighted.

II. MT structure

The presence of MT in fish was first indicated in the goldfish, *Carassius auratus*[51]. Other teleosts reported to contain MT included the copper rockfish, *Sebastodes caurinus*; eel, *Anguilla anguilla;* plaice, *Pleuronectes platessa*; and flounder, *Parophrys vetulus*[14,15,23,59]. The first characterization of fish MT demonstrated that two isoforms existed in eel liver and included the first preliminary amino acid data[58]. It was established that the protein is inducible[59], had an apparent molecular weight close to 7000 Da[49], binds zinc[15], cadmium[52], copper[23] and mercury[14] and had a relatively high cysteine content[58,66]. Since then MT has been isolated and characterized from a number of teleosts.

The first partial MT amino acid sequence was obtained from plaice[67] and revealed that fish and mammalian MT was closely related. In 1987 the first complete sequencing of a teleost MT was accomplished[13]. This demonstrated the presence of two MT isoforms, related to mammalian MT1 and MT2, with a high sequence similarity. The sequences obtained from fish MT show that all fish species so far studied have the cysteines conserved in number and position. Two main differences from mammalian MTs are present (Fig. 1). One is the difference in amino-terminal sequence, the main antigenic epitope, which has been shown to result in low antibody reactivity between distantly related species. The second main difference is the positioning of the cysteine in position 55 in teleosts that is found in position 57 in mammals. The number of cysteines is conserved among fish except for the pupfish (*Cyprinodon nevadensis*) where an extra cysteine has replaced a serine at position 27 (EMBL accession number X97273). The pupfish inhabits desert streams with high daily fluctuations in water temperature. Studies are presently ongoing to determine the effect of this additional cysteine on MT function.

β-Domain (N-terminal)

```
           1       5           10          15          20        25        30
Avian      M D P Q D C X C - - - - - - C X C - - - C X C - - C X C - - C - -
Mammalia   M D P N   C X C - - - - - - C X C - - - C X C - - C X C - - C - -
Piscine    M D P     C X C - - - - - - C X C - - - C X C - - C X C - - C - -
MT III     M D P N T C X C - - - - - - C X C - - - C X C - - C X C - - C - -
MT IV      M D P G E C X C - - - - - - C X C - - - C X C - - C X C - - C - -
```

α-Domain (N-terminal)

```
                 35          40                           55        60
Avian      - C C X C C - - - C X X C - - - C X C - - - - - - - C X C C -
Mammalia   - C C X C C - - - C X X C - - - C X C - - - - - - - C X C C -
Piscine    - C C X C C - - - C X X C - - - C X C - - - - C - - - C C -
MT III     - C C X C C - - - C X X C - - - C X C - - - ∇ - - C X C C -
MT IV      - C C X C C - - - C X X C - - - C X C - - - - - - - C X C C -
```

Fig. 1. Schematic representation of the sequences of MT-I and MT-II isoforms from avian, mammalian and piscine species aligned with that of mammalian MT-III and MT-IV. The arrangement and distribution of the cysteine-containing domain is indicated. The amino acids connecting the cysteine residues are indicated by an X. Amino acids not considered directly involved in metal binding (-). To achieve maximum alignment of the N-terminal domain, gaps in the mammalian and piscine sequences were introduced. The residues are numbered in relation to the avian primary sequence. ∇ represent a variable residue in human, mouse and rat MT-III.

III. MT synthesis and regulation

MT synthesis in fish and mammals is stimulated by a wide variety of agents, including heavy metals, glucocorticoids, inflammatory substances and oxidative stress[24,29,42,47]. MT transcription does not depend on *de novo* protein synthesis and appears to be mainly regulated at the transcriptional level[41]. The use of MT as an indicator of heavy-metal pollution has received a great deal of interest. The relationship between heavy metals and MT was early demonstrated in field studies[60,74]. However, although elevated cadmium levels lead to increased MT levels in different tissues, there are also other compounds and circumstances that can regulate the MT levels in fish.

MT induction by metals has been shown for fish and fish cell lines[38,91]. Other studies have shown that hepatic MT levels are elevated by glucocorticoids, noradrenalin and progesterone[33,38]. MT is endogenously regulated during different developmental life stages[62], during sexual maturation[61] and in response to water temperature[64]. The use of MT as a biomonitoring tool should be performed with caution due to the multitude of circumstances where MT is endogenously regulated.

It has been shown that binding of heavy metals such as Hg and Cd to MT decreases the toxicity of these metals (Fig. 2). Thus toxicity would first occur when the MT pool is saturated with the toxic metals, and the metals start to bind other proteins in the cell. This 'spillover' hypothesis states that toxicological changes coincide with the appearance of toxic metals bound to other proteins than MT[86]. If high amounts of heavy metals accumulate, the basal levels of MT would not be sufficient for protection and *de novo* MT synthesis is required. Therefore, the rate of uptake of the metal and the rate of MT synthesis determine the protection[55]. Several studies have shown that

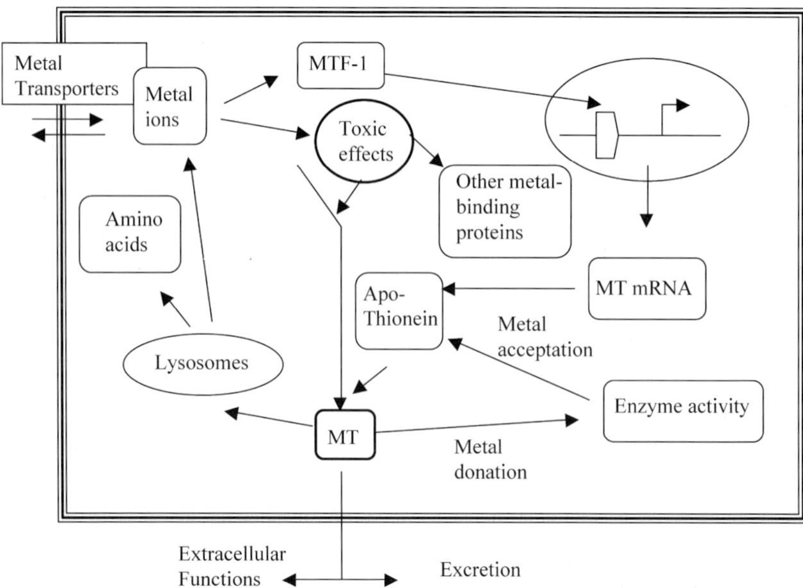

Fig. 2. MTs role in the subcellular distribution of heavy metals. Following uptake, metals activate MTF-1 to initiate transcription at MREs on MT promoters. Raised apo-thionein levels result in reduced toxicity of the heavy metal. Once the metals are bound to MT there is a decrease in free metals and a subsequent MTF-1 inactivation, and lowered MT gene transcription. When the rate of heavy metal accumulation exceeds that of MT synthesis, toxicity may occur. Binding of metals such as Cd and Hg may impair MTs role in regulating the levels of trace elements such as Zn and Cu, essential for maintaining enzyme activity. Other heavy metal binding proteins aid to protect from toxicity.

low MT expressing cell lines[22,68,69] are sensitive to Cd, whereas, excessive expression of MT confers resistance to Cd[9,43]. Moreover, generation of MT-null mice by targeted disruption of MT genes in mice have revealed that those mice become hypersensitive to Cd[54,56]. These results argue in favor of the spillover hypothesis. However, if the main role of MT is to regulate the cellular levels of Zn and Cu, the binding of Hg and Cd would prevent MT from fulfilling its primary function, therefore even low levels would result in toxic effects.

Although Cd and Hg detoxification is a property of MT, this is probably not its primary role. From an evolutionary point of view, it is unlikely that MT was evolved to protect from non-abundant elements. The toxicity of heavy metals such as Cd is more likely to depend on species-specific uptake and other low molecular weight (LMW) heavy metal binding proteins including GSH[7,78].

IV. Metal regulation of MT synthesis

The regulatory regions of mammalian and yeast MTs have been studied in detail, but less is known about fish MT promoter regions. A hallmark of MTs is the rapid transactivation through multiple copies of metal responsive elements (MREs) within their promoters. Comparison of teleost and mammalian MT promoters reveals a high degree of conservation with respect to the MRE consensus sequence 5′ TGC(G/A)CNC 3′ and indicates that the heavy metal response is highly conserved. The rainbow trout MT-B (rtMT-B) gene was the first teleost MT gene to be isolated, and sequencing of that promoter consequently revealed conservation between mammalian and teleost MRE core sequences[90]. A unique feature among teleost MT promoters is the organization of proximal and distal MREs. This has been demonstrated in the rainbow trout[57,63,90], pike (*Esox lucius*) and stone loach (*Noemachelius barbatulus*)[45], Sockeye salmon (*Oncorhynchus nerka*)[18], Antarctic icefish (*Chionodraco hamatus*) MTII[77] and carp (*Cyprinus carpio*) (Genebank accession number AF001983). The MREs in the MTI promoter of Antarctic icefish is an exception among fish since the MREs are only localised within the proximal part of the promoter[77]. Metal-induction of cells transfected with full length promoters correlated well with the number of MREs present in the MT genes, indicating that both proximal and distal MREs are needed to achieve maximum inducibility of MT transcription[63,65,90].

Isolation of the MRE binding transcription factor (MTF-1) from several mammalian species[16,73], Japanese puffer fish, *Fugu rubripes*[5] and zebrafish, *Danio rerio*[19] have revealed a high degree of evolutionary conservation of the heavy metal stress response. MTF-1 belongs to the Cys_2-His_2 family of transcription factors with separate transcriptional activation domains rich in acidic and proline residues[73]. The binding of MTF-1 to MREs is strongly dependent on Zn or agents that mobilize cellular pools of Zn[25,26,69]. The observation that cycloheximide treatment of BHK cells mediated MRE driven vector expression, as well as Zn, led to the speculation that MTF-1 may be under the control of an inhibitor (MTI)[69]. However, more recent data suggest that MTF-1 is the zinc sensor in the cell with a very low binding affinity for Zn. Deletions within recombinant MTF-1 suggests that the Zn finger domain is involved in Zn-dependent activation, where MTF-1 and its Zn finger 1 acts as a sensor for available free Zn[12,25]. The activation of MTF-1 *in vitro* appears to be exclusively dependent on Zn, although MTF-1 from puffer fish appears to be activated by Cd[5]. Interestingly, when comparing the warm and cold-water species such as zebrafish and rainbow trout, it was observed that while the zebrafish contain one MTF-1 isoform, the rainbow trout appears to contain two. While Zn activated one isoform within a wide temperature range, the other isoform was only activated at high temperatures[27]. These adapted MTF-1 isoforms of the rainbow trout allows temperature specific MT regulation in response to metals during the summer and winter season. *In vivo* studies of MTF-1 suggest an important role during development, since MTF-1 knockout mice die *in utero* from hepatic failure[34].

Embryological studies have shown that MTF-1 is strongly and ubiquitously expressed already in the one-cell stage of the zebrafish, whereas MTF-1 is predominantly expressed in the neural parts after hatching, indicating that MTF-1 is important during early development in the zebrafish[19]. The MTF-1 transcription factor is also activated in the zebrafish cell line, ZEM2S, in response to metal treatment[17]. The activity of MTF-1 in the embryological teleost cell line CHSE-214 has been confirmed in three different generations (5, 34 and 59) of the CHSE-214

TABLE 1

Comparison of MRE-dependent MTF-1 activity in three different generations (5, 34 and 59) of the teleost CHSE-214 cell line in response to different concentrations of Zn

Zn (μM)	CHSE-214: 5	CHSE-214: 34	CHSE-214: 59
0	3.1	1.7	1.4
50	3.6	2.4	1.6
100	11	7.1	5.8
150	12	16	15
250	1	1.2	1

CHSE-214: 5, 34 and 59 cells were transfected with a MRE-pEGFP reporter vector prior to the quantitative fluorescent activated cell sorting (FACS) determinations. Relative GFP activity is presented as the mean of six replicas in control or Zn (50, 100 and 150 μM) treated CHSE-214: 5, 34 and 59 cells for 24 h. The GFP activity of CHSE-214: 59 Zn (250 μM) exposed cells, which exhibited the lowest activity, was arbitrarily set to 1.0. All other activities were adjusted accordingly.

cell line. In this experiment there was a correlation between high MTF-1 activity (CHSE-214: 59) and low MT expression following high Zn concentrations. The lower MTF-1 activity observed in CHSE-214: 5 cells may thus be a consequence of the relatively high levels of MT that can sequester Zn (Table 1).

MT was initially considered to be most important for sequestering Cd and thereby important to protect against Cd toxicity. However, more recent data suggest that Zn is the metal primarily sequestered by MT. These novel data were demonstrated in teleost cell lines, which indicate that high MT expression is correlated to Zn resistance but not to Cd resistance[48]. Similar results have also been obtained from studies on mouse, where MTF-1 expression, which is required for normal MT expression, protects against Zn, but not Cd induced toxicity[80].

Metal regulation of MT genes by activation of MTF-1, and its interaction with MRE *cis*-acting sequences, does not appear to be the only mechanism that induces MT transcription by metals. In mammalian species the composite USF/ARE element of the mouse MTI promoter has shown to mediate Cd induction[3]. In addition Cd has also been demonstrated to activate MREs in cells lacking MTF-1[20]. Studies on fish species also suggest other mechanisms for metal-induced MT transcription. Ag has been proposed to act as a primary inducer of MT transcription in teleost cell lines, independently of MTF-1. Transcription was induced at 100-fold lower concentrations of Ag than of Zn in cells transfected with rtMT-A promoter containing six MREs. Thus a redistribution of Zn from cellular Zn pools in response to the low Ag concentrations is unlikely to induce MT transcription and no redistribution of Zn from different cellular compartments was observed in response to Ag. These data strongly support the idea that Ag is a primary inducer of MT transcription in teleosts[93].

V. Free radical regulation of MT synthesis

During respiration, O_2 is reduced to water. However, as a consequence of respiration, toxic by-products such as superoxide and H_2O_2 are produced. These ROS are potentially harmful to macromolecules such as DNA, protein and lipids. To protect from ROS, an organism must be able to regulate its antioxidants accordingly. There are several ways to protect the cellular compartments from oxidative stress, including repair of the oxidative damaged sites or enhancing gene expression of antioxidant genes. The expression can be enhanced either by activating or raising the cellular level of a certain transcription factor. In contrast to eukaryotes which through signal transduction upregulate antioxidant genes, prokaryotes, such as *E. coli*, is dependent on direct oxidative activation of the Oxy R and Sox R factors, leading to upregulation of antioxidant genes[92].

Fish MT is induced in response to factors promoting oxidative stress. These factors include stress and inflammation[6,50] and oxidant generating substances such as hydrogen peroxide[46,47,76]. The genetic basis for MT induction by oxidative stress has recently been investigated in both fish and mammalian species. Mammalian MT genes contain an upstream stimulatory factor (formerly known as the major late transcription factor, MLTF)/antioxidant responsive element (USF/ARE) that directs H_2O_2 responsiveness[24]. The transcription factors involved in regulating the ARE in response to oxidative stress is highly dependent of heterodimerization and subsequent binding of nrf2 and a small maf protein. Cytosolic nrf2 is released from its inhibitor, keap1, and translocated to the nucleus upon kinase phosphorylation[37]. In mammals MTF-1 is activated in response to oxidative stress[26]. Although MTF-1 activation studies have not been performed on fish, the presence of multiple AP-1 elements in the fish gene promoters support the observation that oxidative stress induces MT in fish *via* AP1 elements. AP1 binding occurs at *cis*-acting (5′-TGA G/CTA/C agc-3′) elements sharing homology to the ARE consensus sequence. AP1 was first identified as a transcriptional factor that binds and is required for optimal basal activity of the human MT-IIa promoter[4]. The activity of AP1 is modulated by a multitude of factors including oxidative stress[87]. AP1 consists of dimers from the products of the proto-oncogene families jun and fos. AP1 exists in two forms, as homodimers of jun proteins or heterodimers of jun:fos proteins. The primary role of AP1 is to regulate gene expression in response to mitogenic signals on the cell surface[87]. Therefore, in non-proliferating cells AP1 activity is usually low, consisting mainly of phosphorylated c-Jun homodimers. Like most transcription factors the DNA binding and transactivation domains of fos and jun proteins are physically separated. The activity of both proteins is controlled by phosphorylation. While Jun-N-terminal kinases (JNK) regulate the phosphorylation of c-jun, the kinases involved in c-fos regulation are yet to be identified[30,44,79]. Inducers of oxidative stress, such as UV-irradiation and H_2O_2 lead to AP1 activation and has also been shown to induce the transcription of c-jun and c-fos in several cell lines[70]. Under normal physiological conditions the cellular environment is strongly reducing. It has been observed that proliferating cells contain high levels of MT[82]. The identification of AP1 elements on MT promoters and the observation that proliferating cells contain high levels of MT, raises the question whether MT controls cellular proliferation and differentiation.

Several AP1 sites have been identified in the promoter of rainbow trout, pike, stone loach and Antarctic icefish[45,63,76,77]. In response to oxidative stress, these AP1 elements are involved in the regulation of both rainbow trout MT-A and MT-B genes[63,76]. The rtMT-A promoter contains four AP1 sites located in a proximal

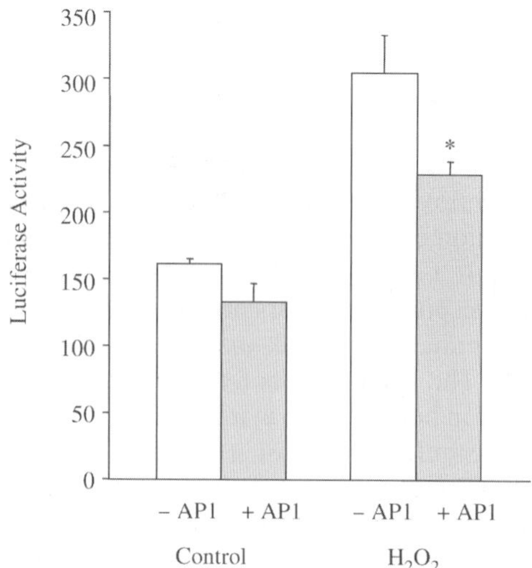

Fig. 3. *In vivo* inhibition of H_2O_2 induced rtMT-A gene activity by AP1 oligonucleotides. Luciferase activities of Hep G-2 cells transfected with a rtMT-A promoter plasmid containing the complete set of AP1 promoter elements alone (−AP1) or together with competing AP1 oligonucleotides (+AP1). Cells were grown in media alone (control) or exposed to 200 μM H_2O_2 for 24 h. The luciferase activities are given in arbitrary units. All activities were normalized for β-galactosidase activity. Significant reduction in H_2O_2 induced rtMT-A gene activity was observed following co-transfection with AP1 oligo as indicated by (*).

and a distal cluster. EMSA analysis and transfection experiments indicate that the AP1 sites are transactivated through the AP1 protein complex and mediate basal level expression of the rtMT-A gene[48]. In addition, it has been demonstrated that rtMT-A induction by oxidative stress is dependent on endogenous AP1 protein. Co-transfections of rtMT-A AP1 oligonucleotides and AP1 containing plasmids demonstrated an inhibition of oxidative induced rtMT-A gene expression (Fig. 3). Unlike mammals, MTF-1 and MREs in fish do not seem to participate in MT regulation during oxidative stress, which indicate that the oxidative stress response is not conserved between fish and mammalian species.

VI. Involvement of MT in protection against oxidative stress

The evolution of antioxidants in aerobic organisms started when photosynthetic bacteria began producing oxygen. Due to its toxicity, oxygen had posed a major threat to life[36]. Teleosts, such as the rainbow trout, inhabit turbid cool rivers, with a high level of dissolved oxygen and this is a potential threat to the rainbow trout. In addition to ROS produced by the respiratory chain, they are also generated in response to UV-irradiation mediated photochemical processes in the water[39]. Hence, fish may have evolved unique regulatory systems, such as MT, for protection from oxygen toxicity. MT is rich in sulfur, and therefore, represents a significant portion of total cellular protein thiols. Thiols are known to be highly reactive towards oxidants and thus MT could be an excellent free radical scavenger. *In vitro* studies have revealed MTs high reactivity towards hydroxyl and superoxide radicals[84]. MT can also protect from DNA strand breaks *in vitro*[1] and *in vivo*[21]. Several studies on aquatic organisms also suggest an important role for MT during oxidative stress. These studies suggest that fish MT are both induced and protected from oxidative stress, and hence suggest a physiological role for MT during free radical production. Inflammation is known to generate excess free radicals and injection of NIP_{11}-LPH antigen in Atlantic salmon results in MT induction[50]. In addition, oxidative stress conditions such as capture stress[6] and paraquat treatment[71] result in MT induction in fish. In the teleost cell line, RTG-2, both MT mRNA and protein were upregulated in response to oxidative stress[47]. Transcriptional induction of MT mRNA in response to oxidative stress was recently reported in the teleost RTH-149 cell line[76]. Pretreatment with Zn or Cd of the rainbow trout cell line, RTG-2, indicate that elevated MT levels protect the cells against oxidative stress, without the interference from GSH mediated protection[46,47]. In addition, the salmonid cell line (CHSE-214), which exhibit low MT expression, rendered resistant to H_2O_2 following over-expression of MT[47], which directly suggest MTs involvement in protecting against oxidative stress (Table 2). In the carp, it was suggested that both elevated GSH and MT afford H_2O_2 protection[88]. Cd-pretreatment of the mussel (*Mytilus galloprovincialis*) result in decreased iron-induced oxyradical formation, which might be the result of Cd-induced MT expression[85]. In addition, the role of MT in protecting from oxyradical-mediated inflammation has also been reported in oyster (*Crassostrea virginica*) hemocytes[2].

Besides the direct interactions of MT with certain oxidants, indirect roles of MT in free radical scavenging have been suggested[83]. Due to the potential toxicological role of heavy metals to alter the cellular and lysosomal membrane leading to lipid peroxidation, MT, as a major metal binding protein, may prevent the formation of heavy metal induced free radicals. Both Cd and Cu can induce lipid peroxidation in the sea bass, *Dicentrarchus labrax*[75]. Hence, MT could inhibit radical formation by binding these metals. It has been suggested that Zn released from MT can stabilize

TABLE 2

Dose-dependent H_2O_2 sensitivity in Zn pre-treated RTG-2 cells and MT over-expressed CHSE-214: 59 cells

H_2O_2 (μM)	RTG-2		CHSE-214: 59	
	Untreated	Zn-treated	pBK-CMV	pBK-CMV-MT
50	46	74	70	101
150	47	60	53	78
300	16	48	43	64
600	5	15	13	30

The relative H_2O_2 tolerance in untreated RTG-2 cells was calculated as the ratio of cells grown in media alone for 72 h, H_2O_2 exposed for 24 h divided by cells grown in media alone for 96 h. The relative H_2O_2 tolerance in Zn pre-treated RTG-2 cells was calculated as the ratio of Zn treated (150 μM, 72 h), H_2O_2 (24 h) exposed cells divided by cells grown in Zn (150 μM) alone for 96 h. The effect of MT over-expression on H_2O_2 sensitivity of CHSE-214: 59 cells was determined by comparing the survival of over-expressed (pBK-CMV-MT) and not over-expressed (pBK-CMV) cells by calculating the ratio of surviving, H_2O_2 exposed cells divided by untreated cells.

the cell membrane, preventing iron-mediated lipid peroxidation[83]. As an important Cu binding protein, MT has a role in preventing Cu-mediated redox cycling. The ability to undergo redox cycling makes Cu a potential mediator of the highly reactive hydroxyl radical. During physiological, reducing conditions a role for MT in preventing Cu(I) mediated lipid peroxidation has been proposed. However, during oxidative stress, Cu is released by MT and enhances the formation of ROS[31]. Thus, oxidant-induced release of Cu from MT may suggest the existence of physiologic redox mechanisms that control the transfer of Cu from MT to Cu-depending enzymes similar to what has been proposed for Zn transfer from MT during minute changes in the cellular redox potential[53].

VII. Summary

Although ample data suggests that MT is involved in the regulation and interactions with metals and oxidants, its physiological role remains elusive. MT knockout models reveal little about the physiological significance of MT, since the animals develop and appear normal. However, it has been observed that MT knockout mice are obese[10]. This suggests a link between MT and energy balance. The animals arc hyperphagic and deposit lipids at a high rate. Thus MT may prevent accumulation of those energy stores. Recently it was suggested that MT could inhibit the respiratory chain by donating Zn[89], resulting in lowered ATP and free radical production. This mechanism may explain how MT prevents obesity, while at the same time being able to regulate the levels of free radicals. Interestingly, defects in energy metabolism have been correlated to longevity in yeast, *C. elegans* and *Drosophila*[81] and also to high MT expression in *C. elegans*[8]. Since aging is correlated to an increased amount of free radicals[11], a possible role for MT is to function as a regulator of energy balance, thereby prolonging life. The correlation between age, energy expenditure, Zn status, free radical production and MT expression are fascinating future prospects.

VIII. References

1. Abel, J. and N. de Ruiter. Inhibition of hydroxyl-radical-generated DNA degradation by metallothionein. *Toxicol. Lett.* 47: 191–196, 1989.
2. Anderson, R.S., K.M. Patel and G. Roesijadi. Oyster metallothionein as an oxyradical scavenger: implications for hemocyte defence responses. *Dev. Comput. Immunol.* 23: 443–449, 1999.
3. Andrews, G.K. Regulation of metallothionein gene expression by oxidative stress and metal ions. *Biochem. Pharmacol.* 59: 95–104, 2000.
4. Angel, P. and M. Karin. The role of Jun, Fos and the AP-1 complex in cell-proliferation and transformation. *Biochim. Biophys. Acta* 1072: 129–157, 1991.
5. Auf der Maur, M.A., T. Belser, G. Elgar, O. Georgiev and S. Schaffner. Characterisation of the transcription factor MTF-1 from Japanese pufferfish (*Fugu rubripes*) reveals evolutionary conservation of heavy metal stress response. *Biol. Chem.* 380: 175–185, 1999.
6. Baer, K.N. and P. Thomas. Influence of capture stress, salinity and reproductive status on zinc, proteins in the liver of three marine teleost species. *Mar. Environ. Res.* 28: 277–287, 1990.
7. Baer, K.N. and P. Thomas. Isolation of novel metal-binding proteins distinct from metallothionein from spotted seatrout (*Cynoscion nebulosus*) and Atlantic croaker (*Micropogonias undulatus*) ovaries. *Mar. Biol.* 108: 31–37, 1991.

8. Barsyte, D., D.A. Lovejoy and G.J. Lithgow. Longevity and heavy metal resistance in daf-2 and age-1 long-lived mutants of *Caenorhabditis elegans*. *FASEB J.* 15: 627–634, 2001.
9. Beach, L.R. and R.D. Palmiter. Amplification of the metallothionein-I gene in cadmium-resistant mouse cells. *Proc. Natl Acad. Sci. USA* 78: 2110–2114, 1981.
10. Beattie, J.H., A.M. Wood, A.M. Newman, I. Bremner, K.H. Choo, A.E. Michalska, J.S. Duncan and P. Trayhurn. Obesity and hyperleptinemia in metallothionein (-I and -II) null mice. *Proc. Natl Acad. USA* 95: 358–363, 1998.
11. Berlett, B.S. and E.R. Stadtman. Protein oxidation in aging, disease and oxidative stress. *J. Biol. Chem.* 272: 20313–20316, 1997.
12. Bittel, D.C., I.V. Smirnova and G.K. Andrews. Functional heterogeneity in the zinc fingers of metalloregulatory protein metal response element-binding transcription factor-1. *J. Biol. Chem.* 275: 37194–37201, 2000.
13. Bonham, K., M. Zafarullah and L. Gedamu. The rainbow trout metallothioneins: molecular cloning and characterization of two distinct cDNA sequences. *DNA* 6: 519–528, 2002.
14. Bouquegneau, J.M., D. Gerday and A. Disteche. Fish mercury binding thionein related to adaptation mechanisms. *FEBS Lett.* 55: 173–177, 1975.
15. Brown, D.A. Increases of Cd and the Cd:Zn ratio in the high molecular-weight protein pool from apparently normal liver of tumor-bearing flounders (*Parophrys vetulus*). *Mar. Biol.* 44: 203–209, 1977.
16. Brugnera, E., O. Georgiev, F. Radke, R. Heuchel, E. Baker, G. Sutherland and W. Schaffner. Cloning, chromosomal mapping and characterisation of the human metal-regulatory transcription factor MTF-1. *Nucleic Acids Res.* 22: 3167–3173, 1994.
17. Carvan, M.J., III, W.A. Solis, G. Lashitew and W.N. Nebert. Activation of transcription factors in zebrafish cell cultures by environmental pollutants. *Arch. Biochem. Biophys.* 376: 320–327, 2000.
18. Chan, W.K. and R.H. Devlin. Polymerase chain reaction amplification and functional characterisation of sockeye salmon histone H3, metallothionein-B, and protamine promoters. *Mol. Mar. Biol. Biotechnol.* 2: 308–318, 1993.
19. Chen, W.-Y., J.A.C. John, C.-H. Lin and C.-Y. Chang. Molecular cloning and developmental expression of zinc finger transcription factor MTF-1 gene in zebrafish, *Danio rerio*. *Biochem. Biophys. Res. Commun.* 291: 798–805, 2002.
20. Chu, W.A., J.D. Moehlenkamp, D. Bittel, G.K. Andrews and J.A. Johnson. Cadmium-mediated activation of the metal response element in human neuroblastoma cells lacking functional metal response element-binding transcription factor-1. *J. Biol. Chem.* 274: 5279–5284, 1999.
21. Chubatsu, L.S. and R. Meneghini. Metallothionein protects DNA from oxidative damage. *Biochem. J.* 291: 193–198, 1993.
22. Compere, S.J. and R.D. Palmiter. DNA methylation controls the inducibility of the mouse metallothionein-I gene lymphoid cells. *Cell* 25: 233–240, 1981.
23. Coombs, T.L. *The Significance of Multielement Analyses in Metal Pollution Studies: Effects of Heavy Metal and Organohalogen Compounds*, edited by A.D. McIntry and F. Mills, New York, Plenum Press, pp. 187–195, 1975.
24. Dalton, T., R.D. Palmiter and G.K. Andrews. Transcriptional induction of the mouse metallothionein-I gene in hydrogen peroxide-treated Hepa cells involves a composite major late transcription factor/antioxidant response element and metal response promoter elements. *Nucleic Acids Res.* 22: 5016–5023, 1994.
25. Dalton, T.P., D. Bittel and G.K. Andrews. Reversible activation of mouse metal response element-binding transcription factor 1 DNA binding involves zinc interaction with the zinc finger domain. *Mol. Cell. Biol.* 17: 2781–2789, 1997.
26. Dalton, T.P., Q. Li, D. Bittel, L. Liang and G.K. Andrews. Oxidative stress activates metal-responsive transcription factor-1 binding activity. Occupancy *in vivo* of metal response elements in the metallothionein-I gene promoter. *J. Biol. Chem.* 271: 26233–26241, 1996.
27. Dalton, T., W. Solis, D. Nebert and M. Carvan, III. Characterisation of the MTF-1 transcription factor from zebrafish and trout cells. *Comp. Biochem. Physiol.* 126B: 325–335, 2000.
28. Davis, S.R. and R.J. Cousins. Metallothionein expression in animals: a physiological perspective on function. *J. Nutr.* 130: 1085–1088, 2000.
29. Durnam, D.M. and R.D. Palmiter. Transcriptional regulation of the mouse metallothionein-I gene by heavy metals. *J. Biol. Chem.* 256: 5712–5716, 1981.
30. Deng, T. and M. Karin. c-Fos transcriptional activity stimulated by H-ras-activated protein kinase distinct from JNK and ERK. *Nature* 371: 171–175, 1994.
31. Fabisiak, J.P., V.A. Tyurin, Y.Y. Tyurina, G.G. Borisenko, A. Korotaeva, B.R. Pitt, J.S. Lazo and V.E. Kagan. Redox regulation of copper–metallothionein. *Arch. Biochem. Biophys.* 363: 171–181, 1999.

32. Friberg, L., M. Piscator, G.F. Nordberg and T. Kjellström (Editors). *Cadmium in the Environment*, 2nd ed., Cleveland, OH, CRC Press, 1974.
33. George, S., D. Burgess, M. Leaver and N. Frerichs. Metallothionein induction in cultured fibroblasts and liver of a marine flatfish, the turbot, *Scopothalmus maximus*. *Fish Physiol. Biochem.* 10: 43–54, 1992.
34. Gunes, C., R. Heuchel, O. Georgiev, K.H. Muller, P. Lichtlen, H. Bluthman, S. Marino, A. Aguzzi and W. Schaffner. Embryonic lethality and liver degeneration in mice lacking the metal-responsive transcriptional activator MTF-1. *EMBO J.* 17: 2846–2854, 1998.
35. Hamer, D.H. Metallothionein. *Annu. Rev. Biochem.* 55: 913–951, 1986.
36. Hassan, H.M. and J.R. Schiavone. The role of oxygen free radicals in biology and evolution. In: *Metazoan Life Without Oxygen*, edited by C. Bryant, London, Chapman and Hall, pp. 17–37, 1991.
37. Huang, H.-C., T. Nguyen and C.B. Pickett. Phosphorylation of Nrf2 at Ser40 by protein kinase C regulates antioxidant response element-mediated transcription. *J. Biol. Chem.* 277: 42769–42774, 2002.
38. Hyllner, S.J., T. Andersson, C. Haux and P.-E. Olsson. Cortisol induction of metallothionein in primary cultures of rainbow trout hepatocytes. *J. Cell. Physiol.* 239: 24–28, 1989.
39. Janssens, B.J., J.J. Childress, F. Baguet and J.-F. Rees. Reduced enzymatic antioxidative defence in deep-sea fish. *J. Exp. Biol.* 203: 3717–3725, 2000.
40. Kägi, J.H.R. and B.L. Vallee. Metallothionein: a cadmium and zinc containing protein from equine renal cortex. *J. Biol. Chem.* 236: 2435–2465, 1960.
41. Karin, M., R. Andersson, E. Slater, K. Smith and H. Herschman. Metallothionein mRNA induction in HeLa cells in response to zinc or dexamethasone is a primary induction response. *Nature* 286: 295–297, 1980.
42. Karin, M., R.J. Imbra, A. Heguy and G. Wong. Interleukin 1 regulates human metallothionein gene expression. *Mol. Cell. Biol.* 5: 2866–2869, 1985.
43. Karin, M., G. Cathala and M.C. Nguyen-Huu. Expression and regulation of a human metallothionein gene carried on an autonomously replicating shuttle vector. *Proc. Natl Acad. Sci. USA* 80: 4040–4044, 1983.
44. Karin, M. The regulation of AP-1 activity by mitogen activated protein kinases. *J. Biol. Chem.* 270: 16483–16486, 1995.
45. Kille, P., J. Kay and G.E. Sweeney. Analysis of regulatory elements flanking metallothionein genes in Cd tolerant fish (pike and stone loach). *Biochim. Biophys. Acta* 1216: 55–64, 1993.
46. Kling, P., L.J. Erkell, P. Kille and P.-E. Olsson. Metallothionein induction in rainbow trout gonadal (RTG-2) cells during free radical exposure. *Mar. Environ. Res.* 42: 33–36, 1996.
47. Kling, P. and P.-E. Olsson. Involvement of differential metallothionein expression in free radical sensitivity of RTG-2 and CHSE-214 cells. *Free Rad. Biol. Med.* 28: 1628–1637, 2000.
48. Kling, P. Metallothionein reguation and function in teleosts during metal- and free radical exposure. Ph.D. Thesis. Department of Molecular Biology, Umeå University, Umeå, pp. 50, 2001.
49. Ley, H.L., M.L. Failla and D.S. Cherry. Isolation and characterization of hepatic metallothionein from rainbow trout (*Salmo gairdneri*). *Comp. Biochem. Physiol.* 74B: 507–513, 1983.
50. Maage, A., R. Waagbo, P.-E. Olsson, K. Julshamn and K. Sandnes. Ascorbate-2 sulphate as a dietary vitamin C source for Atlantic salmon (*Salmo salar*). 2 Effects of dietary levels and immunization on the metabolism of trace elements. *Fish Physiol. Biochem.* 8: 429–436, 1990.
51. Marafante, E., G. Pozzi and P. Scoppa. Detossicazione dei metalli pesanti pesci: isolamento della metallothioneina dal fefato di *Carassius auratus*. *Boll. Soc. Ital. Biol. Sper.* 48: 109, 1972.
52. Marafante, M.. Binding of mercury and zinc to cadmium-binding protein in the liver and kidney of goldfish (*Carassius auratus* L.). *Experientia* 32: 149–150, 1976.
53. Maret, W. The function of zinc metallothionein: a link between cellular zinc and redox state. *J. Nutr.* 130: 1455S–1458S, 2000.
54. Masters, B.A., E.F. Kelly, C.J. Quaife, R.L. Brinster and R.D. Palmiter. Targeted disruption of metallothionein I and II genes increases sensitivity to cadmium. *Proc. Natl Acad. Sci. USA* 91: 584–588, 1994.
55. McCarter, J.A., A.T. Matheson, M. Roch, R.W. Olafson and J.T. Buckley. Chronic exposure of Coho salmon to sublethal concentrations of copper-II. Distribution of copper between high- and low-molecular-weight proteins in liver cytosol and the possible role of metallothionein in detoxification. *Comp. Biochem. Physiol.* 72C: 21–26, 1982.
56. Michalska, A.E. and K.H. Choo. Targeting and germ-line transmission of a null mutation at the metallothionein I and II loci in mouse. *Proc. Natl Acad. Sci. USA* 90: 8088–8092, 1993.
57. Murphy, M., J. Collier, P. Koutz and B. Howard. Nucleotide sequence of the trout metallothionein A gene 5′ regulatory region. *Nucleic Acids Res.* 18: 4622 1990.
58. Noel-Lambot, F., C. Gerday and A. Disteche. Distribution of Cd, Zn and Cu in liver and gills of the eel *Anguilla anguilla* with special reference to metallothioneins. *Comp. Biochem. Physiol.* 61C: 177–187, 1978.

59. Olafson, R.W. and J.A.J. Thompson. Isolation of heavy metal binding proteins from marine vertebrates. *Mar. Biol.* 28: 83–86, 1974.
60. Olsson, P.-E. and C. Haux. Increased hepatic metallothionein content correlates to cadmium accumulation in environmentally exposed perch, *Perca fluviatilis*. *Aquat. Toxicol.* 9: 231–242, 1986.
61. Olsson, P.-E., C. Haux and L. Förlin. Variations in hepatic metallothionein, zinc and copper levels during an annual reproductive cycle in rainbow trout, *Salmo gairdneri*. *Fish Physiol. Biochem.* 3: 39–47, 1987.
62. Olsson, P.-E., M. Zafarullah, R. Foster, T. Hamor and L. Gedamu. Developmental regulation of metallothionein mRNA, zinc and copper levels in rainbow trout, *Salmo gairdneri*. *Eur. J. Biochem.* 193: 229–235, 1990.
63. Olsson, P.-E., P. Kling, L.J. Erkell and P. Kille. Structural and functional analysis of the rainbow trout (*Oncorhynchus mykiss*) metallothionein-A gene. *Eur. J. Biochem.* 230: 344–349, 1995.
64. Olsson, P.-E., Å. Larsson and C. Haux. Influence of seasonal changes in water temperature on cadmium inducibility of hepatic and renal metallothionein in rainbow trout. *Mar. Environ. Res.* 42: 41–44, 1996.
65. Olsson, P.-E. and P. Kille. Functional comparison of the metal-regulated transcriptional control regions of metallothionein genes from cadmium-sensitive and tolerant fish species. *Biochim. Biophys. Acta* 1350: 325–334, 1997.
66. Overnell, J., I.A. Davidson and T.L. Coombs. A cadmium-binding glycoprotein from the liver of the plaice. *Biochem. Soc. Trans.* 5: 267–269, 1977.
67. Overnell, J., C. Berger and K.J. Wilson. Partial amino acid sequence of metallothionein from plaice (*Pleuronectes platessa*). *Biochem. Soc. Trans. 593rd Meet.* : 217–218, 1981.
68. Palmiter, R.D. Molecular biology of metallothionein gene expression. *Experientia Suppl.* 52: 63–80, 1987.
69. Palmiter, R.D. Regulation of metallothionein genes by heavy metals appears to be mediated by a zinc-sensitive inhibitor that interacts with a constitutively active transcription factor, MTF-1. *Proc. Natl Acad. Sci. USA* 91: 1219–1223, 1994.
70. Pahl, H.L. and P.A. Bauerle. Oxygen and the control of gene expression. *Bioessays* 16: 497–502, 1994.
71. Pedrajas, J.R., J. Peinado and J. Lopez-Barea. Oxidative stress in fish exposed to model xenobiotics. Oxidatively modified forms of Cu, Zn-superoxide dismutase as potential biomarkers. *Chem. Biol. Interact.* 98: 267–282, 1995.
72. Piscator, M. Om kadmium I normala människonjurar samt redogörelse för isolering av metallothioein ur lever från kadmium exponerade kaniner. *Nord. Hyg. Tidskr.* 45: 76–82, 1964.
73. Radke, F., R. Heuchel, O. Georgiev, M. Hergersberg, M. Gariglio, Z. Dembic and W. Schaffner. Cloned transcription factor MTF-1 activates the mouse metallothionein I promoter. *EMBO J.* 12: 1355–1362, 1993.
74. Roch, M., J.A. McCarter, A.T. Matheson, M.J.R. Clark and R.W. Olafson. Hepatic metallothionein in rainbow trout (*Salmo gairdneri*) as an indicator of metal pollution in the Campbell river system. *Can. J. Fish. Aquat. Sci.* 39: 1596–1601, 1982.
75. Romeo, M., N. Bennani, M. Gnassia-Barelli, M. Lafaurie and J.P. Girard. Cadmium and copper display different responses towards oxidative stress in the kidney of the sea bass *Dicentrarchus labrax*. *Aquat. Toxicol.* 48: 185–194, 2000.
76. Samson, S.L.-A., W.J. Paramchuk and L. Gedamu. The rainbow trout metallothionein-B gene promoter: contributions of distal promoter elements to metal and oxidant regulation. *Biochim. Biophys. Acta* 1517: 202–211, 2001.
77. Scudiero, R., V. Carginale, C. Capasso, M. Riggio, S. Filosa and E. Parisi. Structural and functional analysis of metal regulatory elements in the promoter region of genes encoding metallothionein isoforms in the Antarctic fish *Chionodraco hamatus* (icefish). *Gene* 274: 199–208, 2001.
78. Schlenk, D. and C.D. Rice. Effect of zinc and cadmium treatment on glutathione, metallothionein and hydrogen peroxide-induced mortality in a teleost hepatoma cell line. *Aquat. Toxicol.* 43: 121–129, 1998.
79. Skinner, M., S. Qu, C. Moore and R. Wisdom. Transcriptional activation and transformation by FosB protein require phosphorylation of the carboxyl-terminal activation domain. *Mol. Cell. Biol.* 17: 2372–2380, 1997.
80. Solis, W.A., N.L. Childs, M.N. Weedon, L. He, D.W. Nebert and T.P. Dalton. Retrovirally expressed metal response element-binding transcripton factor-1 normalizes metallothionein-1 gene expression and protects cells against zinc, but not cadmium, toxicity. *Toxicol. Appl. Pharmacol.* 178: 93–101, 2002.
81. Strauss, E. Growing old together. *Science* 292: 41–42, 2001.
82. Studer, R., C.P. Vogt, M. Cavigelli, P.E. J and R. Kägi. Metallothionein accretion in human hepatic cells is linked to cellular proliferation. *Biochem. J.* 328: 63–67, 1997.
83. Thomas, J.P., G.J. Bachowski and A.W. Girotti. Inhibition of cell membrane lipid peroxidation by cadmium- and zinc-metallothioneins. *Biochim. Biophys. Acta* 884: 448–461, 1986.
84. Thornalley, P.J. and M. Vasak. Possible role for metallothionein in protection against radiation-induced oxidative stress. Kinetics and mechanism of its reaction with superoxide and hydroxyl radicals. *Biochim. Biophys. Acta* 827: 36–44, 1985.

85. Viarengo, A., B. Burlando, M. Cavaletto, B. Marchi, E. Ponzano and J. Blasco. Role of metallothionein against oxidative stress in the mussel *Mytilus galloprovincialis*. *Am. J. Physiol.* 277: R1612–R1619, 1999.
86. Winge, D.R., J. Krasno and A.V. Colucci. Cadmium accumulation in rat liver: correlation between bound metal and pathology, In: *Trace Element Metabolism in Animals*, edited by W.G. Hoekstra, J.W. Suttie, H.E. Ganther and W. Mertz, Baltimore, University Park Press, pp. 500–502, 1974.
87. Wisdom, R. AP-1: one switch for many signals. *Exp. Cell Res.* 253: 180–185, 1999.
88. Wright, J., S. George, E. Martinez-Lara, E. Carpene and M. Kindt. Levels of cellular glutathione and metallothionein affect the toxicity of oxidative stressors in an established carp cell line. *Mar. Environ. Res.* 50: 503–508, 2000.
89. Ye, B., W. Maret and B.L. Vallee. Zinc metallothionein imported into liver mitochondria modulates respiration. *Proc. Natl Acad. Sci. USA* 98: 2317–2322, 2001.
90. Zafarullah, M., K. Bonham and L. Gedamu. Structure of the rainbow trout metallothionein B gene and characterization of its metal-responsive region. *Mol. Cell. Biol.* 8: 4469–4476, 1988.
91. Zafarullah, M., P.-E. Olsson and L. Gedamu. Endogenous and heavy metal induced metallothionein gene expression in salmonid fish tissues and cell lines. *Gene* 183: 85–93, 1989.
92. Zheng, M. and G. Storz. Redox sensing by prokaryotic transcription factors. *Biochem. Pharmacol.* 59: 1–6, 1999.
93. Mayer, G.D., A. Leach, P. Kling, P.-E. Olsson and C. Hogstrand. Activation of the rainbow trout metallothionein-A promoter by silver and zinc. *Comp. Biochem. Physiol. B Biochem. Mol. Biol.* 134: 181–188, 2003.

Biochemistry and Molecular Biology of Fishes, vol. 6
T. P. Mommsen and T. W. Moon (Editors)

CHAPTER 11

Cell death: Investigation and application in fish toxicology

ANTONY W. WOOD*, DAVID M. JANZ** AND GLEN J. VAN DER KRAAK*

**Department of Zoology, University of Guelph, Guelph, ON, Canada N1G 2W1, and **Department of Veterinary Biomedical Sciences, University of Saskatchewan, Saskatoon, SK, S7N 5B4 Canada*

I. Introduction

Cell death is widely recognized to play an important role in the development and function of tissues and organs in all multicellular organisms, including fishes. Cell death can be induced by specific developmental cues that promote active, targeted cell removal ('programmed' cell death)[127] or can be the passive consequence of pathological or toxicological cell injury ('accidental' cell death)[80]. Maintaining an appropriate balance between cell proliferation, cell survival, and cell death is furthermore an important aspect of metazoan tissue homeostasis; disruptions to this balance can compromise tissue integrity, with pathological consequences that may influence an animal's ability to survive or reproduce.

Until relatively recently, most studies of cell death were primarily descriptive in nature. Toxicological investigations were no exception, as cell death in this context was generally assumed to be the passive result of toxicant-induced cell injury. Recognition that cell death can also be an active process under tight genetic and physiological regulation ranks among the most exciting recent developments in modern biology, and has led to a new paradigm in our understanding of tissue dynamics and homeostasis. Unraveling the genetic and biochemical pathways of active cell death is currently the focus of tremendous research activity. From a toxicological perspective, these developments offer many new approaches and experimental models in which to investigate the specific effects of toxicants at cellular and subcellular levels.

While studies of cell death in fishes are in their relative infancy, recognition that many regulatory elements of cell death pathways are evolutionarily conserved makes this field of inquiry ripe for exploration by fish biologists. The use of fish as experimental models in toxicology also offers the potential for novel and informative perspectives on cell death in aquatic vertebrates. Herein, we provide an introduction to recent developments in molecular and biochemical studies of metazoan cell death, and summarize recent applications of this new information to studies of toxicology in fishes.

1. Cell death nomenclature

Pathologists have long recognized that lethal cell injury induces a suite of defined morphological and biochemical changes in affected cells. Severely damaged cells exhibit declining ATP levels, loss of osmoregulatory ability, cell and organelle swelling and vacuolization, and increased membrane permeability. This typically culminates in cell lysis and the release of noxious cytosolic constituents, which induce a localized inflammatory response in adjacent tissues. This passive mode of cell death is generally termed *necrosis*; often afflicting contiguous groups of cells, necrosis is considered indicative of injury or trauma that has exceeded the capacity to maintain local cellular homeostasis.

Necrotic cell death differs markedly from *apoptosis*, a morphologically distinct pathway to cell death under the control of conserved genetic elements (Table 1). Apoptotic cell death (a.k.a. *programmed cell death* or *cell suicide*) is an active cell death mechanism that functions to remove unwanted cells from a tissue in a controlled, orderly fashion. It generally affects cells in isolation, which exhibit a suite of conserved morphological and biochemical features that contrast sharply with those observed during necrosis.

The recent explosive interest in the study of apoptosis has led to the development and promotion of a generalized dichotomy between active (apoptotic) and passive (necrotic) cell death pathways. Unfortunately, this dichotomy becomes problematic

TABLE 1

Biochemical and morphological features of cell death during apoptosis and necrosis

Feature	Apoptosis	Necrosis
Distribution	Single cells	Cell clusters
Cell size	Shrinkage	Swelling
Chromatin	Condensation	No condensation
DNA cleavage	Specific (internucleosomal)	Random
Caspase activity	Generally required	Not required
Endonuclease	Caspase-activated DNase (CAD)	Unknown
RNA synthesis	Required	Not required
Cell membrane	Remains intact	Ruptures
	Encloses apoptotic bodies	No apoptotic bodies
Phosphatidylserine	Translocation to outer membrane	No change
PARP* cleavage	Required	Not required
Tissue response	Phagocytosis by adjacent cells	Inflammation
	Little or no inflammation	Secondary scarring

* PARP, poly (ADP)-ribose polymerase.

in light of recent evidence demonstrating that the terminal phases of apoptotic cell death exhibit necrotic-like degenerative changes (termed *secondary necrosis*)[80,124].

This confusion has led some to suggest the term *oncosis* (from the Greek *onco*-, 'to swell'), in specific reference to the morphological and biochemical features exhibited during the prelethal stages of pathologically induced cell death[80,81,122,124]. According to this convention, necrosis is used specifically in reference to the secondary biochemical and morphological changes that occur after cellular metabolic arrest, regardless of the means by which a cell has died[76]. While these semantic distinctions have potential relevance to toxicological studies of cell death, we have chosen to adopt the conventional approach in this review. Henceforth, we use the term *necrosis* in reference to pathological cell death.

Our efforts in this chapter are devoted primarily to describing the biochemistry and molecular biology of cell death by apoptosis. Space limitations preclude an equivalent review of cell death by necrosis, thus readers are referred elsewhere to recent reviews of this subject[76,80,122,124].

II. Apoptosis

Apoptosis is an active, gene-directed program of cell death that results in the deliberate self-destruction and orderly removal of individual cells from a tissue. This suicidal pathway to cell death plays an instrumental role in many biological processes, including tissue morphogenesis and homeostasis, nervous system development, immune system function, and germ cell selection[56,86,105,127]. The genes associated with apoptosis are highly conserved across diverse taxonomic groups[5,49,59,109], suggesting that all metazoan cells likely possess the intrinsic ability to actively self-destruct. However, the probability of a given cell committing suicide is dependent upon the cell's specific function in the greater context of its multicellular environment.

Cells susceptible to apoptosis during normal development include those destined for deletion during tissue morphogenesis (e.g. tadpole tail resorption), those deemed incapable of performing a specific biological function (e.g. incompetent immune cells), or those that have surpassed their usefulness to tissue function (e.g. senescent cells)[127]. In addition, cells that are infected or that have been stressed or moderately damaged can be targeted for removal by apoptosis[105]. The development and maintenance of healthy tissue structure and function is thus dependent upon appropriate temporal and spatial regulation of cell death by apoptosis; disruptions to this process have been implicated in the progression of many diseases, including cancer, acquired immune deficiency syndrome, and neurodegenerative disorders (e.g. Alzheimer's disease)[109,115].

Although the importance of controlled cell deletion during metazoan development has been recognized for over a century[127], detailed investigations of the phenomenon only began in earnest following recognition that diverse cell types display highly conserved morphological features during demise. This observation led Kerr *et al.*[67] to hypothesize the existence of a universal cell death mechanism; coined from the ancient Greek word for the 'falling off' of leaves or petals, the term *apoptosis* was chosen to reflect the distinct morphological features of cells undergoing active self-destruction.

1. Morphology of apoptosis

The identification of apoptosis initially relied exclusively on morphological criteria, and despite tremendous progress in elucidating its molecular basis, changes to cell and nuclear morphologies remain the gold standard by which to unambiguously identify apoptosis. Cells dying by apoptosis shrink away from their neighboring cells, develop extensive membrane extrusions (blebs), and exhibit condensation (pyknosis) and fragmentation (karyorhexis) of nuclear chromatin. These latter changes yield characteristic 'moon-shaped' chromatin fragments, visible in the periphery of the degenerating nucleus (Fig. 1). Subsequent degenerative changes include fragmentation of the cell into membrane-bound cell fragments (*apoptotic bodies*) and rapid phagocytosis of apoptotic bodies by neighboring cells or macrophages[67]. This sequence of events results in the orderly removal of the dying cell without the inflammatory response typically associated with pathological (necrotic) cell death.

The degenerative changes during apoptosis are conserved among diverse cell types and across phylogenetic groups, a consistency that has been attributed to the underlying genetic control of the apoptotic death program[110]. For this reason, the use of morphological criteria to identify cell death by apoptosis is generally not limited by cell type or species. Morphological criteria have been used in a number of recent studies to identify apoptotic cell death in fish tissues. For example, Drummond *et al.*[36] used cell and nuclear morphological changes to determine the time course of follicle cell apoptosis during postovulatory regression in *Astyanax bimaculatus lacustris*. Genetic screens of zebrafish, *Danio rerio*, revealed mutant strains (*shrunken head* and *yellow head*) exhibiting abnormal retinal development[33]; while the retinae of mutants were observed to develop normally during early embryogenesis, subsequent retinal degeneration exhibited classic morphological features of apoptosis. Wood and Van Der Kraak[133] used morphological criteria to confirm biochemical evidence of theca

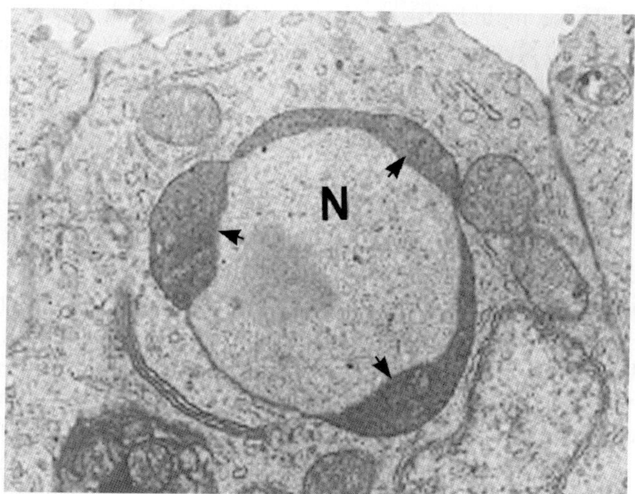

Fig. 1. Electron micrograph demonstrating nuclear morphological changes during apoptosis in postovulatory follicle cell of *Astyanax bimaculatus lacustris*. Arrowheads indicate marginalization of chromatin into 'moon-shaped' fragments in the periphery of the nucleus (N). Micrograph image courtesy of Drummond *et al.*[36].

cell apoptosis in rainbow trout, *Oncorhynchus mykiss*, ovarian follicles, and Rojo and Gonzalez[101,102] used morphological criteria to identify apoptotic cells in the gill epithelium of brown trout, *Salmo trutta*, and rainbow trout.

Importantly, the morphology of apoptotic cell death is not restricted to cells *in situ*. Efforts to develop a Pacific herring *Clupea harengus pallasi* cell line for toxicological studies revealed that density-dependent death of cultured cells was morphologically consistent with an apoptotic mechanism[42]. Similarly, morphological criteria were used to demonstrate increased apoptosis in isolated pituitary cells of adult hybrid tilapia (*Oreochromis niloticus* × *O. aureus*) after exposing the cells *in vitro* to calcium ionophores[38].

Although morphological criteria remain the gold standard by which to identify apoptosis, there are a number of disadvantages with this approach. For example, the requirement for expensive and time-consuming microscopy techniques can be a limiting factor. This is further compounded by the relatively rapid time course of cell death by apoptosis, which typically nears completion in less than a few hours[32]. In addition, the use of morphological criteria to identify apoptosis yields little insight on the biochemical pathways underlying the apoptotic death program. Consequently, the use of morphological criteria to identify apoptosis has been largely displaced by techniques that target biochemical changes associated with earlier stages of the apoptotic death program.

2. *Biochemistry of apoptosis*

Although a single scheme cannot adequately describe the complex biochemical features of apoptosis in all cell types[32], there are generally three stages in the apoptotic death response. These include (i) receipt of a lethal stimulus, by endogenous or exogenous means; (ii) signal transduction, often involving the mitochondria, and a host of second messengers and signal transduction pathways; and (iii) execution, primarily mediated *via* sequential activation of a suite of cysteine protease enzymes called caspases. What follows is a brief summary of the basic elements of each stage in this death pathway; readers are referred elsewhere for more comprehensive reviews of the biochemistry of apoptosis[4,40,48,105,109].

2.1. Stage 1: Lethal stimulus

All cells destined for apoptosis must first receive a lethal stimulus. Lethal stimuli include any molecule, substance, or condition that can induce activation of a cell's genetic death program, but does not directly kill the cell through injurious or pathological damage. As described in more detail below, these stimuli can be derived from within an organism (endogenous lethal stimuli) or from external sources (exogenous lethal stimuli) (Fig. 2). Regardless of origin, lethal stimuli converge on intracellular pathways that initiate the systematic disassembly and removal of the affected cell.

2.1.1. Endogenous lethal stimulus. An endogenous lethal stimulus includes any molecule, substance, or condition originating from within an organism that can trigger the apoptotic death program in selected cells. Endogenous molecular triggers of apoptosis include conventional cell signaling molecules (e.g. hormones, cytokines), intracellular proteins (e.g. protein kinases), cell cycle regulatory proteins (e.g. p53), and free radicals (e.g. reactive oxygen species), all of which have the capacity to extrinsically or intrinsically (see below) induce active cell death. Importantly, reduced or absent cell signaling (cell neglect) can also be an endogenous lethal stimulus;

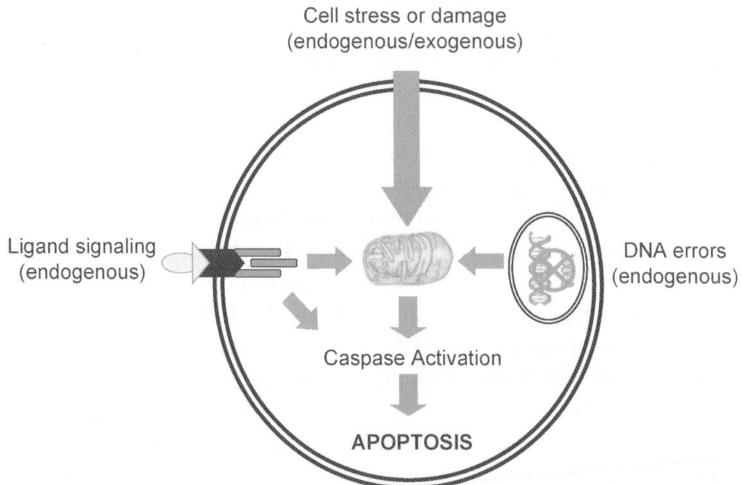

Fig. 2. Schematic representation of endogenous/exogenous origins of lethal stimuli that can trigger apoptosis in a metazoan cell. Lethal stimuli in many instances are integrated through the mitochondria (center), which can function as a central checkpoint in the apoptotic death program.

depriving cells of required stimulatory factors (e.g. matrix- or serum-derived growth factors) or metabolic materials (e.g. oxygen) is a potent death stimulus in many cell types[30,107,125]. Perturbations to cellular homeostasis (e.g. osmotic stress) or genomic integrity (e.g. DNA replication errors) may also serve to endogenously activate the death program in affected cells. Indeed, the diversity of stimuli that can trigger the genetic death program has led to the suggestion that apoptosis may be a default program for metazoan cells, a state of terminal differentiation involving exit from the cell cycle and expression of a fatal phenotype under aberrant conditions[44].

Within the realm of endogenous lethal signaling, two distinct biochemical pathways leading to apoptosis have been identified: the extrinsic (death receptor-mediated) pathway and the intrinsic (mitochondrial) pathway[1]. The extrinsic pathway is mediated through interactions between death-inducing ligands and cell membrane death receptors, whereas the intrinsic pathway involves the mitochondrion as an integrator of multiple pro-apoptotic signals, and as a co-ordinator of cellular execution *via* caspases[45], or as discovered more recently, *via* caspase-independent mechanisms[23].

Extrinsic death pathway. A number of plasma membrane death receptors, belonging to the tumor necrosis factor (TNF) superfamily of cytokine receptors, have been identified in vertebrates. This receptor superfamily is generally associated with cellular inflammatory and immune responses. Fas ligand (FasL), TNFα and TNF-associated apoptosis-inducing ligand (TRAIL) are among the most well-characterized ligands for this receptor family[6,68,89,95,114]. TNF superfamily receptors possess transmembrane domains linked to cytosolic death domains (e.g. TNF-associated death domain (TRADD), Fas-associated death domain (FADD)). Ligand–receptor interaction initiates aggregation and cross-linkage of multiple cytosolic death domains, which initiate the genetic death program by way of caspase activation[6,92] (Fig. 3); this pathway to cell death may or may not involve mitochondria[97].

Receptors and ligands from this gene superfamily have recently been cloned in brook trout, *Salvelinus fontinalis*[18], and rainbow trout[73,94,139]. The brook trout TNF receptor contains an intracellular death domain, suggesting its involvement in activating cell death in response to TNF signaling. In addition, putative TNF decoy receptors have been identified in these species[17,77]; the functions of this receptor type remain enigmatic, though the absence of a death domain suggests their involvement in regulating cell death through decoupling of TNF signaling from intracellular death pathways.

In vertebrates, the extrinsic cell death pathway is generally associated with immune system function. For example, cytotoxic immune cells secrete pro-inflammatory cytokines that induce the destruction and removal of invading pathogens, or infected/tumorigenic host cells, by way of their own apoptotic machinery[89]. Furthermore, termination of an immune response relies upon the cytokine-mediated apoptotic removal of activated lymphocytes[89]. There is relatively little functional information regarding mechanisms of immunocytotoxicity in fish, though emerging evidence suggests that teleost non-specific cytotoxic cells (NCCs) kill target cells by both apoptotic and necrotic mechanisms[46,85,106], mediated by induced secretion

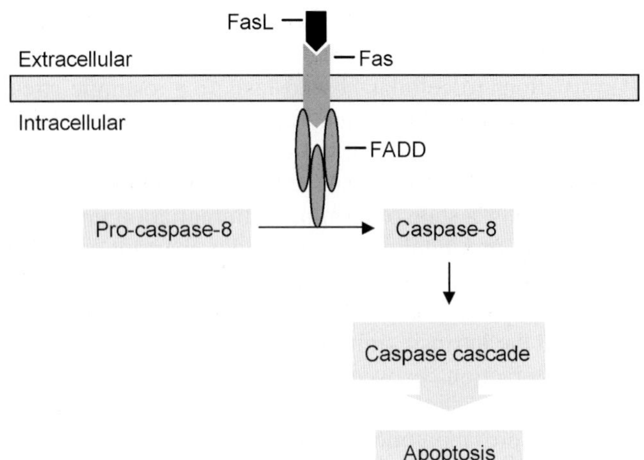

Fig. 3. Schematic representation of extrinsic cell death signaling mediated by Fas ligand (FasL). Binding of FasL to Fas receptor initiates cytosolic aggregation of multiple death domains (e.g. Fas-associated death domains (FADDs)), initiating caspase activation, and subsequent apoptotic death of the cell.

of Fas ligand-like molecules[15,51]. The channel catfish, *Ictalurus punctatus*, and tilapia, *O. niloticus*, are among the more popular models for the study of immunocytotoxicity in fish[14,15,39,91,138].

Intrinsic death pathway. In contrast to that described above, apoptosis induced by conventional cell signaling molecules (e.g. hormones) or stressors (e.g. osmotic stress, DNA damage, oxidative stress, heat shock, xenobiotics) is generally thought to be mediated *via* the mitochondrial (intrinsic) death pathway. This cell death pathway may involve any number of signal transduction elements that converge on the mitochondrion to initiate irreversible commitment to cell death (section II.2.2*). This typically involves the release of cytochrome c* from the mitochondrial intermembrane space, initiating the formation of an apoptosome, leading to caspase activation and cell disassembly (section II.2.3).

A number of endocrine hormones have been shown to induce apoptosis, in a cell type-specific manner, presumably by way of the intrinsic death pathway. For example in rainbow trout, elevated plasma cortisol was shown to induce apoptosis of lymphocytes, pavement cells, mucous cells, and melanocytes[57]. Although initially interpreted as a deleterious effect of elevated cortisol, a more recent study provides evidence for a reinterpretation of these data. Bury *et al.*[22] suggested that cortisol-induced apoptosis in cultured tilapia chloride cells plays a protective role, by attenuating the pathological (necrotic) consequences of toxicant exposure[58]. These findings suggest that glucocorticoid-induced apoptosis may be beneficial in some circumstances, by providing a controlled means by which to remove damaged cells in advance of necrotic degradation.

More recently, Laing *et al.*[72] demonstrated increased expression of caspase-6 mRNA (encoding an apoptosis executioner enzyme) in cultured rainbow trout head kidney leukocytes exposed to cortisol and lipopolysaccharide (LPS). Exposure to either cortisol or LPS alone did not induce increased expression of caspase-6 mRNA in leukocytes, but pretreatment with cortisol dramatically increased leukocyte sensitivity to LPS-induced apoptosis. These data suggest that in some cell types, stress hormones may not directly induce apoptosis, but may indirectly facilitate activation of the genetic death program.

Other endogenous hormones that have been shown to promote apoptosis include androgens, which can induce apoptosis in rat ovarian follicles[12], and gonadotropic hormone releasing hormone (GnRH), which has been shown to induce apoptosis in the mammalian ovary[13], and in the testis of goldfish, *Carassius auratus*[2,3].

Recently, a novel class of apoptosis-inducing flavoproteins (AIPs) was identified in visceral extracts of chub mackerel, *Scomber japonicus*, infected with the larval nematode *Anisakis simplex*[64]. These proteins could potently induce apoptosis in human leukemia (HL-60) cells, but their expression *in vivo* was dependent upon nematode infection and was localized to capsules surrounding nematode larvae in the abdominal cavity of infected fish. AIP-mediated apoptosis of HL-60 cells was Fas/TNF receptor independent, and was abrogated in the presence of free radical scavengers. These data have been interpreted as evidence for the existence of a novel host-defense mechanism involving reactive oxygen-mediated apoptosis of foreign (pathogenic) cells.

Altered expression of cell cycle regulatory genes can also be an important endogenous lethal stimulus. For example, an important role for tumor-suppressor proteins (e.g. p53 family proteins) is to detect mutations or replication errors during the S-phase of the cell cycle. If DNA damage is deemed irreparable, altered expression of tumor-suppressor genes will trigger apoptosis in the damaged cell (Fig. 4). Recognition that a large percentage of cancers involve mutations to conserved regions of p53 has led to widespread interest in the study of p53 signaling in carcinogenesis. Tumor-suppressor gene expression has consequently been targeted as a potential indicator of environmental exposure to genotoxins. To this end, there is genetic sequence information for p53 in a number of model fish species, including chum salmon, *O. keta*; coho salmon, *O. kisutch*; chinook salmon, *O. tshawytscha*; pufferfish, *Tetraodon miurus*; barbel, *Barbus barbus*; and rainbow trout[11,26].

An emerging idea suggests that cross-talk among p53 family members may be important for cell cycle regulation. In support of this, it was recently shown that inhibition of p53 activity by p63 was required for epidermal cell proliferation in embryonic zebrafish[74]. Another p53 family member, p73, was recently reported in the barbel[10], representing the first non-mammalian homologue of this protein to be identified. Further progress in this area will undoubtedly augment our understanding of the molecular basis of cell cycle regulation and chemical carcinogenesis, in fish and other vertebrates.

2.1.2. Exogenous lethal stimulus. Exposure to exogenous factors (e.g. toxins, pathogens) can directly compromise the integrity of a cell's functional, structural, or genetic elements. Irreparable damage that does not cause overt cell lethality will likely result in activation of the intrinsic cell death pathway. This suicidal gesture can serve either of two purposes: (1) removal of a damaged cell whose function has been

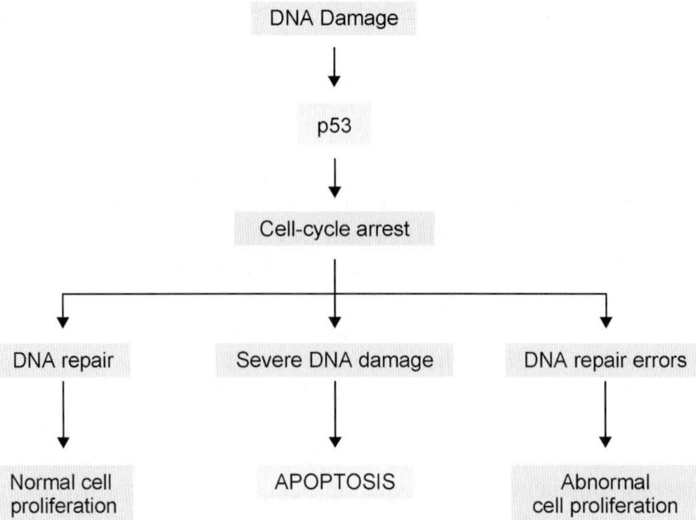

Fig. 4. Flow diagram demonstrating involvement of p53 signaling in life-or-death fate decisions of a metazoan cell. Products of the p53 gene family detect errors in genomic DNA and initiate cell cycle changes required for repair or destruction of the cell.

compromised or (2) removal of a damaged cell that poses a potential risk (e.g. neoplasia) to the organism.

Toxicant exposure is a potentially important exogenous lethal stimulus, and there has consequently been a notable recent increase in the study of cell death in toxicological research. The importance of fish as sentinels of environmental health makes them an interesting and potentially informative model for the study of cell death in toxicology, using both *in vivo* and *in vitro* approaches. We discuss recent developments in this field in more detail below (see section III).

2.2. *Stage 2: Signal transduction*

After receipt of a lethal stimulus, the second stage of apoptosis typically involves the recruitment of intracellular signaling molecules that transduce the lethal stimulus to a genuine cellular response, often referred to as the 'decision to die'. Signal transduction elements include the active intracellular domains of death receptors (e.g. FADD, TRADD), which can directly (*via* caspase activation) or indirectly (*via* mitochondria; see below) initiate cell death. They may also include any number of conventional signal transduction pathways (e.g. MAPK, Erk, Ras) or second messenger molecules (e.g. cAMP, cGMP, ceramide), which transduce the death stimulus to a cellular response. The multiplicity of signal transduction pathways involved in mediating apoptosis precludes an exhaustive review in this chapter.

Elucidation of cell death signal transduction pathways almost invariably requires experimental manipulation of stable cell lines, thus few data have been derived from fish models. However, it was recently shown that ceramide, a sphingolipid-derived second messenger molecule, is up-regulated during apoptosis of cultured fathead minnow (*Pimephales promelas*) tailbud cells. These findings were confirmed *in vivo* using an embryonic zebrafish model[135], demonstrating the potential utility of the fish model as a convenient bridge between *in vitro* and *in vivo* approaches to studying cell death signaling pathways. The increasing availability of stable cell lines from fish species will undoubtedly lead to further progress in this area.

In many model systems, cell death signaling pathways have been shown to converge on the mitochondrion, which can function as a central checkpoint in the apoptotic death program[20,45]. Death signaling induces permeability changes in the mitochondrial membrane, resulting in the release of proteins (e.g. cytochrome *c*) into the cytosol[78]. Binding of these proteins with the scaffolding protein apoptotic protease-activating factor-1 (Apaf-1) induces the formation of an 'apoptosome', an enzyme complex that catalyzes activation of cell death executioner enzymes[137]. This latter sequence of events commits the cell to an irreversible program of proteolytic self-destruction.

The mitochondrial control of cell fate in vertebrates involves interactions among proteins of the B cell lymphoma-2 (Bcl-2) proto-oncogene family, which has been functionally conserved throughout metazoan evolution. Its involvement in programmed cell death was first described in *Caenorhabditis elegans*, where it was shown that expression of *ced-4* was required for cell death activation, whereas *ced-9* overexpression maintained cell viability by suppressing interactions between the products of *ced-4* and downstream cell death executioner proteins[109].

In vertebrates, duplication and divergence of these genes has led to the evolution of a much larger gene family, the products of which play similar roles in regulating cell fate. Homologues of ced-9 (e.g. Bcl-2, Bcl-X_{long}, Bcl-w, Mcl-1) function as inhibitors of apoptosis[49,50,75,108], whereas ced-4 homologues (e.g. Bax, Bid, Bad, Bam, Bok, Bcl-X_{short}) promote cell death[4,7,54,103,112]. Bcl-2 family proteins are characterized by the presence up to four Bcl-2 homology (BH) domains. All anti-apoptotic Bcl-2 family members contain BH1 and BH2 domains; some, but not all, contain BH3 and/or BH4 domains. With the exception of Bad, all pro-apoptotic Bcl-2 family members contain the BH3 domain, which is evidently necessary for dimerization with other Bcl-2 family members[19].

According to the conventional model, anti-apoptotic Bcl-2 proteins inhibit apoptosis by functional suppression of pro-apoptotic Bcl-2 family proteins; the latter promote the release of proteins (e.g. cytochrome *c*, apoptosis-inducing factor (AIF), adenylate kinase-2, HSP60) from the mitochondrial membrane which induce cell death by apoptosis through interactions with Apaf-1 and downstream execution enzymes[50,66,69,137]. A rheostat mechanism has been proposed, in which the balance of expression between pro- and anti-apoptotic Bcl-2 family members ultimately dictates the life or death decisions of a given cell[4].

A number of putative Bcl-2 family members have been identified in fish; all published examples contain at least two BH domains[28,29,52]. The involvement of Bcl-2 family proteins in mediating apoptosis in fish has not been directly investigated. Interestingly, however, one Bcl-2 family member (zfMcl-1a) identified in embryonic zebrafish contains a putative nuclear localization signal[28]. Nuclear localization of zfMcl-1a was subsequently confirmed using a fusion-protein (EGFP) reporter system, providing the first known example of nuclear localization of a Bcl-2 family member.

This finding suggests an alternative role for Bcl-2 family proteins, beyond regulation of mitochondrial membrane permeability[28,29].

A unifying feature of all pro-apoptotic signal transduction pathways is their ability to relay a death stimulus to the caspases, enzymes responsible for the execution phase of apoptosis[37]. As noted previously, however, a caspase-independent mechanism of apoptosis has recently been proposed[23].

2.3. Stage 3: Execution

The execution phase of apoptosis involves the sequential activation of a family of cysteine proteases that initiates irreversible proteolytic disassembly of the cell. These enzymes have been termed caspases (*c*ysteine *asp*artyl prote*ases*), in reference to their targeted cleavage of protein substrates specifically after aspartate residues. Caspases are synthesized as inactive pro-enzymes; during apoptosis, they are activated by proteolytic cleavage that exposes an active site pentapeptide (QACXG, where X is R, Q, or G). Activation may occur by autocatalysis or *via* cleavage by other caspase family members. Sequential activation of multiple caspases initiates an enzymatic cascade that rapidly degrades the cell. The multiplicity of caspase substrates[104,110,116] ensures rapid and complete degradation of the cell's structural components. One family member, caspase-3, triggers activation of an endonuclease (caspase-activated DNase (CAD)) that catalyzes internucleosomal fragmentation of the cell's genomic DNA. The resultant 'ladder' of DNA fragments (visible after electrophoresis; Fig. 5) is a hallmark feature of cell death by apoptosis[87].

The first evidence for the requirement of caspase activity during apoptosis emerged from genetic studies in *C. elegans*, in which programmed cell death was shown to require the expression of proteins encoded by *ced-3*. Sequence comparisons revealed homology between products of *ced-3* and a mammalian cytokine activator, interleukin-1β-converting enzyme (ICE). Though not necessarily required for apoptosis in mammalian cells, further investigations of ICE (subsequently termed caspase-1) led to the identification of an additional 13 mammalian caspase family members (caspases 2–14), some of which are integral to the execution phase of apoptosis (*cf.* Candé *et al.*[23]).

Sequence analyses have revealed that mammalian caspases can be phylogenetically assigned to one of the five lineages[129]: (i) the ICE subfamily (caspases 1, 4, 5, 11–13); (ii) caspase-14; (iii) the caspase-3 subfamily (caspases 3, 6–8, 10); (iv) caspase-2; and (v) caspase-9. Partial or complete sequences for at least nine teleost caspases are currently available in GenBank (http://www.ncbi.nlm.nih.gov/), with more undoubtedly to emerge as genomic sequence data become available. These include homologues of mammalian caspase-1 (*D. rerio*, *Takifugu rubripes*), caspase-2 (*D. rerio*), caspase-3 (*D. rerio*, *T. rubripes*), caspase-4 and -5 (*D. rerio*), caspase-6 (*O. mykiss*, *I. punctatus*, *Sparus aurata*, *D. rerio*), caspase-8 (*T. rubripes*, *D. rerio*), caspase-9 (*S. salar*, *Pleuronectes americanus*, *D. rerio*), and caspase-10 (*T. rubripes*, *Paralichthys olivaceous*).

While few functional studies of caspases have been attempted in fish, the available evidence suggests a generally conserved role for selected caspases in mediating the execution phase of apoptosis. For example, expression of *caspase-6* mRNA in

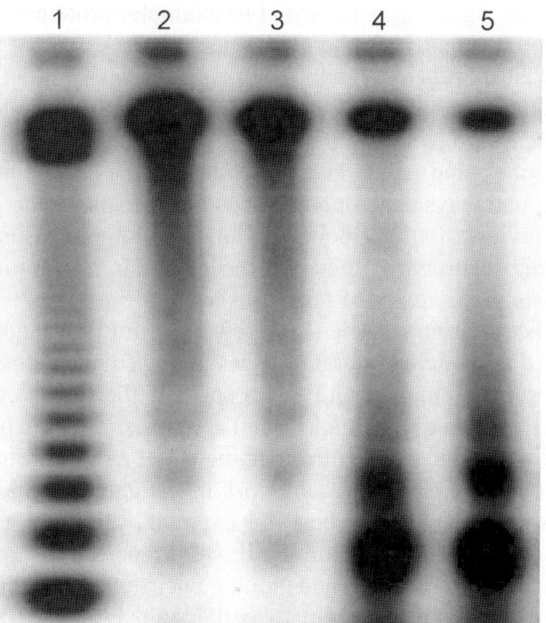

Fig. 5. Autoradiogram demonstrating DNA fragmentation in the 'ladder' pattern characteristic of apoptosis. The image presents genomic DNA from rainbow trout, *Oncorhynchus mykiss*, ovarian follicles 3′ end-labeled with (α-^{32}P)dideoxyATP and electrophoretically fractionated in 2% agarose. Lane 1 is a 123 base-pair (bp) DNA ladder. Lanes 2 and 3 demonstrate constitutive levels of apoptotic DNA fragmentation in vitellogenic follicles; lanes 4 and 5 demonstrate increased oligonucleosomal fragmentation after induction of apoptosis by *in vitro* culture of follicles in serum-free medium. The image is from Wood and Van Der Kraak[131].

rainbow trout leukocytes was increased after combined exposure to cortisol and the bacterial endotoxin lipopolysaccaride[72], while in embryonic zebrafish, caspase-3-like activity was elevated after exposure to heat shock, ultraviolet (UV) and γ-ray irradiation[136]. In overexpression studies, zebrafish caspase-3 induced apoptosis in fathead minnow tailbud cells and embryonic zebrafish[135], and zebrafish caspy and caspy2 (putative homologues of mammalian caspase-1 and -5, respectively) induced apoptosis in mammalian (293T) cells[82]. Interestingly, targeted knockdown of *caspy* in embryonic zebrafish led to abnormal development of the pharyngeal skeleton, coincident with its expression domains, although no changes in the number of apoptotic cells were detected[82]. Further investigations of caspase function(s) in fishes and other vertebrates are clearly warranted. In toxicological research, caspase activity measurements have potential utility as rapid and convenient indicator of stress- or toxin-induced apoptosis[136].

3. Modulators of apoptosis

As discussed previously, cell survival depends on appropriate extracellular signaling, and the absence or depletion of these signals can promote apoptosis through activation of intracellular death signaling pathways. For example, proteins (e.g. integrins) that anchor the extracellular matrix to the cytoskeleton play key roles in cell survival[125]. Similarly, appropriate levels of growth factors are essential for cell survival[96]. Other than these commonalities, hormonal modulators of apoptosis are often specific to cell type, and in many cases, developmental status. In addition, the complex control of cell survival by hormones and other chemical messengers is dependent on multiple cellular signaling pathways that operate in autocrine, paracrine, and/or endocrine manners. Since a review of all hormonal modulators of apoptosis is beyond the scope of this chapter, this section will focus on recent research examining the hormonal control of apoptosis in the ovary of teleost fishes.

The use of intact ovarian follicles cultured in serum-free medium is an *in vitro* approach employed in model vertebrate species to investigate the hormonal control of follicle atresia[123]. In mammalian and avian species, the demise of follicles during atresia is frequently initiated by granulosa cell apoptosis[55,119]. Much of our knowledge of factors controlling apoptosis and atresia in the vertebrate ovary has been elucidated using this approach[55,65], including recent work investigating the hormonal control of apoptosis in the ovary of teleost fishes. For example, salmonid gonadotropin, 17β-estradiol, and epidermal growth factor (EGF) were reported to suppress ovarian cell apoptosis in cultured preovulatory[63] and vitellogenic[133] ovarian follicles from rainbow trout. Unlike in mammals, ovarian somatic cell apoptosis appeared to be restricted to thecal–epithelial cells, rather than granulosa cells, in rainbow trout and goldfish[133,134], suggesting differences between fish and mammals in the constitutive regulation of apoptosis among follicle cell types.

Cell survival is also maintained by an evolutionarily conserved response to cellular stress, *via* expression of heat shock or stress proteins[41]. In general, heat shock proteins (HSPs) maintain cellular homeostasis and provide stress tolerance through regulation of protein–protein interactions. Accumulating evidence indicates that HSPs, particularly the 70 kDa (HSP70) and 27 kDa (HSP27) families, play critical roles in protecting cells from death by apoptosis following cellular stress[8,43]. Importantly, the anti-apoptotic functions of HSP70 and HSP27 have been reported to operate in both intrinsic and extrinsic death pathways, and both upstream and downstream of caspase activation[43]. Although HSPs have been proposed as biochemical indicators of cellular stress following toxicant exposure in aquatic species[60], very little research has addressed their potential anti-apoptotic roles in fishes. However, juvenile channel catfish, *I. punctatus*, exposed acutely to aryl hydrocarbon receptor agonists exhibited a significant inverse relationship between toxicant-induced apoptosis and HSP70 expression in ovarian, but not hepatic, tissues[132], suggesting a tissue-specific relationship between HSP70 expression and apoptosis in fish.

III. Apoptosis as a cellular response to toxicant exposure

There is abundant and accumulating evidence to suggest that apoptosis may be a biologically significant response to toxicant exposure[32,100,118]. The apoptotic removal of cells whose function has been compromised by toxicant exposure is one means by which an organism can minimize the deleterious effects of toxicant exposure. Alternatively, toxicants can disrupt the molecular control elements that regulate constitutive cell death processes. For example, compromising a cell's ability to appropriately self-destruct by apoptosis is one means by which toxicants can promote carcinogenesis. Toxicant-induced disruption to normal cell signaling may also interfere with the external signals regulating cell function, possibly leading to inappropriate up- or down-regulation of the apoptotic death program.

It is currently believed that toxicant-stimulated cell death represents a dose-dependent continuum, with lower exposure doses stimulating apoptosis and higher doses leading to a threshold, after which necrosis predominates[84]. The extent of apoptosis can thus represent an early cellular indicator of toxicity, with potentially greater sensitivity than conventional cytotoxicity assays. Recent work using various fish models has attempted to relate the extent of toxicant-induced apoptosis in specific tissues with functional consequences at higher levels of biological organization, though much work remains to be done in this area.

1. Heavy metals

Heavy metal pollutants have been shown to induce a variety of deleterious effects in aquatic organisms, including fishes[31]. Recently, selected heavy metals were investigated for their specific ability to induce cell death in fish tissues. For example, aqueous copper exposure was shown to induce apoptosis, and at higher concentrations necrosis, in gill tissues of the tropical fish *Prochilodus scrofa*[83]. Similarly, copper exposure caused necrotic and apoptotic cell death in epithelial tissues of carp, *Cyprinus carpio*[58]. Bury *et al.*[22] demonstrated increased chloride cell necrosis in tilapia gill filaments exposed to copper; interestingly, pre-exposure of gill filaments to cortisol attenuated copper-induced necrosis, but increased chloride cell apoptosis.

Chronic dietary cadmium exposure was reported to cause intestinal cell apoptosis in Atlantic salmon, *S. salar*[9]. Similarly, cadmium exposure induced hepatocyte apoptosis in dab, *Limanda limanda*[93], and rainbow trout[99]. *In vitro* exposure of cultured rainbow trout epidermal cells to cadmium (10–50 μM) for 48 h induced apoptosis[79]; exposure to higher concentrations (100 μM) induced necrotic cell death. Importantly, culture conditions were shown to significantly affect the sensitivity of these cells to cadmium exposure, as an absence of serum in the culture medium increased cell death by as much as 25-fold. Lastly, exposure of cultured rainbow trout hepatocytes to the organometallic pesticide tributyltin induced apoptosis in cultured rainbow trout hepatocytes; cell death in this case was reportedly mediated by calcium ions and protein kinase C[98]. More recently, it was shown that tributyltin at concentrations greater than 1 μM led to mitochondrial swelling and alterations to cristae ultrastructure in rainbow trout red blood cells and leukocytes[117]. These responses involved the release of cytochrome *c*, activation of caspases, and fragmentation of DNA, further evidence that tributyltin toxicity is mediated through an apoptotic mechanism.

2. *Organochlorines*

Polychlorinated dibenzo-*p*-dioxins (e.g. 2,3,7,8-tetrachlorodibenzo-*p*-dioxin (TCDD)) are among the most intensively studied organochlorine pollutants. These compounds are potent agonists of the aryl hydrocarbon receptor (AhR). In many vertebrates, including fishes, they cause a variety of toxic effects including cardiovascular dysfunction, edema, hemorrhage, craniofacial malformations, and reduced growth, often with fatal consequences[53]. In fishes, embryonic stages are particularly sensitive to dioxin exposure; 'blue sac disease' is a common manifestation of dioxin exposure in embryonic fishes[128].

Recent studies have begun to investigate the effects of dioxin exposure on cell death in fishes. Exposure of mummichog, *Fundulus heteroclitus*, embryos to TCDD induced apoptosis in a variety of tissues, including brain, eye, gill, kidney, tail, intestine, heart, and the vasculature[120]. Exposure of embryonic medaka, *Oryzias latipes*, to TCDD induced vascular apoptosis, primarily in endothelial cells of the medial yolk vein, which correlated with TCDD dose and the onset of blue sac disease[24,25]. In zebrafish embryos, exposure to TCDD induced apoptosis in cells of the developing dorsal midbrain[34], an effect that was inferred as a secondary consequence of TCDD-induced mesencephalic ischemia[35]. In contrast, a recent study by Hornung *et al.*[53] showed that exposure of rainbow trout yolk sac fry to TCDD induced typical cardiovascular and craniofacial abnormalities, but did not induce apoptosis in the developing vasculature.

In one of the earliest studies investigating organochlorine-induced cell death in fishes, it was shown that sublethal waterborne exposure of rainbow trout to low concentrations of trichloroethylene induced apoptosis in gill, liver, spleen, and kidney. However, exposure to higher concentrations of trichloroethylene induced necrotic cell death in the liver, spleen, and kidney[47].

The effects of organochlorines on thymocyte viability were determined in lake trout, *S. namaycush*[113]. Short-term (6 h) exposure to gamma hexachlorohexane (HCH) (lindane) at a low temperature (6°C) resulted in a dose-dependent increase in thymocyte apoptosis, while exposure to lindane at a higher temperature (17°C), or for a longer duration (24 h), resulted in increased thymocyte necrosis. Among different isomers of HCH, delta-HCH exhibited the most potent cytotoxicity, mainly through the induction of thymocyte necrosis. In the same study, exposure of thymocytes to the commercial polychlorinated biphenyl mixture Aroclor 1254 induced dose-dependent cytotoxicity, through both apoptotic and necrotic mechanisms[113]. These findings demonstrate that toxicant exposure can potentially induce immunosuppression through increased cell death of immunologically important cells, but that the mode of cell death may vary in relation to the intensity and conditions of exposure, and may also vary widely among chemically similar compounds.

3. *Environmental estrogens*

There has recently been much concern regarding xenobiotics with the capacity to alter endocrine homeostasis. Often referred to as endocrine disrupting chemicals

(EDCs), these compounds mimic (agonists) or suppress (antagonists) the actions of endogenous hormones. Recent work in fish has examined the hypothesis that sublethal effects (e.g. reproductive dysfunction) of exposure to environmental estrogens may be mediated by up- or down-regulation of cell death. For example, chronic exposure of medaka to 17α-ethinylestradiol (EE), a potent pharmaceutical found in many aquatic systems, was reported to cause increased cell death (apoptosis or necrosis) in hepatocytes of males and females, in tubule, glomerular and interstitial cells in the head kidney of adult males, and in tubular cells of females[130]. In addition, EE exposure caused an increase in testicular cell death, predominantly in Leydig cells, Sertoli cells, spermatocytes, and spermatids of adult males. These results suggest that impaired breeding capacity of male fish exposed to xenoestrogens may be mechanistically related to increased cell death in testicular tissues.

Alkylphenols are degradation products of alkylphenol ethoxylates, constituents of many household and industrial detergents, and used in the manufacture of paints, plastics, and a variety of pesticides. Alkylphenols have become a source of widespread environmental concern following recognition of their capacity to bind the estrogen receptor in a number of vertebrate species. More recently, studies have demonstrated deleterious effects of alkylphenol exposure in fishes. For example, Toomey *et al.*[121] reported increased apoptosis in brown bullhead catfish (*Ameiurus nebulosus*) hepatocytes exposed to octylphenol *in vitro*. Weber *et al.*[131] reported increased apoptosis of spermatocytes, Sertoli cells, and Leydig cells in medaka testes after *in vivo* exposure to nonylphenol, but similar effects were not observed in the liver, kidney, or intestines of exposed fish. Similarly, exposure of male swordtail, *Xiphophorus helleri*, to nonylphenol and bisphenol A induced increased testicular cell death and reduced spermatogenesis[71].

4. *Pulp mill effluents*

Effluents from pulp and paper mills are complex mixtures containing natural wood sterols, metals, aromatic hydrocarbons, and if chlorine bleaching of pulp occurs, chlorinated hydrocarbons. A characteristic suite of responses in wild fish exposed to these effluents has been observed that includes reductions in egg size, ovary size, fecundity, and circulating levels of sex steroid hormones in females[88,126]. Concomitant with these effects, elevated ovarian cell apoptosis and HSP70 protein expression was reported in prespawning white sucker, *Catostomus commersoni*, exposed to a bleached pulp mill effluent[61]. A subsequent study reported similar effects in vitellogenic white sucker collected downstream of a different bleached pulp mill[62]. Interestingly, in a subsequent study conducted following mill maintenance changes aimed at reducing the discharge of process chemicals, vitellogenic white sucker exhibited no detectable changes in ovary size, ovarian cell apoptosis, or ovarian cell HSP70 expression, which supports the hypothesis that effluent-stimulated ovarian cell death is mechanistically related to reduced ovary size[62].

5. *Irradiation*

Although not a toxicant *per se*, irradiation (and the resultant DNA damage) can potently induce apoptosis in metazoan cells. This has been convincingly demonstrated in mammalian cell lines, and was recently confirmed in a medaka cell line (OCP13), which exhibited a dose-dependent increase in apoptotic cell death after exposure to UV radiation[90]. A teleost embryonic model was furthermore effective in extending these findings *in vivo*: exposure of embryonic zebrafish to UV and γ-irradiation for 3–48 h induced apoptosis in a dose- and time-dependent fashion[136]. These findings have potential implications for feral fish species with pelagic (buoyant) eggs and/or larvae: recent declines in atmospheric ozone levels, and the resultant increase in UV exposure, could have deleterious consequences on developing fish by way of irradiation-induced cell death.

6. *Thermal stress*

Fish are ectothermic vertebrates, and as such are susceptible to fluctuations in temperature of their immediate environment. This has relevance to sewage and industrial effluents, which can be significant sources of thermal pollution in aquatic environments. Recent evidence demonstrating that heat shock can induce apoptosis in a variety of fish tissues[21,111,136] suggests that some deleterious effects of thermal pollution may be mechanistically linked to increased apoptosis of susceptible cell types.

7. *Carcinogens*

Toxicant exposure does not necessarily result in *increased* cell death; exposure to some toxicants has been shown to have the reverse effect. For example, exposure of gulf killifish, *F. grandis*, to the suspected carcinogen *N*-methyl-*N*′-nitro-*N*-nitrosoguanidine (MNNG) resulted in a decreased incidence of hepatocyte apoptosis, concomitant with an increased rate of cell proliferation[16]. The resulting increase in the rate of cell accumulation was postulated to be a potential contributing factor to the suspected carcinogenicity of MNNG.

IV. *Cell death detection*

A variety of approaches are routinely used to qualitatively and/or quantitatively assess toxicant-induced cell death *in vitro*. For example, dye-based cytotoxicity assays are widely employed to assess the viability of cells in culture. While initially developed for use in mammalian cell lines, these approaches have been successfully applied in toxicological studies using fish cell lines[27]. The principle behind many of these assays is the ability of metabolically active cells to either exclude (e.g. trypan blue), incorporate (e.g. neutral red), and/or metabolically transform (e.g. 3-[4,5-dimethylthiazol-3-yl]-2,5-diphenyltetrazolium bromide (MTT)) colorimetric substrates. Other cytotoxicity assays rely upon measurements of ATP or total protein content as indicators of cellular metabolic activity.

While these assays are widely used in toxicology research, they are generally limited in their application to cultured cells. Furthermore, they provide little information on cellular and/or subcellular mechanisms of toxicity, or the specific pathways (apoptosis vs. necrosis) leading to cell death. In addition, standard cytotoxicity assays fail to provide assessments of toxicological effects at higher (e.g. tissue) levels of biological organization, and as such represent relatively coarse indicators of biological toxicity.

By contrast, the complexity of the apoptotic death program offers numerous *in vitro*, *in vivo*, and *in situ* approaches by which to detect active cell death. During the last decade a variety of approaches have been developed, exploiting specific biological events within the cell death program. The gold standard by which to identify apoptosis remains the morphological features of dying cells first identified by Kerr *et al.*[67]. These include cellular shrinkage and detachment, chromatin condensation, and the formation of small, membrane-bound vesicles containing the remnants of dying cells. Despite the proliferation of many sensitive biochemical assays designed to detect apoptosis at different stages of its execution, morphological criteria remain the most reliable means by which to distinguish apoptosis from other cell death modalities (e.g. necrosis). However, the identification of apoptotic features requires a trained eye, appropriate tissue fixation protocols, and recognition that these features are indicative of cells in advanced stages of apoptosis. Furthermore, the morphological features of apoptosis generally persist for less than 9 h, a rapidity that can complicate efforts to detect apoptosis solely on morphological criteria.

During the terminal stages of apoptosis, a Ca^{2+}/Mg^{2+}-dependent endonuclease (CAD) cleaves the dying cell's genomic DNA in the linker regions between nucleosomes[38]. Cleavage by CAD results in the generation of a distinct 'ladder' of oligonucleosomal fragments, in multiples of approximately 200 base-pairs, that is visible after electrophoretic fractionation of genomic DNA. This fragmentation pattern is a hallmark feature of apoptosis, qualitatively distinct from the non-specific degradation of genomic DNA during necrosis. Electrophoretic fractionation of stained or end-labeled genomic DNA remains among the most convenient and widely used methods by which to evaluate the extent of apoptosis. The advantages of this approach are its potential for high sensitivity (particularly if radioisotopic end-labeling is used) and the relative ease of the method. This approach has been successfully used with teleost ovarian tissue to measure apoptosis during normal development[63,133,134] and after exposure to environmental contaminants[61,62,132]. The primary disadvantage with this approach is the inability to identify apoptosis in specific cell types if DNA preparations are derived from homogenates of whole tissue.

An *in situ* approach that relies upon the detection of oligonucleosomal DNA fragmentation and that has been increasingly used by toxicologists is the terminal-deoxynucleotidyl-transferase nick end-labeling (TUNEL) method. This non-isotopic approach permits the detection of DNA fragmentation *in situ*, and thus allows for the identification of cell death at the single cell level. In the teleost ovary, this method was successfully employed to identify cells with fragmented DNA during both normal development[133,134] and after *in vivo* exposure of fish to xenobiotics[130–132]. One important limitation of this approach is its inability to distinguish between apoptotic and necrotic cell deaths, as DNA fragmentation occurs in both circumstances. Complementary biochemical methods are thus required to positively identify the mode of cell death.

Other approaches to detecting apoptosis include colorimetric and fluorometric assays for caspase activity[70]. These methods exploit the specificity of caspases to cleave after aspartate residues during caspase-dependent apoptosis. Cell homogenates are exposed to a synthetic substrate, cleavage of which releases a signal molecule that can be detected colorimetrically or fluorometrically. By targeting relatively early events in the apoptotic cascade, these assays can be a sensitive means by which to detect apoptosis. However, substantial substrate overlap among different caspases can prevent clear identification of specific caspase activities. Commercially available inhibitors of caspase activity (e.g. benzyloxycarbonyl-Val-Ala-Asp-(OMe)-fluoromethylketone (zVAD-fmk)) are frequently used to further discriminate between caspase family members[81].

A cell death detection method used with some degree of success exploits changes in the cell membrane that occur during apoptosis. When a cell becomes irreversibly committed to apoptosis, phosphatidylserine is translocated from the inner to the outer leaflet of the cell membrane, presumably to facilitate recognition of the dying cell by phagocytic macrophages. Phosphatidylserine translocation provides investigators with a convenient means by which to identify apoptotic cells; a number of commercial assay kits have been developed that specifically label cells containing phosphatidylserine in the outer membrane leaflet.

Flow cytometric methods can be used to quantify apoptosis among a population of dispersed cells. Two approaches may be used: (1) detection of apoptotic cells as indicated by ploidy status (apoptotic cells exhibit hypoploidy due to DNA degradation); (2) detection of apoptotic cells after fluorescent tagging of phosphatidylserine residues in the outer leaflet of the plasma membrane. This approach is limited by the necessity to obtain relatively pure populations of dispersed cells, and as such is not amenable to all tissue types.

V. Summary and conclusions

The explosive increase in apoptosis research (e.g. 9721 articles with 'apoptosis' in the Title or Abstract indexed in Medline in 2002) has only recently been expanded to include research in fishes. Furthermore, our overall understanding of underlying mechanisms of apoptotic cell death, particularly with respect to the biochemistry and molecular biology of apoptosis, is far from complete. Thus, great opportunities are currently available to investigate apoptosis in both basic biological (e.g. developmental, molecular, biochemical, physiological) and applied (e.g. toxicological) aspects of fish and fisheries biology. With the genomes of several model fish species anticipated to be sequenced in the near future, there will be considerable interest in investigating functional roles of apoptosis in these and other species. From a toxicological perspective, recognition that stimulation of cell death by xenobiotics may be linked mechanistically to adverse responses at higher levels of biological organization highlights the need for further research in this area. In addition, further toxicological research is needed to develop and refine techniques to reliably assess apoptosis and accurately distinguish between apoptosis and other pathways to cell death.

Acknowledgements. The authors would like to acknowledge the financial support provided by the Natural Sciences and Engineering Research Council of Canada.

VI. References

1. Adrain, C. and S.J. Martin. The mitochondrial apoptosome: a killer unleashed by the cytochrome seas. *Trends Biochem. Sci.* 26: 390–397, 2001.
2. Andreu-Vieyra, C.V. and H.R. Habibi. Effects of salmon GnRH and chicken GnRH-II on testicular apoptosis in goldfish (*Carassius auratus*). *Comp. Biochem. Physiol.* 129B: 483–487, 2001.
3. Andreu-Vieyra, C.V. and H.R. Habibi. Factors controlling ovarian apoptosis. *Can. J. Physiol. Pharamacol.* 78: 1003–1012, 2000.
4. Antonsson, B.. Bax and other pro-apoptotic Bcl-2 family 'killer-proteins' and their victim the mitochondrion. *Cell Tissue Res.* 306: 347–361, 2001.
5. Aravind, L., V.M. Dixit and E.V. Koonin. Apoptotic molecular machinery: vastly increased complexity in vertebrates revealed by genome comparisons. *Science* 291: 1279–1284, 2001.
6. Ashkenazi, A. and V.M. Dixit. Death receptors: signaling and modulation. *Science* 281: 1305–1308, 1998.
7. Bae, J., S.Y. Hsu, C.P. Leo, K. Zell and A.J. Hsueh. Underphosphorylated BAD interacts with diverse antiapoptotic Bcl-2 family proteins to regulate apoptosis. *Apoptosis* 6: 319–330, 2001.
8. Beere, H.M. and D.R. Green. Stress management – heat shock protein-70 and the regulation of apoptosis. *Trends Cell Biol.* 11: 6–10, 2001.
9. Berntssen, M.H., O.O. Aspholm, K. Hylland, S.E. Wendelaar Bonga and A.K. Lundebye. Tissue metallothionein, apoptosis and cell proliferation responses in Atlantic salmon (*Salmo salar* L.) parr fed elevated dietary cadmium. *Comp. Biochem. Physiol.* 128C: 299–310, 2001.
10. Bhaskaran, A., D. May, M. Rand-Weaver and C.R. Tyler. Fish p53 as a possible biomarker for genotoxins in the aquatic environment. *Environ. Mol. Mutagen.* 33: 177–184, 1999.
11. Bhaskaran, A., D. May, M. Rand-Weaver and C.R. Tyler. Molecular characterization of the first non-mammalian p73 cDNA. *Comp. Biochem. Physiol.* 126B: 49–57, 2000.
12. Billig, H., I. Furuta and A.J.W. Hsueh. Estrogens inhibit and androgens enhance ovarian granulosa cell apoptosis. *Endocrinology* 133: 2204–2212, 1993.
13. Billig, H., I. Furuta and A.J.W. Hsueh. Gonadotropin-releasing hormone directly induces apoptotic cell death in the rat ovary: biochemical and *in situ* detection of deoxyribonucleic acid fragmentation in granulosa cells. *Endocrinology* 134: 245–252, 1994.
14. Bishop, G.R., L. Jaso-Friedmann and D.L. Evans. Activation-induced programmed cell death of nonspecific cytotoxic cells and inhibition by apoptosis regulatory factors. *Cell Immunol.* 199: 126–137, 2000.
15. Bishop, G.R., S. Taylor, L. Jaso-Friedmann and D.L. Evans. Mechanisms of nonspecific cytotoxic cell regulation of apoptosis: cytokine-like activity of Fas ligand. *Fish Shellfish Immunol.* 13: 47–67, 2002.
16. Blas-Machado, U., H.W. Taylor and J.C. Means. Apoptosis, PCNA, and p53 in *Fundulus grandis* fish liver after in vivo exposure to *N*-methyl-*N*′-nitro-*N*-nitrosoguanidine and 2-aminofluorene. *Toxicol. Pathol.* 28: 601–609, 2000.
17. Bobe, J. and F.W. Goetz. A tumor necrosis factor decoy receptor homologue is up-regulated in the brook trout (*Salvelinus fontinalis*) ovary at the completion of ovulation. *Biol. Reprod.* 62: 420–426, 2000.
18. Bobe, J. and F.W. Goetz. Molecular cloning and expression of a TNF receptor and two TNF ligands in the fish ovary. *Comp. Biochem. Physiol.* 129B: 475–481, 2001.
19. Borner, C. The Bcl-2 protein family: sensors and checkpoints for life-or-death decisions. *Mol. Immunol.* 39: 615–647, 2003.
20. Brenner, C. and G. Kroemer. Mitochondria – the death signal integrators. *Science* 289: 1150–1151, 2000.

21. Burkhardt-Holm, P., H. Schmidt and W. Meier. Heat shock protein (hsp70) in brown trout epidermis after sudden temperature rise. *Comp. Biochem. Physiol.* 120A: 35–41, 1998.
22. Bury, N., L. Jie, G. Flik, A.C. Lock and S.E. Wendelaar Bonga. Cortisol protects against copper induced necrosis and promotes apoptosis in fish gill chloride cells in vitro. *Aquat. Toxicol.* 40: 193–202, 1998.
23. Candé, C., I. Cohen, E. Daugas, L. Ravagnan, N. Larochette, N. Zamzami and G. Kroemer. Apoptosis-inducing factor (AIF): a novel caspase-independent death effector released from mitochondria. *Biochimie* 84: 215–222, 2002.
24. Cantrell, S.M., J. Joy-Schlezinger, J.J. Stegeman, D.E. Tillitt and M. Hannink. Correlation of 2,3,7,8-tetrachlorodibenzo-*p*-dioxin-induced apoptotic cell death in the embryonic vasculature with embryotoxicity. *Toxicol. Appl. Pharmacol.* 148: 24–34, 1998.
25. Cantrell, S.M., L.H. Lutz, D.E. Tillitt and M. Hannink. Embryotoxicity of 2,3,7,8-tetrachlorodibenzo-*p*-dioxin (TCDD): the embryonic vasculature is a physiological target for TCDD-induced DNA damage and apoptotic cell death in medaka (*Oryzias latipes*). *Toxicol. Appl. Pharmacol.* 141: 23–34, 1996.
26. Caron de Fromentel, C., F. Pakdel, A. Chapus, C. Baney, P. May and T. Soussi. Rainbow trout p53: cDNA cloning and biochemical characterization. *Gene* 112: 241–245, 1992.
27. Castaño, A., N. Bols, T. Braunbeck, P. Dierickx, M. Halder, B. Isomaa, K. Kawahara, L.E.J. Lee, C. Mothersill, P. Pärt, G. Repetto, J.R. Sintes, H. Rufli, R. Smith, C. Wood and H. Segner. The use of fish cells in ecotoxicology. *ATLA* 31: 317–351, 2003.
28. Chen, M.C., H.Y. Gong, C.Y. Cheng, J.P. Wang, J.R. Hong and J.L. Wu. Cloning and characterization of a novel nuclear Bcl-2 family protein, zfMcl-1a, in zebrafish embryo. *Biochem. Biophys. Res. Commun.* 279: 725–731, 2000.
29. Chen, M.C., H.Y. Gong, C.Y. Cheng, J.P. Wang, J.R. Hong and J.L. Wu. Cloning and characterization of zfBLP1, a Bcl-XL homologue from the zebrafish, *Danio rerio*. *Biochim. Biophys. Acta* 1519: 127–133, 2001.
30. Chun, S.-Y. and A.J.W. Hsueh. Paracrine mechanisms of ovarian follicle apoptosis. *J. Reprod. Immunol.* 39: 63–75, 1998.
31. Clearwater, S.J., A.M. Farag and J.S. Meyer. Bioavailability and toxicity of dietborne copper and zinc to fish. *Comp. Biochem. Physiol.* 132C: 269–313, 2002.
32. Corcoran, G.B., L. Fix, D.P. Jones, M.T. Moslen, P. Nicotera, F.A. Oberhammer and R. Buttyan. Apoptosis: molecular control point in toxicity. *Toxicol. Appl. Pharmacol.* 128: 169–181, 1994.
33. Daly, F.J. and J.H. Sandell. Inherited retinal degeneration and apoptosis in mutant zebrafish. *Anat. Rec.* 258: 145–155, 2000.
34. Dong, W., H. Teraoka, S. Kondo and T. Hiraga. 2,3,7,8-Tetrachlorodibenzo-*p*-dioxin induces apoptosis in the dorsal midbrain of zebrafish embryos by activation of arylhydrocarbon receptor. *Neurosci. Lett.* 303: 169–172, 2001.
35. Dong, W., H. Teraoka, K. Yamazaki, S. Tsukiyama, S. Imani, T. Imagawa, J.J. Stegeman, R.E. Peterson and T. Hiraga. 2,3,7,8-Tetrachlorodibenzo-*p*-dioxin toxicity in the zebrafish embryo: local circulation failure in the dorsal midbrain is associated with increased apoptosis. *Toxicol. Sci.* 69: 191–201, 2002.
36. Drummond, C.D., N. Bassoli, E. Rizzo and Y. Sato. Postovulatory follicle: a model for experimental studies of programmed cell death or apoptosis in teleosts. *J. Exp. Zool.* 287: 176–182, 2000.
37. Earnshaw, W.C., L.M. Martins and S.H. Kaufmann. Mammalian caspases: structure, activation, substrates, and functions during apoptosis. *Annu. Rev. Biochem.* 68: 383–424, 1999.
38. Enari, M., H. Sakahira, H. Yokoyama, K. Okawa, A. Iwamatsu and S. Nagata. A caspase-activated DNase that degrades DNA during apoptosis, and its inhibitor ICAD. *Nature* 391: 43–50, 1998.
39. Evans, D.L., M. Oumouna and L. Jaso-Friedmann. Nonradiometric detection of cytotoxicity of teleost nonspecific cytotoxic cells. *Methods Cell Sci.* 22: 233–237, 2000.
40. Fadeel, B., B. Zhivotovsky and S. Orrenius. All along the watchtower: on the regulation of apoptosis regulators. *FASEB J.* 13: 1647–1657, 1999.
41. Feder, M.E. and G.E. Hofmann. Heat-shock proteins, molecular chaperones, and the stress response: evolutionary and ecological physiology. *Annu. Rev. Physiol.* 61: 243–282, 1999.
42. Ganassin, R.C., S.M. Sanders, C.J. Kennedy, E.M. Joyce and N.C. Bols. Development and characterization of a cell line from Pacific herring *Clupea harengus pallasi*, sensitive to both naphthalene cytotoxicity and infection by viral hemorrhagic septicemia virus. *Cell Biol. Toxicol.* 15: 299–309, 1999.
43. Garrido, C., S. Gurbuxani, L. Ravagnan and G. Kroemer. Heat shock proteins: endogenous modulators of apoptotic cell death. *Biochem. Biophys. Res. Commun.* 286: 433–442, 2001.
44. Gerschenson, L.E. and R.J. Rotello. Apoptosis: a different type of cell death. *FASEB J.* 6: 2450–2455, 1992.
45. Green, D.R. and J.C. Reed. Mitochondria and apoptosis. *Science* 281: 1309–1312, 1998.
46. Greenlee, A.R., R.A. Brown and S.S. Ristow. Nonspecific cytotoxic cells of rainbow trout (*Oncorhynchus mykiss*) kill YAC-1 targets by both necrotic and apoptotic mechanisms. *Dev. Comp. Immunol.* 15: 153–164, 1991.

47. Heining, P. and R.W. Hoffmann. Light- and electron microscopical studies on the prolonged toxicity of trichloroethylene on rainbow trout (*Oncorhynchus mykiss*). *Exp. Toxicol. Pathol.* 45: 167–176, 1993.
48. Hengartner, M.O. The biochemistry of apoptosis. *Nature* 407: 770–776, 2000.
49. Hengartner, M.O. and H.R. Horvitz. *C. elegans* cell survival gene *ced-9* encodes a functional homolog of the mammalian proto-oncogene *bcl-2*. *Cell* 76: 665–676, 1994.
50. Hockenbery, D.M., Z.N. Oltvai, X.-M. Yin, C.L. Milliman and S.J. Korsmeyer. Bcl-2 functions in an antioxidant pathway to prevent apoptosis. *Cell* 75: 241–251, 1993.
51. Hogan, R.J., W.R. Taylor, M.A. Cuchens, J.P. Naftel, L.W. Clem, N.W. Miller and V.G. Chinchar. Induction of target cell apoptosis by channel catfish cytotoxic cells. *Cell Immunol.* 195: 110–118, 1999.
52. Hong, J.R., Y.L. Hsu and J.L. Wu. Infectious pancreatic necrosis induces apoptosis due to down-regulation of survival factor MCL-1 protein expression in a fish cell line. *Virus Res.* 63: 75–83, 1999.
53. Hornung, M.W., J.M. Spitsbergen and R.E. Peterson. 2,3,7,8-Tetrachlorodibenzo-*p*-dioxin alters cardiovascular and craniofacial development and function in sac fry of rainbow trout (*Oncorhynchus mykiss*). *Toxicol. Sci.* 47: 40–51, 1999.
54. Hsu, S.-Y., A. Kaipa, E. McGee, M. Lomeli and A.J.W. Hsueh. Bok is a pro-apoptotic Bcl-2 protein with restricted expression in reproductive tissues and heterodimerizes with selective anti-apoptotic Bcl-2 family members. *Proc. Natl Acad. Sci. USA* 94: 12401–12406, 1997.
55. Hsueh, A.J.W., H. Billig and A. Tsafriri. Ovarian follicle atresia: a hormonally controlled apoptotic process. *Endocr. Rev.* 15: 707–724, 1994.
56. Hsueh, A.J.W., K. Eisenhauer, S.-Y. Chun, S.-Y. Hsu and H. Billig. Gonadal cell apoptosis. *Recent Prog. Horm. Res.* 51: 433–456, 1996.
57. Iger, Y., P.H. Balm, H.A. Jenner and S.E. Wendelaar Bonga. Cortisol induces stress-related changes in the skin of rainbow trout (*Oncorhynchus mykiss*). *Gen. Comp. Endocrinol.* 97: 188–198, 1995.
58. Iger, Y., R.A.C. Lock, J.A. Jenner and S.E. Wendelaar Bonga. Cellular responses in the skin of carp (*Cyprinus carpio*) exposed to copper. *Aquat. Toxicol.* 29: 49–64, 1994.
59. Inohara, N. and G. Nuñez. Genes with homology to mammalian apoptosis regulators identified in zebrafish. *Cell Death Differ.* 7: 509–510, 2000.
60. Iwama, G.K., P.T. Thomas, R.B. Forsyth and M.M. Vijayan. Heat shock protein expression in fish. *Rev. Fish Biol. Fish.* 8: 35–56, 1998.
61. Janz, D.M., M.E. McMaster, K.R. Munkittrick and G. Van Der Kraak. Elevated ovarian follicular apoptosis and heat shock protein-70 expression in white sucker exposed to bleached Kraft pulp mill effluent. *Toxicol. Appl. Pharmacol.* 147: 391–398, 1997.
62. Janz, D.M., M.E. McMaster, L.P. Weber, K.R. Munkittrick and G. Van Der Kraak. Recovery of ovary size, follicle cell apoptosis, and HSP70 expression in fish exposed to bleached pulp mill effluent. *Can. J. Fish. Aquat. Sci.* 58: 620–625, 2001.
63. Janz, D.M. and G. Van Der Kraak. Suppression of apoptosis by gonadotropin, 17β-estradiol, and epidermal growth factor in rainbow trout preovulatory ovarian follicles. *Gen. Comp. Endocrinol.* 105: 186–193, 1997.
64. Jung, S.K., A. Mai, M. Iwamoto, N. Arizono, D. Fujimoto, K. Sakamaki and S. Yonehara. Purification and cloning of an apoptosis-inducing protein derived from fish infected with *Anisakis simplex*, a causative nematode of human anisakiasis. *J. Immunol.* 165: 1491–1497, 2000.
65. Kaipia, A. and A.J. Hsueh. Regulation of ovarian follicle atresia. *Annu. Rev. Physiol.* 59: 349–363, 1997.
66. Kane, D.J., T.A. Sarafian, R. Anton, H. Hahn, E.B. Gralla, J.S. Valentine, T. Örd and D.E. Bredesen. Bcl-2 inhibition of neural death: decreased generation of reactive oxygen species. *Science* 262: 1274–1276, 1993.
67. Kerr, J.F., A.H. Wyllie and A.R. Currie. Apoptosis: a basic biological phenomenon with wide-ranging implications in tissue kinetics. *Br. J. Cancer* 26: 239–257, 1972.
68. Kim, J.-M., D.L. Boone, A. Auyeung and B.K. Tsang. Granulosa cell apoptosis induced at the penultimate stage of follicular development is associated with increased levels of Fas and Fas ligand in the rat ovary. *Biol. Reprod.* 58: 1170–1176, 1998.
69. Kluck, R.M., E. Bossy-Wetzel, D.R. Green and D.D. Newmeyer. The release of cytochrome c from mitochondria: a primary site for bcl-2 regulation of apoptosis. *Science* 275: 1132–1136, 1997.
70. Kohler, C., S. Orrenius and B. Zhivotovsky. Evaluation of caspase activity in apoptotic cells. *J. Immunol. Methods* 265: 97–110, 2002.
71. Kwak, H.I., M.O. Bae, M.H. Lee, Y.S. Lee, B.J. Lee, K.S. Kang, C.H. Chae, H.J. Sung, J.S. Shin, J.H. Kim, W.C. Mar, Y.Y. Sheen and M.H. Cho. Effects of nonylphenol, bisphenol A, and their mixture on the viviparous swordtail fish (*Xiphophorus helleri*). *Environ. Toxicol. Chem.* 20: 787–795, 2001.
72. Laing, K.J., J. Holland, S. Bonilla, C. Cunningham and C.J. Secombes. Cloning and sequencing of caspase 6 in rainbow trout, *Oncorhynchus mykiss*, and analysis of its expression under conditions known to induce apoptosis. *Dev. Comp. Immunol.* 25: 303–312, 2001.

73. Laing, K.J., T. Wang, J. Zou, J. Holland, S. Hong, N. Bols, I. Hirono, T. Aoki and C.J. Secombes. Cloning and expression analysis of rainbow trout *Oncorhynchus mykiss* tumor necrosis factor-alpha. *Eur. J. Biochem.* 268: 1315–1322, 2001.
74. Lee, H. and D. Kimelman. A dominant-negative form of p63 is required for epidermal proliferation in zebrafish. *Dev. Cell* 2: 607–616, 2002.
75. Leo, C.P., S.-Y. Hsu, S.-Y. Chun, H.-W. Bae and A.J.W. Hsueh. Characterization of the antiapoptotic Bcl-2 family member myeloid cell leukemia (Mcl-1) and the stimulation of its message by gonadotropins in the rat ovary. *Endocrinology* 140: 5469–5477, 1999.
76. Levin, S. Apoptosis, necrosis, or oncosis: what is your diagnosis? A report from the Cell Death Nomenclature Committee of the Society of Toxicologic Pathologists. *Toxicol. Sci.* 41: 155–156, 1998.
77. Liu, L., K. Fujiki, B. Dixon and R.S. Sundick. Cloning of a novel rainbow trout (*Oncorhynchus mykiss*) CC chemokine with a fractalkine-like stalk and a TNF decoy receptor using cDNA fragments containing AU-rich elements. *Cytokine* 17: 71–81, 2002.
78. Liu, X., C.N. Kim, J. Yang, R. Jemmerson and X. Wang. Induction of apoptotic program in cell-free extracts: requirement for dATP and cytochrome *c*. *Cell* 86: 147–157, 1996.
79. Lyons-Alcantara, M., R. Mooney, F. Lyng, D. Cottell and C. Mothersill. The effects of cadmium exposure on the cytology and function of primary cultures from rainbow trout. *Cell Biochem. Funct.* 16: 1–13, 1998.
80. Majno, G. and I. Joris. Apoptosis, oncosis, and necrosis. An overview of cell death. *Am. J. Pathol.* 146: 3–15, 1995.
81. Majno, G. and I. Joris. Commentary: on the misuse of the term 'necrosis': a step in the right direction. *Toxicol. Pathol.* 27: 494 1999.
82. Masumoto, J., W. Zhou, F.F. Chen, F. Su, J.Y. Kuwada, E. Hidaka, T. Katsuyama, J. Sagara, S. Taniguchi, P. Ngo-Hazelett, J.H. Postlethwait, G. Nunez and N. Inohara. Caspy, a zebrafish caspase, activated by ASC oligomerization is required for pharyngeal arch development. *J. Biol. Chem.* 278: 4268–4276, 2003.
83. Mazon, A.F., C.C. Cerqueira and M.N. Fernandes. Gill cellular changes induced by copper exposure in the South American tropical freshwater fish *Prochilodus scrofa*. *Environ. Res.* 88: 52–63, 2002.
84. McConkey, D.J. Biochemical determinants of apoptosis and necrosis. *Toxicol. Lett.* 99: 157–168, 1998.
85. Meseguer, J., M.A. Esteban and V. Mulero. Nonspecific cell-mediated cytotoxicity in the seawater teleosts (*Sparus aurata* and *Dicentrarchus labrax*): ultrastructural study of target cell death mechanisms. *Anat. Rec.* 244: 499–505, 1996.
86. Morita, Y. and J.L. Tilly. Oocyte apoptosis: like sand through an hourglass. *Dev. Biol.* 213: 1–17, 1999.
87. Mukae, N., M. Enari, H. Sakahira, Y. Fukuda, J. Inazawa, H. Toh and S. Nagata. Molecular cloning and characterization of human caspase-activated DNase. *Proc. Natl Acad. Sci. USA* 95: 9123–9128, 1998.
88. Munkittrick, K.R., M.E. McMaster, L.H. McCarthy, M.R. Servos and G.J. Van Der Kraak. An overview of recent studies on the potential of pulp-mill effluents to alter reproductive parameters in fish. *J. Toxicol. Environ. Health B Crit. Rev.* 1: 347–371, 1998.
89. Nagata, S. and P. Golstein. The Fas death factor. *Science* 267: 1449–1456, 1995.
90. Nishigaki, R., H. Mitani, N. Tsuchida and A. Shima. Effect of cyclobutane pyrimidine dimers on apoptosis induced by different wavelengths of UV. *Photochem. Photobiol.* 70: 228–235, 1999.
91. Oumouna, M., L. Jaso-Friedmann and D.L. Evans. Flow cytometry-based assay for determination of teleost cytotoxic cell lysis of target cells. *Cytometry* 45: 259–266, 2001.
92. Perez, G.I., C.M. Knudson, L. Leykin, S.J. Korsmeyer and J.L. Tilly. Apoptosis-associated signaling pathways are required for chemotherapy-mediated female germ cell destruction. *Nat. Med.* 3: 1228–1232, 1997.
93. Piechotta, G., M. Lacorn, T. Lang, U. Kammann, T. Simat, H.S. Jenke and H. Steinhart. Apoptosis in dab (*Limanda limanda*) as possible new biomarker for anthropogenic stress. *Ecotoxicol. Environ. Saf.* 42: 50–56, 1999.
94. Qin, Q.W., M. Ototake, K. Noguchi, G. Soma, Y. Yokomizo and T. Nakanishi. Tumor necrosis factor alpha (TNFalpha)-like factor produced by macrophages in rainbow trout, *Oncorhynchus mykiss*. *Fish Shellfish Immunol.* 11: 245–256, 2001.
95. Quirk, S.M., R.G. Cowan and S.J. Huber. Fas antigen-mediated apoptosis of ovarian surface epithelial cells. *Endocrinology* 138: 4558–4566, 1997.
96. Raff, M.C. Social controls on cell survival and cell death. *Nature* 356: 397–400, 1992.
97. Ravagnan, L., T. Roumier and G. Kroemer. Mitochondria, the killer organelles and their weapons. *J. Cell Physiol.* 192: 131–137, 2002.
98. Reader, S., V. Moutardier and F. Denizeau. Tributyltin triggers apoptosis in trout hepatocytes: the role of Ca^{2+}, protein kinase C and proteases. *Biochim. Biophys. Acta* 1448: 473–485, 1999.
99. Risso-de Faverney, C., A. Devaux, M. Lafaurie, J.P. Girard, B. Bailly and R. Rahmani. Cadmium induces apoptosis and genotoxicity in rainbow trout hepatocytes through generation of reactive oxygen species. *Aquat. Toxicol.* 53: 65–76, 2001.

100. Roberts, R.A., D.W. Nebert, J.A. Hickman, J.H. Richburg and T.L. Goldsworthy. Perturbation of the mitosis/apoptosis balance: a fundamental mechanism in toxicology. *Fundam. Appl. Toxicol.* 38: 107–115, 1997.
101. Rojo, C. and E. González. Ontogeny and apoptosis of chloride cells in the gill epithelium of newly hatched rainbow trout. *Acta Zool.* 80 1999.
102. Rojo, M.C. and M.E. Gonzalez. In situ detection of apoptotic cells by TUNEL in the gill epithelium of the developing brown trout (*Salmo trutta*). *J. Anat.* 193: 391–398, 1998.
103. Russell, L.D., H. Chiarini-Garcia, S.J. Korsmeyer and C.M. Knudson. *Bax*-dependent spermatogonia apoptosis is required for testicular development and spermatogenesis. *Biol. Reprod.* 66: 950–958, 2002.
104. Salvesen, G.S. and V.M. Dixit. Caspases: intracellular signaling by proteolysis. *Cell* 91: 443–446, 1997.
105. Schwartzman, R.A. and J.A. Cidlowski. Apoptosis: the biochemistry and molecular biology of programmed cell death. *Endocr. Rev.* 14: 133–151, 1993.
106. Shen, L., T.B. Stuge, H. Zhou, M. Khayat, K.S. Barker, S.M. Quiniou, M. Wilson, E. Bengten, V.G. Chinchar, L.W. Clem and N.W. Miller. Channel catfish cytotoxic cells: a mini-review. *Dev. Comp. Immunol.* 26: 141–149, 2002.
107. Shi, Y.B., Q. Li, S. Damjanovski, T. Amano and A. Ishizuya-Oka. Regulation of apoptosis during development: input from the extracellular matrix (review). *Int. J. Mol. Med.* 2: 273–282, 1998.
108. Shimizu, S., Y. Eguchi, H. Kosaka, W. Kamiike, H. Matsuda and Y. Tsuhimoto. Prevention of hypoxia-induced cell death by Bcl-2 and Bcl-xL. *Nature* 374: 811–813, 1995.
109. Steller, H. Mechanisms and genes of cellular suicide. *Science* 267: 1445–1449, 1995.
110. Stroh, C. and K. Schulze-Osthoff. Death by a thousand cuts: an ever increasing list of caspase substrates. *Cell Death Differ.* 5: 997–1000, 1998.
111. Strussmann, C.A., T. Saito and F. Takashima. Heat-induced germ cell deficiency in the teleosts *Odontesthes bonariensis* and *Patagonina hatcheri*. *Comp. Biochem. Physiol.* 119A: 637–644, 1998.
112. Susin, S.A., H.K. Lorenzo, N. Zanzami, I. Marzo, B.E. Snow, G.M. Brothers, J. Mangion, E. Jacotot, P. Costantini, M. Loeffler, N. Larochette, D.R. Goodlett, R. Aebersold, D.P. Siderovski, J.M. Penninger and G. Kroemer. Molecular characterization of mitochondrial apoptosis-inducing factor. *Nature* 397: 441–446, 1999.
113. Sweet, L.I., D.R. Passino-Reader, P.G. Meier and G.M. Omann. Fish thymocyte viability, apoptosis and necrosis: in-vitro effects of organochlorine contaminants. *Fish Shellfish Immunol.* 8: 77–90, 1998.
114. Taylor, M.F., M. De Boer-Brouwer, I. Woolveridge, K.J. Teerds and I.D. Morris. Leydig cell apoptosis after the administration of ethane dimethanesulfonate to the adult male rat is a Fas-mediated process. *Endocrinology* 140: 3797–3804, 1999.
115. Thompson, C.B. Apoptosis in the pathogenesis and treatment of disease. *Science* 267: 1456–1462, 1995.
116. Thornberry, N.A. and Y. Lazebnik. Caspases: enemies within. *Science* 281: 1312–1316, 1998.
117. Tiano, L., D. Fedeli, G. Santoni, I. Davies and G. Falcioni. Effect of tributyltin on trout blood cells: changes in mitochondrial morphology and functionality. *Biochim. Biophys. Acta* 1640: 105–112, 2003.
118. Tilly, J.L. and G.I. Perez. Mechanisms and genes of physiological cell death: a new direction for toxicological risk assessments? In: *Comprehensive Toxicology*, edited by I.G. Sipes, C.A. McQueen and A.J. Gandolfi, Oxford, Elsevier Press, pp. 389–395, 1997.
119. Tilly, J.L., K.I. Kowalski, A.L. Johnson and A.J.W. Hsueh. Involvement of apoptosis in ovarian follicular atresia and postovulatory regression. *Endocrinology* 129: 2799–2801, 1991.
120. Toomey, B.H., S. Bello, M.E. Hahn, S. Cantrell, P. Wright, D.E. Tillitt and R.T. Di Giulio. 2,3,7,8-Tetrachlorodibenzo-*p*-dioxin induces apoptotic cell death and cytochrome P4501A expression in developing *Fundulus heteroclitus* embryos. *Aquat. Toxicol.* 53: 127–138, 2001.
121. Toomey, B.H., G.H. Monteverdi and R.T. Di Giulio. Octylphenol induces vitellogenin production and cell death in fish hepatocytes. *Environ. Toxicol. Chem.* 18: 734–739, 1999.
122. Trump, B.F., I.K. Berezesky, S.H. Chang and P.C. Phelps. The pathways of cell death: oncosis, apoptosis, and necrosis. *Toxicol. Pathol.* 25: 82–88, 1997.
123. Tsafriri, A. and R.H. Braw. Experimental approaches to atresia in mammals. *Oxf. Rev. Reprod. Biol.* 6: 226–265, 1984.
124. Van Cruchten, S. and W. Van Den Broeck. Morphological and biochemical aspects of apoptosis, oncosis and necrosis. *Anat. Histol. Embryol.* 31: 214–223, 2002.
125. van der Flier, A. and A. Sonnenberg. Function and interactions of integrins. *Cell Tissue Res.* 305: 285–298, 2001.
126. Van Der Kraak, G.J., K.R. Munkittrick, M.E. McMaster, C.B. Portt and J.P. Chang. Exposure to bleached Kraft pulp mill effluent disrupts the pituitary-gonadal axis of white sucker at multiple sites. *Toxicol. Appl. Pharmacol.* 115: 224–233, 1992.
127. Vaux, D.L. and S.J. Korsmeyer. Cell death in development. *Cell* 96: 245–254, 1999.

128. Walker, M.K. and R.E. Peterson. Aquatic toxicity of dioxins and related chemicals. In: *Dioxins and Health*, edited by A. Schecter, New York, Plenum Press, pp. 347–387, 1994.
129. Wang, Y. and X. Gu. Functional divergence in the caspase gene family and altered functional constraints: statistical analysis and prediction. *Genetics* 158: 1311–1320, 2001.
130. Weber, L.P., G.C. Balch, C.D. Metcalfe and D.M. Janz. Increased kidney and testicular cell death after developmental exposure to 17α-ethinylestradiol in medaka (*Oryzias latipes*). *Environ. Toxicol. Chem.* 23: 792–797, 2004.
131. Weber, L.P., Y. Kiparissis, G.S. Hwang, A.J. Niimi, D.M. Janz and C.D. Metcalfe. Increased cellular apoptosis after chronic aqueous exposure to nonylphenol and quercetin in adult medaka (*Oryzias latipes*). *Comp. Biochem. Physiol.* 131C: 51–59, 2002.
132. Weber, L.P. and D.M. Janz. Effects of beta-naphthoflavone and dimethylbenz[*a*]anthracene on apoptosis and HSP70 expression in juvenile channel catfish (*Ictalurus punctatus*) ovary. *Aquat. Toxicol.* 54: 39–50, 2001.
133. Wood, A.W. and G. Van Der Kraak. Inhibition of apoptosis in vitellogenic ovarian follicles of rainbow trout (*Oncorhynchus mykiss*) by salmon gonadotropin, epidermal growth factor, and 17β-estradiol. *Mol. Reprod. Dev.* 61: 511–518, 2002.
134. Wood, A.W. and G.J. Van Der Kraak. Apoptosis and ovarian function: novel perspectives from the teleosts. *Biol. Reprod.* 64: 264–271, 2001.
135. Yabu, T., S. Kishi, T. Okazaki and M. Yamashita. Characterization of zebrafish caspase-3 and induction of apoptosis through ceramide generation in fish fathead minnow tailbud cells and zebrafish embryo. *Biochem. J.* 360: 39–47, 2001.
136. Yabu, T., S. Todoriki and M. Yamashita. Stress-induced apoptosis by heat shock, UV and γ-ray irradiation in zebrafish embryos detected by increased caspase activity and whole-mount TUNEL staining. *Fish. Sci.* 67: 333–340, 2001.
137. Yang, J., X. Liu, K. Bhalla, C.N. Kim, A.M. Ibrado, J. Cai, T.-I. Peng, D.P. Jones and X. Wang. Prevention of apoptosis by bcl-2: release of cytochrome c from mitochondria blocked. *Science* 275: 1129–1132, 1997.
138. Zhou, H., T.B. Stuge, N.W. Miller, E. Bengten, J.P. Naftel, J.M. Bernanke, V.G. Chinchar, L.W. Clem and M. Wilson. Heterogeneity of channel catfish CTL with respect to target recognition and cytotoxic mechanisms employed. *J. Immunol.* 167: 1325–1332, 2001.
139. Zou, J., T. Wang, I. Hirono, T. Aoki, H. Inagawa, T. Honda, G.I. Soma, M. Ototake, T. Nakanishi, A.E. Ellis and C.J. Secombes. Differential expression of two tumor necrosis factor genes in rainbow trout, *Oncorhynchus mykiss*. *Dev. Comp. Immunol.* 26: 161–172, 2002.

Endocrine Aspects
a. Stress and HPI axis

Biochemistry and Molecular Biology of Fishes, vol. 6
T. P. Mommsen and T. W. Moon (Editors)

CHAPTER 12

Adrenal toxicology: Environmental pollutants and the HPI axis

ALICE HONTELA

Department of Biological Sciences, Water Institute for Semi-arid Ecosystems (WISE), University of Lethbridge, Lethbridge, AB, Canada T1K 3M4

I. Introduction

The capacity of animals to recognize and react to stressful stimuli or stressors that may be chemical, physical or psychological in nature, is a fundamental physiological response that occurs in all vertebrates, from fish to humans. The response of the organism, referred to as the stress response[125], activates the hypothalamo–pituitary–adrenal (HPA) axis and results in the release of catecholamines (epinephrine, norepinephrine) and adrenal steroid hormones from the adrenal tissue. The term 'interrenal' rather than 'adrenal' has been traditionally used in teleosts to describe the location of the steroid- and catecholamine-producing tissue within the head kidney rather than superior to the kidney as in mammals. Cortisol and corticosterone are the major corticosteroid (CS) hormones in vertebrates – these hormonal effectors of the neuro-endocrine stress response induce changes in a suite of physiological processes. In teleosts, cortisol is the major corticosteroid, it has a role in osmoregulation and maintenance of electrolyte balance, regulation of metabolism[89], modulation of immune function[58] and exerts antigonadal effects[94]. Cortisol seems to favor the maintenance of internal homeostasis and diminish the stress caused by stressful stimuli, although as in mammals and human[105], many issues in fish stress physiology remain unresolved. There are still controversies in the literature concerning the specific role of the CS hormones in the physiological stress response. In a recent article published in Endocrine Reviews[105], Sapolsky and colleagues revisit the role of CSs in the stress response. In one view, CSs actively stimulate and/or have a permissive role in the various facets of the physiological stress response and the increased plasma levels mediate the response. In a different view, CSs suppress the stress response and help to prevent a pathological overactivation of its various components. Despite this ongoing controversy, there is strong evidence that several important processes in life cycles of animals are dependent on corticosteroid hormones that often facilitate actions of other hormones such as thyroid hormones or growth hormone[10]. Smoltification of salmonids, metamorphosis in amphibians and larval development of fishes are examples of corticosteroid-dependent events[4,29,101,116]. The existence of the adrenal tissue or its homologue in all the vertebrates examined[10,91] suggests that this system is important for survival of the organism and that impairment of the synthesis of SGCs will reduce the physiological competence of the organism.

The presence in the aquatic environment of numerous chemical stressors including pollutants, that elicit an adverse physiological response in the exposed organisms, necessitates a clear understanding of the physiological, biochemical and molecular processes that fish depend upon for adaptation to, elimination or avoidance of these potentially harmful chemicals they encounter. Moreover, the responses to short-term (acute) and long-term (chronic) exposures to chemical stressors may differ depending on the concentration/dose of the stressor and the exact duration of the exposure[5,6,50,131]. In this chapter, new findings dealing with the physiological and endocrine responses of fish to pollutants will be reviewed, to integrate the understanding of the mechanisms at different levels of organization, from molecular events to tissue level responses leading to deleterious effects. There is substantial evidence to indicate that limits and quantifiable thresholds exist in the capacity of fish to tolerate some pollutants[17,18,119]. The mechanistic understanding of the link between the presence of these chemicals in surface waters and adverse effects on the physiological fitness of the organisms is essential for establishing cause–effect relationships. Unambiguous causal links, based on a mechanistic understanding, between exposure and effects are important in setting up of acceptable environmental norms (e.g. NOEL, No Observed Effect Level; NOAEL, No Observed Adverse Effects Levels; PNEC, Predicted No Effect Concentration) for concentrations of pollutants from anthropogenic sources that are protective of organisms exposed in their environment[17]. Physiologists, together with toxicologists, can make a very significant contribution by providing data that establish the causal links and define a mechanistic framework for the toxicological effects. More than ever, given the demands on the environment and the capacity to generate new chemicals, there is an urgent need for a mechanism-driven decision-making process[17,119].

This chapter will review the recent studies dealing with toxicology of adrenocortical tissue and corticosteroid hormones (cortisol, corticosterone). Since only few toxicological studies focus on the hypothalamus and corticotrophin-releasing hormone (CRH) secretion, and on the pituitary and adrenocorticotropic hormone (ACTH) secretion, data dealing with these components of the HPA axis will be only briefly reviewed. Similarly, investigations of the chromaffin cells and the adrenal medulla will be dealt with briefly since virtually no toxicological studies are reported in teleostean systems. Many of the findings relating to the interactions of pollutants with cellular components of the adrenocortical cells have general significance for all endocrine cells, in particular other steroidogenic cells, since similarities exist among different endocrine cells in the signal transduction and signaling pathways, the homeostasis of intracellular calcium and the processes regulating cell oxidant–antioxidant equilibrium. Many of the findings and mechanisms presented in this chapter on the hypothalamo-pituitary-interrenal/adrenal (HPI/HPA) axis are relevant to other hormonal systems not only in teleost fishes, but also in other vertebrates.

II. Structural and functional organization of teleost interrenals

1. Anatomy

The adrenal glands of all vertebrates comprise chromaffin cells secreting catecholamines (epinephrine and norepinephrine) and steroidogenic cells secreting, depending on the animal species, steroid hormones such as cortisol, corticosterone, aldosterone, as well as androgens. As mentioned earlier, the nomenclature used for adrenals in different vertebrates varies – in teleosts the term interrenals is often used to designate the head kidney tissue where both chromaffin and steroidogenic cells are located. However, recently Norris[91] proposed to use the term 'adrenocortical' to identify in teleosts and other non-mammalian vertebrates the cells homologous to those found in the adrenal cortex of mammals. Since the toxicological literature, in particular studies pertaining to cellular mechanisms of action of pollutants in endocrine cells, is relevant across vertebrate groups because of the similarities of the cellular constituents, the term 'adrenocortical' will be used here in discussion of fish steroidogenic cells which secrete cortisol. Moreover, since many of the toxicological studies dealing with adrenals are mammalian, the comparison of teleost data with data from mammalian studies will be often made in this chapter.

The degree of anatomical connection between chromaffin cells and adrenocortical cells varies throughout the vertebrate groups[10,91], to evolve into a completely distinct arrangement of the medulla and cortex in mammals. In mammals, the adrenal gland is a compact gland, with a medulla composed of chromaffin cells and a cortex made up of three distinct layers secreting various steroids. The outer zone of the mammalian cortex, the *zona glomerulosa*, secretes the mineralocorticoid aldosterone, the *zona fasciculata* in the middle secretes mainly corticosteroids (cortisol, corticosterone), and the inner zone, *zona reticularis*, secretes weak androgens. The secretion of corticosteroids is dependent on pituitary ACTH. Hypophysectomy leads to atrophy of *zona fasciculata*, as does excess corticosteroids. Treatment with metyrapone (2-methyl-1,2-di-3-pyridyl-1-propanone), a pharmacological blocker of 11β-hydroxylase, key enzyme in the synthesis of corticosteroids, leads to overstimulation of secretion of pituitary ACTH[59], causing hypertrophy and hyperplasia of the adrenal cortex in mammals.

In teleosts, islets of steroidogenic cells with interspersed catecholamine cells are associated with sinuses of the dorsal posterior cardinal veins within the head kidney tissue, easily visible as the most anterior part of the kidney. The head kidney does not contain any renal tubules, instead, besides its secretory function, it is an important organ of the immune system of teleosts. The adrenocortical islets are in fact embedded in a matrix of lymphoid tissue, composed of many small cells important for the production of red blood cells, with small dark packets of melanomacrophages[51,52,58]. It has not been determined thus far whether more than one type of adrenocortical cells exists in teleosts or in other non-mammalian vertebrates[91] and there is no evidence for aldosterone synthesis in teleosts[60]. The close association of chromaffin cells and adrenocortical cells in teleosts, and the absence of a cortex, has important implications for toxicological investigations. Unlike mammals, where the specific layers of the adrenal can be separated, it is difficult in teleosts to isolate the adrenocortical cells from other cells in the head kidney, and most *in vitro* studies aimed at understanding the effects of pollutants on teleost adrenocortical cells and the synthesis of corticosteroid hormones, use head kidney preparations made up of all the cells in the head kidney[77,99,100].

2. *Synthesis of corticosteroid hormones*

Secretion of corticosteroids in teleosts is under the control of the hypothalamo-pituitary axis, through the actions of CRH synthesized by neurons that enter the adenohypophysis and stimulate the secretion of ACTH from pituitary corticotropes. Although other trophic factors have been identified[1,2], ACTH is the most potent hormone to stimulate synthesis of cortisol in teleosts. The signal transduction pathways have been well characterized in mammals[27,129], yet relatively few studies investigated the processes leading to the activation of the enzymatic pathways of cortisol synthesis in teleost adrenocortical cells. In teleosts, it is known that following the binding of ACTH to its membrane receptor in the steroidogenic cell, adenylyl cyclase generates cAMP that subsequently acts as a second messenger[96]. Protein kinase A does seem to mediate the actions of cAMP while protein kinase C has an inhibitory role in the pathways leading to cortisol[68]. Even though in mammalian adrenal cells the roles of G-protein, adenylyl cyclase, phosphodiesterases and other constituents have been documented, surprisingly little is known about these in teleost adrenocortical cells. Even the role of calcium, calmodulin and inositol trisphosphate, cellular constituents which have been thoroughly characterized in teleost ovarian steroidogenic cells[112,120] and in teleost pituitary somatotrophs and gonadotrophs[61,62], has not been investigated in teleost adrenocortical cells. This constitutes an important knowledge gap from the toxicological point of view, since all these cellular constituents are potential targets for toxicants and environmental pollutants.

Corticosteroids, similar to all steroid hormones, are synthesized from cholesterol in a series of reactions mediated by steroidogenic enzymes. Cholesterol is either provided by the circulating low-density lipoprotein complex, obtained from intracellular storage vesicles, or synthesized *de novo* in the steroidogenic cells. ACTH through cAMP does stimulate entry of cholesterol into the cells and synthesis of cholesterol *de novo*[114]. Cholesterol first enters the mitochondrion in a process facilitated by the steroidogenic acute regulatory (StAR) protein that transfers cholesterol from the outer mitochondrial membrane to the inner membrane[115]. The first enzymatic step of steroidogenesis is the hydrolysis of the side chain of cholesterol by $P\text{-}450_{scc}$ (cholesterol side chain cleaving enzyme) yielding pregnenolone (a C21 steroid), a key precursor of corticosteroids[26,91]. Pregnenolone is then transferred to the smooth endoplasmic reticulum where it is converted to progesterone by 3β-hydroxysteroid-$\Delta5$-steroid dehydrogenase (3β-HSD) and hydroxylated by 17α-hydroxylase ($P\text{-}450_{c17}$) and 21-hydroxylase ($P\text{-}450_{c21}$). Returning to the mitochondrion, 11-deoxycortisol is then further modified by 11β-hydroxylase, also designated as $P\text{-}450_{c11}$, the enzyme that converts 11-deoxycortisol to cortisol, the major corticosteroid hormone in teleosts.

Plasma cortisol levels reflect the clearance rate from plasma and the rate of synthesis by adrenocortical cells since the hormone is not stored in the cells but rather released into the circulation as it is synthesized. In mammals, a large proportion of the circulating corticosteroids is bound in plasma to proteins, the most important one is corticosteroid-binding globulin (CBG). The half-life of cortisol is about 90 min and there is a dynamic equilibrium between the bound and the free hormone. Corticosteroids, as well as other steroids synthesized in the body and xenobiotics (foreign substances) absorbed from the environment, are metabolized mainly in the liver by Phase I and Phase II enzymes. These enzymes increase the solubility of the molecules facilitating their excretion through urine or feces. Cortisol may be oxidized to corotic acid or conjugated as glucuronides in mammals[128]. Phase I and II enzymes have been well characterized in teleosts[113] and using GC/MS and radioimmunoassay, Pottinger and colleagues[132] established that the major conjugated steroids in bile of stressed rainbow trout are, in decreasing order of concentrations, tetrahydrocortisone, tetrahydrocortisol, cortisone, cortisol and cortolone. It is important to note that the enzymes metabolizing corticosteroids are also the enzymes that metabolize other substrates including numerous pollutants and may detoxify or bioactivate the xenobiotics[25].

3. *Physiological functions potentially impacted by a disruption of the HPI axis and corticosteroids*

Pathologies resulting from excess or abnormally low levels of CS hormones (Cushing and Addison disease) are well documented in the clinical literature[93]. Congenital adrenal hyperplasia leading to genital ambiguity particularly in females, is another adrenal human pathology caused by a defect in 21-hydroxylase and the inability to synthesize cortisol, leading to an excess production of adrenal androgens[126]. In teleosts, CS hormones have two major functions: a role in intermediary metabolism and a role in osmoregulation. Cortisol receptors have been characterized in liver, gut, adipose tissue, as well as gonads and osmoregulatory organs[86,87,97,109]. The receptors for corticosteroids are present in the cytosol and following binding of the hormone to its receptor, the hormone–receptor complex is translocated to the nucleus. Metabolism of carbohydrates, protein and lipids is affected by cortisol in teleosts[89], and high plasma cortisol results in elevated plasma glucose levels[121]. However, although the gluconeogenic actions of cortisol are well documented in mammals, the metabolic consequences of elevated plasma cortisol are still under some debate in teleost fishes[89].

Osmoregulatory effects of cortisol are well known. Cortisol regulates sodium flux across membranes of gut, gills and kidney, stimulates sodium transport, both in and out[36,75] in numerous teleost species subjected to various osmotic challenges. It stimulates the activity of Na^+/K^+-ATPase in osmoregulatory tissues[88,110] and it seems to facilitate adaptation of teleosts to both sea and fresh water. Salmon (*Oncorhynchus nerka*) exhibit elevated cortisol levels during migration[20] and it is generally accepted that the increase in cortisol facilitates adaptation to either sea water or fresh water during migration.

Larval development is another life stage during which cortisol seems to play an important role, both in teleost fishes and in amphibians during metamorphosis[10,92]. Significant fluctuations in cortisol levels have been detected in fish eggs and larvae throughout development[3,29]. Corticosteroids also play a role in the regulation of the immune function. Immunosuppression is mediated by corticosteroids in all vertebrates investigated thus far, including fishes[58]. Cortisol also exerts antigonadal effects, as demonstrated in laboratory studies[94]. Thus cortisol appears to have several important functions in teleosts, some with constitutive beneficial effects and some, usually those associated to prolonged elevated cortisol levels, with adverse effects. The precise role of glucocorticosteroids during the stress response is still somewhat controversial, some evidence indicating that these hormones help mediate the response while other data suggesting that their role is to prevent the pathological overactivation of the stress response[105]. The causal link between the increase in plasma cortisol in response to acute exposures to chemical stressors interested toxicologists for a long time, cortisol being used as an indicator of a stressed physiological state of the fish.

III. The role of the HPI axis in responses to pollutants – link to toxicology

1. Responses to toxicants

Numerous studies showed that plasma cortisol levels increase in fish subjected to various stressors such as physical confinement and handling or acute exposures to some toxicants[50,52]. Several reviews have been published listing the type, intensity and duration of stressors with the capacity to elevate plasma cortisol levels in teleosts[5,6,50]. Most of the reviewed studies are laboratory-based and report that many of the pollutants of anthropogenic sources elicit a stress response in fish and lead to elevated plasma cortisol levels. These studies are important since they make it possible to identify pollutants that are stressful to fish and may affect their survival. However, acute exposures to some pollutants lead to lower plasma cortisol levels and an impaired cortisol secretion. Such an action has been reported for *o,p′*-DDD [1-(2-chlorophenyl)-1-(4-chlorophenyl)-2,2-dichloroethane] in rainbow trout, *O. mykiss*, and tilapia, *Oreochromis mossambicus*[7,8,56,57]. Thus, some pollutants have the capacity to impair the adrenocortical tissue upon acute exposures. Moreover, pollutants may also interfere with the physiological response to corticosteroid hormones at the target tissue level. Arsenic (As) is reported to block glucocorticoid receptor (GR) activation at the receptor binding level[64].

The physiologically and toxicologically relevant question that must be asked, does not however concern the responses of fish to acute (hours to days) exposure to near lethal concentrations of pollutants in the laboratory, but rather the responses and effects of chronic exposures to environmental concentrations of pollutants and the mechanisms through which such effects are mediated. The assessment of the impact of pollutant-induced endocrine anomalies on the physiological status and fitness is an important component of Toxicological Risk Assessment[119]. Understanding of the causal mechanistic link between exposure to a pollutant and a physiological, biochemical or molecular anomaly is of fundamental importance in defining environmental guidelines that will protect wildlife species living in the aquatic ecosystems impacted by pollutants. Comparative physiologists and biochemists can make an important contribution to Environmental Toxicology by elucidating the mechanisms of action leading to physiological anomalies, and communicating their findings to experts in Environmental Risk Assessment (ERA) who are involved in the setting up of the environmental guidelines for acceptable levels of pollutants in the environment.

2. HPI axis as a target of pollutants

While the HPI axis, by generating the neuroendocrine stress response, enables the organism to respond to environmental stressors including pollutants, the constituents of the axis are themselves potential targets of toxicants. Ribelin[102] surveyed all the publications reporting chemical-induced lesions in endocrine organs of mammals and identified the adrenal gland, specifically the adrenal cortex, as the organ most frequently damaged. In this literature survey, the testes followed as the second most commonly reported organ lesioned by chemicals, with thyroid gland as third. Such a review has not been done for teleosts, but given the cellular and biochemical similarities between the endocrine organs of mammals and teleosts, it is likely that the vulnerabilities of mammalian and teleost endocrine organs are comparable.

2.1. Characteristics defining the vulnerability of the adrenal tissue to toxicants

The review of Ribelin[102] and the recognition of the vulnerability of the adrenal tissue to chemical-induced morphological and functional lesions raises the question of why is the adrenal tissue so sensitive to toxicants? Colby[23] reviewed recent studies in mammalian adrenal toxicology and noted key characteristics of the adrenal cortex that increase its vulnerability to toxicants. These characteristics are likely to be also present in teleosts. The adrenal cortex of mammals, similar to the steroidogenic interrenal tissue in teleosts, is constituted of cholesterol-rich cells with a high affinity for lipids and lipoproteins, making the tissue highly lipophilic. This characteristic, expressed as a high K_{ow} (octanol:water partitioning coefficient), facilitates the uptake and accumulation of lipophilic contaminants in the adrenal cortex. Several mammalian and avian studies reported that organic pollutants accumulate at higher concentrations in the adrenal cortex than in other organs[13,23,24]. Cell-selective binding of *o,p′*-DDD, an isomer of DDT (dichlorodiphenyltrichloroethane), an organochlorine pesticide, was reported recently in corticosteroid producing interrenal cells in Atlantic cod (*Gadus morhua*) using autoradiography and binding studies[80]. Interrenal tissue burdens of pollutants, specifically metals, were also characterized in yellow perch (*Perca flavescens*) sampled from a series of lakes in a mining region of Northwestern Quebec[70,78,79]. Although the kidney and the liver were the organs that accumulated higher concentrations of metals such as Cd, Zn, Cu and Ni than the interrenals, the latter did accumulate easily detectable levels of metals. The concentrations (in μg/g of tissue) in the interrenals were 10–30% of kidney metal concentrations, most likely due to the absence of renal tubules and glomeruli in the head kidney. Importantly, the concentrations of metals in tissues of perch sampled in lakes situated along a contamination gradient reflected very closely the environmental contamination gradient. Moreover, the gradient of bioaccumulated metals was clearly evident in the adrenocortical tissue of the sampled fish and was linked to an endocrine adrenal dysfunction[70,78]. Investigations providing data on adrenal tissue burdens in fish from polluted sites are important since they demonstrate that the adrenal tissue is a target of specific environmental pollutants and that this tissue could be used, with other target organs such as the liver or kidney, for assessing the exposure of the fish living in the wild.

Colby[23] also noted that the extensive blood supply, in addition to the lipophilicity of the adrenal, contributes to high tissue levels of potential adrenotoxicants. Although blood supply to the head kidney has not been measured in teleosts, the large post-cardinal veins draining the interrenals indicate that the tissue is well vascularized. Thus, any toxicant circulating in the blood will reach the head kidney and has the potential, depending on the K_{ow} and other characteristics defining its affinity, to cross the cell membranes and accumulate in the cells.

The presence of steroidogenic enzymes containing cytochrome P450 is another characteristic that makes the adrenal highly vulnerable to pollutants[23]. Although the liver is the major site of biotransformation reactions that make all toxicants more hydrophilic, and thus more easily excretable and usually less toxic, some of these reactions bioactivate the original compound and make it more toxic. Cytochrome P450 enzymes (Phase I) catalyze many of these reactions. Several of the steroidogenic enzymes mediating the synthesis of corticosteroids contain cytochrome P450 and these enzymes seem to have the capacity to recognize substrates other than substituted cholesterol and have in fact the capacity to bioactivate some specific toxicants. A well-documented mechanism of adrenal toxicity in several mammalian species is the bioactivation of *o,p′*-DDD. The substance is activated to a toxic metabolite by the P450c11 (11β-hydroxylase)[14,84] in the adrenal cells. Bioactivation of spironolactone, a highly adrenotoxic pharmaceutical compound used in the past in treatment of high blood pressure, has been linked to transformation by P450c17 (17α-hydroxylase/17,20-lyase)[66]. No teleost studies have investigated such bioactivation reactions in either the interrenal or gonadal steroidogenic tissue as yet, but evidence available thus far suggests that similar processes may occur in teleosts[7,80]. The presence of the cytochrome P450 catalytical cycle and the associated redox processes in all steroidogenic tissues, including the interrenal, makes these tissues highly vulnerable to reactive oxygen species (ROS). Studies by Hornsby and colleagues[54,55] elucidated the phenomenon of pseudosubstrates, pollutants which bind to the adrenal cytochrome P450s, interfering with the normal function of the catalytical cytochrome P450 cycle, and thus generate potentially damaging oxygen species such as superoxide radical and hydrogen peroxide. As demonstrated in classical toxicological studies[23], a dysfunction of the cytochrome P450 catalytic cycle can generate highly reactive oxygen species. The resulting imbalance between the pro-oxidant and antioxidant molecules in the cells may cause a dysfunction, including peroxidation of membrane lipids and other cellular lesions. Recent studies in teleosts provided evidence that oxidative stress is a mechanism of action mediating the adrenal toxicity of some pesticides[30,31].

2.2. *Types of adrenotoxicants*

The complexity and multiple roles of the adrenocortical tissue and corticosteroids necessitated a classification and differentiation between several types of adrenotoxicants. Harvey[43] reviewed adrenal toxicants in his book titled 'The Adrenal in Toxicology' and proposed a useful approach for classifying adrenal toxicants. The approach he presented is of interest to physiologists mainly by the use of physiological principles in the classification of various chemicals that impair the adrenal gland. Some adrenal toxicants appear to act directly on the adrenal tissue and target the adrenal specifically without effects in other organs, endocrine or non-endocrine. A well-known example is *o,p′*-DDD as mentioned earlier. The chemical specifically targets the adrenal steroidogenic tissue in all vertebrates tested thus far, including fish[8,57], birds[13], and mammals[14,63], probably because of specific vulnerability of the steroidogenic enzymes mediating synthesis of corticosteroids (11β-hydroxylase) and other constituents of the steroidogenic cells. The phenomenon of hormone mimicry by pollutants, well documented for sex steroids (see Chapter 16) and xenoestrogens has not been well characterized for corticosteroids since many pollutants activate the HPI axis and simply act as stressors. Non-specific stressors such as crowding or anoxia may activate the HPI axis and ACTH secretion[6] leading to hypertrophy of the adrenal. DDT-like effects similar to the effects of corticosterone, suggesting that DDT may mimic corticosterone, were reported in an anuran amphibian[45]. Other toxicants seem to be indirect-acting, as they damage the adrenal by compromising other endocrine organs and it is the dysfunction of these organs that eventually induces alterations in the adrenal. Chemical-induced lesions may also occur outside of the endocrine system but influence the adrenal. Effects of chemicals such as phenobarbital on liver Phase I and Phase II enzymes may influence the metabolism of corticosteroids. Such an indirect effect on cortisol metabolism and its half-life has been demonstrated for β-naphthoflavone in rainbow trout[127]. In contrast, glucocorticoid hormones or their analogues can facilitate, through a GR, the induction of Phase I enzymes which in turn metabolize pollutants, modify their toxicity and their half-life[21]. Such activation can enhance or decrease the toxicity of the pollutant, depending on the action of the Phase I enzymes.

The list of adrenotoxicants identified as effective in mammals is extensive, while data concerning teleosts are only now becoming available (Table 1). It is important to note that some adrenotoxicants are species specific while others have a similar effect across species. Cadmium (Cd), a ubiquitous metal of anthropogenic sources in the aquatic environment, disrupts steroidogenesis in testicular tissue of mammals[39,72] and teleost fish[117], as well as steroidogenesis in the adrenocortical tissue of mammals[85] and teleosts[76]. Moreover, some adrenotoxicants act directly without any enzymatic transformation, while others seem to require bioactivation and the metabolites are more toxic than the precursor molecule. Such a transformation was demonstrated for benzo[*a*]anthracene in mammalian adrenals[40] – only following the bioactivation to an epoxide intermediate was the molecule adrenotoxic while no effects of the toxicant were detected when the activity of the enzyme aryl hydrocarbon hydroxylase (AHH) was inhibited. Moreover, metyrapone, a specific inhibitor of 11β-hydroxylase, also provided protection against the toxicity of benzo[*a*]anthracene metabolites[41], suggesting that the steroid hydroxylase uses the toxicant as a suicide substrate. Given the diversity of pollutants, ranging from metals to organochlorinated chemicals and pharmaceuticals that have the capacity to interfere with the normal adrenal function, the limited resources for detection of these pollutants in the aquatic ecosystem and the fact that some of them can be transformed *in vivo*, it is crucial to use functional tests and bioassays for detection of endocrine toxicity in the target species.

IV. Diagnosis of interrenal impairment in teleosts

Despite the plethora of studies over the years describing the cortisol stress response in fish subjected to acute (hours to days) exposures to either pollutants (see reviews 46, 50) or non-chemical stressors such as handling or confinement[5,6], the discovery that chronic environmental exposures to pollutants lead to a functional impairment

TABLE 1

Compounds producing structural or functional lesions in the adrenal tissue of vertebrates

Compound	Teleost species	Other vertebrates
o,p′-DDD	Rainbow trout[7,8,69,76], Tilapia[56,57], cod[80]	Sparrow[14], mink[13], otter[13], seal[13], chicken[63], duck[63], dog[84], human[23]
p,p′-DDT	Rainbow trout[7]	
p,p′-DDD	Rainbow trout[7]	Mink[13], otter[13]
Cadmium	Yellow perch[15,38,70,76,78,130], rainbow trout[51,52,130], brown trout[91]	Rat[9*,47*,48*,85], bullfrog[37], *Xenopus*[37]
Nickel		Rat[9*]
Atrazine		Rat[28*], bullfrog[37], human[104]
Mancozeb	Rainbow trout[11]	
Diazinon	Rainbow trout[11]	
Endosulfan	Rainbow trout[11,30,31,77]	*Xenopus*[37], bullfrog[37]
Carbon tetrachloride		Guinea pig[16,23–25,43]
Spironolactone		Guinea pig[24,43,66]
BKME	Yellow perch[51], northern pike[51], whitefish[71]	
Benz[*a*]anthracene		Rat[40,41,54]
Tricresyl phosphate		Rat[73,74]
Mercury	Rainbow trout[76], northern pike[82]	
Phthalates		Bovine[81]
Polychlorinated biphenyls	Tilapia[99,100], carp[116]	
Naphthoflavone	Rainbow trout[127]	

References include investigations of adrenal medulla and cortex, as well adrenal (interrenal, suprarenal) tissue of non-mammalian vertebrates. Studies using cells of the adrenal medulla are indicated by an asterisk.

of the adrenocortical (interrenal) tissue is relatively recent[53]. Although earlier field studies[82] reported lower plasma cortisol levels in northern pike (*Esox lucius*) sampled in lakes contaminated by mercury discharged by pulp mills using mercury-based slimicides, the link between high adrenal tissue burdens of pollutants and a functional impairment of the cortisol-secreting cells was only made in 1992. In a simple field study[53], yellow perch were collected from recreational fishermen at three sites in the St Lawrence River system; fish were immediately blood sampled and cortisol levels assayed in the laboratory. While fish from the unpolluted reference site had high plasma cortisol levels, fish from the other two sites had significantly lower plasma cortisol. Although the cortisol levels were much higher than those reported in other studies with fish (mainly trout and goldfish) sampled in the laboratory[5,6], the data revealed a pattern suggesting that fish from polluted sites may have a decreased ability to secrete cortisol in response to capture and handling stress. Morphological anomalies in the interrenal tissue as well as the pituitary corticotropes were reported later in fish from contaminated sites[51,91] although the findings were not always in agreement, possibly because of differences in the stress protocol used and fish species sampled. Evidence for a cortisol impairment was also documented recently in whitefish (*Coregonus levaretus*) sampled at sites polluted by pulp and paper effluents[71] and in brown trout (*Salmo trutta*) from metal polluted sites[91], in addition to older studies[82]. Since the ability to activate the neuroendocrine stress response and elevate plasma cortisol when subjected to a stressor is a fundamental physiological reaction displayed by healthy, physiologically competent individuals, it was relevant to further investigate the processes leading to the pollutant-induced impairment of this response. Moreover, the similarity of the transduction pathways and the enzymatic constituents of the adrenal steroidogenic cells secreting corticosteroids and the gonadal cells secreting androgens, estrogens and progestins may make any findings concerning the vulnerability of the adrenal cells to pollutants relevant to other steroidogenic cells.

1. Use of stress challenge protocols

Novel approaches to test the functional integrity of the HPI axis in fish from the wild were developed and validated. To test the hypothesis that the dysfunction in fact occurs in the adrenocortical tissue rather than the pituitary gland and the production of ACTH, yellow perch from reference and contaminated sites were injected with ACTH(1–39) to test the secretory capacity of the steroidogenic cells[34]. Treatment with ACTH (ACTH challenge *in vivo*) was less effective in stimulating cortisol secretion in yellow perch sampled at the contaminated sites than at the reference site. Moreover, the same pattern was observed with interrenal tissues isolated from the fish and stimulated with a standard concentration of ACTH *in vitro*[15,70,99,100]. Standardized stress protocols, specifically for the duration of the stress, time of the day, intensity, etc. were used in these experiments to facilitate comparison between studies. The results indicated that the impaired cortisol stress response was caused by a dysfunction of the steroidogenic cells rather than lower circulating ACTH levels.

2. Use of in vitro bioassays

The anatomical complexity of the interrenals and the difficulties of sampling fish in the field where many environmental factors, including mixtures of poorly characterized pollutants can influence the secretion of cortisol, required development of a cell preparation that could be used in the laboratory to further investigate the mechanisms of action mediating interrenal impairment. Similar preparations were validated in mammalian studies with the adrenal cortex and the cells from *zona glomerulosa*, *fasciculata* or *reticularis*[66,85] and also, in classical pharmacological studies with the pheochromocytoma (PC12) cells, a tumor of the adrenal medulla[28]. A cell preparation was developed using the head kidney of rainbow trout[76]. This simple preparation of enzymatically dispersed head kidney (interrenal) cells was subsequently used in toxicological studies designed to compare and quantify the adrenotoxicity of a series of environmentally relevant pollutants (Table 2), and in endocrinological studies aimed at increasing the understanding of cell function in the teleost steroidogenic adrenocortical cells[68].

Exposure of the head kidney cell suspensions *in vitro* to test toxicants suspected of adrenotoxicity was carried out in several studies done under the same conditions

TABLE 2

Adrenotoxicity of pesticides and metals in trout adrenal cells exposed *in vitro*

Compound	Cell viability LC50 (μM)	Cortisol secretion EC50 (μM)	LC50/EC50
Atrazine	>50,000	>50,000	–
	>100,000 (*Xenopus*)[a]	>10,000 (*Xenopus*)	–
	>10,000 (bullfrog)	11 (bullfrog)	>909
$CdCl_2$	10,800	168	64.29
	3312 (*Xenopus*)	127 (*Xenopus*)	26.07 (*Xenopus*)
	907 (bullfrog)	22 (bullfrog)	41.22 (bullfrog)
$ZnCl_2$	22,800	355	64.22
Mancozeb	>5000	207	24.16
Endosulfan	308	17.3	17.85
	4.7 (*Xenopus*)	3.8 (*Xenopus*)	1.23 (*Xenopus*)
	11 (bullfrog)	3.6 (bullfrog)	3.05 (bullfrog)
CH_3HgCl	1140	116	9.83
$HgCl_2$	199	22.3	8.92
o,p′-DDD	385	130	2.96
Diazinon	184	109	1.68

EC50, concentrations that inhibits 50% of ACTH-stimulated cortisol secretion following a 60 min exposure *in vitro*; LC50, concentrations that kills 50% of the adrenal cells following a 60 min exposure *in vitro*.

[a] Values obtained with adrenal cells of two amphibian species are also shown, with the name of the species in brackets.

to facilitate quantitative comparisons of adrenal toxicity of a series of environmentally relevant pollutants. In the initial experiments, rainbow trout head kidneys were digested with collagenase and dispase, the cell suspensions were plated into microplate wells and preincubated in complete Minimum Essential Medium (MEM) for 120 min to reach basal cortisol secretion. Then the cells were exposed *in vitro* to the test toxicant, and following a wash, they were stimulated with ACTH to evaluate the maximal capacity of the steroidogenic cells to secrete cortisol. This approach, rather than measuring basal levels of cortisol secretion, has been favored in the *in vitro* studies since the capacity to activate the neuroendocrine stress response, including the increase in plasma cortisol levels, is required of healthy fish when they are faced with various stressors in the wild. The bioassay generated data of the EC50s (Effective Concentration of the test toxicant that inhibits cortisol secretion by 50%) and also the LC50s (Lethal Concentrations of the toxicant that kills 50% of the cells). A ratio LC50/EC50 of 1.0 suggests that the test toxicant is highly cytotoxic and the loss of cortisol secretion following exposure to the test chemical is due to cell death. In contrast, a high ratio LC50/EC50 of, for example, 100, suggests that the test toxicant is not cytotoxic as a higher concentration (by a factor of 100) is required to kill the endocrine cells than to disrupt cortisol secretion (Fig. 1). Moreover, since the endocrine effects are detected at a lower concentration of the test toxicant than is cell death, the data are interpreted to conclude that the endocrine toxicity of the chemical is higher than its cytotoxicity. The cytotoxic cell death-inducing effects may be specific to the interrenal tissue or may occur in other tissues as well. Table 2 presents adrenotoxicity data for a series of metals and pesticides, administered individually,

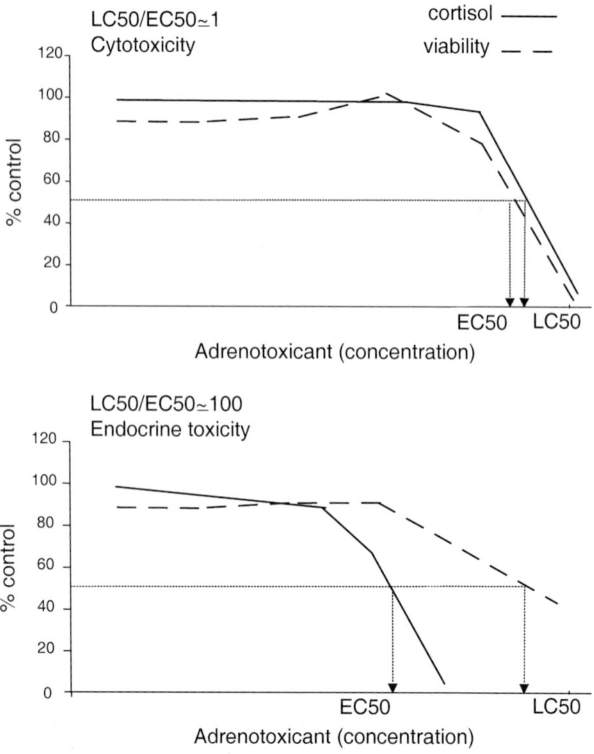

Fig. 1. Adrenotoxicity of a test toxicant assessed *in vitro* using the trout adrenocortical cell suspension[76]. The data are presented as percent of the control, response of cells (cell viability and cortisol secretion) not exposed to the toxicant. The LC50 (concentration that kills 50% of the adrenal cells following an acute *in vitro* exposure to a test toxicant) and EC50 (concentration that inhibits 50% of ACTH-stimulated cortisol secretion following an acute *in vitro* exposure to a test toxicant) are determined statistically and shown on the graphs. A ratio of LC50/EC50 near 1.0 indicates that the loss of cortisol secretion is due to cytotoxicity (cell death). An LC50 greater than the EC50 indicates that the toxicant is a potent endocrine toxicant since it requires a lower concentration (thus higher toxicity) to disrupt hormone synthesis than to kill the cells. The LC50/EC50 ratio for toxicants tested in the trout adrenal bioassay are presented in Table 2.

obtained with this preparation. It shows, for example, that Cd is a toxicant with a high endocrine toxicity to the adrenocortical steroidogenic cells[76], while diazinon, a pesticide, is cytotoxic and the loss of cortisol secretion observed *in vitro* is caused by cell mortality[11] rather than subtle effects within the cell signaling pathways. Although the assay requires some improvements since the exposures used thus far are acute (e.g. 60 min), the concentrations of the toxicants are relatively high compared to environmental concentrations and the cell preparation contains all the cells from the head kidney, the comparisons of the adrenotoxic potential of various pollutants are useful in ranking the chemicals in a rapid, simple and relatively inexpensive endocrine bioassay. There is a need for such assays to facilitate screening of a large number of potentially harmful chemicals prior to more detailed studies which identify the mechanisms of action and the effects of various modifying factors of toxicity, such as age, sex, diet or season.

V. Adrenotoxicants and their mechanisms of action

Given the complexity of the aquatic ecosystem and the multitude of potential endocrine toxicants present in our increasingly industrialized environment, there is a need for a better understanding of the causal relationships between exposure to environmental pollutants and adverse health effects in aquatic species. To elucidate these relationships, it is necessary to understand the mechanisms of action that causes the endocrine or physiological anomaly[119]. Recently significant progress has been made towards the diagnosis of the impairment of cortisol synthesis in teleost steroidogenic cells exposed to adrenotoxicants either accumulated from chronic field exposures in the wild or acute *in vitro* exposure, and also the understanding of the mechanisms of action through which environmental adrenotoxicants disrupt the synthesis of cortisol in teleost steroidogenic cells. Some of the questions asked and methods used in these studies are unique to teleosts, others have been developed for mammalian or avian models and adapted for work with teleost systems. Since teleost adrenal toxicology is a relatively new discipline and only limited data are available thus far, it is important to examine the key findings concerning the mechanisms of action of adrenotoxicants in adrenals of other vertebrates, particularly mammals. Mammalian and human adrenal toxicology is a well-established area of research, using clinical and pharmacological approaches. Since the basic biochemical make-up and functional organization of the steroidogenic cells of mammals and teleosts share many characteristics, some of the findings in the domain of mammalian adrenal toxicology may be relevant to teleosts.

1. Classical mechanisms of action of adrenal toxicants

Colby[23] listed three major mechanisms of action of adrenotoxicants, mechanisms that are similar to actions of other toxicants. These are (1) covalent binding with critical cellular constituents, (2) formation of ROS, and (3) interference with the synthesis or function of cellular constituents such as proteins.

1.1. Covalent interactions with critical macromolecules

One of the keystones in mechanistic cell toxicology and identification of a chemical as harmful to biological constituents, is the formation of covalent links with biomolecules such as DNA, RNA or essential proteins. Chemicals are identified as mutagenic or carcinogenic if they form covalent links with DNA (DNA adducts), considered as potential precursors in the process of abnormal replication of cells. Concentration-dependent formation of covalent links in the guinea pig adrenal cortex was demonstrated for carbon tetrachloride (CCl_4), a well-known organic solvent widely used in industry[25]. In a series of well-executed hypotheses-driven studies, these authors reported that even though CCl_4 is a well-known hepatotoxicant, the numbers of covalent links detected following an exposure to the toxicant is higher in the adrenal cortex than in the liver. The covalent binding decreased if adrenal cells were preincubated *in vitro* with 1-aminobenzotriazole (ABT), a suicide substrate, leading to inactivation of cytochrome P450. Suicide inhibitors of cytochrome P450 are chemicals that following activation by the cytochrome P450 system form metabolites capable of covalently binding to the heme moiety, destroying the heme and inactivating the cytochrome P450. The ABT was used by Colby and colleagues[25] to simply inactivate the cytochrome P450 enzymes. Subsequently, they demonstrated that incubation with antibodies to $P450_{17\alpha}$-hydroxylase and $P450_{c21}$, which inhibit these two specific enzymes, had no effect on the covalent binding of CCl_4. Thus, it was proposed that one of the cytochromes, but not the $P450_{17\alpha}$ or $P450_{c21}$, mediates the formation of covalent links in the cortex, possibly by generating reactive intermediates. In another study investigating covalent binding of adrenotoxicants, avian adrenal cortex was used as a model[63] and adrenal cells of duck and chicken were exposed to two metabolites of DDT, the methylsulfonate-DDE and *o,p′*-DDD. An increase in covalent binding following exposure to the two metabolites was detected in the adrenal cortex of both species. Treatment with metyrapone or SKF 525A, two blockers of cytochrome P450s, prevented the formation of covalent links. Glutathione, an antioxidant, protected against *o,p′*-DDD but not methylsulfonate-DDE. In studies with teleosts, adrenotoxicity of DDT and its isomers was demonstrated *in vivo* and *in vitro* in rainbow trout and tilapia[7,56,69], but covalent links were not measured in these studies. Recently, covalent binding of *o,p′*-DDD has been confirmed in interrenal cells of the cod (*G. morhua*)[80], providing further evidence that covalent binding may be a universal mechanism of adrenal toxicity of this particular organochlorine adrenotoxicant.

Since both CCl_4 and DDT metabolites are organochlorine chemicals, the possibility should be considered that the presence of chlorine (Cl) at specific positions in the molecule increases its toxicity to the adrenal cells. Structure–activity relationships (SAR) and quantitative structure–activity relationships (QSAR), important topics in toxicology and Risk Assessment, are used to facilitate predictions of toxicity and target organ-specific effects, based on the structure of the compounds[12]. The concept of SAR is not new to physiologists and endocrinologists who work with receptors and identify the biologically active parts of hormones and other regulators. However, no SAR studies have been carried out with adrenotoxicants although this area could generate useful data for Risk Assessment in which SARs can be used to speed up the evaluation of a large number of potentially harmful chemicals.

1.2. Generation of highly reactive oxygen species

Extensive evidence conclusively demonstrates that another important mechanism of action of adrenotoxicants is the generation of ROS and the imbalance between pro-oxidants and antioxidants in the cell. The resulting oxidative stress is a mechanism mediating the toxic effects of numerous chemicals in a wide array of cells and organs, in many if not all, animal and plant species. Numerous studies with mammalian models using either *in vitro* exposures of tissues and cells or whole animal studies, provided data to show that following exposure to the toxicant, oxidative stress ensues. It has been reported[16] that exposure of adrenal cortex cells of the guinea pig to CCl_4 (together with NADPH) increases the formation of malondialdehyde, a product of lipid peroxidation. The studies with CCl_4[23] also demonstrated the complexity of the biochemical cell responses to a specific toxicant, responses ranging from the formation of covalent links in the cells[25], to an increase in ROS, generation of oxidative stress and lipid peroxidation[16].

The role of oxidative stress in cortisol impairment in teleost adrenocortical tissue has been investigated recently in rainbow trout and white sucker, *Catostomus commersoni*. Head kidney cells of rainbow trout were exposed *in vitro* to endosulfan, an organochlorine pesticide still extensively used in agriculture[31]. Endosulfan was already identified as an adrenotoxicant in rainbow trout adrenocortical steroidogenic cells[11,77], as well in adrenal cells of two amphibian species, *Xenopus laevis* and *Rana catesbeiana*[37]. The EC50 and LC50 for endosulfan exposure were determined in *in vitro* studies and the adrenotoxic potential was compared in different species of fish and amphibians (Table 2). The recent studies by Dorval[31] identified oxidative stress as one of the mechanisms of adrenotoxicity of endosulfan. Following acute (60 min) *in vitro* exposure to endosulfan, cortisol impairment was diagnosed, as shown previously, and significant concentration-dependent changes in the activities of several key enzymes involved in pro-oxidant–antioxidant balance were detected. Activities of catalase, the enzyme that breaks down H_2O_2, glutathione peroxidase (GPx) which generates oxidized glutathione, as well as the levels of reduced glutathione (GSH) and peroxidation of lipids were measured and changes related to the concentration of endosulfan were shown, along with a loss of the capacity to respond to ACTH. In a follow-up study, the roles of catalase and GSH in defense against endosulfan-generated oxidative stress were confirmed by using several pharmacological agents such as aminotriazole (ATA) that inhibits catalase, *N*-acetylcysteine (NAC) a precursor of glutathione that increases the production of GSH, and L-buthionine sulfoximine (L-BSO) that inhibits GPx[30]. To further investigate the importance of oxidative stress in pesticide-induced interrenal impairment in teleosts, a field study was carried out in the Yamaska River (Quebec), draining an area of intensive agricultural activities. White sucker (*C. commersoni*) was sampled at a reference site and at several sites contaminated by pesticides. The study (Dorval *et al.*, submitted) diagnosed an impairment of cortisol secretion, similar to studies previously carried out with yellow perch in metal-contaminated regions[78,79], and evidence for pesticide-induced oxidative stress was provided.

Thus, oxidative stress and related parameters (levels of antioxidants, enzyme activities, lipid peroxidation, etc.) seem to indicate some of the very early cell responses to toxicant-induced stress and offer a potential for use as early warning markers of cell damage. The generation of ROS may target very specific cellular components within the adrenal steroidogenic cells of teleosts. It was reported[31] that although endosulfan did impair the response to ACTH and dbcAMP, along with the changes in the oxidant status of the cell, incubating the cells with pregnenolone, a cholesterol-like precursor of cortisol, restored the secretory capacity of endosulfan-exposed cells. These results suggest that the intracellular steps downstream from pregnenolone were not disrupted by endosulfan since the cell was able to use pregnenolone for synthesis of cortisol. Toxic effects specific to various steps of the signaling pathways of the endocrine cells are of great interest to toxicologists and endocrinologists alike, since the studies investigating these effects provide important mechanistic data concerning endocrine disruption and toxicity of environmental pollutants. However, the relative vulnerability to ROS of signaling pathways in the adrenal steroidogenic cells and those in other cells and tissues, has not been investigated thus far.

1.3. Interference with protein synthesis and the signaling pathways leading to cortisol synthesis

Although the signaling pathways leading to the synthesis of corticosteroids have been well characterized in mammalian adrenals[95,129], our understanding of the signaling pathways leading to the synthesis of cortisol in teleost adrenocortical cells remains limited. Yet, signaling pathways have been well characterized in teleost gonadal steroidogenic cells[112,120] and in teleost pituitary somatotrophs and gonadotrophs[61,62]. Thus far in teleosts, it is known that cAMP is the key second messenger in cortisol synthesis by the interrenal cells[96], and protein kinase A and C play an important role[68]. However, many steps potentially vulnerable to toxicants within the signaling pathways are yet to be characterized. It is thus necessary to consider teleosts as well as mammalian literature to gain some understanding of the mechanisms through which environmental adrenotoxicants may interfere with essential processes within the steroidogenic cells.

2. Target sites of adrenotoxic chemicals within cortisol-secreting steroidogenic cells

2.1. Effects on adenylyl cyclase and cAMP production

Although the role of adenylyl cyclase in cell function is well documented in mammals[42], limited toxicological data are presently available on this potential intracellular target of adrenotoxicants. It has been reported that some cyclodiene pesticides exert their toxicity by inhibiting intercellular communication through cAMP-dependent processes[103]. Another pesticide, chlorpyrifos, a developmental neurotoxicant in mammalian models, targets the adenylyl cyclase signaling cascade[33,111]. Our recent study investigating the mechanism of action of *o,p'*-DDD on cortisol synthesis in rainbow trout[69] provided some evidence that adenylyl cyclase may be targeted by this organochlorine in teleost adrenocortical cells. Stimulation with ACTH was used to test the functional integrity of the ACTH membrane receptor in cells exposed to the test toxicant *in vitro*, while stimulation with dibutyryl-cAMP (dbcAMP, an analog of cAMP) assessed the integrity of the post-membrane signaling steps and stimulation with pregnenolone tested the steps downstream from this important precursor of cortisol (Fig. 2). Further work is required to characterize the sensitivity and the importance of disrupted cAMP production as a mechanism of action leading to impaired cortisol synthesis during exposure to adrenotoxicants (Fig. 3), and to other types of endocrine disruption. Other techniques, including molecular biology techniques, will provide

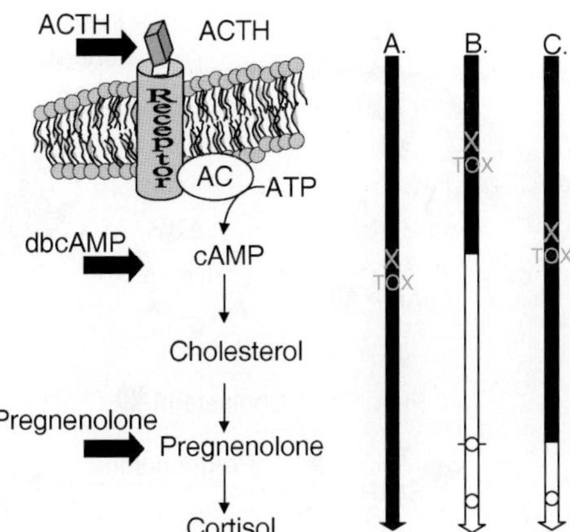

Fig. 2. Schematic representation of tests using exogenous ACTH, dbcAMP and pregnenolone (indicated by black horizontal arrows) of the functional integrity of teleost adrenocortical cells exposed to an adrenotoxicant (TOX). A lack of secretory response to ACTH (pattern A, black vertical arrow with X to show disrupted pathways) indicates a general dysfunction of the secretory pathways, possibly involving the ACTH receptor. No response to ACTH but a secretory response to dbcAMP, an analog of cAMP (pattern B), indicates that the steps downstream from cAMP are functional (white arrow bar) but steps upstream are disrupted by the toxicant. No response to ACTH or dbcAMP with a response to pregnenolone (pattern C) indicates that steps downstream from this precursor of cortisol are functional. Note that the concentrations of ACTH, dbcAMP and pregnenolone used in these functional *in vitro* tests should be physiological rather than pharmacological.

new tools to increase the understanding of mechanisms underlying endocrine and adrenal toxicity.

2.2. *Effects on intracellular calcium homeostasis*

Calcium is an essential messenger within the signaling pathways of most, if not all, cells including the corticosteroidogenic cells[22]. It is necessary for binding of ACTH to its receptor and is required to optimize steroid synthesis and secretion in the adrenal gland. There is extensive evidence that numerous environmental pollutants interfere with the Ca^{2+} signaling processes in cells[65]. Mammalian and teleost studies alike demonstrated that divalent heavy metals such as Cd^{2+}, Pb^{2+} or Hg^{2+} alter the intracellular pools of Ca^{2+} and target Ca^{2+}-dependent processes within various cell types, including hepatocytes, Leydig cells, astrocytes and adrenal cells[35,85,118]. Cadmium (Cd^{2+}), as a divalent heavy metal, competes with and displaces Ca^{2+} at many plasma membrane binding sites, such as Ca^{2+} channels or cell surface receptors. These interactions seem to activate the release of Ca^{2+} stored in the endoplasmic reticulum or in mitochondria[85]. It has been demonstrated that Cd ions pass through the cell membranes of pheochromocytoma cells, a mammalian adrenal neurosecretory cell line, by Ca^{2+} channels[48] and agonist-stimulated Ca^{2+} influx is affected by Cd and Ni[9]. In a rat pituitary cell line, nimodipine, a Ca^{2+} channel antagonist protected against Cd toxicity, suggesting that Cd may enter the cell by voltage-dependent Ca^{2+} channels while BAY K8644, a Ca^{2+} channel agonist, increased Cd toxicity[47].

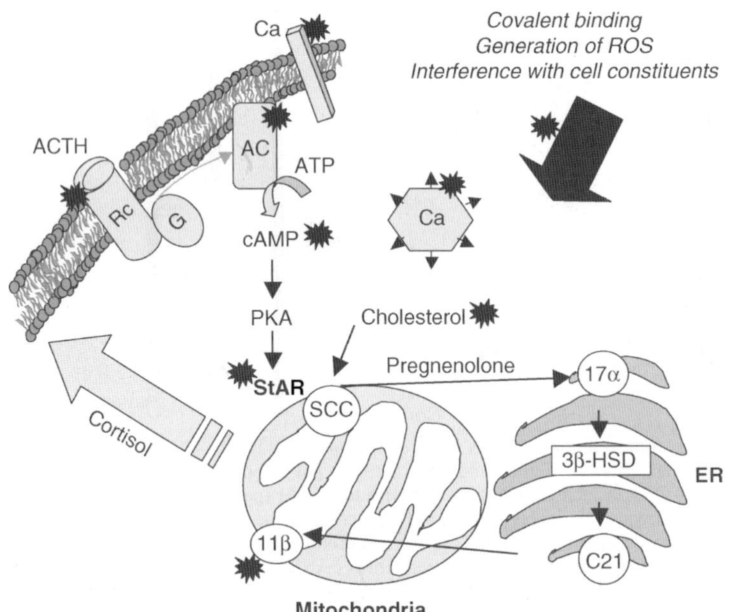

Fig. 3. Summary of key mechanisms of action through which a model adrenotoxicant (indicated by a black star) could disrupt the synthesis of corticosteroids. References presenting data in support of this model are given in the text. ACTH, adrenocorticotropic hormone; Rc, receptor; G, G-protein; AC, adenylyl cyclase; Ca, calcium; ATP, adenosine triphosphate; cAMP, cyclic adenosine monophosphate; PKA, protein kinase A; StAR, Steroid acute regulatory protein; SCC, $P450_{scc}$, cholesterol side chain cleaving enzyme; 11β, 11β-hydroxylase; 17α, 17α-hydroxylase; 3β-HSD, 3β-hydroxysteroid-5Δ-steroid dehydrogenase; C21, 21-hydroxylase; ER, endoplasmic reticulum.

In a study of adrenal steroidogenesis and the effects of Cd in Y-1 mouse adrenal tumor cells, inhibition of ACTH-stimulated steroidogenesis was reversed by using Ca^{2+}-supplemented media[85]. An increase in intracellular Ca^{2+} concentration by the ionophore A23187 provoked an increase in basal and stimulated steroid synthesis by Cd^{2+}-treated adrenal cells. These results suggested that extracellular Ca^{2+} may protect against Cd^{2+} effects on cAMP synthesis and on basal cholesterol metabolism by mitochondria. Measurement by fluorescence of Ca^{2+} entry into the cell suggested that Cd^{2+}, concurrently with Ca^{2+}, freely enters the cell under basal conditions and that its entry is accelerated by ACTH stimulation. However, these results should be interpreted with caution because Ca and Cd signals may overlap in fluorescence measurements[9]. There is strong evidence, however, that Ca^{2+} competes with metals for many intracellular sites. Calmodulin (CaM) may be another cellular target of divalent heavy metals; it has been shown that Cd can replace Ca^{2+} and thus interfere with the CaM function in human trophoblast cells[98]. These experiments provided new data on the cellular targets of Cd in endocrine cells.

In teleosts, the protective effects of Ca^{2+} against Cd^{2+} toxicity in fish exposed through water is well documented *in vivo*[49]. Recent studies in teleost adrenocortical cells demonstrated for the first time that Cd does interfere with Ca channels (Fig. 3) involved in cortisol secretion (Lacroix and Hontela, submitted). Using pharmacological blockers and agonists of calcium channels (Bay K8644, nicardipine, thapsigargin), it was shown that cadmium enters the adrenocortical steroidogenic cells in rainbow trout through Ca^{2+} channels.

It is important to note that there are many elegant clinical pharmacological studies dealing with the disruption of intracellular homeostasis of calcium by xenobiotics using the pheochromocytoma (PC12) cells prepared from tumors isolated from mammalian adrenal medulla or normal chromaffin cells[9]. These studies contributed substantial new knowledge concerning the mechanisms of action of toxicants on these catecholamine-secreting cells[28,81]. Such cell preparations are not available for teleosts, probably because the chromaffin and steroidogenic cells are not arranged into a distinct medulla and cortex, which can be dissected apart, as in mammals. However, pharmacological and toxicological studies using pheochromocytoma cells represent interesting examples of approaches used to identify the intracellular site(s) of action of toxicants within the signaling pathways of endocrine cells.

2.3. Cholesterol metabolism in steroidogenic cells

Steroidogenic cells require cholesterol as substrate for the synthesis of various steroid hormones. Cholesterol is transported by lipoproteins in the blood and eventually, following binding of the lipoproteins (VDL, HDL) to the membranes of the steroidogenic cells, cholesterol is internalized. Cholesterol can also be synthesized *de novo* from acetyl CoA. Cholesterol (free form) is either used for steroidogenesis and shuttled between the mitochondrion and endoplasmic reticulum or it is esterified and the cholesterol esters are stored in lipid vesicles. Key enzymes controlling cholesterol metabolism in the steroidogenic cells are the cholesterol ester hydrolase (CEH) and acyl coenzyme A: cholesterol acyl transferase (ACAT). Latendresse *et al.*[73,74] reported an interesting phenomenon of adrenal toxicity for tricresyl phosphate, a chemical used as an additive to industrial lubricants and oils. Rats exposed *in vivo* to this chemical, or the parent compound butyl cresyl phosphate, exhibited increased weights of the adrenal cortex and gonadal tissue, as well as decreased synthesis of steroids. Histological examination revealed that cells of the steroidogenic tissues were filled with lipid vacuoles, swelling the cells. A biochemical analysis demonstrated that the key enzymes regulating intracellular cholesterol metabolism, particularly the CEH and ACAT activities were significantly lower in rats exposed to the phosphate cresyl. It was proposed that the inability of the enzymatic processes to use cholesterol led to the accumulation of lipids in the vacuoles, distorting the cells, and to the loss of the capacity to synthesize steroids. Thus pollutant-induced anomalies in cholesterol use by the steroidogenic cells disrupted steroidogenesis and resulted in impaired synthesis of key steroids, as well as morphological anomalies. Abnormal cholesterol mobilization was also demonstrated in rat testicular steroidogenic cells exposed to tetrachlorodibenzo-*p*-dioxin (TCDD) *in vitro*[90]. Similarly, studies in teleost gonadal tissue provided evidence that chemicals present in pulp and paper effluents have the capacity to disrupt cholesterol metabolism following *in vivo* exposures[83]. Thus, in addition to oxidative stress and targeted disruption of the signaling pathways, the interference by the pollutants with the metabolic pathways controlling the availability of cholesterol for synthesis of steroids is another important mechanism of action of adrenotoxicants (Fig. 3).

2.4. StAR protein

The steroid acute regulatory (StAR) protein plays a key role in the synthesis of steroid hormones, including corticosteroids. The protein is situated within the mitochondrial membrane and regulates the transfer of cholesterol from the outer mitochondrial membrane to the inner membrane[115]. It is closely linked to cytochrome P450scc, the enzyme that cleaves the side chain from cholesterol. Congenital dysfunction of the StAR protein in humans has been causally linked to a well-characterized pathology, adrenal lipoid hyperplasia[19], a disease that is clinically characterized by sterility, problems with hypotension, symptoms all linked to abnormally low synthesis of key steroids. Steroidogenic tissues of the patients exhibit cells with lipid-filled cytoplasm accumulating cholesterol, and a diminished capacity to complete the synthesis of the hormones. The symptoms of exposure to tricresyl phosphate, described by Latendresse *et al.*[73,74], in rat gonadal and adrenal steroidogenic tissues are similar to those of adrenal lipoid hyperplasia. Recently, it has been reported that exposure to pesticides such as Roundup or lindane disrupts the transcription and expression of the StAR protein gene, leading to low levels of steroids in mammalian models[122–124]. Disruption of the function of this key steroidogenic protein is another mechanism of action of environmental adrenotoxicants (Fig. 3). Toxicological investigations have not yet been completed in fish but StAR protein has been recently characterized[67] in trout head kidney and several other tissues.

2.5. Effects on steroidogenic enzymes

Even though transformation of a xenobiotic into toxic metabolites does often occur in the liver, the major organ of biotransformation, there are many examples of toxic metabolites forming directly in various target organs. The half-lives of most organic pollutants are short and organ or tissue-specific metabolic capabilities, in terms of enzymes, cofactors, pH and other factors that may activate the metabolites, determine the vulnerability of the target organ. Colby and colleagues[24] reported that the adrenotoxicity of spironolactone, a clinically used aldosterone antagonist, is dependent on the bioactivation of spironolactone into adrenotoxic metabolites by adrenal cytochrome P450s. Cortisol secretion was impaired by spironolactone in a concentration-dependent pattern *in vitro*, but the adrenotoxic effects could be completely blocked by treatment with SU-10′603, a cytochrome P450 inhibitor. The example of spironolactone and its toxicity illustrates the bioactivation capacities of the adrenal cells generating toxicants that are deleterious to the cells.

Another classical example of a tissue-specific toxicity is the effect of *o,p′*-DDD on the 11β-hydroxylase, a key steroidogenic enzyme present only in the adrenal tissue. In this particular case, the enzymes effecting steroid production are the specific targets for adrenal toxicants. Steroid hydroxylases contain hemoproteins, constituents that make them vulnerable to toxicants and oxidative stress[54]. The adrenotoxic effects of DDT and its derivatives, and the sensitivity of the 11β-hydroxylase to DDT have been detected in numerous species, including humans. A derivative of DDT, under the name mitotane, was in fact used clinically in chemotherapy treatments for tumors in the adrenal cortex[93] as it specifically destroys adrenal cells. The use of mitotane is similar in its organ-specific toxicity, to the use of radioactive iodide in treatment of hyperthyroidism and cancer of the thyroid gland. In teleosts *o,p'*-DDD is adrenotoxic, but there is evidence that the mechanisms of action may be slightly different from those identified in mammals. Beside the enzymatic targets, the site of action of *o,p'*-DDD may also be situated at steps above the production of cAMP within the signaling pathways leading to cortisol synthesis since stimulation with dbcAMP, the analog of cAMP, restored cortisol synthesis in *o,p'*-DDD exposed cells[7,56]. However, the restoration of cortisol synthesis in *o,p'*-DDD treated cells with dbcAMP was not complete, suggesting that other intracellular sites, possibly the steroidogenic enzymes, may be disrupted as well[69]. Stimulation with pregnenolone, a precursor of cortisol, was used to test the functional integrity of the steroidogenic pathway within the adrenal cells (Fig. 2). Since pregnenolone at physiological concentrations fully restored cortisol secretion in *o,p'*-DDD exposed cells, the enzymes mediating the hydroxylation of pregnenolone appeared functional[69]. Restoration of cortisol secretion by providing pregnenolone *in vitro* has also been observed in both mammalian and teleost adrenal steroidogenic models exposed *in vitro* to cadmium[85,130]. These studies suggest that very specific cellular processes within the steroidogenic cells may be targeted by pollutants. Aromatase, the rate-limiting enzyme in the conversion of androgens to estrogens, is another cellular target of specific endocrine toxicants. Sanderson *et al.*[104] demonstrated in human adrenocortical cells that triazine herbicides such as atrazine, simazine and propazine induced aromatase activity in a dose-dependent pattern. Similar findings have been reported by Hayes *et al.*[44] in amphibians where atrazine also induced aromatase activity, but in gonadal tissue, with major adverse consequences on the reproductive capacity of frogs.

The mechanistic studies reviewed in this chapter are essential for understanding the link between exposure to environmental pollutants and physiological dysfunctions, including endocrine anomalies. Some of the mechanisms may be specific to the adrenal/interrenal tissue, others may be more general. Mechanism-based investigations of target organ-specific toxicities are an important part of toxicological assessment and there is a paucity of endocrine toxicity data, particularly in fish, assessing vulnerabilities of endocrine cells, e.g. different types of steroidogenic cells, to a particular toxicant. Moreover, there are virtually no studies comparing sensitivity to endocrine toxicants in different animal species. Comparative physiologists can greatly contribute to such mechanistic toxicological studies. The challenge that environmental toxicologists face is to predict the adverse effects and set norms that will protect wild species exposed in the field. The elucidation of mechanisms of action does increase the understanding of specific interactions of the toxicant with the cellular targets and our ability to predict effects. There is, however, an urgent need to provide toxicological, physiological and molecular data on wild species. The literature is focused on salmonids, particularly rainbow trout, with only limited information on other teleost species. However, as salmonids are the most sensitive fish species to many pollutants, using this group as a reference to set norms and guideline in environmental toxicology may protect other more tolerant species.

VI. Comparative endocrine toxicology of the HPI axis

Comparative toxicology is a well-established discipline even though the term itself, i.e. the term comparative, is not routinely used. The principle of comparative toxicology is, however, an integral part of toxicology because one of the main goals of a toxicological assessment is the extrapolation of toxicity data between species. Toxicity tests are often done with test species such as rat, guinea pig, mouse or rabbit, and the findings relating to toxicity of specific chemicals are extrapolated to human or other species of interest, using well-established methodology including application of safety factors[32]. Toxicity tests with fish are often carried out with rainbow trout, *O. mykiss*, or minnows, *Pimephales promelas*. There is a lack of data for other fish species in terms of toxicity data (LC50, NOEL, NOAEL, PNEC) and there is even a greater lack of mechanistic data for wild fish species. Comparative approaches and methodologies could fill the knowledge gaps in the area of comparative toxicology and ecotoxicology.

Recently, our laboratory completed experiments designed to provide comparative data on endocrine toxicity of a series of environmental pollutants using the *in vitro* adrenal bioassay with cells from two species of teleost fish and two species of amphibians. Sensitivity of adrenocortical cells isolated from young fish and adults was also compared in trout and perch. In the first series of experiments, the endocrine toxicity of Cd was compared *in vitro* in adrenocortical cells from adult rainbow trout and yellow perch[38,130]. The hypothesis that we tested was simple: given the fact that rainbow trout are more sensitive to Cd *in vivo* than yellow perch (LC50 96 h for trout $<$ perch), are adrenocortical cells of rainbow trout more vulnerable to Cd *in vitro* than cells of perch? If there were no differences between the two fish species in sensitivity to Cd *in vitro*, it could be hypothesized that the difference *in vivo* is determined by rates of uptake of the metal, detoxification capacity and/or other homeostatic mechanisms. If trout cells were in fact more sensitive to Cd *in vitro* than perch cells, it could indicate that the differences between fish species are already expressed at the cell and molecular levels. The adrenal bioassay[76] was used to test the hypothesis in the laboratory. The experiments showed that rainbow trout cells are more sensitive to Cd *in vitro*, as shown by EC50 values (concentration that inhibits cortisol secretion by 50%) but that cell viability (LC50) was similar in the two species. Similar approaches were used to compare young trout and perch, and also young fish with adults[38]. The data showed that adrenocortical cells of young fish are more sensitive to Cd than cells of adults, and provided additional evidence that the endocrine pathways of cells from trout are more vulnerable to the disrupting effects of Cd than cells from perch. Thus the vulnerability to Cd is evident already at the cellular level, even though other differences at tissue and organ levels cannot be excluded.

The adrenotoxic potency of a series of environmental pollutants were also compared in two amphibian species[37], using an adrenal cell preparation similar to the one used in the fish study. Cells of *X. laevis*, the South African clawed frog, and *R. catesbeiana*, the bullfrog, were compared for their sensitivity to Cd, endosulfan (an organochlorine pesticide) and atrazine (a herbicide). The most striking finding concerned the adrenal toxicity of atrazine, a herbicide widely used on corn fields throughout North America: the herbicide was not toxic to adrenal cells of *X. laevis* or rainbow trout[11,37], but was highly toxic to cells of *R. catesbeiana*, the indigenous frog species (Fig. 4). These findings illustrate the difficulty of extrapolating endocrine toxicity (or other toxicity) data across species. Even though *X. laevis* and *R. catesbeiana* are both amphibians, *Rana* seems highly vulnerable to atrazine while *Xenopus* was not. Mechanistic understanding for this difference is limited at present, but would be very important to illustrate the danger of using this herbicide in *Rana* habitats. It has been suggested by others that the use of atrazine poses a significant threat to North American amphibian populations[44].

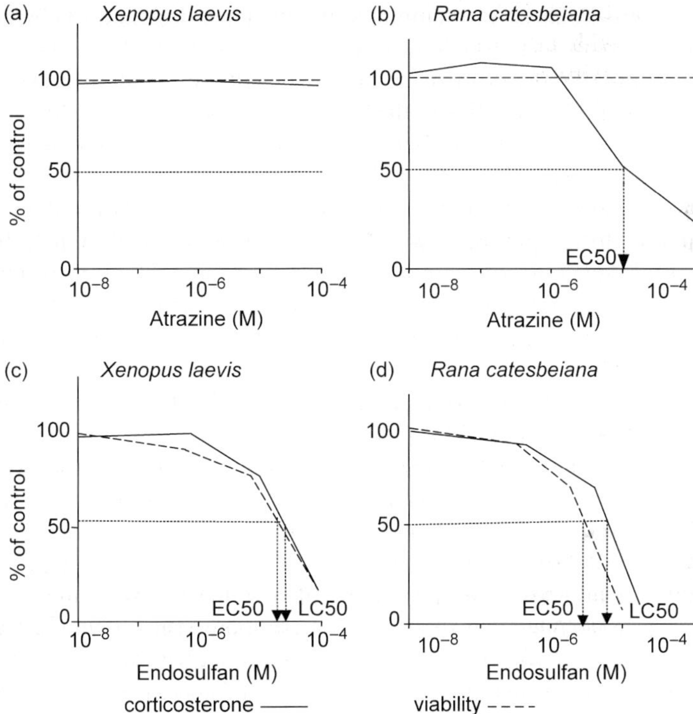

Fig. 4. Adrenotoxicity of atrazine (a herbicide) and endosulfan (an insecticide) extensively used in agriculture in North America, in adrenal cells of *Xenopus laevis* and the bullfrog, *Rana catesbeiana* (modified from Goulet and Hontela[37]). Corticosterone secretion is indicated by solid line, cell viability by broken line. While endosulfan has a similar toxicity in both species, atrazine is a highly potent endocrine toxicant with no cytotoxicity (EC50 $\ll$ LC50) only in *Rana*. The significant difference between the two species in their response to atrazine illustrates the magnitude of species differences in vulnerability to pollutants.

VII. *Effects of impaired corticosteroidogenesis on physiological fitness of fish*

The understanding of the mechanisms mediating adrenal toxicity of environmental pollutants in fish has increased significantly over the last few years. It is now possible to assess adrenal toxicity of any environmental chemical, using *in vitro* methods[76,99,100], as well as *in vivo* exposures in the laboratory or in the field[71,78,79]. Despite these advancements in knowledge, it remains difficult to clearly associate the impaired corticosteroid secretion with specific physiological dysfunctions. Two factors make this linkage difficult. One is the complexity of the field situation where many environmental variables, in addition to pollutant exposure, vary at the same time. The second difficulty is the non-specific role of cortisol, a role with multiple and permissive actions, as are those of some other hormones such as thyroid hormones. In contrast to sex steroids, hormones which affect relatively easily measurable endpoints (gonadal development, fertility), cortisol has multiple functions in fish, including osmoregulation, intermediary metabolism, immune function, development and the acute stress response[50,125]. Despite these shortcomings, field studies designed to characterize the physiological status of fish chronically exposed to pollutants did provide substantial data linking morphological, biochemical and physiological anomalies to pollutant exposures in the field. Anomalies in the use of energy reserves and in the associated enzymes controlling the flow of energy substrates within the organism have been documented in yellow perch sampled in lakes contaminated by metals[79]. Moreover, anomalies in cortisol secretion (lower capacity to secrete cortisol in contaminated fish), thyroid function (lower plasma thyroid levels and damaged thyroid follicles) and delayed gonadal development were also detected in these fish[78]. Sherwood *et al.*[108] examined the bioenergetic capacity of perch from metal-polluted lakes. It was demonstrated using radioisotopic signatures (^{137}Cs) that fish from the polluted lakes had a low conversion efficiency, i.e. the capacity to convert consumed food into biomass. These findings are in agreement with the metabolic impairment reported previously for perch from the metal-polluted lakes[79].

To illustrate some of the difficulties in establishing causal links between exposure to pollutants and physiological effects in the target species, let us consider recent ecophysiological studies by Sherwood and colleagues. These studies[106,107] demonstrated that in addition to the bioenergetic and metabolic impairment linked to environmental metal exposure, fish were also subjected to impoverished food resources. It was concluded that some of the growth impairment was probably due to poor nutrition, since fish from the polluted lakes consumed a diet missing some key prey items, probably eliminated by the pollutants. Thus, despite extensive field studies with fish from the metal-contaminated lakes, studies combining physiological, ecological and toxicological approaches[15,38,70,78,79,106–108], it was not possible to conclusively determine whether all the anomalies diagnosed in the contaminated fish were directly caused by metals or indirectly by effects on food chain structures. Both direct and indirect metal effects[17] can, and probably do, influence the physiological status of fish exposed in the wild. Controlled laboratory studies, using well-characterized exposures and feeding regimes, are required to differentiate between direct and indirect effects of pollutants.

VIII. Importance of mechanism-based toxicological evaluation in Environmental Risk Assessment

Studies reviewed in this chapter illustrate some of the contributions to new knowledge in the area of Comparative Adrenal Toxicology, and the challenges of this relatively new discipline. Most importantly, the contributions, past and future, of comparative biochemistry and physiology, as well as molecular biology, to this discipline have been reviewed. An important challenge to comparative physiologists and toxicologists is to provide data to elucidate the cellular or molecular basis for the differences in vulnerability of various wildlife species to environmental pollutants. There is an overwhelming emphasis on human health in toxicological risk assessment carried out by various government agencies, and a lack of resources leads to very limited, compared to human health risk assessment, investigations of impact of pollutants on wildlife species. Since Environmental Risk Assessment (ERA) does constitute an important mandate for agencies such as US EPA and Environment Canada, the assessment is often done with data obtained for a limited number of species and exposures, with modeling and extrapolations used instead of data for the indigenous species. Although this approach may be acceptable in many situations, there is an opportunity for comparative biochemists and physiologists to become involved in these toxicological studies and provide physiologically meaningful data. Our work in comparative adrenal toxicology, although still limited in scope, did already provide data useful for ERA[17].

Comparative physiologists can make a significant contribution to ERA in several key areas:

- provide physiological data to predict and explain, in terms of biochemical, molecular and physiological interactions at the tissue, organ, cell and gene levels of organization, the effects of pollutants in various aquatic species living in the wild, in seasonally changing environments;
- provide data to explain the differences between species in vulnerability to environmental pollutants;
- elucidate the mechanisms of action of pollutants in a variety of species;
- use experimental design and methodologies traditionally used in comparative physiology to provide and interpret toxicological data for a variety of species;
- provide methods for extrapolation, based on physiological mechanisms, of toxicological data between species;
- provide data that will be used to redefine the environmental norms to ensure protection of wildlife species threatened by multiple stressors, including pollutants.

IX. References

1. Balm, P.H.M., P. Pepels, S. Helfrich, M.L.M. Hovens and S.E. Wendelaar Bonga. Adrenocorticotropic hormone in relation to interrenal function during stress in Tilapia (*Oreochromis mossambicus*). *Gen. Comp. Endocrinol.* 96: 347–360, 1994.
2. Balm, P.H.M. and T.G. Pottinger. Corticotrope and melanotrope POMC-derived peptides in relation to interrenal function during stress in rainbow trout (*Oncorhynchus mykiss*). *Gen. Comp. Endocrinol.* 98: 279–288, 1995.
3. Barry, P.B., J.A. Malison, J.A. Held and J.J. Parrish. Ontogeny of the cortisol stress response in larval rainbow trout. *Gen. Comp. Endocrinol.* 97: 57–65, 1995.
4. Barry, T.P., M. Ochial and J.A. Malison. *In vitro* effects of ACTH on interrenal corticosteroidogenesis during early larval development in rainbow trout. *Gen. Comp. Endocrinol.* 99: 382–387, 1995.
5. Barton, B.A. Stress in finfish: past, present and future – a historical perspective. In: *Fish Stress and Health in Aquaculture*, edited by G.K. Iwama, C.B. Schreck, A.D. Pickering and J. Sumpter, Cambridge, UK, Cambridge University Press, pp. 1–33, 1996.
6. Barton, B.A. and G.K. Iwama. Physiological changes in fish from stress in aquaculture with emphasis on the response and effects of corticosteroids. *Annu. Rev. Fish Dis.* 1: 3–26, 1991.
7. Benguira, S. and A. Hontela. Adrenocorticotrophin- and cyclic adenosine 3′-5′-monophosphate-stimulated cortisol secretion in interrenal tissue of rainbow trout exposed *in vitro* to DDT compounds. *Environ. Toxicol. Chem.* 19: 842–847, 2000.
8. Benguira, S., V.S. Leblond, J.-P. Weber and A. Hontela. Loss of capacity to elevate plasma cortisol in rainbow trout (*Oncorhynchus mykiss*) treated with a single injection of *o,p*′-dichlorodiphenyldichloroethane. *Environ. Toxicol. Chem.* 21: 1753–1756, 2002.
9. Benters, J., T. Schäfer, D. Beyersmann and S. Hechtenberg. Agonist-stimulated calcium transients in PC12 cells are affected differentially by cadmium and nickel. *Cell Calcium* 20: 441–446, 1996.
10. Bentley, P.J. *Comparative Vertebrate Endocrinology*, Cambridge, UK, Cambridge University Press, 1998, 526 pp.
11. Bisson, M. and A. Hontela. Cytotoxic and endocrine-disrupting potential of atrazine, diazinon, endosulfan and mancozeb in adrenocortical steroidogenic cells of rainbow trout exposed *in vitro*. *Toxicol. Appl. Pharmacol.* 180: 110–117, 2002.
12. Bradbury, S.P., O.G. Mekenyan and G.T. Ankley. Quantitative structure–activity relationships for polychlorinated hydroxybiphenyl estrogen receptor binding affinity: an assessment of conformational flexibility. *Environ. Toxicol. Chem.* 15: 1945–1954, 1996.
13. Brandt, I., C.-J. Jönsson and B.-O. Lund. Comparative studies on adrenocorticolytic DDT-metabolites. *Ambio* 21: 602–605, 1992.
14. Breuner, C.W., D.H. Jennings, M.C. Moore and M. Orchinik. Pharmacological adrenalectomy with mitotane. *Gen. Comp. Endocrinol.* 120: 27–34, 2000.
15. Brodeur, J.C., G. Sherwood, J.B. Rasmussen and A. Hontela. Impaired cortisol secretion in yellow perch (*Perca flavescens*) from lakes contaminated by heavy metals: *in vivo* and *in vitro* assessment. *Can. J. Fish. Aquat. Sci.* 54: 2752–2758, 1997.
16. Brogan, W.C.I., P.I. Eacho, D.E. Hinton and H.D. Colby. Effects of carbon tetrachloride on adrenocortical structure and function in guinea pigs. *Toxicol. Appl. Pharmacol.* 75: 118–127, 1984.
17. Campbell, P.G.C., A. Hontela, J.B. Rasmussen, A. Giguère, A. Gravel, L. Kraemer, J. Kovecses, A. Lacroix, H. Levesque and G. Sherwood. Differentiating between direct (physiological) and food-chain mediated (bioenergetic) effects on fish in metal-impacted lakes. *Hum. Ecol. Risk Assess.* 9: 847–866, 2003.
18. Campbell, P.G.C. and A. Tessier. Ecotoxicology of metals in the aquatic environment: geochemical aspects. In: *Ecotoxicology: A Hierarchical Treatment*, edited by M.C. Newman and C.H. Jagoe, Boca Raton, FL, CRC Press, pp. 11–58, 1996.
19. Caron, K.M., S. Soo, W.C. Wetsel, D.M. Stocco, B.J. Clark and K.L. Parker. Targeted disruption of the mouse gene encoding steroidogenic acute regulatory protein provides insights into congenital lipoid adrenal hyperplasia. *Proc. Natl Acad. Sci. USA* 94: 11540–11545, 1997.
20. Carruth, L.L., R.M. Dores, T.A. Maldonado, D.O. Norris, T. Ruth and R.E. Jones. Elevation of plasma cortisol during the spawning migration of landlocked kokanee salmon (*Oncorhynchus nerka kennerlyi*). *Comp. Biochem. Physiol.* 127C: 123–131, 2000.
21. Celander, M., M.E. Bremer, M.E. Hahn and J.J. Stegeman. Glucocorticoid–xenobiotic interactions: dexamethasone-mediated potentiation of cytochrome P4501A induction by β-naphthoflavone in a fish hepatoma cell line (PLHC-1). *Environ. Toxicol. Chem.* 16: 900–907, 1997.
22. Clapham, D.E. Calcium signaling. *Cell Biol. Toxicol.* 80: 259–268, 1995.

23. Colby, H.D. The adrenal cortex as a toxicological target organ. In: *The Adrenal in Toxicology*, edited by P.W. Harvey, London, Taylor & Francis, pp. 131–163, 1996.
24. Colby, H.D. and P.A. Longhurst. Toxicology of the adrenal gland. In: *Endocrine Toxicology*, edited by C.K. Atterwill and J.D. Flack, Cambridge, UK, Cambridge University Press, pp. 243–281, 1992.
25. Colby, H.D., H. Purcelli, S. Kominami, S. Takemori and D.C. Kossor. Adrenal activation of carbon tetrachloride: role of microsomal P450 isozymes. *Toxicology* 94: 31–40, 1994.
26. Conley, A.J. and I.M. Bird. The role of cytochrome P450 17α-hydroxylase and 3β-hydroxylase and 3β-hydroxysteroid dehydrogenase in the integration of gonadal and adrenal steroidogenesis via the $\Delta 5$ and $\Delta 4$ pathways of steroidogenesis in mammals. *Biol. Reprod.* 56: 789–799, 1997.
27. Cooke, B.A. Signal transduction involving cyclic AMP-dependent and cyclic AMP-independent mechanisms in the control of steroidogenesis. *Mol. Cell. Endocrinol.* 151: 25–35, 1999.
28. Das, P.C., W.K. McElroy and R.L. Cooper. Differential modulation of catecholamines by chlorotriazine herbicides in pheochromocytoma (PC12) cells *in vitro*. *Toxicol. Sci.* 56: 324–331, 2000.
29. De Jesus, E.G., T. Hirano and Y. Inui. Changes in cortisol and thyroid hormone concentrations during early development and metamorphosis in the Japanese flounder, *Paralichthys olivaceus*. *Gen. Comp. Endocrinol.* 82: 369–376, 1991.
30. Dorval, J. and A. Hontela. Role of glutathione redox cycle and catalase in defense against oxidative stress induced by endosulfan in adrenocortical cells of rainbow trout (*Oncorhynchus mykiss*). *Toxicol. Appl. Pharmacol.* 192: 191–200, 2003.
31. Dorval, J., V.S. Leblond and A. Hontela. Oxidative stress and loss of cortisol secretion in adrenocortical cells of rainbow trout (*Oncorhynchus mykiss*) exposed *in vitro* to endosulfan, an organochlorine pesticide. *Aquat. Toxicol.* 63: 229–241, 2003.
32. Faustman, E.M. and G.S. Omenn. Risk assessment. In: *Cassarett & Doull's Toxicology: The Basic Science of Poisons*, edited by C.D. Klaassen, New York, USA, McGraw-Hill, pp. 75–88, 1996.
33. Garcia, S.J., F.J. Seidler, T.L. Crumpton and T.A. Slotkin. Does the developmental neurotoxicity of chlorpyrifos involve glial targets? Macromolecule synthesis, adenylyl cyclase signaling, nuclear transcription factors, and formation of reactive oxygen in C6 glioma cells. *Brain Res.* 891: 54–68, 2001.
34. Girard, C., J.C. Brodeur and A. Hontela. Responsiveness of the interrenal tissue of yellow perch (*Perca flavescens*) from contaminated sites to an ACTH challenge test *in vivo*. *Can. J. Fish. Aquat. Sci.* 55: 438–450, 1998.
35. Goldstein, G.W. Evidence that lead acts as a calcium substitute in second messenger metabolism. *Neurotoxicology* 14: 97–102, 1993.
36. Goss, G., S. Perry and P. Laurent. Ultrastructural and morphometric studies on ion and acid–base transport processes in freshwater fish. In: *Cellular and Molecular Approaches to Fish Ionic Regulation*, edited by C.M. Wood and T.J. Shuttleworth, New York, Academic Press, pp. 257–284, 1995.
37. Goulet, B. and A. Hontela. Toxicity of cadmium, endosulfan, and atrazine in adrenal steroidogenic cells of two amphibian species, *Xenopus laevis* and *Rana catesbeiana*. *Environ. Toxicol. Chem.* 22: 2106–2113, 2003.
38. Gravel, A., P.G.C. Campbell and A. Hontela. Disruption of the hypothalamo-pituitary-interrenal axis by metals in the 1^+ yellow perch. *Can. J. Fish. Aquat. Sci.* 2004, in press.
39. Gunnarsson, D., G. Nordberg, P. Lundgren and G. Selstam. Cadmium-induced decrement of the LH receptor expression and cAMP levels in the testis of rats. *Toxicology* 183: 57–63, 2003.
40. Hallberg, E. and J. Rydström. Metabolism and toxic effects of 7,12-dimethyl-benz[*a*]anthracene in isolated rat adrenal cells. *Toxicology* 29: 49–59, 1983.
41. Hallberg, E. and J. Rydström. Toxicity of 7,12-dimethylbenz[*a*]anthracene and 7-hydroxymethyl-12-methylbenz[*a*]anthracene and its prevention in cultured rat adrenal cells. Evidence for a peroxidative mechanism of action. *Toxicology* 47: 259–275, 1987.
42. Hanoune, J., Y. Pouille, E. Tzavara, T. Shen, L. Lipskaya, N. Miyamoto, Y. Suzuki, and N. Defer. Adenylyl cyclases: structure, regulation and function in an enzyme superfamily. *Mol. Cell. Endocrinol.* 128: 179–194, 1997.
43. Harvey, P.W. An overview of adrenal gland involvement in toxicology: from target organ to stress and glucocorticoid modulation of toxicity. In: *The Adrenal in Toxicology*, edited by P.W. Harvey, London, Taylor & Francis, pp. 3–19, 1996.
44. Hayes, T.B., A. Collins, M. Lee, M. Mendoza, N. Noriega, A.A. Stuart and A. Vonk. Hermaphroditic, demasculinized frogs after exposure to the herbicide atrazine at low ecologically relevant doses. *Proc. Natl Acad. Sci. USA* 99: 5476–5480, 2002.
45. Hayes, T.B., T.H. Wu and T.N. Gill. DDT-like effects as a result of corticosterone treatment in an anuran amphibian: is DDT a corticoid mimic or a stressor? *Environ. Toxicol. Chem.* 16: 1948–1953, 1997.

46. Heath, A.G. *Water Pollution and Fish Physiology*, 2nd ed. West Palm Beach, FL, CRC Press, Lewis Publishers, 1995, 359 pp.
47. Hinkle, P.M., P.A. Kinsella and K.C. Osterhoudt. Cadmium uptake and toxicity via voltage-sensitive calcium channels. *J. Biol. Chem.* 262: 16333–16337, 1987.
48. Hinkle, P.M. and M.E. Osborne. Cadmium toxicity in rat pheochromocytoma cells: studies on the mechanism of uptake. *Toxicol. Appl. Pharmacol.* 124: 91–98, 1994.
49. Hollis, L., J.C. McGeer, D.G. McDonald and C.M. Wood. Protective effects of calcium against chronic waterborne cadmium exposure to juvenile rainbow trout. *Environ. Toxicol. Chem.* 19: 2725–2734, 2000.
50. Hontela, A. Endocrine and physiological responses of fish to xenobiotics: role of glucocorticosteroid hormones. *Rev. Toxicol.* 1: 1–46, 1997.
51. Hontela, A., C. Daniel and J.B. Rasmussen. Structural and functional impairment of the hypothalamo-pituitary-interrenal axis in fish exposed to bleached kraft mill effluent in the St Maurice River, Québec. *Ecotoxicology* 6: 1–12, 1997.
52. Hontela, A., C. Daniel and A.C. Ricard. Effects of acute and subacute exposures to cadmium on the interrenal and thyroid function in rainbow trout (*Oncorhynchus mykiss*). *Aquat. Toxicol.* 35: 171–182, 1996.
53. Hontela, A., J.B. Rasmussen, C. Audet and G. Chevalier. Impaired cortisol stress response in fish from environments polluted by PAHs, PCBs, and mercury. *Arch. Environ. Contam. Toxicol.* 22: 278–283, 1992.
54. Hornsby, P.J. Steroid and xenobiotic effects on the adrenal cortex: mediation by oxidative and other mechanisms. *Free Radic. Biol. Med.* 6: 103–115, 1989.
55. Hornsby, P.J. and J.F. Crivello. The role of lipid peroxidation and biological antioxidants in the function of the adrenal cortex. Part 2. *Mol. Cell. Endocrinol.* 30: 123–147, 1983.
56. Ilan, Z. and Z. Yaron. Suppression by organochlorines of the response to adrenocorticotropin of the interrenal tissue in *Sarotherodon Aureus* (Teleostei). *J. Endocrinol.* 87: 185–193, 1980.
57. Ilan, Z. and Z. Yaron. Interference of *o,p′*DDD with interrenal function and cortisol metabolism in *Sarotherodon aureus* (Steindacher). *J. Fish Biol.* 22: 657–669, 1983.
58. Iwama, G.K. and T. Nakanishi. *The Fish Immune System*, New York, Academic Press, 380 pp, 1996.
59. Jain, M.R., F.A. Khan, N.S.R. Krishna and N. Subhedar. Intracranial metyrapone stimulates CRF-ACTH axis in the teleost, *Clarias batrachus*: possible role of neurosteroids. *Neuroendocrinology* 5: 2093–2096, 1994.
60. Jiang, J., G. Young, T. Kobayashi and Y. Nagahama. Eel (*Anguilla japonica*) testis 11*B*-hydroxylase gene is expressed in interrenal tissue and its product lacks aldosterone synthesizing activity. *Mol. Cell. Endocrinol.* 146: 207–211, 1998.
61. Johnson, J.D. and J.P. Chang. Function- and agonist-specific Ca^{2+} signalling: the requirement for and mechanism of spatial and temporal complexity in Ca^{2+} signals. *Biochem. Cell Biol.* 78: 1–24, 2000.
62. Johnson, J.D. and J.P. Chang. Novel, thapsigargin-insensitive intracellular Ca^{2+} stores control growth hormone release from goldfish pituitary cells. *Mol. Cell. Endocrinol.* 165: 139–150, 2000.
63. Jönsson, C.-J., B.O. Lund, B. Brunström and I. Brandt. Toxicity and irreversible binding of two DDT metabolites 3-methylsulfonyl-DDE and *o,p′*-DDD in adrenal interrenal cells in birds. *Environ. Toxicol. Chem.* 13: 1303–1310, 1994.
64. Kaltreider, R.C., A.M. Davis, J.P. Lariviere and J.W. Hamilton. Arsenic alters the function of the glucocorticoid receptor as a transcription factor. *Environ. Health Perspect.* 109: 245–251, 2001.
65. Kass, G.E.N. and S. Orrenius. Calcium signaling and cytotoxicity. *Environ. Health Perspect.* 107: 25–32, 1999.
66. Kossor, D.C., S. Kominami, S. Takemori and H.D. Colby. Role of the steroid 17α-hydroxylase in spironolactone-mediated destruction of adrenal cytochrome P450. *Mol. Pharmacol.* 40: 321–325, 1991.
67. Kusakabe, M., T. Todo, H.J. McQuillan, F.W. Goetz and G. Young. Characterization and expression of steroidogenic acute regulatory protein and MLN64 cDNAs in trout. *Endocrinology* 143: 2062–2070, 2002.
68. Lacroix, M. and A. Hontela. Regulation of acute cortisol synthesis by cAMP-dependant protein kinase and protein kinase C in a teleost species, the rainbow trout (*Oncorhynchus mykiss*). *J. Endocrinol.* 169: 71–78, 2001.
69. Lacroix, M. and A. Hontela. The organochlorine *o,p′*-DDD disrupts the adrenal steroidogenic signaling pathway in rainbow trout (*Oncorhynchus mykiss*). *Toxicol. Appl. Pharmacol.* 190: 197–205, 2003.
70. Laflamme, J.-S., Y. Couillard, P.G.C. Campbell and A. Hontela. Interrenal metallothionein and cortisol secretion in relation to Cd, Cu, and Zn exposure in yellow perch, *Perca flavescens*, from Abitibi lakes. *Can. J. Fish. Aquat. Sci.* 57: 1692–1700, 2000.
71. Lappivaara, J. and A. Oikari. Altered challenge response in whitefish subchronically exposed in areas polluted by bleached kraft mill effluents. *Ecotoxicol. Environ. Saf.* 43: 212–222, 1999.
72. Laskey, J.W. and P.V. Phelps. Effect of cadmium and other metal cations on *in vitro* Leydig cell testosterone production. *Toxicol. Appl. Pharmacol.* 108: 296–306, 1991.

73. Latendresse, J.R., S. Azhar, C.L. Brooks and C.C. Capen. Pathogenesis of cholesteryl lipidosis of adrenocortical and ovarian interstitial cells in F344 rats caused by tricresyl phosphate and butylated triphenyl phosphate. *Toxicol. Appl. Pharmacol.* 122: 281–289, 1993.
74. Latendresse, J.R., C.L. Brooks and C.C. Capen. Pathological effects of butylated triphenyl phosphate-based hydraulic fluid and tricresyl phosphate on the adrenal gland, ovary, and testis in the Fisher-344 rat. *Toxicol. Pathol.* 22: 341–352, 1994.
75. Laurent, P. and S.F. Perry. Effects of cortisol on gill chloride cell morphology and ionic uptake in the freshwater trout, *Salmo gairdneri. Cell Tissue Res.* 259: 429–442, 1990.
76. Leblond, V. and A. Hontela. Effects of *in vitro* exposures to cadmium, mercury, zinc, and 1-(2-chlorophenyl)-1-(4-chlorophenyl)-2,2-dichloroethane on steroidogenesis by dispersed interrenal cells of rainbow trout (*Oncorhynchus mykiss*). *Toxicol. Appl. Pharmacol.* 157: 16–22, 1999.
77. Leblond, V.S., M. Bisson and A. Hontela. Inhibition of cortisol secretion in dispersed head kidney cells of rainbow trout (*Oncorhynchus mykiss*) by endosulfan, an organochlorine pesticide. *Gen. Comp. Endocrinol.* 121: 48–56, 2001.
78. Levesque, H.M., J. Dorval, G.J. Van Der Kraak, P.G.C. Campbell and A. Hontela. Hormonal, morphological, and physiological responses of yellow perch (*Perca flavescens*) to chronic environmental exposures. *J. Toxicol. Environ. Health* 66: 657–676, 2003.
79. Levesque, H.M., T.W. Moon, P.G.C. Campbell and A. Hontela. Seasonal variation in carbohydrate and lipid metabolism of yellow perch (*Perca flavescens*) chronically exposed to metals in the field. *Aquat. Toxicol.* 60: 257–267, 2002.
80. Lindhe, O., I. Brandt, J.S. Christiansen and K. Ingebrigtsen. Irreversible binding of *o,p′*-DDD in interrenal cells of Atlantic cod (*Gadus morhua*). *Chemosphere* 50: 1249–1253, 2003.
81. Liu, P. and C. Lin. Phthalates suppress the calcium signaling of nicotinic acetylcholine receptors in bovine adrenal chromaffin cells. *Toxicol. Appl. Pharmacol.* 183: 92–98, 2002.
82. Lockhart, W.L., J.F. Uthe, A.R. Kenney and P.M. Mehrle. Methylmercury in Northern pike (*Esox lucius*): distribution, elimination, and some biochemical characteristics of contaminated fish. *Can. J. Fish. Aquat. Sci.* 29: 1519–1523, 1972.
83. MacLatchy, D., L. Peters, J. Nickle and G. Van Der Kraak. Exposure to β-sitosterol alters the endocrine status of goldfish differently than 17β-estradiol. *Environ. Toxicol. Chem.* 16: 1895–1904, 1997.
84. Martz, F. and J.A. Straw. Metabolism and covalent binding of 1-(*o*-chlorophenyl)-1-(*p*-chlorophenyl)-2,2-dichloroethane (*o,p′*-DDD). *Drug Metab. Dipos.* 8: 127–130, 1980.
85. Mathias, S.A., O.P. Mgbonyebi, J.R. Motley and J.J. Mrotek. Modulation of adrenal cells functions by cadmium salts. 4. Ca^{2+}-dependant sites affected by $CdCl_2$ during basal and ACTH-stimulated steroid synthesis. *Cell Biol. Toxicol.* 14: 225–236, 1998.
86. Maule, A.G. and C.B. Schreck. Glucorticoid receptors in leucocytes and gill of juvenile coho salmon (*Oncorhynchus kisutch*). *Gen. Comp. Endocrinol.* 77: 448–455, 1990.
87. Maule, A.G. and C.B. Schreck. Stress and cortisol treatment changed affinity and number of glucocorticoid receptors in leukocytes and gill of Coho salmon. *Gen. Comp. Endocrinol.* 84: 83–93, 1991.
88. McCormick, S.. Hormonal control of gill Na^+, K^+-ATPase and chloride cell function. In: *Cellular and Molecular Approaches to Fish Ionic Regulation*, edited by C.M. Wood and T.J. Shuttleworth, New York, Academic Press, pp. 285–316, 1995.
89. Mommsen, T.P., M.M. Vijayan and T.W. Moon. Cortisol in teleosts; dynamics, mechanisms of action, and metabolic regulation. *Rev. Fish Biol. Fish.* 9: 211–268, 1999.
90. Moore, R.W., C.R. Jeffcoate and R.E. Peterson. 2,3,7,8 Tetrachloribenzo-*p*-dioxin inhibits steroidogenesis in the rat testis by inhibiting the mobilization of cholesterol to cytochrome P450 scc. *Toxicol. Appl. Pharmacol.* 109: 85–97, 1991.
91. Norris, D.O. Comparative aspects of vertebrate adrenals. In: *Vertebrate Endocrinology*, edited by D.O. Norris, New York, Academic Press, pp. 329–354, 1997.
92. Norris, D.O. *Vertebrate Endocrinology*, 3rd ed. New York, Academic Press, 634 pp. 1997.
93. Orth, D.N. Cushing's syndrome. *N. Engl. J. Med.* 332: 791–803, 1995.
94. Pankhurst, N.W. and G. Van Der Kraak. Evidence that acute stress inhibits ovarian steroidogenesis in rainbow trout *in vivo*, through actions of cortisol. *Gen. Comp. Endocrinol.* 117: 225–237, 2000.
95. Papadopoulos, V., E.P. Widmaier and P.F. Hall. The role of calmodulin in the response to adrenocorticotropin of plasma membranes from adrenal cells. *Endocrinology* 126: 2465–2473, 1990.
96. Patiño, R., C.S. Bradford and C.B. Schreck. Adenylate cyclase activators and inhibitors, cyclic nucleotide analogs, and phosphatidylinositol: effects on interrenal function of Coho salmon (*Oncorhynchus kisutch*) *in vitro. Gen. Comp. Endocrinol.* 63: 230–235, 1986.

97. Pottinger, T.G., F.R. Knudsen and J. Wilson. Stress-induced changes in the affinity and abundance of cytosolic cortisol-binding sites in the liver of rainbow trout, *Oncorhynchus mykiss* (Walbaum), are not accompanied by changes in measurable nuclear binding. *Fish Physiol. Biochem.* 12: 499–511, 1994.
98. Powlin, S., P.C. Keng and R.K. Miller. Toxicity of cadmium in human trophoblast cells (JAr choriocarcinoma): role of calmodulin and the calmodulin inhibitor, zaldaride maleate. *Toxicol. Appl. Pharmacol.* 144: 225–234, 1997.
99. Quabius, E.S., P.H.M. Balm and S.E. Wendelaar Bonga. Interrenal stress responsiveness of tilapia (*Oreochromis mossambicus*) is impaired by dietary exposure to PCB126. *Gen. Comp. Endocrinol.* 108: 472–482, 1997.
100. Quabius, E.S., D.T. Nolan, C.J. Allin and S.E. Wendelaar Bonga. Influence of dietary exposure to polychlorinated biphenyl 126 and nutritional state on stress response in *Tilapia* (*Oreochromis mossambicus*) and rainbow trout (*Oncorhynchus mykiss*). *Environ. Toxicol. Chem.* 19: 2892–2899, 2000.
101. Redding, J.M., R. Patino and C.B. Schreck. Cortisol effects on plasma electrolytes and thyroid hormones during smoltification in Coho salmon *Oncorhynchus kisutch*. *Gen. Comp. Endocrinol.* 81: 373–382, 1991.
102. Ribelin, W.E.. The effects of drugs and chemicals upon the structure of the adrenal gland. *Fundam. Appl. Toxicol.* 4: 105–119, 1984.
103. Ruch, R.J., R. Fransson, S. Flodstrom, L. Warngard and J.E. Klaunig. Inhibition of hepatocyte gap junctional intercellular communication by endosulfan, chlordane and heptachlor. *Carcinogenesis* 11: 1097–1101, 1990.
104. Sanderson, J.T., W. Seinen, J.P. Giesy and M. van den Berg. 2-Chloro-s-triazine herbicides induce aromatase (CYP19) activity in H295R human adrenocortical carcinoma cells: a novel mechanism for estrogenicity. *Toxicol. Sci.* 54: 121–127, 2000.
105. Sapolsky, R.M., L.M. Romero and A.U. Munck. How do glucocorticoids influence stress responses? Integrating permissive, suppressive, stimulatory, and preparative actions. *Endocr. Rev.* 21: 55–89, 2000.
106. Sherwood, G.D., J. Kovecses, A. Hontela and J.B. Rasmussen. Simplified food webs lead to energetic bottlenecks in polluted lakes. *Can. J. Fish. Aquat. Sci.* 59: 1–5, 2002.
107. Sherwood, G.D., I. Pazzio, A. Moeser, A. Hontela and J.B. Rasmussen. Shifting gears: enzymatic evidence for the energetic advantage of switching diet in wild-living fish. *Can. J. Fish. Aquat. Sci.* 59: 229–241, 2002.
108. Sherwood, G.D., J.B. Rasmussen, D.J. Rowan, J. Brodeur and A. Hontela. Bioenergetic costs of heavy metal exposure in yellow perch (*Perca flavescens*): in situ estimates with a radiotracer (^{137}Cs) technique. *Can. J. Fish. Aquat. Sci.* 57: 441–450, 2000.
109. Shrimpton, J.M. and S.D. McCormick. Regulation of gill cytosolic corticosteroid receptors in juvenile Atlantic salmon: interaction effects of growth hormone with prolactin and triiodothyronine. *Gen. Comp. Endocrinol.* 112: 262–274, 1998.
110. Shrimpton, J.M. and S.D. McCormick. Responsiveness of gill Na^+/K^+-ATPase to cortisol is related to gill corticosteroid receptor concentration in juvenile rainbow trout. *J. Exp. Biol.* 202: 987–995, 1999.
111. Song, X., F.J. Seidler, J.L. Saleh, J. Zhang, S. Padilla and T.A. Slotkin. Cellular mechanisms for developmental toxicity of chlorpyrifos targeting the adenylyl cyclase signaling cascade. *Toxicol. Appl. Pharmacol.* 145: 158–174, 1997.
112. Srivastava, R.K. and G. Van Der Kraak. Effects of activators of different intracellular signaling pathways on steroid production by goldfish vitellogenic ovarian follicles. *Gen. Comp. Endocrinol.* 93: 181–191, 1994.
113. Stegeman, J.J. and M.E. Hahn. Biochemistry and molecular biology of monooxygenases: current perspectives on forms, functions, and regulation of cytochrome P450 in aquatic species. In: *Aquatic Toxicology*, edited by D.C. Malins and G.K. Ostrander, Boca Raton, FL, CRC Press, pp. 87–206, 1994.
114. Stevens, V.L. and J.D. Lambeth. Cholesterol trafficking in steroidogenic cells reversible cycloheximide-dependent accumulation of cholesterol in a pre-steroidogenic pool. *J. Biochem.* 216: 557–563, 1993.
115. Stocco, D.M. Intramitochondrial cholesterol transfer. *Biochim. Biophys. Acta* 1486: 184–197, 2000.
116. Stouthart, A.J.H.X., M.A.J. Huijbregst, P.H.M. Balm, R.A.C. Lock and S.E. Wendelaar Bonga. Endocrine stress response and abnormal development in carp (*Cyprinus carpio*) larvae after exposure of the embryos to PCB 126. *Fish Physiol. Biochem.* 18: 321–329, 1998.
117. Thomas, P. and H.W. Wofford. Effects of cadmium and Aroclor 1254 on lipid peroxidation, glutathione peroxidase activity, and selected antioxidants in Atlantic croaker tissues. *Aquat. Toxicol.* 27: 159–178, 1993.
118. Thoreux-Manlay, A., C. Le Goascogne, D. Segretain, B. Jégou and G. Pinon-Lataillade. Lead affects steroidogenesis in rat Leydig cells *in vivo* and *in vitro*. *Toxicology* 103: 53–62, 1995.
119. Van Der Kraak, G., M. Hewitt, A. Lister, M.E. McMaster and K.R. Munkittrick. Endocrine toxicants and reproductive success in fish. *Hum. Ecol. Risk Assess.* 7: 1017–1025, 2001.

120. Van Der Kraak, G.J. Mechanisms by which calcium ionophore and phorbol ester modulate steroid production by goldfish preovulatory ovarian follicles. *J. Exp. Zool.* 262: 271–278, 1992.
121. Vijayan, M.M., C. Pereira, E.G. Grau and G.K. Iwama. Metabolic responses associated with confinement stress in tilapia: the role of cortisol. *Comp. Biochem. Physiol.* 116C: 89–95, 1997.
122. Walsh, L.P., C. McCormick, C. Martin and D.M. Stocco. Roundup inhibits steroidogenesis by disrupting steroidogenic acute regulatory (StAR) protein expression. *Environ. Health Perspect.* 108: 769–776, 2000.
123. Walsh, L.P. and D.M. Stocco. Effects of lindane on steroidogenesis and steroidogenic acute regulatory protein expression. *Biol. Reprod.* 63: 1024–1033, 2000.
124. Walsh, L.P., D.R. Webster and D.M. Stocco. Dimethoate inhibits steroidogenesis by disrupting transcription of the steroidogenic acute regulatory (StAR) gene. *J. Endocrinol.* 167: 253–263, 2000.
125. Wendelaar Bonga, S.E. The stress response in fish. *Physiol. Rev.* 77: 591–625, 1997.
126. White, P.C. and P.W. Speiser. Congenital adrenal hyperplasia due to 21-hydroxylase deficiency. *Endocrinol. Rev.* 21: 245–291, 2000.
127. Wilson, J.M., M.M. Vijayan, C.M. Kennedy and G.K. Iwama. β-Naphthoflavone abolishes interrenal sensitivity to ACTH stimulation in rainbow trout. *J. Endocrinol.* 157: 63–70, 1998.
128. Yeoh, C., C.B. Schreck, G.W. Feist and M.S. Fitzpatrick. Endogenous steroid metabolism is indicated by fluctuations of endogenous steroid and steroid glucuronide levels in early development of the steelhead trout (*Oncorhynchus mykiss*). *Gen. Comp. Endocrinol.* 103: 107–114, 1996.
129. Yoshida, T., M. Mio and K. Tasaka. Ca^{2+}-induced cortisol secretion from permeabilized bovine adrenocortical cells: the roles of calmodulin, protein kinase C and cyclic AMP. *Pharmacol. Toxicol.* 46: 181–192, 1993.
130. Lacroix, A. and A. Hontela. A comparative assessment of the adrenotoxic effects of cadmium in two teleost species, rainbow trout, *Oncorhynchus mykiss*, and yellow perch, *Perca flavescens*. *Aquat. Toxicol.* 30: 13–21, 2004.
131. Hontela, A. Interrenal dysfunction in fish from contaminated sites: *in vivo* and *in vitro* assessment. *Environ. Toxicol. Chem.* 17: 44–48, 1998.
132. Pottinger, T.G., T.A. Moran and P.A. Cranwell. The biliary accumulation of corticosteroids in rainbow trout, *Oncorhynchus mykiss*, during acute and chronic stress. *Fish Physiol. Biochem.* 10: 55–66, 1992.

Biochemistry and Molecular Biology of Fishes, vol. 6
T. P. Mommsen and T. W. Moon (Editors)

CHAPTER 13

Xenobiotic impact on corticosteroid signaling

Mathilakath M. Vijayan*, Patrick Prunet** and Adrienne N. Boone*

** Department of Biology, University of Waterloo, Waterloo, ON, Canada N2L 3G1, and ** INRA/SCRIBE, Group on Fish Physiology of Adaptation and Stress, Campus de Beaulieu, Rennes Cedex, France*

I. Introduction

In bony fishes, corticosteroids are secreted from the interrenal tissues (analogous to the adrenal cortex) located in the head kidney region[121,182]. Cortisol is the major steroid released from these cells in response to neuroendocrine stimulation emanating from the hypothalamus and the pituitary (see Chapter 12). The release of cortisol in response to stress has been characterized in several species and also in response to different stressors and this topic has been extensively reviewed[10,11,70,182]. Cortisol stimulation is implicated in a wide array of whole animal changes including osmo- and iono-regulation, respiration, immune responses, reproduction, growth and metabolism (see ref. 121). The wide-ranging effects of this steroid on animal physiology make this hormone signaling pathway a prime target for endocrine disruptors affecting animal performance.

Although there is a consistent literature for fish indicating that xenobiotics such as polychlorinated biphenyls (PCBs) may affect the hypothalamus–pituitary–interrenal (HPI) axis (see Chapter 12), very little is known regarding the impact of toxicants on cortisol action in fish. This might be partly due to the fact that studies in fish on corticosteroid signaling and mechanism(s) of action are still in their infancy. However, with the recent molecular characterization of this receptor in fish, one may anticipate that in the near future more studies will clarify possible effects on endocrine disruptors on glucocorticoid receptor (GR). Consequently, our review is not comprehensive as the literature on toxicant impact on this important stress signaling pathway in fish is limited. Nonetheless, it is abundantly clear that there are mechanism(s) in place to impact steroid signaling, especially based on the mode of action of toxicants in other vertebrates. Here, we have tried to step back and identify the potential sites of action of toxicants on corticosteroid signaling, mostly based on information gathered from literature on mammalian models. We hope the materials provided, especially the potential sites of xenobiotic impact, can serve as a launching pad for researchers attempting to formulate hypotheses on the disruption of this steroid signaling pathway in fish. The major assumption for this chapter, based on the majority of published work, is that most corticosteroid signaling is genomic and will ultimately result in the synthesis of new proteins (see Fig. 1). However, we cannot discount the fact that non-genomic signaling may also contribute to the biological action of corticosteroids, even in fish[21–23], but pathways and responses specific to corticosteroid action remain to be clearly identified. Our review will (1) present molecular and biochemical information on the fish corticosteroid receptor (CR), which, overall, shows characteristics very similar to mammalian CR; (2) summarize information on CR signaling, again mostly from mammalian literature, and identify key steps that may be sensitive to direct toxicant impact; and (3) identify possible pathways/molecules that may be indirectly involved in CR signaling and sensitive to contaminant impact.

II. Fish corticosteroid receptor

The corticosteroid receptor in bony fishes is a ligand-dependent transcription factor and recent studies have identified two major classes of CRs, the glucocorticoid receptor and the mineralocorticoid receptor (MR), based on molecular sequences (see below). Here, we summarize studies that have established the receptor characteristics of both these important classes of CRs in fish.

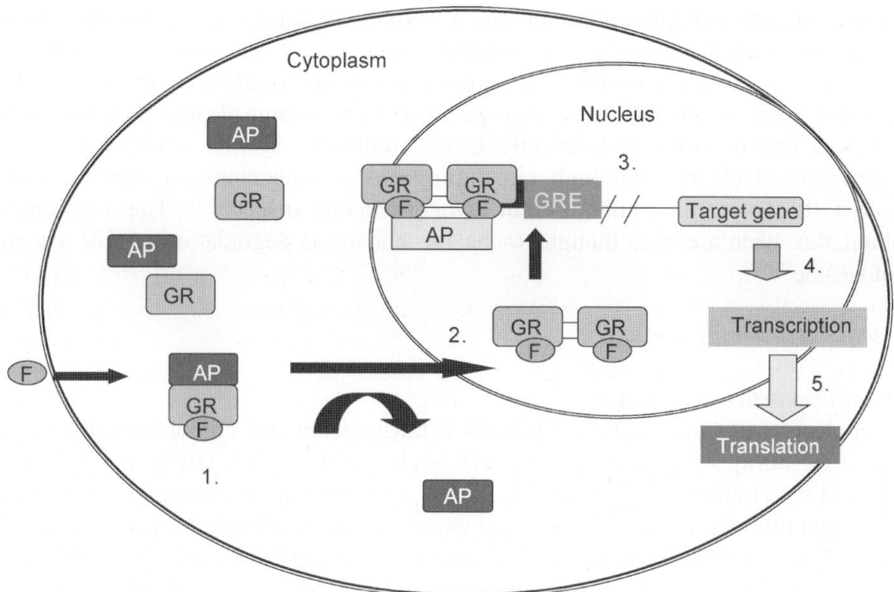

Fig. 1. Glucocorticoid signaling. The mechanism of action has been split into five steps. *Step 1*: binding of cortisol (F) to mature glucocorticoid receptor (GR). The accessory proteins (AP) are crucial for the receptor to be in a ligand-binding conformation and for receptor stability. *Step 2*: translocation of the GR–ligand heterocomplex to the nucleus. The role of accessory proteins in this translocation is not clear, however, cytoskeletal network is thought to be involved. *Step 3*: the GR–ligand complex forms a homodimer and binds to the glucocorticoid-response element (GRE) on the promoter of glucocorticoid-responsive genes. Several proteins are thought to modulate this step, including transactivation of genes. *Steps 4 and 5*: involves the transcription and translation of specific genes by glucocorticoid signaling (see text for details).

1. Binding studies

Early evidence for the presence of CRs in bony fishes came from binding studies using radiolabeled ligands[121]. Most of these studies focused on the GR because an endogenous ligand for the MR, aldosterone in higher vertebrates, was undetected in bony fishes[121]. Piscine GR displayed high affinities for selective agonists, the highest being for triamcinolone acetonide (TA)[54,94,112,160] and dexamethasone[35,102,137,138]. Using either one of these two synthetic glucocorticoids, studies indicated the existence of a single class of high-affinity, low-capacity binding site for GR not only in the cytosol from fish tissues, including gill, intestine, liver, brain, hypothalamus, leukocytes and erythrocytes[4,54,102,112,136,138,154], but also in nuclear extracts from gill[32,110]. While some variability in the measured K_d values in these tissues could be observed, overall these studies indicated that synthetic glucocorticoids (TA or dexamethasone) have higher affinity than cortisol for GR and that dissociation constants reported for teleosts are consistent with accepted dissociation constants for GR measured in other vertebrates, including mammals. The majority of these studies showed changes in GR affinity and/or number of binding sites in response to experimental manipulations, including cortisol treatment, a variety of stressor challenges, seawater transfer and smoltification[111,114,121]. In conclusion, fish GR characteristics closely resemble their mammalian counterparts and, specifically, the GR capacity is modulated to a large extent by circulating plasma cortisol levels. This was further confirmed recently using antibodies against rainbow trout GR which showed clearly that high circulating cortisol concentration downregulated tissue GR content in rainbow trout (*Oncorhynchus mykiss*)[176]. The mechanism behind this decrease was thought to be the enhanced degradation of GR *via* the proteasome[20,156].

2. Molecular characterization

2.1. Glucocorticoid receptor

The molecular characterization of fish GR is fairly recent and, chronologically, began with the cloning of a rainbow trout (*O. mykiss*) GR cDNA (rtGR1)[56]. This was followed by cloning of a partial GR cDNA from tilapia (*Oreochromis mossambicus*)[164] and full length GR cDNAs from Japanese flounder (*Paralichthys olivaceus*)[169], rainbow trout (a second GR; rtGR2)[27] and multiple GRs (HbGR1, HbGR2a, HbGR2b) from a cichlid fish (*Haplochromis burtoni*)[73]. The alignment of fish GR polypeptide sequences with that of mammalian GR showed very high identity with the C domain (DNA-binding domain or DBD) and the E domain (hormone-binding domain or HBD). Within the C domain, there was particularly high homology with the characteristic two zinc-fingers of the human GR[157]. However, the fish GRs also had some distinctive features. Specifically, the amino-terminal region (A/B domain), containing a powerful transcriptional transactivation domain in human GR[157], showed little homology with the fish GR. Moreover, the DBD of fish GR contained an additional nine amino acid sequence between the two zinc-finger regions which was strikingly different from the mammalian GR[56,73,169]. Interestingly, structural analysis of the mammalian GR revealed that these two zinc-finger regions were precisely structured to bind to the specific palindromic sequence of the glucocorticoid-responsive element (GRE) in the vicinity of promoters for genes activated by glucocorticoids[16,67]. However, this additional nine amino acid sequence was not present in all fish GRs: for instance there were only four additional amino acids in rtGR2[27] and none in HbGR2a[73]. In spite of these distinctive features, analysis of the transactivation activity of the fish GR in standard transfection assays indicated that the receptor effectively bound to the mammalian GRE and activated transcription of the reporter gene. Moreover, the specificity of the binding of various steroids to GR and its steroid-binding domain appeared to be essentially the same as for mammalian GR[56,73,165], except for rtGR2[27].

The rtGR2 showed high sequence homology with rtGR1 and phylogenetically these two rtGRs appeared to be the result of a gene duplication common to most teleost fish[144]. Using rtGR2 as a probe, a unique 7.3 kb transcript was observed in various tissues indicating the same tissue distribution as for rtGR1. However, when comparing *in vitro* transactivation activities of both receptors, rtGR2 transcriptional activity was induced at far lower concentrations of either dexamethasone or cortisol compared with rtGR1. Moreover, rtGR2 was less sensitive than rtGR1 to the classical GR antagonist RU486, suggesting that these two receptors have distinct functional specificity[27]. Based on this observation, the authors hypothesized that cortisol preferentially binds to rtGR2 in unstressed fish or upon exposure to low-intensity stressors (low circulating cortisol levels), whereas both rtGR1 and rtGR2 are occupied at times of exposure to high-intensity stressors (high circulating cortisol levels). This duality in receptor response appears specific to fish and has never been observed in mammals.

Expression of trout GR transcript has been found in a number of tissues, but surprisingly, unexpectedly low levels of GR mRNA were detected in trout liver[56,165]. This low GR mRNA transcript may be due to the high GR protein concentration in this tissue as recent studies clearly indicate that GR protein content regulates mRNA abundance *via* a negative feedback loop in rainbow trout hepatocytes[156]. This mismatch between GR mRNA abundance and GR protein content was also seen in trout liver *in vivo*[176]. The development of polyclonal antibodies to trout GR was instrumental in the recent cellular and tissue-specific localization of GR in fish[156,172]. Fish GR showed a wide distribution in the brain[166,167], and, in the gill, GR was detected both in chloride cells of chum salmon (*O. keta*)[173] and rainbow trout[47] and pavement cells of rainbow trout[47]. In trout hepatocytes, GR was distributed both in the cytosol and the nucleus, with the nuclear localization being modulated by cytochalasin D and colchicine suggesting a role for the cytoskeletal network in the regulation of GR localization in trout hepatocytes (A. Boone and M.M. Vijayan, unpublished).

2.2. *Mineralocorticoid receptor*

In mammals and non-mammalian vertebrates, including amphibians, the major mineralocorticoid is aldosterone, while the general consensus is that most fish do not produce this steroid[8,155]. However, recently a partial cDNA sequence of a mineralocorticoid receptor was identified in trout (rtMR)[42] and a full-length cDNA sequence was reported from the cichlid *H. burtoni* (HbMR)[73]. This fish MR bound cortisol with a higher affinity than aldosterone, a feature that is characteristic of mammalian MR[7]. However, transcriptional activity measured using standard *in vitro* tests – which gives a more accurate picture of the functionality of the receptor than binding studies – indicated a higher activity in the presence of aldosterone than with cortisol, suggesting the possibility that an endogenous aldosterone-like steroid could be the major ligand for this MR[27,73]. Moreover, a recent study using specific antagonists suggested that MR activation may be involved in gill chloride cell proliferation associated with acclimation to ion-deficient water in rainbow trout[162]. The recent cloning of a trout 11-beta hydroxysteroid dehydrogenase (11β-HSD), a homolog of mammalian 11β-HSD-2 and involved in the inactivation of cortisol to cortisone, and the wide tissue distribution of the transcript that included gills, lends further credence to the possibility that a mammalian type cortisol inactivation is possible in MR-responsive cells[100].

Overall, these results suggest that control of energy metabolism and hydromineral balance in fish is probably more similar to what is known in other vertebrates, especially mammals, than anticipated before and involves two independent systems using glucocorticoid and mineralocorticoid receptors. However, the precise nature of this regulation, including endogenous ligands and signaling pathways remains to be resolved. In *H. burtoni*, the three CRs showed differential expression in various tissues, but overlapped in the gills[73]. The splice variant HbGR2b was preferentially expressed in gill and liver relative to HbGR2, while the latter was preferentially expressed in kidney and spleen. These results coupled with the different transactivating properties of the multiple GRs suggests to a more subtle control of the corticosteroid effects in fish than seen in mammals.

III. CR heterocomplex and receptor signaling

Despite the recent molecular characterization of fish CRs, little is known about CR signaling in fish and much of the pathways mentioned here are based on mammalian models. However, knowing the conservation in CR sequence between fish and mammals, one could make the assumption that the signaling processes are also very similar, even though empirical evidence is lacking in fish. A number of recent reviews have already detailed the sequence of events, including cofactors involved in GR stability and ligand binding to GR, leading to the binding of GR to a specific DNA sequence in the vicinity of the promoters of glucocorticoid-responsive genes[9,139,157] (see Fig. 1).

1. Cofactors/coactivators

The accessory proteins and the various steps important in the signaling process are identified here only for the purpose of suggesting potential sites of action for toxicants. The identities of the accessory proteins and other factors involved in GR signaling have not been characterized in fish to the same extent as in mammals. The mammalian list involves a suite of cofactors, including hsp70, hsp90, p60/Hop (hsp organizer protein), hsp40, p23, BAG-1, calmodulin, immunophilins and the cytoskeleton, and is explained in great detail in a number of recent reviews[41,84,142]. Briefly, hsp70, ATP hydrolysis, and two molecules of hsp90 are required to open the hormone-binding cleft of GR and the rate is increased by the Hop[123]. Hsp90 has been shown to be crucial for GR maturation, stability and signaling because in the absence of hsp90 binding to GR, the receptor is unable to bind the ligand and also GR is degraded rapidly[50,158,183]. In fish, the chaperones hsp70 and hsp90 have been identified to play a role in GR signaling[14,156]. Also, hsp90 has been shown to be critical for the transcriptional regulation of GR as well as GR protein stability in trout hepatocytes[156].

Hsp40 and p23 (and possibly other proteins) also stabilize the GR complex that is involved in hormone binding[123]. A recent study has suggested that p23 may be a limiting, but essential component of GR complex formation[122]. In addition, immunophilins are part of the complex and coordinate the movement of the GR complex from the cytoplasm to the nucleus along the cytoskeleton network[139,140,142]. Binding of the hormone to the receptor causes both the movement of the receptor to the nucleus and the dissociation of the initial hormone-binding complex. Once inside the nucleus, the hormone-bound GR dimerizes with itself (homodimer) and the two zinc-finger domains in each of the two GRs interact with GRE in the promoter regions of glucocorticoid-responsive genes. Once the GR dimer is bound, additional factors are recruited to the promoter region to enhance transcription by RNA polymerase II[9,139,157]. While the precise roles and the degree of interaction of these different proteins on GR signaling have not been ascertained, it is safe to assume that any of these proteins may be potential candidates for toxicant impact on GR signaling in fish.

2. *GR phosphorylation*

Like many other receptors, GR is also regulated by phosphorylation. Seven phosphorylation sites are present in the amino-terminal domain of the mouse glucocorticoid receptor and many of these sites are conserved and are important in other species[17]. Rat GR is phosphorylated at Thr171 and Ser246 by mitogen-activated protein kinase (MAPK) and by cyclin-dependent kinase (cdk) at Ser224 and Ser232[99]. Interestingly, phosphorylation by MAPK decreases GR-mediated transcription while cdk has the opposite effect[99]. GR-mediated transcription is also inhibited by phosphorylation of Thr171 by glycogen synthase kinase-3[146] and by phosphorylation of Ser246 with c-Jun N-terminal kinase (JNK), a stress-activated kinase[146].

When human GR is phosphorylated by JNK, the GR is rapidly exported from the nucleus, decreasing GR-mediated transcription[79]. Human GR is rapidly phosphorylated at two sites (human Ser203 and Ser211 corresponding to mouse Ser224 and Ser232, respectively) in response to dexamethasone[178]. Biochemical fractionation and immunofluorescence studies found GR that was phosphorylated at Ser203 to be mainly cytoplasmic, while GR phosphorylated at Ser211 was mostly nuclear[178]. Mutation studies of mouse GR suggested that phosphorylation sites may also play a role in receptor protein turnover[181]. Clearly, the phosphorylation status will regulate GR function at multiple levels. Consequently, any factors that activate or interfere with relevant protein kinases or phosphatases (see Table 1) acting on CR will impact corticosteroid signaling.

IV. *Toxicant impact on CR signaling*

From the sequence of events mentioned above for the transcription of corticosteroid-responsive genes, a few potential sites of interaction for toxicants can be envisaged (Table 1). In brief, the sites where toxicants affect corticosteroid signaling may include (i) direct impact on CR protein, including CR content, ligand binding, CR binding to DNA and CR-responsive transactivation processes and (ii) indirect impact on CR signaling, including alteration in accessory proteins, kinases, cytoskeleton and the proteasome (see Fig. 2).

TABLE 1

Impact of xenobiotics on glucocorticoid receptor (GR) signaling

Toxicant	Response	Observed end-point	Cell/tissue type	References
Polychlorinated biphenyls (PCBs)				
PCBs	↓	Hormone-binding capacity	Rat skeletal muscle, human, mouse liver	113
	nc	GR-mediated transcription	Various cell lines	81
	nc	GR-mediated transcription	Reporter assay	135
PCB methyl-sulphones	↓	Dex binding to GR	Mouse liver cytosol and human GR	86
	↓	Dex-induced activation		
TCDD	↑↓	GR protein content	Rat ovary (*in vivo*)	120
	↓	Hormone-binding capacity	Rat liver	163
	↑	Hormone/receptor nuclear uptake	Rat liver	163
	↑ ↑	GR mRNA expression, GR protein	Human palate *in vitro*	2
	↑ ↑	GR mRNA expression, GR protein	Mouse palate *in vivo*	2
Metals				
Cadmium	↓	Hormone/receptor-binding capacity to DNA	Rat liver (*in vivo*)	57, 58, 62
	↓	GR-binding capacity	Rat liver (*in vitro*)	58
	nc	GR-mediated transactivation	Mouse mammary tumor cell line	51
Copper	nc	GR-mediated transactivation	Mouse mammary tumor cell line	51
Mercury	↑	GR-mediated transactivation	Mouse mammary tumor cell line	51
Zinc	↑	GR-mediated transactivation	Mouse mammary tumor cell line	51
Zinc	↑	GR zinc-finger domain binding to DNA	Mouse mammary tumor cell line	51
Arsenite	↓	GR-mediated transcription	Plasmid expression studies	89
	↑	Nuclear translocation of hormone-free GR	Mouse L929 and CHO cells	152
	↑	GR-mediated gene expression	CHO cells	159
Lead	↓	GR-mediated transcription, substitute for Ca^{2+} binding to calmodulin		Reviewed by Bouton and Pevsner[24]
Lead		Improper secondary structure and aberrant transcription (zinc-finger domains)		Reviewed by Bouton and Pevsner[24]

TABLE 1. Continued

Toxicant	Response	Observed end-point	Cell/tissue type	References
Molybdate	↓	Nuclear shuttling	GrH2 rat hepatoma and COS-1 cells	185
Vanadate	↑	GR-mediated transactivation	Reporter cell system	105
Others				
AFB1, DMN, etc.	↓	Hormone/receptor-binding capacity to DNA	Rat liver nuclei	90
Okadaic acid (protein phosphatase inhibitor)	↓	GR nuclear shuttling	3T3 cells	68
Pifithrin α (P53 inhibitor)	↓	GR binding to DNA and transactivation	Mouse fibroblast cell line	96
Genistein (tyrosine kinase inhibitor)	↓	GR-mediated transactivation	Human keratinocytes	72
Glycyrrhizin	↓↑	GR affinity	Rat hepatocytes	186
	nc	GR content	Rat hepatocytes	186
	↑	GR-mediated TAT transcription	Rat hepatocytes	186
Radicicol (anti-cancer drugs)	↓	GR-mediated transactivation	Reporter assay	147
Geldanamycin (hsp90 inhibitor)	↓	GR-mediated transactivation	Reporter assay	147
Herbimycin A (tyrosine kinase inhibitor)	↓	High affinity GR sites, GR protein content, GR-dependent TAT activity	Primary rat hepatocytes	126
Several antidepressants	↑	GR translocation	Human lymphocytes	130
	↓	GR-mediated transcription		
Chlorpromazine (anti-psychotic drug)	↓	Corticosterone-induced gene transcription	Mouse mammary tumor cell line	12
Wortmannin (inhibitor of DNA-dependent protein kinase)	↑	Dex-mediated transactivation	Mouse L929 cells	105
Phospholipase C and L-type Ca^{2+} channel antagonist	↓	GR-mediated transcription	Mouse L929 cells	105
Ca^{2+} and A23187 (Ca^{2+} ionophore)	↓	GR-binding capacity, but not affinity	Mouse L929 cells	105
Tolylfluanid and ketoconazole (fungicides)	↓	GR-mediated transactivation	Reporter assay	85
Ginsenoside-Rg1 (*Panax ginseng* extract)		Binds to GR and activates transcription	Rat hepatocytes	40,103

The observed responses are shown as either an increase (↑), decrease (↓) or no change (nc) compared to the controls. Dex: dexamethasone; TAT: tyrosine aminotransferase.

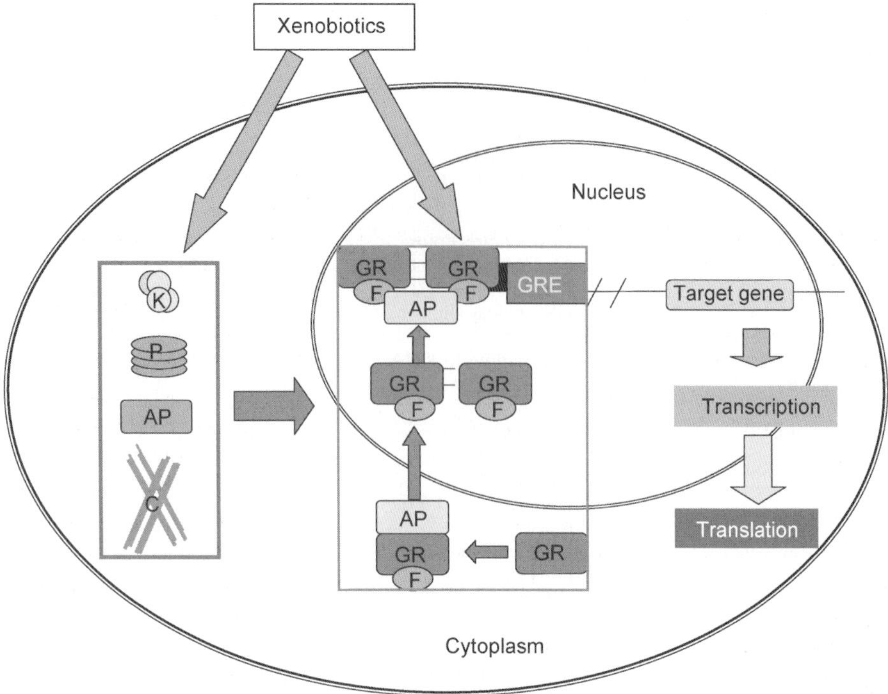

Fig. 2. Xenobiotic impact on glucocorticoid signaling. Xenobiotics impact on GR signaling could include either a direct effect on the GR signaling pathway and/or an indirect effect mediated by changes in other proteins/pathways. The direct effect encompasses impact on cortisol (F) binding to GR, their translocation, binding to GRE and initiation of transcription, whereas the indirect effect may include impact on kinases and phosphatases (K), accessory proteins (AP), the cytoskeletal network (C) and the proteasomal pathways (P), all of which are involved in GR signaling (see text for details).

1. Direct impact on CR function

1.1. Polychlorinated biphenyls

Several chemicals, including PCBs and metals are known to compete with endogenous/synthetic ligands for GR-binding sites (Table 1). For instance, *in vitro* incubation of human GR extract or mouse liver cytosol with a mixture of methylsulfonyl PCBs (mPCBs) resulted in the competitive displacement of dexamethasone[86]. However, mPCBs did not activate a cell system containing GR and GRE-linked reporter constructs[86], and mPCBs also blocked the dexamethasone-mediated transactivation suggesting that these chemicals possess GR antagonistic properties[86]. Although the potency for the mPCBs are ~200 times lower than for RU486 (a competitive GR receptor antagonist), the persistent nature and wide distribution of these compounds in the environment[85] pose a major problem to animal health. Also, the additive effects on GR transactivation seen in co-exposure studies with mPCBs and pesticides suggest that the potency of these compounds may be dictated to a large extent by the chemical mixtures[85], which is most often the case in the aquatic environment.

A decrease in GR binding capacity was observed in rat liver[163] and skeletal muscle[113] following exposure to PCBs. However, liver GR from rats exposed to 2,3,7,8-tetrachlorodibenzo-*p*-dioxin (TCDD) showed increased nuclear uptake of the steroid-bound receptor, despite decreased GR binding capacity[163]. Also, different mixtures of chlorinated PCBs (Aroclor) or chlorinated PCB congeners did not interfere with GR transcriptional activation in an *in vitro* transcription-based reporter assay system[135]. These results tend to suggest that the metabolites of PCBs may be more sensitive to GR function compared with the parent compound, but further studies, especially direct ligand displacement and binding studies, are necessary to identify the mechanisms for the altered transactivational capacity with PCBs.

The xenobiotic response to PCBs is mediated to a large extent by the ligand-dependent transcription factor belonging to the basic helix–loop–helix Per-ARNT-Sim (bHLH-PAS) family of proteins, the aryl hydrocarbon receptor (AhR) (see Chapter 7). AhR and GR have very similar biochemical properties including size, isoelectric point, sedimentation, Stokes radius and axial ratio[45]. For ligand binding and nuclear translocation, AhR requires accessory proteins similar to that of GR, including hsp90[29,125,143]. Once inside the nucleus, AhR binds to the AhR nuclear translocator (arnt) and this heterocomplex binds to dioxin/xenobiotic response elements (XRE or DRE) and activates transcription (see Chapter 7). As both GR and AhR share a similar complement of accessory proteins for their signaling, it is possible that high traffic through one pathway could occupy and deplete the free pool of accessory proteins for the other pathway. In other words, in the presence of large amounts of dioxins, GR signaling may be compromised by a lack of accessory proteins. This idea of regulation by limiting accessory proteins has been explored previously as a possible mechanism for cross-talk between hypoxia-inducing factor (HIF) and AhR[33], but similar studies on GR are lacking. However, indirect evidence points to this possibility. For instance, TCDD impact on mice development showed an increase in GR content with a corresponding downregulation of AhR content, while the opposite occurred with cortisol treatment[1,2]. It is tempting to speculate that this inverse correlation between the two receptors may involve the interaction of accessory proteins, including hsp90, which are crucial for maintaining the ligand-binding conformation and stability of both AhR and GR. Recently, steroid receptor signaling studies using reporter gene expression vectors showed that AhR and steroid receptors, including androgen and estrogen receptors (ER), negatively regulate each other's signaling[82]. This includes TCDD–AhR blocking of ER binding to estrogen response elements as well as ER blocking binding of AhR to XREs[91]. While a similar cross-talk mechanism has yet to be reported for AhR and GR signaling, the positive regulation of AhR and GR content with cortisol and TCDD, respectively[1,2], implies a similar cross-regulation.

In fish, very few studies have actually examined the direct effect of PCBs on GR function (Table 2). Cortisol treatment, both *in vivo* and *in vitro*[31,48,52] modulated CYP1A activation by AhR agonists clearly hinting at possible interactions between

TABLE 2

Xenobiotic impact on glucocorticoid receptor (GR) function in teleost fish

Toxicant	Response	Fish	Observed end-point	Cell/tissue type	References
PCBs					
Aroclor 1254	↓	*S. alpinus*	GR protein expression	Brain	5a
Aroclor 1254	↑	*O. mykiss*	GR protein expression	Liver	Unpublished
Aroclor 1254	↑	*S. alpinus*	GR protein expression	Liver	Unpublished
BNF	↑	*O. mykiss*	GR protein expression	Liver	5b
TCDD	↑	*O. mykiss*	GR protein expression	Hepatocytes in primary culture	Unpublished
Metals					
Copper in water	↓	*O. mykiss*	GR immuno-reactivity	Gills *in vivo*	47
Copper	↑	*O. mykiss*	GR protein expression	Gill filaments *in vitro*	Unpublished
Cadmium in food	nc	*Salmo salar*	GR immuno-reactivity	Gills *in vivo*	46
Arsenite	↓	*O. mykiss*	GR protein expression	Hepatocyte	18
Others					
Alkylphenols and phthalates	nc	*O. mykiss*	GR binding	Liver and brain	95

The observed responses are shown as either an increase (↑), decrease (↓) or no change (nc) compared to the controls. BNF: β-naphthoflavone; TCDD: 2,3,7,8-tetrachlorodibenzo-*p*-dioxin. Unpublished work is from Vijayan's laboratory.

the two signaling pathways. Also, recently we found that Aroclor 1254 treatment downregulated GR content in the brain of fasted Arctic charr (*Salvelinus alpinus*[5a]). Based on this result, it was proposed that the impact of PCBs on the cortisol response to stress in this species[87] may be due to alterations in the negative feedback loop, specifically due to perturbation in cortisol sensing as a consequence of decreased GR content in Arctic charr[5a]. Although the mechanism for PCB downregulation of brain GR is not clear, studies in fish do tend to suggest the presence of AhR in brain as evidenced by the higher ethoxyresorufin *O*-deethylase (EROD) activity in the presence of the AhR agonist β-naphthoflavone[6] and TCDD[132]. A recent study in rainbow trout showed a role for AhR in the GR signaling process[5b]. Trout treated with β-naphthoflavone showed higher liver GR protein content and this effect was further enhanced by co-treating with α-naphthoflavone (an AhR antagonist)[5b] suggesting that these pharmacological agents may impact GR signaling. However, the lack of a corresponding increase in glucocorticoid responsiveness, despite elevated plasma cortisol levels in the β-naphthoflavone-treated fish, lead us to propose that the drug affected cortisol sensitivity perhaps by preventing the entry of cortisol into the cells[5b].

1.2. Metals

With heavy metals, the major concerns are substitution for metals with biological roles (e.g. calcium and zinc), the induction of specific metal-responsive proteins, and the interference/oxidation of necessary structural and active site sulfhydryl groups leading to denatured and aggregated proteins. A number of metals, including cadmium, copper, mercury, zinc, lead, molybdate and vanadate, impact GR signaling in the mammalian system (Table 1). GR has 20 cysteine residues with five in the steroid-binding domain and nine within the DBD; thus GR could potentially be highly susceptible to metals[116]. Recent data have shown that non-toxic concentrations of arsenic can selectively inhibit GR-mediated transcription through altered nuclear function rather than by a decrease in hormone-induced GR transactivation[89]. Interestingly, arsenite, cadmium and selenite interact with the cysteines of the ligand-binding domain of GR and inhibit binding of dexamethasone to the receptor[161]. By using concentrations of cadmium lower than those required to bind all the GR cysteine residues, it was suggested that the metals were actually binding vicinal dithiols (i.e. two thiols very close together) and this was able to reduce steroid binding to the receptor[161]. GR hormone-binding capacity and hormone/receptor binding to DNA were both reduced in livers from rats exposed to cadmium[57,59]. Dithiothreitol (DTT) was able to reverse both of these inhibitory effects of cadmium, further suggesting that cadmium was interfering with GR's thiol groups[59]. In plasmid expression studies, arsenic was able to reduce the hormone/receptor binding to DNA, but steroid-binding capacity did not appear to be altered[89].

Metal substitution in the zinc-finger domains may also be a mechanism for GR disruption. For instance, many transcription factors, including GR and heat shock factor (HSF) interact with DNA through their zinc-finger domains. As their name implies, zinc-finger domains require zinc to maintain the domain in a DNA-binding conformation. When other metals (e.g. lead or cadmium) substitute for the zinc, the secondary structure of the domain is altered[134] and the DNA-binding ability is decreased or lost altogether[145]. Alternatively, incubation with zinc can increase transcription factor binding as observed with the GR in a mouse mammary tumor cell line[51].

In fish, studies on the impact of metals on GR function is limited (Table 2). We have shown that arsenite affects GR signaling and one mechanism(s) involves the downregulation of GR content in trout hepatocytes[18]. Recently, we have seen that copper upregulates GR protein content in a dose–response fashion in isolated trout gill filaments *in vitro* (R. Weil and M.M. Vijayan, unpublished). The mechanisms for this upregulation are not clear, but studies in mouse cell lines suggest a possible role for zinc–metallothionein (MT) in modulating glucocorticoid responsiveness[51]. A review of the metallothionein expression in fish is dealt with in Chapter 10 of this volume and will not be covered in any great detail except with respect to GR interaction. It remains to be seen whether MTs are involved in the GR response to copper exposure in trout gill cells.

Our results are at odds with those of Dang *et al.*[47] showing that copper exposure for 5 days *in vivo* reduced GR-immunoreactivity (ir) in rainbow trout branchial cells. The decrease in GR-ir was inversely related to MT-ir in gills. The difference in response may be attributed to the different experimental protocols, especially as *in vivo* copper exposure also results in elevated plasma cortisol levels[47], which downregulates GR[176], whereas *in vitro* exposure is not affected by plasma cortisol levels. Also, in our *in vitro* study with gill filaments, the GR upregulation with copper was attenuated in the presence of cortisol (R. Weil and M.M. Vijayan, unpublished). Although the mechanism is not known, it appears likely that cortisol increases the breakdown of GR through the proteasomal pathway as shown recently in trout hepatocytes[156].

Interestingly, longer term cadmium exposure did not affect GR-ir in branchial cells of Atlantic salmon, despite an upregulation of MT-ir[46]. However, plasma cortisol levels were not significantly affected by cadmium treatment in this study suggesting that plasma steroid levels in response to metal exposure may be an important regulator of GR content. The opposing response of GR protein to arsenite and copper exposure in the trout study *in vitro* clearly points to mechanisms independent of plasma cortisol levels in modulating GR response to metal toxicity. It is not clear if this is either a metal-specific or tissue-specific response as the arsenite study involved hepatocytes, whereas the copper study was with gill filaments. Together, these findings suggest that heavy metals interact at the level of the hormone receptor to affect GR signaling. The precise nature of this regulation is not clear, but the induction of metal-binding proteins in response to metal exposure may be playing a key role in the modulation of GR responsiveness. Certainly other heavy metals present in the natural environment can now be considered as potential candidates for GR disruption. Future studies characterizing the role of MTs and other metal-binding proteins on GR response and corticosteroid signaling may shed some insight into the mechanisms of metal toxicity on this important signaling pathway.

1.3. Others

Although PCBs and metals are the most commonly studied xenobiotics in the aquatic environment, other chemicals, including pharmaceuticals, can also potentially impact corticosteroid signaling (see Table 1). Mammalian studies have shown that ionophores and inhibitors of kinases, phosphatases and p53, as well as several antidepressants affect GR signaling (Table 1). While the influence of these drugs on GR signaling varies, it is clear that such drugs in the aquatic environment can alter GR signaling in fish. This may include impact on GR affinity and capacity as well as nuclear transport and transactivational activity of corticosteroid-responsive genes (Table 1). Most of these studies with mammalian systems utilized gene expression reporter constructs, and, consequently, extrapolation of the responses to the piscine system needs to be verified experimentally. Also, approaches of correlating biochemical or physiological responses to toxicants as mediated by known endocrine signaling mechanisms may not always hold true. For instance, based on observations that atrazine, a commonly used herbicide, impacts cytokine production by human blood mononuclear cells *in vitro*, Devos and colleagues[53] hypothesized that this compound was interacting with the GR signaling pathway. This hypothesis was also consistent with the lipophilic properties of atrazine that could potentially bind to nuclear receptors as a target. However, the effects of atrazine on cytokine production and expression of a reporter gene in mammalian cell lines was not blocked by RU486 suggesting pathways independent of GR[53]. One way to get around this problem is to directly examine the capacity of chemicals to displace native ligands from their receptors. One such study in fish examined the interaction of representative alkylphenols and phthalates with steroid receptors and showed that these compounds displaced estrogen from estrogen receptors, but not androgen or cortisol from their respective receptors[95]. However, despite the absence of direct binding to the receptor, one cannot rule out the possibility that these chemicals may affect CR signaling indirectly as is the case with GR suppression of ER signaling in fish[104]. In conclusion, one cannot exclude the possibility that the potential exists for chemicals other than PCBs and metals to impact GR signaling in fish. Further, since PCBs and metals also interact with these other pathways, including phosphatases and kinases and ion channels, the interactive effect of these xenobiotics in complex mixtures may have a profound influence on GR signaling and toxicity in fish.

2. *Indirect impact on GR function*

In this review, for sake of clarity, any effect that was not directly mediated by changes in GR structure and function will be categorized as indirect. Several protein families are known to modulate GR function and could be potential sites for xenobiotic impact, including accessory proteins that act as coactivators/cofactors in GR signaling[41,84,142], stress-activated kinases[79,99,146,181] and the proteasome[20,156,177]. While limited information is available regarding xenobiotic impact on these proteins in fish, the heat shock family of proteins has received considerable attention, but mainly as an indicator of cellular stress in fish[64,80]. Only recently have studies started to explore the interaction of hsps with GR in fish[14,156]. Here, we will outline a few potential sites for xenobiotic impact again based on mammalian models.

2.1. *Calmodulin*

Calmodulin, a 17 kDa protein with four calcium-binding sites, is an important intracellular calcium receptor that regulates a number of key enzymes, but is also thought to be an important accessory protein for GR signaling. Immunoaffinity experiments have suggested that Ca^{2+}/calmodulin may interact directly with hsp90 and together they are part of a larger complex with GR, hsp70 and hsp56[127]. Antagonists of calmodulin do not prevent hsp90/GR association or loss of hsp70 or hsp56 from the complex, but inhibit hormone/GR association and prevent translocation to the nucleus, suggesting a role for calmodulin in the GR signaling process[128]. Calmodulin mRNA can be induced in a dose-dependent manner by glucocorticoids and has been linked to glucocorticoid-induced apoptosis[55]. As calmodulin is affected by environmental chemicals, including PCBs and metals, this protein may be a key player in the disruption of GR signaling.

Metals with a similar ionic radius to calcium ($\sim 1 \pm 0.2$ Å, e.g. cadmium, lead, mercury), are able to substitute for calcium in calmodulin[34,36,43]. One main consequence of this substitution is that the metal-binding sites of calmodulin become filled with metals other than calcium and thus less calcium is required to activate calmodulin and the calcium sensing of the cell is disrupted[36]. With calmodulin being involved in so many important aspects of the cell including regulation of intracellular calcium concentration, inappropriate activation by metals can have serious effects. Interestingly, some metals (e.g. mercury and zinc) can have a biphasic influence on calmodulin and activate at low concentrations and inhibit at higher metal concentrations[25,34,43]. Also, xenobiotics such as DDT and its metabolite DDE potently inhibit calmodulin[74,108]. While the mechanism of inhibition has not been fully elucidated, it has been suggested that interaction of PCBs with calmodulin may alter calmodulin's conformation, and reduce its ability to interact with other proteins[131]. DDE has been shown to disrupt progesterone signaling by inhibiting calmodulin[108]. Therefore, a similar mechanism may also be relevant for GR disruption, but remains to be elucidated.

2.2. *Cytoskeleton*

In addition to accessory proteins crucial for GR maturation, stability and nuclear translocation and transactivation, the cytoskeletal network is crucial for GR signaling. Movement of GR between the cytoplasm and the nucleus is believed to be along the cytoskeletal network in an hsp90-dependent manner[69]. Experiments with chemicals that disrupt the cytoskeleton have suggested that the GR complex can also move to the nucleus in the absence of hsp90 and cytoskeleton by diffusion. However, under normal conditions with an intact cytoskeleton, the large size of the GR complex makes it difficult to diffuse to the nucleus easily and must rely on hsp90-dependent movement along the cytoskeleton[69]. Therefore, any toxicant-mediated changes in the cytoskeleton could potentially influence GR signaling. In cultured Swiss 3T3 mouse cells, one characteristic of metal exposure is reorganization of the cytoskeleton[38,106]. Cytoskeletal alterations were observed with arsenite, cadmium, cobalt (II), nickel (II) and chromium (VI) and the extent and type of reorganization varied with the identity, duration of exposure and quantity of metal[38]. Cytoskeletal reorganization has also been observed in human tumor cell lines[30,107]. Therefore, this might be another potential site for xenobiotic impact on the GR signaling process.

2.3. *Stress-activated kinases*

Stress-activated kinases are important in GR regulation and, therefore, any impact on these kinases will limit GR signaling. For instance, in human bronchial epithelial cells, the induction of MAPK, ERK and JNK is observed in response to arsenic, vanadium, zinc, chromium and copper[151]. Also, stress-activated JNKs are activated in mussel hemocytes following exposure to PCBs[28]. As these kinases are thought to be crucial for GR-mediated transcription[79,99,146], it appears likely that xenobiotic impact on steroid disruption may also include this important class of signaling molecules. Examples of the impact of cell signaling inhibitors on GR function are listed in Table 1. A very recent paper has demonstrated that GR has a JNK docking site and binding of JNK to glucocorticoid-bound GR can reduce the free active pool of JNK and can thus reduce JNK cell signaling[26]. The relationship between JNK-mediated phosphorylation of GR and JNK docking with GR has yet to be addressed. However, it is becoming apparent that these stress-activated kinases are involved in all aspects of GR signaling, including GR stability and maturation, translocation and transactivational capacity of GR, and may be an important control point for endocrine disruption. Certainly, more work is required to clearly establish a link between GR disruption associated with xenobiotic impact on stress kinases.

2.4. Heat shock proteins

Hsps are molecular chaperones that play an important role as accessory proteins for GR signaling and have been reviewed in detail before[139]. In the following, we will highlight the impact of toxicants on these important chaperone proteins and potential downstream effects on GR signaling. A number of reviews have comprehensively addressed the cellular stress response, specifically the role of hsps in allowing cells to cope with stress, including xenobiotics[80,133,153]. Briefly, the induction of hsps are thought to be an adaptive response to combat proteotoxicity[76]. Within individual cells, mRNAs for stress proteins are induced during the first few minutes and new protein synthesis follows rapidly. Hsps are a highly conserved family of proteins that function as molecular chaperones in both unstressed and stressed cells. Hsps have many crucial roles: (1) aiding in the folding of new proteins; (2) refolding of incorrectly folded proteins; (3) reducing protein aggregates; (4) presenting proteins in a conformation suitable for degradation by the proteasome; (5) presenting steroid receptors in a ligand-binding conformation; (6) intracellular localization; (7) regulation; and (8) secretion[64]. The hsps are named based on their molecular mass (kDa) and some examples of major families are hsp90, hsp70, hsp60 and the small hsps. Some common stressors that induce hsps in fish are temperature, salinity, hypoxia, radiation, pH changes, oxidation, heavy metals and environmental toxins[15,80,149]. As detailed in Table 3, hsps (70, 90, 60 and small) are induced in fish cells exposed to contaminants, including metals and PCBs[15,80,148,149,174,175]. The recent finding that cells with high hsp70 showed a decreased responsiveness to glucocorticoid stimulation, and the fact that this was due to downregulation of GR protein content argues for a interaction of hsps with GR signaling[18]. One possibility is that high hsp induction affects other protein synthetic pathways, including GR resulting in lower GR content. Alternatively, the high hsps may be binding to aberrant proteins and fewer hsps are available for GR binding resulting in GR degradation. Also, since hsp90 is an important molecular chaperone for maintaining the ligand-binding properties and signaling of both GR and AhR, it remains to be tested whether the hsp90 protein concentration impacts GR and AhR signaling.

One scenario may involve the increased demand for hsp90 in response to the stimulation of one receptor perhaps results in the instability of the other resulting in receptor downregulation. This notion finds support, albeit indirect, from work in mice showing that TCDD and cortisol stimulation negatively cross-regulates GR and AhR content[2], although hsps were not determined in that study. The lack of hsp90 binding to GR has been proposed to downregulate GR content and reduce GR mRNA abundance in trout hepatocytes[156], clearly implicating a crucial role for hsp90 in GR function. Furthermore, the lower hsp90 in the brain of PCB-treated fish correlates positively with GR content, implicating a role for this molecular chaperone in the PCB-impact on the HPI axis in Arctic charr[5a]. Future studies should be aimed

TABLE 3

Impact of xenobiotics on heat shock protein (Hsps) expression in teleosts

Toxicant	Response	Fish type	Cell/tissue type	References
Hsp70				
Arsenic	↑	*O. mykiss*	Primary hepatocytes	19
	↑	*O. mykiss*	RTG-2 cell line	97, 98
	↑	*Pimephales promelas*	Gill and muscle (*in vivo*)	61
Cadmium	↑	*O. mykiss*	Primary hepatocytes	19
	↑	*O. mykiss*	Epidermal cell culture	109
	↑	*O. tshawytscha*	Embryonic (CHSE) cell line	75, 117
	nc	*Trematomus bernacchi*	Primary hepatocytes	77
Chromium	↑	*P. promelas*	Gill and muscle *in vivo*	60
Copper	↑	*O. mykiss*	Primary hepatocytes	19, 65
	↑	*O. mykiss*	Gill filaments	Unpublished
	↑	*Cyprinus carpio*	Liver, gill, erythrocytes (*in vivo*)	49
	↓	*C. carpio*	Kidney, brain (*in vivo*)	49
Zinc	↑	*O. mykiss*	Hepatoma (RTH) cell line	117
	↑	*O. tshawytscha*	Embryonic (CHSE) cell line	75, 117
BNF	↑	*O. mykiss*	Liver (*in vivo*)	174
	↑	*O. mykiss*	Liver and gill (*in vivo*)	13
	↑	*O. mykiss*	Liver (*in vivo*)	179
	↓	*Ictalurus punctatus*	Ovary (*in vivo*)	180
DMBA	↓	*I. punctatus*	Ovary *in vivo* (IP inject)	180
BKME	↑	*O. mykiss*	Liver *in vivo*	175
	↑	*Catostomus commersoni*	Ovarian follicle cells	83
Hsp90				
Arsenic	↑	*O. mykiss*	RTG-2 cell line	97
	↑	*P. promelas*	Muscle *in vivo*	61
Cadmium	↑	*O. tshawytscha*	Embryonic (CHSE) cell	37, 75, 117
Copper	↑	*O. tshawytscha*	Embryonic (CHSE) cell	37
Zinc	↑	*O. mykiss*	Hepatoma (RTH) cell line	117
	↑	*O. tshawytscha*	Embryonic (CHSE) cell line	117
Hsc70				
Arsenic (protein)	nc	*O. mykiss*	Primary hepatocyte	19
Cadmium (protein)	nc	*O. mykiss*	Primary hepatocyte	19
Cadmium (mRNA)	nc	*O. tshawytscha*	Embryonic (CHSE) cell line	188
Copper (protein)	nc	*O. mykiss*	Primary hepatocyte	19
Zinc (mRNA)	nc	*O. tshawytscha*	Embryonic (CHSE) cell line	188
	nc	*O. mykiss*	Red blood cells	44

TABLE 3. Continued

Toxicant	Response	Fish type	Cell/tissue type	References
Hsp60 (^{35}S only)				
Arsenic	↑	*O. mykiss*	RTG-2 cell line	97
Cadmium	↑	*O. tshawytscha*	Embryonic (CHSE) cell line	75
Zinc	↑	*O. tshawytscha*	Embryonic (CHSE) cell line	75
Small hsps				
Arsenic	↑	*O. mykiss*	RTG-2 cell line	97
	↑	*P. promelas*	Gill and muscle *in vivo*	61
Cadmium	↑	*O. tshawytscha*	Embryonic (CHSE) cell line	75, 117
Chromium	↑	*P. promelas*	Gill and muscle *in vivo*	60
Zinc	↑	*O. tshawytscha*	Embryonic (CHSE) cell line	75, 117
	↑	*O. tshawytscha*	Embryonic (CHSE) cell line	117

The observed responses are shown as either an increase (↑), decrease (↓) or no change (nc) compared to the controls. DMBA: dimethylbenz[*a*]anthracene; BNF: β-naphthoflavone; BKME: bleached-kraft pulp mill effluent; RTG: rainbow trout gonadal; CHSE: Chinook salmon embryonic; RTH: rainbow trout hepatoma. Unpublished work is from Vijayan's laboratory.

at understanding the interaction of hsps, particularly the ratio of free to bound hsps, in the regulation of GR in fish tissues. The absence of GR signaling in the presence of geldanamycin (a known blocker of hsp90 binding), despite an increase in hsp90 content, suggests an important role for bound to free pool of hsps in GR signaling[124]. This is especially relevant because most of the adaptive responses to cellular stress, including xenobiotic exposure, involve the induction of hsps, but it is not clear if this heat shock response is at the expense of other critical proteins, including GR, that are crucial for cellular homeostasis. Our recent finding that cells with higher hsp70 content associated with arsenite exposure also showed a decreased responsiveness to glucocorticoid stimulation, and the fact that this was due to downregulation of GR protein content supports this notion[18].

2.5. Heat shock factor and heat shock element

Heat shock elements (HSE) are a family of highly conserved *cis*-acting elements found in the promoters of a variety of proteins including hsps, heme oxygenase, copper/zinc superoxide dismutase, antioxidant protein-1, human multidrug resistance-1, c-fos and yeast metallothionein[63,118,119,187]. For hsps, the active HSF trimer is the transcription factor that binds to the HSE and activates transcription. In unstressed cells, HSF is maintained in a monomeric state by a variety of binding proteins including hsps and HSF-binding protein (HSBP1). When cells are exposed to environmental stressors (e.g. heat, oxidants, metals), damaged proteins accumulate, the hsps holding the HSF in a monomeric state are needed for refolding the damaged proteins, and HSF trimerizes and is activated. In HeLa cells, HSE-binding activity is induced following incubation with zinc, cadmium, silver, copper and heat shock[124]. HSF1 is believed to be the member of the HSF family most influenced by environmental stressors. Two tissue- and temperature-dependent HSF1 isoforms have been identified in zebrafish and the two isoforms are formed by alternative splicing of RNA transcribed from one gene[141]. These two isoforms are differentially expressed in response to stressors, including temperature, oxygen limitation and metal toxicity[3]. The significance of the role of these isoforms is not clear although it is proposed that the ratio of these two forms is indicative of stressor sensitivity of the cells[3]. Regardless, the impact of chemicals on HSF and HSE may be a potential site for the altered heat shock response which would contribute indirectly to alter GR signaling.

2.6. Proteasome

Degradation of critical proteins and accumulation of proteins due to inhibition of the protein degradative pathway are key processes contributing to toxicity. In normal cases, cellular protein degradation is an important homeostatic process necessary for degradation of proteins that are abnormally folded, short-lived, long-lived and endocytosed[71]. Under most conditions in cultured mammalian cells, 80–90% of the protein breakdown occurs *via* the proteasomal pathway[71] and the remainder is *via* the lysosomal pathway[101]. For proteins to be degraded by the proteasomal pathway, the protein must usually first be tagged with multiple copies of the small protein ubiquitin. Once ubiquitinated, the protein must be in a conformation suitable for access to the active cleavage sites of the proteasome and this conformation is facilitated by hsps and other chaperones[71]. Finally, the proteasome degrades the ubiquitinated protein by proteolysis and recycles the ubiquitin molecules. At least three proteins are required for each ubiquitin addition, and the proteasome is comprised of multiple subunits, suggesting the potential existence of multiple sites for inhibition/activation by xenobiotics. As the proteasomal pathway is thought to play a role in GR signaling[177], including fish GR[156], it seems logical that the proteasome may be a key player in the xenobiotic-induced disruption of the GR signaling process, although very little is known at present.

Studies with mammalian cell systems have shown that xenobiotic impact on the proteasome may be both general and targeted. For instance, specific degradation of HIF-I and Na^+/K^+-ATPase has been observed with cadmium[39,168], while TCDD specifically induces degradation of AhR and the estrogen-α receptor[184]. In addition to specific effects, some of the general effects of xenobiotics on protein breakdown include enhancement of cellular ubiquitinylation[66,168] and proteasomal degradation[88] by heavy metals. Also, high levels of arsenic inhibited proteasomal degradation, including depression of E2 proteasomal subunit activity[92,93]. This activation and/or inhibition of the proteasomal pathway by heavy metals are thought to be due in part to the interaction of heavy metals with the key cysteine residues and disulphide bridges resulting in protein denaturation. Consequently, the denatured proteins activate the proteasomal pathway. However, with high concentrations of heavy metals, the cells cannot degrade proteins quickly enough, resulting in aggregated proteins which inhibit the proteasome[99]. This inhibition of the proteasome could perhaps be a mechanism for the higher GR content seen in fish gills exposed to copper (R. Weil and

M.M. Vijayan, unpublished), but this remains to be tested. We have shown in trout that cortisol-induced downregulation of GR involves the proteasome[20,156]. Also, inhibition of the proteasome clearly abolished the attenuation of the hsp70 response to heat shock by cortisol alluding to a key role for this protein breakdown pathway in GR signaling[20], including transcriptional activation of GR[156]. While this field of research is still in its infancy in the piscine system, studies have shown the presence of ubiquitin in many fish species[129] either as free ubiquitin or ubiquitin conjugates[150]. The 26S proteasome has also been examined during goldfish (*Carassius auratus*) oocyte maturation and egg activation[170,171] and has been purified from chum salmon (*O. keta*) sperm[78]. However, to our knowledge no study has yet examined the impact of xenobiotics on the proteasome in fish. This may be an important area of research because any disruption of the proteasomal pathway affects protein homeostasis[71], including GR stability and signaling[20,156,177], responses crucial for maintaining homeostasis.

V. Conclusions and future directions

In the past, significant attention has focused on environmental chemicals that affect estrogen homeostasis (e.g. by binding to the estrogen receptor as agonists or antagonists) and, thereby, affecting fetal development, sex differentiation or reproduction (see review by McLachlan[115]). Many such studies support the idea that chemicals showing such endocrine-disrupting effects do not solely interact with estradiol receptor but certainly also interact with other members of the steroid/thyroid/retinoic nuclear receptor family. This is also the case in fish, but despite that, the focus has been on the reproductive axis perhaps mainly because of the ease of quantifying the responses to xenoestrogens (see Chapter 16). Very little research focuses on CR, possibly because the CR and their signaling pathways are not well characterized in these animals. However, with the recent development of molecular tools for CR research in bony fishes, this animal model may allow for the detection and persistence, as well as the mechanistic analysis of toxicity, associated with CR disruptors in the aquatic environment.

While limited work exists regarding the direct role of toxicants on CR signaling dysfunction in fish, indirect evidence based mostly on the mode of action of toxicants, clearly imply several potential routes for CR toxicity (Table 1; Fig. 2). Many of the described experiments in the literature utilized mammalian cell lines and reporter assay as an *in vitro* system to identify the mechanisms and generate working models for the action of toxicants on CR signaling. In many cases, we have taken these models as representative of the mode of action in fish, although the experiments used to generate these models have not been confirmed in fish. While compiling the literature to write this review article, we realized that CR and the associated signaling pathways are sensitive to toxicant attack at multiple levels (Fig. 2). As some of the components of the signaling pathways, including molecular chaperones, are not just limited to hormonal signaling, the impact of xenobiotics on CR signaling may extend beyond the realm of 'endocrine disruption' and could encompass 'homeostatic disruption'. This is underscored by the fact that CR signaling is involved in all aspects of animal performance, including stress response, growth and metabolism, iono- and osmo-regulation, immune function and reproduction in fish[121]. Nevertheless, it is abundantly clear that we are only skimming the surface and many interesting areas of research, especially mechanistically oriented, still remain to be addressed. Targeted disruption of CR mRNA using novel techniques, including small interfering RNA (siRNA), should reveal the involvement of this important transcription factor in the homeostatic process, including cellular adjustments to xenobiotic insults. The multifaceted responses associated with activation of CR makes this steroid receptor a key target for characterization of the potency of endocrine disruptors on animal performance.

Acknowledgements. Preparation of this review and some of the unpublished work reported here were supported by funding from the Natural Sciences and Engineering Research Council (NSERC), Canada, discovery grant to MMV. The authors thank Alison McGuire, University of Waterloo, for assistance with the literature search.

VI. *References*

1. Abbott, B.D. Review of the interaction between TCDD and glucocorticoids in embryonic palate. *Toxicology* 105: 365–373, 1995.
2. Abbott, B.D., M.R. Probst, G.H. Perdew and A.R. Buckalew. Ah receptor, ARNT, glucocorticoid receptor, EGF receptor, EGF, TGF alpha, TGF beta 1, TGF beta 2, and TGF beta 3 expression in human embryonic palate, and effects of 2,3,7,8-tetrachlorodibenzo-*p*-dioxin (TCDD). *Teratology* 58: 30–43, 1998.
3. Airaksinen, S. Environmental factors regulation gene expression in fish – heat shock response. Dissertation, University of Turku, 2003.
4. Allison, C.M. and R.J. Omeljaniuk. Specific binding sites for [^{3}H]dexamethasone in the hypothalamus of juvenile rainbow trout, *Oncorhynchus mykiss*. *Gen. Comp. Endocrinol.* 110: 2–10, 1998.

5a. Aluru, N., E.H. Jorgensen, A.G. Maule and M.M. Vijayan. PCB disruption of the hypothalamus–pituitary–interrenal axis involves brain glucocorticoid receptor downregulation in anadromous Artic charr. *Am. J. Physiol. Regul. Integr. Comp. Physiol.* 287: R787–R793, 2004.

5b. Aluru, N. and M.M. Vijayan. β-Naphthoflavone disrupts cortisol production and liver glucocorticoid responsiveness in rainbow trout. *Aquat. Toxicol.* 67: 273–285, 2004.

6. Andersson, T. and A. Goksøyr. Distribution and induction of cytochrome P4501A1 in the rainbow trout brain. *Fish Physiol. Biochem.* 13: 335–342, 1994.
7. Baker, M.E. Evolution of glucocorticoid and mineralocorticoid responses: go fish. *Endocrinology* 144: 4223–4225, 2003.
8. Balm, P.H., J.D. Lambert and S.E. Wendelaar Bonga. Corticosteroid biosynthesis in the interrenal cells of the teleost fish, *Oreochromis mossambicus*. *Gen. Comp. Endocrinol.* 76: 53–62, 1989.
9. Bamberger, C.M., H.M. Schulte and G.P. Chrousos. Molecular determinants of glucocorticoid receptor function and tissue sensitivity to glucocorticoids. *Endocr. Rev.* 17: 245–261, 1996.
10. Barton, B.A. and G.K. Iwama. Physiological changes in fish from stress in aquaculture with emphasis on the response and effects of corticosteroids. *Annu. Rev. Fish Dis.* 1: 3–26, 1991.
11. Barton, B.A., J.D. Morgan and M.M. Vijayan. Physiological and condition-related indicators of environmental stress in fish. In: *Biological Indicators of Aquatic Ecosystem Health*, edited by S.M. Adams, Bethesda, MD, American Fisheries Society, pp. 111–148, 2002.
12. Basta-Kaim, A., B. Budziszewska, L. Jaworska-Feil, M. Tetich, M. Leskiewicz, M. Kubera and W. Lason. Chlorpromazine inhibits the glucocorticoid receptor-mediated gene transcription in a calcium-dependent manner. *Neuropharmacology* 43: 1035–1043, 2002.
13. Basu, N., C.J. Kennedy, P.V. Hodson and G.K. Iwama. Altered stress responses in rainbow trout following a dietary administration of cortisol and β-napthoflavone. *Fish Physiol. Biochem.* 25: 131–140, 2002.

14. Basu, N., C.J. Kennedy and G.K. Iwama. The effects of stress on the association between hsp70 and the glucocorticoid receptor in rainbow trout. *Comp. Biochem. Physiol.* 134A: 655–663, 2003.
15. Basu, N., A.E. Todgham, P.A. Ackerman, M.R. Bibeau, K. Nakano, P.M. Schulte and G.K. Iwama. Heat shock protein genes and their functional significance in fish. *Gene* 295: 173–183, 2002.
16. Baumann, H., K. Paulsen, H. Kovacs, H. Berglund, A.P.H. Wright, J.A. Gustafsson and T. Hart. Refined solution structure of the glucocorticoid receptor DNA-binding domain. *Biochemistry* 32: 13463–13471, 1993.
17. Bodwell, J.E., J.C. Webster, C.M. Jewell, J.A. Cidlowski, J.M. Hu and A. Munck. Glucocorticoid receptor phosphorylation: overview, function and cell cycle-dependence. *J. Steroid Biochem. Mol. Biol.* 65: 91–99, 1998.
18. Boone, A.N., B. Ducouret and M.M. Vijayan. Glucocorticoid-induced glucose release is abolished in trout hepatocytes with elevated hsp70 content. *J. Endocrinol.* 172: R1–R5, 2002.
19. Boone, A.N. and M.M. Vijayan. Constitutive heat shock protein 70 (HSC70) expression in rainbow trout hepatocytes: effect of heat shock and heavy metal exposure. *Comp. Biochem. Physiol.* 132C: 223–233, 2002.
20. Boone, A.N. and M.M. Vijayan. Glucocorticoid-mediated attenuation of the hsp70 response in trout hepatocytes involves the proteasome. *Am. J. Physiol. Integr. Comp. Physiol.* 283: R680–R687, 2002.
21. Borski, R.J. Nongenomic membrane actions of glucocorticoids in vertebrates. *Trends Endocrinol. Metab.* 11: 427–436, 2000.
22. Borski, R.J., G.N. Hyde and S. Fruchtman. Signal transduction mechanisms mediating rapid, nongenomic effects of cortisol on prolactin release. *Steroids* 67: 539–548, 2002.
23. Borski, R.J., G.N. Hyde, S. Fruchtman and W.S. Tsai. Cortisol suppresses prolactin release through a non-genomic mechanism involving interactions with the plasma membrane. *Comp. Biochem. Physiol.* 129B: 533–541, 2001.
24. Bouton, C.M. and J. Pevsner. Effects of lead on gene expression. *Neurotoxicology* 21: 1045–1055, 2000.
25. Brewer, G.J., J.C. Aster, C.A. Knutsen and W.C. Kruckeberg. Zinc inhibition of calmodulin: a proposed molecular mechanism of zinc action on cellular functions. *Am. J. Hematol.* 7: 53–60, 1979.
26. Bruna, A., M. Nicolas, A. Munoz, J.M. Kyriakis and C. Caelles. Glucocorticoid receptor–JNK interaction mediates inhibition of the JNK pathway by glucocorticoids. *EMBO J.* 22: 6035–6044, 2003.
27. Bury, N.R., A. Sturm, P. Le Rouzic, C. Lethimonier, B. Ducouret, Y. Guiguen, M. Robinson-Rechavi, V. Laudet, M.E. Rafestin-Oblin and P. Prunet. Evidence for two distinct functional glucocorticoid receptors in teleost fish. *J. Mol. Endocrinol.* 31: 141–156, 2003.
28. Canesi, L., C. Ciacci, M. Betti, A. Scarpato, B. Citterio, C. Pruzzo and G. Gallo. Effects of PCB congeners on the immune function of *Mytilus* hemocytes: alterations of tyrosine kinase-mediated cell signaling. *Aquat. Toxicol.* 63: 293–306, 2003.
29. Carlson, D.B. and G.H. Perdew. A dynamic role for the Ah receptor in cell signaling? Insights from a diverse group of Ah receptor interacting proteins. *J. Biochem. Mol. Toxicol.* 16: 317–325, 2002.
30. Carre, M., G. Carles, N. Andre, S. Douillard, J. Ciccolini, C. Briand and D. Braguer. Involvement of microtubules and mitochondria in the antagonism of arsenic trioxide on paclitaxel-induced apoptosis. *Biochem. Pharmacol.* 63: 1831–1842, 2002.
31. Celander, M., M.E. Hahn and J.J. Stegeman. Cytochromes P450 (CYP) in the *Poeciliopsis lucida* hepatocellular carcinoma cell line (PLHC-1). *Arch. Biochem. Biophys.* 329: 113–122, 1996.
32. Chakraborti, P.K., M. Weisbart and A. Chakraborti. The presence of corticosteroid receptor activity in the gills of the brook trout, *Salvelinus fontinalis*. *Gen. Comp. Endocrinol.* 66: 323–332, 1987.
33. Chan, W.K., G. Yao, Y.Z. Gu and C.A. Bradfield. Cross-talk between the aryl hydrocarbon receptor and hypoxia inducible factor signaling pathways. Demonstration of competition and compensation. *J. Biol. Chem.* 274: 12115–12123, 1999.
34. Chao, S.H., Y. Suzuki, J.R. Zysk and W.Y. Cheung. Activation of calmodulin by various metal cations as a function of ionic radius. *Mol. Pharmacol.* 26: 75–82, 1984.
35. Chen, K.M., W.K. Chan and A.D. Munro. Dexamethasone receptors and their distribution in the brain of the red tilapia. *Fish Physiol. Biochem.* 16: 171–179, 1997.
36. Cheung, W.Y. Calmodulin: its potential role in cell proliferation and heavy metal toxicity. *Fed. Proc.* 43: 2995–2999, 1984.
37. Cho, W.J., S.J. Cha, J.W. Do, J.Y. Choi, J.Y. Lee, C.S. Jeong, K.J. Cho, W.S. Choi, H.S. Kang, H.D. Kim and J.W. Park. A novel 90 kDa stress protein induced in fish cells by fish Rhabdovirus infection. *Biochem. Biophys. Res. Commun.* 233: 316–319, 1997.

38. Chou, I.N. Distinct cytoskeletal injuries induced by As, Cd, Co, Cr, and Ni compounds. *Biomed. Environ. Sci.* 2: 358–365, 1989.
39. Chun, Y.S., E. Choi, G.T. Kim, H. Choi, C.H. Kim, M.J. Lee, M.S. Kim and J.W. Park. Cadmium blocks hypoxia-inducible factor (HIF)-1-mediated response to hypoxia by stimulating the proteasome-dependent degradation of HIF-1α. *Eur. J. Biochem.* 267: 4198–4204, 2000.
40. Chung, E., K.Y. Lee, Y.J. Lee, Y.H. Lee and S.K. Lee. Ginsenoside Rg1 down-regulates glucocorticoid receptor and displays synergistic effects with cAMP. *Steroids* 63: 421–424, 1998.
41. Collingwood, T.N., F.D. Urnov and A.P. Wolffe. Nuclear receptors: coactivators, corepressors and chromatin remodeling in the control of transcription. *J. Mol. Endocrinol.* 23: 255–275, 1999.
42. Colombe, L., A. Fostier, N. Bury, F. Pakdel and Y. Guiguen. A mineralocorticoid-like receptor in the rainbow trout, *Oncorhynchus mykiss*: cloning and characterization of its steroid binding domain. *Steroids* 65: 319–328, 2000.
43. Cox, J.L. and S.D. Harrison, Jr. Correlation of metal toxicity with *in vitro* calmodulin inhibition. *Biochem. Biophys. Res. Commun.* 115: 106–111, 1983.
44. Currie, S., B.L. Tufts and C.D. Moyes. Influence of bioenergetic stress on heat shock protein gene expression in nucleated red blood cells of fish. *Am. J. Physiol.* 276: R990–R996, 1999.
45. Cuthill, S., A. Wilhelmsson, G.G. Mason, M. Gillner, L. Poellinger and J.A. Gustafsson. The dioxin receptor: a comparison with the glucocorticoid receptor. *J. Steroid Biochem.* 30: 277–280, 1988.
46. Dang, Z.C., M.H. Berntssen, A.K. Lundebye, G. Flik, S.E. Wendelaar Bonga and R.A. Lock. Metallothionein and cortisol receptor expression in gills of Atlantic salmon, *Salmo salar*, exposed to dietary cadmium. *Aquat. Toxicol.* 53: 91–101, 2001.
47. Dang, Z.C., G. Flik, B. Ducouret, C. Hogstrand, S.E. Wendelaar Bonga and R.A. Lock. Effects of copper on cortisol receptor and metallothionein expression in gills of *Oncorhynchus mykiss*. *Aquat. Toxicol.* 51: 45–54, 2000.
48. Dasmahapatra, A.K. and P.C. Lee. Down regulation of CYP 1A1 by glucocorticoids in trout hepatocytes *in vitro*. *In Vitro Cell Dev. Biol. Anim.* 29A: 643–648, 1993.
49. De Boeck, G., B. De Wachter, A. Vlaeminck and R. Blust. Effect of cortisol treatment and/or sublethal copper exposure on copper uptake and heat shock protein levels in common carp, *Cyprinus carpio*. *Environ. Toxicol. Chem.* 22: 1122–1126, 2003.
50. DeFranco, D.B., C. Ramakrishnan and Y. Tang. Molecular chaperones and subcellular trafficking of steroid receptors. *J. Steroid Biochem. Mol. Biol.* 65: 51–58, 1998.
51. DeMoor, J.M., W.A. Kennette, O.M. Collins and J. Koropatnick. Zinc-metallothionein levels are correlated with enhanced glucocorticoid responsiveness in mouse cells exposed to $ZnCl_2$, $HgCl_2$, and heat shock. *Toxicol. Sci.* 64: 67–76, 2001.
52. Devaux, A., M. Pesonen, G. Monod and T. Andersson. Glucocorticoid-mediated potentiation of P450 induction in primary culture of rainbow trout hepatocytes. *Biochem. Pharmacol.* 43: 898–901, 1992.
53. Devos, S., K. De Bosscher, B. Staels, E. Bauer, F. Roels, W. Berghe, G. Haegeman, R. Hooghe and E.L. Hooghe-Peters. Inhibition of cytokine production by the herbicide atrazine. Search for nuclear receptor targets. *Biochem. Pharmacol.* 65: 303–308, 2003.
54. DiBattista, J.A., A.Z. Mehdi and T. Sandor. Intestinal triamcinolone acetonide receptors of the eel (*Anguilla rostrata*). *Gen. Comp. Endocrinol.* 51: 228–238, 1983.
55. Dowd, D.R., P.N. MacDonald, B.S. Komm, M.R. Haussler and R. Miesfeld. Evidence for early induction of calmodulin gene expression in lymphocytes undergoing glucocorticoid-mediated apoptosis. *J. Biol. Chem.* 266: 18423–18426, 1991.
56. Ducouret, B., M. Tujague, J. Ashraf, N. Mouchel, N. Servel, Y. Valotaire and E.B. Thompson. Cloning of a teleost fish glucocorticoid receptor shows that it contains a deoxyribonucleic acid-binding domain different from that of mammals. *Endocrinology* 136: 3774–3783, 1995.
57. Dundjerski, J., B. Butorovic, J. Kipic, D. Trajkovic and G. Matic. Cadmium affects the activity of rat liver tyrosine aminotransferase and its induction by dexamethasone. *Arch. Toxicol.* 70: 390–395, 1996.
58. Dundjerski, J., T. Kovac, N. Pavkovic, A. Cvoro and G. Matic. Glucocorticoid receptor–Hsp90 interaction in the liver cytosol of cadmium-intoxicated rats. *Cell Biol. Toxicol.* 16: 375–383, 2000.
59. Dundjerski, J., J. Stanosevic, B. Ristic, D. Trajkovic and G. Matic. *In vivo* effects of cadmium on rat liver glucocorticoid receptor functional properties. *Int. J. Biochem.* 24: 1065–1072, 1992.
60. Dyer, S.D., K.L. Dickson and E.G. Zimmerman. A laboratory evaluation of the use of stress proteins in fish to detect changes in water quality. In: *Environmental Toxicology and Risk Assessment*, edited by W.G. Landis, J.J. Hughes and M.A. Lewis, Philadelphia, ASTM Publishers, 273 pp., 1993.
61. Dyer, S.D., K.L. Dickson and E.G. Zimmerman. Synthesis and accumulation of stress proteins in tissues of arsenite exposed fathead minnow (*Pimephales promelas*). *Environ. Toxicol. Chem.* 12: 913–924, 1993.

62. Elez, D., J. Dundjerski and G. Matic. Cadmium affects the redox state of rat liver glucocorticoid receptor. *Cell Biol. Toxicol.* 17: 169–177, 2001.
63. Fatma, N., D.P. Singh, T. Shinohara and L.T. Chylack, Jr. Transcriptional regulation of the antioxidant protein 2 gene, a thiol-specific antioxidant, by lens epithelium-derived growth factor to protect cells from oxidative stress. *J. Biol. Chem.* 276: 48899–48907, 2001.
64. Feder, M.E. and G.E. Hofmann. Heat-shock proteins, molecular chaperones, and the stress response: evolutionary and ecological physiology. *Annu. Rev. Physiol.* 61: 243–282, 1999.
65. Feng, Q., A.N. Boone and M.M. Vijayan. Copper impact on heat shock protein 70 expression and apoptosis in rainbow trout hepatocytes. *Comp. Biochem. Physiol.* 135C: 345–355, 2003.
66. Figueiredo-Pereira, M.E. and G. Cohen. The ubiquitin/proteasome pathway: friend or foe in zinc-, cadmium-, and H_2O_2-induced neuronal oxidative stress. *Mol. Biol. Rep.* 26: 65–69, 1999.
67. Freedman, L.P., B.F. Lusis, Z.R. Korszun, R. Basavappa, P.B. Sigler and K.R. Yamamoto. The function and structure of the metal coordination sites within the glucocorticoid receptor DNA binding domain. *Nature* 334: 543–546, 1988.
68. Galigniana, M.D., P.R. Housley, D.B. DeFranco and W.B. Pratt. Inhibition of glucocorticoid receptor nucleocytoplasmic shuttling by okadaic acid requires intact cytoskeleton. *J. Biol. Chem.* 274: 16222–16227, 1999.
69. Galigniana, M.D., J.L. Scruggs, J. Herrington, M.J. Welsh, C. Carter-Su, P.R. Housley and W.B. Pratt. Heat shock protein 90-dependent (geldanamycin-inhibited) movement of the glucocorticoid receptor through the cytoplasm to the nucleus requires intact cytoskeleton. *Mol. Endocrinol.* 12: 1903–1913, 1998.
70. Gamperl, A.K., M. Wilkinson and R.G. Boutilier. Beta-adrenoreceptors in the trout (*Oncorhynchus mykiss*) heart: characterization, quantification, and effects of repeated catecholamine exposure. *Gen. Comp. Endocrinol.* 95: 259–272, 1994.
71. Glickman, M.H. and A. Ciechanover. The ubiquitin–proteasome proteolytic pathway: destruction for the sake of construction. *Physiol. Rev.* 82: 373–428, 2002.
72. Gradin, K., M.L. Whitelaw, R. Toftgard, L. Poellinger and A. Berghard. A tyrosine kinase-dependent pathway regulates ligand-dependent activation of the dioxin receptor in human keratinocytes. *J. Biol. Chem.* 269: 23800–23807, 1994.
73. Greenwood, A.K., P.C. Butler, R.B. White, U. DeMarco, D. Pearce and R.D. Fernald. Multiple corticosteroid receptors in a teleost fish: distinct sequences, expression patterns, and transcriptional activities. *Endocrinology* 144: 4226–4236, 2003.
74. Hagmann, J. Inhibition of calmodulin-stimulated cyclic nucleotide phosphodiesterase by the insecticide DDT. *FEBS Lett.* 143: 52–54, 1982.
75. Heikkila, J.J., G.A. Schultz, K. Iatrou and L. Gedamu. Expression of a set of fish genes following heat or metal ion exposure. *J. Biol. Chem.* 257: 12000–12005, 1982.
76. Hightower, L.E. Heat shock, stress proteins, chaperones, and proteotoxicity. *Cell* 66: 191–197, 1991.
77. Hofmann, G.E., B.A. Buckley, S. Airaksinen, J.E. Keen and G.N. Somero. Heat-shock protein expression is absent in the Antarctic fish *Trematomus bernacchii* (family Nototheniidae). *J. Exp. Biol.* 203: 2331–2339, 2000.
78. Inaba, K., Y. Akazome and M. Morisawa. Purification of proteasomes from salmonid fish sperm and their localization along sperm flagella. *J. Cell Sci.* 104: 907–915, 1993.
79. Itoh, M., M. Adachi, H. Yasui, M. Takekawa, H. Tanaka and K. Imai. Nuclear export of glucocorticoid receptor is enhanced by c-Jun N-terminal kinase-mediated phosphorylation. *Mol. Endocrinol.* 16: 2382–2392, 2002.
80. Iwama, G.K., P.T. Thomas, R.B. Forsyth and M.M. Vijayan. Heat shock protein expression in fish. *Rev. Fish Biol. Fish.* 8: 35–56, 1998.
81. Iwasaki, T., W. Miyazaki, A. Takeshita, Y. Kuroda and N. Koibuchi. Polychlorinated biphenyls suppress thyroid hormone-induced transactivation. *Biochem. Biophys. Res. Commun.* 299: 384–388, 2002.
82. Jana, N.R., S. Sarkar, M. Ishizuka, J. Yonemoto, C. Tohyama and H. Sone. Comparative effects of 2,3,7,8-tetrachlorodibenzo-*p*-dioxin on MCF-7, RL95-2, and LNCaP cells: role of target steroid hormones in cellular responsiveness to CYP1A1 induction. *Mol. Cell Biol. Res. Commun.* 4: 174–180, 2000.
83. Janz, D.M., M.E. McMaster, K.R. Munkittrick and G. Van der Kraak. Elevated ovarian follicular apoptosis and heat shock protein-70 expression in white sucker exposed to bleached kraft pulp mill effluent. *Toxicol. Appl. Pharmacol.* 147: 391–398, 1997.
84. Jenkins, B.D., C.B. Pullen and B.D. Darimont. Novel glucocorticoid receptor coactivator effector mechanisms. *Trends Endocrinol. Metab.* 12: 122–126, 2001.

85. Johansson, M. Interaction of xenobiotics with the glucocorticoid hormone system *in vitro*. Dissertation, Uppsala University, 2002.
86. Johansson, M., S. Nilsson and B.O. Lund. Interactions between methylsulfonyl PCBs and the glucocorticoid receptor. *Environ. Health Perspect.* 106: 769–772, 1998.
87. Jorgensen, E.H., M.M. Vijayan, N. Aluru and A.G. Maule. Fasting modifies Aroclor 1254 impact on plasma cortisol, glucose and lactate responses to a handling disturbance in Arctic charr. *Comp. Biochem. Physiol.* 132C: 235–245, 2002.
88. Jungmann, J., H.A. Reins, C. Schobert and S. Jentsch. Resistance to cadmium mediated by ubiquitin-dependent proteolysis. *Nature* 361: 369–371, 1993.
89. Kaltreider, R.C., A.M. Davis, J.P. Lariviere and J.W. Hamilton. Arsenic alters the function of the glucocorticoid receptor as a transcription factor. *Environ. Health Perspect.* 109: 245–251, 2001.
90. Kensler, T.W., W.F. Busby, Jr., N.E. Davidson and G.N. Wogan. Effect of hepatocarcinogens on the binding of glucocorticoid-receptor complex in rat liver nuclei. *Cancer Res.* 36: 4647–4651, 1976.
91. Kharat, I. and F. Saatcioglu. Antiestrogenic effects of 2,3,7,8-tetrachlorodibenzo-*p*- dioxin are mediated by direct transcriptional interference with the liganded estrogen receptor. Cross-talk between aryl hydrocarbon- and estrogen-mediated signaling. *J. Biol. Chem.* 271: 10533–10537, 1996.
92. Klemperer, N.S., E.S. Berleth and C.M. Pickart. A novel, arsenite-sensitive E2 of the ubiquitin pathway: purification and properties. *Biochemistry* 28: 6035–6041, 1989.
93. Klemperer, N.S. and C.M. Pickart. Arsenite inhibits two steps in the ubiquitin-dependent proteolytic pathway. *J. Biol. Chem.* 264: 19245–19252, 1989.
94. Knoebl, I., M.S. Fitzpatrick and C.B. Schreck. Characterization of a glucocorticoid receptor in the brains of Chinook salmon, *Oncorhynchus tshawytscha*. *Gen. Comp. Endocrinol.* 101: 195–204, 1996.
95. Knudsen, F.R. and T.G. Pottinger. Interaction of endocrine disrupting chemicals, singly and in combination, with estrogen-, androgen-, and corticosteroid-binding sites in rainbow trout (*Oncorhynchus mykiss*). *Aquat. Toxicol.* 44: 159–170, 1999.
96. Komarova, E.A., N. Neznanov, P.G. Komarov, M.V. Chernov, K. Wang and A.V. Gudkov. p53 inhibitor pifithrin α can suppress heat shock and glucocorticoid signaling pathways. *J. Biol. Chem.* 278: 15465–15468, 2003.
97. Kothary, R.K. and E.P. Candido. Induction of a novel set of polypeptides by heat shock or sodium arsenite in cultured cells of rainbow trout, *Salmo gairdnerii*. *Can. J. Biochem.* 60: 347–355, 1982.
98. Kothary, R.K., D. Jones and E.P. Candido. 70-Kilodalton heat shock polypeptides from rainbow trout: characterization of cDNA sequences. *Mol. Cell. Biol.* 4: 1785–1791, 1984.
99. Krstic, M.D., I. Rogatsky, K.R. Yamamoto and M.J. Garabedian. Mitogen-activated and cyclin-dependent protein kinases selectively and differentially modulate transcriptional enhancement by the glucocorticoid receptor. *Mol. Cell. Biol.* 17: 3947–3954, 1997.
100. Kusakabe, M., I. Nakamura and G. Young. 11β-Hydroxysteroid dehydrogenase complementary deoxyribonucleic acid in rainbow trout: cloning, sites of expression, and seasonal changes in gonads. *Endocrinology* 144: 2534–2545, 2003.
101. Lee, D.H. and A.L. Goldberg. Proteasome inhibitors: valuable new tools for cell biologists. *Trends Cell Biol.* 8: 397–403, 1998.
102. Lee, P.C. and M. Struve. Unsaturated fatty acids inhibit glucocorticoid receptor binding of trout hepatic cytosol. *Comp. Biochem. Physiol.* 102B: 707–711, 1992.
103. Lee, Y.J., E. Chung, K.Y. Lee, Y.H. Lee, B. Huh and S.K. Lee. Ginsenoside-Rg1, one of the major active molecules from *Panax ginseng*, is a functional ligand of glucocorticoid receptor. *Mol. Cell. Endocrinol.* 133: 135–140, 1997.
104. Lethimonier, C., G. Flouriot, O. Kah and B. Ducouret. The glucocorticoid receptor represses the positive autoregulation of the trout estrogen receptor gene by preventing the enhancer effect of a C/EBPbeta-like protein. *Endocrinology* 143: 2961–2974, 2002.
105. Li, D.P., S. Periyasamy, T.J. Jones and E.R. Sanchez. Heat and chemical shock potentiation of glucocorticoid receptor transactivation requires heat shock factor (HSF) activity. Modulation of HSF by vanadate and wortmannin. *J. Biol. Chem.* 275: 26058–26065, 2000.
106. Li, W. and I.N. Chou. Effects of sodium arsenite on the cytoskeleton and cellular glutathione levels in cultured cells. *Toxicol. Appl. Pharmacol.* 114: 132–139, 1992.
107. Ling, Y.H., J.D. Jiang, J.F. Holland and R. Perez-Soler. Arsenic trioxide produces polymerization of microtubules and mitotic arrest before apoptosis in human tumor cell lines. *Mol. Pharmacol.* 62: 529–538, 2002.
108. Lundholm, C.E.. The effects of DDE, PCB and chlordane on the binding of progesterone to its cytoplasmic receptor in the eggshell gland mucosa of birds and the endometrium of mammalian uterus. *Comp. Biochem. Physiol.* 89C: 361–368, 1988.

109. Lyons-Alcantara, M., R. Mooney, F. Lyng, D. Cottell and C. Mothersill. The effects of cadmium exposure on the cytology and function of primary cultures from rainbow trout. *Cell Biochem. Funct.* 16: 1–13, 1998.
110. Marsigliante, S., S. Barker, E. Jimenez and C. Storelli. Glucocorticoid receptors in the euryhaline teleost. *Anguilla anguilla. Mol. Cell. Endocrinol.* 162: 193–201, 2000.
111. Marsigliante, S., A. Muscella, S. Vilella and C. Storelli. Dexamethasone modulates the activity of the eel branchial Na^+/K^+ATPase in both chloride and pavement cells. *Life Sci.* 66: 1663–1673, 2000.
112. Maule, A.G. and C.B. Schreck. Stress and cortisol treatment changed affinity and number of glucocorticoid receptors in leukocytes and gill of coho salmon. *Gen. Comp. Endocrinol.* 84: 83–93, 1991.
113. Max, S.R. and E.K. Silbergeld. Skeletal muscle glucocorticoid receptor and glutamine synthetase activity in the wasting syndrome in rats treated with 2,3,7,8- tetrachlorodibenzo-*p*-dioxin. *Toxicol. Appl. Pharmacol.* 87: 523–527, 1987.
114. McCormick, S.D. Hormonal control of gill Na^+,K^+-ATPase and chloride cell function. In: *Cellular and Molecular Approaches to Fish Ionic Regulation*, edited by C.M. Wood and T.J. Shuttleworth, San Diego, Academic Press, pp. 285–315, 1995.
115. McLachlan, J.A.. Environmental signaling: what embryos and evolution teach us about endocrine disrupting chemicals. *Endocr. Rev.* 22: 319–341, 2001.
116. Miesfeld, R., S. Rusconi, P.J. Godowski, B.A. Maler, S. Okret, A.C. Wikstrom, J.A. Gustafsson and K.R. Yamamoto. Genetic complementation of a glucocorticoid receptor deficiency by expression of cloned receptor cDNA. *Cell* 46: 389–399, 1986.
117. Misra, S., M. Zafarullah, J. Price-Haughey and L. Gedamu. Analysis of stress-induced gene expression in fish cell lines exposed to heavy metals and heat shock. *Biochim. Biophys. Acta* 1007: 325–333, 1989.
118. Mitani, K., H. Fujita, S. Sassa and A. Kappas. A heat-inducible nuclear factor that binds to the heat-shock element of the human haem oxygenase gene. *Biochem. J.* 277: 895–897, 1991.
119. Miyazaki, M., K. Kohno, T. Uchiumi, H. Tanimura, K. Matsuo, M. Nasu and M. Kuwano. Activation of human multidrug resistance-1 gene promoter in response to heat shock stress. *Biochem. Biophys. Res. Commun.* 187: 677–684, 1992.
120. Mizuyachi, K., D.S. Son, K.K. Rozman and P.F. Terranova. Alteration in ovarian gene expression in response to 2,3,7,8- tetrachlorodibenzo-*p*-dioxin: reduction of cyclooxygenase-2 in the blockage of ovulation. *Reprod. Toxicol.* 16: 299–307, 2002.
121. Mommsen, T.P., M.M. Vijayan and T.W. Moon. Cortisol in teleosts: dynamics, mechanisms of action, and metabolic regulation. *Rev. Fish Biol. Fish.* 9: 211–268, 1999.
122. Morishima, Y., K.C. Kanelakis, P.J. Murphy, E.R. Lowe, G.J. Jenkins, Y. Osawa, R.K. Sunahara and W.B. Pratt. The hsp90 cochaperone p23 is the limiting component of the multiprotein hsp90/hsp70-based chaperone system *in vivo* where it acts to stabilize the client protein: hsp90 complex. *J. Biol. Chem.* 278: 48754–48763, 2003.
123. Morishima, Y., K.C. Kanelakis, A.M. Silverstein, K.D. Dittmar, L. Estrada and W.B. Pratt. The Hsp organizer protein hop enhances the rate of but is not essential for glucocorticoid receptor folding by the multiprotein Hsp90-based chaperone system. *J. Biol. Chem.* 275: 6894–6900, 2000.
124. Murata, M., P. Gong, K. Suzuki and S. Koizumi. Differential metal response and regulation of human heavy metal-inducible genes. *J. Cell. Physiol.* 180: 105–113, 1999.
125. Nair, S.C., E.J. Toran, R.A. Rimerman, S. Hjermstad, T.E. Smithgall and D.F. Smith. A pathway of multi-chaperone interactions common to diverse regulatory proteins: estrogen receptor, Fes tyrosine kinase, heat shock transcription factor Hsf1, and the aryl hydrocarbon receptor. *Cell Stress Chaperones* 1: 237–250, 1996.
126. Niimi, S., T. Yamaguchi and T. Hayakawa. Regulation of glucocorticoid receptor by the tyrosine kinase inhibitor herbimycin A in the cytosolic fraction of primary cultured rat hepatocytes. *J. Steroid Biochem. Mol. Biol.* 61: 65–71, 1997.
127. Ning, Y.M. and E.R. Sanchez. Evidence for a functional interaction between calmodulin and the glucocorticoid receptor. *Biochem. Biophys. Res. Commun.* 208: 48–54, 1995.
128. Ning, Y.M. and E.R. Sanchez. *In vivo* evidence for the generation of a glucocorticoid receptor-heat shock protein-90 complex incapable of binding hormone by the calmodulin antagonist phenoxybenzamine. *Mol. Endocrinol.* 10: 14–23, 1996.
129. Okubo, K., K. Yamano, Q. Qin, K. Aoyagi, M. Ototake, T. Nakanishi, H. Fukuda and J.M. Dijkstra. Ubiquitin genes in rainbow trout (*Oncorhynchus mykiss*). *Fish Shellfish Immunol.* 12: 335–351, 2002.
130. Okuyama-Tamura, M., M. Mikuni and I. Kojima. Modulation of the human glucocorticoid receptor function by antidepressive compounds. *Neurosci. Lett.* 342: 206–210, 2003.
131. Olivero, J. and P.E. Ganey. Participation of Ca^{2+}/calmodulin during activation of rat neutrophils by polychlorinated biphenyls. *Biochem. Pharmacol.* 62: 1125–1132, 2001.

132. Ortiz-Delgado, J.B., C. Sarasquete, A. Behrens, M.L. Gonzalez de Canales and H. Segner. Expression, cellular distribution and induction of cytochrome P4501A (CYP1A) in gilthead seabream, *Sparus aurata*, brain. *Aquat. Toxicol.* 60: 269–283, 2002.
133. Parsell, D.A. and S. Lindquist. The function of heat-shock proteins in stress tolerance: degradation and reactivation of damaged proteins. *Annu. Rev. Genet.* 27: 437–496, 1993.
134. Payne, J.C., M.L. Horst and H.A. Godwin. Lead fingers: Pb^{2+} binding to structural zinc-binding domains determined directly by monitoring lead-thiolate charge-transfer bands. *J. Am. Chem. Soc.* 121: 6850–6855, 1999.
135. Portigal, C.L., S.P. Cowell, M.N. Fedoruk, C.M. Butler, P.S. Rennie and C.C. Nelson. Polychlorinated biphenyls interfere with androgen-induced transcriptional activation and hormone binding. *Toxicol. Appl. Pharmacol.* 179: 185–194, 2002.
136. Pottinger, T. and I.I. Brierley. A putative cortisol receptor in the rainbow trout erythrocyte: stress prevents starvation-induced increases in specific binding of cortisol. *J. Exp. Biol.* 200: 2035–2043, 1997.
137. Pottinger, T.G. The effect of stress and exogenous cortisol on receptor-like binding of cortisol in the liver of rainbow trout, *Oncorhynchus mykiss*. *Gen. Comp. Endocrinol.* 78: 194–203, 1990.
138. Pottinger, T.G., F.R. Knudsen and J. Wilson. Stress-induced changes in the affinity and abundance of cytosolic cortisol-binding sites in the liver of rainbow trout, *Oncorhynchus mykiss*, are not accompanied by changes in measurable nuclear binding. *Fish Physiol. Biochem.* 12: 499–511, 1994.
139. Pratt, W.B., A.M. Silverstein and M.D. Galigniana. A model for the cytoplasmic trafficking of signalling proteins involving the hsp90-binding immunophilins and p50cdc37. *Cell. Signal.* 11: 839–851, 1999.
140. Pratt, W.B. and D.O. Toft. Steroid receptor interactions with heat shock protein and immunophilin chaperones. *Endocrinol. Rev.* 18: 306–360, 1997.
141. Rabergh, C.M., S. Airaksinen, A. Soitamo, H.V. Bjorklund, T. Johansson, M. Nikinmaa and L. Sistonen. Tissue-specific expression of zebrafish (*Danio rerio*) heat shock factor 1 mRNAs in response to heat stress. *J. Exp. Biol.* 203: 1817–1824, 2000.
142. Ratajczak, T., B.K. Ward and R.F. Minchin. Immunophilin chaperones in steroid receptor signalling. *Curr. Top. Med. Chem.* 3: 1348–1357, 2003.
143. Richter, K. and J. Buchner. Hsp90: chaperoning signal transduction. *J. Cell. Physiol.* 188: 281–290, 2001.
144. Robinson-Rechavi, M., O. Marchand, H. Escriva, P.L. Bardet, D. Zelus, S. Hughes and V. Laudet. Euteleost fish genomes are characterized by expansion of gene families. *Genome Res.* 11: 781–788, 2001.
145. Roesijadi, G., R. Bogumil, M. Vasak and J.H. Kagi. Modulation of DNA binding of a tramtrack zinc finger peptide by the metallothionein–thionein conjugate pair. *J. Biol. Chem.* 273: 17425–17432, 1998.
146. Rogatsky, I., C.L. Waase and M.J. Garabedian. Phosphorylation and inhibition of rat glucocorticoid receptor transcriptional activation by glycogen synthase kinase-3 (GSK-3). Species-specific differences between human and rat glucocorticoid receptor signaling as revealed through GSK-3 phosphorylation. *J. Biol. Chem.* 273: 14315–14321, 1998.
147. Rosenhagen, M.C., J.C. Young, G.M. Wochnik, A.S. Herr, U. Schmidt, F.U. Hartl, F. Holsboer and T. Rein. Synergistic inhibition of the glucocorticoid receptor by radicicol and benzoquinone ansamycins. *Biol. Chem.* 382: 499–504, 2001.
148. Ryan, J.A. and L.E. Hightower. Evaluation of heavy-metal ion toxicity in fish cells using a combined stress protein and cytotoxicity assay. *Environ. Toxicol. Chem.* 13: 1231–1240, 1994.
149. Ryan, J.A. and L.E. Hightower. Stress proteins as molecular biomarkers for environmental toxicology. *EXS* 77: 411–424, 1996.
150. Ryan, S.N., N.W. Pankhurst and R.M.G. Wells. A possible role for ubiquitin in the stress response of the teleost fish blue mao mao (*Scorpis violaceus*). *Physiol. Zool.* 68: 1077–1092, 1995.
151. Samet, J.M., L.M. Graves, J. Quay, L.A. Dailey, R.B. Devlin, A.J. Ghio, W. Wu, P.A. Bromberg and W. Reed. Activation of MAPKs in human bronchial epithelial cells exposed to metals. *Am. J. Physiol. Lung Cell. Mol. Physiol.* 275: L551–L558, 1998.
152. Sanchez, E.R.. Heat shock induces translocation to the nucleus of the unliganded glucocorticoid receptor. *J. Biol. Chem.* 267: 17–20, 1992.
153. Sanders, B.M.. Stress proteins in aquatic organisms: an environmental perspective. *Crit. Rev. Toxicol.* 23: 49–75, 1993.
154. Sandor, T., J.A. DiBattista and A.Z. Mehdi. Glucocorticoid receptors in the gill tissue of fish. *Gen. Comp. Endocrinol.* 53: 353–364, 1984.
155. Sangalang, G.B. and J.F. Uthe. Corticosteroid activity, *in vitro*, in interrenal tissue of Atlantic salmon (*Salmo salar*) parr. 1. Synthetic profiles. *Gen. Comp. Endocrinol.* 95: 273–285, 1994.
156. Sathiyaa, R. and M.M. Vijayan. Autoregulation of glucocorticoid receptor by cortisol in rainbow trout hepatocytes. *Am. J. Physiol. Cell Physiol.* 284: C1508–C1515, 2003.

157. Schaaf, M.J. and J.A. Cidlowski. Molecular determinants of glucocorticoid receptor mobility in living cells: the importance of ligand affinity. *Mol. Cell. Biol.* 23: 1922–1934, 2003.
158. Segnitz, B. and U. Gehring. The function of steroid hormone receptors is inhibited by the hsp90-specific compound geldanamycin. *J. Biol. Chem.* 272: 18694–18701, 1997.
159. Shen, P., Z.J. Xie, H. Li and E.R. Sanchez. Glucocorticoid receptor conversion to high affinity nuclear binding and transcription enhancement activity in Chinese hamster ovary cells subjected to heat and chemical stress. *J. Steroid Biochem. Mol. Biol.* 47: 55–64, 1993.
160. Shrimpton, J.M. and D.J. Randall. Downregulation of corticosteroid receptors in gills of coho salmon due to stress and cortisol treatment. *Am. J. Physiol.* 267: R432–R438, 1994.
161. Simons, S.S., Jr., P.K. Chakraborti and A.H. Cavanaugh. Arsenite and cadmium(II) as probes of glucocorticoid receptor structure and function. *J. Biol. Chem.* 265: 1938–1945, 1990.
162. Sloman, K.A., P.R. Desforges and K.M. Gilmour. Evidence for a mineralocorticoid-like receptor linked to branchial chloride cell proliferation in freshwater rainbow trout. *J. Exp. Biol.* 204: 3953–3961, 2001.
163. Sunahara, G.I., G.W. Lucier, Z. McCoy, E.H. Bresnick, E.R. Sanchez and K.G. Nelson. Characterization of 2,3,7,8-tetrachlorodibenzo-*p*-dioxin-mediated decreases in dexamethasone binding to rat hepatic cytosolic glucocorticoid receptor. *Mol. Pharmacol.* 36: 239–247, 1989.
164. Tagawa, M., H. Hagiwara, A. Takemura, S. Hirose and T. Hirano. Partial cloning of the hormone-binding domain of the cortisol receptor in tilapia, *Oreochromis mossambicus*, and changes in the mRNA level during embryonic development. *Gen. Comp. Endocrinol.* 108: 132–140, 1997.
165. Takeo, J., J. Hata, C. Segawa, H. Toyohara and S. Yamashita. Fish glucocorticoid receptor with splicing variants in the DNA binding domain. *FEBS Lett.* 389: 244–248, 1996.
166. Teitsma, C.A., I. Anglade, C. Lethimonier, G. Le Drean, D. Saligaut, B. Ducouret and O. Kah. Glucocorticoid receptor immunoreactivity in neurons and pituitary cells implicated in reproductive functions in rainbow trout: a double immunohistochemical study. *Biol. Reprod.* 60: 642–650, 1999.
167. Teitsma, C.A., I. Anglade, G. Toutirais, J.A. Munoz-Cueto, D. Saligaut, B. Ducouret and O. Kah. Immunohistochemical localization of glucocorticoid receptors in the forebrain of the rainbow trout (*Oncorhynchus mykiss*). *J. Comp. Neurol.* 401: 395–410, 1998.
168. Thevenod, F. and J.M. Friedmann. Cadmium-mediated oxidative stress in kidney proximal tubule cells induces degradation of Na^+/K^+-ATPase through proteasomal and endo-/lysosomal proteolytic pathways. *FASEB J.* 13: 1751–1761, 1999.
169. Tokuda, Y., K. Touhata, M. Kinoshita, H. Toyohara, M. Sakaguchi, Y. Yokoyama, T. Ichikawa and S. Yamashita. Sequence and expression of a cDNA encoding Japanese flounder glucocorticoid receptor. *Fish. Sci. (Tokyo)* 3: 466–471, 1999.
170. Tokumoto, M., R. Horiguchi, M. Yamashita, Y. Nagahama and T. Tokumoto. Involvement of 26S proteasome in oocyte maturation of goldfish *Carassius auratus*. *Zool. Sci.* 14: 347–351, 1997.
171. Tokumoto, T., M. Yamashita and Y. Nagahama. Proteasome during oocyte maturation and egg activation in goldfish, *Carassius auratus*. *Zool. Sci.* 9: 1162 1992.
172. Tujague, M., D. Saligaut, C. Teitsma, O. Kah, Y. Valotaire and B. Ducouret. Rainbow trout glucocorticoid receptor overexpression in *Escherichia coli*: production of antibodies for Western blotting and immunohistochemistry. *Gen. Comp. Endocrinol.* 110: 201–211, 1998.
173. Uchida, K., T. Kaneko, M. Tagawa and T. Hirano. Localization of cortisol receptor in branchial chloride cells in chum salmon fry. *Gen. Comp. Endocrinol.* 109: 175–185, 1998.
174. Vijayan, M.M., C. Pereira, R.B. Forsyth, C.J. Kennedy and G.K. Iwama. Handling stress does not affect the expression of hepatic heat shock protein 70 and conjugation enzymes in rainbow trout treated with β-naphthoflavone. *Life Sci.* 61: 117–127, 1997.
175. Vijayan, M.M., C. Pereira, G. Kruzynski and G.K. Iwama. Sublethal concentrations of contaminant induce the expression of hepatic heat shock protein 70 in two salmonids. *Aquat. Toxicol.* 40: 101–108, 1998.
176. Vijayan, M.M., S. Raptis and R. Sathiyaa. Cortisol treatment affects glucocorticoid receptor and glucocorticoid-responsive genes in the liver of rainbow trout. *Gen. Comp. Endocrinol.* 132: 256–263, 2003.
177. Wallace, A.D. and J.A. Cidlowski. Proteasome-mediated glucocorticoid receptor degradation restricts transcriptional signaling by glucocorticoids. *J. Biol. Chem.* 276: 42714–42721, 2001.
178. Wang, Z., J. Frederick and M.J. Garabedian. Deciphering the phosphorylation "code" of the glucocorticoid receptor *in vivo*. *J. Biol. Chem.* 277: 26573–26580, 2002.
179. Weber, L.P., S.L. Diamond, S.M. Bandiera and D.M. Janz. Expression of HSP70 and CYP1A protein in ovary and liver of juvenile rainbow trout exposed to β-naphthoflavone. *Comp. Biochem. Physiol.* 131C: 387–394, 2002.
180. Weber, L.P. and D.M. Janz. Effect of β-naphthoflavone and dimethylbenz[*a*]anthracene on apoptosis and HSP70 expression in juvenile channel catfish (*Ictalurus punctatus*) ovary. *Aquat. Toxicol.* 54: 39–50, 2001.

181. Webster, J.C., C.M. Jewell, J.E. Bodwell, A. Munck, M. Sar and J.A. Cidlowski. Mouse glucocorticoid receptor phosphorylation status influences multiple functions of the receptor protein. *J. Biol. Chem.* 272: 9287–9293, 1997.
182. Wendelaar Bonga, S.E.. The stress response in fish. *Physiol. Rev.* 77: 591–625, 1997.
183. Whitesell, L. and P. Cook. Stable and specific binding of heat shock protein 90 by geldanamycin disrupts glucocorticoid receptor function in intact cells. *Mol. Endocrinol.* 10: 705–712, 1996.
184. Wormke, M., M. Stoner, B. Saville and S. Safe. Crosstalk between estrogen receptor α and the aryl hydrocarbon receptor in breast cancer cells involves unidirectional activation of proteasomes. *FEBS Lett.* 478: 109–112, 2000.
185. Yang, J. and D.B. DeFranco. Assessment of glucocorticoid receptor–heat shock protein 90 interactions *in vivo* during nucleocytoplasmic trafficking. *Mol. Endocrinol.* 10: 3–13, 1996.
186. Yoh, T., T. Nakashima, Y. Sumida, Y. Kakisaka, Y. Nakajima, H. Ishikawa, Y. Sakamoto, T. Okanoue and H. Mitsuyoshi. Effects of glycyrrhizin on glucocorticoid signaling pathway in hepatocytes. *Dig. Dis. Sci.* 47: 1775–1781, 2002.
187. Yoo, H.Y., M.S. Chang and H.M. Rho. The activation of the rat copper/zinc superoxide dismutase gene by hydrogen peroxide through the hydrogen peroxide-responsive element and by paraquat and heat shock through the same heat shock element. *J. Biol. Chem.* 274: 23887–23892, 1999.
188. Zafarullah, M., J. Wisniewski, N.W. Shworak, S. Schieman, S. Misra and L. Gedamu. Molecular cloning and characterization of a constitutively expressed heat-shock-cognate hsc71 gene from rainbow trout. *Eur. J. Biochem.* 204: 893–900, 1992.

b. Thyroid and Retinoids

Biochemistry and Molecular Biology of Fishes, vol. 6
T. P. Mommsen and T. W. Moon (Editors)

CHAPTER 14

Thyroid hormones

J.G. Eales* and S.B. Brown**

**Department of Zoology, University of Manitoba, Winnipeg, MB, Canada R3T 3B8, and **Environment Canada, P.O. Box 5050, Burlington, ON, Canada L7R 4A6*

I. Introduction

It is well established that the vertebrate thyroid system can be influenced by environmental chemicals identified as ecotoxicants (xenobiotics). Much of the literature explores the responses of the mammalian thyroid (see reviews in refs. 18,25,76). However, there is also a substantial fish literature showing that a variety of thyroid endpoints can be altered by acute or chronic exposure to a broad spectrum of potential or proven contaminants.

Although this literature has been reviewed extensively[16,36,47,54–56,65] it is still not understood precisely how the fish thyroid system responds to any given xenobiotic and what the biological consequences might be. In light of this we have focused on possible mechanisms of fish thyroid disruption. We have commenced with brief overviews of the properties of thyroid hormones (TH) and the thyroid 'cascade' and have then considered the problems inherent in evaluating fish thyroidal responses.

However, we have devoted most of the review to a step-by-step analysis of the thyroid 'cascade', focusing on laboratory studies of chemicals likely to have direct effects at a given step.

II. Thyroid hormone structures and properties

The TH are found in all vertebrates but are phylogenetically ancient molecules synthesized in protochordates and with reported actions in several invertebrate taxa[34]. L-thyroxine (T4), the parent TH, is made from two diiodinated tyrosines which form the unusual T4 amino acid, not coded by DNA and containing four iodine atoms equally distributed on two ether-linked aromatic rings (Fig. 1). T4 is highly hydrophobic (and lipophilic) and in water at pH 7.0 has a maximum solubility of only 2.3 μM[45]. As a result, most T4 in aqueous biological systems are not free in solution but reversibly (non-covalently) bound to water-soluble proteins. Many TH properties depend on the presence of specific iodine atoms and on the ionization state of the 4′-OH (phenol) group. Its p*K* for T4 is 6.73 and, at pH 7.0, 80% of the T4 molecules are ionized at this site. In contrast, the T4 derivative, 3,5,3′-triiodo-L-thyronine (T3), has a 4′-OH p*K* of 8.45 and at pH 7.0 only 10% of its molecules are ionized[45]. Thus T3 is less hydrophobic and more lipophilic than T4. The 4′-OH group is also involved in the reversible binding of TH to receptors and transport proteins and it engages in enzyme-regulated reactions to form conjugates with physical and chemical properties different from those of the parent TH. All these reactions are strongly influenced by the position of specific iodines on the inner and outer aromatic rings.

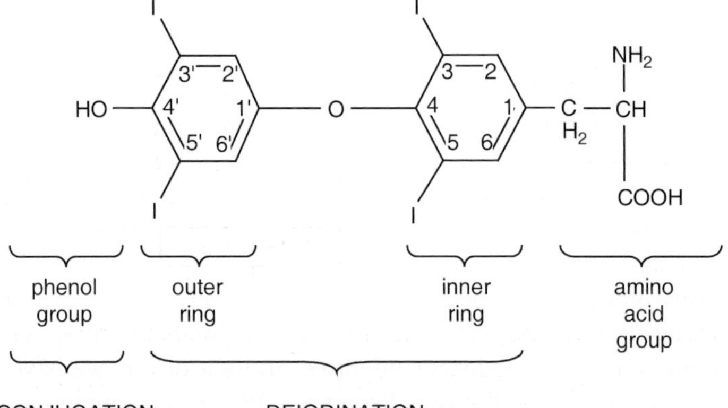

Fig. 1. Structure of thyroxine (T4) showing the outer and inner aromatic rings and the 4′-OH involved in sulfate or glucuronide conjugation. T3 is formed by loss of the 5′ iodine and rT3 by loss of the 5 iodine.

III. *Overview of the fish thyroid 'cascade'*

Figure 2 represents a working model of the fish thyroid 'cascade' based mainly on salmonid species. The thyroid is the endogenous source of TH but this may be supplemented with variable exogenous dietary sources[34]. Iodide (I^-), a scarce environmental element, is transported from the plasma into thyroid follicles which synthesize and store TH. The main hormone secreted by the thyroid is T4. It circulates in plasma bound mainly to proteins with $< 1\%$ in the free form and available for tissue exchange. The brain–pituitary axis regulates T4 secretion and through negative feedback maintains plasma T4 at a set-point appropriate for a given physiologic state. This represents *central regulation*. However, T4 is enzymatically monodeiodinated to T3 by liver (Fig. 3), resulting in plasma T3 levels often exceeding those of T4. Since T3 is more potent than T4 this is considered as an activation step. Extrahepatic tissues may also deiodinate T4 to T3 but mainly for their own local use. Thus a target tissue may use either systemic T3 or T3 made in that tissue. Some tissues may also degrade T3 by deiodination to presumed inactive 3,3′-diiodothyronine (T2) and so adjust their T3 levels. Thus deiodination pathways are pivotal in the control of thyroidal status. However, other secondary TH metabolic conversions occur. TH and their metabolites may be conjugated enzymatically forming less active and excretion-prone

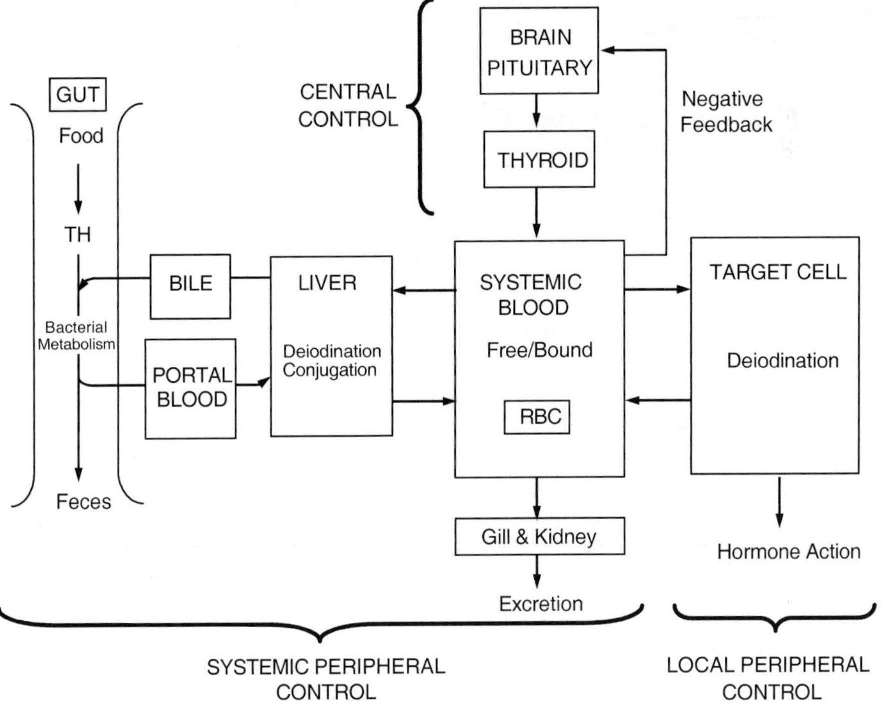

Fig. 2. The thyroid cascade showing the sites for *central* control and for *systemic* and *local peripheral* control. Also shown is the possibility of an exogenous TH supply from food.

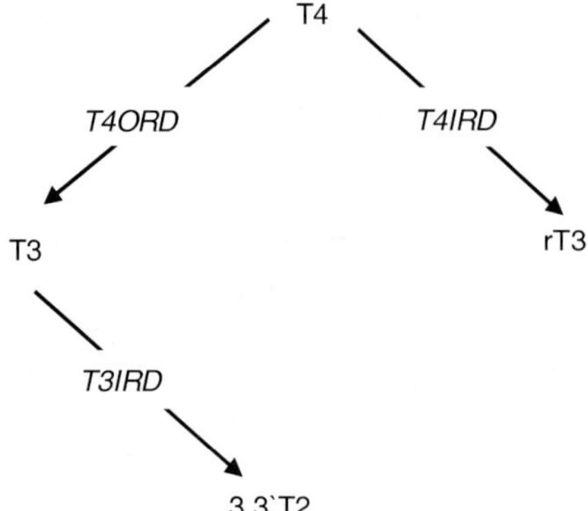

Fig. 3. The main pathways for TH monodeiodination. ORD, outer-ring deiodination; IRD, inner-ring deiodination.

glucuronide or sulfate metabolites (Fig. 3). In summary, several extrathyroidal pathways exist to activate or inactivate TH. Collectively these comprise the *peripheral regulation* (both *systemic* and *local*) of bioactive T3 supply to target tissue receptors. From the few teleost species studied to date, peripheral regulation is especially prominent in determining thyroidal status. Indeed, peripheral regulation may have preceded central regulation in the evolving thyroid system[32,34,35].

IV. Problems in interpreting the effects of ecotoxicants on the fish thyroid

Diverse chemicals have been reported to affect measured endpoints of fish thyroidal status. These chemicals include aromatic hydrocarbons, planar halogenated aromatic hydrocarbons (dioxans, furans, coplanar PCBs), organochlorine, organophosphorus and carbamate pesticides, chlorinated paraffins, cyanide compounds, methyl bromide, phenol, ammonia, metals (aluminum, arsenic, cadmium, lead and mercury), low pH conditions, environmental steroids and a variety of pharmaceutical agents. For the following reasons their modes of action appear complex and are poorly understood[16].

There are many instances where the results of one study have not been confirmed by the results of another study using the same chemical. Such disturbing discrepancies may be explained in part by differences in protocol for chemical administration (food, ambient water or injection) or dose or duration of the treatment. Furthermore, over 30 different model species from marine, freshwater, tropical, temperate and cold-water environments have been used. Also, fish have rarely been at a comparable physiologic state or stage of development or maturation and have been held under diverse environmental conditions. If indeed, as seems to be the case, variations in thyroidal responses to ecotoxicants can be explained in part by variations in species, environmental conditions and physiologic state, then selection of a standardized species and protocol to evaluate the effects of ecotoxicants on thyroidal status will continue to be problematic.

The interpretation of thyroidal responses to ecotoxicants is also confounded by several inherent properties of the thyroid system. As shown in Fig. 2, there are numerous steps, many of them enzyme- or ligand-controlled, between the initial uptake of environmentally scarce I^- from the environment to the eventual excretion/degradation of the TH or their interaction with target tissue receptors. Virtually each of these steps has the potential to be developed as a thyroid index, resulting in many endpoints. Indeed there is probably no other endocrine system where the choice of method of measurement is so great[36]. Thus some discrepancies may be explained by use of endpoints responding differently to an ecotoxicant. A suite of endpoints has been recommended to minimize this type of problem[36]. These include endpoints of central control (plasma total and free T4 levels and thyroid histological appearance), peripheral control (plasma and muscle T3 levels and deiodination activity) and post-receptor biologic actions (as yet not well established).

Thyroid autoregulation also creates problems. Possibly because of its long evolution under conditions of potentially low I^- availability or temporary dietary TH excess, the thyroid 'cascade' has developed considerable capacity for compensation (thyroid homeostasis) under otherwise compromising conditions. Compensation can occur at several levels. First, the thyroid is unique amongst endocrine tissues in storing relatively massive quantities of TH and I-containing precursors in the extracellular follicle lumen. This large store protects the system against short-term I^- deficiency or interference with TH synthesis. Secondly, should plasma T4 levels tend to deviate from the norm, the altered plasma-free T4 level feeds back on the brain–pituitary axis and modifies thyroid stimulating hormone (TSH) secretion, signaling the thyroid to reinstate the free T4 level. Thus the thyroid may appear hyper- (or hypo-) active based on its histological appearance but may merely be maintaining the plasma T4 euthyroid state. Thirdly, T3 homeostasis can be achieved independently of T4 homeostasis by adjusting the activities of the various peripheral deiodination pathways regulating T3 production or degradation. In summary, thyroidal status may still be normal regardless of major changes in certain thyroid endpoints.

However, despite its considerable compensatory abilities the thyroid system may eventually become exhausted due to agents that block TH synthesis/storage. There have been few studies conducted over a sufficient period to determine the point when this occurs. Consequently most studies on the fish thyroid system have likely provided *biomarkers of exposure* to an ecotoxicant rather than *measures of its effect* on thyroid-mediated events. The latter are more difficult to determine, not only on account of the requirement for a long-term study but also because for fish there are, as yet, few reliable physiological actions of TH that can be used as TH-specific endpoints.

The final point is perhaps the most critical as well as the most overlooked. Even under normal circumstances the fish thyroid cascade responds to a wide array of environmental variables and changes in whole-animal physiologic state. In fact, much fish thyroid literature (unrelated to ecotoxicants) documents thyroidal responses to variables such as temperature, stressors, nutritional availability and sexual state. The thyroidal system is also pivotal in the total endocrine network. Consequently, a contaminant inducing a stress response, reducing appetite or even changing some other endocrine system could *secondarily*, or *indirectly*, modify thyroidal status. Therefore carefully designed experiments, often involving *in vitro* methods, are needed to determine whether an ecotoxicant has actually acted directly at some step in the thyroid cascade or if change occurred indirectly due to some other metabolic change. Most studies have not addressed this fundamental distinction.

Because of the problems in interpreting the numerous exposure studies, we have not attempted an exhaustive review of the entire literature on the effects of ecocontaminants on fish thyroid systems. Previous reviews have comprehensively documented individual case histories of ecotoxicant effects on the fish thyroid[16,36,54,55]. They have been valuable in summarizing previously scattered information and in revealing the above complications in interpreting ecotoxicant effects on the fish thyroid system. We believe that the most fruitful next approach is *to consider each major step in the fish thyroid cascade from I^- acquisition to TH action and to evaluate the effects of exogenous chemicals on that step*.This has the advantage of determining if any chemically diverse ecotoxicants have common effects at a given step. It also draws attention to steps in the fish cascade that may be particularly susceptible to ecotoxicants or to steps in the cascade that have potential susceptibility but have hitherto escaped critical study.

V. Step-by-step analysis of the thyroid 'cascade' and the effects of potential ecotoxicants

1. Iodide acquisition

Iodine is essential for TH synthesis. It occurs in natural environments and body fluids mainly as I^- rather than elemental iodine. It is a scarce element. This is particularly true in freshwater but even in freshwater naturally occurring iodine-deficiency goiters are rarely encountered. Fish acquire I^- from food (gastrointestinal tract) or ambient water (gills). At both sites I^- uptake depends on active Na^+ transport. Plasma I^- levels range from 0.5 to 3000 μg/100 ml, depending on species and I^- availability. Some of the highest plasma I^- levels occur in the order Clupeiformes, that includes the salmonids, due to their possession of a prealbumin that reversibly binds I^- and transports it in plasma. I^- may be recycled following its salvaging by TH degradative deiodination. In order to participate in TH synthesis the I^- must be taken up by the epithelial cells forming the thyroid follicle. Assuming a system comparable to that of mammals, free I^- is symported with Na^+ across the basal membrane of the thyroid epithelial cells. Once inside the follicle cell, I^- follows its electrochemical gradient to the apical membrane and to the luminal colloid.

Clinical studies show the following ions competitively inhibit the mammalian thyroid I^- transport: ClO_4^- (perchlorate) > ReO_4^- (perrhenate) > BF_4^- (fluoroborate) > $SeCN^-$ (selenocyanate) > SO_3F^- (fluorosulfite) > SCN^- (thiocyanate) > I^- > NO_3^- (nitrate) > OCN^- (cyanate) > Br^- (bromide)[2]. They are all monovalent anions with a partial molal ionic volume of 25–50 ml/mole. Most of them are spherical or tetrahedral and as well as competing with I^- are transported into the thyroid. However, the pseudohalides, $SeCN^-$ and SCN^- are more linear in shape and are not transported into the thyroid. Of the true halides, only Br^- and the heaviest halide, astidide (At^-), which occurs only as a radionuclide, are transported by the I^- carrier. Extrathyroidal I^- transporters are affected by these anions, since in mammals the same gene is responsible for the transporter at both thyroidal and extrathyroidal sites. Thus, if present in the environment, for fish some of these ions could compromise both I^- acquisition from the environment and thyroidal I^- uptake from plasma. Even low levels of these competing monovalent anions may be effective due to their low K_i and the low ambient I^- in some natural waters.

However, few of these anions have been tested on fish. Injected SCN^- depressed thyroid $^{125}I^-$ uptake in rainbow trout, *Oncorhynchus mykiss*, but did not depress plasma TH levels[38]. White suckers, *Catastomus commersoni*, captured in SCN^--contaminated water did not show any histological evidence of goiter and did not have detectable levels of SCN^- in their plasma[12]. This may be due to the inability of the competing SCN^- to be transported by the I^- pump. Thiocyanate may also be derived indirectly by kidney rhodanese conversion from cyanide (CN^-)[7] or from ingested cyanogenic glucosides common in a variety of plants. Consistent with this, depressed plasma T4 and T3 levels, thyroid hyperplasia and colloid depletion occurred in CN^--treated rainbow trout or fathead minnows, *Pimephales promelas*[50–52,66].

Chemicals other than monovalent anions may also influence fish thyroid I^- transport. Endrin (an organochlorine pesticide)[69], malathion[62], lead[46] and mercury[48] all depressed thyroid I^- uptake in various Indian teleosts. However, it is not clear if these represent *direct* actions on thyroid I^- uptake since it can also be indirectly affected by I^- use in TH biosynthesis. Therefore an *in vitro* assay is needed to determine if a particular test substance is directly affecting I^- transport. The most useful assay is likely to be one that measures solely I^- transport. The extrathyroidal acquisition of ambient I^- across the fish gill has the potential to be developed into such an assay. Since both extrathyroidal and thyroidal I^- acquisition are crucial for normal thyroid function and may be influenced by a wide range of materials, development of a reliable I^- transport assay should be a high priority.

2. TH biosynthesis and storage

The thyroid epithelial cells concentrate I^- and also synthesize two key proteins: a large glycoprotein, thyroglobulin (TG, MW of 660 kDa) and an iodoperoxidase (IP). The TG is packaged by the Golgi into vesicles whose membranes contain the IP. The vesicles move towards and fuse with the apical membrane and discharge TG to create the colloid, while IP becomes part of the apical membrane with its active site facing the colloid. IP activates I^- so that it replaces 3 and 5 hydrogens on tyrosine residues of TG to form diiodotyrosyl (DIT) residues. IP also couples strategic DIT pairs to form tetraiodinated T4 still joined by peptide bonds to other TG amino acids. T4 is thus stored in an immense TG molecule trapped in the lumen.

Functional IP is essential for TH synthesis but is inhibited by numerous chemicals, some of which are prescribed for treatment of hyperthyroidism. They include thionamides (thiourea, thiouracil, propylthiouracil, methimazole, carbimazole, ethionamide, 6-mercaptopurine and metahexamide), aniline derivatives (sulfonamides, carbutamide, *para*-aminobenzoic acid, *para*-aminosalicylic acid, amphenone B and tobutamide), substituted phenols (resorcinol, phloroglucinol and 2,4-dihydroxybenzoic acid) and other diverse chemicals (aminotriazole, tricyanoaminoprene, antipyrene (phenazone) and phenylbutazone)[39,53,71,72]. Many of these chemicals are unlikely to be present in the environment at concentrations likely to affect IP. However, they are listed to show the diversity of chemical compounds to which IP is sensitive and to highlight IP as a potential lesion site for xenobiotic action.

Several thionamides induce a histological goitrous state in fish[33,61], suggesting a similar susceptibility of fish IP. Under certain conditions the potential environmental xenobiotics, malathion, carbaryl, ammonia and $HgCl_2$ also depress fish IP[6,19,70,75]. However, there are several reports of enhanced IP activity after treatment of fish with 3-methylcolanthrene, endosulfan, malathion, carbofuran or phenol[19,68,70,75]. Unless these compounds duplicate natural ligands that enhance IP activity, the most reasonable explanation is that they compromised thyroid function in some other way and secondarily caused a compensatory increase in IP activity. Bhattacharya and co-workers[5] reported that Endrin inhibited TH synthesis in a head-kidney preparation of the teleost, *Anabas testudineus*, while Yadav and Singh[75] found that other pesticides inhibited IP activity *in vitro*, suggesting direct effects.

The thyroid is unusual amongst endocrine tissues in containing large hormone reserves in the colloid. The ability of the fish thyroid to endure exposure to an ecotoxicant that compromises TH synthesis may thus depend on the extent to which thyroidal stores of iodine/TH are depleted. In theory this can be determined by measuring the amounts of TH stored in the thyroid. The size of the reserve is not known for fish but in humans it may last for up to 3 months. Endrin and malathion decrease thyroid TH levels[5,75] but levels were increased by 3-methylcolanthrene[68]. TH stores may also relate to the quality of the luminal colloid. Changes in colloid histological appearance have been noted by several workers after exposure to cadmium, lead and other chemicals (reviewed in ref. 16), but may reflect indirect actions.

3. TH secretion

Release of T4 from the thyroid requires endocytosis of colloid (TG) across the apical membrane and subsequent proteolysis of the TG within the ingestion vacuole by lysosomal enzymes. The liberated T4 then passes across the basal membrane to the circulation. The thyroid may also secrete some T3. This may be formed either by linkage of a monoiodinated tyrosyl residue with a DIT residue within the TG or by intrathyroidal enzymatic 5′-monodeiodination of liberated T4. The extent of T3 secretion from the fish thyroid is still somewhat controversial and likely species-dependent. Lithium (Li^+) inhibits thyroid secretion in humans[39]. No inhibitor with a direct action on thyroidal secretion has been identified for fish.

The thyroid secretion rate (pmoles of T4 secreted/h/unit body mass) is potentially useful in detecting thyroid disruption but is difficult to measure directly in most fish due to their diffuse thyroid. However, assuming no changes in TH balance over the period of measurement, the secretion rate is equal to the TH degradation rate (DR) described below.

4. Control of thyroid secretion

Constitutive secretion of TH from the thyroid is augmented by secretion regulated by thyroid stimulating hormone (TSH = thyrotropin) produced by pituitary thyrotrope cells. The fish thyrotropes receive direct innervation from hypothalamic nuclei that either stimulate or inhibit TSH secretion. In fish the net influence of the hypothalamus is to inhibit thyrotrope activity. Thus the pituitary–thyroid axis is largely free-running unless inhibited by the hypothalamus[37]. As in higher vertebrates, the pituitary and hypothalamus are influenced by plasma TH levels to create a negative feedback loop. This enables plasma T4 levels to be held at a set-point appropriate for a given physiologic state. Plasma T4 can change rapidly during the day and in response to stress and feeding presumably due to change in the set-point. At least in some fish, T4 is the primary TH exerting negative feedback with little influence by T3. Thus the brain–pituitary axis represents the central control of the thyroid system and seems particularly important in regulating plasma T4 levels.

There is no specific information on the effects of ecotoxicants on the fish brain–pituitary axis. Assays for fish TSH are not yet routine. The only current practical measure of thyrotrope activity is from their histological appearance but this is difficult to interpret since it may represent adjustments in negative feedback to maintain plasma free T4 levels.

5. TH in the circulation and tissues

T4 and T3 are the main circulating TH. Their levels vary between species and physiologic states but usually fall within the range <1 to >50 ng/ml[45]. Typically over 99% of TH is reversibly (non-covalently) bound to plasma proteins. A transthyretin (TTR)-like prealbumin has been described for some fish. In the small number of fish species studied to date T3 tends to be bound to a greater extent than T4. TH also exchange across the red blood cell (RBC) membrane and T3 flux is especially rapid suggesting that RBC may supplement plasma sites in providing a dynamic circulating store of T3. Changes in the properties of TH-binding plasma proteins can influence total plasma TH levels without necessarily altering thyroidal status. This is because the plasma-free TH concentration determines tissue TH uptake and feedback action on the brain–pituitary axis. Total plasma T4 and T3 levels also represent the net result of their addition and removal from the circulating pool. Thus total plasma TH levels, the most commonly used indices of thyroidal status, cannot be interpreted fully without knowledge of the properties of plasma TH-binding proteins and the TH disappearance rate (DR). Because of TH homeostasis, plasma TH levels may be maintained despite effects elsewhere in the thyroid cascade. Therefore while numerous ecotoxicants exert either acute or chronic effects on fish total plasma TH levels it is often difficult to draw firm conclusions of their effects on thyroidal status.

Total plasma T4 and T3 levels may independently increase, decrease or remain unchanged in response to a broad spectrum of ecotoxicants, precluding any generalizations[16].

Limited attempts have been made by toxicologists to supplement measures of total plasma TH levels with measures of plasma TH-binding properties[9,10]. Certain PCB derivatives can displace TH from mammalian TH-binding protein sites. However, Besselink *et al.*[4] reported that TCDD had no influence on free T4 levels in flounder, *Platichthys flesus*, while Adams *et al.*[1] found that PCB 77, PCB 126 and several hydroxylated PCB derivatives present in high concentrations *in vitro* did not alter plasma-free TH levels in American plaice, *Hippoglossoides platessoides*. Several other chemicals can competitively inhibit TH binding to plasma proteins of mammals[11] and have the potential to do this in fish. In mammals, estrogens increase the plasma levels of T4-binding globulin. Injected estradiol (E2) increases the levels of TH-binding plasma proteins in rainbow trout[27,29,42].

Although lipophilic and able to diffuse across the cell membrane TH can enter cells more rapidly by facilitated transport systems that tend to be specific for T4 or T3. Based on studies on isolated hepatocytes and RBCs, TH uptake in fish can be blocked by inhibitors of protein binding (phloretin, bromosulphothalein, 5,5′-diphenylhydantoin and 8-anilino-1-naphthalene sulphonic acid) and by certain sulphydryl blockers (*p*-hydroxymercuribenzoate and *N*-ethylmaleimide)[59,63,64].

All tissues probably contain some T4 and T3 but muscle on account of its bulk (up to 60% of body weight) may contain up to 80% of total carcass T3. A combination of low ambient pH and Al^{3+} depleted these presumably mobile reserves by 80%[44]. The underlying mechanism was not explored. Orally administered PCB 126 depressed muscle thyroid hormone levels in juvenile rainbow trout[14].

6. *TH metabolism*

The main pathways for the metabolism of TH are deiodination and conjugation. The deiodination pathways demonstrated for fish (Fig. 3) involve either outer-ring deiodination (ORD) or inner-ring deiodination (IRD). They have been designated as T4ORD to convert T4 to active T3, T4IRD to convert T4 to inactive reverse T3 (rT3), T3IRD to convert T3 to inactive T2 and rT3ORD to degrade rT3 to T2. In mammals these conversions are accomplished by a suite of three deiodinases (Types I, II and III) typically found in the endoplasmic reticulum (microsomes). While there is considerable structural homology between the fish and mammalian enzymes (all appear to have selenium at the active site), they do differ in several physiological properties. In fish the activities of all these deiodination pathways may be independently adjusted in different tissues[36]. This provides the powerful potential of site-specific regulation of T3 availability not only by altering T3 formation from T4 but by altering the rate of degradation of T4 or T3 in relation to T4 or T3 systemic supplies. Deiodination also serves to salvage I^- from TH prior to their excretion.

Mammalian studies show that deiodinases in general are inhibited by iodoacetate and gold thioglucose. Heavy metals with a single positive charge, such as gold, inhibit deiodination by interaction with the negatively charged selenium atom which forms a stable Se–S complex and blocks regeneration of the active enzyme[53]. Type I deiodinase is also inhibited by thionamides (thiourylenes) including thiouracil and propylthiouracil (PTU). Iopanoic acid and its derivatives also depress deiodination[39].

In fish, iodoacetate and aurothioglucose (ATG) are effective inhibitors of all rainbow trout deiodination pathways *in vitro*[41] and iopodate is effective *in vivo*[26], but all the fish deiodinases tested so far are negligibly sensitive to PTU administered *in vitro*. E2 depresses T4ORD in rainbow trout[30,43] and E2-like substances (e.g. nonylphenol) tested at environmentally relevant levels lower hepatic deiodination activities in Atlantic salmon, *Salmo salar*, at parr-smolt transformation[15]. In contrast androgens may increase hepatic T4ORD activity[58]. Exposure of rainbow trout to low pH and Al^{3+} depressed *in vitro* T4ORD activity by increasing K_m and decreasing V_{max}[13]; Al^{3+} may have acted directly on the deiodinase. However, a 5-day exposure of brown trout, *S. trutta*, to low levels of Al^{3+} in acidic soft water (pH 5.0) increased hepatic T4ORD activity and liver TH content[73,74]. Studies in mammals have shown that deiodination is decreased by treatment with heavy metals[22,24] and there is some indirect evidence for inhibition of deiodination in some teleosts[21,23]. Adams *et al.*[1] injected American plaice with PCB 77 and observed an increase in hepatic T4ORD activity accompanied by a decrease or no change in plasma T3. The enhanced T4ORD activity was likely compensation for an independent PCB action to depress plasma T3. This was consistent with the inability of PCB 77, PCB 126 or selected hydroxylated derivatives to modify directly deiodination activities when added *in vitro*.

TH may be excreted in the urine, bile and gastrointestinal route or across the gills. The biliary route predominates. Prior to biliary loss a high proportion of the TH and their metabolites are enzymatically conjugated (mainly at the 4′-OH) to glucuronic acid to form TH-glucuronides (TH-G) or to sulfate to form TH-sulfates (TH-S). In trout, as in mammals, there appear to be separate microsomal glucuronyltransferase (GT) systems responsible for T3G and T4G formation[40]. These enzyme systems also conjugate non-thyroidal substances including toxins or foreign substances, which if present in sufficient amounts, can induce GT activity[49]. Thus increased conjugation of TH could stem from GT induction by an ecotoxicant. This could be the case in lake trout treated with PCB 126 where there was an increase in T4G but not T3G formation due to selective induction of the GT system that happened to be specific for T4 but not for T3[17]. TH-sulfate conjugation occurs mainly in the cytoplasm and one enzyme may conjugate both T4 and T3 in trout. There have been no studies of the effects of ecotoxicants on TH sulfation in fish. Since the liver is one of the first lines of defense against foreign substances and since conjugation is an important process in their inactivation and elimination from the body, these secondary metabolic steps are likely crucial in interpreting the effects of xenobiotics on the fish thyroid system.

7. TH degradation rate (DR)

The plasma clearance of an injected labeled TH can be used to calculate both the exchange of TH between plasma and major tissue compartments and the irreversible plasma clearance of TH which can be used to determine the TH DR[67]. TH plasma clearance and TH DR are powerful tools that have been used to study effects of potential ecotoxicants on the fish thyroid cascade. Cyr and Eales[28] showed that E2 consistently depressed T4 DR and depressed plasma T3 with no change in T3 DR. This was consistent with the marked E2 inhibition of hepatic T4ORD activity[30]. Brown *et al.*[13] found that low pH and Al^{3+} had no effect on the T4 DR but increased the loss of T3 from plasma and lowered the plasma T3 with no decrease in hepatic T4ORD. Thus the reduced plasma T3 was presumably due to an effect of low pH and Al^{3+} to promote T3 degradation or excretion. Brown *et al.*[17] chronically treated lake trout, *Salvelinus namaycush*, with PCB 126 by oral gavage and found an increased T4 DR with no change in the plasma T4 or T3 levels. However, this could be explained by an increase in hepatic T4 glucuronosyltransferase activity to form T4G for excretion in bile.

8. TH actions

Although TH can exert rapid non-genomic actions[31], TH are believed to achieve most of their effects by T3 binding to nuclear thyroid receptors (TRs) that then bind in pairs (dimers) to adjacent thyroid response elements of the DNA. The resulting complex can then act as a transcription factor and regulate transcription of an adjacent gene and alter specific protein synthesis. The steps between T3 binding to nuclear TR and the eventual cellular response have not been examined in any fish species but are assumed to correspond to those established for other vertebrates. Almost all the substances that inhibit T3 binding to TRs are competitive TH analogues that, depending on their affinity for TRs, also have thyromimetic action. Only amirodone[39] and the recently discovered NH-3[57] bind to TRs antagonistically. There is also the possibility that a substance could up- or down-regulate TR numbers. Iopanoic acid reduced trout hepatic TR abundance but this was likely an indirect effect[8].

Experiments involving either the administration of TH or the inhibition of their production or action have led to a bewilderingly diverse array of claimed TH effects generally relating to development, growth and aspects of reproduction of fish. This probably reflects the permissive mode of action of TH, which may be required at minimal levels for overt effects of other hormones. Thus it is challenging to identify TH-specific responses even at the cellular level. TH have been shown recently to have significant roles in the early stages of fish development and metamorphosis, including a window of a few weeks during parr-smolt transformation. However, there is yet no reliable fish-based bioassay for bioactive TH.

If indeed the fish system is similar to that of other vertebrates then the TH–TR complex will also have the potential to form heterodimers by combining with other ligand–receptor complexes and in the process modify cellular responses. Potential ligands with these properties are the retinoids which associated with their receptors form heterodimers with the TH–TR complex. While literature on the role of retinoids in fish is rapidly expanding (see Chapter 15), there are few studies where aspects of thyroid and retinoid functions have been studied simultaneously for fish[3,4,14,60] and we are unaware of any published study of the mechanism of their combined regulation of any physiologic response.

VI. Summary and conclusions

Observations under controlled laboratory conditions and some field studies on 30 + species of teleost fish show that a broad spectrum of chemical contaminants can modify, or have the potential to modify, thyroid endpoints. Numerous thyroid endpoints exist since the thyroid 'cascade' involves many steps from acquisition of environmentally scarce iodide to TH binding to target receptors and expression of physiologic responses. Our present scrutiny of foreign chemical action at each of the steps in the 'cascade' points to several potential lesion sites, but for the most part fails to establish a precise mechanism of action for the majority of substances believed to influence the thyroid system. One underlying problem is the failure to distinguish direct and indirect thyroid responses. However, interpretation is further complicated by the inherent capacity of the thyroid system to compensate at both central and peripheral levels for deficiencies or excesses in TH availability. Thus some changes in thyroid endpoints may merely represent an autoregulation, without alteration of thyroidal status or disruption of TH-dependent physiologic processes. While deficiencies in TH production may ultimately lead to diminished thyroidal status this will depend on the extent of thyroidal and extrathyroidal hormone stores which are poorly understood for any fish species. Consequently, most studies may not have been sufficiently prolonged to reach the deficiency state. Because TH likely have permissive roles in fish in modifying the actions of other hormones, little progress has been made in identifying physiologic endpoints characteristic for TH. However, changes in early development and metamorphosis show considerable promise in this regard. One group of ligands shown in higher vertebrates to have major interactions with TH in regulating development are the retinoids but we are unaware of any published fish study simultaneously analyzing the effects of TH and retinoids in this area. We suggest that more attention be given to at least some of the above points if significant progress is to be made in unraveling the actions of foreign chemicals on the fish thyroid system.

VII. References

1. Adams, B.A., D.G. Cyr and J.G. Eales. Thyroid hormone deiodination in tissues of American plaice, *Hippoglossoides platessoides*: characterization and short-term responses to polychlorinated biphenyls (PCBs) 77 and 126. *Comp. Biochem. Physiol.* 127C: 367–378, 2000.
2. Bastomsky, C.H. Thyroid iodide transport, Handbook of Physiology, Vol. III, edited by R.O. Greep and E.B. Astwood, Washington, DC, American Physiological Society, 1974, Section 7.
3. Besselink, H.T., S. van Beusekom, E. Roex, A.D. Vethaak, J.H. Koeman and A. Brouwer. Low hepatic 7-ethoxyresoufin-*o*-deethylase (EROD) activity and minor alterations in retinoid and thyroid hormone levels in flounder (*Platichthys flesus*) exposed to the polychlorinated biphenyl (PCB) mixture, Clophen A50. *Environ. Pollut.* 92: 267–274, 1996.
4. Besselink, H.T., E. van Santen, W. Vorstman, A.D. Vethaak, J.H. Koeman and A. Brouwer. High induction of cytochrome P4501A activity without changes in retinoid and thyroid hormone levels in flounder (*Platichthys flesus*) exposed to 2,3,7,8-tetrachlorodibenzo-*p*-dioxin. *Environ. Toxicol. Chem.* 16: 816–823, 1997.
5. Bhattacharya, S., D. Kumar and R.H. Das. Inhibition of thyroid hormone formation by Endrin in the head kidney of a teleost, *Anabas testudineus* (Bloch). *Indian J. Exp. Biol.* 16: 1310–1312, 1978.

6. Bhattacharya, T., S. Bhattacharya, A.K. Ray and S. Dey. Influence of industrial pollutants on thyroid function in *Channa punctatus* (Bloch). *Indian J. Exp. Biol.* 27: 65–68, 1989.
7. Bourdoux, P., F. Delange, M. Gerard, M. Mafuta, A. Hanson and A.M. Ermans. Evidence that cassava ingestion increases thiocyanate formation: a possible etiologic factor in endemic goiter. *J. Clin. Endocrinol. Metab.* 46: 613–621, 1978.
8. Bres, O., D.G. Cyr and J.G. Eales. Factors influencing the properties of saturable T_3-binding sites in hepatic nuclei of rainbow trout, *Salmo gairdneri*. *J. Exp. Zool.* 254: 63–71, 1990.
9. Brouwer, A. The role of biotransformation in PCB-induced alterations in vitamin A and thyroid hormone metabolism in laboratory and wildlife species. *Biochem. Soc. Trans.* 19: 731–737, 1991.
10. Brouwer, A., W.S. Blaner, A. Kukler and K.J. Van Den Berg. Study on the mechanism of interference of 3,4, 3′,4′-tetrachlorobiphenyl with the plasma retinol-binding proteins in rodents. *Chem.-Biol. Interact.* 68: 203–217, 1988.
11. Brouwer, A., D.C. Morse, M.C. Lans, A.G. Schuur, A.J. Murk, E. Klasson-Wehler, A. Bergman and T.J. Visser. Interactions of persistent environmental organohalogens with the thyroid hormone system: mechanisms and possible consequences for animal and human health. *Toxicol. Ind. Health* 14: 59–84, 1998.
12. Brown, D.G., R.P. Lanno, M.R. Van Den Heuvel and D.G. Dixon. HPLC determination of plasma thiocyanate concentrations in fish blood: application to laboratory pharmacokinetic and field-monitoring studies. *Ecotoxicol. Environ. Saf.* 30: 302–308, 1995.
13. Brown, S.B., D.L. MacLatchy, T. Hara and J.G. Eales. Effects of low ambient pH and aluminum on plasma kinetics of cortisol, T_3 and T_4 in rainbow trout (*Oncorhynchus mykiss*). *Can. J. Zool.* 68: 1537–1543, 1990.
14. Brown, S.B., A.T. Fisk, M. Brown, M. Villella, M. Cooley, R.E. Evans, D.A. Metner, A.L. Lockhart and D.C.G. Muir. Dietary accumulation and biochemical responses of juvenile rainbow trout (*Oncorhynchus mykiss*) to 3,3′,4,4′,5-pentachlorobiphenyl (PCB 126). *Aquat. Toxicol.* 59: 139–152, 2002.
15. Brown, S.B., W. Fairchild, K. Haya, J.G. Eales, D. MacLatchy, J. Sherry, D. Bennie, K. Burnison and R. Evans. The effects of endocrine disrupters on sea water adaptability, growth and survival of salmon smolts. Final Project Report (TSRI #173), 51pp., June 2002.
16. Brown, S.B., B.A. Adams, D.G. Cyr and J.G. Eales. The ecotoxicology of the teleost fish thyroid. *Environ. Toxicol. Chem.* 23: 1680–1701, 2004.
17. Brown, S.B., R.E. Evans, L. Vandenbyllardt, K.W. Finnson, V.P. Palace, A.S. Kane, A.Y. Yarechewski and D.C.G. Muir. Altered thyroid status in lake trout (*Salvelinus namaycush*) exposed to co-planar 3,3′,4,4′,5-pentachlorobiphenyl. *Aquat. Toxicol.* 67: 75–85, 2004.
18. Capen, C.C. Mechanistic data and risk assessment of selected toxic endpoints of the thyroid gland. *Toxicol. Pathol.* 25: 39–48, 1997.
19. Chatterjee, S. and S. Bhattacharya. Response of climbing perch, *Anabas testudineus* to industrial pollutants: head kidney peroxidase, iodide peroxidase, and blood thyroid profiles. *Water Air Soil Pollut.* 25: 161–174, 1985.
20. Chatterjee, S., N.B. Jash and S. Bhattacharya. *In vivo* and *in vitro* activation of peroxidase in the head kidney of *Channa punctatus* (Bloch) by sublethal doses of carbofuran. Comp. *Physiol. Ecol.* 7: 56–58, 1982.
21. Chaurasia, S.S., P. Gupta, A. Kar and P.K. Maiti. Lead-induced thyroid dysfunction in the fish *Clarias batrachus* with special reference to hepatic type I – 5′-monodeiodinase activity. *Bull. Environ. Contam. Toxicol.* 56: 649–654, 1996.
22. Chaurasia, S.S., P. Gupta, A. Kar and P.K. Maiti. Free-radical mediated membrane perturbation and inhibition of type-I iodothyronine 5′-monodeiodinase activity by lead and cadmium in rat-liver homogenate. *Biochem. Mol. Biol. Int.* 39: 765–770, 1996.
23. Chaurasia, S.S. and A. Kar. An oxidative mechanism for the inhibition of iodothyronine 5′-monodeiodinase activity by lead nitrate in the fish, *Heteropneustes fossilis*. *Water Air Soil Pollut.* 111: 417–423, 1999.
24. Chaurasia, S.S. and A. Kar. Influence of lead on type-I iodothyronine 5’-monodeiodinase activity in male-mouse. *Horm. Metab. Res.* 29: 532–533, 1997.
25. Colbourn, T.. Clues from wildlife to create an assay for thyroid hormone disruption. *Environ. Health Perspect.* 110(Suppl. 30): 363–367, 2002.
26. Cyr, D.G. and J.G. Eales. The effects of sodium ipodate (ORAgrafin) on thyroid function in rainbow trout, *Salmo gairdneri*. *Gen. Comp. Endocrinol.* 63: 86–92, 1986.
27. Cyr, D.G. and J.G. Eales. The effects of short-term 17β-estradiol treatment on the binding properties of T_3-binding proteins in plasma of immature rainbow trout, *Salmo gairdneri*. *J. Exp. Zool.* 252: 245–251, 1989.
28. Cyr, D.G. and J.G. Eales. The effects of short-term estradiol treatment on plasma thyroid hormone kinetics in rainbow trout, *Salmo gairdneri*. *J. Fish Biol.* 36: 391–400, 1990.
29. Cyr, D.G. and J.G. Eales. The effects of short-term 17β-estradiol treatment on plasma thyroxine-binding proteins in rainbow trout, *Salmo gairdneri*. *J. Exp. Zool.* 262: 441–446, 1992.

30. Cyr, D.G., D.L. Maclatchy and J.G. Eales. The influence of short-term 17β-estradiol on plasma T_3 levels and *in vivo* hepatic T_4 5′-monodeiodinase activity in rainbow trout. *Gen. Comp. Endocrinol.* 69: 481–489, 1988.
31. Davis, P.J. and F.B. Davis. Nongenomic actions of thyroid hormone. *Thyroid* 6: 497–504, 1996.
32. Dickhoff, W.W. and D.S. Darling. Evolution of thyroid function and its control in lower vertebrates. *Am. Zool.* 23: 697–707, 1983.
33. Eales, J.G. Thyroid functions in cyclostomes and fishes, In: *Hormones and Evolution*, Vol. 1, edited by E.J.W. Barrington, New York, Academic Press, pp. 341–436, 1979.
34. Eales, J.G. Iodine metabolism and thyroid-related functions in organisms lacking thyroid follicles: are thyroid hormones also vitamins? *Proc. Soc. Exp. Biol. Med.* 214: 302–317, 1997.
35. Eales, J.G. and S.B. Brown. Measurement and regulation of thyroidal status in teleost fish. *Rev. Fish Biol. Fish.* 3: 299–347, 1993.
36 Eales, J.G., S.B. Brown, D.G. Cyr, B.A. Adams and K.R. Finnson. Deiodination as an index of chemical disruption of thyroid hormone homeostasis and thyroidal status in fish. In: *Environmental Toxicology and Risk Assessment: Standardization of Biomarkers for Endocrine Disruption and Environmental Assessment: Eighth Volume*, ASTM STP 1364, edited by D.S. Henshel, M.C. Black and M.C. Harrass, West Conshohocken, PA, American Society for Testing and Materials, pp. 136–164, 1999.
37. Eales, J.G. and B.A. Himick. Regulation of hormone release from the thyroid of teleost fish. *Adv. Comp. Endocrinol.* 1: 33–43, 1991.
38. Eales, J.G. and S. Shostak. Influence of potassium thiocyanate on thyroid function of rainbow trout, *Salmo gairdneri. Gen. Comp. Endocrinol.* 51: 39–43, 1983.
39. Farwell, A.P. and L.E. Braverman. Thyroid and antithyroid drugs. In: *Goodman and Gillman's the Pharmacological Basis of Medical Practice, 10th edition*, edited by J.G. Hardman and L.E. Limbird, New York, McGraw-Hill, pp. 1563–1596, 2001.
40. Finnson, K.W. and J.G. Eales. Glucuronidation of thyroxine and 3,5,3′-triiodothyronine by hepatic microsomes in rainbow trout, *Oncorhynchus mykiss. Comp. Biochem. Physiol.* 120C: 415–420, 1997.
41. Finnson, K.W., J.M. McLeese and J.G. Eales. Deiodination and deconjugation of thyroid hormone conjugates and type I deiodination in liver of rainbow trout, *Oncorhynchus mykiss. Gen. Comp. Endocrinol.* 115: 387–397, 1999.
42. Flett, P.A. and J.F. Leatherland. Dose-related effects of 17β-estradiol (E_2) on liver weight, plasma E_2, protein, calcium, and measurement of the binding of thyroid hormones to vitellogenin in rainbow trout *Salmo gairdneri. J. Fish Biol.* 34: 515–527, 1989.
43. Flett, P.A. and J.F. Leatherland. *In vitro* hepatic monodeiodination of L-thyroxine and the temporal effect of 17β-estradiol on deiodination of rainbow trout (*Salmo gairdneri Richardson*). *Fish Physiol. Biochem.* 6: 129–138, 1989.
44. Fok, P., J.G. Eales and S.B. Brown. Determination of 3,5,3′-triiodo-L-thyronine (T_3) levels in tissues of rainbow trout (*Salmo gairdneri*) and the effect of low ambient pH and aluminum. *Fish Physiol. Biochem.* 8: 281–290, 1990.
45. Hulbert, A.J. Thyroid hormones and their effects: a new perspective. *Biol. Rev.* 75: 519–631, 2000.
46. Katti, S.R. and A.G. Sathyanesan. Lead nitrate-induced changes in the thyroid physiology of the catfish, *Clarias batrachus* (L). *Ecotoxicol. Environ. Saf.* 13: 1–6, 1987.
47. Kime, D.E.. *Endocrine Disruption in Fish*, Dordrecht, The Netherlands, Kluwer Academic Publishers, 1998.
48. Kirubagaran, R. and K.P. Joy. Effects of short-term exposure to methylmercury chloride and its withdrawal on serum levels of thyroid hormones in the catfish, *Clarias batrachus* (L). *Bull. Environ. Contam. Toxicol.* 53: 166–170, 1994.
49. Kohn, M.C.. Effects of TCDD on thyroid hormone homeostasis in the rat. *Drug Chem. Toxicol.* 1: 259–277, 2000.
50. Lanno, R.P. and D.G. Dixon. Chronic toxicity of waterborne thiocyanate to the fathead minnow (*Pimephales promelas*): a partial life-cycle study. *Environ. Toxicol. Chem.* 13: 1423–1432, 1994.
51. Lanno, R.P. and D.G. Dixon. Chronic toxicity of waterborne thiocyanate to rainbow trout (*Oncorhynchus mykiss*). *Can. J. Fish. Aquat. Sci.* 53: 2137–2146, 1996.
52. Lanno, R.P. and D.G. Dixon. The comparative chronic toxicity of thiocyanate and cyanide to rainbow trout. *Aquat. Toxicol.* 36: 177–187, 1996.
53. Larsen, P.R., T.F. Davies and I.D. Hay. The thyroid gland. In: *Williams Textbook of Endocrinology, 9th edition*, edited by J.D. Wilson, D.W. Foster, H.M. Kronenberg and P.R. Larsen, Philadelphia, W.B. Saunders, pp. 389–515, 1998.
54. Leatherland, J.F. Field observations on reproductive and developmental dysfunction in introduced and native salmonids from the Great Lakes. *Great Lakes Res.* 19: 737–751, 1993.
55. Leatherland, J.F. *Reflections on the Thyroidology of Fish: from Molecules to Humankind, Guelph Ichthyological Reviews*, Vol. 2, Neptune City, NJ, T.F.H. Publications, 1994.

56. Leatherland, J.F. Contaminant altered thyroid function in wildlife. In: *Environmental Endocrine Disruptors: An Evolutionary Perspective*, edited by L.J. Guillette and D.A. Crain, New York, Taylor & Francis, pp. 155–181, 2000.
57. Lim, W., N.-H. Nguyen, H.Y. Yang, T.S. Scanlan and J.D. Furlow. A thyroid hormone antagonist that inhibits thyroid hormone action *in vivo*. *J. Biol. Chem.* 277: 35664–35670, 2002.
58. MacLatchy, D.L. and J.G. Eales. Short-term treatment with testosterone increases plasma T_3 and hepatic T_4 5′-monodeiodinase levels in Arctic char, *Salvelinus alpinus*. *Gen. Comp. Endocrinol.* 71: 10–16, 1988.
59. McLeese, J.M. and J.G. Eales. Characteristics of the uptake of 3,5,3′-triiodo-L-thyronine and L-thyroxine into red blood cells of rainbow trout, *Oncorhynchus mykiss*. *Gen. Comp. Endocrinol.* 103: 200–208, 1996.
60. Palace, V.P., S.M. Allen-Gill, S.B. Brown, R.E. Evans, D.A. Metner, D.H. Landers, L.R. Curtis, J.F. Klaverkamp, C.L. Baron and W.L. Lockhart. Vitamin and thyroid status in Arctic grayling (*Thymallus arcticus*) exposed to doses of 3,3′,4,4′-tetrachlorobiphenyl that induce the phase I enzyme system. *Chemosphere* 45: 185–193, 2001.
61. Pickford, G.E. and J.W. Atz. *Physiology of the Pituitary Gland of Fishes*, New York, New York Zoological Society, 1957.
62. Ram, R.N., K.P. Joy and A.G. Sathyanesan. Cythion-induced histophysiological changes in thyroid and thyrotrophs of the teleost fish, *Clarias punctatus* (Bloch). *Ecotoxicol. Environ. Saf.* 17: 272–278, 1989.
63. Riley, W.W. and J.G. Eales. Characterization of L-thyroxine transport into hepatocytes isolated from juvenile rainbow trout (*Oncorhynchus mykiss*). *Gen. Comp. Endocrinol.* 90: 31–42, 1993.
64. Riley, W.W. and J.G. Eales. Characterization of 3,5,3′-triiodo-L-thyronine transport into hepatocytes isolated from juvenile rainbow trout (*Oncorhynchus mykiss*), and comparison with L-thyroxine transport. *Gen. Comp. Endocrinol.* 90: 31–42, 1994.
65. Rolland, R.M. A review of chemically-induced alterations in thyroid and vitamin A status from field studies of wildlife and fish. *J. Wildl. Dis.* 36: 615–635, 2000.
66. Ruby, S.M., D.R. Idler and Y.P. So. Plasma vitellogenin, 17β-estradiol, T_3 and T_4 levels in sexually maturing rainbow trout *Oncorhynchus mykiss* following sublethal HCN exposure. *Aquat. Toxicol.* 26: 91–102, 1993.
67. Sefkow, A.J., J.J. Distefano, III, B.A. Himick, S.B. Brown and J.G. Eales. Kinetic analysis of thyroid hormone secretion and interconversion in the 5-day fasted rainbow trout, *Oncorhynchus mykiss*. *Gen. Comp. Endocrinol.* 101: 123–138, 1996.
68. Singh, T.P. Interaction of xenobiotics with reproductive endocrine functions in a protogynous teleost, *Monopterus albus*. *Mar. Environ. Res.* 28: 285–289, 1989.
69. Singh, H. and T.P. Singh. Thyroid activity and TSH potency of the pituitary gland and blood serum in response to cythion and hexadrin treatment in the freshwater catfish, *Heteropneustes fossilis* (Bloch). *Environ. Rev.* 22: 184–189, 1980.
70. Sinha, N., B. Lal and T.P. Singh. Effect of endosulfan on thyroid physiology in the freshwater catfish, *Clarias batrachus*. *Toxicology* 67: 187–197, 1991.
71. Taurog, A. Hormone synthesis, In: *Endocrinology*, Vol. 1, edited by L.J. Degroot, New York, Grune and Stratton, pp. 331–342, 1979.
72. Taurog, A. Hormone synthesis: thyroid iodine metabolism. In: *Werner and Ingbar's the Thyroid, 8th edition*, edited by L.E. Braverman and R.D. Utiger, New York, Lippincott, Williams and Wilkins, pp. 61–84, 2000.
73. Waring, C.P. and J.A. Brown. Plasma and tissue thyroxine and triiodothyronine contents in sublethally stressed, aluminum-exposed brown trout (*Salmo trutta*). *Gen. Comp. Endocrinol.* 106: 120–126, 1997.
74. Waring, C.P., J.A. Brown, J.E. Collins and P. Prunet. Plasma prolactin, cortisol and thyroid responses of brown trout (*Salmo trutta*) exposed to lethal and sublethal aluminum in acidic soft waters. *Gen. Comp. Endocrinol.* 102: 337–385, 1996.
75. Yadav, A.K. and T.P. Singh. Effect of pesticides on thyroid peroxidase activity in *Heteropneustes fossilis* (Teleostei, Clariidae): an *in vitro* study. *Zool. Jb. Physiol.* 91: 333–336, 1987.
76. Zoeller, T.R. Thyroid hormone, brain development and the environment. *Environ. Health Perspect.* 110(Suppl. 3): 355–361, 2002.

Biochemistry and Molecular Biology of Fishes, vol. 6
T. P. Mommsen and T. W. Moon (Editors)

CHAPTER 15

The biology and toxicology of retinoids in fish

D. ALSOP*, GLEN J. VAN DER KRAAK*, S.B. BROWN** AND J.G. EALES†

**Department of Zoology, University of Guelph, Guelph, ON, Canada N1G 2W1, **Environment Canada, P.O. Box 5050, Burlington, ON, Canada L7R 4A6, and †Department of Zoology, University of Manitoba, Winnipeg, MB, Canada R3T 3B8*

I. Introduction

The retinoids originate from carotenoids obtained exclusively from the diet as vitamins and are best known as visual pigments involved in signal transduction in the retina[39,103]. More recently, retinoids such as retinoic acid have been shown to regulate gene expression, through binding to nuclear receptors, thereby controlling aspects of growth, reproduction and development[44].

Previous reviews on retinoid metabolism and action have focused almost exclusively on studies with mammals[16,30,31,44,58,60,62–65,104,109]. Our understanding of the retinoid system in fish has grown immensely over the past decade and while there are many similarities with what has been shown in mammals, some subtle differences exist. There is also growing evidence that the retinoid system of many vertebrates is sensitive to toxicant exposure. The effects of toxicants on the retinoid system in mammals have been previously reviewed[68,93,108]. Additionally, Rolland[87] has summarized cases of retinoid and thyroid disruptions in wildlife that include information about retinoids in feral fish from polluted areas.

Our first objective of this chapter is to summarize the available information about the retinoid system in fish and to highlight the components that set it apart from the mammalian system. Our second objective is to review the effects of toxicant exposure on the fish retinoid system, as documented in both the field and lab studies.

II. Structure and metabolism

Retinoids are lipophilic, polyisoprenoids with a cyclohexane ring (Fig. 1). The term retinoid refers to a broad grouping of compounds from animal sources including retinol, retinal and retinoic acid (RA), as well as synthetic forms developed for pharmaceutical purposes.

Retinoids are essential to fish and other vertebrates. Predators acquire them in the diet, either as retinol or retinyl esters; they may also be acquired through the conversion of dietary carotenoids produced by plants (Fig. 2). In fish, carotenoids may be metabolized to retinol (vitamin A_1) or didehydroretinol (vitamin A_2) (Fig. 1). For example, fish can convert canthaxanthin to β-carotene[47], which is a dimer of retinal that is readily converted to retinol (Figs. 1 and 2). Didehydroretinol (Fig. 1) is formed

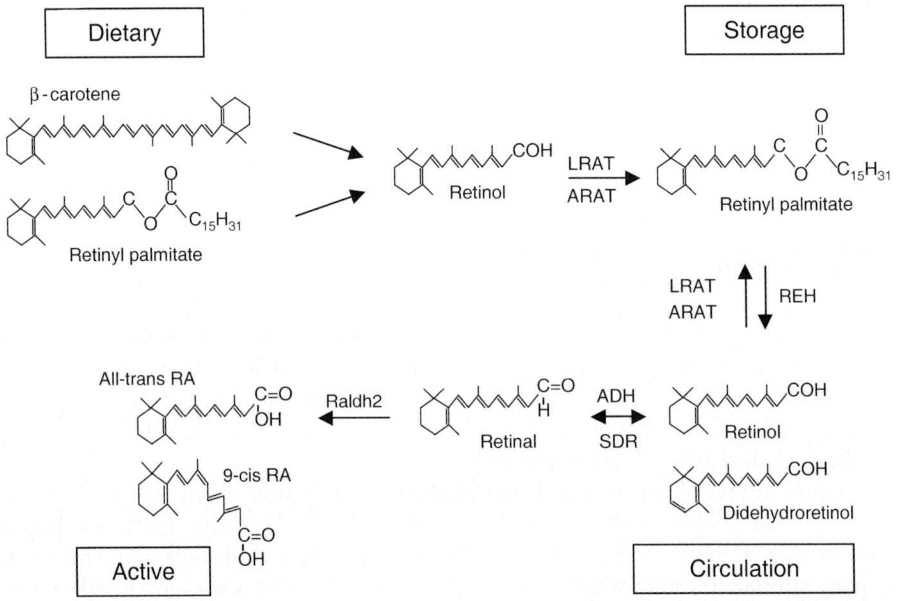

Fig. 1. The structures of key retinoids and their precursors. Fish convert retinyl esters (e.g. retinyl palmitate (RP)) and carotenoids (e.g. β-carotene) to retinol in the gut lumen prior to intestinal absorption. Retinyl esters (e.g. RP) stored in the liver are synthesized from retinol by lecithin:retinol acyltransferase (LRAT) and acyl CoA:retinol acyltransferase (ARAT). The retinyl esters are mobilized through their conversion to retinol by retinyl ester hydrolase (REH), which is then transported in the circulation to various sites in the body. Retinol is further metabolized within specific tissues to retinal by alcohol dehydrogenases (ADH) or short-chain dehydrogenase/reductase. Retinal is converted to the two major biologically active forms of retinoic acid (RA) (all-*trans* and 9-*cis* RA). Retinaldehyde dehydrogenase 2 (Raldh2) synthesizes all-*trans* RA from all-*trans* precursors and 9-*cis* RA form 9-*cis* precursors.

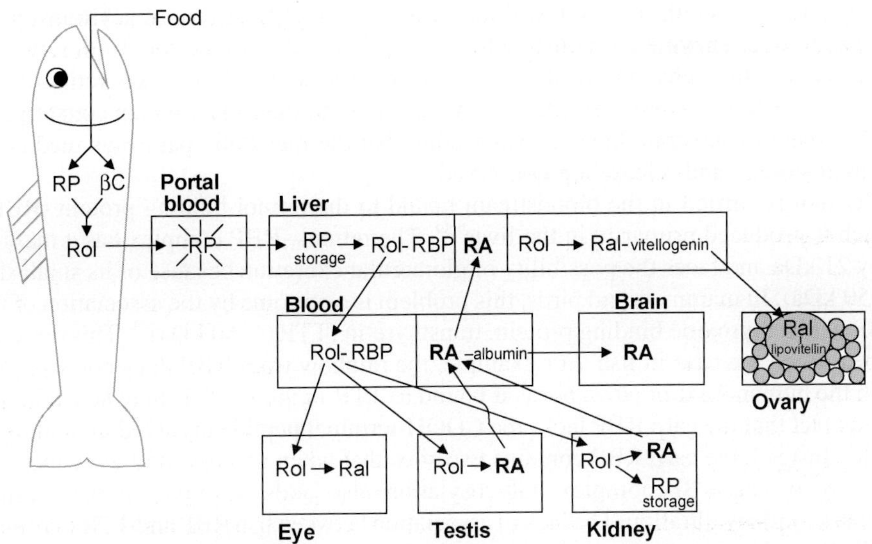

Fig. 2. Tissue distribution and metabolism of retinoids in fish. Dietary carotenoids (e.g. β-carotene (βC)) and retinyl esters (e.g. retinyl palmitate (RP)) are converted into retinol (Rol) in the lumen of the gut. Retinol is then re-esterified and packaged into chylomicrons and transported to the portal circulation. When required elsewhere, stored retinyl esters (e.g. RP) in the liver are hydrolyzed to retinol and transported in the blood bound to the retinol-binding protein (RBP). Retinol is converted in target tissues to RA, RP or retinal (Ral). RA may exert its effects locally, or be returned to the circulation and transported throughout the body bound to albumin. RA can then be sequestered in other tissues.

by the oxidative conversion of retinol[90,91] or by the direct conversion of carotenoids such as astaxanthin or zeaxanthin[53]. Rainbow trout (*Oncorhynchus mykiss*) require a minimum of about 0.75 μg retinol equivalents/g diet, with juvenile fish requiring more than adults[54]. Below this limit, fish show signs of deficiency, including increased mortality, decreased growth, decreased hepatosomatic index, fading of body color, anemia and hemorrhaging[54]. Similar symptoms develop in adult cherry salmon (*O. masou*) when fed with a retinoid-free diet for 15 weeks[98].

Little is known about the intestinal metabolism and uptake of retinoids in fish. This is surprising since these processes have been well described in mammals (see reviews [16,64,88]) and dietary carotenoids (used to enhance muscle color) can amount to 20% of the total cost of diets used in aquaculture. Few studies have considered the intestinal absorption of retinoids, although a study with juvenile sunshine bass (*Morone chrysops x M. saxatilis*) reported that retinol uptake appeared to be carrier mediated[28], as is the case with mammals[36].

Most dietary retinoids are stored in the liver as esters of long-chain fatty acids (e.g. palmitate, stearate, oleate). Fish also store retinyl esters in their kidneys[71]. Measurement of hepatic retinyl ester concentrations provides an indication of total body reserves. However, with both retinoid and didehydroretinoid forms mediating physiological functions[17,44], all varieties require consideration when assessing the retinoid status of fish. A study with brook trout (*Salvelinus fontinalis*) characterized the activities of lecithin:retinol acyltransferase and acyl CoA:retinol acyltransferase which are liver enzymes responsible for the synthesis of retinyl esters from retinol[67]. Additionally, they characterized retinyl ester hydrolase, which releases retinol from the retinyl esters[67]. These enzymes behave similarly to their mammalian counterparts with respect to substrate kinetics, suggesting that the metabolic pathways mediating retinoid storage and release are conserved.

Retinol is carried in the bloodstream bound to the retinol-binding protein (RBP), which is produced primarily in the liver[9,89]. The retinol–RBP complex has a mass of only 21 kDa and faces the possibility of glomerular excretion because of its small size (< 50 kDa). In mammals and birds, this problem is overcome by the association of the RBP with a thyroxine binding protein, transthyretin (TTR: ~60 kDa)[43]. This does not appear to be the case in fish. For example, the rainbow trout RBP does not associate with the human TTR *in vitro*, nor is it bound to TTR *in vivo*[10]. This may be explained by the fact that the carp RBP lacks the COOH-terminal peptide involved in binding to TTR[9]. Instead, the carp RBP contains two sites that when glycosylated, may increase the size of the RBP complex. Glycosylation also adds negative charges, which decreases kidney filtration. The lack of association between fish RBP and TTR may have toxicological consequences. Certain polychlorinated biphenyl (PCB) congeners have been shown to interfere with the association of RBP and TTR in mammals and it has been postulated that this may be the reason for the decreased levels of circulating thyroxine and retinol in PCB-exposed animals[20,21]. Obviously, this toxicological mechanism may not affect fish due to the normal lack of association of these proteins.

A RBP receptor has been characterized in mammals and is thought to mediate retinol uptake into specific tissues[7,94]. The possible involvement of a RBP receptor in retinol uptake into fish tissues is yet to be studied. Cellular pools of retinol can be used in multiple ways. In the eye, retinol can be converted to 11-*cis* retinal that is used in the signal transduction processes associated with vision[33,39,103]. Retinol and β-carotene have antioxidant activity and may be used in this respect in a variety of tissues[78,85,86]. During gonad maturation in female fish, hepatic retinol is converted to retinal and bound to vitellogenin (vg) in the liver. The retinal–vg complex is secreted into the circulation and sequestered in the developing oocytes[50]. Retinal and didehydroretinal are the predominant forms of retinoids in the eggs of marine and freshwater teleosts including herring (*Clupea harengus*), plaice (*Pleuronectes platessa*), cod (*Gadus callarias*), turbot (*Scophthalmus maximus*)[83,84], white sucker (*Catostomus commersoni*)[19], chum salmon (*Onchorhynchus keta*), black porgy (*Acanthopagrus schlegeli*), marbled flounder (*Pleuronectes yokohamae*) and stingfish (*Inimicus japonicus*)[50]. However, the eggs of lamprey and marine elasmobranchs appear to contain more retinyl esters and retinol than retinal[84]. The evolutionary significance or physiological consequence of this difference is unknown. Some fish eggs can also contain considerable amounts of carotenoids[84].

The cellular metabolism of retinoids involves the interaction of several enzymes together with a number of retinoid-binding proteins[44,64] (Fig. 3). While these binding proteins have been well described in mammals, little is known about their properties in fish, although orthologues of the mammalian proteins do exist[29,55]. Retinol can be metabolized to retinal and, in turn, be irreversibly converted to RA in tissues that

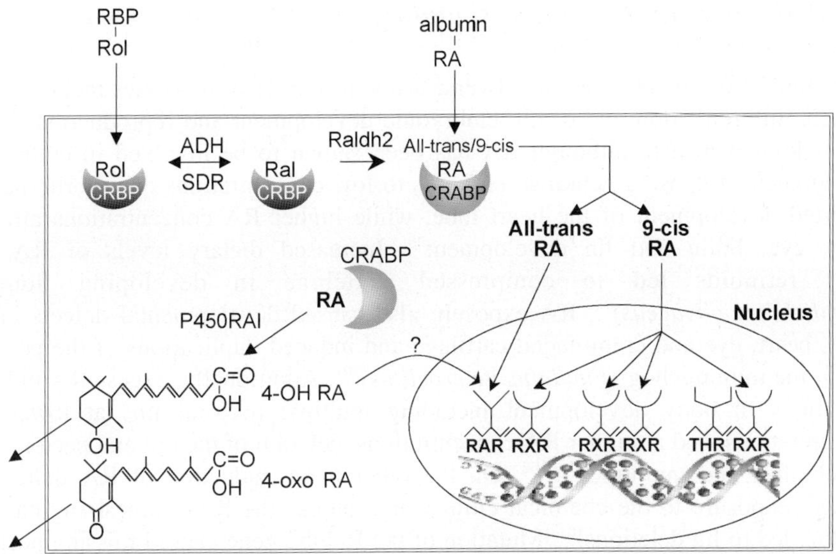

Fig. 3. Cellular retinoid pathways are mediated by a complex interaction between retinoids, retinoid binding proteins, enzymes and RA receptors. Retinol (Rol) is sequestered in target tissues and bound to cellular retinol-binding proteins (CRBP). Retinol is converted through the actions of alcohol dehydrogenase (ADH) or short-chain dehydrogenase/reductase (SDR) to retinal (Ral), which in turn is irreversibly converted to RA and bound by cellular RA binding protein (CRABP). RA interacts with two classes of nuclear receptors, the retinoic acid receptors (RARs) and retinoid X receptors (RXRs). All-*trans* RA binds to the RARs, while 9-*cis* RA binds to both the RARs and RXRs. The RXRs dimerize and form transcriptional complexes with RARs, RXRs, thyroid hormone receptors and vitamin D receptors to regulate gene transcription. RA is also further metabolized by P450RAI to 4-OH and 4-oxo RA to facilitate excretion, though 4-OH RA can bind to fish RARs with high affinity.

express the enzyme retinal dehydrogenase 2 (Raldh2) (Figs. 1 and 3). Raldh2 expression, and presumably RA itself, is first detected in the zebrafish (*Brachydanio rerio*) embryo at 30% epiboly[46]. During later stages of development, Raldh2 is expressed in the eye, branchial arch primordium, pronephric duct, posterior to the fin bud and in different regions of the brain[46]. RA may have actions in the tissues where it is formed, or be returned to the circulation and transported to other target tissues[56]. In mammals, RA is carried in the bloodstream bound to albumin[14]. Little is known of RA synthesis, transport and uptake in fish.

The all-*trans* and 9-*cis* isomers of RA (Fig. 1) have historically received most attention due to their biological activity. However, the pathways that lead to the formation of each isomer are not entirely understood. In mammals, it appears that Raldh2 is responsible for the synthesis of not only all-*trans* RA from all-*trans* precursors (e.g. all-*trans* retinal) but can also synthesize 9-*cis* RA from 9-*cis* precursors (e.g. 9-*cis* retinal)[65]. RA can be further metabolized to more polar compounds to facilitate their excretion. These include 4-hydroxy RA (4-OH RA) and 4-oxo RA which are formed by certain P450 enzymes, including the RA-specific P450RAI described in zebrafish (also known as CYP26)[105] (Fig. 3).

III. Retinoic acid action and signaling

In mammals, RA is involved in a diverse array of biological processes including cell growth, differentiation and death, embryonic development and reproduction. Much less is known in fish, although RA has been shown to be involved in embryonic development. Exposing zebrafish embryos to low concentrations of waterborne RA distorted development of the heart tube, while higher RA concentrations affected heart, eye, brain and fin development[96]. Increased dietary levels of RA and other retinoids led to compressed vertebrae in developing flounder (*Paralichthys olivaceus*)[97]. RA exposure also caused developmental defects in the brain, heart, eye and craniofacial cartilage and induced duplications of the pectoral fins in the mummichog (*Fundulus heteroclitus*)[101]. Additionally, excess RA induced alterations in body development including multiple pectoral fins at lower RA concentrations, and at higher RA concentrations, deletion of the tail and pectoral fins in zebrafish and mummichog[102]. On the other hand, induction of RA deficiency through exposure to the chemical citral which blocks the RA-synthesizing enzyme Raldh2, led to fin deletions[102]. Mutation of the Raldh2 gene caused missing pectoral fins as well as hindbrain and neural crest defects in zebrafish[8].

RA regulates biological processes through binding to RA receptors which in turn gain the ability to bind to RA response elements on the DNA and initiate gene transcription. The RA receptors belong to the superfamily of ligand-activated transcriptional regulators that also include the steroid hormone receptors, vitamin D receptors and thyroid hormone receptors[30]. Two classes of RA receptors have been described in mammals and fish. These include the retinoic acid receptors (RARs) and the retinoid X receptors (RXRs), and each class contains multiple subtypes. RA receptors appear to be ubiquitously distributed since at least one subtype has been found in all vertebrate tissues examined, including those from zebrafish and rainbow trout[1,2,44,52]. RARs form heterodimers with RXRs and bind to specific RA response elements within the DNA and modify gene transcription[59]. In contrast, RXRs can either form homodimers with other RXRs or form heterodimers with other receptors including the thyroid hormone receptor, the vitamin D receptor and the peroxisome proliferator-activated receptor and these in turn bind to unique response elements[59].

RARs and RXRs are also distinguished on the basis of their ability to bind different isomers of RA. In fish and mammals, the RARs bind both the all-*trans* and 9-*cis* isomers of RA (Figs. 1 and 3), while the RXRs only bind 9-*cis* RA[30,51,52]. The binding affinity of 9-*cis* RA to RARs and RXRs is much lower than that for all-*trans* RA binding to RARs in rainbow trout (K_d of all-*trans* RA binding to RARs = 1.0 nM, 9-*cis* RA binding to the RARs = 7.0 nM and RXRs = 51 nM)[2]. This suggests that compared to all-*trans* RA, higher concentrations of endogenous 9-*cis* RA would need to be present to activate the RXRs. This is perplexing in that researchers have been unable to detect the 9-*cis* isomer in zebrafish embryos and adults[32]. It may be that a high-affinity ligand for the RXRs is yet to be discovered.

Other forms of RA may also serve as RA receptor ligands. The 13-*cis* RA isomer was detected in zebrafish[32], although the biological significance of this isomer is not fully understood[15]. Studies have shown that 13-*cis* RA binds weakly to rainbow trout RARs and RXRs[2]. In mammals, the 4-OH and 4-oxo RA metabolites possess little biological activity, as they fail to induce differentiation of a RA-sensitive mammalian cell line[42]. However, it was later found that 4-oxo RA binds to African clawed frog (*Xenopus laevis*) RARs and was biologically active since it disrupted normal embryonic development[82]. In other studies, Idres *et al.*[49] found that 4-OH RA bound to mammalian RARs, albeit with 85-fold weaker affinity than all-*trans* RA. In sharp contrast, Alsop *et al.*[3] found that 4-OH RA bound to rainbow trout RARs with the same affinity as all-*trans* RA. The apparent contrast in these activities may be due to differences in the RARs between mammals and fish/frogs, and raises questions on the roles of different RA metabolites across species.

Compared with mammals, little is known about the physiology of RA receptors in fish, including their spatial and temporal expression. Joore *et al.*[52] identified the cDNAs encoding the zebrafish homologues of the mammalian RARα and RARγ. While no zebrafish orthologue of RARβ was found, both RARα and RARγ were highly expressed during gastrulation, neurulation and somitogenesis. In the adult, low levels of the RARs were found in the liver, intestine and stomach, while RARγ was strongly expressed in the ovary and RARα and RARγ were both expressed in the brain and muscle. Jones *et al.*[51] described zebrafish RXRα and RXRγ, as well as two additional subtypes RXRδ and RXRε which do not exist in mammals. Unlike RXRα and γ, RXRδ and ε did not bind 9-*cis* RA, or any other ligands that were tested, despite a high degree of homology with the RXRα and γ in the ligand-binding domain. However, like RXRα and γ, RXRδ and ε did form homodimers and heterodimers with other receptors suggesting they are active in the retinoid system[51]. Further research is needed to identify RA receptors, their ligands and their roles in the expression of biological responses in fish.

IV. Pollutant effects on the retinoid system

There is growing evidence that toxicants can disrupt the retinoid system in fish. The most commonly reported response is a reduction of hepatic retinoid stores, which is emerging as a sensitive marker of organic pollutant exposures. Research on the mechanisms of retinoid reduction is more limited but suggests that planar halogenated aromatic hydrocarbons (PHAHs) may accelerate retinoid metabolism. Recent research has also demonstrated that pulp mill effluents contain ligands that bind to fish RA receptors. However, little is known about the consequences of decreased retinoid stores or the presence of RA receptor ligands on physiological parameters. All these topics will be discussed below.

It is well established that exposure to planar organic chemicals (dioxins, coplanar PCBs, and furans) severely disturbs retinoid metabolism in mammals[22,68]. The PHAHs bind to the aryl hydrocarbon receptor (AhR) and induce CYP1A gene transcription[69], along with a variety of toxic responses including thymic involution, teratogenicity, changes in thyroid function and, in the context of this section, altered retinoid metabolism.

1. Reduction of retinoid stores: Field studies

The first studies of retinoid status in fish were published in 1992 and 1995 and showed that wild white sucker from the polluted Des Prairies River near Montrèal, QC had approximately 90% less retinol and retinyl palmitate than fish from an unpolluted site[19,95]. Similar results were obtained for lake sturgeon (*Acipenser fulvescens*) from Lake St Louis, QC[37], where there is a high level of PCB contamination[79].

Subsequently, there have been other reports of retinoid depletion in fish collected from organochlorine contaminated sites. These include silver carp (*Hypophthalmichys molitrix*), from Ya-Er Lake, China, where lower hepatic retinol and retinyl palmitate were associated with elevated hepatic CYP1A activity and glucuronidation activities and increased organochlorine pollution[106,107]. Nacci *et al.*[61] reported modest but consistent reductions in liver and egg retinoid concentrations in mummichogs collected near a PCB-contaminated site. Arcand-Hoy and Metcalfe[4] found markedly depressed concentrations of hepatic retinoids in brown bullheads (*Ameiurus nebulosus*) collected from contaminated sites in the Great Lakes relative to reference locations. As noted in other species, the lower hepatic retinoid concentrations were associated with elevated CYP1A activity, suggesting that these responses are mediated by compounds that act through the AhR.

Brown and Vandenbyllardt[24] reported lower levels of retinyl palmitate in longnose suckers (*Catostomus catostomus*) from the Athabasca River downstream of a pulp mill located at Hinton, AB. Other investigations showed greatly diminished retinoid stores in Winnipeg River white suckers exposed to effluents from the pulp mill at Pine Falls[25,41]. In Atlantic tomcod (*Microgadus tomcod*), hepatic retinoids were negatively correlated with the tissue concentrations of organic contaminants typical of exposure to pulp mill effluents[3,5,40]. Alsop *et al.*[3] confirmed that white suckers residing downstream of a pulp mill site on the Mattagami River in northeastern Ontario had dramatic reductions of stored hepatic retinoids (retinol, didehydroretinol, total retinyl esters, total didehydroretinyl esters). This effect was specific to the retinoids, as tocopherol levels (vitamin E, a fat-soluble antioxidant) were unaffected.

Most retinoid reductions have been reported in response to organic contaminants, but other pollutants may affect storage levels. Lake trout (*Salvelinus namaycush*) inhabiting an area contaminated with iron-ore mine tailings had hepatic retinol and retinyl palmitate levels reduced by approximately 95%[80]. These fish also exhibited oxidative damage, providing indirect evidence that the loss of retinoid stores may be related to increased oxidative stress.

2. Reduction of retinoid stores: Experimental studies

Several PHAHs, 2,3,7,8-tetrachlorodibenzo-*p*-dioxin (TCDD), co-planar PCBs (PCB 77 and PCB 126), 2,3,4,7,8-pentachlorodibenzofuran (PCDF), and 2,3,7,8-tetra-chlordibenzofuran (TCDF) affect the retinoid status of fish. Short-term exposures (<4 weeks) do not appear to deplete retinoid stores. Besselink *et al.*[12] exposed flounder (*Platichthys flesus*) to TCDD by oral gavage and found few changes in plasma and hepatic retinoids after 10 d, but there was a 27-fold increase of CYP1A activity.

Also examination of flounder 10 d after an intraperitoneal injection of the PCB mixture Clophen A50 revealed only minor changes in plasma, hepatic and kidney retinoid concentrations[11]. Exposure (<4 weeks) of lake sturgeon to TCDF did not affect hepatic retinoid stores[72]. In longer term experiments (>4 weeks), single dietary or injected doses of coplanar PCBs altered tissue retinoid status in lake trout[71,73], brook trout[18,66], rainbow trout[27] and Arctic grayling (*Thymallus arcticus*)[75]. Male rainbow trout exposed to a single intraperitoneal injection of PCDF for an entire reproductive cycle had depleted hepatic retinoid stores[26].

Delorme[35] collected adult lake trout and white sucker from the Experimental Lakes Area (ELA) in northwestern Ontario and after a single intraperitoneal injection of PCDF, the fish were released into the lake. Subsequent long-term sampling showed sustained induction of hepatic CYP1A activity in both the species. While hepatic vitamins were unaffected in white suckers, hepatic retinol and retinyl palmitate were much lower in PCDF-treated lake trout. Chronic exposure (3 years) of flounder (*P. flesus*) to contaminated harbor sludge and sludge wash-off containing PCBs, polyaromatic hydrocarbons and metals lowered hepatic retinol and retinyl palmitate[13]. In all of the long-term studies, hepatic retinol showed a negative relationship to CYP1A activity[13,23,35,71–73].

Earlier studies have shown that estrogens elevated circulating retinol levels 25 to 40-fold in the African clawed frog[6]. Fish may be exposed to estrogenic contaminants including ethynylestradiol (EE_2) and estrone from municipal sewage discharges [99,100]. Palace *et al.*[76] injected lake trout with 17β-estradiol and after 21 d showed that hepatic concentrations of retinol were lowered but that retinyl esters were unaffected. However, in juvenile lake sturgeon, waterborne EE_2 produced minor reductions in liver retinyl esters, and larger reductions in kidney retinol levels[77].

3. *Effects on RA receptors*

Few studies have tested whether toxicants can bind to the RARs or RXRs, and most of these have only considered binding to human or mouse RA receptors[48,70,92]. Because mammalian and fish RA receptors have different binding specificities for compounds such as 4-OH RA[3], the use of mammalian systems may not be a suitable surrogate for studying responses in fish.

A radioligand competitive binding assay was used to test whether pulp mill effluent extracts bind to rainbow trout RARs and RXRs[3]. Extracts of the final effluent from six of eleven pulp mills displaced greater than 25% of the [^{3}H]RA bound to the RARs and RXRs (Fig. 4). There is evidence to suggest that the receptor ligands present in the effluent may originate from the wood furnish itself. Schoff and Ankley[92] reported that the effluent from a separate pulp mill, contained ligands that acted as RAR antagonists in a mammalian receptor–reporter gene bioassay.

4. *Retinoid toxicology: Uncertainties and unanswered questions*

All toxicology studies to date measuring retinoid levels in fish have focused on the storage forms as opposed to RA, which is the active form binding to the nuclear

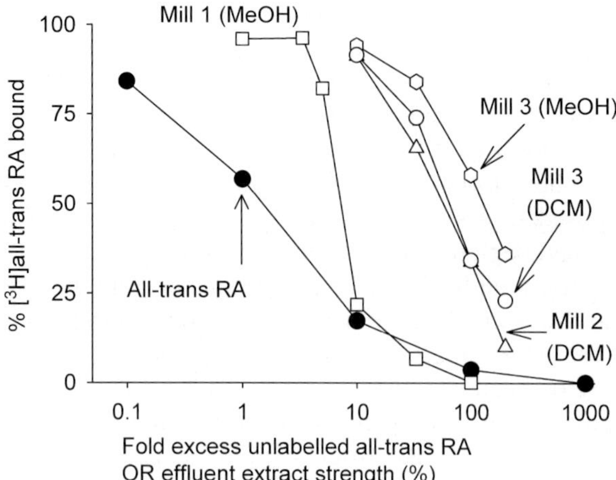

Fig. 4. Effects of effluent extracts from three pulp mills on retinoic acid (RA) binding to fish RA receptors (RARs) (adapted from Alsop *et al.*[3]). Displacement of [^{3}H]-all-*trans* RA bound to rainbow trout RARs by unlabelled all-*trans* RA (filled circles), MeOH extract from mill 1 (open squares), dichloromethane (DCM) extract from mill 2 (open triangles), MeOH extract from mill 3 (open hexagons) and DCM extract from mill 3 (open circles).

receptors. This probably reflects the difficulty in measuring the low concentrations of RA by conventional HPLC methods routinely used for other retinoids. Changes in the levels of storage forms may not be accompanied by overt changes in physiological performance of the individual. For example, zebrafish fed with a retinoid-free diet for 175 d exhibited no apparent changes in health, even though whole body retinyl esters and retinol were depleted by 97 and 88%, respectively (with most of the depletion occurring in the first 70 d; D. Alsop, unpublished observations). By comparison, levels of retinal, localized to the eye, did not change over the last 100 d of the experiment and may reflect sparing of retinoids for vision. However, there is certain to be some level of retinoid depletion that will induce adverse health effects, given the two previous studies in which fed fish retinoid deficient diets exhibited a variety of adverse responses (see above)[54,98]. In the future, RA-mediated gene expression assays and germ-line transgenic zebrafish, which serve as biosensors for chemicals that affect the RA receptor mediated regulatory pathways, may aid in evaluating disruptions in retinoid status[57,81].

Little is known of the mechanisms by which environmental chemicals deplete retinoid stores. In many field studies there is evidence of increased activity of phase I and II biotransformation enzymes, and both are thought to directly metabolize retinoids. P450 enzymes also produce oxyradicals causing oxidative stress[34,72,73], which may lead to retinoid use as antioxidants. There was evidence of oxidative stress and retinoid depletion in lake trout inhabiting an area contaminated with iron-ore mine tailings[80], yet it is unlikely that this exposure would be associated with induction of phase I or II biotransformation enzymes.

Accelerated metabolism and breakdown of retinoids may be another mechanism responsible for the depletion of retinoid stores following contaminant exposure.

Studies in brook trout showed that injection of PCB 77 decreased hepatic retinyl ester dehydrogenase activity in a dose-dependent manner[67]. In other studies, Palace *et al.*[74] showed greater elimination of hepatic ^{3}H-retinol to the bile of lake trout 12 weeks after a single dietary exposure to PCB 126. Gilbert *et al.*[45] reported that intraperitoneal injections of PCB 77 enhanced the hydroxylation of RA in rainbow trout liver microsomes *in vitro*. In a follow-up study, brook trout exposed to PCB 77 exhibited lowered hepatic retinoid stores, greater production of 4-OH RA and increased RA glucuronidation *in vitro*[18]. Increases in these hydroxylated products may have implications for the fish RA receptor activation (see above)[3]. The rate of RA hydroxylation was also increased in St Lawrence River sturgeon and liver retinoid concentrations exhibited a strong negative correlation with the rate of RA hydroxylation[38]. Collectively, these studies suggest that there are contaminant-induced effects on retinoid metabolism in fish.

V. Conclusions

Although our understanding of the retinoid system in fish is far from complete, it appears similar, but more complex, than that of the mammals. In fish, there are two forms of the retinoids, the 'standard' retinoids as well as the didehydroretinoids. The fish have two additional receptors: RXRδ and RXRε. RA metabolites such as 4-OH RA bind to fish RARs with high affinity, while they do not in mammals. In fish, RBP and TTR circulate independent of one another. While we know that these differences exist, we have little idea of their implications on normal functions. Do the didehydroretinoids have the same functions as the 'standard' retinoids? What are the functions of RXRδ and RXRε and what are their ligands? Do RA metabolites play a role in gene expression or are they simply excreted? These are just some of the questions that need to be addressed in future studies.

The retinoid toxicology of fish is a new subject, with only a decade's worth of research. The reduction of retinoid stores with toxicant exposure has been well documented; however, the mechanism of this reduction and the implications on RA and fish physiology and health have not been determined. This is partially due to the lack of technology that would allow the easy measurement of RA in fish tissues (required in some remarkably small tissue samples compared to mammals!). It is very important for future studies to establish whether reductions of stored retinoids alter RA levels and whether changes in RA are responsible for the effects of a toxicant observed in fish. However, this may be a challenge in that long-term studies (months to years) may be required to induce retinoid deficiency in fish.

Acknowledgements. Unpublished research cited in this report was supported by grants from NSERC and the Canadian Network of Toxicology Centers. The comments provided on an earlier draft of this manuscript by three reviewers were very helpful and appreciated.

VI. References

1. Allegretto, E. Detection of RARs and RXRs in cells and tissues using specific ligand-binding assays and ligand-binding immunoprecipitation techniques. In: *Retinoid Protocols*, edited by C.P.F. Redfern, Totowa, NJ, Humana Press, pp. 219–232, 1998.
2. Alsop, D., S. Brown and G. Van Der Kraak. Development of a retinoic acid receptor-binding assay with rainbow trout tissue: characterization of retinoic acid binding, receptor tissue distribution, and developmental changes. *Gen. Comp. Endocrinol.* 123: 254–267, 2001.
3. Alsop, D., M. Hewitt, M. Kohli, S. Brown and G. Van Der Kraak. Constituents within pulp mill effluent deplete retinoid stores in white sucker and bind to rainbow trout retinoic acid receptors and retinoid X receptors. *Environ. Toxicol. Chem.* 22: 2969–2976, 2003.
4. Arcand-Hoy, L.D. and C.D. Metcalfe. Biomarkers of exposure of brown bullheads (*Ameiurus nebulosus*) to contaminants in the Lower Great Lakes, North America. *Environ. Toxicol. Chem.* 18: 740–749, 1999.
5. Arsenault, J.T., W.L. Fairchild, S.B. Brown and D.C.G. Muir. Comparison of organic contaminants and hepatic retinoid and tocopherol concentrations in Atlantic tomcod from two estuaries in the Gulf of St. Lawrence. 21st Annual Aquatic Toxicity Workshop, Sarnia, ON, 1994. *Can. Tech. Rep. Fish. Aquat. Sci.* 2050: 149, 1995.
6. Azuma, M., T. Irie and T. Seki. Retinals and retinols induced by estrogen in the blood plasma of *Xenopus laevis*. *J. Exp. Biol.* 178: 89–96, 1993.
7. Bavik, C.O., A. Peterson and U. Eriksson. Retinol-binding protein mediates uptake of retinol to cultured human keratinocytes. *Exp. Cell Res.* 216: 358–362, 1995.
8. Begemann, G., T. Schiling, G.J. Rauch, R. Geisler and P. Ingham. The zebrafish *neckless* mutation reveals a requirement for Raldh2 in mesodermal signals that pattern the hindbrain. *Development* 128: 3081–3094, 2001.
9. Bellovino, D., T. Morimoto, E. Mengheri, G. Perozzi, I. Garaguso, F. Nobili and S. Gaerani. Unique biochemical nature of carp retinol-binding protein. *J. Biol. Chem.* 276: 13949–13956, 2001.
10. Berni, R., M. Stoppini and M.C. Zapponi. The piscine plasma retinol-binding protein. Purification, partial amino acid sequence and interaction with mammalian transthyretin. *Eur. J. Biochem.* 204: 99–106, 1992.
11. Besselink, H.T., S. van Beusekom, E. Roex, A.D. Vethaak, J.H. Koeman and A. Brouwer. Low hepatic 7-ethoxyresoufin-o-deethylase (EROD) activity and minor alterations in retinoid and thyroid hormone levels in flounder (*Platichthys flesus*) exposed to the polychlorinated biphenyl (PCB) mixture, Clophen A50. *Environ. Pollut.* 92: 267–274, 1996.
12. Besselink, H.T., E. van Santen, W. Vorstman, A.D. Vethaak, J.H. Koeman and A. Brouwer. High induction of cytochrome P4501A activity without changes in retinoid and thyroid hormone levels in flounder (*Platichthys flesus*) exposed to 2,3,7,8-tetrachlorodibenzo-*p*-dioxin. *Environ. Toxicol. Chem.* 16: 816–823, 1997.
13. Besselink, H.T., E.M.T.E. Flipsen, M.L. Eggens, A.D. Vethaak, J.H. Koeman and A. Brouwer. Alterations in plasma and hepatic retinoid levels in flounder (*Platichthys flesus*) after chronic exposure to harbour sludge in a mesocosm study. *Aquat. Toxicol.* 42: 271–285, 1998.
14. Blaner, W. and J. Olsen. Retinol and retinoic acid metabolism. In: *The Retinoids, Biology, Chemistry, and Medicine*, 2nd ed., edited by M.B. Sporn, A.B. Roberts and D.S. Goodman, New York, Raven Press, pp. 229–255, 1994.
15. Blaner, W.S. Cellular metabolism and actions of 13-*cis*-retinoic acid. *J. Am. Acad. Dermatol.* 45: S129–S135, 2001.
16. Blomhoff, R., M.H. Green, J.B. Green, T. Berg and K.R. Norum. Vitamin A metabolism: new perspectives on absorption, transport, and storage. *Physiol. Rev.* 71: 951–990, 1991.
17. Bowmaker, J.K. Visual pigments of fishes. In: *The Visual System of Fish*, edited by R.H. Douglas and M.B.A. Djamgoz, New York, Chapman & Hall, pp. 81–107, 1990.
18. Boyer, P.M., A. Ndayibagira and P.A. Spear. Dose-dependant stimulation of hepatic retinoic acid hydroxylation/oxidation and glucuronidation in brook trout, *Salvelinus fontinalis*, after exposure to 3,3′,4,4′-tetrachlorobiphyenyl. *Environ. Toxicol. Chem.* 19: 700–705, 2000.
19. Branchaud, A., A. Gendron, R. Fortin, P. Anderson and P.A. Spear. Vitamin A stores, teratogenesis, and EROD activity in white sucker, *Catostomus commersoni*, from Rivière des Prairies near Montrèal and a reference site. *Can. J. Fish. Aquat. Sci.* 52: 1703–1713, 1995.
20. Brouwer, A. and K.J. Van Den Berg. Binding of a metabolite of 3,4,3′,4′-tetrachlorobiphenyl to transthyretin reduces serum vitamin A transport by inhibiting the formation of the protein complex carrying both retinol and thyroxin. *Toxicol. Appl. Pharmacol.* 85: 301–312, 1986.

21. Brouwer, A., W.S. Blaner, A. Kukler and K.J. Van Den Berg. Study on the mechanism of interference of 3,4,3′,4′-tetrachlorobiphenyl with the plasma retinol-binding proteins in rodents. *Chem. Biol. Interact.* 68: 203–217, 1988.
22. Brouwer, A. The role of biotransformation in PCB-induced alterations in vitamin A and thyroid hormone metabolism in laboratory and wildlife species. *Biochem. Soc. Trans.* 19: 731–737, 1991.
23. Brown, S.B. and V.P. Palace. Vitamin dynamics in fish exposed to contaminated environments. *39th Conference of the International Association for Great Lakes Research*, Erindale, ON, 1996.
24. Brown, S.B. and L. Vandenbyllardt. Analysis of dehydroretinol, retinol, retinyl palmitate and tocopherol in fish, Peace, Athabasca and Slave River basins, September to December, 1994. *Northern River Basins Study Project Report No. 90*, Northern River Basins Study, Edmonton, Alta., Canada, 47pp., 1996.
25. Brown, S.B., K. Munkittrick, C. Bezte, W.L. Lockhart and L. Vandenbyllardt. Retinoid status in fish exposed to pulp mill effluent. *18th Annual Society of Environmental Toxicology and Chemistry (SETAC) Meeting*, San Francisco, CA, 1997.
26. Brown, S.B., P.D. Delorme, R.E. Evans, W.L. Lockhart, D.C.G. Muir and F.J. Ward. Biochemical and histological responses in rainbow trout (*Oncorhynchus mykiss*) exposed to 2,3,4,7,8-pentachlorodibenzofuran. *Environ. Toxicol. Chem.* 17: 915–921, 1998.
27. Brown, S.B., A.T. Fisk, M. Brown, M. Villella, M. Cooley, R.E. Evans, D.A. Metner, W.L. Lockhart and D.C.G. Muir. Dietary accumulation and biochemical responses of juvenile rainbow trout (*Oncorhynchus mykiss*) to 3,3′,4,4′,5-pentachlorobiphenyl (PCB 126). *Aquat. Toxicol.* 59: 139–152, 2002.
28. Buddington, R., K. Buddington, D.F. Deng, G.I. Hemre and R. Wilson. A high retinol dietary intake increases its apical absorption by the proximal small intestine of juvenile sunshine bass (*Morone chrysops x M. saxatilis*). *J. Nutr.* 132: 2713–2716, 2002.
29. Calderone, V., C. Folli, A. Marchesani, R. Berni and G. Zanotti. Identification and structural analysis of a zebrafish apo- and holo-cellular retinol-binding protein. *J. Mol. Biol.* 321: 527–535, 2002.
30. Chambon, P. A decade of molecular biology of retinoic acid receptors. *FASEB J.* 10: 940–954, 1996.
31. Clagett-Dame, M. and H.F. DeLuca. The role of vitamin A in mammalian reproduction and embryonic development. *Annu. Rev. Nutr.* 22: 347–381, 2002.
32. Costaridis, P., C. Horton, J. Zeitlinger, N. Holder and M. Maden. Endogenous retinoids in the zebrafish embryo and adult. *Dev. Dyn.* 205: 41–51, 1996.
33. Crouch, R.K. and J.-X. Ma. The role of vitamin A in visual transduction. In: *Vitamin A and Retinoids: An Update of Biological Aspects and Clinical Applications*, edited by M.A. Livrea, Boston, Birkhauser Verlag, pp. 59–72, 2000.
34. Dalton, T.P., A. Puga and H.G. Shertzer. Induction of cellular oxidative stress by aryl hydrocarbon receptor activation. *Chem. Biol. Interact.* 141: 77–95, 2002.
35. P.D. Delorme. The effects of toxaphene, chlordane and 2,3,4,7,8-pentachlorodibenzofuran on lake trout and white sucker in an ecosystem experiment and the distribution and effects of 2,3,4,7,8-pentachlorodibenzofuran on white suckers and broodstock rainbow trout in laboratory experiments. PhD Thesis, University of Manitoba, Winnipeg, Man., 1995.
36. Dew, S.E. and D.E. Ong. Specificity of the retinol transporter of the rat small intestine brush border. *Biochemistry* 33: 12340–12345, 1994.
37. Doyon, C., S. Boileau, R. Fortin and P. Spear. Rapid HPLC analysis of retinoids and dehydroretinoids stored in fish liver: comparison of two lake sturgeon populations. *J. Fish Biol.* 53: 973–986, 1998.
38. Doyon, C., R. Fortin and P. Spear. Retinoic acid hydroxylation and teratogenesis in lake sturgeon (*Acipenser fulvescens*) from the St Lawrence River and Abitibi region, Quebec. *Can. J. Fish. Aquat. Sci.* 56: 1428–1436, 1999.
39. Duester, G. Families of retinoid dehydrogenases regulating vitamin A function: production of visual pigment and retinoic acid. *Eur. J. Biochem.* 267: 4315–4324, 2000.
40. Fairchild, W.L., J.T. Arsenault, D.C.G. Muir and S.B. Brown, Organic contaminants and retinoids in Atlantic tomcod from two estuaries in the Gulf of St Lawrence. *15th Annual Society of Environmental Toxicology and Chemistry (SETAC) Meeting*, Denver, CO, 1994.
41. Friesen, C., W.L. Lockhart and S.B. Brown, Results from the analysis of Winnipeg River water, sediment and fish. *Report to the Department of Indian Affairs and Northern Development*, 35pp., 1994.
42. Frolik, C., A. Roberts, T. Tavela, P. Roller, D. Newton and M. Sporn. Isolation and identification of 4-hydroxy- and 4-oxoretinoic acid. In vitro metabolites of all-*trans* retinoic acid in hamster trachea and liver. *Biochemistry* 18: 2092–2097, 1979.
43. Gamble, M.V. and W.S. Blaner. Factors affecting blood levels of vitamin A. In: *Vitamin A and Retinoids: An Update of Biological Aspects and Clinical Applications*, edited by M.A. Livrea, Boston, Birkhäuser Verlag, pp. 1–16, 2000.

44. Giguère, V. Retinoic acid receptors and cellular retinoid binding proteins: complex interplay in retinoid signaling. *Endocr. Rev.* 15: 61–79, 1994.
45. Gilbert, N.L., M.J. Cloutier and P.A. Spear. Retinoic acid hydroxylation in rainbow trout (*Oncorhynchus mykiss*) and the effect of a coplanar PCB, 3,3′,4,4′-tetrachlorobiphenyl. *Aquat. Toxicol.* 32: 177–187, 1995.
46. Grandel, H., K. Lun, G.J. Rauch, M. Rhinn, T. Piotrowski, C. Houart, P. Sordino, A. Kuchler, S. Schulte-Merker, R. Geisler, N. Holder, S. Wilson and M. Brand. Retinoic acid signaling in the zebrafish embryo is necessary during presegmentation stages to pattern the anterior-posterior axis of the CNS and to induce a pectoral fin bud. *Development* 129: 2851–2865, 2002.
47. Guillou, A., G. Choubert, T. Storebakken, J. de la Noue and S. Kausik. Bioconversion pathway of astaxanthin into retinol$_2$ in mature rainbow trout (*Salmo gairdneri* Rich.). *Comp. Biochem. Physiol.* 94B: 481–485, 1989.
48. Harmon, M., M.F. Boehm, R.A. Heyman and D.J. Mangelsdorf. Activation of mammalian retinoid X receptors by the insect growth regulator methoprene. *Proc. Natl Acad. Sci. USA* 92: 6157–6160, 1995.
49. Idres, N., J. Marill, M.A. Flexor and G.G. Chabot. Activation of retinoic acid receptor-dependent transcription by all-*trans*-retinoic acid metabolites and isomers. *J. Biol. Chem.* 277: 31491–31498, 2002.
50. Irie, T. and T. Seki. Retinoid composition and retinal localization in the eggs of teleost fishes. *Comp. Biochem. Physiol.* 131B: 209–219, 2002.
51. Jones, B., C. Ohno, G. Allenby, M. Boffa, A. Levin, J. Grippo and M. Petkovich. New retinoid X receptor subtypes in zebrafish (*Danio rerio*) differentially modulate transcription and do not bind 9-*cis* retinoic acid. *Mol. Cell. Biol.* 15: 5226–5234, 1995.
52. Joore, J., G. van der Lans, P. Lanser, J. Vervaart, D. Zivkovic, J. Speksnijder and W. Kruijer. Effects of retinoic acid on the expression of retinoic acid receptors during zebrafish embryogenesis. *Mech. Dev.* 46: 137–150, 1994.
53. Katsuyama, M. and T. Matsuno. Carotinoid and vitamin A and metabolism of carotinoids, β-carotene, canthaxanthin, astaxanthin, zeaxantin, lutein and tunaxanthin in tilapia, *Tilapia nilotica*. *Comp. Biochem. Physiol.* 90B: 133–139, 1988.
54. Kitamura, S., T. Suwa, S. Ohara and K. Nakagawa. Studies on vitamin requirements of rainbow trout – III requirement for vitamin A and deficiency symptoms. *Bull. Jpn. Soc. Sci. Fish* 33: 1126–1131, 1967.
55. Kleinjan, D.A., S. Dekker, J.A. Guy and F.G. Grosveld. Cloning and sequencing of the CRABP-I locus from chicken and pufferfish: analysis of the promoter regions in transgenic mice. *Transgenic Res.* 7: 85–94, 1998.
56. Kurlandsky, S., M. Gamble, R. Ramakrishnan and W. Blaner. Plasma delivery of retinoic acid to tissues in the rat. *J. Biol. Chem.* 270: 17850–17857, 1995.
57. Linney, E.A., K. Yacisin, B. Dobbs-McAuliffe, S. Donerly, Q. Zhao and L. Christopher, Germ-line transgenic zebrafish as biosensors. *23rd Annual Society of Environmental Toxicology and Chemistry (SETAC) Meeting*, Salt Lake City, UT, 2002.
58. Maden, M. Heads or tails? Retinoic acid will decide. *BioEssays* 21: 809–812, 1999.
59. Mangelsdorf, D.J., K. Umesono and R.M. Evans. The retinoid receptors. In: *The Retinoids, Biology, Chemistry, and Medicine*, 2nd ed., edited by M.B. Sporn, A.B. Roberts and D.S. Goodman, New York, Raven Press, pp. 319–349, 1994.
60. McCaffery, P. and U.C. Dräger. Regulation of retinoic acid signaling in the embryonic nervous system: a master differentiation factor. *Cytokine Growth Factor Rev.* 11: 233–249, 2000.
61. Nacci, D., S. Jayaraman and J. Specker. Stored retinoids in populations of estuarine fish *Fundulus heteroclitus* indigenous to PCB-contaminated and reference sites. *Arch. Environ. Contam. Toxicol.* 40: 511–518, 2001.
62. Napoli, J.L. Retinoic acid biosynthesis and metabolism. *FASEB J.* 10: 993–1001, 1996.
63. Napoli, J.L. Biochemical pathways of retinoid transport, metabolism, and signal transduction. *Clin. Immunol. Immunopathol.* 80: S52–S62, 1996.
64. Napoli, J.L. Retinoid binding-proteins redirect retinoid metabolism: biosynthesis and metabolism of retinoic acid. *Semin. Cell Dev. Biol.* 8: 403–415, 1997.
65. Napoli, J.L. Enzymology and biogenesis of retinoic acid. In: *Vitamin A and Retinoids: An Update of Biological Aspects and Clinical Applications*, edited by M.A. Livrea, Boston, Birkhauser Verlag, pp. 17–27, 2000.
66. Ndayibagara, A., M.J. Cloutier, P.D. Anderson and P.A. Spear. Effects of 3,3′,4,4′-tetrachlorobiphenyl on the dynamics of vitamin A in brook trout (*Salvelinus fontinalis*) and intestinal retinoid concentrations in lake sturgeon (*Acipenser fulvescens*). *Can. J. Fish. Aquat. Sci.* 52: 512–520, 1995.
67. Ndayibagara, A. and P.A. Spear. Esterfication and hydrolysis of vitamin A in the liver of brook trout (*Salvelinus fontinalis*) and the influence of a coplanar polychlorinated biphenyl. *Comp. Biochem. Physiol.* 122C: 317–325, 1999.

68. Nilsson, C.B. and H. Håkansson. The retinoid signaling system – a target in dioxin toxicity. *Crit. Rev. Toxicol.* 32: 211–232, 2002.
69. Okino, S.T. and J.P. Whitlock. The aromatic hydrocarbon receptor, transcription, and endocrine aspects of dioxin action. *Vitam. Horm.* 59: 241–264, 2000.
70. Paganetto, G., F. Campi, K. Varani, A. Piffanelli, G. Giovannini and P.A. Borea. Endocrine-disrupting agents on healthy human tissues. *Pharmacol. Toxicol.* 86: 24–29, 2000.
71. Palace, V.P. and S.B. Brown. HPLC determination of tocopherol, retinol, dehydroretinol and retinyl palmitate in tissues of lake char (*Salvelinus namaycush*) exposed to coplanar 3,3′,4,4′,5-pentachlorobiphenyl. *Environ. Toxicol. Chem.* 13: 473–476, 1994.
72. Palace, V.P., T.A. Dick, S.B. Brown, C.L. Baron and J.F. Klaverkamp. Oxidative stress in juvenile lake sturgeon orally exposed to 2,3,7,8-tetrachlordibenzofuran. *Aquat. Toxicol.* 35: 79–82, 1996.
73. Palace, V.P., S.B. Brown, D.A. Metner, W.L. Lockhart, D.C.G. Muir and J.F. Klaverkamp. Mixed function oxidase enzyme activity and oxidative stress in lake trout (*Salvelinus namaycush*) exposed to 3,3′,4,4′,5-pentachlorobiphenyl. *Environ. Toxicol. Chem.* 15: 955–960, 1996.
74. Palace, V.P., J.F. Klaverkamp, C.L. Baron and S.B. Brown. Metabolism of ^{3}H-retinol by lake trout (*Salvelinus namaycush*) pre-exposed to 3,3′,4,4′,5-pentachlorobiphenyl (PCB 126). *Aquat. Toxicol.* 39: 321–332, 1997.
75. Palace, V.P., S.M. Allen-Gill, S.B. Brown, R.E. Evans, D.A. Metner, D.H. Landers, L.R. Curtis, J.F. Klaverkamp, C.L. Baron and W.L. Lockhart. Vitamin and thyroid status in Arctic grayling (*Thymallus arcticus*) exposed to doses of 3,3′,4,4′-tetrachlorobiphenyl that induce the phase I enzyme system. *Chemosphere* 45: 185–193, 2001.
76. Palace, V.P., K. Wautier, R.E. Evans, C. Baron, J. Werner, C. Ranson, J.F. Klaverkamp and K. Kidd. Effects of 17β-estradiol exposure on metallothionein and fat-soluble antioxidant vitamins in juvenile lake trout (*Salvelinus namaycush*). *Bull. Environ. Contam. Toxicol.* 66: 591–596, 2001.
77. Palace, V.P., R.E. Evans, K. Wautier, C. Baron, J. Werner, J.F. Klaverkamp, K. Kidd and T.A. Dick. Altered distribution of lipid-soluble, antioxidant vitamins in juvenile sturgeon exposed to waterborne ethynylestradiol. *Environ. Toxicol. Chem.* 20: 2370–2376, 2001.
78. Palozza, P. and N.I. Krinsky. The inhibition of radical initiated peroxidation of microsomal lipids by both α-tocopherol and β-carotene. *Free Rad. Biol. Med.* 11: 407–414, 1991.
79. Paul, M. and D. Laliberté. *Teneurs en mercure, plom, cadmium, BPC et pesticides organochlorés des sediments et de la chair de poisson du fleuve Saint-Laurent et de la Rivière des Outaouais en 1985*, QEN/QE-86-07. Technical Report, Montréal, Que., Ministère de l'Environment du Québec, 1988.
80. Payne, J., D. Malins, S. Gunselman, A. Rahimtula and P. Yeats. DNA oxidative damage and vitamin A reductions in fish from a large lake system in Labrador, Newfoundland, contaminated with iron-ore mine tailings. *Mar. Environ. Res.* 46: 289–294, 1998.
81. Perz-Edwards, A., N.L. Hardison and E. Linney. Retinoic acid-mediated gene expression in transgenic reporter zebrafish. *Dev. Biol.* 229: 89–101, 2001.
82. Pijnappel, W., H. Hendriks, G. Folkers, C. van den Brink, E. Dekker, C. Edelenbosch, P. van der Saag and A. Durston. The retinoid ligand 4-oxo-retinoic acid is a highly active modulator of positional specification. *Nature* 366: 340–344, 1993.
83. Plack, P.A., S.K. Kon and S.Y. Thompson. Vitamin A_1 aldehyde in the eggs of the herring (*Clupea harengus* L.) and other marine teleosts. *Biochem. J.* 71: 467–476, 1959.
84. Plack, P.A. and S.K. Kon. A comparative survey of the distribution of vitamin A aldehyde in eggs. *Biochem. J.* 81: 561–570, 1961.
85. Ribera, D., J.F. Narbonne, X. Michel, D.R. Livingstone and S.H. O'Hara. Responses of antioxidants and lipid peroxidation in mussels to oxidative damage exposure. *Comp. Biochem. Physiol.* 100C: 177–181, 1991.
86. Roberfroid, M.B. and P.B. Calderon. *Pharmacology of antioxidant molecules: analysis of their mechanism of action*, Free Radicals and Oxidation Phenomena in Biological Systems, New York, Marcel Dekker, pp. 237–252, 1995.
87. Rolland, R.M. A review of chemically-induced alterations in thyroid and vitamin A status from field studies of wildlife and fish. *J. Wildl. Dis.* 36: 615–635, 2000.
88. Ross, A. Overview of retinoid metabolism. *J. Nutr.* 123: S346–S350, 1993.
89. Sammar, M., P.J. Babin, M. Durliat, I. Meiri, I. Zchori, A. Elizur and E. Lubzens. Retinol binding protein in rainbow trout: molecular properties and mRNA expression in tissues. *Gen. Comp. Endocrinol.* 123: 51–61, 2001.
90. Scheidt, K., F.J. Leuenberger, M. Vecchi and E. Glinz. Absorption, retention and metabolic transformations of carotinoids in rainbow trout, salmons and chicken. *Pure Appl. Chem.* 57: 685–692, 1985.

91. Scheidt, K., M. Vecchi and E. Glinz. Astaxanthin and its metabolites in wild rainbow trout (*Salmo gairdneri* R.). *Comp. Biochem. Physiol.* 83B: 9–12, 1986.
92. Schoff, P.K. and G.T. Ankley. Inhibition of retinoid activity by components of a paper mill effluent. *Environ. Pollut.* 119: 1–4, 2002.
93. Simms, W. and P.S. Ross. Vitamin A physiology and its implication as a biomarker of contaminant-related toxicity in marine mammals: a review. *Toxicol. Ind. Health* 16: 291–302, 2000.
94. Sivaprasadarao, A. and J.B. Findlay. The interaction of retinol-binding protein with its plasma-membrane receptor. *Biochem. J.* 255: 561–569, 1988.
95. Spear, P.A., A.Y. Bilodeau and A. Branchaud. Retinoids: from metabolism to environmental monitoring. *Chemosphere* 25: 1733–1738, 1992.
96. Stainier, D. and M. Fishman. Patterning the zebrafish heart tube: acquisition of anterioposterior polarity. *Dev. Biol.* 153: 91–101, 1992.
97. Takeuchi, T., J. Dedi, Y. Haga, T. Seikai and T. Watanabe. Effect of vitamin A compounds on bone deformity in larval Japanese flounder (*Paralichthys olivaceus*). *Aquaculture* 169: 155–165, 1998.
98. Taveekijakarn, P. and T. Miyazaki. Vitamin A deficiency in cherry salmon. *J. Aquat. Anim. Health* 6: 251–259, 1994.
99. Ternes, T.A., H. Anderson, D. Gilberg and M. Bonerz. Determination of estrogens in sludge and sediments by liquid extraction and GC/MS/MS. *Anal. Chem.* 74: 498–504, 2002.
100. Ternes, T.A., M. Stumpf, K. Haberer, R.D. Wilken and M. Servos. Behavior and occurrence of estrogen in municipal sewage treatment plants – I. Investigations in Germany, Canada and Brazil. *Sci. Total Environ.* 225: 81–90, 1999.
101. Vandersea, M., R. McCarthy, P. Fleming and D. Smith. Exogenous retinoic acid during gastrulation induces cartilaginous and other craniofacial defects in *Fundulus heteroclitus*. *Biol. Bull.* 196: 281–296, 1998.
102. Vandersea, M., P. Fleming, R. McCarthy and D. Smith. Fin duplications and deletions induced by disruption of retinoic acid signaling. *Dev. Genes Evol.* 208: 61–68, 1998.
103. Wald, G. The chemistry of rod vision. *Science* 113: 287–291, 1951.
104. Ward, S.J. and G.M. Morriss-Kay. The functional basis of tissue-specific retinoic acid signaling in embryos. *Semin. Cell Dev. Biol.* 8: 429–435, 1997.
105. White, J., Y.D. Guo, K. Baetz, B. Beckett-Jones, J. Bonasoro, K. Hsu, F. Dilworth, G. Jones and M. Petkovich. Identification of the retinoic acid-inducible all-*trans*-retinoic acid 4-hydroxylase. *J. Biol. Chem.* 271: 29922–29927, 1996.
106. Xu, Y., J. Zhang, K.W. Schramm and A. Kettrup. Endocrine effects of sublethal exposure to persistent organic pollutants (POPs) on silver carp (*Hypophthalmicthys molitrix*). *Environ. Pollut.* 120: 683–690, 2002.
107. Zhang, J., Y. Xu, K.W. Schramm and A. Kettrup. Alterations in retinoids, tocopherol and microsomal enzyme activities in the liver of silver carp (*Hypophthalmicthys molitrix*) from Ya-Er Lake, China. *Bull. Environ. Contam. Toxicol.* 68: 660–667, 2002.
108. Zile, M.H. Vitamin A homeostasis endangered by environmental pollutants. *Proc. Soc. Exp. Biol. Med.* 201: 141–153, 1992.
109. Zile, M.H. Function of vitamin A in vertebrate embryonic development. *J. Nutr.* 131: 705–708, 2001.

c. Reproduction

Biochemistry and Molecular Biology of Fishes, vol. 6
T. P. Mommsen and T. W. Moon (Editors)

CHAPTER 16

Vitellogenesis and endocrine disruption

NAOSHI HIRAMATSU *, ANN O. CHEEK **, CRAIG V. SULLIVAN *,
TAKAHIRO MATSUBARA† AND AKIHIKO HARA‡

**Department of Zoology, North Carolina State University, Box 7617, Raleigh, NC 27695-7617, USA, **Department of Biological Sciences, Southeastern Louisiana University, Box 10736, Hammond, LA 70402, USA, †Hokkaido National Fisheries Research Institute, 116, Katsurakoi, Kushiro, Hokkaido 085-0802, Japan, and ‡Division of Marine Biosciences, Graduate School of Fisheries Sciences, Hokkaido University, Hakodate, Hokkaido, 041-8611, Japan*

I. Introduction

Many chemical compounds used in human daily life have the potential to impact the vertebrate neuroendocrine system, which regulates vital processes including development, growth, metabolism, and reproduction. Over the past decade, research on the identification and effects of such compounds has become an important area of human and environmental health science. Chemicals that either mimic or antagonize the actions of naturally occurring hormones are termed endocrine disrupters or endocrine disrupting chemicals (EDCs). These include environmental estrogens, substances that elicit an estrogenic response by mimicking the action of endogenous estradiol-17β (E_2). In addition to natural estrogens, various kinds of industrial, municipal, and agricultural wastes are known or suspected to be estrogenic. Examples include biodegradation products of alkylphenol polyethoxylates, polychlorinated biphenyls (PCBs), and pesticides (e.g. dichlorodiphenyltrichloroethane (DDT), chlordecone, and methoxychlor), as well as synthetic estrogens such as ethinylestradiol (EE_2) and diethylstilbestrol (DES) (reviews: Sumpter[189]; Arcand-Hoy and Benson[7]; Giesy and Snyder[52]; Kime[102]; Pait and Nelson[151]; Arukwe and Goksøyr[9]).

Endogenous sex steroid hormones may have two broad categories of influence on physiological systems, organization and activation[8]. Organizational effects are usually permanent changes in morphology that persist after the hormone stimulus is removed and that affect subsequent function and behavior. Activational effects are usually transient changes in morphology, function, and behavior that disappear when the hormone stimulus is removed. Classically, the organizational effects of sex steroid hormones include gender-specific morphological differentiation of the gonads and brain. Activational effects of sex steroids in vertebrates include seasonal gonadal recrudescence and onset of breeding coloration and behavior[8]. Both categories of action are essential for normal reproduction.

Endocrine disruption can alter both the organizational and the activational effects of reproductive hormones, possibly having a profound effect on an organism's ability to reproduce and therefore on its fitness. EDCs can have organizational effects on sex differentiation and expression of secondary sex characteristics in fish[37,53,84,131]. In addition, numerous studies have documented the activational effects of exogenous chemicals on levels of sex steroid hormones and on hormone-mediated protein synthesis in fish[3,4,41,70,71,95,120,130,134–137,187]. Activational effects of EDCs, such as alterations in hormone and protein levels, are effective biomarkers of exposure to an estrogenic compound. Only recently have studies addressed the relationship between these activational effects and potential organizational effects such as altered gonad morphology and reproductive function.

Assessment of the estrogenic impacts of EDCs requires development of sensitive, simple, and accurate assay systems for evaluating their bioactivity. Several estrogen-regulated proteins have been utilized as biomarkers of exposure of fish and wildlife to EDCs. The yolk precursor protein, vitellogenin (Vg), has been used most frequently for such assessments. In oviparous vertebrates, Vg is produced by the liver of maturing females in response to E_2, secreted into the bloodstream, and then taken up by growing oocytes to be processed into a suite of yolk proteins that are stored in the ooplasm. The current status of our understanding of teleost Vg and vitellogenesis was recently reviewed[78,158,186]. Vitellogenin has served as an ideal marker for detecting the onset of puberty and progression of maturation in female fishes, especially in aquaculture. The following characteristics make Vg a functional biomarker for exposure of fishes to estrogenic compounds in the aquatic environment[154,197]: (1) induction of vitellogenesis is a specific physiological response to estrogen or estrogenic chemicals, (2) mechanisms of Vg production have been extensively studied as a model for steroid hormone action, (3) within broad limits, induction of Vg synthesis by estrogen proceeds in a dose-dependent manner, (4) Vg naturally appears in the blood of maturing females but not in the blood of males or immature fish, (5) Vg is inducible in male and juvenile fish by exogenous estrogen or estrogenic EDCs, and, therefore, (6) the presence of Vg in the blood of male or juvenile fish can be used as an indicator of exposure to xenobiotic estrogens.

Sensitive and specific assays for measuring teleost Vg have been, and will continue to be, valuable tools for identifying environmental estrogens. However, most of these Vg assays were developed, and their results were interpreted, based on the classical 'single Vg model' of teleost oogenesis. This simplistic model ignores the fact that some species produce multiple Vgs with different physiological functions. Recent gene cloning and immuno-biochemical analyses confirmed that the presence of multiple forms of Vg is the norm in fishes, leading to the adoption of a new 'multiple Vg model' of teleost oocyte growth (reviews: Hiramatsu *et al.*[78]; Patiño and Sullivan[158]; Matsubara *et al.*[127]). The new model includes unique functions of individual Vgs and their yolk protein derivatives during oocyte maturation, early embryogenesis, and later larval development. Exposure to EDCs has the potential to impair fish egg quality by disturbing the coordinated induction of functionally discrete forms of Vg because the responsiveness of the various Vg genes to estrogen may differ within and between species. Yolk protein derivatives of specific Vgs have been used to raise antisera employed in various immunoassays. However, the relationship of these yolk proteins to their parent Vg molecule has rarely been verified. Exposure of fishes to EDCs could also disrupt endocrine regulation of the stepwise proteolysis of individual Vgs and their product yolk proteins during oocyte growth and maturation, although practical studies of this phenomenon are needed. Choriogenins (Chgs), which are protein precursors of the teleost egg envelope (chorion), have potential in some species to be an even more useful biomarker of estrogen exposure than Vgs[9]. However, expression of these chorion precursors may not always be restricted to the liver and tightly controlled by estrogen, indicating that choriogenins may not be as universally applicable as Vgs for detecting environmental estrogens. Research directed at discovering new biomarkers, aside from Vg and Chg, that could be useful for surveying exposure of fishes to estrogenic EDCs has only just begun.

It is now apparent that teleost vitellogenesis involves a suite of distinct Vgs and product yolk proteins with specific functions during oocyte maturation and development of embryos and larvae. Thus vitellogenesis is a far more complicated and interesting process than was previously thought. These complications in the biology of vitellogenesis should not preclude use of Vg-based bioassays for simple and effective detection of animal exposure to EDCs. On the contrary, the new knowledge of Vg diversity in form and function should open the door to a better interpretation of bioassay data and to the discovery of some reproductive effects of EDCs that have previously been overlooked. Accordingly, this review will focus on our current understanding of the extent and function of Vg multiplicity in teleosts as it relates to potential improvements in development and interpretation of Vg-based bioassays of fish exposure to EDCs. We also provide practical examples and evaluations of recent surveys in which Vg or other estrogen-regulated proteins (e.g. Chgs) were utilized as biomarkers of fish exposure to environmental estrogens or other EDCs.

II. Teleostean vitellogenesis and choriogenesis

1. Discovery of vitellogenin

In oviparous animals, vitellogenesis is associated with a number of changes in the profile of plasma proteins seen in females. These changes are under hormonal control and can be induced in immature animals of either sex or in castrates by exogenous estrogens. Nearly a century ago, Uhlenhuth and Kodama[199] found a serum protein component that specifically appeared in the blood of sexually maturing carp and named it 'ovumin'. Laskowski[111] and Roepke and Hughes[167] later found that serum from the laying fowl contained a phosphoprotein, that they named serum 'vitellin', which was not present in non-laying female or male birds and could not be found in mammalian sera. In early studies of fishes, estrogen injections resulted in increasing levels of total serum protein and protein-bound phosphorus, as well as the *de novo* appearance of egg proteins in the blood[10,83,200,201]. Specific proteins observed in the serum of vitellogenic female or estrogen-treated fish were initially given various names such as serum vitellin[205], Sm-antigen[105], HM-factor[201], serum lipovitellin[5], and female-specific plasma protein (FSPP)[2].

The term 'vitellogenin' was first used by Pan *et al.*[155] to describe a female-specific protein in the hemolymph of the *Cecropia* moth. It has since become the generally accepted name for all the vertebrate serum proteins mentioned above. In teleost fishes, Hara and Hirai[66] discovered an iron-binding female-specific serum protein (FSSP) in chum salmon (*Oncorhynchus keta*) and rainbow trout (*O. mykiss*). Emmersen and Petersen[38] found a similar female-specific lipophosphoprotein in the serum of vitellogenic flounder (*Platichtys flesus*) and partially purified the protein. The FSSP was later purified from rainbow trout and was immunologically identified as Vg for the first time in any teleostean species[67]. By definition, Vg is produced by somatic organs or tissues aside from the ovary (e.g. liver) and is the major circulating precursor to the egg yolk proteins. To date, Vg has been purified and identified in scores of teleosts by various biochemical and immunological procedures (see reviews: Wallace[207]; Mommsen and Walsh[133]; Specker and Sullivan[186]). Recent advances in purification, identification, and characterization of multiple forms of teleost Vg will be discussed later in this section.

2. *Vitellogenesis and oocyte growth*

Ovarian follicle growth in fishes can be broadly classified into previtellogenic and vitellogenic stages. As noted, vitellogenic growth is characterized by extensive accumulation of exogenously synthesized yolk proteins within the oocyte[158] but, by common usage, the term 'vitellogenesis' includes various other processes related to oocyte growth.

The yolk or vitellus of an ovulated egg largely consists of a material that has been deposited in the growing oocyte to be later utilized by the nascent embryo as a source of nutrition to support development. Accordingly, 'yolk' includes a diverse mixture of endogenously synthesized or maternally derived materials necessary for embryo development, such as proteins, sugars, lipids, vitamins, and minerals. In addition, teleost eggs contain regulatory compounds such as sex steroids[39] and other lipophilic hormones[15,186], antibodies such as immunoglobulin M[48], and maternal mRNAs encoding vitellogenin receptors, yolk-processing enzymes, and other bioactive molecules[158]. In salmonid species, yolk proteins derived from Vg can account for 80–90% of the dry mass of ovulated eggs[179,198]. However, in some marine or anadromous fishes that spawn heavily lipidated eggs, while Vg is an important carrier of yolk proteins and lipids into oocytes, it is perhaps not the major source of egg yolk substances, especially of lipids[158].

Endocrine regulation of the induction of Vg synthesis by the liver has been extensively studied both *in vivo* and *in vitro* (reviews: Wallace[207]; Mommsen and Walsh[133]; Specker and Sullivan[186]). Briefly, hepatic Vg synthesis naturally ensues from the activation of the hypothalamus–pituitary–gonad neuroendocrine axis by environmental and endogenous signals. Data obtained for some teleosts (e.g. salmonids) indicate that increased levels of follicle-stimulating hormone (FSH) in the blood induces follicular production of E_2, which triggers hepatic synthesis of Vg[186]. In other species, ovarian E_2 production may be regulated by luteinizing hormone (LH)[148]. Other estrogens, such as estrone, may contribute to the induction of vitellogenesis by priming the liver to respond more strongly to E_2[202].

A major focus of a number of recent investigations is oocyte uptake and deposition of Vg. The process of teleost vitellogenesis and related oocyte growth is summarized in Fig. 1 (reviews: Wallace[207]; Mommsen and Walsh[133]; Specker and Sullivan[186]; Hiramatsu *et al.*[78]; Patiño and Sullivan[158]). After synthesis, Vg is released into the bloodstream, taken up by the developing oocytes, and enzymatically processed into yolk proteins that are stored throughout the ooplasm in yolk granules or globules. Growing oocytes selectively accumulate Vg *via* receptor-mediated endocytosis. A membrane receptor on the oocyte surface with a high affinity for Vg, called the Vg receptor (VgR), mediates the endocytotic process. Vg bound to the VgR is clustered in clathrin-coated pits that invaginate to form coated vesicles. These vesicles fuse with lysosomes in the peripheral ooplasm, forming multivesicular bodies. The lysosomes contain cathepsin D and possibly other enzymes that process Vg into its product yolk proteins.

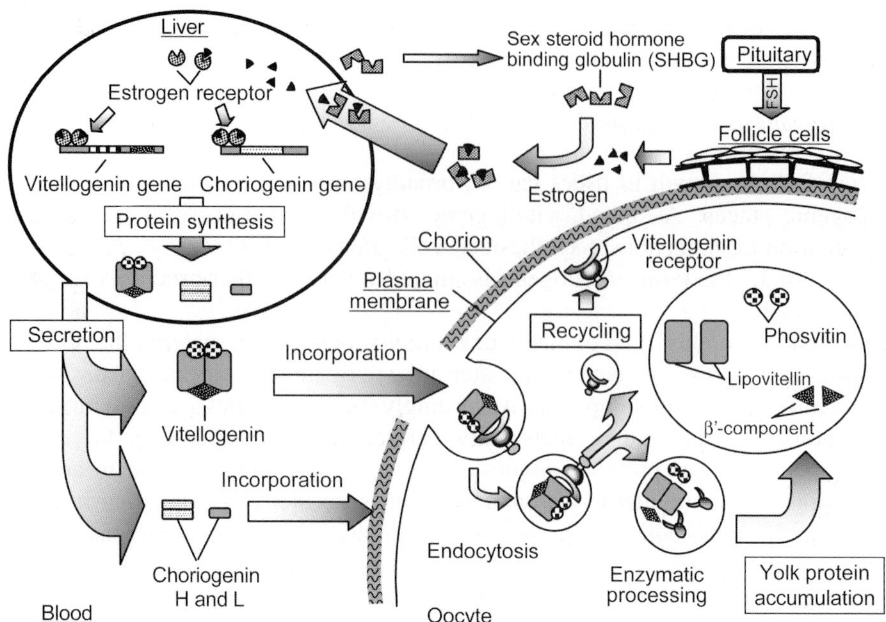

Fig. 1. Outline of the processes of vitellogenesis and choriogenesis in teleost fish. For discussion of vitellogenesis, choriogenesis, and yolk protein formation, see section II. Although only a single form of Vg is illustrated in this figure, multiple forms of vitellogenin (Vg) and choriogenin (Chg) have been discovered in several teleost species (see Fig. 2). Choriogenins produced by the liver include high molecular weight choriogenin (choriogenin H) and low molecular weight choriogenin (choriogenin L). In some teleosts, Chgs of ovarian origin (homologues of mammalian zona pellucida proteins) also are involved in the formation of chorion.

3. *Relationship between vitellogenin and yolk proteins*

Vitellogenin is a large glycolipophosphoprotein that usually circulates in blood as a homodimer of ~350–600 kDa. In the chicken and in *Xenopus laevis*, as well as in teleost fishes, after being taken up by growing oocytes, Vg gives rise to two major yolk proteins, lipovitellin (Lv) and phosvitin (Pv)[13,29,67,207]. Lipovitellin is a heavily lipidated protein carrying as much as 20% lipid by mass. It is the largest Vg-derived yolk product and consists of two polypeptides, a heavy chain (named LvI or LvH) and a light chain (LvII or LvL). The main function of Lv appears to be as a nutritional source of amino acids and lipids needed for embryonic development. Phosvitin consists largely (~50%) of serine residues to which phosphate is covalently linked and to which calcium is ionically bound. Thus, Pv delivers minerals needed by developing embryos for skeletal development and metabolic functions. By virtue of its heavily phosphorylated nature, Pv has some unique characteristics among Vg-derived yolk proteins, such as limited antigenicity and poor stainability with commonly used biochemical dyes[76].

Additional proteins derived from Vg, the yolk plasma glycoproteins (YGPs), were recently discovered in the chicken[214]. Based on recent molecular biological analyses, the coding sequence of the chicken Vg gene appears to be arranged in linear fashion with respect to its yolk protein domains as follows: NH_2–LvH–Pv–LvL–YGP–COOH. The YGPs contain a unique motif of repeated cysteine residues whose coding sequence is highly conserved among Vg genes from the chicken, *X. laevis*, and rainbow trout[142,214]. In teleosts, Jared and Wallace[92] identified another protein in rainbow trout yolk fractionated by TEAE-cellulose column chromatography. Unlike Lv and Pv, this yolk protein contained neither lipid nor phosphorus, and it was named β-component after the nominal chromatographic fraction in which it was found. Later, Markert and Vanstone[123] recovered a similar yolk protein from coho salmon (*O. kisutch*) eggs by gel filtration after water and ammonium sulfate precipitation. This protein, named β'-component (β'-c), was later immunologically confirmed to be a *bona fide* Vg-derived yolk protein[16,76]. In addition to salmonids, β'-c has been identified in teleosts from diverse taxa, such as barfin flounder (*Verasper moseri*) and white perch (*Morone americana*)[79,125], suggesting that this yolk protein is generally present in teleost eggs.

Using the information on the N-terminal peptide sequence and mass of purified teleost yolk proteins, these proteins were mapped to available complete amino acid sequences for teleost Vgs (deduced from cDNA). These analyses revealed the primary structure of teleost Vg, with respect to its major yolk protein domains, to be as follows: NH_2-LvH-Pv-LvL-β'-component-C-terminal coding sequence-COOH[78,79,126]. This model implied that there was a YGP-like, cysteine-rich, teleost yolk protein derived from the C-terminus of Vg downstream from its β'-c domain which remained to be discovered. Recently, the existence of the C-terminal domain of Vg as a fourth teleost yolk protein was confirmed for barfin flounder by Western blotting using an antiserum raised against a recombinant C-terminal Vg polypeptide[127]. With respect to embryonic nutrition or physiology, the functions of the β'-c and the yolk protein derived from the C-terminus of Vg are unknown[79]. Interestingly, we found that β'-c from fish eggs was one of the most powerful egg allergens to humans (Kubo *et al.*, unpublished), suggesting that β'-c has remarkable antigenicity and good potential for induction of Vg-specific antibodies.

4. *Proteolysis of vitellogenin*

It is becoming apparent that the ovarian enzyme responsible for specific conversion of Vg into its constituent yolk proteins in vertebrates is cathepsin D[141,166]. Both cathepsin D and Vg are known to exist together in the multivesicular bodies of oocytes from vitellogenic rainbow trout[184]. We recently demonstrated for white spotted charr (*Salvelinus leucomaenis*) that Vg can be converted into its product yolk proteins *in vitro* by bovine cathepsin D or by a pepstatin-sensitive, water-soluble fraction of vitellogenic charr ovaries[77]. Carnevali *et al.*[20] later purified cathepsin D and another lysosomal cathepsin (cathepsin L) from the ovary of sea bream (*Sparus aurata*) and suggested that cathepsin D was responsible for the generation of yolk proteins from Vg in this species. Subsequently, a pepstatin A-sensitive enzyme involved in Vg processing was purified from the masu salmon (*O. masou*) ovary[80] and identified as a cathepsin D-like protein. Activity of cathepsin D and another lysosomal cathepsin (cathepsin B) was high in ovaries from sea bream undergoing early vitellogenesis, indicating that both cathepsins may be responsible for Vg processing in this species[20]. Interestingly, cathepsin D and cathepsin L were found to be potential markers of egg quality in sea bream or sea bass (*Dicentrarchus labrax*)[19]. For example, eggs of sea bass are identified as being of good vs. poor quality by virtue of their tendency to float or sink, respectively, in seawater. Activity of cathepsin D was highest in sinking sea bass eggs, while cathepsin L's activity was higher in floating eggs. Since cathepsin D is an estrogen-regulated protein, at least in a human breast cancer cell line[21], and appears to be involved in egg quality as described above, EDCs may affect fish egg quality by disrupting activation and/or expression of cathepsin D during oocyte development and maturation. Utilization of cathepsin D as a new biomarker for the presence of EDCs in the environment was recently suggested by Carnevali and Maradonna[18] (see also section VII).

5. *Multiplicity of vitellogenin*

While only a single class of Vg gene or protein had been confirmed in teleosts up until the early 1990s, multiple Vg genes or cDNAs encoding Vg had been described for the chicken[183,204] and for *X. laevis*[40,206]. Recently, evidence has been accumulating that for fishes multiplicity of Vg genes and proteins exists across diverse taxa[78]. This multiplicity is not entirely surprising because of the apparent genome duplication that occurred early during the evolution of *Actinopterygii* as well as the subsequent additional genome duplications which are evident in many lineages of teleosts[192,203].

Full-length cDNAs encoding two clearly distinct types of Vg were obtained from mummichog, *Fundulus heteroclitus*[107,108], and haddock, *Melanogrammus aeglefinus*[165]. For rainbow trout, Trichet *et al.*[196] revealed 20 complete Vg genes and 10 pseudogenes per haploid genome, although these genes differed from one another by less than 3% at the nucleotide level and likely produce indistinguishable protein products. On the other hand, Wang *et al.*[210] showed that the zebrafish (*Danio rerio*) has at least seven different functional Vg genes including one encoding a novel form of Vg (Vg3) that is characterized by a missing Pv domain (phosvitin-less Vg). With regard to Vg proteins, at least two classes of Vgs or their yolk protein products have been detected in several teleost species (review: Hiramatsu *et al.*[78]), including tilapia (genus *Oreochromis*), barfin flounder, haddock, and medaka (*Oryzias latipes*). We recently purified and characterized three forms of Vg (VgA, VgB, and a putative phosvitin-less Vg (VgC)) from the plasma of estradiol-treated white perch[81]. To purify the three perch Vgs, we utilized several types of ion-exchange chromatography including a novel, fast flow, strong anion-exchanger (POROS media), followed by gel filtration. Each Vg appeared to be distinct in its profile of antigenicity. Endoprotease digestion of each Vg followed by N-terminal peptide sequencing and mapping of the peptides to published Vg sequences supported the immunological evidence that each white perch Vg is a distinct protein and not the degraded product of any other Vg protein.

Because structural and functional multiplicity of Vgs is a relatively new concept for the research on teleosts, naming and classification of dual or multiple Vg proteins and their corresponding genes was somewhat confusing, even for individual species. Therefore, we proposed an interim model for classification of multiple Vgs, including the three perch Vgs discovered in our study, as follows. Vitellogenins that possess a complete yolk protein domain structure (complete Vg: NH_2-LvH-Pv-LvL-β'-component-C-terminal coding sequence-COOH) can be divided into two types, VgA and VgB. These complete Vgs usually elute in fractions at relatively high NaCl concentration during anion-exchange chromatography. The VgA type of complete Vg is structurally homologous to mummichog VgI, barfin flounder VgA, haddock VgA, and medaka Vg1, and its constituent LvH (LvI) is heavily degraded during oocyte hydration associated with final maturation (see section III). The VgB type of complete Vg is structurally homologous to mummichog VgII, barfin flounder VgB, haddock VgB, and medaka Vg2, and its constituent LvH (LvI) is either not degraded or only partially hydrolyzed during oocyte maturation. An incomplete Vg lacking a phosvitin domain, or having a greatly shortened phosvitin domain, and that is most homologous to zebrafish Vg3, insect Vgs, or chicken VgIII should be considered to be a C type of Vg. Structural and biochemical features for classification of a VgC could include a lower content of phosphorus or serine residues, considerably lower molecular mass than VgAs or VgBs, and elution in fractions at lower NaCl concentration than the other types of Vg during anion-exchange chromatography.

Three classes of Vg proteins were also detected in mosquitofish (*Gambusia affinis*) and their corresponding complete cDNA sequences (VgA, VgB, and phosvitin-less Vg or VgC) were determined[172]. Further investigations of the extent of Vg multiplicity among teleost species drawn from diverse phylogenetic taxa are now in progress using several degenerate PCR primer sets targeted to distinctive regions of each type of Vg transcript. Preliminary results from these trials are shown in Fig. 2. More than one form of Vg transcript was detected in all species examined. In addition to mosquitofish and white perch, all three forms of Vg (A–C) were detected in red sea bream (*Pagrus major*), white-edged rockfish (*Sebastes taczanowskii*), mummichog, and striped mullet (*Mugil cephalus*). It appears that, in general, members of higher teleost taxa (*Paracanthopterygii* and *Acanthopterygii*) express both VgA and VgB. Moreover, phosvitin-less Vg (Vg-C) seems to be widely present among teleosts, supporting the concept that a primitive Vg in the ancestor of both invertebrates and vertebrates was likely a phosvitin-less Vg[210].

6. *Induction of multiple vitellogenins by estrogen*

In fishes possessing two or more Vg proteins, differential responses of the liver to estrogen with regard to production of each distinct Vg are just beginning to be elucidated. For example, Takemura and Kim[191] developed sandwich ELISAs for two distinct Vgs (VTG210 and VTG140) in a tilapia species (*O. mossambicus*).

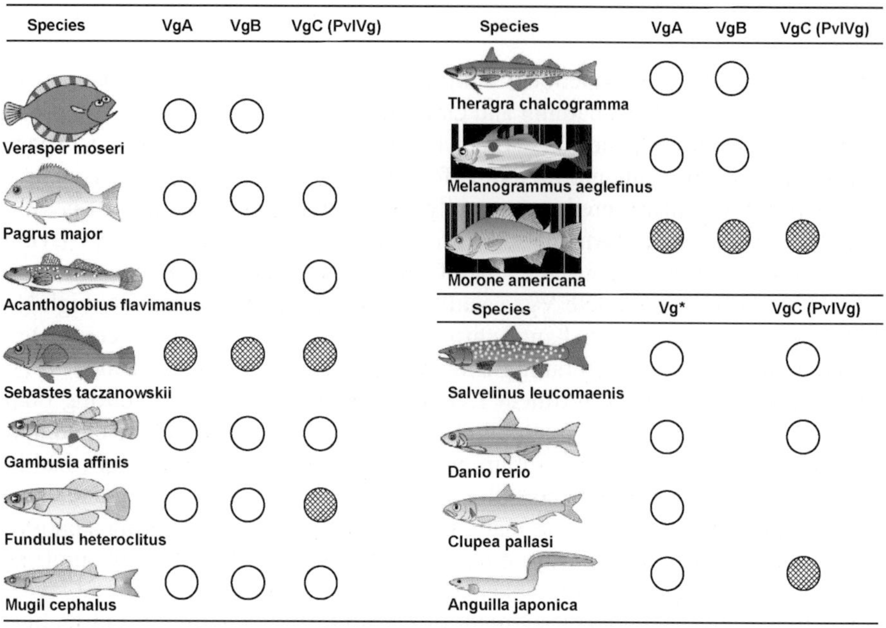

Species	VgA	VgB	VgC (PvlVg)
Verasper moseri	○	○	
Pagrus major	○	○	○
Acanthogobius flavimanus	○		○
Sebastes taczanowskii	●	●	●
Gambusia affinis	○	○	○
Fundulus heteroclitus	○	○	●
Mugil cephalus	○	○	○
Theragra chalcogramma	○	○	
Melanogrammus aeglefinus	○	○	
Morone americana	●	●	●

Species	Vg*	VgC (PvlVg)
Salvelinus leucomaenis	○	○
Danio rerio	○	○
Clupea pallasi	○	
Anguilla japonica	○	●

*Vitellogenin which does not belong to A, B, or C (PvlVg: phosvitin-less vitellogenin) group.

Fig. 2. Multiple vitellogenin (Vg) mRNA transcripts in various teleosts. Full-length (open circle) or partial (shaded circle) cDNA sequences encoding Vg mRNA were obtained for all but three species in our recent studies (unpublished). The three Vgs present in mosquitofish (*Gambusia affinis*) and white perch (*Morone americana*) also have been characterized as purified proteins[81,172]. Complete cDNA sequences encoding mummichog (*Fundulus heteroclitus*), zebrafish (*Danio rerio*), and haddock (*Melanogrammus aeglefinus*) Vgs were obtained from GenBank (accession numbers T43141 and AAB17152 for mummichog VgA and VgB; AAK94945 and AAG30407 for zebrafish Vg and VgC; AAK15158 and AAK15157 for haddock VgA and VgB, respectively). See section II.5 for explanation of the classification of complete Vgs (VgA and VgB) and the incomplete, phosvitin-less Vg (PvlVg or VgC).

The VTG140 (a putative phosvitin-less VgC in this species) exhibited clearly lower sensitivity to E_2 induction *in vitro*. The VTG210 was induced in primary hepatocyte cultures exposed to E_2 for 2 days at doses of 1×10^{-7} M or above, while VTG140 was only induced at E_2 doses of 5×10^{-7} M or above. After a single injection of different doses of E_2 into male tilapia *in vivo*, plasma levels of VTG140 were always lower than those of VTG210. For example, 24 h after injection of E_2 at a dose of 10 μg/g body mass, VTG140 levels reached only ~300 μg/ml whereas VTG210 levels reached ~2300 μg/ml. Very similar results were obtained for the dual Vgs (Vg-530 and Vg-320) of Japanese common goby (*Acanthogobius flavimanus*)[146]. The goby Vg-320, thought to be phosvitin-less VgC in this species, was much less responsive to E_2 induction than the 'complete' Vg-530, as was the case in tilapia. It remains to be verified whether the comparatively weak induction by estrogen of VgCs is common in other fishes. With regard to their inducibility by estrogen, differences between teleost A-type and B-type Vgs remain to be explored.

7. Choriogenins and choriogenesis

Recently, other estrogen-inducible proteins synthesized by the liver, choriogenins (Chgs), have been used as new biomarkers of exposure of fishes to estrogenic EDCs (review: Arukwe and Goksøyr[9]). Teleost oocytes and eggs are surrounded by an acellular egg envelope (also named the chorion, zona radiata, or vitelline envelope (VE)), which is generally composed of a thin outer layer and a thick inner layer. Usually, most of the vitelline envelope of teleost species is composed of the inner layer, which is constructed of three or four major subunit proteins, the Chgs. Considerable growth in the thickness of this inner layer has been observed to occur simultaneously with vitellogenesis[179].

Elucidation of the fundamental mechanisms of formation of the teleost VE, known as choriogenesis, has been difficult and often confusing. Based on histological data, the origin of the material comprising the inner VE layer was originally considered to be the oocyte (review: Dumont and Brummett[36]). However, a series of studies of medaka demonstrated that two precursor proteins of the VE matrix are synthesized in the liver under the influence of E_2, secreted into the bloodstream, and then incorporated into the inner layer of the VE of growing oocytes[63,138]. Recently, LaFleur *et al.*[107] and Yamagami[213] proposed the general name 'choriogenin' for these circulating, liver-derived precursors of fish egg envelope proteins. Subsequently, there have been numerous reports on immunological detection of circulating Chgs in several teleost species[85–87,110,150]. These reports are inconsistent with the scenario that VE proteins are exclusively derived from the oocyte itself, as indicated by the earlier histological studies.

The results of recent molecular biological studies shed new light on the process of choriogenesis in teleosts. Murata *et al.*[139,140] isolated cDNA clones encoding the two known Chgs of medaka, and analyses of the cDNAs revealed that these two medaka Chgs are homologs of two mammalian zona pellucida (ZP) proteins (ZPB and ZPC). Subsequently, a liver cDNA encoding a third medaka Chg (ChgH minor) was identified[188]. More recently, Kanamori[99] isolated cDNAs encoding protein components of the outer layer of the VE from ovaries of immature medaka. The deduced protein products were identified as homologs of mammalian ZPA, ZPB, and ZPC. Accordingly, it seems that at least six ZP homologs, three of which are liver-derived and three of which are synthesized by the ovary, are involved in choriogenesis in the medaka.

Identification of the complete spectrum of Chgs contributing to the growth of VE in other teleost species remains to be accomplished. Three cDNA clones encoding choriogenins have been isolated from the liver of rainbow trout[88]. On the other hand, cDNAs encoding major components of the VE (ZPB and ZPC) have been isolated from the ovary of carp (*Cyprinus carpio*) and zebrafish. Expression of transcripts encoding the carp ZP homologs was exclusively found in the ovary[24,25,209].

To date, thorough purification of liver-derived Chgs has been reported for only three species of teleosts, medaka[62,138], Sakhalin taimen, *Hucho perryi*[181], and masu salmon, *O. masou*[47]. In these species, two distinct Chgs (high molecular weight and low molecular weight Chgs: ChgH and ChgL) have been identified. The ChgH is constructed of a single subunit, of molecular mass ranging from 53 to 76 kDa, and it has a high content of glutamic acid and proline. The ChgL is constructed of a single subunit (46–47 kDa) rich in asparatic acid and glutamic acid. Murata *et al.*[140] reported the existence of a proline-X-Y repeat sequence, where X is mostly glutamine, encoded by a cDNA clone of medaka ChgH. This proline-X-Y repeat domain is a common feature of fish VE proteins[118]. The high content of glutamic acid in Chgs and ZP proteins may be related to the mechanism of transglutaminase-associated hardening of the VE after spawning or fertilization, as reported for the cod, *Gadus morhua*[149] and rainbow trout[61]. Transglutaminase is known to catalyze the formation of an amido bond between glutamine and lysine.

III. Roles of multiple vitellogenins in oocyte maturation

1. Genomic vs. cytoplasmic maturation

In a broad sense, a major function of Vg is to invest growing oocytes with nutrients essential for developing embryos. However, the recent discovery of multiple teleost Vgs whose yolk protein products have different proteolytic fates during oogenesis led us to explore potential differences among these proteins with respect to physiological function.

The term ‘ovarian follicle maturation’ is defined as the suite of LH-induced changes in the follicle that are necessary for the resumption of meiosis by growing oocytes[159]. The process of ovarian follicle maturation includes cytoplasmic changes in the oocyte that are required for successful fertilization and embryonic development, in addition to several structural and functional changes in the follicle that activate oocyte meiosis (nuclear or genomic maturation). These changes in the cytoplasm are referred to as ‘cytoplasmic maturation’. Our current knowledge of ovarian cytoplasmic maturation in teleosts was recently reviewed with a special attention to roles of dual or multiple Vgs in this process[78,158].

Typical morphological changes that occur in oocytes undergoing cytoplasmic maturation include extensive hydration and clarification of the ooplasm, which is accompanied in some species by coalescence of the major lipid droplets. With respect to the involvement of Vg and its product yolk proteins in oocyte cytoplasmic maturation, limited proteolysis of Vg-derived yolk proteins during ovarian follicle maturation was first discovered in *F. heteroclitus*[208]. This phenomenon was also found in other marine or brackish water fishes that exhibit remarkable hydration of their oocytes during follicle maturation[58,125]. The increase in free amino acid (FAA) content of the ooplasm resulting from selective proteolysis of yolk proteins is thought to play a significant role in generating osmotic effectors that drive water influx during oocyte maturation and hydration[59,194]. When further studies were conducted to identify proteolytic fates of yolk proteins derived from distinct Vgs (VgA and VgB) in barfin flounder, the results strongly supported the concept that proteolysis of Vg-derived yolk proteins is a major cause of oocyte hydration in teleosts spawning pelagic eggs[124,125].

2. Dual vitellogenin model

We recently discovered the duality of Vg in barfin flounder and proposed a 'dual Vg model' in which the two forms of Lv (LvA and LvB) derived from two distinct Vgs (VgA and VgB) play different physiological roles during cytoplasmic maturation and early vs. late embryonic and larval development[126,145]. A diagram of our current dual Vg model for this species is shown in Fig. 3. Both forms of Lv, as well as the other yolk proteins derived from each separate type of Vg (Pvs, β'-cs, and C-terminal components), are present in oocytes from vitellogenic (v) or fully grown follicles that have not yet initiated maturation. Each Lv appears to be comprised of distinct heavy chains (vLvHA and vLvHB) and light chains (vLvLA and vLvLB). During cytoplasmic maturation, the native dimeric LvB (vLvB; ~400 kDa) dissociates into the native monomer found in ovulated (o) eggs (oLvB, ~170 kDa). This large oLvB appears to be used as a major nutrient source at later stages of embryonic and larval development[145]. At the same time, the vLvA and other classes of Vg-derived yolk proteins are nearly completely degraded into FAAs. Only a small fragment of vLvA remains in ovulated eggs as the apoprotein component of a large lipoprotein consumed by late-stage embryos and larvae. The resulting large pool of Vg-derived FAA in the ooplasm of maturing oocytes likely acts as a potent osmotic effector driving oocyte hydration, and also functions as a source of diffusible nutrients during early embryogenesis[145]. These results suggest that the quantitative ratio of VgA vs. VgB in post-vitellogenic oocytes may control yields of FAAs, regulating oocyte hydration and egg buoyancy. Furthermore, the new dual Vg model explains for the first time how different protein and lipid components of Vg might be targeted for use at specific stages of embryonic and larval life. A similar pattern of yolk proteolysis during cytoplasmic maturation has been observed in oocytes of the haddock[165], suggesting that this type of dual Vg system may be widely conserved among marine teleosts spawning pelagic eggs.

3. Regulation of yolk protein proteolysis during cytoplasmic maturation

The specific enzymes involved in proteolysis of Vg-derived yolk proteins during oocyte cytoplasmic maturation have not been identified. In the sea bream, cathepsin L was implicated as the protease responsible for this process[20]. We recently developed similar evidence for the involvement of a cathepsin B-like protease in proteolysis of Lvs during oocyte maturation in barfin flounder[127]. Thus, there may be differences between species in the specific type(s) of cathepsin responsible for the maturation-associated proteolysis of Lv. The endocrine regulation of yolk protein proteolysis associated with cytoplasmic maturation is also largely unexplored. Morphological changes during cytoplasmic maturation, such as ooplasm hydration and clarification, clearly depend on a maturation-inducing hormone (MIH) in numbers of teleosts[215]. Our recent studies of oocyte cytoplasmic maturation in *Morone* species also support the scenario that proteolysis of yolk proteins is a MIH-dependent process[78]. In black

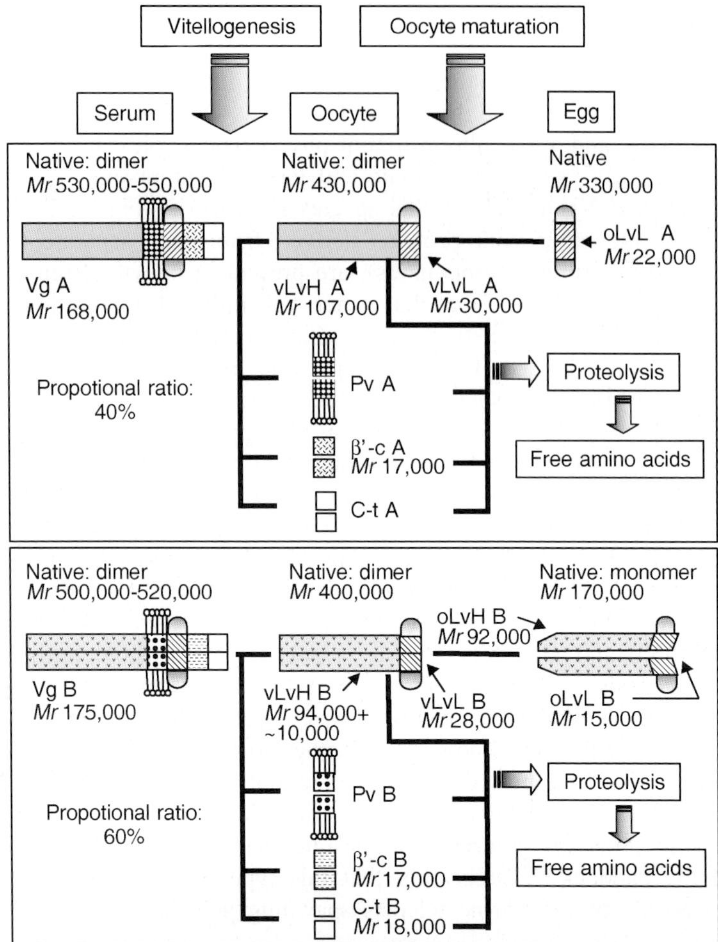

Fig. 3. Diagram of the dual vitellogenin (VgA and VgB) system of barfin flounder (*Verasper moseri*) illustrating Vg processing during oocyte maturation in this species. The molecular structures of the two Vgs and their product yolk proteins in serum from vitellogenic fish, growing oocytes, and ovulated eggs are shown. The yolk proteins include lipovitellin (Lv), phosvitin (Pv), β'-component (β'-c), and the newly discovered C-terminal component (C-t). The origin of each form of Lv chain or yolk protein (in VgA or VgB) is indicated by the upper case letter (A, B) following the abbreviation for that molecule. The abbreviations vLvH and vLvL refer to Lv heavy chains and light chains, respectively, present in growing (vitellogenic) oocytes. The abbreviations oLvH and oLvL refer to portions of Lv heavy and light chains, respectively, remaining in ovulated eggs. The percentage of total Vg comprised by VgA and VgB in fully grown oocytes is indicated on the left side of the figure (proportional ratio). Numbers indicate the relative molecular mass (M_r) of each Vg, yolk protein, or yolk protein fragment. Lipids bound to Vg, intact Lv, or Lv fragments are indicated by the darkly shaded half-ovals. See section III.2 for a detailed discussion of the dual Vg system of barfin flounder.

sea bass (*Centropristis striata*), this MIH-induced yolk protein proteolysis, as well as oocyte hydration, can be inhibited by bafilomycin A1, suggesting that vacuolar ATPase-dependent yolk acidification is required for activation of the specific cathepsins involved in these processes[180].

IV. Vitellogenin assay systems

1. Detection and measurement of vitellogenin

Vitellogenin has served as an ideal marker of the onset of puberty and the progression of maturity in female fishes, especially for farmed species. Assay systems for Vg have evolved mainly to support basic and applied research on fish reproduction. Because Vg binds substantial quantities of calcium and phosphorus, blood Vg levels were originally estimated in fish indirectly, by measuring calcium or alkali-labile protein phosphorus[38,143]. Immunological methods, such as immunodiffusion[67,163,201] or immunoelectrophoresis[30,32,54,121] were later adopted for direct detection and quantification of Vg. These early immunoassays had limited sensitivity, being able to detect only several μg of Vg per ml of plasma or serum. However, these techniques, especially radial immunodiffusion, are simple in operation, and their Vg detection limits are generally sufficient to measure the protein in blood from naturally maturing females.

In the late 1970s, a precise radioimmunoassay (RIA) was developed for fish Vg, achieving sensitivity on the order of several ng of Vg per ml of serum[16,89]. Although such RIAs are sensitive, they have several disadvantages, including a short working life of radiolabeled Vg-tracer and the need for specialized equipment and radioactive waste disposal. Beginning in the late 1980s, enzyme-linked immunosorbent assays (ELISAs) were developed to detect Vg[122,185]. These ELISAs eliminate disadvantages associated with RIA and have become the most popular method for measuring Vg in teleosts. We have utilized a sandwich ELISA for detecting Vg in salmonid species (e.g. Kwon *et al.*[106]; Matsubara *et al.*[128]; Hiramatsu *et al.*[82]) and a competitive ELISA for measuring Vg in some other teleosts (e.g. Heppell *et al.*[74]; Heppell and Sullivan[75]; Parks *et al.*[157]). Assay sensitivities and working ranges of these ELISA systems are generally equal to or better than those achieved using RIAs. The sandwich ELISA tends to require more primary antibody (anti-Vg) and has more procedural steps, while the competitive ELISA requires more purified Vg to coat the ELISA plate for antibody capture.

In addition to detection and measurement of Vg proteins, recent studies have utilized Vg mRNA expression as a marker for assaying EDCs. Such measurements have been conducted by conventional or real-time quantitative RT-PCR or by advanced macroarray technology[22,45,109]. Expression of Vg mRNA is becoming relatively easy to measure due to recent advances in molecular biology procedures and instrumentation. However, since multiplicity of Vg transcripts is apparent in a number of teleosts (Fig. 2) and certain Vg variants may be relatively insensitive to induction by estrogens or EDCs (e.g. VgCs), investigators evaluating Vg mRNA expression will need to develop a rationale and justification for targeting specific Vg transcripts for assay. Researchers measuring expression of Vg proteins must also face this challenge of Vg multiplicity (see section IV.3). Additionally, after stimulation with E_2 or EDCs, Vg mRNA is cleared far more quickly from hepatic cells than Vg proteins are cleared from the blood plasma *in vivo*; therefore, expression of Vg mRNA may reflect only very recent (up to several days) exposure to EDCs in male fish (see section VI.2).

2. Advanced immunoassays for vitellogenin

Current and future investigations of fish will require more sensitive Vg assays. As examples, measurement of Vg for detection of compounds with a weak estrogenic activity, and functional studies of minor Vg isoforms, will require assays sensitive enough to detect a few pg of Vg per ml of plasma or serum. Such highly sensitive assays will be especially valuable when quantities of plasma or serum are limited, as is often the case for small species like medaka, mosquitofish, or zebrafish, or when very low levels of Vg are present in the samples.

We recently developed an ultra-sensitive chemiluminescent immunoassay (CLIA) using anti-Vg labeled with acridinium ester to measure Vg in salmonids and carp[50,51]. The CLIA was applicable to five salmonid species and could detect Vg at levels as low as 30 pg/ml serum while requiring less time to complete than RIAs or ELISAs. The carp Vg assay is a simultaneous CLIA, meaning that both sample and labeled antibody are added to the reaction mixture at the same time (Fig. 4). An advantage of the simultaneous CLIA is its simplicity, achieved by omitting one step from the general sandwich CLIA technique, although the sensitivity of the simultaneous CLIA is generally lower ($\sim$2 ng Vg/ml sample). These CLIA techniques have the potential to achieve Vg detection limits within the fg/ml range, as demonstrated by the CLIA we developed for salmon growth hormone[49].

Since serum Vg concentrations can change widely and quickly in teleosts either during their natural reproductive cycle or in response to administration of estrogen, simple and fast semi-quantitative Vg assays are preferable as tools for initial screening, before more complex, quantitative Vg assays are applied. In addition, assessments of the exposure of fishes to EDCs conducted in the field would benefit from availability of rapid and simple Vg detection systems. Accordingly, we developed simple immunochromatography kits for detecting Vg in several fish species (Fig. 5). These kits, which resemble those marketed as the human 'Early Pregnancy Test', can be used successfully in field studies, even by relatively unskilled personnel. They are capable of detecting Vg at ng/ml levels within 15 min and can even be designed to specifically detect different Vg isoforms (A. Hara, unpublished).

3. Standard vitellogenin proteins and antibody generation

Purified Vg is not always necessary to generate a specific antiserum or antibody directed against Vg. Serum (or plasma) from naturally vitellogenic females, estrogen-treated fish, or extracts of vitellogenic ovaries can be used as antigens to immunize rabbits. The resulting antiserum can be adsorbed by male plasma to obtain an antiserum that can be used to detect serum proteins specifically present in females. These adsorbed antisera strongly recognize Vg, although their specificity for Vg must be confirmed because in some species they may react with female-specific proteins other than Vg, such as the choriogenins (see section II.7). To avoid cross-reactivity

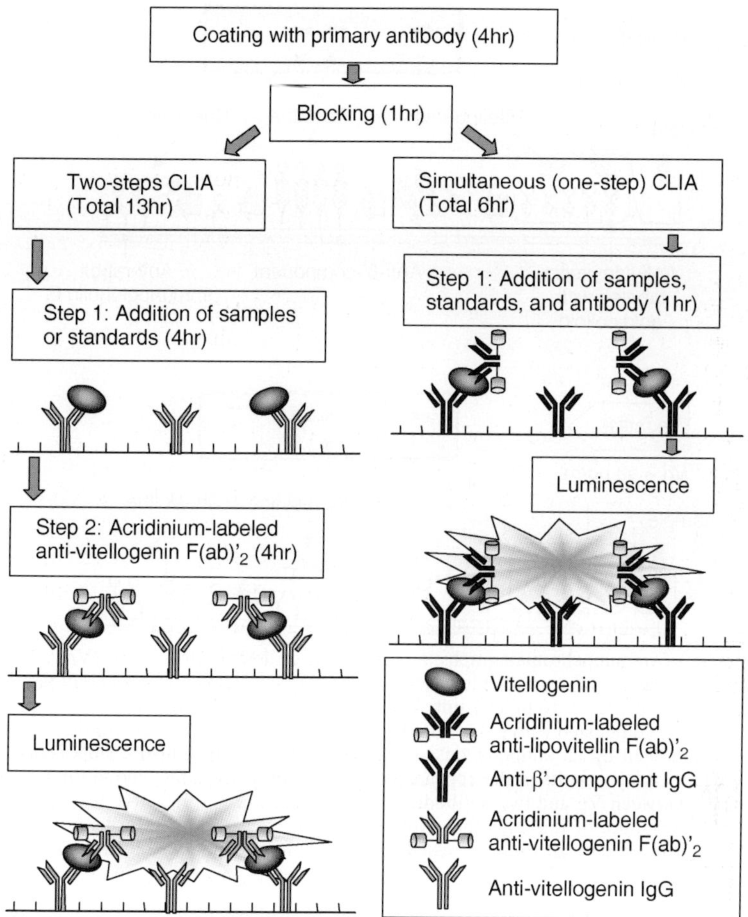

Fig. 4. Principle of chemiluminescent immunoassay (CLIA) of vitellogenin (Vg). Polyclonal antibodies raised against two distinct classes of Vg-derived yolk proteins (anti-β'-component and anti-lipovitellin), which are derived from different parts of the same parent Vg molecule, can be utilized to develop a simultaneous one-step CLIA (right side of the figure). Alternatively, a single anti-Vg antibody can be used in a two-step CLIA (left side of the figure). Advantages of each type of CLIA are discussed in detail in section VI.2. IgG, immunoglobulin G.

problems, it is generally preferable to use purified Vg as an antigen, if it is available. In addition to Vg, its product yolk proteins, Lv and β'-c, can be used as antigens to raise Vg-specific antibodies. As mentioned earlier, β'-c seems to possess particularly strong antigenicity, while purification of Lv is relatively easy because it is generally the dominant protein present in fish ovarian extracts. Native Pv is not antigenic[76]. Development of two separate antisera, one directed against Lv and the other against β'-c, has one obvious advantage as compared to raising a single antiserum against Vg. Since Lv and β'-c are both components of the Vg molecule, but are antigenically distinct from one another, generation of anti-Lv and anti-β'-c antisera allows

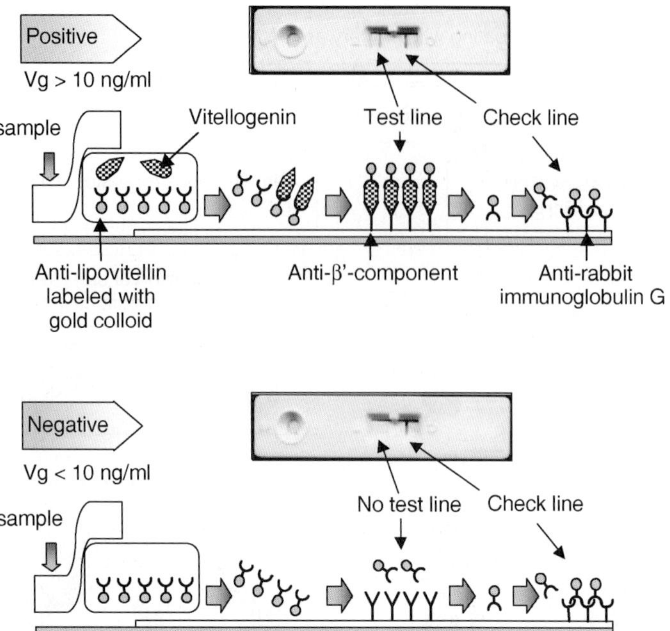

Fig. 5. Principle of immunochromatography as performed to detect vitellogenin (Vg). The procedure illustrated involves the use of a simple kit similar to that marketed as the human 'early pregnancy test'. Vitellogenin in the sample (>10 ng/ml) initially binds to a polyclonal rabbit antibody raised against lipovitellin (anti-Lv) conjugated with gold colloid and then diffuses to the position of the test line window where it is immobilized by an antibody raised against β'-component (anti-β'-component). A positive reaction (existence of Vg) results in the appearance of a line in this window due to the formation of an immunocomplex between Vg and two antibodies, while absence of Vg (negative reaction) produces no test line due to the failure of this immunocomplex to form. Excess anti-Lv forms a line in the check window where it is trapped by an antibody raised against rabbit immunoglobulin G, verifying the validity of assay results.

development of a simultaneous two-site CLIA or ELISA in which each antiserum is utilized at different steps of the procedure. Trials have been conducted to obtain a 'universal Vg antibody' by screening monoclonal antibodies with various heterologous Vgs, or by using a synthetic consensus peptide representing the conserved N-terminal amino acid sequence of Vg as an antigen[42,73]. Results of these trials indicated that it is feasible to generate antibodies capable of recognizing Vg without regard to species and that development of a universal Vg assay is an achievable goal, although a 'universal' Vg standard protein, necessary for quantitative assay, is still lacking.

Vitellogenin is known to be a proteolytically unstable protein[178]. The instability of Vg is probably related to its role as a precursor for shorter yolk polypeptides. Therefore, use of suitable protease inhibitors is necessary during blood sampling, purification, and storage in order to avoid degradation of Vg. Among these, several serine protease inhibitors, such as aprotinin or phenylmethanesulfonyl fluoride, are often used in Vg purification procedures. A 170 kDa, trypsin-like, serine protease that possibly degrades Vg during purification was identified in the plasma of the tilapia, *O. niloticus*[90]. It follows that, for quantitative Vg assays, suitable quality control procedures must be in place to ensure that each batch of Vg standard is stored under identical conditions that prevent proteolytic breakdown.

Vitellogenin immunoassay procedures developed for environmental toxicology must ensure that resulting Vg values are consistent within and between assays or studies. At the very least, complete assay validations (e.g. intra- and inter-assay variation, etc.) are necessary. Ideally, the extent of Vg multiplicity in the target species also should be considered. Investigators should verify how many distinct Vg proteins exist in the species under investigation or at least should identify the major form of Vg in that species. Without this information, the Vg immunoassay employed could possibly be directed against only a single class of minor Vg proteins, potentially resulting in an insensitive assessment or measurement of a Vg that is relatively resistant to induction by estrogen or estrogenic substances. If a Vg immunoassay is based on an antiserum directed against an antigen mixture that includes all forms of Vg (total Vg) in the target species, the composition of each Vg (e.g. the ratio of VgA:VgB:VgC) in the standard Vg solution and samples must be fairly constant if assay results are to be highly reproducible. In the future, researchers should endeavor to prove the constancy of Vg ratios in an induction experiment with various doses of estrogen because distinct Vgs show different sensitivities to estrogen in terms of dose–response kinetics and maximal Vg levels produced (see section II). For some species, it may eventually prove advantageous to utilize an antiserum and assay system targeting a specific, major estrogen-sensitive Vg rather than an assay that measures all forms of Vg (total Vg). However, although the multiplicity of Vg has been confirmed for several teleost species, procedures for purifying each separate form of Vg have been developed for only a few species. Advanced purification methods, such as our procedures for isolating the three Vgs of white perch[81], need to be developed in order to isolate multiple Vgs from other species for use as antigens and immunoassay standards.

V. Applications of vitellogenin as a biomarker for detection of estrogenic activity

1. Laboratory vitellogenin assays

A large number of laboratory studies of fishes have utilized the appearance of Vg, either in the bloodstream (*in vivo* Vg induction assay) or in liver cell cultures (*in vitro* Vg induction assay), as an indicator of the estrogenic bioactivity of various natural or man-made compounds (reviews: Palmer and Selcer[154]; Sumpter[189]; Kime *et al.*[103]; Pait and Nelson[151]; Arukwe and Goksøyr[9]). Pelissero *et al.*[162] first proposed the use of an *in vitro* Vg induction assay for rainbow trout to test the estrogenic activity of several compounds, including phytoestrogens. Using these *in vivo* and *in vitro* assay procedures, numerous types of chemicals were found to possess estrogenic activity based on their ability to induce Vg production in a dose-dependent manner.

An investigation by Jobling and Sumpter[93], as well as a considerable number of later studies (review: Pait and Nelson[151]), revealed that exposure of male fish to nonylphenol (NP) elevated plasma Vg levels or hepatic Vg mRNA expression and inhibited testicular growth and spermatogenesis in rainbow trout, eelpout (*Zoarces viviparus*), sheepshead minnow (*Cyprinodon variegatus*), and carp. Certain pesticides (DDT and its metabolites, and methoxychlor) and a number of industrial chemicals (bisphenol-A, PCBs, and octylphenol) also showed estrogenic bioactivity in fish as evidenced by their ability to stimulate vitellogenesis. Other industrial chemicals, including PCBs, polyaromatic hydrocarbons (benzo[*a*]pyrene and β-naphthoflavone), and insecticides (Mirex and endosulfan) inhibit estrogen-mediated processes, including vitellogenin synthesis[6,11,23,28].

2. Field surveys using vitellogenin assays

It was quite a shock to environmentalists when it was first discovered that the waters of many rivers in the United Kingdom and Europe that receive sewage effluents have a strong estrogenic potency as judged from the induction of vitellogenesis in male fish caged and immersed in river water[69,70,164,190]. Further investigations have shown that wild male fish living in streams receiving sewage effluent are morphologically feminized as well; males have an intersex gonad in which the testis contains ovarian tissue[95]. The major estrogenic chemicals that have been identified in sewage effluents include natural estrogens (E_2 and estrone), 17β-ethinylestradiol (EE_2), and alkylphenols[34,93,94,170,212]. Similar induction of Vg in caged and wild fish in freshwater streams and rivers receiving sewage effluent has also been reported in the United States[41,43,55,195].

Sewage and industrial wastes also enter estuaries and the sea, both directly and as a consequence of river discharges. A somewhat lesser estrogenic activity of estuarine and seawaters might be expected based on the greater dilution factors involved when sewage or industrial effluents enter these environments. However, feminization of marine teleosts such as flounder (*P. flesus*) has been observed in the United Kingdom and the Netherlands; feminization of the male flounder was confirmed with respect to the atypical formation of an ovotestis coinciding with extremely high serum Vg levels[3,4,91,115–117,129]. Moreover, similar biochemical feminization has been reported in male marbled sole (*Pleuronectes yokohamae*) in Japan[71] and in swordfish (*Xiphias gladius*) and bluefin tuna (*Thunnus thynnus*) in Italy[46]. Morphological feminization of males, involving either intersex testis or intermediate male secondary sex characteristics, has also been observed in marbled sole in Japan[71] and sand goby (*Pomatoschistus minutus* and *P. lozanoi*) in the UK[104].

Recently, we developed sandwich ELISAs for each of the two distinct Vgs (Vg-530 and Vg-320) of the Japanese common goby (*A. flavimanus*) and used the two Vg assays to survey, for the first time, the impact of environmental estrogens on production of more than one distinct Vg molecule by a fish species[146,147]. These field studies were conducted in various coastal waters of Japan. The Japanese common goby is a suitable model species for monitoring chemical pollutants in coastal waters because of their broad distribution in brackish water and seawater environments throughout Japan. In a preliminary experiment, male gobies were exposed to water containing E_2, nonylphenol (NP), and bisphenol-A (BPA). Among these compounds, E_2 showed the strongest estrogenic activity, inducing production of both goby Vgs at E_2 concentrations of 10.5 ng/l seawater. Exposure of the male fish to NP resulted in a significant induction of both Vgs at an NP concentration of 19.0 μg/l, while fish exposed to 133.7 μg/l BPA produced neither type of Vg. A subsequent field survey of serum Vg levels in male goby revealed that the mean Vg levels in serum of fish sampled from some estuaries near urban areas of Japan were similar to values from the fish exposed to water containing 10.5 ng/l E_2. However, the actual concentrations of E_2 in the environments where the fish were sampled were considerably lower than this value (see section V.3). While the appearance of Vg in field-collected male goby may indicate that these fish are currently exposed to exogenous estrogens or estrogenic compounds, and could be taken as a warning of potential reproductive damage, these assumptions still need to be rigorously evaluated, as described below for laboratory studies of medaka.

3. Bioaccumulation and other enhancers

Levels of E_2 in the serum of male fish exposed to water containing E_2 can be much higher than those levels in the surrounding water. For example, serum E_2 levels in male goby (*A. flavimanus*) exposed to water containing 100 pg/ml E_2 reached $\sim$2 ng/ml[146]. While it is possible that exposure to low levels of estrogen may stimulate estrogen biosynthesis in males as it does in female fish[153], it is also possible that this phenomenon is due to selective bioaccumulation of environmental E_2. However, the specific mechanisms of this type of bioaccumulation remain to be verified. Sex steroid hormone binding globulin (SHBG) or sex steroid binding proteins (SBPs) have been found in the serum of several teleost species (Fig. 1) and possess a strong binding affinity to sex steroids, including E_2 and testosterone[182]. Accordingly, SHBG and SBPs are believed to be important modulators of endogenous steroid hormone action[65]. Because the blood serum comes into extraordinarily close proximity with the environment as it courses through the vascular network underlying the respiratory epithelium, it is possible that these and other serum proteins may play a role in bioaccumulation of environmental E_2 by creating a localized positive transepithelial concentration gradient for estrogen. The highly lipophilic nature of most estrogens also likely facilitates uptake of these compounds.

In reality, fish are exposed to a spectrum of various kinds of estrogenic chemicals, rather than to any single EDC, in polluted aquatic environments. Thus it is important to know how fish respond to mixtures of estrogenic compounds rather than to any individual estrogen. As mentioned earlier, levels of E_2 and NP in seawater from urban areas of Japan were measured and judged to be insufficient to independently induce Vg production by male goby, even though male fish with appreciable serum Vg levels were observed in these areas[147]. After consideration of these results, we concluded that the estrogenic activity of estuarine and seawaters around urban areas of Japan depends not only on the impacts of E_2 or NP but also on the total synergistic effects of other environmental estrogens present in these areas. Synergistic effects of mixtures of estrogenic compounds on the induction of vitellogenesis in male or juvenile fishes have been described in some detail[190].

VI. Interpretation of vitellogenin presence

1. Background in control males

Interpreting Vg production by domesticated male fish that are caged and transferred to field sites being surveyed for the presence of EDCs is relatively easy. Resulting Vg levels can simply be compared to initial values in the same fish prior to transfer. On the other hand, when wild male fish are used as sentinels for the presence of estrogenic chemicals in the environment, it is important to acquire detailed knowledge of the particular species and its environment so that reasonable conclusions can be made about the reasons for Vg induction when it is observed. For example, Vg may be synthesized by the liver of male fish in response to estrogenic substances in food or forage, to low levels of endogenous estrogen, or to estradiol excreted by females living in close proximity. Some prepared fish diets contain phytoestrogens that can elicit an elevation of Vg levels in the blood of male fish[12,112–114,160–162], and natural forage obviously has the potential to produce the same effect. It is also possible that leaky expression of Vg genes may occur in males in the absence of an estrogenic stimulus.

The appearance of low levels of Vg in the serum of naturally maturing males has been observed in a variety of species including roach (*Rutilus rutilus*)[96], fathead minnow (*Pimephales promelas*)[68,156], cutthroat trout (*O. clarki*)[51], rainbow trout[174], mummichog[119], largemouth bass (*Micropterus salmoides*)[33], longear sunfish (*Lepomis megalotis*; A.O. Cheek, unpublished data), and cunner (*Tautogolabrus adspersus*)[132]. The levels of Vg measured ranged from several ng/ml in roach, fathead minnow, and rainbow trout to hundreds of μg/ml in sunfish and cunner (Table 1). Vg may cycle seasonally as it does in male longear sunfish, largemouth bass, and cutthroat trout. In longear sunfish, Vg is present at very low levels throughout the year, but increases during the non-spawning season. In bass and cutthroat trout, Vg only appears in the circulation of males just before or during the spawning season. In other species, Vg may be present at low, unchanging concentrations (5–100 ng/ml) as appears to be the case in male roach and fathead minnows. In cutthroat trout and largemouth bass, elevation of E_2 levels in male serum from basal values ($\sim$200 pg/ml) was relatively minor, reaching only $\sim$700 pg/ml during the spawning season. It seems reasonable to surmise that Vg could be induced by low levels of endogenous E_2 in males of these species, probably due to limited conversion of testosterone to E_2 (aromatase activity) by the testes or an unspecified somatic tissue.

To provide background information needed for interpretation of Vg levels in male fish exposed to putative EDCs, we verified that serum Vg levels in normal control males of salmonid species (in the presence or absence of conspecific females) never exceeded 10 μg/ml during the natural reproductive cycle and that this value for serum Vg levels can be utilized as a baseline for males in *in vivo* bioassays of EDCs[51,64]. The appearance of Vg in the serum of wild or cultured male salmonids at levels exceeding

TABLE 1

Levels of vitellogenin (Vg) measured in control males of example species

Classification	Vg concentration	Type of antibody	Fish source[a]
Cypriniformes			
Roach (*Rutilus rutilus*)[96]	5–100 ng/ml	Carp polyclonal	Wild
Fathead minnow (*Pimephales promelas*)[68,156]	5–60 ng/ml	Carp polyclonal	Hatchery
Salmoniformes			
Cutthroat trout (*Oncorhynchus clarki*)[51]	< 10 μg/ml	*Oncorhynchus* polyclonal	Hatchery
Rainbow trout (*Oncorhynchus mykiss*)[174]	30–50 ng/ml	Rainbow trout polyclonal	Hatchery
Cyprinodontiformes			
Mummichog (*Fundulus heteroclitus*)[119,152]	0[152], 100–200[119] μg/ml	*Fundulus* monoclonal[b]	Wild
Sheepshead minnow (*Cyprinodon variegatus*)[72]	0 mg/ml	*Cyprinodon* monoclonal	Wild
Perciformes			
Longear sunfish (*Lepomis megalotis*)[c]	100–350 μg/ml	*Lepomis* polyclonal	Wild
Cunner (*Tautogolabrus adspersus*)[132]	8–1500 μg/ml	Swordfish (*Xiphias*) monoclonal	Wild

[a] Wild caught or hatchery-reared fish.
[b] Monoclonal antibodies from different sources.
[c] Cheek, unpublished data.

10 μg/ml can be taken as evidence that the fish have been exposed to estrogenic chemicals in their environment. The source of the great variability in Vg concentrations measured in male fish of non-salmonid species is unclear. This variability could be due to differences in the types of antibodies used to develop ELISAs, monoclonal or polyclonal; in fish source, wild caught or hatchery-reared; or between species in phylogenetic position. The only clear pattern that emerges from Table 1 is that wild male perciforms appear to have high levels of Vg, regardless of antibody type.

2. *Signal duration*

Evaluating the presence of Vg in wild male fish also requires some knowledge of the persistence of Vg in the circulation. The half-life of plasma Vg seems to depend somewhat on the temperature to which a particular species is adapted. In rainbow trout, a cold-water species, the half-life of plasma Vg induced by EE_2 exposure ranged from 50 to 145 days, depending upon the EE_2 dose[173]. In sheepshead minnow, a warm water species, the half-life of plasma Vg induced by E_2 or nonylphenol exposure was approximately 14 days[72]. A similar half-life has been reported for flounder (*P. flesus*)[3]. In fathead minnow, roach, sheepshead minnow, and mummichog, several months are required for plasma Vg levels to return to baseline after stimulation with E_2, sewage effluent, or alkylphenols[72,152,156,169]. In these species, elevated plasma Vg levels appear to indicate that exposure to an estrogen mimic has occurred within the previous 4–5 months[72].

Interpretation of Vg mRNA levels is entirely different. In *Xenopus*, Vg mRNA has a half-life of 16–30 h in the absence of exogenous E_2 stimulation or 500 h in the presence of exogenous E_2[14]. In sheepshead minnows, Vg mRNA has a similarly short half-life of approximately 24 h, even in the presence of E_2 stimulation[72]. Clearly, one should measure both Vg mRNA and plasma Vg if there is a need to distinguish between very recent (within 24–36 h) or relatively recent (within weeks to months) exposure of wild fish to estrogen or EDCs.

3. *Species-specific variation in sensitivity*

The magnitude of the vitellogenic response to estrogenic stimulation varies between species and probably depends upon several factors. First, the capacity for maximal Vg synthesis may differ between species. Some fishes exhibit plasma Vg levels of only a few mg/ml Vg (e.g. carp) during the reproductive season, while others have plasma levels of Vg exceeding 50 mg/ml (e.g. rainbow trout)[98]. Second, the sensitivity of the estrogen receptor to endogenous estrogen and various xenobiotics differs between species[31]. Third, the temperature during a laboratory exposure may affect the amount of Vg produced, depending upon whether the temperature is close to what the fish experiences during the natural reproductive season[98]. Fourth, the life history stage at which exposure occurs also plays an important role in the apparent sensitivity of a species to estrogen with regard to vitellogenic response. This issue is addressed in section VI.4. A few recent laboratory studies have directly addressed the issue of species-specific responses. When the responses of immature male carp and rainbow trout to the same doses of estrogenic sewage effluent at the same test temperature (7–11°C) and exposure duration were compared, trout appeared to be more sensitive than carp. However, trout naturally undergo vitellogenesis and spawn at cooler temperatures (7–11°C) than carp (15–20°C)[98]. When the Vg response of adult male medaka and channel catfish (*Ictalurus punctatus*) to the same doses of E_2 at the same test temperature and duration were compared, both species were equally sensitive[193]. These two species live and reproduce at similar warm temperatures.

4. *Relationship between vitellogenin production and fitness*

Elevated production of Vg in males or immature fish may be associated with two different types of consequences for an individual: decreased survival or impaired reproduction. First, extremely high levels of Vg in the circulation of males (tens of mg/ml) may cause renal failure and therefore pose a threat to survival, as has been demonstrated in summer flounder, *Paralichthys dentatus*[44]. Second, high levels of Vg in males or juveniles may be produced in response to an EDC exposure that also alters sexual development and causes impaired fertility. In the first case, Vg overproduction causes an overt toxicity problem; in the second, it serves as an indicator of a more subtle reproductive problem. The association between elevated plasma Vg and other physiological parameters such as gonad morphology, sperm motility, and fertility is one of the most important issues of endocrine disruption in fishes. What does abnormally high Vg mean in terms of individual reproduction, population stability, and community structure?

A variety of recent studies have begun to explore the reproductive and developmental correlates of abnormally high plasma Vg concentrations. The association between Vg level, gonad morphology, and fertility has been most thoroughly documented in the roach, a cyprinid fish. The intersex condition, in which a male's testis contains variable amounts of ovarian tissue, occurs in 4–100% of wild male roach in the UK[97]. The background level of intersexuality in wild roach appears to be approximately 4%, occurring in populations living in lakes receiving no industrial or agricultural runoff, but the proportion of phenotypic males with intersex gonads varies from 9% to as high as 100% in certain rivers, depending largely upon the amount and type of sewage effluent entering the environment[95,97]. Rigorous field surveys have shown that the proportion of ovarian tissue in the intersex gonad is positively correlated with plasma Vg concentration[95,96]. In addition, these intersex fish have levels of circulating E_2 intermediate between those seen in normal males and females[96]. Laboratory studies of spawning by roach caught in the field also have shown that intersex fish have significantly reduced sperm motility and fertility[96]. In addition, Rodgers-Gray *et al.*[169] demonstrated that concentrations of sewage effluent that induce vitellogenesis in adult males are capable of causing juveniles to develop an intersex gonad structure: testes with an ovarian cavity. However, no development of oocytes in the testis was observed, begging the question of what causes development of ovarian follicles in the testis of wild intersex roach. Because the relationship between plasma Vg concentration, gonad morphology, and fertility has been explicitly catalogued in wild roach and the background level of Vg in males from reference sites is known, measurement of high Vg concentrations in male roach is a reliable indicator not only of exposure to an environmental estrogen but also of probable adverse effects on the reproductive function of males in the population being analyzed.

A relationship between Vg induction by exogenous estrogens or EDCs and impairment of reproductive function has not been as rigorously demonstrated for wild fishes of other species. However, this relationship has been evaluated in the laboratory, both with regard to effects on adult males and on developing juveniles. Adult male medaka[60,100,175], cunner (*T. adspersus*)[132], fathead minnows[68], and rainbow trout[174] have been examined in this regard. In medaka, circulating Vg concentration was negatively correlated with fertility[60], while in cunner, Vg levels and fertility were unrelated, but increasing Vg concentrations in males were associated with decreased daily egg production by their female mates[132]. In adult male medaka and fathead minnows, 3-week exposures of breeding pairs to nonylphenol or EE_2 showed that induction of Vg and of testis–ova (intersex gonads) occurred at lower estrogen or EDC concentrations than did alterations in fecundity and fertility[68,100,175]. Adult male roach exposed to sewage effluent had elevated Vg levels, but intersex gonads were not induced by a 4-month exposure[168]. These data suggest that in adults, activational responses, such as induction of vitellogenesis, are more sensitive to EDC exposure than are organizational responses, such as alterations of reproductive function.

The period of sex differentiation is likely the life history stage most sensitive to endocrine disruption. The issue of how abnormal Vg levels are related to sex differentiation and subsequent fertility has been most thoroughly examined by exposing developing medaka and rainbow trout to EDCs and steroidal estrogens at dosages that induce vitellogenesis. A variety of studies have established that exposure of developing medaka to E_2, alkylphenols, *o,p′*-DDT, methoxychlor, and β-hexachlorocyclohexane causes development of testis–ova[37,56,57,131,144,211]. In contrast, none of the compounds or doses examined to date, including nonylphenol, octylphenol, *o,p′*-2,2-bis(chloro-phenyl)-1,1-dichloroethylene (*o,p′*-DDE), *p,p′*-DDE, or chlordecone cause development of intersex gonads (testis–ova) in rainbow trout, even in studies in which juveniles were exposed from the eyed-egg stage through 1 year of age[1,17].

Recent work has specifically examined whether intersexuality, impaired fertility, and induction of vitellogenesis each require the same or different degrees of fish exposure to estrogenic EDCs. In a short-term exposure of medaka fry to 0.2–4.3 μg/l *o,p′*-DDT (for 14 days post-hatch (dph)), no intersex gonads were observed in adults sampled at 124 dph, and Vg induction was not detectable during the exposure. However, the fertility (percent of eggs fertilized) of exposed females and the hatching success of their progeny were significantly reduced at the lowest DDT dose[26]. Long-term developmental exposure (for approximately 60 dph) of medaka fry to *o,p′*-DDT[27], 4-nonylphenol, or 4-tert-octylphenol[176] induced significantly elevated Vg concentrations in males at the same doses that caused development of intersex gonads and significant reductions in both male and female fertilities. The hatching success of the progeny of exposed fish was also reduced (DDT only). Unfortunately, the possible correlation between Vg concentration and the proportion of ovarian tissue in the testis was not examined in these studies. Exposure of medaka to 4-tert-pentylphenol for up to 60 dph elevated male Vg levels, induced intersex gonads, and reduced male fertility. However, doses of 4-tert-pentylphenol that altered gonad development and impaired fertility were fourfold higher than those that induced Vg production[177].

Transgenerational and multigenerational exposure of fish to EDCs may elicit different physiological responses than exposures conducted for discrete periods of juvenile or adult life. Exposure of mating pairs of rainbow trout (P_1) to nonylphenol caused elevated Vg levels in the males without any alteration in sperm production or gonad morphology. Although fecundity and fertility were not measured, the physiological status of offspring was assessed. Both male and female offspring exhibited altered steroid hormone profiles and Vg concentrations. Male offspring (F_1) showed no changes in Vg production, but approximately 1% of these fish had intersex gonads (testis with some ovarian tissue). Female offspring (F_1) had elevated Vg levels, and approximately 12% of F_1 fish with male secondary sex characteristics had ovaries, i.e. they appeared to be masculinized females[174]. The absence of Vg induction in the F_1 males and the occurrence of intersex (both feminized and masculinized) F_1 fish are very different from the responses of juvenile rainbow trout to the same doses of nonylphenol. Discrete exposure of juveniles (for 1 year) caused Vg production by males, but no intersex gonads were found in either sex[1].

A similar disparity between the effects of a discrete exposure of juveniles to an EDC and the effects of a continuous transgenerational exposure has been observed in medaka. Exposure of medaka fry to 4-tert-pentylphenol (PP) from 12 h post-fertilization to 101 dph caused elevated Vg in males at doses ≥ 51.1 μg/l, and induced intersex gonads and reduced fertility at doses ≥ 224 μg/l. Offspring (F_1) of these exposed individuals (P_1, termed F_0 in this particular study) were then continually exposed to PP through 61 dph. Vitellogenin levels in the F_1 males were positively correlated with PP dose, but no PP dose caused Vg production that was significantly greater than that of F_1 controls. Intersex gonads were found in F_1 males exposed to ≥ 51.1 μg/l PP, a concentration fourfold lower than that required to induce intersexuality in P_1 males[177]. The limited data from laboratory transgenerational studies strongly suggest that the vitellogenic response to an EDC may be attenuated in the second generation, while sex differentiation becomes more sensitive to disruption by EDC exposure, at least in the case of alkylphenols.

Steroidogenesis also seems to be permanently altered in rainbow trout by transgenerational exposure to nonylphenol. The F_1 male progeny of fish exposed to nonylphenol (P_1) had twice the plasma E_2 levels found in controls, without any associated change in testosterone level. Levels of testosterone in the corresponding F_1 females were 13 times the control values, without an associated change in E_2 levels, but with a significant increase in plasma levels of Vg[174]. These data suggest that exposure of fish to an estrogen mimic during development may permanently alter expression or activity of steroidogenic enzymes in a sex-specific manner. Exposed males may have elevated aromatase activity (conversion of testosterone to E_2), while exposed females may have permanently elevated 17β-hydroxysteroid dehydrogenase activity (conversion of androstenedione or dehydroepiandrosterone to testosterone) and 3β-hydroxysteroid dehydrogenase activity (conversion of 5-androsten-3β, 17β-diol to testosterone)[35]. How elevated testosterone could cause increased vitellogenesis in these females without being converted to E_2 is a conundrum. Perhaps testosterone alters the number, affinity, or function of hepatic estrogen receptors.

Data from wild and domestic roach also support the idea that developmental exposure to an estrogenic EDC can permanently alter steroidogenesis. As noted above, wild intersex roach have plasma E_2 levels intermediate between those seen in normal males and females. In addition, they have higher testosterone levels than males and Vg levels intermediate between those seen in normal males and females[96]. Laboratory exposure of hatchery-reared roach to sewage effluent induced an elevation of E_2 levels in all juveniles (data not separated by sex)[169], again suggesting persistent changes in the mechanisms of steroidogenesis. Elevated Vg levels in adult males may then be a response to a developmental exposure to an EDC that causes permanent increases in production of endogenous estrogen.

What does the presence or level of Vg in the blood or liver of a male fish really tell us? Drawing general conclusions about the relationship between Vg induction by an EDC and the consequences of EDC exposure to fish reproductive function is difficult

TABLE 2

Juvenile vs. adult sensitivity to EDCs

Species	Life history stage exposed	Compound (dose range)	LOEC			F_1 generation	
			Vg induction in males	Intersex male gonad	Fertility	Effect on males	Effect on females
Rainbow trout[174]	Adult mating groups[a]	NP (1 or 10 μg/l)	1 μg/l	X	NE	E_2 increase, 1% feminization	T and Vg increase, 12% masculinization
Rainbow trout[1]	Juvenile[b]	NP (1 or 10 μg/l)	1 μg/l	X	NE	NE	NE
Roach[97]	All[c]	Sewage effluent	NE	+++	− − −	NE	NE
Roach[96]	All[d]	Sewage effluent	+++	+++	NE	NE	NE
Roach[169]	Juvenile[e]	Sewage effluent (12.5–100%)	≥23.6%	≥23.6%[e]	NE	NE	NE
Roach[168]	Adult male[f]	Sewage effluent (9.4–100%)	≥9.4%	X	NE	NE	NE
Medaka[175]	Adult male[g]	EE_2 (32.6–488 ng/l)	≥64 ng/l	≥64 ng/l	488 ng/l	NE	NE
Medaka[100]	Adult male[g]	NP (24.8–184 μg/l)	≥51 μg/l	≥25 μg/l	184 μg/l	NE	NE
Medaka[60]	Adult male[g]	OP (20–230 μg/l)	+++	≥74 μg/l	− − −	NE	NE
Medaka[26]	Juvenile[h]	*o,p′*-DDT (0.2–4 μg/l)	X	X	≥0.2 μg/l	NE	NE
Medaka[26]	Juvenile[i]	*o,p′*-DDT (0.3–5 μg/l)	≥0.3 μg/l	≥1.9 μg/l	≥0.3 μg/l	NE	NE
Medaka[176]	Juvenile[j]	NP (3.1–50 μg/l)	≥12 μg/l	≥12 μg/l	NE	NE	NE

Medaka[176]	Juvenile[j]	OP (6.3–100 μg/l)	≥11 μg/l	≥11 μg/l	NE	NE	NE
Medaka[177]	Juvenile P_1 to adult F_1[k]	PP (51–993 μg/l)	≥51 μg/l	≥224 μg/l	≥224 μg/l	Vg – increase	Vg – no change
Fathead minnow[68]	Adult mating pairs[l]	NP (10 or 100 μg/l)	≥10 μg/l	NE	100 μg/l	NE	NE
Cunner[132]	Adult male[m]	E_2 or EE_2 (various protocols)	+++	NE	– – –	NE	NE

Vitellogenesis is usually induced by exposure to an estrogenic EDC regardless of the life history stage at which treatment occurs, while fertility is significantly decreased at much lower EDC doses when exposure occurs during sexual differentiation. Concentrations given for various effects are the lowest concentration at which a significant effect occurred (LOEC, lowest observed effective concentration). NE, not evaluated; X, no effect at any dose tested; +++, effect positively correlated with dose (no LOEC determined); – – –, effect negatively correlated with dose (no LOEC determined); E_2, estradiol-17β; EE_2, ethinylestradiol; NP, nonylphenol; OP, 4-tert-octylphenol; *o,p′*-DDT, 1,1,1-trichloro-2-(*p*-chlorophenyl)-2-(*o*-chlorophenyl); PP, 4-tert-pentylphenol; T, testosterone; Vg, vitellogenin.

[a] Exposed 10 days per month for 4 months immediately prior to spawning.
[b] Exposed from eyed-egg stage to 1 year old.
[c] Wild male fish collected as adults from several rivers and artificially spawned in the laboratory.
[d] Wild male fish collected as adults from several rivers.
[e] Exposed 50–100 days post-hatch (dph), intersex gonads had ovarian cavity but no oocytes.
[f] Exposed for 1–4 months.
[g] Exposed for 21 days.
[h] Exposed 2–14 dph.
[i] Exposed 2–58 dph.
[j] Exposed 16 h post-fertilization to 60 dph.
[k] P_1 exposed 12 h post-fertilization to 101 dph, F_1 collected and exposed until 61 dph.
[l] Exposed 21 days.
[m] Various exposure protocols.

because of interspecific variation in sensitivity to different chemicals and differences in exposure regimen. To further complicate such interpretations, induction of vitellogenesis in adult males may occur in response to current exposure to an estrogen mimic or in response to elevated endogenous estrogen, which may be the result of a prior (developmental) exposure. One clear pattern emerges, however. Life history stage seems to determine whether induction of abnormal vitellogenesis or reproductive impairment is a more sensitive response to an estrogenic EDC (Table 2). In adults, significant increases in plasma Vg occur at the same or similar doses as do changes in gonad morphology, but fertility is either completely unaffected or impaired only at very high EDC doses. In other words, in adults activational responses such as production of Vg are more sensitive to estrogenic compounds than are organizational responses such as fertility and fecundity. Within broad limits, activational responses and organizational responses of juveniles to estrogenic EDCs seem to be equally sensitive. Significant increases in plasma Vg occur at the same or similar doses of an estrogenic EDC as do changes in sex differentiation, fertility, and fecundity. However, in a few cases of juvenile or transgenerational exposure, Vg production was unaffected by EDC doses that cause significant disruption of sex differentiation and fertility[1,26,174,177]. It can be surmised that elevated Vg levels measured in wild male fish should be considered a warning that juveniles undergoing sex differentiation under the same conditions may experience some reproductive impairment.

Will reproductive impairment in individuals have measurable, significant consequences for populations and communities of fish? One recent study specifically examined the relationship between Vg induction, sediment contamination, and measures of the abundance and diversity of wild flatfish (English sole, *P. vetulus*; bigmouth sole, *Hippoglossina stomata*; and hornyhead turbot, *Pleuronichthys verticalis*) off the California (USA) coast. Interestingly, although the level of plasma and hepatic Vg expression in males was similar to that seen in mature females, no correlation with species abundance or diversity was observed[171]. Clearly, exposure to an estrogen mimic is occurring in this case, but no significant adverse effect on flatfish populations or on the fish community was measurable. Many more studies evaluating population and community responses to EDC exposure in conjunction with assessments of Vg induction will be required to resolve this issue.

VII. New biomarkers: Choriogenins and more

1. Advantage of choriogenins as biomarkers

As noted in a recent review by Arukwe and Goksøyr[9], Chgs have a potential to be even more sensitive biomarkers of fish exposure to estrogenic EDCs than Vgs, based on the greater responsiveness of choriogenesis to estrogenic substances as compared to vitellogenesis in Atlantic salmon (*Salmo salar*) exposed to E_2, NP, and effluent from an oil refinery. We have developed ELISAs for two distinct salmonid Chgs (ChgH and ChgL, see section II.7) and evaluated the effect of E_2 administration on the induction of these Chgs and Vg in male masu salmon (Fujita, T. and A. Hara, unpublished). In male salmon injected with a low concentration of E_2 (10 μg/kg body mass), circulating levels of the Chgs were higher than Vg levels. However, the relationship between circulating levels of Chgs and Vg was reversed in fish injected with high doses (>100 μg/kg body mass) of E_2. These results indicate that Chg may be a better biomarker of fish exposure to estrogenic EDCs when responses to weak estrogens or low levels of estrogenic compounds are being assessed. These preliminary findings support the results of a study where real-time PCR was used to evaluate hepatic production of Chg and Vg transcripts by rainbow trout. Transcription of Chg mRNA was more highly activated than transcription of Vg mRNA in rainbow trout exposed to low doses of E_2 and α-zearalenol[22].

2. Other new biomarkers

In addition to Vg and Chgs, other estrogen-regulated proteins, such as cathepsin D and heat shock protein 70 (HSP70), have been suggested as candidate biomarkers of exposure of fish to estrogenic (e.g. E_2 and NP) and anti-estrogenic (e.g. β-naphthoflavone) compounds[18]. Cathepsin D gene transcription and enzymatic activity were up-regulated by E_2 and NP but down-regulated by β-naphthoflavone in the male goby, *Gobius niger*. In contrast, expression of the HSP70 gene was up-regulated by all three toxicant compounds in the goby. These findings suggest that, in addition to cathepsin D, HSP70 may be a useful biomarker of fish exposure to pollutants present in aquatic environments, especially in light of the numerous and multifaceted physiological roles of HSP70s in vertebrates[101]. As mentioned earlier, cathepsin D is one of the enzymes responsible for the specific proteolysis of Vgs and somehow contributes to egg quality of cultured fishes. Understanding the effects of EDCs on regulation of cathepsin D gene transcription or enzyme activity may be key to understanding impacts of these chemicals on fish reproduction and especially on oocyte growth during vitellogenesis. However, basic research on the regulation of cathepsin D expression (gene and protein) and activation during fish vitellogenesis have been limited so far[18,78,80,82].

VIII. Summary

Over the last decade, much progress has been made with regard to identification of EDCs and evaluation of their estrogenic potency. These developments have included establishment of accurate and reliable assay systems for measuring circulating Vgs and Chgs, as well as identification of other new biomarkers with the potential to detect and evaluate the potency of estrogenic EDCs. Future investigations need to focus on development of specific and sensitive immunoassays for individual Vg and Chg molecules and, in general, far more consideration needs to be given to the constitutive multiplicity of teleost Vgs and Chgs. On the other hand, substantial knowledge of the extent of estrogenic endocrine disruption in fish has accumulated from field surveys. It is becoming apparent that exposure of fishes to these compounds is widespread in marine, brackish, and fresh waters around the world. Combined with results from several laboratory studies of fishes, these observations send the obvious warning that many contemporary aquatic environments possess an estrogenic potency and have the potential to disrupt reproduction of fishes, or even wildlife and humans consuming water or aquatic species from these areas. The extent of estrogenic EDC impact on reproduction of wild fishes, and the proximal causes of such effects, have been and remain largely unknown. Basic research on reproductive processes in fish will reveal critical mechanisms subject to impairment by EDCs. Especially important in this regard will be complete elucidation of the roles of dual or multiple Vgs, cathepsins, and Chgs in ovarian follicle growth and maturation, egg quality, and embryonic and larval development. Such basic studies of the functional mechanisms of oogenesis and embryogenesis will, in turn, provide the next generation of biomarkers for assessing the potential impacts of aquatic EDCs on fishes, wildlife, and human beings.

Acknowledgements. We are deeply indebted to our collaborators, T. Fujita, H. Fukada, M. Shimizu, N. Ohkubo, K. Hiramatsu, S. Sawaguchi, and M. Fujita for their seminal contributions to this work. We also thank T. Mommsen, B.J. Reading, and C.R. Couch for the critical reading of the manuscript. The present study was partly supported by a Grant-in-Aid for Scientific Research from Ministry of Education, Culture, Sports, Science and Technology of Japan. This work was also supported by grants from the US Department of Agriculture (NRICGP Animal Reproduction, 01-35203-11131 to C.V.S) and the National Sea Grant (SG) College Program (North Carolina SG, NA46ROG0087 to C.V.S.), and the US Environmental Protection Agency (STAR EaGLe, R82945801 to A.O.C.).

IX. References

1. Ackermann, G., J. Schwaiger, R. Negele and K. Fent. Effects of long-term nonylphenol exposure on gonadal development and biomarkers of estrogenicity in juvenile rainbow trout (*Oncorhynchus mykiss*). *Aquat. Toxicol.* 60: 203–221, 2002.
2. Aida, K., P.V. Ngan and T. Hibiya. Physiological studies of gonadal maturation of fishes. I. Sexual differences in composition of plasma protein of ayu in relation to gonadal maturation. *Bull. Jpn. Soc. Sci. Fish.* 39: 1107–1115, 1973.
3. Allen, Y., P. Matthiessen, A.P. Scott, S. Haworth, S. Feist and J.E. Thain. The extent of oestrogenic contamination in the UK estuarine and marine environments – further surveys of flounder. *Sci. Total Environ.* 233: 5–20, 1999.
4. Allen, Y., A. Scott, P. Matthiessen, S. Haworth, J. Thain and S. Feist. Survery of estrogenic activity in United Kingdom estuarine and coastal waters and its effects on gonadal development of the flounder *Platichthys flesus*. *Environ. Toxicol. Chem.* 18: 1791–1800, 1999.
5. Amirante, G.A. Immunochemical studies on rainbow trout lipovitellin. *Acta Embryol. Morphol. Exp.* (Suppl.): 373–383, 1972.
6. Anderson, M.J., M.R. Miller and D.E. Hinton. *In vitro* modulation of 17-β-estradiol-induced vitellogenin synthesis: effects of cytochrome P4501A1 inducing compounds on rainbow trout (*Oncorhynchus mykiss*) liver cells. *Aquat. Toxicol.* 34: 327–350, 1996.
7. Arcand-Hoy, L.D. and W.H. Benson. Fish reproduction: an ecologically relevant indicator of endocrine disruption. *Environ. Toxicol. Chem.* 17: 49–57, 1998.

8. Arnold, A.P. and S.M. Breedlove. Organizational and activational effects of sex steroids on brain and behavior: a reanalysis. *Horm. Behav.* 19: 469–498, 1985.
9. Arukwe, A. and A. Goksøyr. Eggshell and egg yolk proteins in fish: hepatic proteins for the next generation: oogenetic, population, and evolutionary implications of endocrine disruption. *Comp. Hepatol.* 2: 1–21, 2003.
10. Bailey, R.E. The effect of estradiol on serum calcium, phosphorus, and protein of goldfish. *J. Exp. Zool.* 136: 455–469, 1957.
11. Barry, T., P. Thomas and G. Callard. Stage-related production of 21-hydroxylated progestins by the dogfish (*Squalus acanthias*) testis. *J. Exp. Zool.* 265: 522–532, 1993.
12. Bennetau-Pelissero, C., B.B. Breton, B. Bennetau, G. Corraze, F. Le Menn, B. Davail-Cuisset, C. Helou and S.J. Kaushik. Effect of genistein-enriched diets on the endocrine process of gametogenesis and on reproduction efficiency of the rainbow trout *Oncorhynchus mykiss*. *Gen. Comp. Endocrinol.* 121: 173–187, 2001.
13. Bergink, E.W. and R.A. Wallace. Precursor–product relationship between amphibian vitellogenin and yolk proteins, lipovitellin and phosvitin. *J. Biol. Chem.* 249: 2897–2903, 1974.
14. Blume, J. and D. Shapiro. Ribosome loading, but not protein synthesis, is required for estrogen stabilization of *Xenopus laevis* vitellogenin mRNA. *Nucleic Acids Res.* 17: 9003–9014, 1989.
15. Brown, C.L. and J.M. Nunez. Hormone actions and applications in embryogenesis. In: *Perspectives in Comparative Endocrinology*, edited by K.G. Davey, R.E. Peter and S.S. Tobe, Ottawa, National Research Council, pp. 333–339, 1994.
16. Campbell, C.M. and D.R. Idler. Characterization of an estradiol-induced protein from rainbow trout as vitellogenin by the cross reactivity to ovarian yolk fractions. *Biol. Reprod.* 22: 605–617, 1980.
17. Carlson, D., L. Curtis and D. Williams. Salmonid sexual development is not consistently altered by embryonic exposure to endocrine-active chemicals. *Environ. Health Perspect.* 108: 249–255, 2000.
18. Carnevali, O. and F. Maradonna. Exposure to xenobiotic compounds: looking for new biomarkers. *Gen. Comp. Endocrinol.* 131: 203–208, 2003.
19. Carnevali, O., G. Mosconi, M. Cardinali, I. Meiri and A. Polzonetti-Magni. Molecular components related to egg viability in the gilthead sea bream, *Sparus aurata*. *Mol. Reprod. Dev.* 58: 330–335, 2001.
20. Carnevali, O., R. Carletta, A. Cambi, A. Vita and N. Bromage. Yolk formation and degradation during oocyte maturation in seabream, *Sparus aurata*: involvement of two lysosomal proteinases. *Biol. Reprod.* 60: 140–146, 1999.
21. Cavailles, V., P. Augereau, M. Garcia and H. Rochefort. Estrogens and growth factors induce the mRNA of the 52K-pro-cathepsin-D secreted by breast cancer cells. *Nucleic Acids Res.* 16: 1903–1919, 1988.
22. Celius, T., J.B. Matthews, J.P. Giesy and T.R. Zacharewski. Quantification of rainbow trout (*Oncorhynchus mykiss*) zona radiata and vitellogenin mRNA levels using real-time PCR after *in vivo* treatment with estradiol-17β or α-zeralenol. *Steroid Biochem. Mol. Biol.* 75: 109–119, 2000.
23. Chakravorty, S., B. Lal and T.P. Singh. Effect of endosulfan (thiodan) on vitellogenesis and its modulation by different hormones in the vitellogenic catfish *Clarias batrachus*. *Toxicology* 75: 191–198, 1992.
24. Chang, Y.S., S.C. Wang, C.C. Tsao and F.L. Huang. Molecular cloning, structural analysis, and expression of carp ZP3 gene. *Mol. Reprod. Dev.* 44: 295–304, 1996.
25. Chang, Y.S., C.C. Hsu, S.C. Wang, C.C. Tsao and F.L. Huang. Molecular cloning, structural analysis, and expression of carp ZP2 gene. *Mol. Reprod. Dev.* 46: 258–267, 1997.
26. Cheek, A., T. Brouwer, S. Carroll, S. Manning, M. Brouwer and J. McLachlan. Developmental exposure to anthracene and estradiol alters reproductive success in medaka (*Oryzias latipes*). *Environ. Sci.* 8: 31–45, 2001.
27. Cheek, A., T. Brouwer, S. Carroll, S. Manning, J. McLachlan and M. Brouwer. Experimental evaluation of vitellogenin as a predictive biomarker of reproductive disruption. *Environ. Health Perspect.* 109: 681–690, 2001.
28. Chen, T.T., P.C. Reid, R. VanBeneden and R.A. Sonstegard. Effect of Aroclor 1254 and Mirex on estradiol-induced vitellogenin production in juvenile rainbow trout (*Salmo gairdneri*). *Can. J. Fish. Aquat. Sci.* 43: 169–173, 1986.
29. Christmann, J.L., M.J. Grayson and R.C.C. Huang. Comparative study of hen yolk phosvitin and plasma vitellogenin. *Biochemistry* 16: 3250–3256, 1977.
30. Craik, J.C.A. Plasma levels of vitellogenin in the elasmobranch *Scyliorhinus canicula* L. (lesser spotted dogfish). *Comp. Biochem. Physiol.* 60B: 9–18, 1978.
31. Crews, D., E. Willingham and J. Skipper. Endocrine disruptors: present issues, future problems. *Q. Rev. Biol.* 75: 243–260, 2000.
32. Crim, L.W. and D.R. Idler. Plasma gonadotrophin, estradiol, and vitellogenin and gonad phosvitin levels in relation to the seasonal reproductive cycles of female brown trout. *Ann. Biol. Anim. Biochem. Biophys.* 18: 1001–1005, 1978.

33. Denslow, N., M. Chow, K. Kroll, C. Weiser, J. Wiebe, B. Johnson, T. Shoeb and T. Gross. Determination of baseline seasonal information on vitellogenin production in female and male largemouth bass collected from lakes in central Florida. SETAC 18th Annual Meeting, San Francisco, CA, 16–20 November, 1997, pp. 256–257, 1997.
34. Desbrow, C., J. Routledge, G.C. Brighty, J.P. Sumpter and M. Waldock. Identification of estrogenic chemicals in STW effluent. 1. Chemical fractionation and in vitro biological screening. *Environ. Sci. Technol.* 32: 1549–1558, 1998.
35. Devlin, R. and Y. Nagahama. Sex determination and sex differentiation in fish: an overview of genetic, physiological, and environmental influences. *Aquaculture* 208: 191–364, 2002.
36. Dumont, J.N. and A.R. Brummett. Egg envelope in vertebrates. In: *Developmental Biology*, Vol. 1, edited by L.W. Browder, New York, Plenum Press, pp. 235–288, 1985.
37. Edmunds, J.S.G., R.A. McCarthy and J.S. Ramsdell. Permanent and functional male-to-female sex reversal in d-rR strain medaka (*Oryzias latipes*) following egg microinjection of *o,p'*-DDT. *Environ. Health Perspect.* 108: 219–224, 2000.
38. Emmersen, B.K. and I.M. Petersen. Natural occurrence and experimental induction by estradiol-17β, of a lipophosphoprotein (vitellogenin) in flounder (*Platichthys flesus*, L.). *Comp. Biochem. Physiol.* 54B: 443–446, 1976.
39. Feist, G., C.B. Schreck, M.S. Fitzpatrick and J.M. Redding. Sex steroid profiles of coho salmon (*Oncorhynchus kisutch*) during early development and sexual differentiation. *Gen. Comp. Endocrinol.* 80: 299–313, 1990.
40. Felber, B.K., S. Maurhofer, R.B. Jaggi, T. Wyler, W. Wahli, G.U. Ryffel and R. Weber. Isolation and translation in vitro of four related vitellogenin mRNAs of estrogen-stimulated *Xenopus laevis*. *Eur. J. Biochem.* 105: 17–24, 1980.
41. Folmar, L.C., N.D. Denslow, V. Rao, M. Chow, D.A. Crain, J. Enblom, J. Marcino and L.J. Guillette, Jr. Vitellogenin induction and reduced serum testosterone concentrations in feral male carp (*Cyprinus carpio*) captured near a major metropolitan sewage treatment plant. *Environ. Health Perspect.* 104: 1096–1101, 1996.
42. Folmar, L.C., N.D. Denslow, R.A. Wallace, G. LaFleur, T.S. Gross, S. Bonomelli and C.V. Sullivan. A highly conserved N-terminal sequence for teleost vitellogenin with potential value to the biochemistry, molecular biology and pathology of vitellogenesis. *J. Fish Biol.* 46: 255–263, 1995.
43. Folmar, L.C., N.D. Denslow, K. Kroll, E.F. Orlando, J. Enblom, J. Marcino, C. Metcalfe and L.J. Guillette, Jr.. Altered serum sex steroids and vitellogenin induction in walleye (*Stizostedion vitreum*) collected near a Metropolitan sewage treatment plant. *Arch. Environ. Contam. Toxicol.* 40: 392–398, 2001.
44. Folmar, L.C., G.R. Gardner, M.P. Schreibman, L. Magliulo-Cepriano, L.J. Mills, G. Zaroogian, R. Gutjahr-Gobell, R. Haebler, D.B. Horowitz and N.D. Denslow. Vitellogenin-induced pathology in male summer flounder (*Paralichthys dentatus*). *Aquat. Toxicol.* 51: 431–441, 2001.
45. Folmar, L.C., M.J. Hemmer, R. Hemmer, C. Bowman, K. Kroll and N.D. Denslow. Comparative estrogenicity of estradiol, ethynyl estradiol and diethylstilbestrol in an in vivo, male sheepshead minnow (*Cyprinodon variegates*), vitellogenin bioassay. *Aquat. Toxicol.* 49: 77–88, 2000.
46. Fossi, M.C., S. Casini, S. Ancora, A. Moscatelli, A. Ausili and G. Notarbartolo-di-Sciara. Do endocrine disrupting chemicals threaten Mediterranean swordfish? Preliminary results of vitellogenin and zona radiata proteins in *Xiphias gladius*. *Mar. Environ. Res.* 52: 447–483, 2001.
47. Fujita, T., M. Shimizu, N. Hiramatsu, H. Fukada and A. Hara. Purification of serum precursor proteins to vitelline envelope (choriogenins) in masu salmon, *Oncorhynchus masou*. *Comp. Biochem. Physiol.* 32B: 599–610, 2002.
48. Fuda, T., A. Hara, F. Yamazaki and K. Kobayashi. A peculiar immunoglobulin M (IgM) identified in eggs of chum salmon (*Oncorhynchus keta*). *Dev. Comp. Immunol.* 16: 415–423, 1992.
49. Fukada, H., M. Ban, H. Chiba and A. Hara. Immune complex transfer two-site chemiluminescent immunoassay for serum growth hormone in alevin chum salmon. *J. Biolumin. Chemilumin.* 13: 107–111, 1998.
50. Fukada, H., Y. Fujiwara, T. Takahashi, N. Hiramatsu, C.V. Sullivan and A. Hara. Carp (*Cyprinus carpio*) vitellogenin: purification and development of a simultaneous chemiluminescent immunoassay. *Comp. Biochem. Physiol.* 134A: 615–623, 2003.
51. Fukada, H., A. Haga, T. Fujita, N. Hiramatsu, C.V. Sullivan and A. Hara. Development and validation of chemi-luminescent immunoassay for vitellogenin in five salmonid species. *Comp. Biochem. Physiol.* 130A: 163–170, 2001.
52. Giesy, J.P. and E.M. Synder. Xenobiotic modulation of endocrine function in fish. In: *Principles and Processes for Evaluating Endocrine Disruption in Wildlife*, edited by R.J. Kendall, R.L. Dickerson, J.P. Giesy and W.A. Suk, Florida, SETAC Press, pp. 155–237, 1998.

53. Gimeno, S., A. Gerritsen, T. Bowmer and H. Komen. Feminization of male carp. *Nature* 384: 221–222, 1996.
54. Goedmakers, A. and B.L. Verboom. Studies on the maturation and fecundity of the pike, *Esox lucius* Linnaeus. *Aquaculture* 4: 3–12, 1974.
55. Goodbred, S.L., R.J. Gillion, T.S. Gross, N.P. Denslow, W.L. Bryant and T.R. Schoeb. Reconnaissance of 17β-estradiol, 11-ketotestosterone, vitellogenin, and gonad histopathology in common carp of United States streams: potential for contaminant-induced endocrine disruption. US Geological Survey Report, pp. 1–47, 1997.
56. Gray, M.A. and C.D. Metcalfe. Induction of testis–ova in Japanese medaka (*Oryzias latipes*) exposed to *p*-nonlyphenol. *Environ. Toxicol. Chem.* 16: 1082–1086, 1997.
57. Gray, M.A., A.J. Niimi and C.D. Metcalfe. Factors affecting the development of testis–ova in medaka, *Oryzias latipes*, exposed to octylphenol. *Environ. Toxicol. Chem.* 18: 1835–1842, 1999.
58. Greeley, M.S., Jr., D.R. Calder and R.A. Wallace. Changes in teleost yolk proteins during oocyte maturation: correlation of yolk proteolysis with oocyte hydration. *Comp. Biochem. Physiol.* 84B: 1–9, 1986.
59. Greeley, M.S., Jr., H. Hols and R.A. Wallace. Changes in size, hydration and low molecular weight osmotic effectors during meiotic maturation of *Fundulus* oocytes *in vivo*. *Comp. Biochem. Physiol.* 100A: 639–647, 1991.
60. Gronen, S., N. Denslow, S. Manning, S. Barnes, D. Barnes and M. Brouwer. Serum vitellogenin levels and reproductive impairment of male Japanese medaka (*Oryzias latipes*) exposed to 4-*tert*-octylphenol. *Environ. Health Perspect.* 107: 385–390, 1999.
61. Ha, C.R. and I. Iuchi. Purification and partial characterization of 76 kDa transglutaminase in the egg envelope (chorion) of rainbow trout, *Oncorhynchus mykiss*. *J. Biochem.* 122: 947–954, 1997.
62. Hamazaki, T.S., I. Iuchi and K. Yamagami. Isolation and partial characterization of a 'spawning female-specific substance' in the teleost, *Oryzias latipes*. *J. Exp. Zool.* 242: 343–349, 1987.
63. Hamazaki, T.S., Y. Nagahama, I. Iuchi and K. Yamagami. A glycoprotein from the liver constitutes the inner layer of the egg envelope (*zona pellucida interna*) of the fish, *Oryzias latipes*. *Dev. Biol.* 133: 101–110, 1989.
64. Haga, A., H. Fukada, T. Fujita, N. Hiramatsu and A. Hara. Estimation of baseline vitellogenin level in male salmonid serum. *Environ. Sci.* 8: 173 2001.
65. Hammond, G.L. and W.P. Bocchinfuso. Sex hormone-binding globulin/androgen-binding protein: steroid-binding and dimerization domains. *J. Steroid Biochem. Mol. Biol.* 53: 543–552, 1995.
66. Hara, A. and H. Hirai. Iron-binding activity of female-specific serum proteins of rainbow trout (*Salmo gairdneri*) and chum salmon (*Oncorhynchus keta*). *Biochim. Biophys. Acta* 437: 549–557, 1976.
67. Hara, A. and H. Hirai. Comparative studies on immunochemical properties of female-specific serum protein and egg yolk proteins in rainbow trout (*Salmo gairdneri*). *Comp. Biochem. Physiol.* 59B: 339–343, 1978.
68. Harries, J., T. Runnalls, E. Hill, C. Harris, S. Maddix, J. Sumpter and C. Tyler. Development of a reproductive performance test for endocrine disrupting chemicals using pair-breeding fathead minnow (*Pimephales promelas*). *Environ. Sci. Technol.* 34: 3003–3011, 2000.
69. Harries, J.E., D.A. Sheahan, S. Jobling, P. Matthiessen, P. Neall, E.J. Routledge, R. Rycroft, J.P. Sumpter and T. Tylor. A survey of estrogenic activity in United Kingdom inland waters. *Environ. Toxicol. Chem.* 15: 1993–2002, 1996.
70. Harries, J.E., D.A. Sheahan, S. Jobling, P. Matthiessen, P. Neall, J.P. Sumpter, T. Tylor and N. Zaman. Estrogenic activity in five United Kingdom rivers detected by measurement of vitellogenesis in caged male trout. *Environ. Toxicol. Chem.* 16: 534–542, 1997.
71. Hashimoto, S., H. Bessho, A. Hara, M. Nakamura, T. Iguchi and K. Fujita. Elevated serum vitellogenin levels and gonadal abnormalities in wild male flounder (*Pleuronectes yokohamae*) from Tokyo Bay, Japan. *Mar. Environ. Res.* 49: 37–53, 2000.
72. Hemmer, M., C. Bowman, B. Hemmer, S. Friedman, D. Marcovich, K. Kroll and N. Denslow. Vitellogenin mRNA regulation and plasma clearance in male sheepshead minnows (*Cyprinodon variegatus*) after cessation of exposure to 17β-estradiol and *p*-nonylphenol. *Aquat. Toxicol.* 58: 99–112, 2002.
73. Heppell, S.A., N.D. Denslow, L.C. Folmar and C.V. Sullivan. Universal assay of vitellogenin as a biomarker for environmental estrogens. *Environ. Health Perspect.* 103: 9–15, 1995.
74. Heppell, S.A., L.F. Jackson, G.M. Weber and C.V. Sullivan. Enzyme-linked immunosorbent assay (ELISA) of vitellogenin in temperate basses (Genus *Morone*): plasma and *in vitro* analysis. *Trans. Am. Fish. Soc.* 128: 532–541, 1999.
75. Heppell, S.A. and C.V. Sullivan. Gag (*Mycteroperca microlepis*) vitellogenin: purification, characterization and use for enzyme-linked immunosorbent assay (ELISA) of female maturity in three species of grouper. *Fish Physiol. Biochem.* 20: 361–374, 1999.

76. Hiramatsu, N. and A. Hara. Relationship between vitellogenin and its related egg yolk proteins in Sakhalin taimen (*Hucho perryi*). *Comp. Biochem. Physiol.* 115A: 243–251, 1996.
77. Hiramatsu, N. and A. Hara. Specific proteolysis of vitellogenin to egg yolk proteins in white spotted-charr *Salvelinus leucomaenis*. *Nippon Suisan Gakkaishi* 63: 701–708, 1997.
78. Hiramatsu, N., T. Matsubara, G.M. Weber, C.V. Sullivan and A. Hara. Vitellogenesis in aquatic animals. *Fish. Sci.* 68(Suppl. I): 694–699, 2002.
79. Hiramatsu, N., A. Hara, K. Hiramatsu, H. Fukada, G.M. Weber, N.D. Denslow and C.V. Sullivan. Vitellogenin-derived yolk proteins of white perch, *Morone americana*: purification, characterization and vitellogenin-receptor binding. *Biol. Reprod.* 67: 655–667, 2002.
80. Hiramatsu, N., N. Ichikawa, H. Fukada, T. Fujita, C.V. Sullivan and A. Hara. Identification and characterization of proteases involved in specific proteolysis of vitellogenin and yolk proteins in salmonids. *J. Exp. Zool.* 292: 11–25, 2002.
81. Hiramatsu, N., T. Matsubara, A. Hara, D.M. Donato, K. Hiramatsu, N.D. Denslow and C.V. Sullivan. Identification, purification and classification of multiple forms of vitellogenin from white perch (*Morone americana*). *Fish Physiol. Biochem.* 26: 355–370, 2002.
82. Hiramatsu, N., M. Shimizu, H. Fukada, M. Kitamura, K. Ura, H. Fuda and A. Hara. Transition of serum vitellogenin cycle in Sakhalin taimen (*Hucho perryi*). *Comp. Biochem. Physiol.* 118C: 149–157, 1997.
83. Ho, C.W. and W.E. Vanstone. Effect of estradiol monobenzoate on some serum constituents of maturing sockeye salmon (*Oncorhynchus nerka*). *J. Fish. Res. Board Can.* 18: 859–864, 1961.
84. Howell, W.M., D.A. Black and S.A. Bortone. Abnormal expression of secondary sex characters in a population of mosquitofish, *Gambusia affinis holbrooki*: evidence for environmentally-induced masculinization. *Copeia* 4: 676–681, 1980.
85. Hyllner, S.J. and C. Haux. Immunochemical detection of the major vitelline envelope proteins in the plasma and oocytes of the maturing female rainbow trout, *Oncorhynchus mykiss*. *J. Endocrinol.* 135: 303–309, 1992.
86. Hyllner, S.J., B. Norberg and C. Haux. Isolation, partial characterization, induction, and the occurrence in plasma of the major vitelline envelope proteins in the Atlantic halibut (*Hippoglossus hippoglossus*) during sexual maturation. *Can. J. Fish. Aquat. Sci.* 51: 1700–1707, 1994.
87. Hyllner, S.J., H.F.-P. Barber, D.G.J. Larsson and C. Haux. Amino acid composition and endocrine control of vitelline envelope proteins in European sea bass (*Dicentrarchus labrax*) and gilthead sea bream (*Sparus aurata*). *Mol. Reprod. Dev.* 41: 339–347, 1995.
88. Hyllner, S.J., L. Westerlund, P.E. Olsson and A. Schopen. Cloning of rainbow trout egg envelope proteins: members of a unique group of structural proteins. *Biol. Reprod.* 64: 805–811, 2001.
89. Idler, D.R., S.J. Hwang and L.W. Crim. Quantification of vitellogenin in Atlantic salmon (*Salmo salar*) plasma by radioimmunoassay. *J. Fish. Res. Board Can.* 36: 574–578, 1979.
90. Inaba, K., C.C. Buerano, F.F. Natividad and M. Morisawa. Degradation of vitellogenins by 170 kDa trypsin-like protease in the plasma of the tilapia, *Oreochromis niloticus*. *Comp. Biochem. Physiol.* 118B: 85–90, 1997.
91. Janssen, P.A.H., J.G.D. Lambert, A.D. Vethaak and H.J.Th. Goos. Environmental pollution caused elevated concentrations of oestradiol and vitellogenin in the female flounder, *Platichthys flesus* (L.). *Aquat. Toxicol.* 39: 195–214, 1997.
92. Jared, D.W. and R.A. Wallace. Comparative chromatography of the yolk proteins of teleosts. *Comp. Biochem. Physiol.* 24: 437–443, 1968.
93. Jobling, S. and J.P. Sumpter. Detergent components in sewage effluent are weakly oestrogenic to fish: an in vitro study using rainbow trout (*Oncorhynchus mykiss*) hepatocytes. *Aquat. Toxicol.* 27: 361–372, 1993.
94. Jobling, S., T. Reynolds, R. White, M.G. Parker and J.P. Sumpter. A variety of environmentally persistent chemicals, including some phthalate plasticizers, are weakly estrogenic. *Environ. Health Perspect.* 103: 582–587, 1995.
95. Jobling, S., M. Nolan, C.R. Tyler, G. Brighty and J.P. Sumpter. Widespread sexual disruption in wild fish. *Environ. Sci. Technol.* 32: 2498–2506, 1998.
96. Jobling, S., N. Beresford, M. Nolan, T. Rodgers-Gray, G. Brighty, J. Sumpter and C. Tyler. Altered sexual maturation and gamete production in wild roach (*Rutilus rutilus*) living in rivers that receive treated sewage effluents. *Biol. Reprod.* 66: 272–281, 2002.
97. Jobling, S., S. Coey, J. Whitmore, D. Kime, K. VanLook, B. McAllister, N. Beresford, A. Henshaw, G. Brighty, C. Tyler and J. Sumpter. Wild intersex roach (*Rutilus rutilus*) have reduced fertility. *Biol. Reprod.* 67: 515–524, 2002.
98. Jobling, S., D. Casey, T. Rodgers-Gray, J. Oehlmann, U. Schulte-Oehlmann, S. Pawlowski, T. Baunbeck, A. Turner and C. Tyler. Comparative responses of molluscs and fish to environmental estrogens and an estrogenic effluent. *Aquat. Toxicol.* 65: 205–220, 2003.

99. Kanamori, A. Systematic identification of genes expressed during early oogenesis in medaka. *Mol. Reprod. Dev.* 55: 31–36, 2000.
100. Kang, I., H. Yokota, Y. Oshima, Y. Tsuruda, T. Hano, M. Maeda, N. Imada, H. Tadokoro and T. Honjo. Effects of 4-nonylphenol on reproduction of Japanese medaka (*Oryzias latipes*). *Environ. Toxicol. Chem.* 22: 2438–2445, 2003.
101. Kiang, J.G. and G.C. Tsokos. Heat shock proteins 70 kDa: molecular biology, biochemistry, and physiology. *Pharmacol. Ther.* 80: 183–201, 1998.
102. Kime, D.E. *Endocrine Disruption in Fish*, Boston, Kluwer Academic Publishers, 416 pp., 1998.
103. Kime, D.E., J.P. Nash and A.P. Scott. Vitellogenesis as a biomarker of reproductive disruption by xenobiotics. *Aquaculture* 177: 345–352, 1999.
104. Kirby, M., J. Bignell, E. Brown, J. Craft, I. Davies, R. Dyer, S. Feist, G. Jones, P. Matthiessen, C. Megginson, F. Robertson and C. Robinson. The presence of morphologically intermediate papilla syndrome in United Kingdom populations of sand goby (*Pomatoschistus* spp): endocrine disruption? *Environ. Toxicol. Chem.* 22: 239–251, 2003.
105. Krauel, K.K. and G.J. Ridgway. Immunoelectrophoretic studies of red salmon (*Oncorhynchus nerka*) serum. *Int. Arch. Allergy* 23: 246–253, 1963.
106. Kwon, H.C., Y. Mugiya, J. Yamada and A. Hara. Enzyme linked-immunosorbent assay (ELISA) of vitellogenin in white-spotted charr, *Salvelinus leucomaenis*. *Bull. Fac. Fish. Hokkaido Univ.* 41: 241–259, 1990.
107. LaFleur, G.J., Jr., B.M. Byrne, J. Kanungo, L.D. Nelson, R.M. Greenberg and R.A. Wallace. *Fundulus heteroclitus* vitellogenin: the deduced primary structure of a piscine precursor to noncrystalline, liquid-phase yolk protein. *J. Mol. Evol.* 41: 505–521, 1995.
108. LaFleur, G.J., Jr., B.M. Byrne, C. Haux, R.M. Greenberg and R.A. Wallace. Liver-derived cDNAs: vitellogenin and vitelline envelope protein precursors (choriogenins). In: *Proceedings of the Fifth International Symposium on the Reproductive Physiology of Fish*, edited by F.W. Goetz and P. Thomas, Austin, University of Texas, pp. 336–338, 1995.
109. Larkin, P., L.C. Folmar, M.J. Hemmer, A.J. Poston and N.D. Denslow. Expression profiling of estrogenic compounds using a sheepshead minnow cDNA macroarray. *Environ. Health Perspect.* 111: 839–846, 2003.
110. Larsson, D.G.J., S.J. Hyllner and C. Haux. Induction of vitelline envelope proteins by estradiol-17β in 10 teleost species. *Gen. Comp. Endocrinol.* 96: 445–450, 1994.
111. Laskowski, M. Über das vorkommen des serumvitellins im blute der wirebeltiere. *Biochem. Z.* 284: 318–3321, 1936.
112. Latonnelle, K., F. Le Menn and C. Bennetau-Pelissero. *In vitro* estrogenic effects of phytoestrogens in rainbow trout and Siberian sturgeon. *Ecotoxicology* 9: 115–125, 2000.
113. Latonnelle, K., F. Le Menn, S.J. Kaushik and C. Bennetau-Pelissero. Effects of dietry phytoestrogens *in vivo* and *in vitro* in rainbow trout and Siberian sturgeon: interests and limits of the *in vitro* studies of interspecies differences. *Gen. Comp. Endocrinol.* 126: 39–51, 2002.
114. Latonnelle, K., A. Fostier, F. Le Menn and C. Bennetau-Pelissero. Binding affinities of hepatic nuclear estrogen receptors for phytoestrogens in rainbow trout (*Oncorhynchus mykiss*) and Siberian sturgeon (*Acipenser baeri*). *Gen. Comp. Endocrinol.* 129: 69–79, 2002.
115. Lye, C.M., C.L.J. Frid and M.E. Gill. Seasonal reproductive health of flounder *Platichthys flesus* exposed to sewage effluent. *Mar. Ecol. Prog. Ser.* 170: 249–260, 1998.
116. Lye, C.M., C.L.J. Frid, M.E. Gill and D. McCormick. Abnormalities in the reproductive health of flounder *Platichthys flesus* exposed to effluent from a sewage treatment works. *Mar. Pollut. Bull.* 34: 34–41, 1997.
117. Lye, C.M., C.L.J. Frid, M.E. Gill, D.W. Cooper and D.M. Jones. Estrogenic alkylphenols in fish tissues, sediments, and waters from the U.K. Tyne and Tees estuaries. *Environ. Sci. Technol.* 33: 1009–1014, 1999.
118. Lyons, C.E., K.L. Payette, J.L. Price and R.C.C. Huang. Expression and structural analysis of a teleost homologue of a mammalian zona pellucida gene. *J. Biol. Chem.* 268: 21351–21358, 1993.
119. MacLatchy, D., S. Courtenay, C. Rice and G. Van Der Kraak. Development of a short-term reproductive endocrine bioassay using steroid hormone and vitellogenin endpoints in the estuarine mummichog (*Fundulus heteroclitus*). *Environ. Toxicol. Chem.* 22: 996–1008, 2003.
120. MacLatchy, D.L. and G.J. Van Der Kraak. The phytoestrogen β-sitosterol alters the reproductive endocrine status of goldfish. *Toxicol. Appl. Pharmacol.* 134: 305–312, 1995.
121. Maitre, J.L., C. Leguellec, S. Derrien, M. Tenniswood and Y. Valotaire. Measurement of vitellogenin from rainbow trout by rocket immunoelectrophoresis – application to the kinetic-analysis of estrogen stimulation in the male. *Can. J. Biochem. Cell Biol.* 63: 982–987, 1985.
122. Maitre, J.L., Y. Valotaire and C. Guguen-Guillouzo. Estradiol-17β stimulation of vitellogenin synthesis in primary culture of male rainbow trout hepatocytes. *In Vitro Cell. Dev. Biol.* 22: 337–343, 1986.

123. Markert, J.P. and W.E. Vanstone. Egg proteins of coho salmon (*Oncorhynchus kisutch*). *J. Fish. Res. Board Can.* 28: 1853–1856, 1971.
124. Matsubara, T. and Y. Koya. Course of proteolytic cleavage in three classes of yolk proteins during oocyte maturation in barfin flounder (*Verasper moseri*). *J. Exp. Zool.* 272: 34–45, 1997.
125. Matsubara, T. and K. Sawano. Proteolytic cleavage of vitellogenin and yolk proteins during vitellogenin uptake and oocyte maturation in barfin flounder (*Verasper moseri*). *J. Exp. Zool.* 272: 34–45, 1995.
126. Matsubara, T., N. Ohkubo, T. Andoh, C.V. Sullivan and A. Hara. Two forms of vitellogenin, yielding two distinct lipovitellins, play different roles during oocyte maturation and early development of barfin flounder, *Verasper moseri*, a marine teleost that spawns pelagic eggs. *Dev. Biol.* 213: 18–32, 1999.
127. Matsubara, T., M. Nagae, N. Ohkubo, T. Andoh, S. Sawaguchi, N. Hiramatsu, C.V. Sullivan and A. Hara. Multiple vitellogenins and their unique roles in marine teleosts. *Fish Physiol. Biochem.* 28: 295–299, 2003.
128. Matsubara, T., T. Wada and A. Hara. Purification and establishment of ELISA for vitellogenin of Japanese sardine (*Sardinops melanostictus*). *Comp. Biochem. Physiol.* 109B: 545–555, 1994.
129. Matthiessen, P., Y.T. Allen, C.R. Allchin, S.W. Feist, M.F. Kirby, R.J. Law, A.P. Scott, J.E. Thain and K.V. Thomas. Oestrogenic endocrine disruption in flounder (*Platichthys flesus* L.) from United Kingdom estuarine and marine waters. *Sci. Ser. Tech. Rep., CEFAS, Lowestoft.* 107: 1–48, 1998.
130. McMaster, M.E., G.J. Van Der Kraak and K.R. Munkittrick. An epidemiological evaluation of the biochemical basis for steroid hormone depressions in fish exposed to industrial wastes. *J. Great Lakes Res.* 22: 153–171, 1996.
131. Metcalfe, T.L., C.D. Metcalfe, Y. Kiparissis, A.J. Niimi, C.M. Foran and W.H. Benson. Gonadal development and endocrine responses in Japanese medaka (*Oryzias latipes*) exposed to *o,p′*-DDT in water or through maternal transfer. *Environ. Toxicol. Chem.* 19: 1893–1900, 2000.
132. Mills, L., R. Gutjahr-Gobell, D. Horowitz, N. Denslow, M. Chow and G. Zaroogian. Relationship between reproductive success and male plasma vitellogenin concentrations in cunner, *Tautogolabrus adspersus*. *Environ. Health Perspect.* 111: 93–99, 2003.
133. Mommsen, T.P. and P.J. Walsh. Vitellogenesis and oocyte assembly. In: *Fish Physiology*, Vol. XIA, edited by W.S. Hoar, D.J. Randall and A.P. Farrell, San Diego, CA, Academic Press, pp. 347–406, 1988.
134. Munkittrick, K.R., C.B. Portt, G.J. Van Der Kraak, I.R. Smith and D.A. Rokosh. Impact of bleached kraft mill effluent on population characteristics, liver MFO activity, and serum steroid levels of a Lake Superior white sucker (*Catostomus commersoni*) population. *Can. J. Fish. Aquat. Sci.* 48: 1371–1380, 1991.
135. Munkittrick, D.R., M.E. McMaster, C.B. Portt, G.J. Van Der Kraak, I.R. Smith and D.G. Dixon. Changes in maturity, plasma sex steroid levels, hepatic mixed function oxygenase activity, and the presence of external lesions in lake whitefish (*Coregonus clupeaformis*) exposed to bleached Kraft mill effluent. *Can. J. Fish. Aquat. Sci.* 49: 1560–1569, 1992.
136. Munkittrick, K.R., G.J. Van Der Kraak, M.E. McMaster and C.B. Portt. Response of hepatic MFO activity and plasma sex steroids to secondary treatment of bleached kraft pulp mill effluent and mill shutdown. *Environ. Toxicol. Chem.* 11: 1427–1439, 1992.
137. Munkittrick, K.R. and G.J. Van Der Kraak. Receiving water environmental effects associated with discharges from Ontario pulp mills. *Pulp Pap. Can.* 95: 57–59, 1994.
138. Murata, K., I. Iuchi and K. Yamagami. Isolation of H-SF substance, the high-molecular-weight precursors of egg envelope proteins, from the ascites accumulated in the oestrogen-treated fish, *Oryzias latipes*. *Zygote* 1: 315–324, 1993.
139. Murata, K., T. Sasaki, S. Yasumasu, I. Iuchi, J. Enami, I. Yasumasu and K. Yamagami. Cloning of cDNAs for the precursor proteins of low-molecular-weight subunit of the inner layer of the egg envelope (chorion) of the fish *Oryzias latipes*. *Dev. Biol.* 167: 9–17, 1995.
140. Murata, K., H. Sugiyama, S. Yasumasu, I. Iuchi, I. Yasumasu and K. Yamagami. Cloning of cDNA and estrogen-induced hepatic gene expression for choriogenin H, a precursor protein of the fish egg envelope (chorion). *Proc. Natl Acad. Sci. USA* 94: 2050–2055, 1997.
141. Nakamura, K., S. Yonezawa and N. Yoshizaki. Vitellogenesis-related ovary cathepsin D from *Xenopus laevis*: purification and properties in comparison with liver cathepsin D. *Comp. Biochem. Physiol.* 113B: 835–840, 1996.
142. Nardelli, D., F.D. van het Schip, S. Gerber-Huber, J.A. Haefliger, M. Gruber, G. Ab and W. Wahli. Comparison of the organization and fine structure of a chicken and a *Xenopus laevis* vitellogenin gene. *J. Biol. Chem.* 262: 15337–15385, 1987.
143. Nath, P. and B.I. Sundararaj. Isolation and identification of female-specific serum lipophosphoprotein (vitellogenin) in the catfish, *Heteropneustes fossilis*. *Gen. Comp. Endocrinol.* 43: 184–190, 1981.
144. Nimrod, A. and W. Benson. Reproduction and development in Japanese medaka following an early life stage exposure to xenoestrogens. *Aquat. Toxicol.* 44: 141–156, 1998.

145. Ohkubo, N. and T. Matsubara. Sequential utilization of free amino acids, yolk proteins and lipids in developing eggs and yolk-sac larvae of barfin flounder *Verasper moseri*. *Mar. Biol.* 140: 187–196, 2002.
146. Ohkubo, N., K. Mochida, S. Adachi, A. Hara, K. Hotta, Y. Nakamura and T. Matsubara. Development of enzyme-linked immunosorbent assays (ELISAs) for two forms of vitellogenin in Japanese common goby (*Acanthogobius flavimanus*). *Gen. Comp. Endocrinol.* 131: 353–364, 2003.
147. Ohkubo, N., K. Mochida, S. Adachi, A. Hara, K. Hotta, Y. Nakamura and T. Matsubara. Estrogenic activity in coastal areas around Japan evaluated by measuring male serum vitellogenins in Japanese common goby (*Acanthogobius flavimanus*). *Fish. Sci.* 69: 1133–1143, 2003.
148. Okuzawa, K. Puberty in teleosts. *Fish Physiol. Biochem.* 26: 31–41, 2002.
149. Oppen-Berntsen, D.O., J.V. Helvik and B.T. Walther. The major structural proteins of cod (*Gadus morhua*) eggshells and protein crosslinking during teleost egg hardening. *Dev. Biol.* 137: 258–265, 1990.
150. Oppen-Berntsen, D.O., S.J. Hyllner, C. Haux, J.V. Helvik and B.T. Walther. Eggshell zona radiata-proteins from cod (*Gadus morhua*): extra-ovarian origin and induction by estradiol-17β. *Int. J. Dev. Biol.* 36: 247–254, 1992.
151. Pait, A.S. and J.O. Nelson. *Endocrine Disruption in Fish: An Assessment of Recent Research and Results*, NOAA Technical Memo. NOS NCCOS CCMA 149, Silver Spring, MD, NOAA, NOS, Center for Coastal Monitoring and Assessment, 55 pp., 2002.
152. Pait, A. and J. Nelson. Vitellogenesis in male *Fundulus heteroclitus* (killifish) induced by selected estrogenic compounds. *Aquat. Toxicol.* 64: 331–342, 2003.
153. Pakdel, F., S. Feon, F. Legac, F. Le Menn and Y. Valotaire. *In vivo* estrogen induction of hepatic estrogen-receptor messenger-RNA and correlation with vitellogenin messenger-RNA in rainbow trout. *Mol. Cell. Endocrinol.* 75: 205–212, 1991.
154. Palmer, B.D. and K.W. Selcer. Vitellogenin as a biomarker for xenobiotic estrogens: a review. In: *Environmental Toxicology and Risk Assessment: Biomarkers and Risk Assessment*, ASTM STP 1306, Vol. 5, edited by D.A. Bengtson and D.S. Henshel, pp. 3–21, 1996.
155. Pan, M.J., W.J. Bell and W.H. Telfer. Vitellogenic blood protein synthesis by insect fat body. *Science* 165: 393–394, 1969.
156. Panter, G., R. Thompson and J. Sumpter. Intermittent exposure of fish to estradiol. *Environ. Sci. Technol.* 34: 2756–2760, 2000.
157. Parks, L.G., A.O. Cheek, N.D. Denslow, S.A. Heppell, J.A. McLachlan, G.A. LeBlanc and C.V. Sullivan. Fathead minnow (*Pimephales promelas*) vitellogenin: purification, characterization and quantitative immunoassay for the detection of estrogenic compounds. *Comp. Biochem. Physiol.* 123C: 113–125, 1999.
158. Patiño, R. and C.V. Sullivan. Ovarian follicle growth, maturation, and ovulation in teleost fish. *Fish Physiol. Biochem.* 26: 57–70, 2002.
159. Patiño, R., G. Yoshizaki, P. Thomas and H. Kagawa. Gonadotropic control of ovarian follicle maturation: the two-stage concept and its mechanisms. *Comp. Biochem. Physiol.* 129B: 427–439, 2001.
160. Pelissero, C., B. Bennetau, B. Babin, F. Le Menn and J. Dunogues. The estrogenic activity of certain phytoestrogens in the Siberian sturgeon *Acipenser baeri*. *J. Steroid Biochem. Mol. Biol.* 38: 293–299, 1991.
161. Pelissero, C., F. Le Menn and S. Kaushik. Estrogenic effect of dietary soya bean meal on vitellogenesis in cultured Siberian sturgeon *Acipenser baeri*. *Gen. Comp. Endocrinol.* 83: 447–457, 1991.
162. Pelissero, C., G. Flouriot, J.L. Foucher, B. Bennetau, J. Dunogues, F. Le Gac and J.P. Sumpter. Vitellogenin synthesis in cultured hepatocytes; an in vitro test for the estrogenic potency of chemicals. *J. Steroid Biochem. Mol. Biol.* 44: 263–272, 1993.
163. Plack, P.A., D.J. Pritchard and N.W. Fraser. Egg proteins in cod serum: natural occurrence and induction by injections of oestradiol 3-benzoate. *Biochem. J.* 121: 847–856, 1971.
164. Purdom, C.E., P.A. Hardiman, V.J. Bye, N.C. Eno, C.R. Tyler and J.P. Sumpter. Estrogenic effects of effluents from sewage treatment works. *Chem. Ecol.* 8: 275–285, 1994.
165. Reith, M., J. Munholland, J. Kelly, R.N. Finn and H.J. Fyhn. Lipovitellins derived from two forms of vitellogenin are differentially processed during oocyte maturation in haddock (*Melanogrammus aeglefinus*). *J. Exp. Zool.* 291: 58–67, 2001.
166. Retzek, H., E. Steyrer, E.J. Sanders, J. Nimpf and W.J. Schneider. Molecular cloning and functional characterization of chicken cathepsin D, a key enzyme for yolk formation. *DNA Cell Biol.* 11: 661–672, 1992.
167. Roepke, R.R. and J.S. Hughes. Phosphorus partition in the blood of laying hens. *J. Biol. Chem.* 108: 79–83, 1935.
168. Rodgers-Gray, T., S. Jobling, S. Morris, C. Kelly, S. Kirby, A. Janbakhsh, J. Harries, M. Waldock, J. Sumpter and C. Tyler. Long-term temporal changes in the estrogenic composition of treated sewage effluent and its biological effects on fish. *Environ. Sci. Technol.* 34: 1521–1528, 2000.

169. Rodgers-Gray, T., S. Jobling, C. Kelly, S. Morris, G. Brighty, M. Waldock, J. Sumpter and C. Tyler. Exposure of juvenile roach (*Rutilus rutilus*) to treated sewage effluent induces dose-dependent and persistent disruption in gonadal duct development. *Environ. Sci. Technol.* 35: 462–470, 2001.
170. Routledge, E.J., D. Sheahan, C. Desbrow, G.C. Brightly, M. Waldock and J.P. Sumpter. Identification of estrogenic chemicals in STW effluent. 2. *In vivo* responses in trout and roach. *Environ. Sci. Technol.* 32: 1559–1565, 1998.
171. Roy, L., J. Armstrong, K. Sakamoto, S. Steinert, E. Perkins, D. Lomax, L. Johnson and D. Schlenk. The relationships of biochemical endpoints to histopathology and population metrics in feral flatfish species collected near the municipal wastewater outfall of Orange County, California, USA. *Environ. Toxicol. Chem.* 22: 1309–1317, 2003.
172. Sawaguchi, S., Y. Koya and T. Matsubara. Primary structures of three types of vitellogenin in mosquitofish (*Gambusia affinis*), a viviparous fish. *Fish Physiol. Biochem.* 28: 363–364, 2003.
173. Schultz, I., G. Orner, J. Merdink and A. Skillman. Dose–response relationships and pharmacokinetics of vitellogenin in rainbow trout after intravascular administration of 17β-ethynylestradiol. *Aquat. Toxicol.* 51: 305–318, 2001.
174. Schwaiger, J., U. Mallow, H. Ferling, S. Knoerr, T. Braunbeck, W. Kalbfus and R. Negele. How estrogenic is nonylphenol? A transgenerational study using rainbow trout (*Oncorhynchus mykiss*) as a test organism. *Aquat. Toxicol.* 59: 177–189, 2002.
175. Seki, M., H. Yokota, H. Matsubara, Y. Tsuruda, M. Maeda, H. Tadokoro and K. Kobayashi. Effect of ethinylestradiol on the reproduction and induction of vitellogenin and testis–ova in medaka (*Oryzias latipes*). *Environ. Toxicol. Chem.* 21: 1692–1698, 2002.
176. Seki, M., H. Yokota, M. Maeda, H. Tadokoro and K. Kobayashi. Effects of 4-nonylphenol and 4-*tert*-octylphenol on sex differentiation and vitellogenin induction in medaka (*Oryzias latipes*). *Environ. Toxicol. Chem.* 22: 1507–1516, 2003.
177. Seki, M., H. Yokota, H. Matsubara, M. Maeda, H. Tadokoro and K. Kobayashi. Fish full life-cycle testing for the weak estrogen 4-*tert*-pentylphenol on medaka (*Oryzias latipes*). *Environ. Toxicol. Chem.* 22: 1487–1496, 2003.
178. Selman, K. and R.A. Wallace. Oogenesis in *Fundulus heteroclitus*. III. Vitellogenesis. *J. Exp. Zool.* 226: 441–457, 1983.
179. Selman, K. and R.A. Wallace. Cellular aspects of oocyte growth in teleost. *Zool. Sci.* 6: 211–231, 1989.
180. Selman, K., R.A. Wallace and J. Cerda. Bafilomycin A1 inhibits proteolytic cleavage and hydration but not yolk crystal disassembly or meiosis during maturation of sea bass oocytes. *J. Exp. Zool.* 290: 265–278, 2001.
181. Shimizu, M., T. Fujita and A. Hara. Purification of the precursors to vitelline envelope proteins from serum of Sakhalin taimen, *Hucho perryi*. *J. Exp. Zool.* 282: 385–395, 1998.
182. Siiteri, P.K., J.T. Murai, G.L. Hammond, J.A. Nisker, W.J. Raymoure and R.W. Kuhn. The serum transport of steroid hormones. *Recent Prog. Horm. Res.* 38: 457–510, 1982.
183. Silva, R., A.H. Fischer and J.B. Burch. The major and minor chicken vitellogenin genes are each adjacent to partially deleted pseudogene copies of the other. *Mol. Cell Biol.* 9: 3557–3562, 1989.
184. Sire, M.F., P.J. Babin and J.M. Vernier. Involvement of the lysosomal system in yolk protein deposit and degradation during vitellogenesis and embryonic development in trout. *J. Exp. Zool.* 269: 69–83, 1994.
185. Specker, J.L. and T.R. Anderson. Developing an ELISA for a model protein – vitellogenin. In: *Biochemistry and Molecular Biology of Fishes*, Vol. 3, Analytical Techniques, edited by P.W. Hochachka and T.P. Mommsen, Amsterdam, Elsevier, pp. 567–578, 1994.
186. Specker, J.L. and C.V. Sullivan. Vitellogenesis in fishes: status and perspectives. In: *Perspectives in Comparative Endocrinology*, edited by K.G. Davey, R.E. Peter and S.S. Tobe, Ottawa, National Research Council, pp. 304–315, 1994.
187. Spies, R. and P. Thomas. Reproductive and endocrine status of female kelp bass from a contaminated site in the Southern California Bight and estrogen receptor binding of DDTs. In: *Chemically Induced Alterations in Functional Development and Reproduction of Fishes*, edited by R.R. Rolland, M. Gilbertson and R.E. Peterson, Pensacola, SETAC Press, pp. 113–134, 1997.
188. Sugiyama, H., S. Yasumasu, K. Murata, I. Iuchi and K. Yamagami. The third egg envelope subunit in fish: cDNA cloning and analysis, and gene expression. *Dev. Growth Differ.* 40: 35–45, 1998.
189. Sumpter, J.P. Environmental control of fish reproduction: a different perspective. *Fish Physiol. Biochem.* 17: 25–31, 1997.
190. Sumpter, J.P. and S. Jobling. Vitellogenesis as a biomarker for estrogenic contamination of the aquatic environment. *Environ. Health Perspect.* 103: 173–178, 1995.
191. Takemura, A. and B.H. Kim. Effects of estradiol-17β treatment on in vitro and in vivo synthesis of two distinct vitellogenins in tilapia. *Comp. Biochem. Physiol.* 129A: 641–651, 2001.

192. Taylor, J.S., Y. Van de Peer, I. Braasch and A. Meyer. Comparative genomics provides evidence for an ancient genome duplication event in fish. *Philos. Trans. R. Soc. Lond. B* 356: 1661–1679, 2001.
193. Thompson, S., F. Tilton, D. Schlenk and W. Benson. Comparative vitellogenic responses in three teleost species: extrapolation to in situ field studies. *Mar. Environ. Res.* 51: 185–189, 2000.
194. Thorsen, A. and H.J. Fyhn. Final oocyte maturation in vivo and in vitro in marine fishes with pelagic eggs: yolk protein hydrolysis and free amino acid content. *J. Fish. Biol.* 48: 1195–1209, 1996.
195. Tilton, F., W. Benson and D. Schlenk. Evaluation of estrogenic activity from a municipal wastewater treatment plant with predominantly domestic input. *Aquat. Toxicol.* 61: 211–224, 2002.
196. Trichet, V., N. Buisine, N. Mouchel, P. Moran, A.M. Pendas, J.P. Le Pennec and J. Wolff. Genomic analysis of the vitellogenin locus in rainbow trout (*Oncorhynchus mykiss*) reveals a complex history of gene amplification and retroposon activity. *Mol. Gen. Genet.* 263: 828–837, 2000.
197. Tyler, C.R., B. van der Eerden, S. Jobling, G. Panter and J.P. Sumpter. Measurement of vitellogenin, a biomarker for exposure to estrogenic chemicals, in a wide variety of cyprinid fish. *J. Comp. Physiol.* 166B: 418–426, 1996.
198. Tyler, C.R., E.M. Santos and F. Prat. Unscrambling the egg – cellular, biochemical, molecular and endocrine advances in oogenesis. In: *Proceedings of the Sixth International Symposium on the Reproductive Physiology of Fish*, edited by B. Norberg, O.S Kjesbu, G.L. Taranger, E. Andersson and S.O. Stefansson, Bergen, Norway, Institute of Marine Research and University of Bergen, pp. 273–280, 1999.
199. Uhlenhuth, P. and T. Kodama. Studien über die Geschlechtsdifferenzierung. *Kokka Igakkai Zassi* 324–335: 385–405 or *Nippon Eiseigaku-Zasshi*, 10: 13–34, 1914.
200. Urist, M.R. and A.O. Schjeide. The partition of calcium and protein in the blood of oviparous vertebrates during estrus. *J. Gen. Physiol.* 44: 743–756, 1961.
201. Utter, F.M. and G.J. Ridgeway. A serologically detected serum factor associated with maturity in the English sole, *Parophrys vetulus* and Pacific halibut, *Hippoglossus stenolepis. US Fish Wildl. Serv. Fish. Bull.* 66: 47–48, 1967.
202. van Bohemen, C.G., J.G.D. Lambert and P.G.W.J. van Oordt. Vitellogenin induction by estradiol in estrone-primed rainbow trout, *Salmo gairdneri. Gen. Comp. Endocrinol.* 46: 136–139, 1982.
203. van de Peer, Y., J.S. Taylor, J. Joseph and A. Meyer. Wanda: a database of duplicated fish genes. *Nucleic Acids Res.* 30: 109–112, 2002.
204. van-het Schip, F.D., J. Samallo, J. Broos, J. Ophuis, M. Mojet, M. Gruber and G. Ab. Nucleotide sequences of a chicken vitellogenin gene and derived amino acid sequence of the enclosed yolk precursor protein. *J. Mol. Biol.* 196: 245–260, 1987.
205. Vanstone, W.E. and F.C.W. Ho. Plasma proteins of coho salmon, *Oncorhynchus kisutch*, as separated by zone electrophoresis. *J. Fish. Res. Board Can.* 18: 393–399, 1961.
206. Wahli, W., I.B. Dawid, T. Wyler, R.B. Jaggi, R. Weber and G.U. Ryffel. Vitellogenin in *Xenopus laevis* is encoded in a small family of genes. *Cell* 16: 535–549, 1979.
207. Wallace, R.A. Vitellogenesis and oocyte growth in non-mammalian vertebrates. In: *Developmental Biology*, Vol. 1, edited by L.W. Browder, New York, Plenum Press, pp. 127–177, 1985.
208. Wallace, R.A. and P.C. Begovac. Phosvitins in *Fundulus* oocytes and eggs. Preliminary chromatographic and electrophoretic analyses together with biological considerations. *J. Biol. Chem.* 260: 11268–11274, 1985.
209. Wang, H. and Z. Gong. Characterization of two zebrafish cDNA clones encoding egg envelope proteins ZP2 and ZP3. *Biochim. Biophys. Acta* 1446: 156–160, 1999.
210. Wang, H., T. Yan, J.T.T. Tan and Z.A. Gong. Zebrafish vitellogenin gene (*vg3*) encodes a novel vitellogenin without a phosvitin domain and may represent a primitive vertebrate vitellogenin gene. *Gene* 256: 303–310, 2000.
211. Wester, P. and J. Canton. Histopathological study of *Oryzias latipes* (medaka) after long-term β-hexachlorocyclohexane exposure. *Aquat. Toxicol.* 9: 21–45, 1986.
212. White, R., S. Jobling, S.A. Hoare, J.P. Sumpter and M.G. Parker. Environmentally persistent alkylphenolic compounds are estrogenic. *Endocrinology* 135: 175–182, 1994.
213. Yamagami, K. Studies on the hatching enzyme (choriolysin) and its substrate, egg envelope, constructed of the precursors (choriogenins) in *Oryzias latipes*: a sequel to the information in 1991/1992. *Zool. Sci.* 13: 331–340, 1996.
214. Yamamura, J.I., T. Adachi, N. Aoki, H. Nakajima, R. Nakamura and T. Matsuda. Precursor–product relationship between chicken vitellogenin and the yolk proteins: the 40 kDa yolk plasma glycoprotein is derived from the C-terminal cysteine-rich domain of vitellogenin II. *Biochim. Biophys. Acta* 1244: 384–394, 1995.
215. Yoshizaki, G., R. Patino, P. Thomas, D. Bolamba and X. Chang. Effects of maturation-inducing hormone on heterologous gap junctional coupling in ovarian follicles of Atlantic croaker. *Gen. Comp. Endocrinol.* 124: 359–366, 2001.

Emerging Areas

Biochemistry and Molecular Biology of Fishes, vol. 6
T. P. Mommsen and T. W. Moon (Editors)

CHAPTER 17

Pharmaceuticals in the environment: Drugged fish?

VANCE L. TRUDEAU*, CHRIS D. METCALFE**, CAROLINE MIMEAULT*
AND THOMAS W. MOON*

** Centre for Advanced Research in Environmental Genomics and Department of Biology, University of Ottawa, Ottawa, ON, Canada K1N 6N5, and ** Environmental & Resource Studies Program, Trent University, 1600 West Bank Drive, Peterborough, ON, Canada K9J 7B8*

I. Introduction

Prescription and non-prescription drugs are produced and used by human populations in quantities that exceed hundreds of metric tonnes annually[22]. Most of these drugs or their metabolites are excreted or discarded into urban wastewaters and eventually make their way to municipal sewage treatment plants (STPs). In a recent review, Daughton and Ternes[13] came to the conclusion that many pharmaceuticals have the potential to impact the environment through release from STPs. The purpose of this chapter is to review what we know about the potential impact of these drugs in the aquatic environment. Few studies have addressed this question, and these studies will be covered here. It should also be mentioned that there are other sources of pharmaceuticals in the environment, such as veterinary drugs or feed additives for livestock, but a discussion of these sources is beyond the scope of this review.

II. Presence of pharmaceuticals in the environment

1. Source and fate

The most direct route of pharmaceuticals and metabolites to aquatic environments is through discharges of STP effluents into streams, rivers, lakes, estuaries and marine coastal zones. In addition, bio-solids containing drugs may be placed in landfills or spread on agricultural land for soil amendment. As a result, these drugs may be transported by run-off into the surrounding surface water and may leach into underlying groundwater increasing the potential for contamination of other areas and municipal water supplies.

The potential for these compounds to enter the environment depends upon the quantities of the drugs sold and utilized by humans, the pharmacokinetic behavior of the drugs in humans and the rates of degradation of the compounds in the environment, including within STPs. Most drugs are taken orally and are absorbed into the body through the gastrointestinal tract. The absorption capacity of orally administered drugs vary from 5 to less than 100%[28] leading in some cases to a large proportion of ingested drugs eliminated in the feces without being biomodified. Substances that are absorbed are usually metabolized by Phase I (oxidation, reduction or hydrolysis) and Phase II (conjugation, glucuronidation, methylation or acetylation) mechanisms prior to excretion in urine. However, excreted drug metabolites are not necessarily biologically inactive. For example, some drugs such as the lipid-regulating agent clofibrate, are designed to be metabolized in the body to an active form, clofibric acid, which is persistent in the environment[7]. In addition, the synthetic estrogen, 17β-ethinylestradiol is excreted by women in an inactive conjugated form, but microbial de-conjugation of the metabolite in STP bio-solids returns this compound to its biologically active form[63,64].

It had been anticipated until relatively recently that drugs would degrade rapidly in STPs, generally through microbial degradation processes under anaerobic or aerobic conditions. However, according to the available data, it appears that efficient degradation of drugs in STPs is more an exception than the rule[22,26,52,60]. There are three principal fates of drugs in STPs: (i) the substance is rapidly degraded and mineralized; (ii) the substance or its metabolite(s) is relatively hydrophilic and persistent and remains dissolved in the aqueous phase of the STP effluent; and (iii) the substance or its metabolite(s) is relatively hydrophobic and persistent and binds to STP bio-solids[22] (i.e. 'sludge'). There are some pharmaceuticals that are degraded relatively rapidly in STPs, such as acetylsalicylic acid[52] (aspirin) but there are other drugs that do not appear to be degraded at all, such as the anti-epileptic, carbamazepine[62]. Lipid regulators were also reported to be quite persistent[13]. Other compounds appear to have intermediate biodegradability in STPs, including ibuprofen[8,62]. Little is known about the fate of drugs applied to agricultural soils as bio-solids, although this is an active area of research. To our knowledge, there are no published data on the transport of drugs from agricultural lands into groundwater or into surrounding surface waters. However, it is clear that groundwater and surface water samples collected in agricultural areas in some areas of the world have been contaminated with drugs that are used by humans[54,55,57].

2. *Concentration and distribution*

Most previous work on the distribution of drugs in STP effluents and the aquatic environment had been conducted in Europe. However, there are now some data on the distribution of drugs in STP effluents and surface waters in North America[40]. Table 1 provides a summary of the classes of prescription and non-prescription drugs detected in effluents and surface water, and selected sources of these data. Anti-inflammatories/analgesics which are either prescription drugs (e.g. naproxen) or non-prescription drugs (e.g. ibuprofen) are found frequently in effluents and surface waters[33,40,55,61,62] probably because of the high rates of usage of these compounds to treat the symptoms of colds, aches and pains or for the treatment of arthritic conditions. A variety of lipid-regulating agents, including clofibric acid and gemfibrozil, have been identified in STP effluents and river water by investigators in both Germany and North America[33,40,55,62]. The anti-epileptic drug, carbamazepine, and several antidepressants used in psychopharmacotherapy (e.g. diazepam and fluoxetine) have also been detected in effluents and/or surface waters[33,55,61,62]. A range of antibiotics have been detected in effluents and surface waters in Europe and North America, including trimethoprim, antibiotics from the fluoroquinoline class (e.g. ciprofloxacin), sulfonamide class (e.g. sulfamethoxazole), the erythromycin class, and the tetracycline class[19,23,27,33]. It is possible that the source of some of these antibiotics is livestock operations where antibiotics are still used as feed additives. The synthetic estrogen, ethinylestradiol, which is used in birth-control pills, has been detected in sewage effluents in Europe and North America[63,64]. The effects and risks of ethinylestradiol and related compounds are well known[2,4] and will not be considered here. A variety of other drugs used to treat heart arrhythmias, hypertension, stomach acidity, asthma and diabetes, as well as drugs used in chemotherapy have been detected occasionally in effluents and/or surface waters[33,34,62].

In general, the levels of drugs in STP effluents range from high ng/l (ppt) to low μg/l (ppb) concentrations. In surface waters, the concentrations of these compounds are rarely above 100 ng/l and are frequently below 10 ng/l. It is anticipated that the highest concentrations of these compounds will occur at sites close to point sources, such as near discharges from STPs. However, the lipid-regulating agent, clofibric

TABLE 1

Classes of drugs detected in sewage treatment plant effluents and surface waters in Europe and North America and their therapeutic applications

Class of drug	Applications	Examples	References
Analgesic/ Anti-inflammatory	*Non-prescription*: treatment of colds, allergies, pain	ASA (aspirin), ibuprofen, indomethacin, acetaminophen	33, 40, 55, 61, 62
	Prescription: treatment of chronic pain, arthritis, migraines, etc.	Naproxen, diclofenac, propyphenazone, codeine	
Lipid-regulating	Reduce blood cholesterol and treat hyperlipidemia	Clofibric acid, gemfibrozil	33, 40, 55, 62
Anti-epileptic	Anti-convulsant	Carbamazepine	55, 61, 62
Psychiatric	Psychopharmacotherapy, sedative	Fluoxetine, diazepam	33, 61, 62
Betablockers	Treatment of heart arrhythmias, hypertension	Propranolol, nadolol	33, 62
β_2-Sympatho-mimetics	Bronchodilator for treatment of asthma	Salbutamol	33, 62
Antacid	Treatment of stomach acidity	Cimetidine	33
Cytostatic	Cancer chemotherapy	Ifosfamide, cyclophosphamide	62
Antibiotics	Treatment of bacterial infections	Sulfamethoxazole, trimethoprim, tetracycline, ciprofloxacin	19, 23, 27, 33
Synthetic estrogens	Birth control, post-menopausal therapy	17β-ethinylestradiol	64

acid, has been detected in Europe in surface waters remote from point sources at concentrations up to 10 ng/l[7].

3. *Biological and environmental concerns*

Aquatic organisms may be exposed to prescription and non-prescription drugs through direct discharges from STPs into receiving waters or through surface run-off of drugs from agricultural fields. Since most pharmaceuticals are designed not to bioaccumulate in human tissues, there is little potential for drugs to bioaccumulate in the tissues of exposed fish but bioconcentration in the blood might be a possibility. In addition, drugs used in human applications usually have low acute toxicity, with the possible exception of chemotherapy agents[34] and mutagenic antibiotics[24]. Therefore, the biological impacts will likely occur as a result of the actions of these drugs on fish through similar mechanisms by which they act in humans. Further, it is likely that these impacts will occur at chronically low concentrations (i.e. ng/l) only if the drugs act through interactions with cellular receptors. For instance, it has been shown that ethinylestradiol can alter the gonadal development of fish at exposure concentrations below 10 ng/l[38], well within the range of concentrations of this compound in surface waters near STPs. It is likely that biological impacts will occur in fish exposed to other pharmaceuticals acting through receptor-mediated mechanisms, such as drugs that bind to steroid, thyroid or peroxisome proliferator-activated receptors (PPARs), or drugs acting upon the neuroendocrine system. In addition, chronic exposure due to the persistence of some of these drugs might lead to multigenerational effects.

Of the many thousands of drugs currently in use, it is difficult to predict possible effects on non-target species in the environment. Fish represent ideal models to study initial impacts of drugs specifically designed to affect vertebrate systems. Such pharmaceutical pollutants are vastly different from what are now well-known endocrine disrupting chemicals. Industrial waste, pesticides and other pollutants were for the most part not designed to be hormone mimics or high-affinity receptor ligands. Pharmaceuticals on the other hand were designed to be potently biologically active with the lowest possible toxicity for the human species. Many hormonal and metabolic systems targeted by pharmaceuticals (i.e. drugs principally for humans, livestock and furry pets and occasionally birds) are conserved in vertebrates, thus drug actions on aquatic animals such as fish and frogs can therefore be expected.

This review will focus on the potential biological impact of two classes of pharmaceuticals prevalent in the environment because of their wide therapeutic use: lipid regulators and central nervous system drugs. This should serve as a foundation for future studies on the mechanism of action of these pharmaceuticals in non-target species and as a warning of the next wave of endocrine disrupting substances that we are adding to our environment continuously.

III. Potential biological effects of pharmaceuticals in the environment

Pharmaceuticals are designed to target specific enzyme or receptor systems. As a result of the functional interactions between metabolic, endocrine and immune systems, a high potential exists for undesirable side effects with many drugs. This is a specific concern in the case of pharmaceuticals in the environment and their effects on non-target aquatic species. We will first briefly review the mode of action of the drugs as well as the physiological effects and finally the potential effects on fish following environmental exposure.

1. Fibrate drugs

Fibrate drugs are a class of lipid regulatory drugs. They are used to regulate blood lipid and cholesterol levels. They include drugs such as gemfibrozil, ciprofibrate, clofibrate, bezafibrate and fenofibrate (Fig. 1). As noted in Table 1, gemfibrozil (GEM) was identified in μg/l concentrations in post-STP effluents collected from

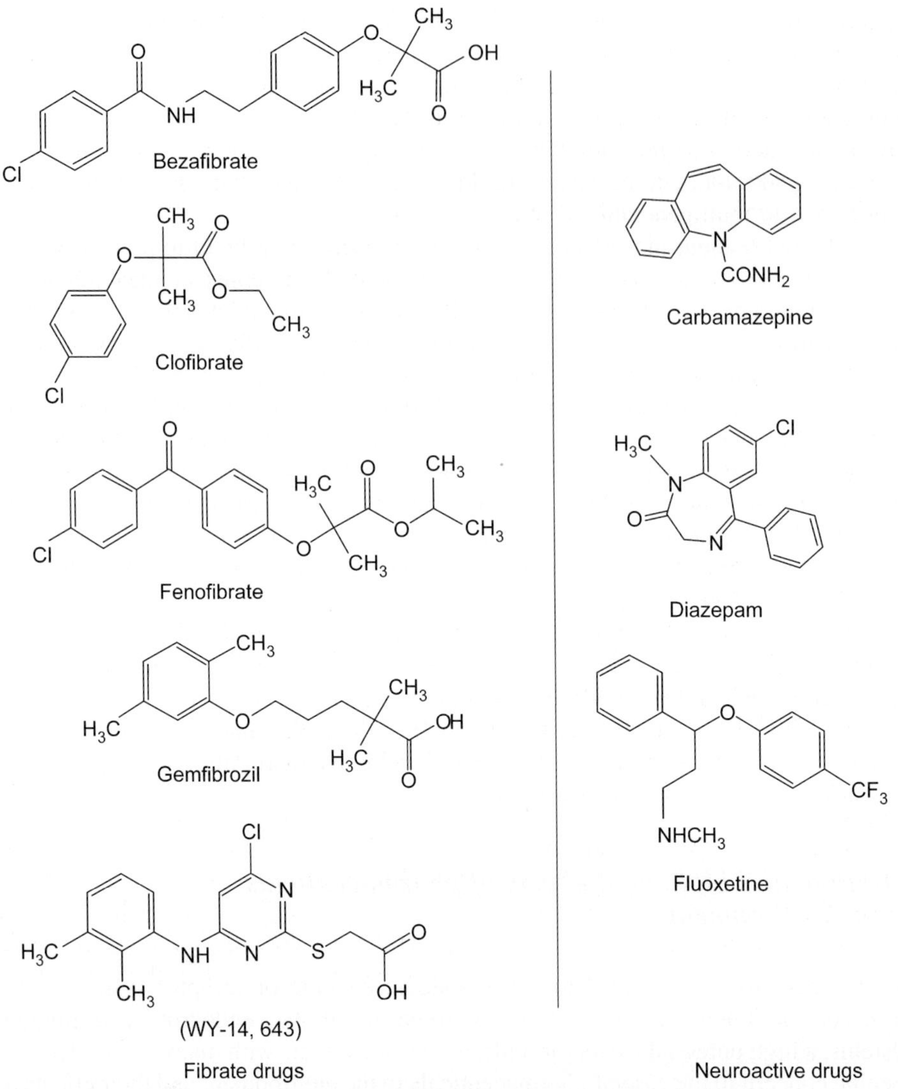

Bezafibrate

Clofibrate

Fenofibrate

Gemfibrozil

(WY-14, 643)

Fibrate drugs

Carbamazepine

Diazepam

Fluoxetine

Neuroactive drugs

Fig. 1. Chemical structures of different fibrate drugs known to be peroxisomal proliferators (PP) and of neuroactive drugs. Note that WY-14,643 is a fibrate drug and a potent PP but is not used clinically.

a number of Canadian urban areas[40] and was therefore chosen to investigate potential impacts on aquatic life.

1.1. Mechanisms of action

Fibrates are classified as peroxisomal proliferators (PPs), as these drugs increase the number and size of cellular peroxisomes not only in the liver but also in many other tissues of susceptible species[17,21]. Peroxisomes are single membrane bound organelles that segregate harmful products, such as H_2O_2, from the rest of the cell while carrying on β-oxidation of very long-chain fatty acids[21]. In addition to fibrate drugs, polyunsaturated fatty acids (PUFA), arachidonic acid and other eicosanoids, some hormones, phthalate ester plasticizers, herbicides and solvents also act as PPs[14,21,49,69]. These PPs are amongst the most widespread non-genotoxic hepatocarcinogens, although the precise mechanism of hepatocarcinogenesis is yet to be fully understood.

The therapeutic role and hepatocarcinogenesis of GEM, and other fibrates are initiated as these PPs bind to and activate a family of transcription factors called the peroxisomal proliferator-activated receptors. The strongest evidence in support of this linkage is that gene-knockout mice lacking PPARα, one PPAR isoform, are resistant to peroxisome proliferation and hepatocarcinogenesis caused by PP-agonists[37]. The activated PPAR moves to the nucleus where it forms a heterodimer with RXRα (Fig. 2), the retinoic acid X receptor[14]. Neither activated PPAR nor RXRα can individually initiate gene transcription. This heterodimer of PPARα:RXRα binds to regulatory regions called the PPAR responsive element (PPRE) within the upstream promoter of PP-sensitive genes[14,35,21,49]. Targeted genes containing a PPRE in their promoter region, mainly encode for proteins involved in lipid and carbohydrate metabolism, transcription, cell cycle and apoptosis[10]. A potential PPRE has also been found in the zebrafish aromatase (CYP192A) promoter[32]. In addition, PPARs are regulated by phosphorylation, hormones (glucocorticoids, insulin, leptin) and by stress and fasting[14,17,35]. These studies implicate PPAR-agonists in the control of numerous metabolic processes.

The first PPAR was reported by Issemann and Green[30] in rodents. To date, three distinct PPARs, encoded by three distinct genes have been identified: PPARα, PPARβ, and PPARγ. These isoforms have different tissue distributions, with PPARα expressed primarily in liver, heart and kidney, PPARγ in brown and white adipose tissues, and PPARβ (also known as δ) is more ubiquitously expressed, but most abundant in the central nervous system[14,17,35]. The role of each isoform is beyond the scope of this review; as fibrate drugs are mainly PPARα-ligands[14], only this form will be examined in detail here. Nevertheless, it should be mentioned that isoforms of the PPAR family have now been cloned from *Xenopus laevis*, a number of mammals[49,50], as well as in some fish species[1].

1.2. Physiological effects

Peroxisome proliferators impact many tissues and metabolic processes. Significant species differences with respect to the effects of PPs are reported, with rodents much more sensitive than primates (including humans). Studies have demonstrated that hepatic expression of PPARα in rodents is very much higher than in other mammals[21] and the human gene is shortened compared with its rodent counterpart, but whether these or some other components of the signaling pathway are involved in species sensitivity is not understood[18,39,69]. Regardless, our greatest understanding of PP action is associated with the effects of PPARα on fatty acid metabolism in the mammalian liver.

Peroxisomes generally occupy about 2% of hepatic cytoplasmic volume in rodents; when exposed to a PP, this volume percent can increase to 18–25%[49].

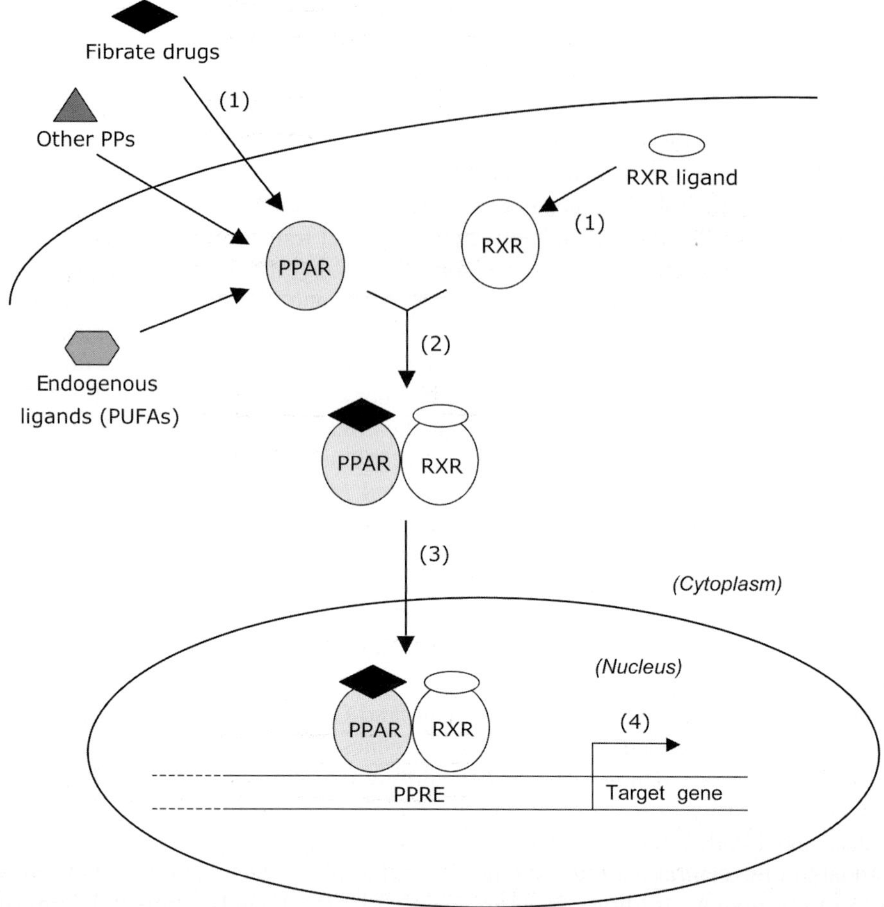

Fig. 2. Summary of the mode of action of peroxisome proliferators (PP) through the nuclear peroxisome proliferator-activated receptor (PPAR). The activation of PPAR requires the retinoic X receptor (RXR). The receptors are found in the cytoplasm and once the ligands bind (1) they undergo a conformational change (2) that allows the complex to enter the nucleus (3). The complex then binds to the peroxisome proliferator response element (PPRE) in the promoter region of target genes to alter transcription (4). Modified from Gervois *et al.*[17]

This hepatomegaly is correlated to PPARα activation of acyl-CoA oxidase (AOX), the first enzyme of peroxisomal β-oxidation of fatty acids and a gene with a PPRE in its promoter region[14]. In addition, hepatic mitochondrial β-oxidation and microsomal ω-oxidation of fatty acids are increased, as a direct result of PPARα activation of mRNA of specific enzymes associated with these pathways (carnitine palmitoyl transferase I and cytochrome P4504A, respectively). Activation of fatty acid oxidation by these three pathways would lead to enhanced fatty acid oxidation, given the appropriate substrate. PPARα has also been shown to enhance delivery of fatty acids to the oxidizing systems (Fig. 3).

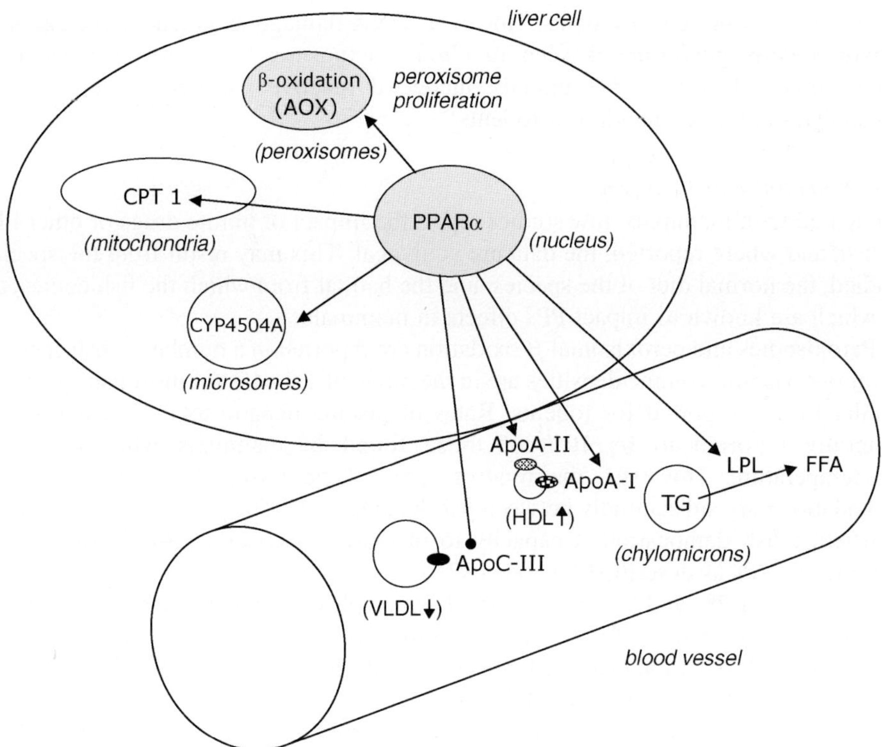

Fig. 3. Effects of PPARα activation by fibrate drugs on lipid metabolism in mammals. Target genes are up- or down-regulated through the activation of PPARα. Shaded areas indicate known effects in fish. In addition to effects in the liver cell, PPARα regulates circulating lipoproteins and membrane fatty acid transporters to enhance delivery of fatty acids to the oxidizing systems. These actions of PPARα lead to enhanced hepatic oxidation of fatty acids, reduced very low density lipoprotein production (VLDL) and increased LDL removal while increasing HDL production by increasing cholesterol movement to the liver. AOX, acyl-CoA oxidase; CPT-1, carnitine palmitoyl transferase-1; PPAR, peroxisome proliferator-activated receptor; HDL, high-density lipoprotein; VLDL, very low density lipoprotein; TG, triglyceride.

The formation of oxidative stress following PP exposure has been proposed as one hypothesis to explain the observed hepatocarcinogenesis in rodent[21]. Hydrogen peroxide (H_2O_2) production increases as AOX donates its electron directly to molecular oxygen rather than to reduced flavin dinucleotide as in mitochondrial fatty acid oxidation. Generally there is a dissociation between AOX and catalase (CAT) activities in response to a PP, with the former increasing by 20–30-fold compared with only 2–3-fold in rats, respectively[21,49], as a result of the presence of a PPRE in AOX promoter region but not in CAT. Thus, increased H_2O_2 production may escape the peroxisome resulting in cellular oxidative damage, leading potentially to hepatic tumor production. Other components of the cellular antioxidant system including glutathione peroxidase and total glutathione are often depressed with chronic administration of PPs[42], but significant changes in oxidative damage are generally not reported in rodents. However, DNA damage as noted by increases in 8-hydroxydeoxyguanosine is seen in chronic exposures[21]. *In vitro* studies have demonstrated that clofibrate directly increased reactive oxygen species in both hepatocytes and mitochondria of rodents[48].

1.3. Potential effects in fish

Compared with mammals, few studies report the impact of fibrate drugs or other PPs in fish, and where reported, the data are equivocal. This may result from the species studied, the normal diet of the species, and the habitat from which the fish comes, all of which are known to impact PPs effects in mammals.

Peroxisomes and peroxisomal β-oxidation are reported in a number of fish species. Liver peroxisome volume densities are in the range of 1.3–4% for most fish species[43], similar to that reported for rodents. Rates of piscine hepatic total β-oxidation are generally reported to be 10% of those found in mammals when corrected for temperature. However, the relative rates of peroxisomal vs. mitochondrial β-oxidation vary substantially between fish species[11,41,50]. The pertinent question here is whether fish demonstrate a capacity to increase peroxisomal β-oxidation in the presence of PPs as described for rodents?

A number of PPs and fish species have been employed, and in general, fish respond to hypolipidemic drugs[15,56,70] by increasing the activity of AOX, the first enzyme of peroxisomal β-oxidation (Table 2). Species sensitivities vary but in those species that do respond, the response is reported to be weaker than in rodents to the same PP[56]; the marine sea bass (*Dicentrarchus labrax*) was reported to be refractory to clofibrate[47] but further investigations are needed as AOX was not measured. The marine flatfish, *Pleuronectes platessa* did respond to a PP by an increased expression of glutathione-*S*-transferase (GST), an enzyme associated with cellular antioxidant status[36]. Furthermore, a 1.7-fold increase in AOX after administration of clofibric acid and bezafibrate was observed in salmon (*Salmo salar*) hepatocyte culture[53]. In addition,

TABLE 2

Summary of observed effects on enzyme activities following acute exposure to peroxisome proliferators in fish

PPs	Species	Enzyme activity	Reference
	In vivo		
Clofibrate	Rainbow trout	↑ ACO	70
		↑ CAT	70
Gemfibrozil	Rainbow trout	↑ ACO	56
	Japanese medaka	– ACO	56
	In vitro		
Gemfibrozil	Rainbow trout hepatocytes	– ACO	15
Ciprofibrate	Rainbow trout hepatocytes	↑ ACO	15
Clofibric acid	Rainbow trout hepatocytes	↑ ACO	15
	Atlantic salmon hepatocytes	↑ ACO	53
Bezafibrate	Atlantic salmon hepatocytes	↑ ACO	53

↑, increase in enzyme activities; –, no change in enzyme activities.

dehydroepiandrosterone, a known hormonal PP in rodents, enhances hepatocarcinogenesis in rainbow trout (*Oncorhynchus mykiss*) through a peroxisomal-independent pathway, acting as a genotoxic carcinogen[44]. These studies suggest that PPs are less effective in the fish species studied to date, but most studies are short term, using a limited dosage range, and with a limited number of fish species. Whether PPs affect fish in an environmental context has yet to be effectively studied.

Our recent studies have found that gemfibrozil (GEM) is bioactive in the goldfish, *Carassius auratus* (C. Mimeault *et al.*, unpublished). GEM injected into goldfish at 10 and 100 μg/g fish doubles blood glucose and halves blood triglycerides. However, this study does not mimic the exposure route of GEM in water. In addition, given the importance of fatty acids in gonadal steroidogenesis and reproduction, the impact of GEM on reproductive processes must be examined.

Limited gene sequence information on fish PPARs is available. Using semi-nested PCR amplifications, Escriva *et al.*[16] identified several PPAR consensus sequences from a shark (*Scyliorhinus canicula*), a hagfish (*Myxine glutinosa*), a lamprey (*Petromyzon marinus*) and the zebrafish (*Danio rerio*). In addition to unclassified PPAR forms, PPARα, PPARβ and PPARγ homologs were found in the artic char, lamprey and zebrafish, and PPARα and PPARγ were found in the shark (GenBank, 2003). Andersen *et al.*[1] amplified from Atlantic salmon (*S. salar*) liver mRNA a PPARγ cDNA with a deduced amino acid sequence of 544 amino acids and 47% overall sequence identity to the mammalian PPARγ. GenBank also contains PPARγ sequences for the plaice (*P. platessa*) and the European flounder (*Platichthys flesus*). We have cloned and sequenced partial cDNAs from the goldfish (Genbank Acc. AY198322) that are homologous to the mammalian PPARα and PPARγ isoforms. Therefore, we can conclude that several PPARs are expressed in fish supporting the observation that these drugs can be bioactive in these non-target species.

2. *Neuroactive drugs*

Drugs that act on the central nervous system include, amongst others, antidepressant, anti-seizure and sedative drugs. Pharmaceuticals from these classes have been found in aquatic environments[13,33] and are of interest because they have the potential to be endocrine disruptors although for the most part this hypothesis has yet to be tested in any aquatic species. To illustrate the possible endocrine disrupting effects, first it will be necessary to briefly describe several aspects of the fish neuroendocrine system.

2.1. Fish neuroendocrine system

Multiple neurotransmitters and neuropeptides from the hypothalamus and pre-optic region of the brain interact to regulate hormone synthesis and release from the pituitary gland. Best described are the gonadotropin-II (GTH-II; also called fish luteinizing hormone) and growth hormone (GH) releasing systems in the goldfish[5,6,65,67,68].

The synthesis and release of GTH-II is regulated by multiple stimulatory systems counteracted by a single potent inhibitory dopaminergic (DA) system (Fig. 4). The principal stimulatory neuropeptide-regulating GTH-II is the decapeptide

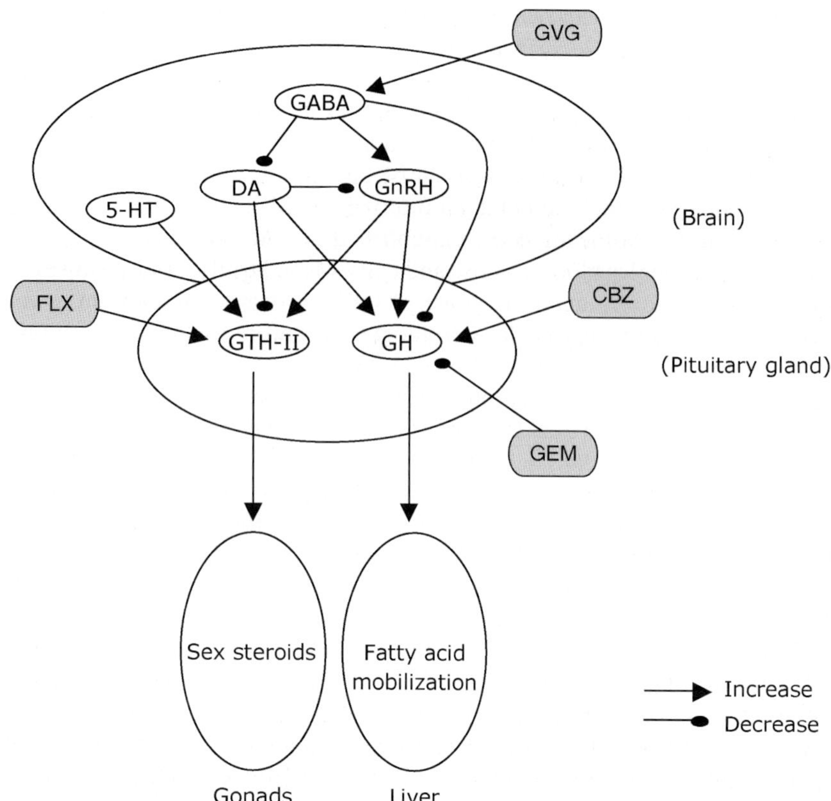

Fig. 4. Observed effects of fibrate and neuroactive drugs on the fish endocrine system. Shown in grey shaded boxes are pharmaceuticals with known biological activity in fish. CBZ, carbamazepine; FLX, fluoxetine; GEM, gemfibrozil; GVG, gamma vinyl-GABA. The potential of these drugs to disrupt endocrine functions are significant given that multiple neuroendocrine interactions are likely. Note that GVG is biologically active in fish but its concentrations in aquatic ecosystems and sewage are not known. Adapted from Trudeau[65].

gonadotropin-releasing hormone (GnRH). Fish have at least two GnRHs, and some fish have three different GnRH forms[20]. Considerable research in several fish species have shown that DA inhibits GTH-II release by multiple mechanisms, both directly at the gonadotroph and indirectly through inhibition of GnRH release and action[9,65].

The GH release system is also regulated by multiple neurotransmitters and neuropeptides and has been reviewed extensively[5,45,65]. Relevant here is the role of gamma-aminobutyric acid (GABA). In contrast to the stimulatory effects of GABA on GTH-II, GABA inhibits GH release in goldfish[68]. This inhibitory effect of GABA on GH varies seasonally and is induced by sex steroids. It is beyond the scope of this review to cover the complete regulation of GH and the more than 20 factors regulating GTH-II release. However, two additional neurotransmitter systems have direct relevance to the endocrine-disrupting potential of pharmaceuticals found in the environment, serotonin (5-HT) and GABA.

Somoza *et al.*[58,59] demonstrated that 5-HT plays an important stimulatory role in the regulation of GTH-II release. Intraperitoneal injections of 5-HT rapidly elevated serum GTH II levels. Serotonin also stimulated GTH-II release from goldfish pituitary fragments *in vitro*. Antagonists of the 5-HT2 receptor type system blocked this *in vitro* effect. Moreover, 5-HT in Atlantic croaker (*Micropogonias undulates*), a marine fish species, is considered a principal stimulator of GTH-II release[31].

Another neurotransmitter system, that is a potential target for psychoactive drugs detected in the environment, is GABA, which is one of the most abundant neurotransmitters in the vertebrate central nervous system and is considered the main inhibitory neurotransmitter. Injections of GABA rapidly increased GTH-II release in goldfish by stimulation of GnRH release and inhibition of the inhibitory DA system[66].

2.2. *Potential effects in fish*

Fluoxetine is a 5-HT re-uptake inhibitor used as an antidepressant and is found in the environment[33] and could potentially disrupt normal GTH-II release. The results of Kahn and Thomas[31] where fluoxetine potentiated 5-HT action on GTH-II release in Atlantic croaker support this idea. In addition, the effects of 5-HT are known to vary seasonally and thus, the biological effects would vary depending on the time of the reproductive cycle when fish are exposed[31].

Anti-epileptic drugs have also been found in the environment, but for the most part their effects in fish have not been assessed. One drug, gamma vinyl-GABA (GVG), is an inhibitor of the GABA-degrading GABA transaminase enzyme and has important biological effects when injected in fish. It is not yet known if it can be detected in sewage effluent, but its actions serve to illustrate possible effects of drugs in this class. Following injection of GVG, GABA levels increase in the brain and pituitary. This causes increased release of GTH-II and also increases the expression of GTH-II β subunit mRNA in the pituitary[67]. Its action in goldfish has been shown to work through the stimulation of GnRH release and inhibition of DA, as for GABA itself[66].

The GABA-A receptor is a complex chloride channel with the GABA ligand-binding site along with multiple binding sites for modulators of GABA function. In particular, benzodiazepines, which are the most important type of sedatives, specifically bind to the GABA-A receptor to potentiate GABA action. The anxiolytic diazepam is a widely prescribed benzodiazepine and is found in the environment (Table 1). Although the effects of benzodiazepines on GTH-II release in fish have not been studied, given the importance of GABA to regulate GTH-II release, and the well-known sedative effects of diazepam, it can be reasonably predicted that benzodiazepines would affect the endocrine system or alter the behavior in fish. Neuroendocrine effects of benzodiazepines have been reported for the human[29] and chicken[25]. Diazepam decreases alertness in a weakly electric fish species[3] (*Gymnotus carapo*) and benzodiazepine-binding sites have been detected in minnow species (*Pimephales promelas*) brain[51]. As described for GTH-II, it is not known if diazepam or other benzodiazepines affect GH release in fish. However, given that benzodiazepines affect GH release in other species, its actions on GH in fish should be investigated.

2.3. Preliminary results

The above speculations suggest that at least a few medical pharmaceuticals are of concern. However, considerable research remains to be conducted to test the hypothesis that pharmaceuticals act as endocrine-disrupting chemicals in aquatic organisms. To this end, we have begun to assess the effects of *in vivo* administration of the anti-epileptic carbamazepine (CBZ) and the fibrate GEM, two drugs found in the environment[40], on pituitary function in goldfish (*C. auratus*). Carbamazepine (5 mg/kg) and GEM (100 mg/kg) were injected every second day for 10 days into male goldfish and the pituitaries were collected 24 h following the last injection for RNA extraction and blood for the analysis of growth hormone levels by radioimmunoassay. GH levels in the blood for control, CBZ and GEM in goldfish were 87 ± 18, 134 ± 15 and 78 ± 49 ng/ml, respectively, demonstrating that CBZ stimulated the release of GH ($p < 0.05$) whereas GEM had no effect on serum levels. In contrast, CBZ had very little effect on GH or secretogranin-II (Sg-II) mRNA expression in the pituitary (Figs. 5 and 6). However, GEM significantly depressed GH and Sg-II gene expression ($p < 0.01$) (Figs. 5 and 6). These data demonstrate that both CBZ and GEM are biologically active modulators of the endocrine system in goldfish when given at doses equivalent to human therapeutic levels.

The results obtained were not predictable based on the pharmacological and endocrinological literature derived from experiments on mammals. The case of GEM best illustrates this point. In mammals, it is well known that GH stimulates fatty acids mobilization from the liver. In turn, fatty acids feedback negatively on GH release[46]. By lowering circulating fatty acid levels with GEM in fish, it was predicted that GH mRNA levels and/or serum GH levels would increase. We observed a decrease in GH mRNA and no effects of GEM on serum GH.

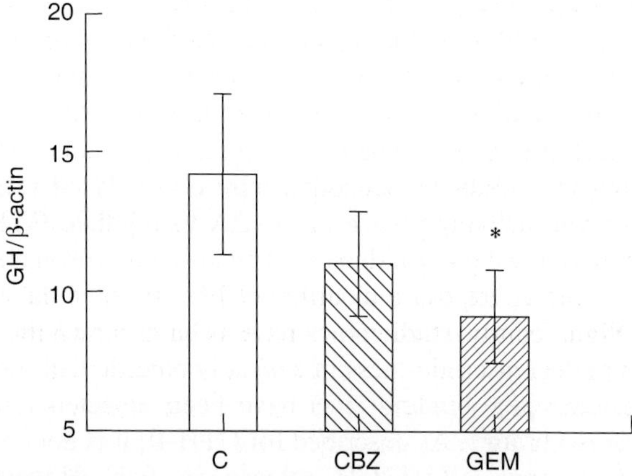

Fig. 5. Effects of carbamazepine (CBZ) and gemfibrozil (GEM) on relative growth hormone (GH) mRNA expression levels in goldfish pituitary. Data represent means ± standard errors with $n = 10$. C, control; $*p < 0.05$ compared with C by one-way ANOVA; Tukey *post hoc* test.

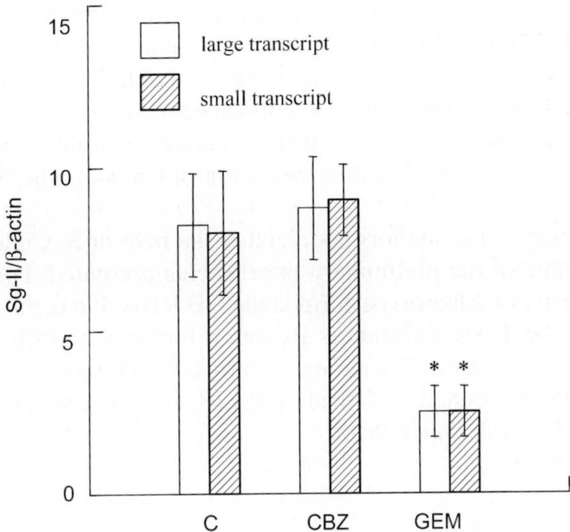

Fig. 6. Effects of carbamazepine (CBZ) and gemfibrozil (GEM) on relative secretoganin-II (SG-II) mRNA expressions levels in goldfish pituitary. There are two Sg-II mRNA transcripts and these are co-suppressed by GEM. Data represent means ± standard errors with $n = 10$. C, control; $*p < 0.05$ compared with C by one-way ANOVA; Tukey *post hoc* test.

IV. Conclusions

Thousands of medical pharmaceuticals are released into the environment. Some are highly biologically active in non-target species and others are highly persistent in STP effluents and in the aquatic environment. As required by law, significant research and development time and funds are invested to determine the mechanisms of action, safety and side effects of pharmaceuticals before a drug can be used clinically. Until recently, there was little concern of the effects of pharmaceuticals in the environment because it was not known that they were persistent. Now that this is known, it is critical that we begin examining the impact of these drugs in the aquatic environment.

Through this review, we examined the potential impacts as well as some preliminary results demonstrating unexpected effects. It should also be mentioned that drugs in the environment will not occur in isolation, but with other toxicants that may be altering other biochemical processes. Additive and/or synergestic effects could also be observed. Furthermore, exposure at different life stages should be examined as effects may vary. Significantly, studies with rodents have shown that exposure to PPs including clofibric acid during early developmental stages alters the binding efficiency of endogenous steroids to estrogen and androgen receptors during later life stages and affects endocrine-mediated responses such as mating behavior[12].

Studies are beginning to assess the effects of some drugs in fish to explore the basic biology of a particular physiological or endocrine system. However, it is not yet known if chronic low-level exposure to environmental levels of GEM, CBZ or any of the thousands of other medical pharmaceuticals would produce a deleterious or endocrine-disrupting effect in fish. Moreover, it cannot always be assumed that a pharmaceutical developed for humans has identical actions in non-target species, even though many receptors and hormones are conserved between species. Clearly, we are only at the beginning of testing whether drugs and other pharmaceuticals and personal care products represent a new threat to the health of fish and other wildlife species.

Acknowledgements. The authors appreciated the help of S. Chiu, B. Hibbert and M. Samia with some of the preliminary experiments presented. Drs E.J. Fraser and N. Stacey (University of Alberta) performed the GH RIAs. Financial support for these studies was from the Toxic Substances Research Initiative (TSRI – government of Canada), Canadian Network of Toxicology Centres (CNTC) and Natural Sciences and Engineering Research Council of Canada (NSERC) Discovery and Strategic Grants programs to V.L.T., C.M. and T.W.M.

V. References

1. Andersen, O., V.G.H. Eijsink and M. Thomassen. Multiple variants of the peroxisome proliferator-activated receptor (PPAR) γ are expressed in the liver of Atlantic salmon (*Salmo salar*). *Gene* 255: 411–418, 2000.
2. Andersson, H.R., A.M. Andersson, S.F. Arnold, H. Autrup, M. Barfoed, N.A. Beresford, P. Bjerragaard, L.B. Christiansen, B. Gissel, R. Hummel, E.B. Jorgensen, B. Korsgaard, R. Le Guevel, H. Leffers, J. McLachlan, A. Moller, J.B. Nielson, N. Olea, O.A. Karasko, F. Pakdel, K.L. Pederson, P. Perez, N.E. Skakkeboek, C. Sonnenschein, A.M. Soto, J.P. Sumpter, S.M. Thorpe and P. Grandjean. Comparison of short-term estrogenicity tests for identification of hormone-disrupting chemicals. *Environ. Health Perspect.* 107: 89–108, 1999.
3. Aparecida, S., L. Correa and A. Hoffmann. Effect of drugs that alter alertness and emotionality on the novelty response of a weak electric fish, *Gymnotus carapo*. *Physiol. Behav.* 65: 863–869, 1999.
4. Arcand-Hoy, L.D. and W.H. Benson. Fish reproduction: an ecologically relevant indicator of endocrine disruption. *Environ. Toxicol. Chem.* 17: 49–57, 1998.
5. Blazquez, M., P.T. Bosma, J.P. Chang, K. Docherty and V.L. Trudeau. Gamma-aminobutyric acid upregulates the expression of a novel secretogranin-II mRNA in the goldfish pituitary. *Endocrinology* 139: 4870–4880, 1998.
6. Blazquez, M., P.T. Bosma, E.J. Fraser, K. Van Look and V.L. Trudeau. Fish as models for neuroendocrine regulation of growth and reproduction. *Comp. Biochem. Physiol.* 119C: 339–364, 1998.
7. Buser, H.-R., M.D. Muller and N. Theobald. Occurrence of the pharmaceutical drug clofibric acid and herbicide mecoprop in various Swiss lakes and in the North Sea. *Environ. Sci. Technol.* 32: 188–192, 1998.
8. Buser, H.-R., T. Poiger and M.D. Muller. Occurrence and environmental behavior of the chiral pharmaceutical drug ibuprofen in surface waters and in wastewater. *Environ. Sci. Technol.* 33: 2529–2535, 1999.
9. Chang, J.P., A.O. Wong and R.E. Peter. Differential actions of dopamine receptor subtypes on gonadotropin and growth hormone release in vitro in goldfish. *Neuroendocrinology* 51: 664–674, 1990.
10. Cherkaoui-Malki, M., K. Meyer, W.-Q. Cao, N. Latruffe, A.V. Yeldandi, M.S. Rao, C.A. Bradfield and J.K. Reddy. Identification of novel peroxisome proliferator-activated receptor α (PPARα) target genes in mouse liver using cDNA microarray analysis. *Gene Express.* 9: 291–304, 2001.
11. Crockett, E.L. and B.C. Sidell. Peroxisomal β-oxidation is a significant pathway for catabolism of fatty acids in a marine teleosts. *Am. J. Physiol.* 264: R1004–R1009, 1993.
12. Csaba, G., A. Inczefi-Gonda, C. Karabelyos and E. Pap. Hormonal imprinting: neonatal treatment of rats with the peroxysome proliferator clofibrate irreversibly affects sexual behaviour. *Physiol. Behav.* 58: 1203–1207, 1995.
13. Daughton, C.G. and T.A. Ternes. Pharmaceuticals and personal care products in the environment: agents of subtle change? *Environ. Health Perspect.* 107: 907–938, 1999.
14. Desvergne, B. and W. Wahli. Peroxisome proliferator-activated receptors: nuclear control of metabolism. *Endocrinol. Rev.* 20: 649–688, 1999.

15. Donohue, M., L.A. Baldwin, D.A. Leonard, P.T. Kostecki and E.J. Calabrese. Effects of hypolipidemic drugs gemfibrozil, ciprofibrate, and clofibric acid on peroxisomal β-oxidation in primary cultures of rainbow trout hepatocytes. *Ecotoxicol. Environ. Saf.* 26: 127–132, 1993.
16. Escriva, H., R. Safi, C. Hanni, M.-C. Langlois, P. Saumitou-Laprade, D. Stehelin, A. Capron, R. Pierce and V. Laudet. Ligand binding was acquired during evolution of nuclear receptors. *Proc. Natl Acad. Sci. USA* 94: 6803–6808, 1997.
17. Gervois, P., I.P. Torra, J.-C. Fruchart and B. Staels. Regulation of lipid and lipoprotein metabolism by PPAR activators. *Clin. Chem. Lab. Med.* 38: 3–11, 2000.
18. Giometti, C.S., S.L. Tollaksen, X. Liang and M.L. Cunningham. A comparison of liver protein changes in mice and hamsters treated with the peroxisome proliferator Wy-14,643. *Electrophoresis* 19: 2498–2505, 1998.
19. Golet, E.M., A.C. Alder, A. Hartmann, T.A. Ternes and W. Giger. Trace determination of fluoroquinolone antibacterial agents in urban wastewater by solid-phase extraction and liquid chromatography with fluorescence detection. *Anal. Chem.* 73: 3632–3638, 2001.
20. Gonzalez-Martinez, D., T. Madigou, N. Zmora, I. Anglade, S. Zanuy, Y. Zohar, Munoz-Cueto and O. Kah. Differential expression of three different prepro-GnRH (gonadotrophin-releasing hormone) messengers in the brain of the European sea bass (*Dicentrarchus labrax*). *J. Comp. Neurol.* 429: 144–155, 2001.
21. Gonzalez, F.J., J.M. Peters and R.C. Cattley. Mechanism of action of the nongenotoxic peroxisome proliferators: role of the peroxisome proliferator-activated receptor α. *J. Natl Cancer Inst.* 90: 1702–1709, 1998.
22. Halling-Sørenson, B., S.N. Nielsen, P.F. Lanzky, F. Ingerslev, H.C. Holten Lutzhoft and S.E. Jorgensen. Occurrence, fate and effects of pharmaceutical substances in the environment – a review. *Chemosphere* 36: 357–393, 1998.
23. Hartig, C., T. Storm and M. Jekel. Detection and identification of sulphonamide drugs in municipal wastewater by liquid chromatography coupled with electrospray ionisation tandem mass spectrometry. *J. Chromatogr.* 854A: 163–173, 1999.
24. Hartmann, A., A.C. Alder, T. Koller and R.M. Widmer. Identification of fluoroquinolone antibiotics as the main source of *umu*C genotoxicity in native hospital wastewater. *Environ. Toxicol. Chem.* 17: 377–382, 1998.
25. Harvey, S. Benzodiazepine antagonism of thyrotropin-releasing hormone receptors: biphasic actions on growth hormone secretion in domestic fowl. *J. Endocrinol.* 137: 35–42, 1993.
26. Henschel, K.P., A. Wenzel, M. Didrich and A. Fleiner. Environmental hazard assessment of pharmaceuticals. *Regul. Toxicol. Pharmacol.* 18: 220–225, 1997.
27. Hirsch, R., T.A. Ternes, K.L. Kratz, K. Haberer, A. Mehlick, F. Ballwanz and K.-L. Kratz. Occurrence of antibiotics in the aquatic environment. *Sci. Total Environ.* 225: 109–118, 1999.
28. Holford, N.H.G. Pharmacokinetics and pharmacodynamics: rational dosing and the time course of drug action. In: *Basic and Clinical Pharmacology*, 8th ed., edited by B.G. Katzung, New York, McGraw-Hill, pp. 35–50, 2001.
29. Humbert, T. Effets neuroendocrines des benzodiazepines. *Ann. Med. Psychol. (Paris)* 152: 161–171, 1994.
30. Issemann, I. and S. Green. Activation of a member of the steroid hormone receptor superfamily by peroxisome proliferators. *Nature* 347: 645–650, 1990.
31. Kahn, A. and P. Thomas. Stimulatory effects of serotonin on maturational gonadotropin release in the Atlantic croaker *Micropogonias undulates*. *Gen. Comp. Endocrinol.* 88: 388–396, 1992.
32. Kazeto, Y., S. Ijiri, A.R. Place, Y. Zohar and J.M. Trant. The 5′-flanking regions of CYP19A1 and CYP19A2 in zebrafish. *Biochem. Biophys. Res. Commun.* 288: 503–508, 2001.
33. Kolpin, D.W., E.T. Furlong, M. Meyer, E.M. Thurman, S.D. Zaugg, L.B. Barber and H.T. Buxton. Pharmaceuticals, hormones, and other organic wastewater contaminants in US streams, 1999–2000: a national reconnaissance. *Environ. Sci. Technol.* 36: 1202–1211, 2002.
34. Kummerer, K., T. Steger-Hartmann and M. Meyer. Biodegradability of the anti-tumour agent ifosfamide and its occurrence in hospital effluents and communal sewage. *Water Res.* 31: 2705–2710, 1997.
35. Latruffe, N., M. Cherkaoui-Malki, V. Nicolas-Frances, B. Jannin, M.-C. Clemencet, F. Hansmannel, P. Passilly-Degrace and J.-P. Berlot. Peroxisome-proliferator-activated receptors as physiological sensors of fatty acid metabolism: molecular regulation in peroxisomes. *Biochem. Soc. Trans.* 29: 305–309, 2001.
36. Leaver, M.J., J. Wright and S.G. George. A peroxisomal proliferator-activated receptor gene from the marine flatfish, the plaice (*Pleuronectes platessa*). *Mar. Environ. Res.* 46: 75–79, 1998.
37. Lee, S.S., T. Pineau, J. Drago, E.J. Lee, J.W. Owens and D.L. Kroetz. Targeted disruption of the α isoform of the peroxisome proliferator-activated receptor gene in mice results in abolishment of the pleitropic effects of peroxisome proliferators. *Mol. Cell. Biol.* 15: 3012–3022, 1995.

38. Metcalfe, C.D., T.L. Metcalfe, Y. Kiparissis, B.G. Koenig, C. Khan, R.J. Hughes, T. Croley, R.E. March and T. Potter. The estrogenic potency of chemicals detected in sewage treatment plant effluents as determined by *in vivo* assays with the Japanese medaka *Oryzias latipes*. *Environ. Toxicol. Chem.* 20: 297–308, 2001.
39. Meyer, K., A. Volkl, H.F. Kuhnle and J. Pill. Species differences in induction of hepatic enzymes by BM 17.0744, an activator of peroxisome proliferator-activated receptor alpha (PPARalpha). *Arch. Toxicol.* 73: 440–450, 1999.
40. Miao, X.-S., B.G. Koenig and C.D. Metcalfe. Analysis of acidic pharmaceutical drugs in the aquatic environment using liquid chromatography–electrospray tandem mass spectrometry. *J. Chromatogr.* 95A: 139–147, 2002.
41. Moyes, C.D., R.K. Suarez, G.S. Brown and P.W. Hochachka. Peroxisomal β-oxidation: insights from comparative biochemistry. *J. Exp. Zool.* 260: 267–273, 1991.
42. O'Brien, M., M.L. Cunningham, B.T. Spear and H.P. Glauert. Effects of peroxisome proliferators on glutathione and glutathione-related enzymes in rats and hamsters. *Toxicol. Appl. Pharmacol.* 171: 27–37, 2001.
43. Orbea, A., K. Beier, A. Volkl, H.D. Fahimi and M.P. Cajaraville. Ultrastructural, immunocytochemical and morphometric characterization of liver peroxisomes in gray mullet, *Mugil caphalus*. *Cell Tissue Res.* 297: 493–502, 1999.
44. Orner, G.A., C. Mathews, J.D. Hendricks, H.M. Carpenter, G.S. Bailey and D.E. Williams. Dehydroepiandrosterone is a complete hepatocarcinogen and potent tumor promoter in the absence of peroxisome proliferation in rainbow trout. *Carcinogenesis* 16: 2893–2898, 1995.
45. Peter, R.E. and T.A. Marchant. The endocrinology of growth in carp and related species. *Aquaculture* 129: 299–321, 1995.
46. Pombo, M., C.M. Pombo, R. Astorga, F. Cordido, V. Popovic, R.V. Garcia-Mayor, C. Dieguez and F.F. Casanueva. Regulation of growth hormone secretion by signals produced by the adipose tissue. *J. Endocrinol. Invest.* 22: 22–26, 1999.
47. Pretti, C., S. Novi, V. Longo and P.G. Gervasi. Effects of clofibrate, a peroxisome proliferator, in sea bass (*Dicentrarchus labrax*), a marine fish. *Environ. Res.* 80A: 294–296, 1999.
48. Qu, B., Q.-T. Li, K.P. Wong, T.M.C. Tan and B. Halliwell. Mechanism of clofibrate hepatotoxicity: mitochondrial damage and oxidative stress in hepatocytes. *Free Radic. Biol. Med.* 31: 659–669, 2001.
49. Reddy, J.K. and G.P. Mannaerts. Peroxisomal β-oxidation. *Annu. Rev. Nutr.* 14: 343–370, 1994.
50. Reddy, J.K. and T. Hashimoto. Peroxisomal β-oxidation and peroxisome proliferator-activated receptor α: an adaptive metabolic system. *Annu. Rev. Nutr.* 21: 193–230, 2001.
51. Rehnberg, B.G., E.H. Bates, R.J. Smith, B.D. Sloley and J.S. Richardson. Brain benzodiazepine receptors in fathead minnows and the behavioral response to alarm pheromone. *Pharmacol. Biochem. Behav.* 33: 435–442, 1989.
52. Richardson, M.L. and J.M. Bowron. The fate of pharmaceutical chemicals in the aquatic environment. *J. Pharm. Pharmacol.* 37: 1–12, 1985.
53. Ruyter, B., O. Andersen, A. Dehli, A.-K. Ostlund Farrants, T. Gjoen and M.S. Thomassen. Peroxisome proliferator receptors in Atlantic salmon (*Salmo salar*): effects on PPAR transcription and acyl-CoA oxidase activity in hepatocytes by peroxisome proliferators and fatty acids. *Biochim. Biophys. Acta* 1348: 331–338, 1997.
54. Sacher, F., E. Lochow, D. Bethmann and H.-J. Brauch. Occurrence of drugs in surface waters. *Vom Wasser* 90: 233–243, 1998.
55. Sacher, F., F.T. Lange, H.-J. Brauch and I. Blankenhorn. Pharmaceuticals in groundwaters: analytical methods and results of a monitoring program in Baden-Wurttemberg, Germany. *J. Chromatogr.* 938A: 199–210, 2001.
56. Scarano, L.J., E.J. Calabrese, P.T. Kostecki, L.A. Baldwin and D.A. Leonard. Evaluation of a rodent peroxisome proliferator in two species of freshwater fish: rainbow trout (*Onchorynchus mykiss*) and Japanese medaka (*Oryzias latipes*). *Ecotoxicol. Environ. Saf.* 29: 13–19, 1994.
57. Seiler, R.L., S.D. Zaug, J.M. Thomas and D.L. Howcroft. Caffeine and pharmaceuticals as indicators of waste water contamination in wells. *Ground Water* 37: 405–410, 1999.
58. Somoza, G.M. and R.E. Peter. Effects of serotonin on gonadotropin and growth hormone release from in vitro perifused goldfish pituitary fragments. *Gen. Comp. Endocrinol.* 82: 103–110, 1991.
59. Somoza, G.M., K.L. Yu and R.E. Peter. Serotonin stimulates gonadotropin release in female and male goldfish *Carassius auratus* L. *Gen. Comp. Endocrinol.* 72: 374–382, 1988.
60. Stuer-Lauridsen, F., M. Birkved, L.P. Hansen, H.-C. Holten Lutzhoft and B. Halling-Sørensen. Environmental risk assessment of human pharmaceuticals in Denmark after normal therapeutic use. *Chemosphere* 40: 783–793, 2000.

61. Ternes, T., M. Bonerz and T. Schmidt. Determination of neutral pharmaceuticals in wastewater and rivers by liquid chromatography–electrospray tandem mass spectrometry. *J. Chromatogr.* 938A: 175–185, 2001.
62. Ternes, T.A. Occurrence of drugs in German sewage treatment plants and rivers. *Water Res.* 12: 3245–3260, 1998.
63. Ternes, T.A., M. Stumpf, J. Muller, K. Haberer, R.-D. Wilken and M. Servos. Behavior and occurrence of estrogens in municipal sewage treatment plants – I. Investigations in Germany and Canada. *Sci. Total Environ.* 225: 81–90, 1999.
64. Ternes, T.A., P. Kreckel and J. Mueller. Behavior and occurrence of estrogens in municipal sewage treatment plants – II. Aerobic batch experiments with activated sludge. *Sci. Total Environ.* 225: 91–101, 1999.
65. Trudeau, V.L. Neuroendocrine regulation of gonadotrophin-II release and gonadal growth in the goldfish *Carassius auratus*. *Rev. Reprod.* 2: 55–68, 1997.
66. Trudeau, V.L., B.D. Sloley and R.E. Peter. GABA stimulates gonadotropin-II secretion in the goldfish: involvement of a GABA-A receptor, dopamine and gonadal steroids. *Am. J. Physiol.* 265: R348–R355, 1993.
67. Trudeau, V.L., D. Spanswick, E.J. Fraser, K. Lariviere, D. Crump, S. Chiu, M. MacMillan and R. Schulz. The role of amino acid neurotransmitters in the regulation of pituitary gonadotropin release in fish. *Biochem. Cell Biol.* 78: 241–259, 2000.
68. Trudeau, V.L., O. Kah, J.P. Chag, B.D. Sloley, P. Dubourg, E.J. Fraser and R.E. Peter. The inhibitory effects of gamma-aminobutyric acid (GABA) on growth hormone secretion in the goldfish are modulated by sex steroids. *J. Exp. Biol.* 203: 1477–1485, 2000.
69. Waxman, D.J. P450 gene induction by structurally diverse xenochemicals: central role of nuclear receptors CAR, RXR, and PPAR. *Arch. Biochem. Biophys.* 369: 11–23, 1999.
70. Yang, J.-H., P.T. Kostecki, E.J. Calabrese and L.A. Baldwin. Induction of peroxisome proliferation in rainbow trout exposed to ciprofibrate. *Toxicol. Appl. Pharmacol.* 104: 476–482, 1990.

Biochemistry and Molecular Biology of Fishes, vol. 6
T. P. Mommsen and T. W. Moon (Editors)

CHAPTER 18

P-glycoproteins and xenobiotic efflux transport in fish

ARMIN STURM* AND HELMUT SEGNER**

* *King's College London, Division of Life Sciences, Franklin-Wilkins Building, 150 Stamford Street, London SE1 9NH, UK, and* ** *Centre for Fish and Wildlife Health, University of Bern, Post Box, CH-3001 Bern, Switzerland*

I. Introduction

Biological membranes are barriers for the uptake, distribution and elimination of xenobiotics in organisms[159]. According to a traditional view in toxicology, the bioaccumulation of organic chemicals occurs mainly passively, driven by the compound's hydrophobicity. Once bioaccumulated, the xenobiotic may undergo phase I and II biotransformation metabolisms, the products of which in general are less toxic than the mother compound. Consequently, hydrophobicity and biotransformation rate have traditionally been considered the main factors determining whether a compound is adequately detoxified. However, there are cases where biotransformation only partly inactivates toxic xenobiotics; in other cases the products of biotransformation may even be more toxic than the mother compound. Hence, the removal of xenobiotics or their biotransformation products from the cell has paramount importance to the detoxification process[218]. The excretion of xenobiotics or of their biotransformation products frequently involves active transport by efflux transporters across the epithelia of excretory organs. The term 'phase III' was coined to denote transport proteins involved in the elimination of xenobiotics[72], stressing the possible cooperation of xenobiotic transport with biotransformation. Apart from their role in the epithelia of excretory organs, xenobiotic efflux transporters can limit the uptake of xenobiotics across epithelia, for instance in the intestine. Similarly, transporters accomplishing xenobiotic efflux from cells in tissues forming permeation barriers, such as the blood–brain barrier, can prevent xenobiotic accumulation in certain body compartments.

In order to achieve a mechanistic understanding of toxic action in ecotoxicology, biotransformation enzymes of non-mammalian species, including fish, have been studied quite extensively. Moreover, some biotransformation enzymes, e.g. cytochrome P4501A, are inducible by certain pollutants and have gained importance as biomarkers of pollutant exposure. Compared to biotransformation enzymes, the research on xenobiotic transporters is still in its infancy, and has largely been limited to mammalian models. The best-studied xenobiotic efflux transporter is probably the multidrug resistance (MDR) pump P-glycoprotein (P-gp)[17,57,58]. This broad-specificity drug transporter is in many cases the molecular basis for the cross-resistance of tumors against a broad spectrum of structurally and functionally unrelated cytostatic drugs. However, P-gp is also expressed in normal tissues, where its localization at the apical cellular pole and its abundance in epithelia of organs with roles in excretion suggests that it is part of the organism's defense against exogenous and endogenous toxicants.

The pioneering work of Kurelec[89,94] pointed out the ecotoxicological significance of xenobiotic transport in aquatic animals. Kurelec demonstrated the presence of a xenobiotic efflux transport mechanism in fish and aquatic invertebrates that resembles the mammalian MDR P-gp (reviewed in refs. 12,92). Increased drug transporter activity and protein levels in aquatic animal populations from heavily polluted habitats, reported in some studies, suggest that elevated levels of drug efflux transporters may indeed be of adaptational value, allowing survival and reproduction in spite of the presence of toxic contaminants. Because of many similarities to MDR P-gp, Kurelec dubbed the invertebrate transporter 'multixenobiotic resistance (MXR) mechanism'. Evidence available today confirms that in most cases indeed P-gp(s) constitute the biochemical basis for the MXR mechanism. Nevertheless, related broad-specificity drug transporters may be further involved in xenobiotic transport in aquatic animals. In some studies, environmental pollutants have been shown to induce piscine and invertebrate MXR protein(s). However, the identity of inducers is unknown, as are species differences in inducibility and the spectrum of chemicals against which a protection may be achieved.

The MXR mechanism in aquatic animals has been the subject of a recent excellent review[12]. However, considerable progress has been achieved in the understanding of the molecular biology and biochemistry of P-gp and related transporters in recent years. Moreover, an increasing number of studies have focused on P-gp in fish. Consequently, this review will deal with these more recent data and more specifically focus on P-gp in fish. Before addressing P-glycoprotein in fish, we will review the general biochemistry of P-gp.

II. General aspects of P-glycoprotein

1. ABC transporters

Many xenobiotic efflux transporters belong to the large gene superfamily of ATP-binding cassette (ABC) transporters[142]. Members of this evolutionary ancient group of proteins can have very different functions and are found in species ranging from bacteria to man[34]. The functional ABC transporter comprises multiple functional units organized in a characteristic fashion[66]. The typical transporter consists of four domains, two transmembrane domains (TMD) and two nucleotide-binding domains (NBD). The TMD is highly hydrophobic and usually comprises six transmembrane helices. The NBD contains several consensus sequences, called the Walker A and B motifs[200] and the ABC family signature[84]. The individual domains of an ABC transporter may exist as separate polypeptides, particularly in prokaryotic species. In eukaryotic species, many ABC transporters combine the above four domains in one polypeptide having the N- to C-terminal order of domains: TMD1–NBD1–TMD2–NBD2.

The Human Genome Project has led to the discovery of a great number of novel ABC transporters[2,84]. To make the great number of ABC proteins manageable, the Human Gene Nomenclature Committee has developed a new nomenclature based on sequence homology in order to replace the numerous (and still popular) trivial names. In the new nomenclature, the ABC superfamily is divided into seven subfamilies, ABCA to ABCG[84]. The individual transporters are identified by the name of the subfamily followed by a number. Information on the new nomenclature can be accessed electronically (http://www.gene.ucl.ac.uk/nomenclature/genefamily/abc.html).

1.1. P-glycoproteins

1.1.1. P-glycoproteins and multidrug resistance. The importance of transmembrane transporters as modulators of the cytotoxic effects of foreign compounds is highlighted by the phenomenon of MDR. MDR is the acquired or intrinsic resistance of cancers against a broad variety of structurally unrelated drugs and causes a major problem in cancer chemotherapy[17,57,58]. The MDR phenotype can be produced *in vitro* by growing drug-sensitive tumor cell lines in the presence of anti-cancer drugs. A main step towards understanding the biochemical basis of MDR was achieved when a 170 kDa glycoprotein was discovered that is highly expressed in the plasma membrane of MDR cell lines and tumors, but not in drug-sensitive cell lines. The protein, dubbed P-glycoprotein because its expression correlates with a decreased drug permeability of the cells, was shown to be an ATP-dependent drug pump that actively extrudes drugs from the cell, thereby reducing the accumulation of cytotoxic drugs in MDR cells. While P-gp probably represents the most common biochemical factor causing MDR in cancer, it must be emphasized that MDR is a multifactorial phenomenon and can also involve, and even be solely based on, cellular factors other than P-gp. Such factors include the expression of other drug transporters, changes in drug metabolism and defective apoptotic pathways[57].

1.1.2. Isoforms of P-glycoprotein. P-glycoproteins are ABC transporters encoded by *mdr* genes. P-gps are phosphoglycoproteins located in the cytoplasmic membrane and consist of one polypeptide chain of about 1300 amino acids. The molecular mass of the mature protein may vary from 130 to 180 kDa, reflecting differences in glycosylation. P-gp shows the domain architecture of a prototypical eukaryotic ABC transporter, TMD1–NBD1–TMD2–NBD2, consisting of two homologous halves arranged in a tandem-like manner. Each half is composed of one TMD and one NBD[17,56].

mdr1-type P-gps (also called class I P-gps) are ATP-dependent drug transporters capable of extruding a broad variety of hydrophobic organic chemicals from the cell. The transfection of *mdr1*-type cDNAs confers the MDR phenotype to drug-sensitive cell lines[26,62]. By contrast, mdr2-type (or class II) P-gps are phosphatidylcholine transporters[137,168] and are not able to bind or transport most mdr1-type substrates. As this review focuses primarily on xenobiotic transport, 'P-gp' will, from here on, refer to mdr1-type P-gps, except where stated otherwise.

In humans, one mdr1-type P-gp exists, MDR1 (ABCB1) (Table 1). In rodents, two genes encoding mdr1-type P-gps have been identified, *mdr1a* (*abcb1a*, also called *mdr3* in mice or *pgp1* in rat and hamster) and *mdr1b* (*abcb1b*, also called *mdr1* in mice or *pgp2* in rat and hamster)[35,36,39,61,70,122,158]. In man, high levels of MDR1 are found in adrenal cortex, renal proximal tubules, the canalicular pole of hepatocytes, small and large intestinal mucosal cells and pancreatic ductules[48,179]. Lower levels of MDR1 have been detected in other tissues, including the capillary endothelia of the brain, testis and placenta[180,211]. Moreover, certain types of leukocytes also show MDR1 expression[86,112]. In rodents, the *mdr1a* and *mdr1b* genes appear to share the function of the human *MDR1* gene, as together they show a similar distribution among tissues as human *MDR1*[191].

mdr2-type P-gps are encoded by the human *MDR2* gene (*ABCB4*, also called *MDR3*) and by the rodent *mdr2* gene (*abcb4*, also called *pgp3*)[22,35,63,122,136,194]. The *MDR2* gene product shows a high expression only in the liver where it is located at the canalicular pole of hepatocytes[160] (Table 1). Despite their different functions, MDR2/mdr2 show high sequence homology to MDR1/mdr1. For instance a 78% amino acid identity is observed for the two human transporters[217].

1.2. ABC-transporters other than P-glycoprotein

A gene with high homology to *MDR1* and *MDR3* has been identified in the human and initially named 'sister of P-glycoprotein' (*SPGP*)[27]. *SPGP* was later shown to encode the canalicular bile salt export pump (*BSEP*, *ABCB11*)[55]. Expression of *BSEP* is largely limited to the liver[182] (Table 1).

TABLE 1

Overview on selected human ABC transporters

Trivial name	Systematic name	Tissues showing major expression	Substrates	Localization in polarized cells	References
MDR1, P-gp	ABCB1	Intestine, liver, kidney, placenta, blood–brain barrier	Neutral and cationic organic compounds, including many common drugs	Apical	17,56,58
MDR2	ABCB4	Liver	Phosphatidylcholine	Apical	137,160,168
BSEP, SPGP	ABCB11	Liver	Bile salts	Apical	27,55,182
MRP1	ABCC1	All tissues	Glutathione and other conjugates, organic anions, leukotriene C4	Basolateral	19,80,131
MRP2	ABCC2	Liver, kidney, intestine	Similar to MRP1	Apical	19,80,131
MRP3	ABCC3	Pancreas, kidney, intestine, liver, adrenal	Glucuronate and glutathione conjugates, bile acids	Basolateral	19,131
MRP4	ABCC4	Prostate, testis, ovary, intestine, pancreas, lung	Nucleotide analogs, organic anions	?	19,131,142
MRP5	ABCC5	Most tissues	Nucleotide analogs, cyclic nucleotides, organic anions	Basolateral	19,131
MRP6	ABCC6	Liver, kidney	Anionic cyclic pentapeptide	Basolateral	19,131
BCRP, MXR, ABC-P	ABCG2	Placenta, intestine, breast, liver	Organic compounds including many common drugs	Apical	142

Modified from Gottesman *et al.*[57] and Schinkel and Jonker[142].

The multidrug resistance-associated protein (MRP1, ABCC1) is an ABC transporter isolated from tumors that showed the MDR phenotype without expressing P-gps[28,108,213]. Meanwhile, at least six isoforms of MRP are known (MRP1 to 6, ABCC1 to 6)[19] (Table 1). MRP1, MRP2, MRP3 and MRP6 differ from the general ABC transporter architecture (TMD1–NBD1–TMD2–NBD2) by having an additional N-terminal TMD[19]. This additional domain is lacking in MRP4 and MRP5[19]. The isoforms associated with the MDR phenotype – MRP1, MRP2 and MRP3 – provide resistance against a similar spectrum of drugs as MDR1. However, MRPs show only a low degree of homology to P-gp (e.g. 20% amino acid identity between MRP1 and MDR1[131]). In contrast to P-gp that transports unconjugated organic compounds, MRP1 transports glutathione-, glucuronate- and sulfate-conjugates[80,212]. Additionally, MRP1 can co-transport certain unconjugated xenobiotics with reduced glutathione[101]. In normal tissues, MRP1 and MRP5 show a ubiquitous tissue distribution[19]. MRP2 is highly expressed in liver and also found in small intestine and kidney. MRP3 and MRP6 are expressed in liver and kidney, with MRP3 also being present in adrenals, pancreas and gut[19]. MRP4 is found in the prostate, lung, muscle pancreas, gonads, bladder and gallbladder[19]. MRP2 was shown to be identical to the canalicular multispecific anion transporter, cMOAT[80]. MRP2 and MRP4 display an apical localization in polarized cells, while the localization of the remaining MRP isoforms is basolateral or, in the case of MRP4, still debated[142].

The breast cancer resistance protein (BCRP) was cloned from a drug-resistant breast cancer cell line[38]. Different groups have obtained the cDNA sequence of BCRP independently, so that alternative names of the protein are mitoxantrone-resistance protein (MXR)[121] and placental ABC protein (ABCP)[3] (Table 1). BCRP mediates resistance against a similar spectrum of cytotoxic drugs as MDR1[142]. BCRP differs from the ABC transporters discussed so far in that it comprises only one NBD and one TMD[142]. ABC transporters of this domain architecture are called half-transporters, because of the general assumption that they form dimers to constitute the functional transporter.

2. *Functional and structural characteristics of P-glycoprotein*

2.1. *Structural features of substrates and inhibitors of P-glycoprotein*

The most astonishing characteristic of P-gp is that it transports structurally and functionally very diverse substrates. Transport substrates of P-gp (Table 2) include anticancer drugs such as the vinca alkaloids, anthracyclines, epipodophyllotoxins and taxanes[187], as well as numerous other classes of compounds, e.g. cytotoxic agents, cyclic and linear peptides, steroid hormones, lipids, HIV protease inhibitors and pesticides[172]. Substances that can revert the MDR phenotype by inhibiting P-gp are called reversal agents, chemosensitizers, resistance modifiers or MDR modulators[49,172]. Just like the transport substrates, reversal agents also comprise compounds of diverse structure and pharmacological characteristics, including calcium channel blockers, calmodulin antagonists, steroids, detergents, cyclic peptides and pesticides (Table 3)[11,49].

TABLE 2

Transport substrates of P-gp

Class	Example	References
Anticancer drugs	Vinca alkaloids (vincristine, vinblastine)	7
	Anthracyclines (doxorubicin, daunorubicin)	7
	Epipodophyllotoxins	7
	Taxanes (paclitaxel)	7
Other cytotoxic agents	Colchicine	7
	Ethidium bromide	7
Cyclic and linear peptides	Gramicidin D	46
	Opioid peptides	124
	N-acetyl-leucyl-leucyl-norleucinal	156
Steroid hormones	Cortisol	186
	Aldosterone	186
	Estriol	187
Lipids	Phospholipids	135, 195
HIV protease inhibitors	Indinavir	7
Pesticides	Ivermectin	151
	Endosulfan	10
Other compounds	Benzo(*a*)pyrene	210
	Nonylphenol ethoxylates	110
	Hoechst 33342	7
	Rhodamine 123	7
	Calcein-AM	7
	Triton X-100	151
	Yohimbine	151
	Morphine	151

Studies aimed at the identification of substrates and inhibitors of P-gp have mostly had a pharmacological perspective, related to the role of P-gp in MDR. Few of the compounds studied in this context have environmental relevance. However, the unusually broad spectrum of structurally unrelated molecules interacting with P-gp suggests that P-gp substrates may also be present among environmental pollutants. When 38 pesticides were screened for interaction with P-gp, many of the organophosphates and organochlorines inhibited doxorubicin efflux from an MDR cell line[10] (Table 3). In contrast, carbamate and pyrethroid insecticides showed little interaction with P-gp[10]. P-gp has been reported to transport benzo(*a*)pyrene and different nonylphenol ethoxylates, but not nonylphenol or the PCB 3,3′,4,4′-tetrachlorobiphenyl[51,110,210].

Many studies have aimed at unraveling the structural features that predispose chemicals to interact with P-gp (for review, see refs. 172,215). Known mdr1-type substrates are relatively large organic compounds (M_r between 300 and 2000[187]) and are, at physiological pH, neutral or positively charged. A certain minimum size is required (molar refractivity > 9.7). It has been suggested that compounds containing one basic nitrogen atom and two planar aromatic domains are MDR substrates[214]. However, certain mdr1-type substrates, e.g. cyclosporins, lack these features.

A structure–activity analysis of 609 compounds[87] revealed different substructural features ('biophores') related to MDR reversal activity. The biophore most significantly related to activity had the generic form C–C–X–C–C, where X = N,

TABLE 3

Reversal agents of P-gp

Class	Compound	References
Calcium channel blockers	Verapamil	157
	Azidopine	157
Calmodulin antagonists	Trifluoperazine	157
	Chlorpromazine	157
Steroid hormone	Progesterone	187
	Deoxycorticosterone	187
Detergents and amphiphiles	Cremophor EL	157
	Tween 80	157
Cyclic peptides	Cyclosporin A	157
	SDZ PCS 833	157
	Valinomycin	157
Pesticides	Parathion	10
	Chlorpyrifos	10
	Permethrin	10
	Clotrimazol	10
Miscellaneous	Quinidine	157
	Chloroquine	157
	Reserpine	157
	Green tea polyphenols	73
	Rosemary extract	127
	Ginsenoside Rg_3	81

NH or O (preferably a tertiary nitrogen), linked to two unsubstituted alkyl groups. Activity was further enhanced by the presence of (di)methoxyphenyl groups. In contrast, the presence of a stable quaternary ammonium salt, a carboxylic, a phenol or an aniline group was found to impede MDR reversal.

Following her analysis of 100 known P-gp transport substrates, Seelig[151] defined units of structural elements required for an interaction with P-gp. Type I units of such recognition elements consist of two electron donor groups with a spatial separation of 2.5 ± 0.3 Å. Type II units contain either two groups with a spatial separation of 4.6 ± 0.6 Å or three groups with the outer two having a spatial separation of 4.6 ± 0.6 Å. Molecules containing at least one type I unit or one type II unit are predicted to be transport substrates of MDR1.

Bain and co-workers[11] investigated 44 compounds, including P-gp transport substrates, reversal agents and non-interacting compounds. Transport substrates were differentiated by molecular characteristics, including several molecular size/shape parameters, lipophilicity and hydrogen bonding potential. Transport substrates had a significantly higher molecular mass (724 vs. 408 Da) and a lower log K_{ow} (-0.34 vs. 4.35) than reversal agents. Electrostatic features differentiated reversal agents from non-interacting compounds.

In conclusion, attempts to find characteristic features of P-gp substrates and inhibitors have so far only led to the identification of rather general features of the two classes of compounds that interact with P-gp. Given the broad range of compounds interacting with P-gp, this finding is not unexpected.

2.2. Nature of the drug-binding site(s) of P-glycoprotein

Different experimental approaches have been adopted to elucidate the nature and number of drug-interacting sites of P-gp, the most important being photoaffinity labeling of P-gp with analogs of MDR drugs, studies of the kinetics of drug transport, genetic analyses of natural or targeted MDR mutants showing deviant substrate affinity and analyses of recombinant chimeras between different MDR isoforms. However, the number and exact localization of drug-binding sites in P-gp cannot be derived directly from such studies and remain controversial. The proposition of at least two distinct drug-binding sites[197] opposes the suggestion of one large drug-binding surface encompassing different sub-sites[157].

Radiolabeled analogs of MDR drugs or reversal agents carrying photoactivable moieties have been used in photoaffinity labeling studies of P-gp[138]. After reversible binding, the photoactive probe is converted to a highly reactive intermediate by irradiation, and binds covalently to P-gp at or near the binding site. The labeled regions are then identified by protein mapping. In general, photoactivable probes simultaneously label two regions of the P-gp polypeptide, one located in the N-terminal TMD within or close to TM (transmembrane domain) 4–6, the other in the C-terminal TMD within or close to TM11–12[56,138]. The binding regions are not identical for different drugs, but overlap. An excess of non-photoactive substrates generally inhibits the labeling. However, certain combinations of drugs do not behave competitively, e.g. anthracyclines were poor inhibitors of labeling with a photoactivable vinblastine analog[138].

Evidence for the presence of multiple drug-binding sites comes also from analyzes of substrate kinetics. Shapiro and Ling[154] found a mutual stimulation of transport between the substrates rhodamine 123 and Hoechst 33342. Further substrates tested for their interaction with transport of rhodamine 123 and Hoechst 33342 could be divided into three groups, i.e. compounds stimulating rhodamine 123 transport and inhibiting Hoechst 33342 transport (colchicines, quercetin), compounds stimulating Hoechst 33342 transport and inhibiting rhodamine 123 transport (anthracyclines) and compounds inhibiting transport of both (vinblastine, actinomycin D, etoposide). Based on these results, Shapiro and Ling[154] proposed a functional model of P-gp with two positively cooperative drug-binding sites, called the R and H sites (for rhodamine 123 and Hoechst 33342, respectively). In the model, drugs preferentially binding to the same site will show mutual transport inhibition, while drugs clearly preferring one of the sites will enhance transport of drugs preferring the other site[154]. In a subsequent study, prazosin and progesterone were shown to enhance the transport of both rhodamine 123 and Hoechst 33342[153]. This was interpreted as indicative of a third, allosteric drug-binding site[153].

Mutational analyses showed that single amino acid mutations can greatly alter the activity of P-gp in a substrate-dependent manner (for review, see refs. 7,56). While amino acids whose mutation has a marked effect on substrate selectivity of P-gp are found in all the regions of the polypeptide, they cluster in TMs 5 and 6 of the N-terminal TMD and the corresponding TMs 11 and 12 of the C-terminal TMD of P-gp[7], suggesting a role for these regions of the P-gp polypeptide in substrate recognition.

A role of TM6 and TM11–12 of P-gp in substrate recognition is also supported by studies with recombinant MDR1–MDR2 chimeras[216,217]. While MDR1 and MDR2 show 78% amino acid identity, the phosphatidylcholine transporter MDR2 fails to transport most mdr1-type substrates. Chimeras derived from MDR1, where the N-terminal part of MDR1 up to amino acid 394 was replaced by the homologous segment of MDR2 (MDR2(1–394)–MDR1), lost the ability to confer drug resistance to colchicine, daunomycin and vincristine, and to transport rhodamine 123 and daunomycin. In contrast, this chimera's ability to confer bisanthrene resistance was greatly reduced[217]. To identify structural features responsible for this functional loss, selected MDR1 amino acids were reintroduced into MDR2(1–394)–MDR1. The simultaneous introduction of three amino acids located in the TM6 and conserved among mdr1-type P-gps (Q330, V331, L332) into the chimera partially restored drug transport and MDR-conferring properties of the mutated P-gp[217]. A subsequent study confirmed that large parts of the N-terminal half of MDR1 can be replaced by the homologous segments of MDR2 without a marked loss of mdr1-type activity, as long as residues of TM6 were not touched[216]. However, corresponding segments of the C-terminal half of the transporter do not show this degree of the exchangeability. MDR2 mutants where residues of both TM6 and TM11–12 had been replaced by MDR1 amino acid residues were able to bind mdr1-type substrates, but could no longer transport them[216].

2.3. ATP hydrolysis and transport activity

The transport activity of P-gp is ATP-dependent[33], and the polypeptide sequence of P-gp shows two nucleotide-binding sites[26]. Photolabeling studies with analogs of ATP confirmed that P-gp binds ATP[56]. Most mutations of amino acids at and around the two nucleotide-binding sites abolish transport activity, suggesting that nucleotide binding is necessary for P-gp activity[7,56]. The mutation of one of the two ATP-binding sites is sufficient to abrogate the functional integrity of P-gp, indicating cooperation between the two ATP-binding sites and the two homologous halves of the transporter[7,56].

The availability of P-gp-enriched membrane preparations and of reconstituted purified P-gp has greatly facilitated studies on the mechanisms of the catalytic activity of P-gp[7,157]. Membrane vesicle preparations of P-gp are characterized by a high constitutive ATPase activity that can be stimulated by substrate addition[157]. Most substrates show a biphasic profile of stimulation of *in vitro* ATPase activity, with inhibition at high doses[104]. However, some substrates transported *in vivo* act solely as inhibitors *in vitro*, e.g. cyclosporin A[104,206]. Moreover, many reversal agents, for which no net transport is seen *in vivo*, are particularly potent in stimulating ATPase activity *in vitro*[157].

Different approaches have shown that the two halves of P-gp work in an interdependent, cooperative and alternating manner (reviewed in ref. 152). The inhibitor *N*-ethylmaleimide (NEM) binds covalently to the ATP-binding sites of P-gp[152]. The blocking of either of the two sites with NEM results in complete inhibition of activity, but can be prevented by Mg-ATP with similar efficacy for both sites[152]. The inhibition of P-gp by *ortho*-vanadate results in trapping the nucleotide in a catalytic site after ATP hydrolysis[188]. Vanadate trapping of a photoactivable ATP analog showed that trapping occurred in both nucleotide-binding sites, and that trapping of nucleotide in either catalytic site completely blocked hydrolysis at both sites[189]. Loo and Clarke[109] expressed 'half-molecules' of P-gp that comprised either the N-terminal TMD and NBD or the C-terminal TMD and NBD separately in Sf6 cells. Both half-molecules showed basal ATPase activity when expressed singly, indicating that both NBDs can potentially bind and hydrolyze ATP. However, stimulation of basal ATPase activity by drugs was not observed except when both half molecules were co-expressed.

The stoichiometry of ATP hydrolysis and drug transport is important for interpreting the physiological relevance of P-gp as a drug transporter[7]. Early estimates were in the range of 50 ATP molecules per transported drug molecule, a value that appeared too high to account for the observed resistance conferred by P-gp, and too high by comparison with other ABC transporters[7]. With the substrates $^{86}Rb^+$-complexed valinomycin[45], rhodamine 123[155] and vinblastine[6], three recent studies estimated that 1–3 ATP molecules are hydrolyzed per substrate molecule transported, which compares well with known ion-translocating pumps[7].

2.4. Mechanism of transport

The mechanism by which P-gp affects drug exclusion remains unknown. For a long time, drug pump models, assuming a direct transport of substrates, have opposed models in which P-gp indirectly affects transport by building up transmembrane gradients (e.g. of ions) that in turn are linked to substrate efflux. The most recent and refined version of the latter group of models is the altered partitioning model[133], where P-gp is proposed to change the intracellular pH and/or the plasma membrane potential. As most P-gp substrates are charged at physiological pH, one could imagine that their accumulation could be influenced by changed intracellular pH and/or membrane potential. However, most available experimental evidence strongly favors a direct transport of substrates by P-gp (reviewed by Stein[171]). Conflicting with the altered partitioning model, a correlation between P-gp-dependent transport and membrane potential, or transport and intra- or extracellular pH, has not been shown[4,111,171]. Moreover, as reviewed above, evidence for a direct, specific interaction between P-gp and its substrates is provided by photoaffinity labeling studies and mutational analyses.

According to the drug pump model, P-gp directly pumps its substrates out of the cell in an ATP-dependent mechanism[33]. However, assuming the transport of substrates from the cytoplasm to the extracellular space leads to some contradictions. As P-gp shows very broad substrate specificity, it is likely to display low substrate affinities. Because most substrates of P-gp are hydrophobic, they show a strong tendency to partition into cellular membranes and other cellular sinks. This should result in low cytosolic concentrations of substrates, and low-affinity transporter should not work efficiently under these conditions. To resolve this contradiction, the initial drug pump model was modified by proposing that P-gp extracts its substrates from the plasma membrane rather than from the cytosol. Due to the partitioning of hydrophobic drugs into the plasma membrane phase, the expected drug concentrations in the membrane are high enough to be reconciled with the transport by a low-affinity transporter. Two variants of the modified drug pump model exist today. In the hydrophobic vacuum cleaner model[59], the MDR protein extracts its drug substrates from the lipid bilayer and transports them directly into the external water phase. According to the flippase model[67], P-gp recognizes the substrate in the inner leaflet of the plasma membrane and transports it to the outer leaflet of the plasma membrane, from where it diffuses to the external water phase.

2.5. Mechanism of MDR reversal

Different mechanisms for the action of reversal agents are conceivable. MDR reversing activity could result from blocking either substrate recognition, binding of ATP, ATP hydrolysis or the coupling of ATP hydrolysis to substrate transport[7]. The behavior of reversal agents is puzzling in many aspects. Most reversal agents act as competitive or non-competitive inhibitors and compete with transport substrates for binding to drug interaction sites[7]. In contrast, many drugs against which P-gp confers MDR do not inhibit P-gp strongly and exert no effect on drug uptake in MDR cells[16,44]. Certain reversal agents, such as verapamil and cyclosporin A, are transport substrates of P-gp, but MDR cells are not resistant to them[157]. For other reversal agents, no transport by P-gp can be measured. Many reversal agents parallel the behavior of substrates in that they stimulate *in vitro* ATPase activity, with the stimulation often being much greater than that by transport substrates[157]. Other reversal agents inhibit *in vitro* ATPase activity[206].

Eytan and coworkers[46] provide a model explaining some of the above contradictory observations. In their study, rates of passive movement across an artificial membrane bilayer were compared between MDR drugs and reversal agents. Five reversal agents showed much faster transmembrane movement rates than five MDR drugs. Based on the flippase model, where P-gp transports its substrates from the inner to the outer leaflet of the plasma membrane, the authors proposed that MDR drugs are excluded from the cell because the flipping rate by P-gp exceeds the drug's rate of equilibration between the two membrane leaflets[46]. Consequently, a concentration gradient is built up. In contrast, reversal agents show rates of equilibration between the two leaflets of the membrane bilayer that exceed or are close to the flipping rates by the P-gp pump. As a result, transport of reversal agents by P-gp is futile because they re-enter the inner leaflet of the membrane as fast as they have been pumped out[46].

The kinetics of P-gp inhibition are complex and appear to reflect allosteric interactions between at least two distinct drug-binding sites. In their analysis of the effects of reversal agents on daunomycin accumulation by MDR P388 leukemia cells, Ayesh and coworkers[9] found that the effects of the reversal agents cyclosporin, verapamil and trifluoperazine could be described adequately assuming a Hill coefficient of unity, suggesting that these compounds interact with P-gp as single molecules. By contrast, vinblastine, mefloquine, dipyridamol, tamoxifen and quinidine showed Hill numbers close to 2, indicating that these molecules probably act as pairs in affecting P-gp inhibition[9]. With pairs of reversal agents, examples of both competitive and non-competitive behaviors were found[9]. This finding was interpreted as indicative of two modulatory sites in the P-gp pump, with mefloquine, and vinblastine preferentially binding to the first. Verapamil, dipyridamol, trifluoperazine and quinidine bind to the second site, while cyclosporin interacts with both sites[9]. That reversal agents may act cooperatively, and interact with each other either competitively or non-competitively, was later confirmed by different experimental approaches, including measurement of the transport of fluorescent P-gp substrates in MDR cells[203,205] and of ATPase activity in P-gp-rich membrane vesicles[105,126,204].

3. Physiological function of P-glycoprotein

Yeh and coworkers[210] suggested that P-gp might be part of the organism's defense against toxicants. Such a function is indicated both by the broad substrate specificity of P-gp to hydrophobic organic compounds and by its tissue distribution and cellular localization[179,180]. The apical expression of P-gp in intestinal epithelial cells supports its role in limiting the uptake of toxic substrates from the intestinal lumen[183]. Moreover, the apical localization of P-gp in several epithelia that form boundaries within the body, e.g. the placental epithelium or the capillary endothelia in brain and testis[69,175,211], further supports its function in preventing entry of toxic substrates into 'sanctuary sites' of the body. Finally, the presence of P-gp in the liver and kidney suggests that P-gp has a role in excretion by driving the secretion of substrates into bile and urine across the canalicular membrane of hepatocytes[60] and the luminal membrane of epithelial cells in the kidney tubule[129], respectively.

The role of P-gp in the defense against xenobiotics is today largely accepted. It has received major support from studies experimentally deactivating P-gps, either using reversal agents (drug interaction studies) or animals in which *mdr1* gene(s) are either experimentally disrupted (transgenic 'knockout' mice) or non-functional as a consequence of natural mutations. These studies will be reviewed in some detail below. In addition, it has been suggested that P-gp accomplishes the transmembrane transport of numerous endogenous substrates, thus linking P-gp to a plethora of physiological processes[20,74]. Indeed, some sites of physiological expression of P-gp cannot be explained by its assumed role in the protection against toxicants. Why is P-gp expressed at high concentrations in the adrenal[48,71]? Why do natural killer (NK)-cells and subpopulations of CD8 + T-cells have relatively high levels of P-gp expression[86,112]? *In vitro*, P-gp transports steroids[186,208], small peptides[124,156] and lipids[135,195]. Accordingly, it has been hypothesized, for instance, that P-gp could have roles in steroid synthesis, secretion and/or the regulation of cellular steroid access and secretion of neuropeptides and cytokines. Moreover, it has been proposed that P-gp regulates the activity of an unidentified Cl^- channel, plays a role in the cytotoxic activity of lymphocytes or in apoptosis (for reviews, see refs. 20,74). Most proposed roles of P-gp remain controversial. Yet some evidence exists for a role of P-gp in steroid secretion and/or regulation of steroid access to cells. The targeted disruption of the mdr1b gene in mouse Y1 adrenal cells leads to decreased steroid secretion in cells with the null mutation[5]. However, the lack of gross pathologies in mdr-knockout mice shows that steroid secretion does not completely depend on P-gp.

Recently, a role of P-gp in the defense against viruses has been proposed, which could explain the aforementioned P-gp expression in certain subpopulations of peripheral blood mononuclear cells. The invasion of enveloped viruses into the host cell is facilitated by the insertion of hydrophobic peptides of the virus envelope into the lipid domain of the host cell membrane. Because P-gp extrudes hydrophobic peptides from the plasma membrane, Raviv *et al.*[130] investigated the effect of P-gp overexpression in MDR cells on their susceptibility to virus infection. MDR cells were resistant to infection by envelope viruses that invade cells by fusion with the plasma membrane, but not to infection by pH-dependent viruses that enter cells by fusion with endocytotic vesicles[130]. In a different study, the overexpression of P-gp in a CD4 + T-cell line greatly reduced HIV-1 virus production[100]. Interestingly, the reversal agents quinidine and PSC833 only partly overcame the effect of P-gp overexpression, and non-functional mutants of P-gp were also effective in decreasing HIV-1 infectivity[100], suggesting that in addition to pumping activity, other unknown mechanisms may have played a role.

3.1. Evidence for the role of P-glycoprotein in the defense against organic toxicants

3.1.1. mdr knockout mice. The mouse, the standard model for targeted gene disruption, possesses two genes coding for mdr1-type P-gps, *mdr1a* and *mdr1b*. In mouse brain, liver and intestine, the *mdr1a* product is the predominant form, while in pregnant uterus and kidney the *mdr1b* gene product predominates[20]. Moreover, the mouse has one gene encoding an mdr2-type P-gp, *mdr2*. The initially generated *mdr1a* knockout (*mdr1a* $^{-/-}$) mouse shows no detectable P-gp in brain or intestine, but *mdr1b* expression in liver and kidney appears up-regulated compared with wild-type mice[20]. Later *mdr1b* knockout and *mdr1ab* double-knockout (*mdr1ab* $^{-/-}$) mouse strains were generated – the latter completely lacking detectable mdr1-type P-gp[20]. The three mdr1-type knockout mice strains are viable and fertile, and show no discernible pathologies[143,144]. By contrast, the disruption of the mdr2 gene leads to the complete blocking of biliary phospholipid secretion and severe progressive liver disease[161]. However, mdr1-type knockout mice are characterized by a markedly increased susceptibility to certain toxicants[196]. This supports the suggested function of *mdr1* genes in the defense against xenobiotics. Should the *mdr1* gene products have other roles, these appear at least not essential for survival in the laboratory environment.

3.1.2. Role of P-gp in the intestine. Studies with *mdr1a* $^{-/-}$ mice collectively suggest a dual role of intestinal P-gp by first limiting xenobiotic uptake from the intestinal lumen and, second, mediating secretion of xenobiotic into the gut lumen[103]. This is illustrated by studies with paclitaxel[170]. In wild-type mice, a significant part of the dose was excreted as the unchanged compound in feces (90% after oral, or ~40% after i.v. administration), but fecal excretion of unchanged compound was insignificant in *mdr1a* knockout mice (<3%). In mice with a cannulated gall bladder and dosed by i.v. injection, the quantity of unchanged paclitaxel in the intestinal content was higher in the wild type than in the *mdr1a* $^{-/-}$ strain (10.8 vs. 2.5%), demonstrating the direct excretion of paclitaxel by intestinal P-gp[170]. Similarly, 90 min after i.v. administration of digoxin the direct intestinal excretion was appreciable in wild-type mice (16% of dose), but insignificant in *mdr1a* $^{-/-}$ mice (<2%)[114]. The intestinal excretion of digoxin in wild-type mice could be blocked completely by the reversal agent PSC833[114].

mdr1a $^{-/-}$ mice show an increased risk to develop severe spontaneous intestinal inflammation, pointing to a potential role of P-gp in preventing entry of bacterial toxins into the gut wall mucosa[125]. It has been hypothesized that impaired P-gp function in the intestine could play a role in the etiology of inflammatory bowel diseases, such as ulcerative colitis[50]. In support of this hypothesis, it has recently been reported that the *MDR1* polymorphism C3435T associated with decreased intestinal P-gp expression occurs more frequently in patients with ulcerative colitis than in healthy subjects[150].

3.1.3. Role of P-gp in the blood–brain barrier. The most striking change in drug pharmacokinetics in *mdr1a* $^{-/-}$ and *mdr1ab* $^{-/-}$ mice is their markedly increased brain accumulation of hydrophobic drugs[143,144]. Normally, the entry of drugs into the brain is strongly restricted by the blood–brain barrier. The endothelial cells in the brain capillaries that form this barrier show tight junctions at cell boundaries, a lack of intercellular clefts and a low degree of fenestration and pinocytosis[141]. In mammals with functional *mdr* genes, a high concentration of P-gp is present in the luminal (apical) membrane of the endothelial cells in the brain[141], resulting in drug transport from the cytoplasm into the blood capillary lumen[118,123]. The deficiency of the blood–brain barrier in mdr1-type knockout mice was discovered when the pesticide ivermectin, given as a routine treatment against mite infestation, unexpectedly killed almost all *mdr1a* $^{-/-}$ mice[144]. *mdr1a* $^{-/-}$ mice are almost 100-fold more sensitive to ivermectin than the wild-type strain, and showed 87-fold increased brain concentrations of the pesticide upon treatment, although their ivermectin plasma concentrations were increased by less than threefold[144]. An increased brain accumulation (7- to 35-fold) in *mdr1a* $^{-/-}$ mice is also observed for various other drugs known to be substrates of P-gp[103]. The role of P-gp in the blood–brain barrier was corroborated in studies with reversal agents that markedly increased the accumulation or toxicity of MDR drugs in the brain of wild-type mice[113,114].

3.1.4. Role of P-gp in biliary and renal excretion. Studies with *mdr1* knockout mice also provide evidence for a role of P-gp in the biliary excretion of drugs, although the overall effects of blocking P-gp pharmacologically or by gene disruption tend to be smaller than in the intestine or the brain. This could reflect a shift between alternative detoxification pathways for a xenobiotic in the liver, an organ rich in biotransformation enzymes and one expressing multiple xenobiotic transporters.

In *mdr1a* $^{-/-}$ mice with a cannulated gall bladder, the biliary excretion of doxorubicin and vinblastine, respectively, was 5.5- and 2.9-fold decreased compared with wild-type mice[190]. Similarly, the biliary clearance of digoxin is significantly smaller in the *mdr1a* $^{-/-}$ than in the wild-type mouse[78]. However, *mdr1a* $^{-/-}$ mice show an up-regulation of *mdr1b* in liver and kidney[144], so that the contribution of P-gp to biliary excretion is likely to be underestimated in this model. This is suggested by a comparison of biliary drug excretion between *mdr1ab* $^{-/-}$ and *mdr1a* $^{-/-}$ mice[163,164]. While biliary excretion of three drugs was at least reduced by 45% in *mdr1a* $^{-/-}$ mice[163], it was reduced by at least 70% in *mdr1ab* $^{-/-}$ mice when compared to the wild type[164].

Results on renal excretion in mdr1 knockout mice are less clear. The renal clearance of digoxin was threefold decreased in *mdr1a* $^{-/-}$ mice compared with wild type[78]. In contrast, the renal excretion of unchanged doxorubicin was higher in *mdr1a* $^{-/-}$ mice than in wild-type mice (15 vs. 10% of dose)[191]. Similarly, a renal clearance exceeding that in wild-type mice was also found in *mdr1ab* $^{-/-}$ mice for three compounds[164]. There is no apparent explanation for these findings that may point to hitherto unidentified transporters being up-regulated after mdr1 gene disruption.

3.1.5. Role of P-gp in the placenta. The role of P-gp in the placenta was studied in the CF1 mouse in which a natural *mdr1a* mutation occurs[99]. *mdr1a* heterozygous females (+/−) were mated to homozygous negative (−/−) and positive (+/+) males and treated with the teratogen L-652,280 on gestation days 6–15. Following this treatment, 63% of the fetuses developed a cleft palate. All the fetuses of homozygous negative (−/−) and ~30% of the fetuses of heterozygous (+/−) matings showed the malformation, while none of the homozygous positive (+/+) fetuses were affected[99]. In another study with mdr double-knockout mice[162], mdr1a$^{+/-}$/1b$^{+/-}$ females were mated with mdr1a$^{+/-}$/1b$^{+/-}$ males to obtain three genotypes (mdr1a$^{+/+}$/1b$^{+/+}$; mdr1a$^{+/-}$/1b$^{+/-}$; mdr1a$^{-/-}$/1b$^{-/-}$) in a single mother. When the mdr1-type drugs digoxin, saquinavir or paclitaxel were given i.v. to pregnant dams, 2.4-, 7- or 16-fold more drug, respectively, entered mdr1a$^{-/-}$/1b$^{-/-}$ than wild-type fetuses[162]. These findings suggest that placental P-gp has an important role in limiting the fetal penetration of toxic compounds.

3.2. A role for P-gp in the resistance to inorganic toxicants?

Recently, a role of mammalian P-gp has been proposed in the toxicity of arsenic and cadmium[40,106,107]. Upon arsenic selection, a rat liver epithelial cell line underwent malignant transformation and acquired arsenic tolerance[106]. The arsenic resistant cells showed an up-regulation of glutathione-*S*-transferases, MDR1, MDR2, MRP1 and MRP2[106]. The resistance could be reversed by an alternative treatment of the cells with either glutathione-depleting agents, the reversal agent PSC833 or the MRP-inhibitor MK571[106]. *mdr1ab* double-knockout mice are more sensitive to arsenic and accumulate more arsenic than wild-type mice[107]. In cultured renal epithelial cells, P-gp expression was negatively correlated with cadmium accumulation, and cadmium accumulation could be increased by pretreatment of cells with reversal agents[40].

While these studies suggest a role of P-gp as a factor in the toxicity of inorganic chemicals, the study on rat liver cells[106] suggests that different biochemical factors act cooperatively in conferring arsenic tolerance. A direct transport of heavy metals appears difficult to reconcile with current models of P-gp function. However, one could hypothesize that inorganic chemicals are transported as complexes with hydrophobic endogenous compounds. Interestingly, certain P-gp homologs in invertebrates and protozoans are known to mediate heavy metal and/or arsenic tolerance[21,23,24].

4. Regulation of P-gp expression and activity

The growing albeit still fragmentary conceptual understanding of the transcriptional regulation of the MDR1 gene has been the subject of several excellent reviews[76,98,134]. Transcription of the human *MDR1* gene appears to be regulated at two levels. First, for a given cell type, epigenetic mechanisms decide whether the *MDR1* promoter has basal activity and MDR1 is expressed. Second, in cells expressing *MDR1*, *MDR1* promoter activity can be up-regulated by different agents and treatments, such as drugs, hormones, DNA damage, heat shock, serum starvation and UV irradiation[76,98]. The *MDR1* promoter lacks a TATA box, shows an initiator element and at least two GC boxes. This suggests that *MDR1* is a housekeeping gene. However, going beyond a simple housekeeping gene concept, the regulation of the *MDR1* gene is complex and involves multiple signal transduction pathways[76,98]. Accordingly, the *MDR1* promoter contains consensus sequences for several transcription factors, including a CAAT and a CCAAT box, AP-1 and YB-1 recognition sites and a putative heat shock element[76].

Chemicals that induce P-gp expression in mammals are numerous and include polycyclic aromatic hydrocarbons such as 3-methylcholanthrene[52], cytotoxic drugs, e.g. mitoxantrone[146], reversal agents such as cyclosporin A[102], antibiotics such as rifampicin[148], and steroids such as dexamethasone[47]. It is noteworthy that drugs inducing P-gp in mammals often also induce cytochrome P4503A (CYP3A). In a study of 15 drugs, the induction of P-gp and CYP3A was correlated with an *r*-value of 0.7[147]. P-gp and CYP3A also show broadly overlapping substrate and inhibitor specificities[81,201,202].

Recently, the pregnane X receptor (PXR, also called the steroid and xenobiotic receptor) was identified as a key regulator of the expression of P4503A genes in mammals[18,75,139,149]. The PXR is a ligand-inducible transcription factor that binds both endogenous compounds, such as steroids or bile acids, and organic xenobiotics. Following heterodimer formation with the retinoic acid receptor (RXR), the liganded PXR can regulate the transcription of genes after binding to consensus sequences in their regulatory regions[18]. Large differences in the ligand selectivity of PXRs of different mammalian species have been observed that correlate well with interspecific differences in the responsiveness to CYP3A-inducing agents[75,209]. It was recently reported that the PXR coordinately regulates P4503A isoforms and the MDR1 drug pump[177]. Moreover, an upstream enhancer was discovered in the promoter of MDR1, the sequence of which ligand-dependently binds the PXR–RXR dimer[53], suggesting that the PXR may also participate in the regulation of P-gp.

In addition to regulation at the transcriptional level, expression of P-gp may be regulated post-transcriptionally, e.g. at the level of mRNA stability[103]. Moreover, in certain cell types such as hepatocytes, intracellular pools of ABC transporters exist that can be transported to the plasma membrane on demand[83]. Furthermore, a potential role of phosphorylation in the regulation of P-gp activity has been studied extensively, but remains controversial (reviewed in ref. 56). Many studies have used inhibitors of protein kinases, namely of protein kinases A and C. In many reports, the inhibition of protein kinases was accompanied by a reduced phosphorylation and activity of P-gp, and it has often been suggested that phosphorylation of P-gp increases its activity[56]. However, apart from phosphorylating specific amino acids in the transporter polypeptide, protein kinase inhibitors may act directly as inhibitors of P-gp or affect P-gp expression[56]. Moreover, directed mutation of amino acid residues that are the target of phosphorylation in P-gp has no marked effects on transport activity[56].

III. P-glycoproteins in aquatic animals

1. Multixenobiotic resistance and P-gp in aquatic animals

Kurelec and coworkers were the first to demonstrate that alterations in xenobiotic transport are significant strategies adopted by aquatic organisms to adapt to organic pollution in their habitats (reviewed in refs. 12,41,92). Certain marine and brackish water animals are able to survive and reproduce in heavily polluted habitats, and often show surprisingly low accumulation of pollutants in body tissues[96]. As mentioned elsewhere, Kurelec coined the term multixenobiotic resistance for this phenotype[96], and searched to identify the biochemical basis for this adaptational capacity. Most of the studies on MXR deal with marine and aquatic invertebrates and will be reviewed first, followed by studies on fish.

In two studies intended to identify the biochemical basis for pollution resistance in the bivalves *Anodonta cygnea* and *Mytilus galloprovincialis*, a first interesting finding was that these animals had comparatively high glutathione-*S*-transferase and glutathione-peroxidase activities[96,97]. Moreover, membrane fractions from tissues of both species showed a high potential to bind 2-acetylaminofluorene, with the binding being sensitive to verapamil and trypsin[96,97]. Co-exposure of *M. galloprovincialis* to verapamil increased the accumulation of 2-acetylaminofluorene[97]. These latter findings suggested the presence of a drug transporter resembling P-gp in both bivalves, which Kurelec later termed the 'multixenobiotic resistance mechanism'.

Invertebrate P-gp-like proteins were characterized in some depth in marine sponges (Porifera)[94] and an echiuran worm (*Urechis caupo*)[181]. In two sponge species[94], a membrane protein fraction of 125 kDa cross-reacted with antiserum raised against mammalian mdr1. Treatment of membrane vesicles with endoglycosidase F decreased the apparent molecular mass of the immunoreactive protein fraction to 105 kDa, confirming that it consisted of glycoproteins. Immunolabeling of viable sponge cells localized the immunoreactive proteins to the cell surface[94]. In Northern analysis, a cDNA probe derived from human *MDR1* recognized an mRNA species of 4.2 kb in the sponge *Geodia cydonium*. Membrane fractions prepared from sponge cells bound the mdr1-type substrates, 2-acetylaminofluorene and vincristine, in a saturable and verapamil-sensitive fashion[94]. In the eggs of the marine worm *U. caupo*, different antibodies raised against mammalian P-gp detected membrane protein(s) of about 140 kDa in immunoblots[181]. A photoactive forskolin derivative labeled different proteins in *U. caupo* egg membranes, of which a major 145 kDa protein is immunoprecipitated by an antibody directed against mammalian P-gp[181]. The accumulation of rhodamine B and forskolin by *U. caupo* embryos was markedly increased in the presence of different reversal agents[181]. Collectively, these results suggest that the two species of sponges and the echiuran worm express one or several proteins strongly resembling mammalian P-gp.

Immunochemical evidence for invertebrate P-gp-like proteins has also been reported for aquatic mollusks and insects. In Western blots, antibodies raised against mammalian P-gp recognized immunoreactive protein bands of a molecular mass within the expected range of P-gps (130–170 kDa) in different mollusks, including the bivalves *M. galloprovincialis*, *M. californianus* and *Corbicula fluminea*, and the marine gastropod *Monodonta turbinata*[32,93,95,199]. Immunohistochemistry using the antibody C219, which recognizes a conserved epitope of P-gp, resulted in a strong positive staining of anal papillae[128] in larvae of the midge *Chironomus riparius* (Diptera). These immunochemical findings suggesting the presence of P-gp-like proteins in marine and aquatic invertebrates were corroborated by data on transporter function. Different reversal agents increased the accumulation of the fluorescent mdr1-type substrate rhodamine B by *M. californianus* gills[32]. Similarly, verapamil increased the accumulation of rhodamine B and forskolin, and potentiated cytotoxic effects of vinblastine and emetine in *U. caupo* embryos[181]. The accumulation of vincristine by *M. turbinata*, *C. fluminea* and *M. galloprovincialis* maintained under control conditions or collected from pristine sites increased in the presence of the reversal agent verapamil[93,95,199].

M. turbinata and *M. galloprovincialis* collected at polluted sites showed a markedly decreased accumulation of vincristine when compared to control animals. The same result was observed for *M. turbinata* snails after 2 days of exposure to Diesel-2 oil[93,95], suggesting an induction of the MXR mechanism. Interestingly, in animals from polluted sites showing decreased accumulation of vincristine, reversal agents had only minor effects on vincristine accumulation[93,95]. Eufemia and Epel studied factors inducing P-gp in the mussel *M. californianus*[42,43]. P-gp-dependent transport activity, measured in a rhodamine B exclusion assay, and P-gp protein levels were induced not only by organic chemicals such as the pesticide chlorthal, the PCB arochlor 1254, pentachlorophenol and DDE, but also by cadmium, arsenic and heat shock[42,43]. The effect of seasonal factors and pollution on P-gp expression was studied in oysters (*Crassostrea virginica*)[79]. P-gp expression tended to be higher in the warmer months and lower in the colder months[79]. There was no clear relationship between P-gp expression and pollution, although P-gp protein levels tended to decrease at polluted sites in the warmer months[79].

Some information on environmentally relevant reversal agents is available from aquatic invertebrates. P-gp-related dye exclusion in the mussel *M. californianus* was inhibited by the moderately hydrophobic pesticides pentachlorophenol, dacthal, chlorbenside and sulfallate, but not by DDT, DDD, DDE or the PCB Arochlor 1254[32]. Similarly, Diesel-2 oil water-extracts increased the accumulation of vincristine in the gills of *M. galloprovincialis*[91]. In the freshwater mussel *Dreissena polymorpha*, water from the polluted Sava River increased the accumulation of fluorescent P-gp substrates in gill tissue to a similar extent as 20 μM verapamil[165]. Water from two relatively pristine rivers also increased dye accumulation by the mussel gills, but to a lesser extent[165]. In another study, the increase in vincristine accumulation by *M. galloprovincialis* gills was similar after exposure to water and sediment XAD-2 concentrates from the polluted Sava River, and from a non-polluted reference river[91]. While these studies suggest the presence of compounds with reversal agent activity in natural waters, not all of these compounds may be of anthropogenic origin. Secondary metabolites of animals and plants can have reversal agent activity. For instance, a cyanobacterial toxin and a cyclic peptide produced by a marine tunicate have been found to inhibit P-gp[169,207]. Chemosensitizing activity has further been reported for many plant secondary metabolites, including green tea polyphenols and compounds contained in medical plants such as ginseng and rosemary[73,82,127].

2. *Evidence for fish genes coding for P-gp or related transporters*

Evidence for genes in fish homologous to *mdr* genes in mammals has been presented for several teleostean species. Sequence information available in GenBank at the date of this review is summarized in Table 4.

Screening a genomic DNA library of winter flounder, *Pleuronectes americanus*[25], identified two genes with sequence homologies to mammalian mdr genes and produced partial sequences termed the flounder genes fP-gp A and fP-gp B[25]. However, it was later shown that P-gp A is a BSEP homolog[27]. In contrast, fP-gp B shows moderate homology to both *mdr1* and *mdr2* genes, but cannot be attributed to either class (*mdr1*-type or *mdr2*-type) of mammalian *mdr* genes.

A similar pattern to winter flounder has been found in other fish species. In a number of teleosts, the search for *mdr*-homologs by RT-PCR techniques has resulted in the identification of one BSEP homolog, and one mdr homolog that could not be attributed to the *mdr1/mdr2* dichotomy. From the liver of killifish, *Fundulus heteroclitus*[29], 3 kb of a cDNA was amplified with similarities to mammalian BSEP. The authors further amplified 2.7 kb of another cDNA from killifish intestine and liver, and this fragment was similar to both mammalian *mdr1* and *mdr2*. From the liver of turbot, *Scophthalmus maximus*, Tutundjian and coworkers[184] cloned a 473 bp partial cDNA that shares up to 73% identity with mammalian mdr sequences, but is equally distant to *mdr1* and *mdr2*. Semiquantitative estimates of P-gp expression levels in various tissues by RT-PCR were high in brain, intestine, kidney and esophagus, intermediate in liver, heart and gill, and low in skeletal muscle[184].

It has been speculated that the mammalian *mdr1* and *mdr2* genes arose from gene duplication after the separation of teleosts and tetrapod vertebrates. Consequently, fish would possess only one class of *mdr* genes[29,184]. Apart from the question about the origin of the unclassifiable *mdr* genes identified in fish, the central question is what the function of their protein products might be. *In vitro* expression of fish *mdr* genes in suitable cellular systems would be the method of choice to identify the spectrum of substrates of the transporter. However, up to now, no complete coding cDNA sequence of a fish *mdr* gene is available, making this approach unfeasible. The presence of mdr1-like transport in teleost tissues (reviewed below) makes it appear likely that

TABLE 4

Sequences of piscine P-gp and BSEP available in GenBank

Species	Gene	Accession number	Type of sequence	Length	References
Fundulus heteroclitus	mdr homolog	AF099732	Partial cds	2752 bp	Cooper, P.S., P.A. van Veld and K.S. Reece, unpublished
Fundulus heteroclitus	BSEP homolog	AF135793	Partial cds	3004 bp	Cooper, P.S., P.A. van Veld and K.S. Reece, unpublished
Pseudopleuronectes americanus	mdr homolog	AY053461	Partial cds	2454 bp	Adilakshmi, T. and R.O. Laine, unpublished
Pseudopleuronectes americanus	mdr homolog	X72068	Partial gene sequence	379 bp	Chan *et al.*[25]
Pseudopleuronectes americanus	mdr homolog	X72069	Partial gene sequence	2301 bp	Chan *et al.*[25]
Pseudopleuronectes americanus	BSEP	X72066	Partial gene sequence	564 bp	Chan *et al.*[25]
Pseudopleuronectes americanus	BSEP	X72067	Partial gene sequence	1536 bp	Chan *et al.*[25]
Takifugu rubripes	mdr homolog	AAO20901	Predicted protein sequence	1292 aa	Liu, Y., X. Zhang, J. Yu and H. Yang, unpublished
Takifugu rubripes	mdr homolog	AAO20902	Predicted protein sequence	1271 aa	Liu, Y., X. Zhang, J. Yu and H. Yang unpublished
Platichthys flesus	BSEP	AF179307	Partial cds	432 bp	Alpermann, T.J., A. Luedeking and A. Koehler, unpublished
Danio rerio	mdr homolog	BQ262340	EST	606 bp	Clark, M., S.L. Johnson, H. Lehrach, *et al.*, unpublished
Danio rerio	mdr homolog	BE016688	EST	651 bp	Clark, M., S.L. Johnson, H. Lehrach, *et al.*, unpublished
Danio rerio	mdr homolog	BG302675	EST	661 bp	Sugano, S., K. Kawakami, S. Johnson, *et al.*, unpublished
Oncorhynchus mykiss	mdr homolog	CA366489	EST	704 bp	Rexroad, C.E. and J.W. Keele, unpublished
Oncorhynchus mykiss	mdr homolog	CA363431	EST	755 bp	Rexroad, C.E. and J.W. Keele, unpublished
Oncorhynchus mykiss	BSEP	CA363965	EST	696 bp	Rexroad, C.E. and J.W. Keele, unpublished

the products of teleostean mdr-homologs mediate this transport; however, at present, this remains a hypothesis. The number of *mdr*-genes in fish also remains unknown. Genomic research activities have led to the identification of several expressed sequence tags encoding mdr-homologs in rainbow trout (*Oncorhynchus mykiss*) and zebrafish (*Danio rerio*) (Table 4). It remains to be elucidated whether the small differences between the sequences of different clones within either species are due to cloning artifacts, allelic differences or different genes. The prevalence of gene duplications in teleosts has provoked the controversial hypothesis that a whole-genome duplication might have occurred before the radiation of teleosts[8,132,178]. In addition, lineage-specific duplications exist in groups that underwent tetraploidization[178]. Hence, it would not be surprising if multiple *mdr*-homologs were found in fish. Interestingly, the analysis of the pufferfish (*Tagifugu rubripes*) genome led to the identification of two possible *mdr* homologs (Table 4).

3. *Immunochemical evidence for P-gp in fish*

Homologous antibodies against teleostean P-gp are not yet available. Using antibodies raised against mammalian P-gp, the presence of immunoreactive proteins could be demonstrated for a number of teleostean species. The antibody most frequently used in immunochemical studies on P-gp expression in fish is the monoclonal antibody C219. This antibody recognizes an epitope common to all known P-gps, including the mdr2-type P-gps and BSEP[27,54,193]. Hence, the C219 antibody indicates whether a specific tissue expresses P-gps, but not which P-gp isoform is present.

Western blot protocols and immunohistochemical detection of P-gp-like proteins in fish tissues are available[30,64]. Using immunoblot techniques, putative P-gp proteins with molecular masses of approximately 170 kDa were found in channel catfish (*Ictalurus punctatus*)[37], rainbow trout[174], high cockscomb prickleback (*Anoplarchus purpurescens*)[15] and killifish[1,13,31]. A molecular mass of 170 kDa agrees well with the molecular masses reported for mammalian P-gps. The apparent molecular mass of mammalian P-gp may vary from 130 to 180 kDa due to differences in the degree of glycosylation[56]. A contrasting result was reported for turbot where a P-gp-immunoreactive protein of 83 kDa was found[184]. A likely explanation for this unexpected finding is that the 83 kDa protein represents a proteolytic breakdown product of P-gp[184]. This interpretation is supported by Doi and coworkers[37] who found both an 80 and a 170 kDa P-gp-immunoreactive protein in channel catfish. After changes in the composition of the sampling buffer and inclusion of protease inhibitors, exclusively the 170 kDa protein was detectable[37]. The P-gp protein may be prone to proteolytic breakdown at the linkage region, which reportedly divides the molecule into two mirror halves[58].

Hemmer and coworkers[64] compared the influence of three different fixatives on immunoreactivity of P-gp antigen in tissue sections and reported that fixation conditions can strongly influence the staining results. Generally, Bouin's fixative was found to be inferior to Dietrich's or Lillie's fixative. Formalin was not considered in that study[64], but in our hands it results in a better conservation of P-gp antigenicity for immunohistochemical staining than Bouin's fixative.

In an immunohistochemical study on P-gp distribution in the guppy, *Poecilia reticulata*, Hemmer and coworkers[65] used the monoclonal antibody C219, which recognizes an epitope in both MDR1 and MDR3, the monoclonal antibody C494, which specifically recognizes an epitope in the C-terminal half of MDR1, but not MDR3, and the monoclonal antibody JSB-1, which recognizes a conserved cytoplasmic epitope of MDR. They also used the polyclonal antibody mdr(Ab-1) raised against a synthetic peptide encircling the epitope sequence of C219. All four antibodies stained positive on intestine, kidney and exocrine pancreatic tissues, albeit with varying intensities. Hepatic bile canaliculi, pseudobranch and gill chondrocytes were positive for C219, JSB-1 and mdr(Ab-1) but not for C494. Tissues staining positive only with C494 and JSB-1, but not with C219 and mdr(Ab-1) included the interrenal, the branchial transverse septum and the blood vessels. The described staining responses are not consistent with expectations based on the epitope specificity of these antibodies in mammals. Because the epitope recognized by C219 is present in both mdr1-type and mdr2-type P-gps and in BSEP, there should not be a tissue negative for C219, but positive for any of the other antibodies. Such deviations from the expected staining response may indicate the presence of different P-gp proteins in fish, but it may also reflect technical effects related to conservation of the antigenic epitopes during tissue fixation and processing. Based on current data, without fish-specific homologous probes available, a tissue-specific distribution of specific P-gp isoforms cannot be demonstrated in fish.

3.1. Immunohistochemical studies on tissue distribution of P-gp

Immunohistochemical studies on tissue-specific distribution of P-gp in fish revealed a pattern similar to that described in mammals, i.e. immunoreactive P-gp occurs in epithelial tissues involved in secretion, absorption or serving a barrier function. Further, P-gp-like proteins were observed at the bile canaliculi of the liver, in the proximal tubules of the excretory kidney, at the endothelia of brain capillaries, in the intestinal tract, exocrine pancreas, operculi, gills, pseudobranch, gas gland, interrenal tissue and skeletal muscle (Table 5). This tissue distribution is in agreement with the assumed role of P-gp in protection against xenobiotics.

Hemmer and coworkers[65] made a direct comparison of the tissue distribution of P-gp immunoreactivity as obtained with both C219 and JSB-1 in either guppy (*P. reticulata*) or man (*Homo sapiens*): the only major difference found between the two species was the observation that adrenal cortical cells were strongly positive in man, while the adrenal cortical homolog in guppy, the interrenal cells, was negative with C219 but (weakly) positive with JSB-1.

Findings on P-gp tissue distribution in different fish species are generally in agreement, with only a few contradictory results. One example is the localization of P-gp in trunk kidney. In guppy and sheephead minnow (*Cyprinodon variegatus*) the tubular epithelia, but not the kidney interstitium, showed positive staining for P-gp[64,65], while in channel catfish the interstitium, but not the proximal tubular epithelial cells, stained positive for P-gp[85].

TABLE 5

Tissue distribution of immunoreactive staining in immunohistochemical studies of fish P-gp

Tissue and cell type showing positive reaction	Fish species	Antibody	References
Liver			
Biliary pole of hepatocytes	*Poecilia reticulata*	C219, JSB-1, mdr(Ab-1)	65
	Fundulus heteroclitus	C219	1, 13, 30, 3164
	Cyprinodon variegatus	C219, JSB-1, mdr(Ab-1)	37, 85
	Ictalurus punctatus	C219	88
	Platichthys flesus	C219	174
	Oncorhynchus mykiss	C219	15
	Anoplarchus purpurescens	C219	
Kidney			
Luminar membrane of tubular epithelial cells	*Poecilia reticulata*	C219, JSB-1, C494, mdr(Ab-1)	65
Tubular cells and renal corpuscles	*Pleuronectes americanus*	C219	176
	Cyprinodon variegatus	C219	64
Extratubular tissue (no staining of tubuli and corpuscles)	*Ictalurus punctatus*	C219	85
Brain			
Endothelia of capillaries	*Fundulus heteroclitus*	C219	116
	Squalus acanthus		
Intestine			
Luminal surface of enterocytes	*Poecilia reticulata*	C219, JSB-1, C494, mdr(Ab-1)	65
Surface and cytoplasm of enterocytes	*Ictalurus punctatus*	C219	85
	Cyprinodon variegates	C219, JSB-1, C494, mdr(Ab-1)	64
Epithelial cells of posterior intestine	*Fundulus heteroclitus*	C219	13
Gills			
Chondrocytes, transverse septum	*Poecilia reticulata*	C219, C494, JSB-1	65
Exocrine pancreas	*Poecilia reticulata*	C219, JSB-1, C494, mdr(Ab-1)	198
	Cyprinodon variegatus		
Interrenals	*Poecilia reticulata*	C494, JSB-1, C219	198
	Cyprinodon variegatus	JSB-1, C494, mdr(Ab-1)	191
Gas gland	*Poecilia reticulata*	C219	198
Pseudobranch	*Poecilia reticulata*	C219, JSB-1, mdr(Ab-1)	198
Hematopoietic tissue	*Cyprinodon variegatus*	C494	191

4. Evidence for mdr1 P-glycoprotein-like transport in fish tissues

4.1. Liver

In the mammalian liver, P-gps are an important component of the ensemble of canalicular transmembrane transporters. In the liver of fish, immunohistochemical staining indicated that P-gp immunoreactive proteins are localized to the bile canalicular pole of the hepatocytes[1,30,65,174]. Studies with isolated fish hepatocytes demonstrated that teleostean liver cells possess transport activities that display classic characteristics of P-gp-mediated transport[1,174,185]. The accumulation of P-gp substrates in teleost hepatocytes or the efflux of these substrates from hepatocytes is sensitive to reversal agents. In rainbow trout hepatocytes, both the accumulation and the efflux of the fluorescent P-gp substrate rhodamine 123 were modified by the reversal agents verapamil, cyclosporin A and vinblastine, the P-gp substrate doxorubicin and the ATPase inhibitor vanadate[174]. In contrast, tetraethylammonium chloride, a substrate for type I sinusoidal organic cation uptake systems and electroneutral canalicular H^+/organic cation antiporter, had no effect on rhodamine 123 transport[174]. Similarly, in isolated hepatocytes of killifish and turbot as well as in carp (*Cyprinus carpio*) *in vivo*, uptake of fluorescent P-gp substrates was verapamil-sensitive[1,167,185].

4.2. Kidney

Besides the liver, the kidney is a major site for the excretion of drugs and toxicants from the body. In the mammalian kidney, the proximal tubule is the major site of P-gp-mediated transport[192]. The situation is likely to be the same in teleost fish. A number of studies characterized the transport of fluorescent P-gp substrates in isolated killifish (*F. heteroclitus*) renal proximal tubules[115,119,145]. This *in vitro* model takes advantage of the fact that fish kidneys contain a high proportion of proximal tubules which can be dissected out and remain viable in culture. During culture, the broken ends of the isolated tubules spontaneously reseal to form a closed, fluid-filled luminal compartment that communicates with the medium only through the tubular epithelium. Thus, the *in vitro* preparation retains the polarity of the kidney tubules *in vivo*. When fluorochrome P-gp substrate was added to the culture medium, fluorescence accumulated both in tubular cells and the tubular lumen, with luminal fluorescence at steady state being 2–4 times higher than cellular fluorescence, depending on the compound used[115,119,145]. Luminal accumulation of fluorescent P-gp substrates in killifish kidney tubules could be inhibited by resistance modifiers such as verapamil and cyclosporin A, the ATPase inhibitor vanadate and by metabolic inhibitors such as cyanide[145]. In contrast, prototypical substrates of cation and anion transporters other than P-gp such as tetraethylammonium and *p*-aminohippurate did not affect the luminal accumulation of P-gp substrates[145]. Using primary cultures of winter flounder (*P. americanus*) renal tubular epithelia mounted in Ussing chambers, Sussman-Turner and Renfro[176] described a transepithelial transport for daunomycin that was not only sensitive to P-gp inhibitors, but also partly sensitive to tetraethylammonium. Mild heat shock of primary cultures of winter flounder renal cells led to elevated P-gp-mediated drug transport[176], suggesting that P-gp might be part of a general cellular stress response.

4.3. Other tissues

Single studies are available that have investigated P-gp-mediated transport in fish tissues other than liver and kidney. In particular, data on P-gp function are available for the intestine[37], the operculum[77] and brain capillaries[116]. In isolated brain capillaries from killifish (*F. heteroclitus*) and spiny dogfish shark, *Squalus acanthius*, an uphill transport of fluorescent P-gp substrates from incubation bath medium to capillary lumen was observed. The transport could be inhibited by KCN, a metabolic inhibitor, and by P-gp substrates and inhibitors[116]. These findings strongly argue for the presence of a P-gp-mediated efflux pump in the endothelia of brain capillaries, which, together with the tight junctions, contributes to the barrier function of the non-fenestrated endothelium against the entry of hydrophobic xenobiotics into the brain.

5. Occurrence of P-glycoprotein in tumors of fish

The initial interest to study P-gp was because it plays a role in the MDR of cancers. An intriguing question is whether fish tumor cells express P-gp and at what level. Surprisingly, this issue has received little attention. To our knowledge, to date only three studies have analyzed P-gp expression in fish tumors: one study on winter flounder, *P. americanus*[14], one on European flounder *Platichthys flesus*[88], and one on killifish *F. heteroclitus*[31]. Bard and co-workers[14] examined tumors of winter flounder from Boston Harbor. This area is strongly contaminated with carcinogenic compounds, and hosts a flatfish population showing an enhanced prevalence of cholangiocellular neoplasms. Among the individuals with neoplastic changes, only 13% showed immunohistochemically detectable P-gp expression in the tumors, while the normal, unaltered liver tissue exhibited enhanced P-gp expression. The authors hypothesized that this somewhat unexpected finding could reflect an influence of the neoplastic tissue on the surrounding, unaltered liver parenchyma to increase P-gp expression as a putative defense mechanism. In another study, a creosote-resistant killifish population showed an increased prevalence of adenomas and hepatocellular carcinomas[198]. Normal and neoplastic livers of fish from the resistant population displayed two to threefold higher levels of P-gp-like protein in C219 immunoblots than livers of fish from a reference site. The cellular localization of P-gp in normal livers of creosote-resistant fish was like that in livers of control fish, i.e. exclusively at the bile canalicular surface; in neoplastic livers, however, P-gp was found along the whole cell surface as well as in the cytoplasm of the hepatocytes from the non-neoplastic tissue areas, while within the tumor areas, a strong overexpression of P-gp could be observed. Interestingly, P-gp overexpression was only present in the adenomas and carcinomas, while early neoplastic foci showed no elevated P-gp expression. The authors opined that P-gp overexpression in the more progressed neoplastic lesions is related to tumor progression and oncogene activation rather than selection of expression early in carcinogenesis. This interpretation implies that enhanced P-gp levels in neoplastic and non-neoplastic liver cells of creosote-resistant killifish probably result from different causes. In a third study on P-gp expression in fish tumors, flounders collected from polluted sites of the North Sea were examined[88]. While eosinophilic foci showed no P-gp immunostaining, P-gp became overexpressed with the transition to the basophilic cell type (for a description of the progression of chemically induced carcinogenesis in fish, see ref. 68). The described results suggest that fish tumor cells can exhibit elevated P-gp, but that this is not always the case. For a more thorough understanding of under what conditions or in what type of neoplastic alteration P-gp may be overexpressed, we clearly need more studies relating P-gp expression to tumor type or progression stage of neoplastic changes.

6. *P-gp expression and activity in fish exposed to xenobiotics*

Exposure of fish to environmental contaminants may modulate the P-gp transporter mainly in two ways: first, the xenobiotic can directly interact with the transporter either as substrate or as inhibitor (reversal agent) and second, the xenobiotic modulates P-gp expression levels.

6.1. *Xenobiotics as substrates or inhibitors of P-glycoprotein in fish*

Few studies have examined the interaction of environmental contaminants as substrates or inhibitors with P-gp transporters in fish. Indirect evidence that P-gp is able to modulate the accumulation of xenobiotics in fish is provided by two experiments of Kurelec's group. In one experiment, golden ides, *Leuciscus idus melanotus*, were exposed to the mutagenic compound 2-aminoanthracene resulting in the induction of DNA single strand breaks. Co-exposure of fish to 2-aminoanthracene and verapamil led to a significant increase of DNA damage compared with exposure to 2-aminoanthracene alone[89]. This could reflect increased bioaccumulation of 2-aminoanthracene due to the inhibition of P-gp activity by verapamil[89]. In another study, a 3-day aqueous exposure of carp to low levels of hydrocarbon-rich Diesel-2 oil resulted in no alteration of hepatic CYP1A activity[90]. However, when verapamil was added to the exposure water together with Diesel-2 oil, hepatic CYP1A activity was induced. A possible explanation for this finding was that inhibition of P-gp by verapamil led to an elevation of Diesel-2 oil tissue burdens above the threshold for CYP1A induction.

In an *in vivo* study with carp, Smital and Sauerborn[167] noticed that the reversal agents verapamil and cyclosporin A increased the hepatic accumulation of the P-gp substrate rhodamine B by 54 and 170%, respectively. The organophosphorus insecticide malathion (5 and 20 μM) also increased rhodamine B accumulation in the liver and was comparable in potency to verapamil. XAD-7 concentrates from a polluted river, but not concentrates from a non-polluted river, increased hepatic rhodamine B accumulation in carp. This suggests that malathion and unidentified compounds from the river water acted as reversal agents of hepatic P-gp in carp[167]. Inhibitory effects of environmentally relevant compounds have also been observed in a study on P-gp-mediated transport in cultured rainbow trout hepatocytes[173]. Nonylphenol diethoxylate and prochloraz (an imidazole derivative) significantly reduced the efflux of rhodamine 123 from trout hepatocytes. However, the P-gp substrates, vinblastine and rhodamine 123, and the reversal agent verapamil did not significantly influence the hepatocellular accumulation or efflux of [^{14}C]-prochloraz radioactivity, suggesting that prochloraz is not transported by trout hepatic P-gp.

6.2. Xenobiotics as inducers of P-glycoprotein in fish

Studies that attempted to identify xenobiotics that up-regulate P-gp expression in fish are yet to provide a consistent picture on the identity of P-gp inducers in fish. Sturm and coworkers[173] did not find induction of C219-immunoreactive P-gp protein levels after *in vivo* exposure of rainbow trout to prochloraz or nonylphenol diethoxylate. Similarly, dietary exposure of channel catfish to the xenobiotics β-naphthoflavone, benzo(*a*)pyrene or 3,4,3′,4′-tetrachlorobiphenyl did not affect the levels of C219-immunoreactive proteins or P-gp activity in intestine or liver[37]. In the liver of killifish, the expression of C219-immunoreactive P-gp protein was significantly enhanced, both *in vitro* and *in vivo*, after exposure to chlorpyrifos oxon and the carcinogen, *N*-nitrosodiethylamine[1]. A study by Bard and coworkers[15] on intertidal blennies (*A. purpurescens*) suggests that the relationship between xenobiotic exposure and P-gp expression in fish is complex. Hepatic P-gp immunostaining was significantly elevated in blennies collected from sites downstream of pulp mills compared with control fish held for 4 weeks in clean water in the laboratory. However, hepatic P-gp expression of blennies in the field showed no apparent correlation with the proximity of sampling sites to the pulp mill. In the laboratory, exposure of blennies to oiled sediments significantly elevated hepatic P-gp levels, whereas exposure to sediments from the pulp mill sites or injection of β-naphthoflavone had no effect on hepatic P-gp levels[15].

Two interesting studies addressed P-gp expression in pollutant-resistant killifish populations[13,31]. Cooper and coworkers investigated a creosote-resistant killifish population from a polluted site in the Elizabeth River, Virginia (USA)[31]. Adult fish from this population show a high prevalence of hepatic tumors[198]. In immunoblots with the C219 antibody, both normal and neoplastic livers of killifish from the polluted site contained significantly higher levels of P-gp-like proteins than the livers of killifish from a reference site[31]. This finding may be taken as an indication that increased P-gp expression in the liver could contribute to creosote resistance, since biliary excretion is a major elimination route for PAH metabolites. Observations contrasting the findings on the killifish from the Elizabeth River were found in a study on killifish from New Bedford Harbor (MA, USA) which show acquired resistance to planar halogenated aromatic hydrocarbons (PHAH)[13]. In this case, the contaminant-resistant fish showed lower levels of hepatic C219-immunoreactive protein than killifish from a relatively unpolluted site[13]. However, hepatic P-gp levels in fish from both sites decreased within 8 days of maintenance in clean laboratory water, suggesting the presence of inducers at both sites. In about half the killifish from New Bedford Harbor, intestinal P-gp protein reached detectable levels, whereas it was rarely detectable in fish from the control site or fish maintained in clean laboratory water[13]. While both the Elizabeth River and the New Bedford Harbor studies suggest induction of P-gp by environmental contaminants, the identity of inducers remains elusive. The differences in the tissue pattern of induction among sites could be explained if different inducing compounds with different routes of exposure had been present. The PHAH 2,3,7,8-tetrachlorodibenzofuran, known to be present in New Bedford Harbor, did not affect intestinal or hepatic P-gp expression in killifish[13].

Similarly, 3-methylcholanthrene, a PAH contained in creosote, did not affect hepatic levels of P-gp[31].

Overall, thc database on the relation between xenobiotics and P-gp in fish is small and dispersed. The published reports do not yield a consistent pattern that may be explained in part by between-study differences in species, xenobiotics and experimental design. However, it may also be due to technical limitations such as the lack of fish-specific probes for P-gp and to the insufficient knowledge on basic parameters of P-gp regulation in fish. From mammalian models, evidence exists that P-gp expression can be induced as part of a general response to cellular stress, and further be specifically regulated by ligands of the PXR. The latter way of regulation also applies to an important group of mammalian biotransformation enzymes, CYP3A. Furthermore, mammalian P-gp and CYP1A share some inducers, albeit they are not regulated by the same pathways. For fish, published evidence on PXR is lacking, and the understanding of piscine CYP3A is still fragmentary. In several studies with fish it was attempted to provoke a co-induction of CYP1A and P-gp. Cooper and co-workers exposed killifish (*F. heteroclitus*) to the AhR ligand 3-methylcholanthrene and observed an induction of CYP1A but not of P-gp expression[31]. Similarly, in the study of Bard and co-workers, treatment of killifish with the AhR ligand 2,3,7,8-tetrachlorodibenzofuran resulted in the induction of CYP1A but did not affect P-gp[13]. In rainbow trout, prochloraz induced hepatic CYP1A but not hepatic P-gp[173]. A correlation between CYP1A and P-gp levels was found in blennies collected at polluted field sites, however, when blennies were treated in the laboratory with the AhR ligand β-naphthoflavone, only CYP1A but not P-gp was elevated[15]. In flounders collected at three sites with differing pollution level in the Seine Bay, CYP1A and P-gp levels showed exactly opposite trends[120]. These data indicate that P-gp and CYP1A are not coordinately regulated in fish, however, these two proteins may act in a complementary way in cellular detoxicification.

7. *Toxicological implications of P-gp in fish*

Kurelec[89] postulated that the efflux transport of toxic compounds mediated by P-gp is a major biochemical mechanism behind the phenotype of MXR found in aquatic organisms of highly polluted habitats. In this concept, P-gp-dependent transport is considered as a first line of defense against harmful environmental substances[41,89]. Evidence for the toxicological relevance of P-gp in marine and aquatic invertebrates comes from considerable amount of studies. By comparison, a toxicological relevance of P-gp in teleost fish is at present less well documented. A protective role of fish P-gp is suggested by certain laboratory studies that have used model compounds, e.g. ref. 89. Moreover, the observation of enhanced P-gp expression in resistant fish populations from highly polluted field sites points to a protective function of P-gp[13,31]. However, there are also findings of a reduced P-gp expression in fish from polluted sites when compared with reference sites[13,120]. An interpretation of the present data is further complicated by the lack of understanding of which xenobiotics are inducers of P-gp in fish. Consequently, the toxicological implications of P-gps in fish are at present difficult to judge.

In the environment, fish are exposed to chemical mixtures. Environmental substances with low or no toxicity inhibiting P-gp, i.e. environmental chemosensitizers, might block the P-gp-dependent efflux transport of potent toxicants from fish tissues and thus indirectly lead to toxicity[166]. While a number of studies have addressed the effects of environmentally relevant chemosensitizers in fish, data to date remain fragmentary and are restricted to laboratory observations. In conclusion, it remains uncertain to which extent environmental chemosensitizers have ecotoxicological relevance in fish.

IV. Perspectives

There is clear evidence for the existence of P-gp in fish on the genetic, the immunochemical and the functional levels. However, the present knowledge on fish P-gp is rudimentary. Progress in research on piscine P-gp is hampered by a number of factors. For example, the exact number of P-gp isoforms in fish remains unknown. Isoform-specific probes for piscine *mdr* genes and *mdr*-encoded proteins are not available. Since to date no complete coding sequence of a fish *mdr* gene is available, it is not possible to express fish *mdr* genes in suitable cellular systems *in vitro* to identify the spectrum of substrates of the transporter.

Apart from the need to further develop basic tools for research on piscine P-gp, progress is needed in understanding the physiological and toxicological functions of P-gp in fish. To date, it remains unknown which classes of environmental toxicants can induce P-gp expression in fish, which natural factors may modulate P-gp induction and to what extent P-gp transport activity protects fish from the toxic action of xenobiotics. Concerning this latter point, the only data in fish available concern the accumulation of P-gp model substrates such as rhodamine, however, almost no information exists on how P-gp activity modulates body and tissue burdens of environmental contaminants. An important aspect to be considered in this context will be the life stage-specific expression of P-gp in fish. In both mammals and amphibia, it has been shown that the *mdr* gene is developmentally regulated[140,219]. Also, species-specific differences in P-gp expression should be studied in more detail, since they may offer competitive advantages in a particular dietary or environmental niche. Up to now, only single species studies have been performed, which do not allow concluding whether reported differences between fish species are methodologically caused or whether they represent true species differences. Finally, efflux transporters other than P-gp may significantly contribute to toxicant resistance, and certainly merit investigation. In mammals, the toxicological relevance of the MRPs has attracted much attention during recent years[101]. Only one study on MRP is available for fish and provides evidence that this transporter exists in elasmobranchs[117]. In conclusion, a number of important questions on basic traits and xenobiotic interaction of piscine P-gp remain to be answered before the toxicological and ecotoxicological relevances of P-gp in fish can be appropriately assessed.

V. *References*

1. Albertus, J.A. and R.O. Laine. Enhanced xenobiotic transporter expression in normal teleost hepatocytes: response to environmental and chemotherapeutic toxins. *J. Exp. Biol.* 204: 217–227, 2001.
2. Allikmets, R., B. Gerrard, A. Hutchinson and M. Dean. Characterization of the human ABC superfamily: isolation and mapping of 21 new genes using the expressed sequence tags database. *Hum. Mol. Genet.* 5: 1649–1655, 1996.
3. Allikmets, R., L.M. Schriml, A. Hutchinson, V. Romano-Spica and M. Dean. A human placenta-specific ATP-binding cassette gene (ABCP) on chromosome 4q22 that is involved in multidrug resistance. *Cancer Res.* 58: 5337–5339, 1998.
4. Altenberg, G.A., G. Young, J.K. Horton, D. Glass, J.A. Belli and L. Reuss. Changes in intra- or extracellular pH do not mediate P-glycoprotein-dependent multidrug resistance. *Proc. Natl Acad. Sci. USA* 90: 9735–9738, 1993.
5. Altuvia, S., W.D. Stein, S. Goldenberg, S.E. Kane, I. Pastan and M.M. Gottesman. Targeted disruption of the mouse mdr1b gene reveals that steroid hormones enhance mdr gene expression. *J. Biol. Chem.* 268: 27127–27132, 1993.
6. Ambudkar, S.V., C.O. Cardarelli, I. Pashinsky and W.D. Stein. Relation between the turnover number for vinblastine transport and for vinblastine-stimulated ATP hydrolysis by human P-glycoprotein. *J. Biol. Chem.* 272: 21160–21166, 1997.
7. Ambudkar, S.V., S. Dey, C.A. Hrycyna, M. Ramachandra, I. Pastan and M.M. Gottesman. Biochemical, cellular, and pharmacological aspects of the multidrug transporter. *Annu. Rev. Pharmacol. Toxicol.* 39: 361–398, 1999.
8. Amores, A., A. Force, Y.L. Yan, L. Joly, C. Amemiya, A. Fritz, R.K. Ho, J. Langeland, V. Prince, Y.L. Wang, M. Westerfield, M. Ekker and J.H. Postlethwait. Zebrafish hox clusters and vertebrate genome evolution. *Science* 282: 1711–1714, 1998.
9. Ayesh, S., Y.-M. Shao and W.D. Stein. Co-operative, competitive and non-competitive interactions between modulators of P-glycoprotein. *Biochim. Biophys. Acta* 1316: 8–18, 1996.
10. Bain, L.J. and G.A. LeBlanc. Interaction of structurally diverse pesticides with the human MDR1 gene product P-glycoprotein. *Toxicol. Appl. Pharmacol.* 141: 288–298, 1996.
11. Bain, L.J., J.B. McLachlan and G.A. LeBlanc. Structure–activity relationships for xenobiotic transport substrates and inhibitory ligands of P-glycoprotein. *Environ. Health Perspect.* 105: 812–818, 1997.
12. Bard, S.M. Multixenobiotic resistance as a cellular defense mechanism in aquatic organisms. *Aquat. Toxicol.* 48: 357–389, 2000.
13. Bard, S.M., S.M. Bello, M.E. Hahn and J.J. Stegeman. Expression of P-glycoprotein in killifish (*Fundulus heteroclitus*) exposed to environmental xenobiotics. *Aquat. Toxicol.* 59: 237–251, 2002.
14. Bard, S.M., M. Moore and J.J. Stegeman. Immunohistochemical study of multixenobiotic resistance and cytochrome P4501A expression in winter flounder liver tumours. 18th Annual Meeting of the Society of Environmental Toxicology and Chemistry, San Francisco, CA, Abstract #398, 1997.
15. Bard, S.M., B.R. Woodin and J.J. Stegeman. Expression of P-glycoprotein and cytochrome p450 1A in intertidal fish (*Anoplarchus purpurescens*) exposed to environmental contaminants. *Aquat. Toxicol.* 60: 17–32, 2002.
16. Barecki-Roach, M., E.J. Wang and W.W. Johnson. Many P-glycoprotein substrates do not inhibit the transport process across cell membranes. *Xenobiotica* 33: 131–140, 2003.
17. Bellamy, W.T. P-glycoproteins and multidrug resistance. *Annu. Rev. Pharmacol. Toxicol.* 36: 161–183, 1996.
18. Blumberg, B., W. Sabbagh, H. Juguilon, J. Bolado, C.M. van Meter, E.S. Ong and R.M. Evans. SXR, a novel steroid and xenobiotic-sensing nuclear receptor. *Genes Dev.* 12: 3195–3205, 1998.
19. Borst, P., R. Evers, M. Kool and J. Wijnholds. The multidrug resistance protein family. *Biochim. Biophys. Acta* 1461: 347–357, 1999.
20. Borst, P. and A.H. Schinkel. What have we learnt thus far from mice with disrupted P-glycoprotein genes? *Eur. J. Cancer* 32A: 985–990, 1996.
21. Broeks, A., B. Gerrard, R. Allikmets, M. Dean and R.H. Plasterk. Homologues of the human multidrug resistance genes MRP and MDR contribute to heavy metal resistance in the soil nematode *Caenorhabditis elegans. EMBO J.* 15: 6132–6143, 1996.
22. Brown, P.C., S.S. Thorgeirsson and J.A. Silverman. Cloning and regulation of the rat *mdr2* gene. *Nucleic Acids Res.* 21: 3885–3891, 1993.
23. Callaghan, A. and N. Denny. Evidence for an interaction between p-glycoprotein and cadmium toxicity in cadmium-resistant and -susceptible strains of *Drosophila melanogaster. Ecotoxicol. Environ. Saf.* 52: 211–213, 2002.

24. Callahan, H.L. and S.M. Beverley. Heavy metal resistance: a new role for P-glycoproteins in *Leishmania. J. Biol. Chem.* 266: 18427–18430, 1991.
25. Chan, K.M., P.L. Davies, S. Childs, L. Veinot and V. Ling. P-glycoprotein genes in the winter flounder, *Pleuronectes americanus*: isolation of two types of genomic clones carrying 3′ terminal exons. *Biochim. Biophys. Acta* 1171: 65–72, 1992.
26. Chen, C.J., J.E. Chin, K. Ueda, D.P. Clarkl, I. Pastan, M.M. Gottesman and I.B. Roninson. Internal duplication and homology with bacterial transport proteins in the mdr 1 (P-glycoprotein) gene from multidrug-resistant human cells. *Cell* 47: 381–389, 1986.
27. Childs, S., R.L. Yeh, E. Georges and V. Ling. Identification of a sister gene to P-glycoprotein. *Cancer Res.* 55: 2029–2034, 1995.
28. Cole, S.P., G. Bhardwaj, J.H. Gerlach, J.E. Mackie, C.E. Grant, K.C. Almquist, A.J. Stewart, E.U. Kurz, A.M. Duncan and R.G. Deeley. Overexpression of a transporter gene in a multidrug-resistant human lung cancer cell line. *Science* 258: 1650–1654, 1992.
29. Cooper, P.S., K.S. Reece, W.K. Vogelbein and P.A. van Veld. P-glycoprotein and sister P-glycoprotein sequences from mummichog, *Fundulus heteroclitus*, liver and intestine. *Mar. Environ. Res.* 50: 335 2000.
30. Cooper, P.S., W.K. Vogelbein and P.A. Van Veld. Immunohistochemical and immunoblot detection of P-glycoprotein in normal and neoplastic fish liver. In; *Techniques in Aquatic Toxicology*, edited by G.K. Ostrander, Boca Raton, FL, CRC Press, pp. 307–325, 1996.
31. Cooper, P.S., W.K. Vogelbein and P.A. Van Veld. Altered expression of the xenobiotic transporter P-glycoprotein in liver and liver tumours of mummichog (*Fundulus heteroclitus*) from a creosote-contaminated environment. *Biomarkers* 4: 48–58, 1999.
32. Cornwall, R., B.H. Toomey, S. Bard, C. Bacon, W.M. Jarman and D. Epel. Characterization of the multixenobiotic/multidrug transport in the gills of the mussel *Mytilus californianus* and identification of environmental substrates. *Aquat. Toxicol.* 31: 277–296, 1995.
33. Dano, K. Active outward transport of daunomycin in resistant Ehrlich ascites tumor cells. *Biochim. Biophys. Acta* 323: 466–483, 1973.
34. Dassa, E. and P. Bouige. The ABC of ABCs: a phylogenetic and functional classification of ABC systems in living organisms. *Res. Microbiol.* 152: 211–229, 2001.
35. Deuchars, K.L., M. Duthie and V. Ling. Identification of distinct P-glycoprotein gene sequences in rat. *Biochim. Biophys. Acta* 1130: 157–165, 1992.
36. Devault, A. and P. Gros. Two members of the mouse mdr gene family confer multidrug resistance with overlapping but distinct drug specificities. *Mol. Cell. Biol.* 10: 1652–1663, 1990.
37. Doi, A.M., E. Holmes and K.M. Kleinow. P-glycoprotein in the catfish intestine: inducibility by xenobiotics and functional properties. *Aquat. Toxicol.* 55: 157–170, 2001.
38. Doyle, L.A., W. Yang, L.V. Abruzzo, T. Krogmann, Y. Gao, A.K. Rishi and D.D. Ross. A multidrug resistance transporter from human MCF-7 breast cancer cells. *Proc. Natl Acad. Sci. USA* 95: 15665–15670, 1998.
39. Endicott, J.A., P.F. Juranka, F. Sarangi, J.H. Gerlach, K.L. Deuchars and V. Ling. Simultaneous expression of two P-glycoprotein genes in drug-sensitive Chinese hamster ovary cells. *Mol. Cell. Biol.* 7: 4075–4081, 1987.
40. Endo, T., O. Kimura and M. Sakata. Effects of P-glycoprotein inhibitors on cadmium accumulation in cultured renal epithelial cells, LLC-PK1, and OK. *Toxicol. Appl. Pharmacol.* 185: 166–171, 2002.
41. Epel, D. Use of multidrug transporters as first lines of defense against toxins in aquatic organisms. *Comp. Biochem. Physiol.* 120A: 23–28, 1998.
42. Eufemia, N.A. and D. Epel. The multixenobiotic defense mechanism in mussels is induced by substrates and non-substrates: implications for a general stress response. *Mar. Environ. Res.* 46: 401–405, 1998.
43. Eufemia, N.A. and D. Epel. Induction of the multixenobiotic defense mechanism (MXR), P-glycoprotein, in the mussel *Mytilus californianus* as a general cellular response to environmental stresses. *Aquat. Toxicol.* 49: 89–100, 2000.
44. Eytan, G.D. and P.W. Kuchel. Mechanism of action of P-glycoprotein in relation to passive membrane permeation. *Int. Rev. Cytol.* 190: 175–250, 1999.
45. Eytan, G.D., R. Regev and Y.G. Assaraf. Functional reconstitution of P-glycoprotein reveals an apparent near stoichiometric drug transport to ATP hydrolysis. *J. Biol. Chem.* 271: 3172–3178, 1996.
46. Eytan, G.D., R. Regev, G. Oren and Y.G. Assaraf. The role of passive transbilayer drug movement in multidrug resistance and its modulation. *J. Biol. Chem.* 271: 12897–12902, 1996.
47. Fardel, O., V. Lecureur and A. Guillouzo. Regulation by dexamethasone of P-glycoprotein expression in cultured rat hepatocytes. *FEBS Lett.* 327: 189–193, 1993.
48. Fojo, A.T., K. Ueda, D.J. Slamon, D.G. Poplack, M.M. Gottesman and I. Pastan. Expression of a multidrug-resistance gene in human tumors and tissues. *Proc. Natl Acad. Sci. USA* 84: 265–269, 1987.

49. Ford, J. and W.N. Hait. Pharmacology of drugs that alter multidrug resistance in cancer. *Pharmacol. Rev.* 42: 155–199, 1990.
50. Fromm, M.F. The influence of MDR1 polymorphisms on P-glycoprotein expression and function in humans. *Adv. Drug Delivery Rev.* 54: 1295–1310, 2002.
51. Fujise, H., T. Annoura, S. Sasawatari, T. Ikeda and K. Ueda. Transepithelial transport and cellular accumulation of steroid hormones and polychlorobiphenyl in porcine kidney cells expressed with human P-glycoprotein. *Chemosphere* 46: 1505–1511, 2002.
52. Gant, T.W., J.A. Silverman, H.C. Bisgaart, R.K. Burt, P.A. Marino and S.S. Thorgeirsson. Regulation of 2-acetylaminofluorene and 3-methylcholanthrene-mediated induction of multidrug resistance and cytochrome P450IA gene family expression in primary hepatocyte cultures and rat liver. *Mol. Carcinog.* 4: 499–509, 1991.
53. Geick, A., M. Eichelbaum and O. Burk. Nuclear receptor response elements mediate induction of intestinal MDR1 by rifampin. *J. Biol. Chem.* 276: 14581–14587, 2001.
54. Georges, E., G. Bradley, J. Gariepy and V. Ling. Detection of P-glycoprotein by gene-specific monoclonal antibodies. *Proc. Natl Acad. Sci. USA* 87: 152–156, 1990.
55. Gerloff, T., B. Stieger, B. Hagenbuch, J. Madon, L. Landmann, J. Roth, A.F. Hofmann and P.J. Meier. The sister of P-glycoprotein represents the canalicular bile salt export pump of mammalian liver. *J. Biol. Chem.* 273: 10046–10050, 1998.
56. Germann, U.A.. P-glycoprotein – a mediator of multidrug resistance in tumour cells. *Eur. J. Cancer* 32A: 927–944, 1996.
57. Gottesman, M.M., T. Fojo and S.E. Bates. Multidrug resistance in cancer: role of ATP-dependent transporters. *Nat. Rev. Cancer* 2: 48–58, 2002.
58. Gottesman, M.M. and I. Pastan. Biochemistry of multidrug resistance mediated by the multidrug transporter. *Annu. Rev. Biochem.* 62: 385–427, 1993.
59. Gottesman, M.M., P.V. Schoenlein, S.J. Currier, E.P. Bruggemann and I. Pastan. Biochemical basis for multidrug resistance in cancer. In: *Biochemical Aspects of Selected Cancers*, edited by T. Pretlow and T. Pretlow, San Diego, CA, Academic Press, pp. 339–379, 1991.
60. Groothuis, G.M.M. and D.K.F. Meijer. Drug traffic in the hepatobiliary system. *J. Hepatol.* 24(Suppl. 1): 3–28, 1996.
61. Gros, P., J. Croop and D. Housman. Mammalian multidrug resistance gene: complete cDNA sequence indicates strong homology to bacterial transport proteins. *Cell* 47: 371–380, 1986.
62. Gros, P., Y.B. Neriah, J.M. Croop and D.E. Housman. Isolation and expression of a complementary DNA that confers multidrug resistance. *Nature* 323: 728–731, 1986.
63. Gros, P., M. Raymond, J. Bell and D. Housman. Cloning and characterization of a second member of the mouse mdr gene family. *Mol. Cell. Biol.* 8: 2770–2778, 1988.
64. Hemmer, M.J., L.A. Courtney and W.H. Benson. Comparison of three histological fixatives on the immunoreactivity of mammalian P-glycoprotein antibodies in the sheepshead minnow, *Cyprinodon variegatus. J. Exp. Zool.* 281: 251–259, 1998.
65. Hemmer, M.J., L.A. Courtney and L.S. Ortego. Immunohistochemical detection of P-glycoprotein in teleost tissues using mammalian polyclonal and monoclonal antibodies. *J. Exp. Zool.* 272: 69–77, 1995.
66. Higgins, C.F. ABC transporters: from microorganism to man. *Annu. Rev. Cell Biol.* 8: 67–113, 1992.
67. Higgins, C.F. and M.M. Gottesman. Is the multidrug transporter a flippase? *Trends Biochem. Sci.* 17: 18–21, 1992.
68. Hinton, D.E., H. Segner and T. Braunbeck. Toxic responses of the liver. In: *Target Organ Toxicity in Marine and Freshwater Teleosts*, edited by D. Schlenk and W.H. Benson, London, UK, Taylor & Francis, pp. 224–268, 2001.
69. Holash, J.A., S.I. Harik, G. Perry and P.A. Stewart. Barrier properties of testis microvessels. *Proc. Natl Acad. Sci. USA* 90: 11069–11073, 1993.
70. Hsu, S.I.-H., L. Lothstein and S.B. Horwitz. Differential overexpression of three mdr gene family members in multidrug-resistant J774.2 mouse cells. *J. Biol. Chem.* 264: 12053–12062, 1989.
71. Ichikawa-Haraguchi, M., T. Sumizawa, A. Yoshimura, T. Furukawa, S. Hiramoto, M. Sugita and S. Akiyama. Progesterone and its metabolites: the potent inhibitors of the transporting activity of P-glycoprotein in the adrenal gland. *Biochim. Biophys. Acta* 1158: 201–208, 1993.
72. Ishikawa, T. The ATP-dependent glutathione S-conjugate export pump. *Trends Biochem. Sci.* 17: 463–468, 1992.
73. Jodoin, J., M. Demeule and R. Beliveau. Inhibition of the multidrug resistance P-glycoprotein activity by green tea polyphenols. *Biochim. Biophys. Acta* 1542: 149–159, 2002.
74. Johnstone, R.W., A.A. Ruefli and M.J. Smyth. Multiple physiological functions for multidrug transporter P-glycoprotein? *Trends Biochem. Sci.* 25: 1–6, 2000.

75. Jones, S.A., L.B. Moore, J.L. Shenk, G.B. Wisely, G.A. Hamilton, D.D. McKee, N.C.O. Tomkinson, E.L. LeCluyse, M.L. Lambert, T.M. Willson, S.A. Kliewer and J.T. Moore. The pregnane X receptor: a promiscuous xenobiotic receptor that has diverged during evolution. *Mol. Endocrinol.* 14: 27–39, 2000.
76. Kantharidis, P., S. El-Osta, M. de Silva, G. Lee, X.F. Hu and J. Zalcberg. Regulation of MDRI gene expression: emerging concepts. *Drug Resist. Updates* 3: 99–108, 2000.
77. Karnaky, K.J., K. Sluggs, S. French and S.C. Willingham. Evidence for the multidrug transporter, P-glycoprotein, in the killifish, *Fundulus heteroclitus. Bull. Mt Desert Isl. Biol. Lab.* 32: 61–62, 1993.
78. Kawahara, M., A. Sakata, T. Miyashita, I. Tamai and A. Tsuji. Physiologically based pharmacokinetics of digoxin in mdr1a knockout mice. *J. Pharm. Sci.* 88: 1281–1287, 1999.
79. Keppler, C. and A.H. Ringwood. Expression of P-glycoprotein in the gills of oysters, *Crassostrea virginica*: seasonal and pollutant related effects. *Aquat. Toxicol.* 54: 195–204, 2001.
80. Keppler, D., I. Leier and G. Jedlitschky. Transport of glutathione conjugates and glucuronides by the multidrug proteins MRP1 and MRP2. *Biol. Chem.* 378: 787–791, 1997.
81. Kim, R.B., C. Wandel, B. Leake, M. Cvetkovic, M.F. Fromm, P.J. Dempsey, M.M. Roden, F. Belas, A.K. Chaudhary, D.M. Roden, A.J. Wood and G.R. Wilkinson. Interrelationship between substrates and inhibitors of human CYP3A and P-glycoprotein. *Pharm. Res.* 16: 408–414, 1999.
82. Kim, S.W., H.Y. Kwon, D.W. Chi, J.H. Shim, J.D. Park, Y.H. Lee, S. Pyo and D.K. Rhee. Reversal of P-glycoprotein-mediated multidrug resistance by ginsenoside Rg(3). *Biochem. Pharmacol.* 65: 75–82, 2003.
83. Kipp, H., N. Pichetshote and I.M. Arias. Transporters on demand: intrahepatic pools of canalicular ATP binding cassette transporters in rat liver. *J. Biol. Chem.* 276: 7218–7224, 2001.
84. Klein, I., B. Sarkadi and A. Váradi. An inventory of the human ABC proteins. *Biochim. Biophys. Acta* 1461: 237–262, 1999.
85. Kleinow, K.M., A.M. Doi and A.A. Smith. Distribution and inducibility of P-glycoprotein in the catfish: immunohistochemical detection using the mammalian C-219 monoclonal. *Mar. Environ. Res.* 50: 313–317, 2000.
86. Klimecki, W.T., B.W. Futscher, T.M. Grogan and W.S. Dalton. P-glycoprotein expression and function in circulating blood cells from normal volunteers. *Blood* 83: 2451–2458, 1994.
87. Klopman, G., L.M. Shi and A. Ramu. Quantitative structure–activity relationship of multidrug resistance reversal agents. *Mol. Pharmacol.* 52: 323–334, 1997.
88. Köhler, A., B. Lauritzen, S. Bahns, S.G. George, L. Förlin and C.J.F. Van Noorden. Clonal adaptation of cancer cells in flatfish liver to environmental contamination by changes in expression of P-gp related MXR, CYP450, GST-A and G6PDH activity. *Mar. Environ. Res.* 46: 191–195, 1998.
89. Kurelec, B. The multixenobiotic resistance mechanism in aquatic organisms. *Crit. Rev. Toxicol.* 22: 23–43, 1992.
90. Kurelec, B. Inhibition of the multixenobiotic resistance mechanism: ecotoxic consequences. *Sci. Total Environ.* 171: 197–204, 1995.
91. Kurelec, B. Reversion of the multixenobiotic resistance mechanism in gills of a marine mussel *Mytilus galloprovincialis* by a model inhibitor and environmental modulators of P170-glycoprotein. *Aquat. Toxicol.* 33: 93–103, 1995.
92. Kurelec, B. Pumping-out: the first line cellular defence to water pollutants in aquatic organisms. *Toxicol. Ecotoxicol. News* 4: 104–109, 1997.
93. Kurelec, B., S. Krca and D. Lucic. Expression of a multixenobiotic resistance mechanism in a marine mussel *Mytilus galloprovincialis* as a biomarker of exposure to polluted environments. *Comp. Biochem. Physiol.* 113C: 282–289, 1996.
94. Kurelec, B., S. Krca, B. Pivcevic, D. Ugarcovic, M. Bachmann and W.E.G. Müller. Expression of P-glycoprotein gene in marine sponges. Identification and characterization of the 125-kDa drug-binding glycoprotein. *Carcinogenesis* 13: 69–72, 1992.
95. Kurelec, B., D. Lucic, B. Pivcevic and S. Krca. Induction and reversion of multixenobiotic resistance in the marine snail *Monodonta turbinata. Mar. Biol.* 123: 305–312, 1995.
96. Kurelec, B. and B. Pivcevic. Distinct glutathione-dependent enzyme activities and a verapamil-sensitive binding of xenobiotics in a fresh water mussel *Anodonta cygnea. Biochem. Biophys. Res. Commun.* 164: 934–940, 1989.
97. Kurelec, B. and B. Pivcevic. Evidence for a multixenobiotic resistance mechanism in the mussel, *Mytilus galloprovincialis. Aquat. Toxicol.* 19: 291–302, 1991.
98. Labialle, S., L. Gayet, E. Marthinet, D. Rigal and L.G. Baggetto. Transcriptional regulators of the human multidrug resistance 1 gene: recent views. *Biochem. Pharmacol.* 64: 943–948, 2002.

99. Lankas, G.R., L.D. Wise, M.E. Cartwright, T. Pippert and D.R. Umbenhauer. Placental P-glycoprotein deficiency enhances susceptibility to chemically induced birth defects in mice. *Reprod. Toxicol.* 12: 457–463, 1998.
100. Lee, C.G., M. Ramachandra, K.T. Jeang, M.A. Martin, I. Pastan and M.M. Gottesman. Effect of ABC transporters on HIV-1 infection: inhibition of virus production by the MDR1 transporter. *FASEB J.* 14: 516–522, 2000.
101. Leslie, E.M., R.G. Deeley and S.P.C. Cole. Toxicological relevance of the multidrug resistance protein 1, MRP1 (ABCC1) and related transporters. *Toxicology* 167: 3–23, 2001.
102. Lette, L., E. Beaulieu, J.-M. Leclerc and R. Beliveau. Cyclosporin A treatment induces overexpression of P-glycoprotein in the kidney and other tissues. *Am. J. Physiol.* 270: F756–F765, 1996.
103. Lin, J.H. Drug–drug interaction mediated by inhibition and induction of P-glycoprotein. *Adv. Drug Delivery Rev.* 55: 53–81, 2003.
104. Litman, T., T. Zeuthen, T. Skosgaard and W.D. Stein. Structure–activity relationships of P-glycoprotein interacting drugs: kinetic characterization of their effects on ATP-ase activity. *Biochim. Biophys. Acta* 1361: 159–168, 1997.
105. Litman, T., T. Zeuthen, T. Skovsgaard and W.D. Stein. Competitive, non-competitive and cooperative interactions between substrates of P-glycoprotein as measured by its ATPase activity. *Biochim. Biophys. Acta* 1361: 169–176, 1997.
106. Liu, J., H. Chen, D.S. Miller, J.E. Saavedra, L.K. Keefer, D.R. Johnson, C.D. Klaassen and M.P. Waalkes. Overexpression of glutathione S-transferase II and multidrug resistance transport proteins is associated with acquired tolerance to inorganic arsenic. *Mol. Pharmacol.* 60: 302–309, 2001.
107. Liu, J., Y. Liu, D.A. Powell, M.P. Waalkes and C.D. Klaassen. Multidrug-resistance mdr1a/1b double knockout mice are more sensitive than wild type mice to acute arsenic toxicity, with higher arsenic accumulation in tissues. *Toxicology* 170: 55–62, 2002.
108. Loe, D.W., R.G. Deeley and S.P.C. Cole. Biology of the multidrug resistance-associated protein, MRP. *Eur. J. Cancer* 32A: 945–957, 1996.
109. Loo, T.W. and D.M. Clarke. Reconstitution of drug-stimulated ATPase activity following co-expression of each half of human P-glycoprotein as separate polypeptides. *J. Biol. Chem.* 269: 7750–7755, 1994.
110. Loo, T.W. and D.M. Clarke. Nonylphenol ethoxylates, but not nonylphenol, are substrates of the human multidrug resistance P-glycoprotein. *Biochem. Biophys. Res. Commun.* 247: 478–480, 1998.
111. Luker, G.D., T.P. Flagg, Q. Sha, K.E. Luker, C.M. Pica, C.G. Nichols and D. Piwnica-Worms. MDR1 P-glycoprotein reduces influx of substrates without affecting membrane potential. *J. Biol. Chem.* 276: 49053–49060, 2001.
112. Malorni, W., M.B. Lucia, G. Rainaldi, R. Cauda, M. Cianfriglia, G. Donelli and L. Ortona. Intracellular expression of P-170 glycoprotein in peripheral blood mononuclear cell subsets from healthy donors and HIV-infected patients. *Haematologica* 83: 13–20, 1998.
113. Marques-Santos, L.F., R.R. Bernardo, E.F. de Paula and V.M. Rumjanek. Cyclosporin A and trifluoperazine, two resistance-modulating agents, increase ivermectin neurotoxicity in mice. *Pharmacol. Toxicol.* 84: 125–129, 1999.
114. Mayer, U., E. Wagenaar, B. Dorobek, J.H. Beijnen, P. Borst and A.H. Schinkel. Full blockade of intestinal P-glycoprotein and extensive inhibition of blood–brain barrier P-glycoprotein by oral treatment of mice with PSC833. *J. Clin. Invest.* 100: 2430–2436, 1997.
115. Miller, D.S. Daunomycin secretion by killifish proximal tubules. *Am. J. Physiol.* 269: R370–R379, 1995.
116. Miller, D.S., C. Graeff, L. Droulle, S. Fricker and G. Fricker. Xenobiotic efflux pumps in isolated fish brain capillaries. *Am. J. Physiol. Regul. Integr. Comp. Physiol.* 282: R191–R198, 2002.
117. Miller, D.S., R. Masereeuw, J. Henson and K.J. Karnaky. Excretory transport of xenobiotics by dogfish rectal gland tubules. *Am. J. Physiol.* 275: R697–R705, 1998.
118. Miller, D.S., S.N. Nobmann, H. Gutmann, M. Toeroek, J. Drewe and G. Fricker. Xenobiotic transport across isolated brain microvessels studied by confocal microscopy. *Mol. Pharmacol.* 58: 1357–1367, 2000.
119. Miller, D.S., C.R. Sussman and J.L. Renfro. Protein kinase C regulation of p-glycoprotein-mediated xenobiotic secretion in renal proximal tubule. *Am. J. Physiol.* 275: F785–F795, 1998.
120. Minier, C.F., F. Levy, D. Rabel, G. Bocquené, D. Godefroy, T. Burgeot and F. Leboulenger. Flounder health status in the Seine Bay. A multibiomarker study. *Mar. Environ. Res.* 50: 373–377, 2000.
121. Miyake, K., L. Mickley, T. Litman, Z. Zhan, R. Robey, B. Cristensen, M. Brangi, L. Greenberger, M. Dean, T. Fojo and S.E. Bates. Molecular cloning of cDNAs which are highly overexpressed in mitoxantrone-resistant cells: demonstration of homology to ABC transport genes. *Cancer Res.* 59: 8–13, 1999.
122. Ng, W.F., F. Sarangi, R.L. Zastawny, L. Veinot-Drebot and V. Ling. Identification of members of the P-glycoprotein multigene family. *Mol. Cell. Biol.* 9: 1224–1232, 1989.

123. Nobmann, S., B. Bauer and G. Fricker. Ivermectin excretion by isolated functionally intact brain endothelial capillaries. *Br. J. Pharmacol.* 132: 722–728, 2001.
124. Oude Elferink, R.P.J. and J. Zadina. MDR1 P-glycoprotein transports endogenous opioid peptides. *Peptides* 22: 2015–2020, 2001.
125. Panwala, C.M., J.C. Jones and J.L. Viney. A novel model of inflammatory bowel disease: mice deficient for the multiple drug resistance gene, mdr1a, spontaneously develop colitis. *J. Immunol.* 161: 5733–5744, 1998.
126. Pascaud, C., M. Garrigos and S. Orlowski. Multidrug resistance transporter P-glycoprotein has distinct but interacting binding sites for cytotoxic drugs and reversing agents. *Biochem. J.* 333: 351–358, 1998.
127. Plouzek, C.A., H.P. Ciolino, R. Clarke and G.C. Yeh. Inhibition of P-glycoprotein activity and reversal of multidrug resistance *in vitro* by rosemary extract. *Eur. J. Cancer* 35: 1541–1545, 1999.
128. Podsiadlowski, L., V. Matha and A. Vilcinskas. Detection of a P-glycoprotein related pump in *Chironomus* larvae and its inhibition by verapamil and cyclosporin A. *Comp. Biochem. Physiol.* 121B: 443–450, 1998.
129. Pritchard, J.B. and D.S. Miller. Mechanisms mediating renal secretion of organic anions and cations. *Physiol. Rev.* 73: 765–796, 1993.
130. Raviv, Y., A. Puri and R. Blumenthal. P-glycoprotein-overexpressing multidrug-resistant cells are resistant to infection by enveloped viruses that enter via the plasma membrane. *FASEB J.* 14: 511–515, 2000.
131. Renes, J., E.G. de Vries, P.L. Jansen and M. Muller. The (patho)physiological functions of the MRP family. *Drug Resist. Updates* 3: 289–302, 2000.
132. Robinson-Rechavi, M., O. Marchand, H. Escriva and V. Laudet. An ancestral whole-genome duplication may not have been responsible for the abundance of duplicated fish genes. *Curr. Biol.* 11: R458–R459, 2001.
133. Roepe, P.D. The role of the MDR protein in altered drug translocation across tumor cell membranes. *Biochim. Biophys. Acta* 1241: 385–406, 1995.
134. Rohlff, C. and R.I. Glazer. Regulation of multidrug resistance through the cAMP and EGF signalling pathways. *Cell. Signal.* 7: 431–443, 1995.
135. Romsicki, Y. and F.J. Sharom. Phospholipid flippase activity of the reconstituted P-glycoprotein multidrug transporter. *Biochemistry* 40: 6937–6947, 2001.
136. Roninson, I.B., J.E. Chin, K. Choi, P. Gros, D.E. Housman, A. Fojo, D.-W. Shen, M.M. Gottesman and I. Pastan. Isolation of human mdr DNA sequences amplified in multidrug resistant KB carcinoma cells. *Proc. Natl Acad. Sci. USA* 83: 4538–4542, 1986.
137. Ruetz, S. and P. Gros. Phosphatidylcholine translocase: a physiological role for the *mdr2* gene. *Cell* 77: 564–571, 1994.
138. Safa, A.R. Photoaffinity analogs for multidrug resistance-related transporters and their use in identifying chemosensitizers. *Drug Resist. Updates* 2: 371–381, 1999.
139. Savas, Ü., K.J. Griffin and E.J. Johnson. Molecular mechanisms of cytochrome-P450 induction by xenobiotics: an expanded role for nuclear hormone receptors. *Mol. Pharmacol.* 56: 851–857, 1999.
140. Schiengold, M., L. Schwantes, G. Schwartsmann, J.A. Chies and N.B. Nardi. Multidrug resistance gene expression during the murine ontogeny. *Mech. Ageing Dev.* 122: 255–270, 2001.
141. Schinkel, A.H. P-glycoprotein, a gatekeeper in the blood–brain barrier. *Adv. Drug Delivery Rev.* 36: 179–194, 1999.
142. Schinkel, A.H. and J.W. Jonker. Mammalian drug efflux transporters of the ATP-binding cassette (ABC) family: an overview. *Adv. Drug Delivery Rev.* 55: 3–29, 2003.
143. Schinkel, A.H., U. Mayer, E. Wagenaar, C.A. Mol, L. van Deemter, J.J. Smit, M.A. van der Valk, A.C. Voordouw, H. Spits, O. van Telingen, J.M. Zijlmans, W.E. Fibbe and P. Borst. Normal viability and altered pharmacokinetics in mice lacking mdr1-type (drug-transporting) P-glycoproteins. *Proc. Natl Acad. Sci. USA* 94: 4028–4033, 1997.
144. Schinkel, A.H., J.J. Smit, O. van Tellingen, J.H. Beijnen, E. Wagenaar, L. van Deemter, C.A. Mol, M.A. van der Valk, E.C. Robanus-Maandag, H.P. te Riele, A.J.M. Berns and P. Borst. Disruption of the mouse mdr1a P-glycoprotein gene leads to a deficiency in the blood–brain barrier and to increased sensitivity to drugs. *Cell* 77: 491–502, 1994.
145. Schramm, U., G. Fricker, R. Wenger and D.S. Miller. P-glycoprotein-mediated secretion of a fluorescent cyclosporin analogue by teleost renal proximal tubules. *Am. J. Physiol.* 268: F46–F52, 1995.
146. Schrenk, D., A. Michalke, T.W. Gant, P.C. Brown, J.A. Silverman and S.S. Thorgeirsson. Multidrug resistance gene expression in rodents and rodent hepatocytes treated with mitoxantrone. *Biochem. Pharmacol.* 52: 1453–1460, 1996.
147. Schuetz, E.G., W.T. Beck and J.D. Schuetz. Modulators and substrates of P-glycoprotein and cytochrome P4503A coordinately up-regulate these proteins in human colon carcinoma cells. *Mol. Pharmacol.* 49: 311–318, 1996.

148. Schuetz, E.G., A.H. Schinkel, M.V. Relling and J.D. Schuetz. P-glycoprotein: a major determinant of rifampicin-inducible expression of cytochrome P4503A in mice and humans. *Proc. Natl Acad. Sci. USA* 93: 4001–4005, 1996.
149. Schuetz, E.G., S. Strom, K. Yasuda, V. Lecureur, M. Assem, C. Brimer, J. Lamba, R.B. Kim, V. Ramachandran, B.J. Komoroski, R. Venkataramanan, H. Cai, C.J. Sinal, F.J. Gonzalez and J.D. Schuetz. Disrupted bile acid homeostasis reveals an unexpected interaction among nuclear hormone receptors, transporters, and cytochrome P450. *J. Biol. Chem.* 276: 39411–39418, 2001.
150. Schwab, M., E. Schaeffeler, C. Marx, M.F. Fromm, B. Kaskas, J. Metzler, E. Stange, H. Herfarth, J. Schoelmerich, M. Gregor, S. Walker, I. Cascorbi, I. Roots, U. Brinkmann, U.M. Zanger and M. Eichelbaum. Association between the C3435T MDR1 gene polymorphism and susceptibility for ulcerative colitis. *Gastroenterology* 124: 26–33, 2003.
151. Seelig, A. A general pattern for substrate recognition by P-glycoprotein. *Eur. J. Biochem.* 251: 252–261, 1998.
152. Senior, A.E., M.K. al-Shawi and I.L. Urbatsch. The catalytic cycle of P-glycoprotein. *FEBS Lett.* 377: 285–289, 1995.
153. Shapiro, A.B., K. Fox, P. Lam and V. Ling. Stimulation of P-glycoprotein-mediated drug transport by prazosin and progesterone. Evidence for a third drug-binding site. *Eur. J. Biochem.* 259: 841–850, 1999.
154. Shapiro, A.B. and V. Ling. Positively cooperative sites for drug transport by P-glycoprotein with distinct drug specificities. *Eur. J. Biochem.* 250: 130–137, 1997.
155. Shapiro, A.B. and V. Ling. Stoichiometry of coupling of rhodamine 123 transport to ATP hydrolysis by P-glycoprotein. *Eur. J. Biochem.* 254: 189–193, 1998.
156. Sharma, R.C., S. Inoue, J. Roitelman, R.T. Schimke and R.D. Simoni. Peptide transport by the multidrug resistance pump. *J. Biol. Chem.* 267: 5731–5734, 1992.
157. Sharom, F.J. The P-glycoprotein efflux pump: how does it transport drugs. *J. Membr. Biol.* 160: 161–175, 1997.
158. Silverman, J.A., H. Raunio, T.W. Gant and S.S. Thorgeirsson. Cloning and characterization of a member of the rat multidrug resistance (*mdr*) gene family. *Gene* 106: 229–236, 1991.
159. Simkiss, K.. Ecotoxicants at the cell membrane barrier. In: *Ecotoxicology: A Hierarchical Treatment*, edited by M. Newman and C.H. Jagoe, Boca Raton, FL, Lewis Publishers, pp. 59–83, 1995.
160. Smit, J.J., A.H. Schinkel, C.A. Mol, D. Majoor, W.J. Mooi, A.P. Jongsma, C.R. Lincke and P. Borst. Tissue distribution of the human MDR3 P-glycoprotein. *Lab. Invest.* 71: 638–649, 1994.
161. Smit, J.J.M., A.H. Schinkel, R.P.J. Oude Elferink, A.K. Groen, E. Wagenaar, L. van Deemter, C.A.A.M. Mol, R. Ottenfoff, N.M.T. van der Ligt, M.A. van Roon, G.J.A. Offerhaus, A.J.M. Berns and P. Borst. Homozygous disruption of the murine mdr2 P-glycoprotein gene leads to a complete absence of phospholipid from the bile and to liver disease. *Cell* 75: 451–462, 1993.
162. Smit, J.W., M.T. Huisman, O. van Tellingen, H.R. Wiltshire and A.H. Schinkel. Absence or pharmacological blocking of placental P-glycoprotein profoundly increases fetal drug exposure. *J. Clin. Invest.* 104: 1441–1447, 1999.
163. Smit, J.W., A.H. Schinkel, M. Muller, B. Weert and D.K. Meijer. Contribution of the murine mdr1a P-glycoprotein to hepatobiliary and intestinal elimination of cationic drugs as measured in mice with an mdr1a gene disruption. *Hepatology* 27: 1056–1063, 1998.
164. Smit, J.W., A.H. Schinkel, B. Weert and D.K. Meijer. Hepatobiliary and intestinal clearance of amphiphilic cationic drugs in mice in which both mdr1a and mdr1b genes have been disrupted. *Br. J. Pharmacol.* 124: 416–424, 1998.
165. Smital, T. and B. Kurelec. Inhibitors of the multixenobiotic resistance mechanism in natural waters: in vivo demonstration of their effects. *Environ. Toxicol. Chem.* 16: 2164–2170, 1997.
166. Smital, T. and B. Kurelec. The chemosensitizers of multixenobiotic resistance mechanism in aquatic invertebrates: a new class of pollutants. *Mutat. Res.* 399: 43–53, 1998.
167. Smital, T. and R. Sauerborn. Measurement of the activity of multixenobiotic resistance mechanism in the common carp *Cyprinus carpio*. *Mar. Environ. Res.* 54: 449–453, 2002.
168. Smith, A.J., J.L.P.M. Timmermanns-Hereijgers, B. Roelefsen, K.W.A. Wirtz, W.J. Van Blitterswijk, J.J.M. Smit, A.H. Schinkel and P. Borst. The human MDR3 P-glycoprotein promotes translocation of phosphatidylcholine through the plasma membrane of fibroblasts from transgenic mice. *FEBS Lett.* 354: 263–266, 1994.
169. Smith, C.D., M.R. Prinsep, F.R. Caplan, R.E. Moore and G.M. Patterson. Reversal of multiple drug resistance by tolyporphin, a novel cyanobacterial natural product. *Oncol. Res.* 6: 211–218, 1994.
170. Sparreboom, A., J. van Asperen, U. Mayer, A.H. Schinkel, J.W. Smit, D.K. Meijer, P. Borst, W.J. Nooijen, J.H. Beijnen and O. van Tellingen. Limited oral bioavailability and active epithelial excretion of paclitaxel (Taxol) caused by P-glycoprotein in the intestine. *Proc. Natl Acad. Sci. USA* 94: 2031–2035, 1997.

171. Stein, W.D.. Kinetics of the multidrug transporter (P-glycoprotein) and its reversal. *Physiol. Rev.* 77: 545–590, 1997.
172. Stouch, T.R. and O. Gudmundsson. Progress in understanding the structure–activity relationships of P-glycoprotein. *Adv. Drug Delivery Rev.* 54: 315–328, 2002.
173. Sturm, A., J.-P. Cravedi and H. Segner. Prochloraz and nonylphenol diethoxylate inhibit an mdr1-like activity in vitro, but do not alter hepatic levels of P-glycoprotein in trout exposed in vivo. *Aquat. Toxicol.* 53: 215–228, 2001.
174. Sturm, A., C. Ziemann, K. Hirsch-Ernst and H. Segner. Expression and functional activity of P-glycoprotein in cultured hepatocytes from *Oncorhynchus mykiss*. *Am. J. Physiol. Regul. I.* 281: R1119–R1126, 2001.
175. Sun, H., H. Dai, N. Shaik and W.F. Elmquist. Drug efflux transporters in the CNS. *Adv. Drug Delivery Rev.* 55: 83–105, 2003.
176. Sussman-Turner, C. and J.L. Renfro. Heat-shock stimulated transepithelial daunomycin secretion by flounder renal proximal tubule primary cultures. *Am. J. Physiol.* 267: F135–F144, 1995.
177. Synold, T.W., I. Dussault and B.M. Forman. The orphan nuclear receptor SXR coordinately regulates drug metabolism and efflux. *Nat. Med.* 7: 584–590, 2001.
178. Taylor, J.S., Y. Van de Peer and A. Meyer. Revisiting recent challenges to the ancient fish-specific genome duplication hypothesis. *Curr. Biol.* 11: R1005–R1008, 2001.
179. Thiebaut, F., T. Tsuruo, H. Hamada, M.M. Gottesman and I. Pastan. Cellular localization of the multidrug-resistance gene product in normal human tissues. *Proc. Natl Acad. Sci. USA* 84: 7735–7738, 1987.
180. Thiebaut, F., T. Tsuruo, H. Haqmada, M.M. Gottesmann, I. Pastan and M.C. Willingham. Immunohistochemical localization in normal tissues of different epitopes in the multidrug transport protein P170: evidence for localization in brain capillaries and cross-reactivity of one antibody with a muscle protein. *J. Histochem. Cytochem.* 37: 159–164, 1989.
181. Toomey, B.H. and D. Epel. Multixenobiotic resistance in *Urechis caupo* embryos: protection from environmental toxins. *Biol. Bull.* 185: 355–364, 1993.
182. Török, M., H. Gutmann, G. Frickert and J. Drewe. Sister of P-glycoprotein expression in different tissues. *Biochem. Pharmacol.* 57: 833–835, 1999.
183. Tsuji, A. and I. Tamai. Carrier-mediated intestinal transport of drugs. *Pharmocol. Res.* 13: 963–977, 1996.
184. Tutundjian, R., J. Cachot, F. Leboulenger and C. Minier. Genetic and immunological characterisation of a multixenobiotic resistance system in the turbot (*Scophthalmus maximus*). *Comp. Biochem. Physiol.* 132B: 463–471, 2002.
185. Tutundjian, R., C. Minier, F. Le Foll and F. Leboulenger. Rhodamine exclusion activity in primary cultured turbot (*Scophthalmus maximus*) hepatocytes. *Mar. Environ. Res.* 54: 443–447, 2002.
186. Ueda, K., N. Okamura, M. Hirai, Y. Tanigawara, T. Saeki, N. Kioka, T. Komano and R. Hori. Human P-glycoprotein transports cortisol, aldosterone, and dexamethasone, but not progesterone. *J. Biol. Chem.* 267: 24248–24252, 1992.
187. Ueda, K., Y. Taguchi and M. Morishima. How does P-glycoprotein recognize its substrates? *Semin. Cancer Biol.* 8: 151–159, 1997.
188. Urbatsch, I.L., B. Sankaran, S. Bhagat and A.E. Senior. Both P-glycoprotein nucleotide-binding sites are catalytically active. *J. Biol. Chem.* 270: 26956–26961, 1995.
189. Urbatsch, I.L., B. Sankaran, J. Weber and A.E. Senior. P-glycoprotein is stably inhibited by vanadate-induced trapping of nucleotide at a single catalytic site. *J. Biol. Chem.* 270: 19383–19390, 1995.
190. van Asperen, J., O. van Tellingen and J.H. Beijnen. The role of mdr1a P-glycoprotein in the biliary and intestinal secretion of doxorubicin and vinblastine in mice. *Drug Metab. Dispos.* 28: 264–267, 2000.
191. van Asperen, J., O. van Tellingen, F. Tijssen, A.H. Schinkel and J.H. Beijnen. Increased accumulation of doxorubicin and doxorubicinol in cardiac tissue of mice lacking mdr1a P-glycoprotein. *Br. J. Cancer* 79: 108–113, 1999.
192. van Aubel, R.A., R. Masereeuw and F.G. Russel. Molecular pharmacology of renal organic anion transporters. *Am. J. Physiol. Renal Physiol.* 279: F216–F232, 2000.
193. van den Elsen, J.M.H., D.A. Kuntz, F.J. Hoedemaeker and D.R. Rose. Antibody C219 recognizes an α-helical epitope on P-glycoprotein. *Proc. Natl Acad. Sci. USA* 96: 13679–13684, 1999.
194. van der Bliek, A.M., P.M. Kooiman, C. Schneider and P. Borst. Sequence of mdr3 cDNA encoding a human P-glycoprotein. *Gene* 71: 401–411, 1988.
195. van Helvoort, A., A.J. Smith, H. Sprong, I. Fritzsche, A.H. Schinkel, P. Borst and G. van Meer. MDR1 P-glycoprotein is a lipid translocase of broad specificity, while MDR3 P-glycoprotein specifically translocates phosphatidylcholine. *Cell* 87: 507–517, 1996.
196. van Tellingen, O. The importance of drug-transporting P-glycoproteins in toxicology. *Toxicol. Lett.* 120: 31–41, 2001.

197. van Veen, H.W., C.F. Higgins and W.N. Konings. Multidrug transport by ATP binding cassette transporters: a proposed two-cylinder engine mechanism. *Res. Microbiol.* 152: 365–374, 2001.
198. Vogelbein, W.K., J.W. Fournie, P.A. van Veld and R.J. Huggett. Hepatic neoplasms in the mummichog *Fundulus heteroclitus* from a creosote-contaminated site. *Cancer Res.* 50: 5978–5986, 1990.
199. Waldmann, P., B. Pivcevic, W.E.G. Müller, R.K. Zahn and B. Kurelec. Increased genotoxicity of acetylaminofluorene by modulators of multixenobiotic resistance mechanism: studies with the fresh water clam *Corbicula fluminea. Mutat. Res.* 342: 113–123, 1995.
200. Walker, J.E., M. Saraste, M.J. Runswick and N.J. Gay. Distantly related sequences in the α- and β-subunits of ATP synthase, myosin, kinases and other ATP requiring enzymes and a common nucleotide binding fold. *EMBO J.* 1: 945–951, 1982.
201. Wandel, C., R.B. Kim, S. Kajiji, P. Guengerich, G.R. Wilkinson and A.J. Wood. P-glycoprotein and cytochrome P-450 3A inhibition: dissociation of inhibitory potencies. *Cancer Res.* 59: 3944–3948, 1999.
202. Wang, E., K. Lew, M. Barecki, C.N. Casciano, R.P. Clement and W.W. Johnson. Quantitative distinctions of active site molecular recognition by P-glycoprotein and cytochrome P450 3A4. *Chem. Res. Toxicol.* 14: 1596–1603, 2001.
203. Wang, E.J., C.N. Casciano, R.P. Clement and W.W. Johnson. Cooperativity in the inhibition of P-glycoprotein-mediated daunorubicin transport: evidence for half-of-the-sites reactivity. *Arch. Biochem. Biophys.* 383: 91–98, 2000.
204. Wang, E.J., C.N. Casciano, R.P. Clement and W.W. Johnson. Two transport binding sites of P-glycoprotein are unequal yet contingent: initial rate kinetic analysis by ATP hydrolysis demonstrates intersite dependence. *Biochim. Biophys. Acta* 1481: 63–74, 2000.
205. Wang, E.J., C.N. Casciano, R.P. Clement and W.W. Johnson. Active transport of fluorescent P-glycoprotein substrates: evaluation as markers and interaction with inhibitors. *Biochem. Biophys. Res. Commun.* 289: 580–585, 2001.
206. Watanabe, T., N. Kokubu, S.B. Charnick, M. Naito, T. Tsuruo and D. Cohen. Interaction of cyclosporin derivatives with the ATPase activity of human P-glycoprotein. *Br. J. Pharmacol.* 122: 241–248, 1997.
207. Williams, A.B. and R.S. Jacobs. A marine natural product, patellamide D, reverses multidrug resistance in a human leukemic cell line. *Cancer Lett.* 71: 97–102, 1993.
208. Wolf, D.C. and S.B. Horwitz. P-glycoprotein transports corticosterone and is photoaffinity-labelled by the steroid. *Int. J. Cancer* 52: 141–146, 1992.
209. Xie, W., J.L. Barwick, M. Downes, B. Blumberg, C.M. Simon, M.C. Nelson, B.A. Neuschwander-Tetri, E.M. Brunt, P.S. Guzelian and R.M. Evans. Humanized xenobiotic response in mice expressing nuclear receptor SXR. *Nature* 406: 435–439, 2000.
210. Yeh, G.C., J. Lopaczynska, C.M. Poore and J.M. Phang. A new functional role for P-glycoprotein: efflux pump for benzo(*a*)pyrene in human breast cancer MCF-7 cells. *Cancer Res.* 52: 6692–6695, 1992.
211. Young, A.M., C.E. Allen and K.L. Audus. Efflux transporters of the human placenta. *Adv. Drug Delivery Rev.* 55: 125–132, 2003.
212. Zaman, G.J., J. Lankelma, O. van Tellingen, J. Beijnen, H. Dekker, C. Paulusma, R.P. Oude Elferink, F. Baas and P. Borst. Role of glutathione in the export of compounds from cells by the multidrug-resistance-associated protein. *Proc. Natl Acad. Sci. USA* 92: 7690–7694, 1995.
213. Zaman, G.J.R., M.J. Flens, M.R. Van Leusden, M. De Haas, H.S. Mülder, J. Lankelma, H.M. Pinedo, R.J. Scheper, H.J. Brosterman and P. Borst. The human multidrug resistance-associated protein MRP is a plasma membrane drug-efflux pump. *Proc. Natl Acad. Sci. USA* 91: 8822–8826, 1994.
214. Zamora, J.M., H.L. Pearce and W.T. Beck. Physical–chemical properties shared by compounds that modulate multidrug resistance in human leukemic cells. *Mol. Pharmacol.* 33: 454–462, 1988.
215. Zhang, E.Y., M.A. Phelps, C. Cheng, S. Ekins and P.W. Swaan. Modeling of active transport systems. *Adv. Drug Delivery Rev.* 54: 329–354, 2002.
216. Zhou, Y., M.M. Gottesman and I. Pastan. Domain exchangeability between the multidrug transporter (MDR1) and phosphatidylcholine flippase (MDR2). *Mol. Pharmacol.* 56: 997–1004, 1999.
217. Zhou, Y., M.M. Gottesman and I. Pastan. Studies of human MDR1–MDR2 chimeras demonstrate the functional exchangeability of a major transmembrane segment of the multidrug transporter and phosphatidylcholine flippase. *Mol. Cell. Biol.* 19: 1450–1459, 1999.
218. Zimniak, P., S. Pikula, J. Bandorowicz-Pikula and Y.C. Awasthi. Mechanisms for xenobiotic transport in biological membranes. *Toxicol. Lett.* 106: 107–118, 1999.
219. Zucker, S.N., G. Castillo and S.B. Horwitz. Down-regulation of the mdr gene by thyroid hormone during *Xenopus laevis* development. *Mol. Cell. Endocrinol.* 129: 73–81, 1997.

Species Index

Subject Index